The McGraw-Hill Reader

Issues Across the Disciplines

NINTH EDITION

Gilbert H. Muller
The City University of New York
LaGuardia

Boston   Burr Ridge, IL   Dubuque, IA   Madison, WI   New York
San Francisco   St. Louis   Bangkok   Bogotá   Caracas   Kuala Lumpur
Lisbon   London   Madrid   Mexico City   Milan   Montreal   New Delhi
Santiago   Seoul   Singapore   Sydney   Taipei   Toronto

THE MCGRAW-HILL READER: ISSUES ACROSS THE DISCIPLINES

Published by McGraw-Hill, a business unit of The McGraw-Hill Companies, Inc., 1221 Avenue of the Americas, New York, NY, 10020. Copyright © 2006, 2003, 2000, 1997, 1994, 1991, 1988, 1985, 1982, by The McGraw-Hill Companies, Inc. All rights reserved. No part of this publication may be reproduced or distributed in any form or by any means, or stored in a database or retrieval system, without the prior written consent of The McGraw-Hill Companies, Inc., including, but not limited to, in any network or other electronic storage or transmission, or broadcast for distance learning.

Some ancillaries, including electronic and print components, may not be available to customers outside the United States.

This book is printed on acid-free paper.

1 2 3 4 5 6 7 8 9 0 DOC/DOC 0 9 8 7 6 5

Student Edition ISBN: 0-07-296295-X
Instructor's Examination Copy ISBN: 0-07-321747-6

Editor in Chief: *Emily Barrosse*
Publisher: *Lisa Moore*
Sponsoring Editor: *Christopher Bennem*
Developmental Editor II: *Bennett Morrison*
Marketing Manager: *Lori DeShazo*
Managing Editor: *Jean Dal Porto*
Project Manager: *Becky Komro*
Lead and Cover Designer: *Gino Cieslik*

Photo Research Coordinator: *Natalia C. Peschiera*
Cover Credit: *Getty Images*
Senior Production Supervisor: *Carol A. Bielski*
Permissions Editor: *Marty Granahan*
Composition: *10/12 Palatino, by Cenveo*
Printing: *45# New Era Matte, R.R. Donnelley/ Crawfordsville, IN*

Credits: The credits section for this book begins on page C and is considered an extension of the copyright page.

Library of Congress Cataloging-in-Publication Data

The McGraw-Hill reader : issues across the disciplines / [edited by] Gilbert H. Muller.—
   9th ed.
      p. cm.
   Includes bibliographical references and index.
   ISBN 0-07-296295-X (softcover : alk. paper)
      1. College readers. 2. Interdisciplinary approach in education—Problems, exercises, etc.
3. English language—Rhetoric—Problems, exercises, etc. 4. Academic writing—Problems, exercises, etc.   I. Muller, Gilbert H., 1941–
PE1417.M44 2006
808'.0427—dc22

                                                                          2005049233

The Internet addresses listed in the text were accurate at the time of publication. The inclusion of a website does not indicate an endorsement by the authors of McGraw-Hill, and McGraw-Hill does not guarantee the accuracy of the information presented at these sites.

www.mhhe.com

About the Author

GILBERT H. MULLER, who received a PhD in English and American literature from Stanford University, is professor emeritus of English at the LaGuardia campus of the City University of New York. He has also taught at Stanford University, Vassar College, and several universities overseas. Dr. Muller is the author of the award-winning *Nightmares and Visions: Flannery O'Connor and the Catholic Grotesque; Chester Himes; New Strangers in Paradise: The Immigrant Experience and Contemporary American Fiction;* and other critical studies. His essays and reviews have appeared in *The New York Times, The New Republic, The Nation, The Sewanee Review, The Georgia Review,* and elsewhere. He is also a noted author and editor of textbooks in English and composition, including *The Short Prose Reader* with Harvey Wiener, and, with John A. Williams, *The McGraw-Hill Introduction to Literature, Bridges: Literature across Cultures,* and *Ways In: Reading and Writing about Literature and Film.* Among Dr. Muller's awards are fellowships from the National Endowment for the Humanities, the Fulbright Commission, the Ford Foundation, and the Mellon Foundation.

To Parisa and Darius
My favorite readers

Brief Contents

Contents

PART 1
AN OVERVIEW OF COLLEGE WRITING

PART 2
ISSUES ACROSS THE DISCIPLINES

Contents of Essays
by Rhetorical Mode

NARRATION

DESCRIPTION

ILLUSTRATION

COMPARISON AND CONTRAST

ANALOGY

DEFINITION

CLASSIFICATION

PROCESS ANALYSIS

CAUSAL ANALYSIS

ARGUMENT AND PERSUASION

HUMOR, IRONY, AND SATIRE

Preface

Through eight previous editions, *The McGraw-Hill Reader* has presented the finest classic and contemporary essays, works that span various ages, cultures, and disciplines, providing students with a range of quality prose works. Eudora Welty speaks of reading as "a sweet devouring." This anthology alerts students to the vast and varied pleasures of reading and writing, while offering them opportunities to experience numerous perspectives on academic discourse.

Addressing the abiding national interest in core liberal arts programs, interdisciplinary issues, and multicultural perspectives, this ninth edition continues to offer students and instructors a full range of quality prose models important to writing courses, reading sequences, and key undergraduate disciplines. All of the selections have been chosen for their significance, vitality, and technical precision. With the high quality of its essays, its consistent humanistic emphases, and its clear organization, *The McGraw-Hill Reader* is a lively, sophisticated, and eminently flexible text for college composition and reading programs.

Organization and Proven Features

Composed of 14 chapters, *The McGraw-Hill Reader* covers the major modes of writing and most of the disciplines that college students will encounter as undergraduates. Chapter 1 presents an extensive overview of the critical thinking, reading, and writing processes. Chapter 2 provides extensive coverage of argument and persuasion. Chapter 3 is a concise guide to research and documentation in the electronic age. Chapters 4 through 14 cover core liberal arts disciplines, including education, the social sciences, business and economics, the humanities, and the sciences. Each chapter asks a key question drawn from the disciplines it represents and designed to elicit constructive class discussion and sound critical writing. These disciplinary chapters offer prose models that allow students to practice skills they will need throughout college, including analysis, criticism, argumentation, and persuasion.

Throughout its previous editions, instructors and students have appreciated the following features of *The McGraw-Hill Reader:*

- **A rich selection of readings:** A distinct strength of *The McGraw-Hill Reader*—perhaps the primary one for teachers who prefer to create their own approaches to composition and reading courses—is the wide range of material and the varied constituencies represented in the text. The essays in this book have been selected carefully to embrace a rich assortment of authors, to achieve balance among constituencies, to cover major historical periods, and to provide prose models and styles for class analysis, discussion, and imitation. The authors in this text—whether Plato or Julia Alvarez, Jonathan Swift or Amy Tan—have high visibility as writers and thinkers of value. Some of these authors are represented by more than one essay. All the authors—writing from such vantage points as literature, journalism, anthropology, sociology, art history, biology, and philosophy—start from the perspective that ideas exist in the world, that we should be alert to them, and that we should be able to deal with them in our own discourse.

- **A text that works with a wide variety of levels and approaches:** Because the selections range from very simple essays to the most abstract and complex modes of prose, teachers and students will be able to use *The McGraw-Hill Reader* at virtually all levels of a program. Containing 118 complete essays, *The McGraw-Hill Reader* thus is a flexible companion for composition courses. It can be used with any of the major pedagogical perspectives common to the practice of composition today: as a writing-across-the-curricula text; as the basis for a rhetorically focused course; as a thematic reader; as a multicultural anthology, as an in-depth reader. An alternate table of contents, listing carefully selected essays in 11 rhetorical categories, also makes *The McGraw-Hill Reader* adaptable to an approach based on the rhetorical patterns. Above all, teachers can develop their own sequences of essays that will contribute not only to their students' reading and writing proficiency but also to their growing intellectual power.

- **Chapter introductions that encourage students to reflect on major issues in the discipline:** The introduction to each disciplinary chapter gives students a broad perspective on the field at hand by putting major issues and concerns in context. Each introduction ends with a previewing section that alerts students to strategies for reading, discussion, and writing.

- **Uniform apparatus that reinforces critical reading and writing:** Another major strength of *The McGraw-Hill Reader* is in the uniform apparatus that accompanies every essay. Much can be learned from any well-written essay, especially if the apparatus is systematic in design. Each selection in this text is preceded by a brief introduction that offers biographical information about the author. The questions that follow each essay are organized in a consistent format created to reinforce essential reading, writing, and oral communication skills. Arranged in three categories—Comprehension, Rhetoric, and Writing—these questions reflect current compositional theory as they move students from audience analysis to various modes and processes of composition. All specialized terms used in the questions are defined for students in an extensive Glossary of Terms at the end of the text. The integrated design of these questions makes each essay—simple or complex, short or long, old or new—accessible to college students who possess varied reading and writing abilities.

- **"Connections for Critical Thinking" sections:** The essay topics listed at the end of each chapter help students make comparative assessments of various groups of essays and use Internet resources.
- **A guide to research and documentation:** Chapter 3 offers guidance on the most current research writing processes and the documentation styles recommended by the Modern Language Association and the American Psychological Association.

Highlights of the Ninth Edition

Informed by the comments and suggestions of more than sixty instructors from across the country who reviewed the previous edition, this ninth edition of *The McGraw-Hill Reader* offers a number of new and significant features.

- **A new plan of organization:** Part 1, "An Overview of College Writing," presents three chapters introducing students to critical reading, writing, and research, with an emphasis on argumentation. Part 2, "Issues across the Disciplines," offers 11 chapters covering the humanities, social sciences, sciences, and professions.
- **Three new portfolios of essays in Part 1:** Portfolios on *writing and communication, argumentation,* and *professional research papers*—almost two dozen selections in all—make Part 1 more interactive for teachers and students and give students an introduction to forms of academic and persuasive writing.
- **Three new chapters in Part 2:** Exciting chapters on *Media and Popular Culture, Literature and the Arts,* and *Health and Medicine,* all of high interest to students, expand the coverage of academic disciplines.
- **Forty new selections:** Fresh essays on topics of current and enduring interest such as literature, religion, medicine, biotechnology, globalization, and terrorism will elicit provocative student writing. New readings to this edition include essays by Ishmael Reed, Julia Alvarez, Thomas Friedman, Leslie Marmon Silko, David Sedaris, Joan Didion, Salman Rushdie, Barbara Ehrenreich, Dinesh D'Souza, Barbara Kingsolver, and other prominent writers and thinkers.
- **Expanded treatment of argument, plagiarism, summary and précis, and electronic research:** The three chapters in Part 1 provide integrated guidelines on these topics. Additionally, argumentation is stressed throughout the text, with the last writing assignment following all selections asking students to compose an argumentative essay.

- **A four-color insert on Advertising and Culture:** This insert encourages students to respond comparatively to advertisements as visual texts. New classic and contemporary visuals also appear at the beginning of each chapter. These visuals include photographs, paintings, cartoons, and other varieties of visual texts. Students are engaged by visual texts, and these provocative, high-quality images, along with the accompanying "Using a Critical Perspective" questions, will serve to interest them in the topic of the chapter and get them thinking and writing.

Useful Supplements

The following supplements are designed to help instructors and students derive the full benefit from *The McGraw-Hill Reader:*

- **A powerful Online Learning Center powered by Catalyst 2.0.** The Web site for *The McGraw-Hill Reader* (www.mhhe.com/mhreader) features a wealth of online resources including additional resources that complement the readings in *The McGraw-Hill Reader,* tutorials for document design and visual rhetoric, guides for avoiding plagiarism and evaluating sources, Bibliomaker software for MLA, APA, *Chicago,* and CSE styles of documentation, and many more tools that support students with their writing at every stage of the process. Catalyst 2.0 features a full classroom management system that includes an online grade book and collaborative peer review environment.
- **A *Guide to the McGraw-Hill Reader,*** by Gilbert H. Muller and David Pecan, offers well-considered strategies for teaching individual essays, sample rhetorical analyses, answers to questions, additional thought-provoking questions, comparative essay discussion formats, and tips for prewriting and guided writing activities. There is also a bibliography of criticism and research on the teaching of composition. This guide is available on the Online Learning Center at www.mhhe.com/mhreader.
- **Teaching Composition Faculty Listserv (www.mhhe.com/tcomp).** Moderated by Chris Anson at University of North Carolina, Raleigh, and offered by McGraw-Hill as a service to the composition community, this listserv brings together senior members of the college composition community with newer members—junior faculty, adjuncts, and teaching assistants—in an online newsletter and accompanying discussion group to address issues of pedagogy, in theory and in practice.

Acknowledgments

It is a pleasure to acknowledge the support, assistance, and guidance of numerous individuals who helped create *The McGraw-Hill Reader.* I want to thank the excellent McGraw-Hill family of assistants, editors, and executives who participated enthusiastically in the project from the outset and who encouraged me at every step. My editor, Christopher Bennem, has been an enthusiastic supporter of *The McGraw-Hill Reader*. Meg Botteon assisted in the revision of the chapter on research writing. David Pecan helped with the preparation of the instructor's manual. Above all, I want to thank Bennett Morrison, my development editor, a helpful friend and full partner through the revision process.

The final content and design of *The McGraw-Hill Reader*, Ninth Edition, reflects the expertise and advice offered by college instructors across the country who gave generously of their time when asked to review the text. These include Cora Agatucci, Central Oregon Community College; Lise Buranen, California State University; Andrea Dace, Thomas Nelson Community College; Ed Eleazer, Francis Marion University; Yuri Horner, San Jacinto College South; David Judkins, University of Houston; Jim Kenkel, Eastern Kentucky University; Terry Long, Boston College; Walter Lowe, Green River Community College; Susan J. Miller, Santa Fe Community College; RoseAnn Morgan, Middlesex Community College; Barbara Mueller, Cerritos College; Larry Rochelle, Johnson County Community College; William Ryder, College of the Canyons; Jeremy Schneider, University of Nebraska at Kearney; John Schaffer, Blinn College; Michel Small, Shasta College; Dan Thorpe, College of DuPage; Patrick Vecchio, St. Bonaventure University; Stephen H. Wells, Community College of Allegheny College; Lana A. Whited, Ferrum College; Steve Yarborough, Bellevue Community College. In addition, the following instructors replied thoughtfully to a survey administered by McGraw-Hill: Cora Agatucci, Central Oregon Community College; Leisa Belleau, University of Southern Indiana; Sandra Coyle, College of St. Joseph in Vermont; Gay Degani, Pasadena City College; Michael Dittman, Butler County Community College; William Feeler, Midland College; Mary Haley, Austin Community College-Northridge; Marianna Hofer, University of Findlay; Beth Impson, Bryan College; Josh H. Knight, Midland College; Carrie Leverenz, Texas Christian University; Joe Lostracco, Austin Community College; Walter Lowe, Green River Community College, Auburn; John Marsden, Indiana University of Pennsylvania; Elizabeth Mandrell, Morehead State University; Floyd Moos, College of the Canyons; Rebecca Neagle, Louisburg College; Eric Nelson, Green River Community College, Auburn; Ron Reed, Hazard Community College; Connie Smith, College of St. Joseph in Vermont; Wes Spradley, University of Texas at San Antonio; Steve Swellander, University of Texas-San Antonio; Elyce Wakerman, California State University, Northridge; Laviece Ward, Wake Tech Community College.

I am pleased to acknowledge support from the Mellon Foundation, the Graduate Center of The City University of New York, and the United States Department of Education (Title III and Title IV) that enabled me to develop this text.

Gilbert H. Muller

part *1*

An Overview of College Writing

chapter *1*

Critical Thinking, Reading, and Writing

This book will help you read and write critically for college courses. Although we have entered the electronic information age, where mastery of computer skills and visual literacy seem to be essential for understanding and maintaining our lives and careers, most college work still requires an ability to understand and reflect intelligently on written texts and, subsequently, to respond in writing to them. College courses typically involve the reading of challenging texts. As a college student, you will need to approach these texts with skills that go beyond those of casual reading, that is, the type of reading you may do for leisure, for pleasure or for escapism, or simply to pass time. Even in courses where a preponderance of work is in learning forms of knowledge and new technologies, such as computers, mathematics, and science, you are sure to find a healthy amount of reading that will supplement any other work done in the classroom or laboratory.

The reading and writing skills you develop during your college years will also help you in your future profession. Think of a lawyer reviewing legal history or preparing a legal brief, a doctor reviewing current literature on medical innovations or writing an article for a professional journal, or an environmental scientist reading and writing about issues regarding pollution and global warming. All these activities require the ability to think, read, and write about complex material. Learning the tools of critical reading and writing not only teaches you the "what" of an issue, but also helps you think about and respond intelligently to the relative strength of the writer's opinions, ideas, and theories. Critical thinking, reading, and writing enable you to distinguish between informed ideas and pure speculation, rational arguments and emotional ones, and organized essays and structurally deficient ones.

As you hone your critical thinking, reading, and writing skills by tackling the essays in this anthology, you should soon understand how the written word is still the primary medium with which thinkers transmit the intricacies of controversial issues involving the family, society, politics, work, gender, and class. You will encounter complex texts that require you to extract maximum meaning from them, compare your own views with those of the authors you read, and respond to what you read in an informed and coherent manner. The reading selections in this textbook have been chosen specifically to assist you in developing such skills. As you tackle these texts, you will realize that sound reading habits will permit you to understand the fine points of logic, reasoning, analysis, argumentation, and evaluation.

STEPS TO READING CRITICALLY AND ACTIVELY

You can find numerous reasons to rationalize a failure to read carefully and critically. You have a headache. You're hungry. The material is boring. The writer puts you to sleep. Your roommates are talking. You have a date. In short, there are many internal and external barriers to critical reading. Fortunately, there are techniques—a critical reading process—to guide you through this maze of distractions. Consider these five strategies:

1. *Develop an attitude of "critical consciousness."* In other words, do not be passive, uncritical, or alienated from the writer or the text. Instead, be active, critical, and engaged with the writer and his or her text.
2. *Read attentively.* Give your full attention to the text in order to understand it. Do not let your mind wander.
3. *Paraphrase.* Periodically restate what you read. Learn to process bits of key information. Keep a running inventory of highlights. Take mental or actual notes on the text's main points. (More information on paraphrase and summary appears in this chapter and in Chapter 3.)
4. *Ask questions.* If for any reason you are uncertain about any aspect of the text, pose a question about it and try to answer it yourself. You might seek immediate help from a friend or classmate. If you are unable to answer your question, ask for clarification from the instructor.
5. *Control your biases.* You must both control and correct any prejudices that might interfere with the claim, information, or tone of a text. You might, for example, have misgivings about a liberal or conservative writer, about a feminist or a creationist, but such strong emotions can erode your ability to keep an open mind and your power to think critically about a subject or issue.

These five strategies will help you begin to overcome the barriers to critical reading.

One way to view critical reading is through the concept of active reading. Active reading suggests that you, as a reader, have an obligation to yourself and the author to bring an alert, critical, and responsive perspective to your encounter with the written word. Active reading means learning to annotate (a strategy discussed later in this chapter), to reflect on what you read, and to develop personal responses in order to prepare yourself for writing assignments that your instructor will present to you during the term. This process—reading critically in order to write critically—is not merely an "academic" exercise. It is a skill that can enrich you as a person throughout your life and career. It will teach you to respond critically to the admonitions of politicians or to the seductions of advertisements and, if you choose, to participate intelligently in the "national conversation," which can lead to a rewarding life and responsible citizenship.

When you read an essay or any other type of text, you create meaning out of the material the author has presented. If the essay is relatively simple, clear, and concise, the experience that you construct from your reading may be very similar to what the author intended. Nevertheless, the way that you interact with even the most comprehensible texts will never be identical to the way another reader interacts.

Consider an essay that you will encounter in this anthology, Langston Hughes's "Salvation." A chapter from his autobiography, *The Big Sea* (1940), this essay tells of a childhood incident in which the young Hughes's faith was tested. The essay focuses on a church revival meeting that Hughes was taken to and the increasing pressure he sensed at the meeting to "testify" to the presence of Jesus in his life. At first the young Hughes holds out against the fervor of the congregation, but ultimately he pretends to be converted, or "saved." That night, however, he weeps and then testifies to something entirely unexpected: the loss of faith he experienced because Jesus did not "save" him in a time of need.

As your class reads this essay, individuals among you may be struck by the compressed energy of the narration and the description of the event; by the swift characterization and revealing dialogue; or by the conflict and mounting tension. Moreover, the heightened personal and spiritual conflict will force class members to consider the sad irony inherent in the title "Salvation."

Even if your class arrives at a broad consensus on the intentions of the author, individual reader responses to the text will vary. Readers who have attended revival meetings will respond differently from those who have not. Evangelical Christians will see the text from a different perspective than will Catholics, Muslims, or Jews. African American readers (Hughes was black) may respond differently than white readers. Women may respond differently than men, and so on.

In this brief assessment of possible reader responses, we are trying to establish meaning from a shifting series of critical perspectives. Although we can establish a consensus of meaning over what Hughes probably intended, our own interpretation and evaluation of the text will be conditioned by our personal experiences, backgrounds, attitudes, biases, and beliefs. In other words, even as the class attempts to construct a common reading, each member of the class is also constructing a somewhat different meaning, one based on the individual's own interaction with the text.

PREPARING TO READ

This textbook contains many essays covering a variety of subjects by writers from a wealth of backgrounds and historical periods. You may be familiar with some, unfamiliar with others. All, however, have something to say and a way of saying it that others have found significant. Hence, many have stood the test of time, whether a year, a decade, or centuries. Essays are a recognized genre, or form of literature, and the finest essays have staying power. As Ezra Pound said, "Literature is news that stays news," and the best examples of the essay convey this sense of permanent value. Thus, you have an obligation to be an active and critical reader to do justice to the work that was put into these texts. Most were written with care, over extended periods of time, and by people who themselves studied the art of writing and the topics of their discourse. During your first week of class, you may wish to read some of their brief biographies to understand these authors' personal and educational backgrounds, their beliefs and credos, and some of the significant moments of their lives. You will often find that there are logical connections between the stories of their lives and the topics they have written about.

Sharpening your reading skills will be important because you may not be able to personally choose the essays from the text. You may find some topics and essays more interesting than others. But if you are prepared to read critically, you will be able to bring the same set of skills to any selection your instructor assigns. With this principle in mind, we present an overview of the active reading process, which will culminate in a case study using this process with an essay in this chapter—"The Cult of Ethnicity," by Arthur M. Schlesinger Jr.

When you are given a reading assignment from the textbook, a good strategy in preparing to read is to locate the text as well as possible within its broader context. Read the biographical notes about the author. Focus on the title of the essay. What can you infer from the title? What is the length of the essay? Although many students delight at the thought of reading shorter texts rather than longer ones, you may find that this variable is not always the deciding one in determining how easily you "get through" the essay. Short essays can be intricate and difficult; long ones can be more transparent and simpler. A long essay on a topic in which you are interested may be more rewarding than a short essay that you find lacking in relevance. Other basic prereading activities can include noting whether there are section breaks in the essay, whether there are subheadings, whether the author has used footnotes, and if so, how extensive they are. Other preliminary questions to answer could be, What is the date of the original publication of the essay? Is the essay a fully contained work or is it an excerpt from a larger text? Are there visual or mathematical aids, such as graphs, charts, diagrams, or lists? Because authors often use typographical signals to highlight things or to help organize what they have written, you might ask, Does the author use quotation marks to "signal" certain words? Is italic type used, and if so, what is its purpose? Are other books and authors cited in the essay? Does the author use organizational tools such as Arabic or Roman numerals? Once you have answered these questions regarding mechanics, you will be prepared to deal more substantively with the essay as a unit of meaning and communication.

Preparation for reading also means understanding that you bring your own knowledge, opinions, experiences, and attitudes to the text. You are not an empty glass to be filled with the knowledge and opinions of the authors, but rather a learner who can bring to bear your own reflections on what you read even if you think your knowledge is minimal. Often we do not know just how much ability we have in thinking about a topic until we actively respond to what others confront us with in their writings. By tackling the reading assignments in the text, you will not only learn new information and confront opinions that may challenge your own, but find that reading frees up your ability to express your own opinions. For this reason, most English teachers look upon reading as a two-way process: an exchange between writer and reader.

Although the credentials and experience of a professional writer may seem impressive, they should not deter you from considering your own critical talents as you read. But first, you must find a way to harness those abilities.

www.mhhe.com/
mhreader

For a student sample of preparing to read and an interactive previewing exercise, go to:
More Resources > Ch. 1 Reading & Writing

CRITICAL READING

It should be evident to you by now that you are not a mere recipient of information who passively accepts what the writer conveys. Instead, you should feel comfortable about engaging the author as you might a friend in a lively conversation or argument. And just as a talk with a friend involves active listening, rebuttal, use of facts, and logic, the interaction between yourself and the author needs to be a dynamic one as well. Active reading is so important in the learning process that one of America's most popular philosophers, Mortimer Adler, wrote an article that has become a classic on this topic. It is entitled "How to Mark a Book," and appears beginning on page 46.

Among the essential elements of your close reading are annotating, note taking, and questioning the text.

Annotating

Annotating refers to marking your text by making content notes, by using symbols such as question marks and exclamation points, and by recording personal reactions. Annotating is not, however, mere underlining or highlighting. These latter two methods often serve little purpose in helping you comprehend a text.

Most likely, when you return to passages you've marked with these simple procedures, you will have forgotten why you felt they were important in the first place. If you do underline or highlight, you should be sure to link your marking with a note in the margin. Simply drawing attention to someone else's words does little in the way of expanding your own thoughts on a topic. Learning is best accomplished by restating ideas in your own words.

Note Taking

Many essays in your anthology will require more than jotting down marginal notes in order to comprehend them fully or to respond to them in depth. Just as you might take down notes during a classroom lecture, you may find it useful to take notes to supplement your annotations. You may wish, for example, to write down quotations so that you can see them together. Or you may wish to summarize the essay by outlining its key points, a reversal of the process you would use to develop your own essay, wherein you begin with an outline and expand it into paragraphs. By collapsing an essay into an outline, you have a handy reference of the author's thesis (main idea) and supporting points, and the methods used to develop them. Another function of note taking is to overcome the simple habit most of us have of thinking we will remember things without jotting them down, only to find out later we cannot recall significant information from memory. You will appreciate the benefits of taking notes when you tackle lengthy essays, which may run 15 or 20 pages in length.

Questioning the Text

Posing key questions about a text and then answering them to the best of your ability is a helpful means of understanding more cogently an essay's substance and structure. Certain basic questions are salient for nearly any text you confront, and answering them for yourself can be a powerful means of enhancing your comprehension. As you read your text, such questions help you spot the significant issues that lie within most essays, regardless of their form or length. It is a good habit to have these questions in mind as you read, and then to return to them once you've thought through your reading. They serve as guideposts along the way of your reading experience and assist you in focusing on those issues that are most important to a text. When you become comfortable with them, you will probably find that your mind automatically poses them as you read, making your comprehension of difficult texts easier.

- What is the thesis or main point of the text?
- What methods does the author use to support these points, for instance, illustration, example, citing authorities, citing studies or statistics, description, personal experience, or history?
- What value position, if any, does the author present? In other words, is the author either directly or indirectly presenting her or his moral framework on an issue, or is she or he summarizing or describing an issue?

- Does the author use any special terms or expressions that need to be eluci-dated to understand the essay? You will find that authors, when address-ing innovative or revolutionary ideas within the context of their times, must use vocabulary that often needs to be defined. Take for example, the term *multiculturalism*. Exactly what does an author mean by that word?
- What is the level of discourse of the essay? Or what is the audience's level of educational attainment the author presumes?
- Who is the implied audience for the essay? Is it written for a specialized profession (such as scientists or educators); is it written for individuals with a focus on their particular role in society, for example, as parents or consumers or citizens?

The following essay, "The Cult of Ethnicity," by the influential historian Arthur M. Schlesinger Jr., has been annotated to demonstrate how a student might respond to it. Schlesinger's essay also will be used to explain aspects of the reading and writing process as we move through this section.

The history of the world has been in great part the history of the mixing of peoples. Modern communication and transport accel-erate mass migrations from one continent to another. Ethnic and racial diversity are more than ever a salient fact of the age.

But what happens when people of different origins, speak-ing different languages and professing different religions, inhabit the same locality and live under the same political sovereignty? Ethnic and racial conflict—far more than ideological conflict—is the explosive problem of our times.

This seems like the thesis. Where are his supports? Or is it the thesis?

On every side today ethnicity is breaking up nations. The Soviet Union, India, Yugoslavia, Ethiopia, are all in crisis. Ethnic tensions disturb and divide Sri Lanka, Burma, Indonesia, Iraq, Cyprus, Nigeria, Angola, Lebanon, Guyana, Trinidad—you name it. Even nations as stable and civilized as Britain and France, Belgium and Spain, face growing ethnic troubles. Is there any large multiethnic state that can be made to work?

Look these up. Demonstrates knowledge on the part of the author.

The answer to that question has been, until recently, the United States. "No other nation," Margaret Thatcher has said, "has so successfully combined people of different races and na-tions within a single culture." How have Americans succeeded in pulling off this almost unprecedented trick?

We have always been a multiethnic country. Hector St. John de Crevecoeur, who came from France in the 18th century, marveled at the astonishing diversity of the settlers—"a mixture of English, Scotch, Irish, French, Dutch, Germans and Swedes . . . this promiscuous breed." He propounded a famous question: "What then is the American, this new man?" And he gave a fa-mous answer: "Here individuals of all nations are melted into a new race of men." *E pluribus unum*.

Historical figure— who was he?

The United States escaped the divisiveness of a multiethnic society by a brilliant solution: the creation of a brand-new national identity. The point of America was not to preserve old cultures but to *forge a new, American culture. "By an intermixture with our people," President George Washington told Vice President John Adams, immigrants will "get assimilated to our customs, measures and laws: in a word, soon become one people." This was the ideal that a century later Israel Zangwill crystallized in the title of his popular 1908 play *The Melting Pot.* And no institution was more potent in molding Crevecoeur's "promiscuous breed" into Washington's "one people" than the American public school.

The new American nationality was inescapably English in language, ideas, and institutions. The pot did not melt everybody, not even all the white immigrants; deeply bred racism put black Americans, yellow Americans, red Americans and brown Americans well outside the pale. Still, the infusion of other stocks, even of nonwhite stocks, and the experience of the New World reconfigured the British legacy and made the United States, as we all know, a very different country from Britain.

In the 20th century, new immigration laws altered the composition of the American people, and a cult of ethnicity erupted both among non-Anglo whites and among nonwhite minorities. This had many healthy consequences. The American culture at last began to give shamefully overdue recognition to the achievements of groups subordinated and spurned during the high noon of Anglo dominance, and it began to acknowledge the great swirling world beyond Europe. Americans acquired a more complex and invigorating sense of their world—and of themselves.

But, pressed too far, the cult of ethnicity has unhealthy consequences. It gives rise, for example, to the conception of the United States as a nation composed not of individuals making their own choices but of inviolable ethnic and racial groups. It rejects the historic American goals of assimilation and integration.

And, in an excess of zeal, well-intentioned people seek to transform our system of education from a means of creating "one people" into a means of promoting, celebrating and perpetuating separate ethnic origins and identities. The balance is shifting from *unum* to *pluribus.*

That is the issue that lies behind the hullabaloo over "multiculturalism" and "political correctness," the attack on the "Eurocentric" curriculum and the rise of the notion that history and literature should be taught not as disciplines but as therapies whose function is to raise minority self-esteem. Group separatism crystallizes the differences, magnifies tensions, intensifies hostilities. Europe—the unique source of the liberating ideas of democracy, civil liberties and human rights—is portrayed as the root of all evil, and non-European cultures, their own many crimes deleted, are presented as the means of redemption.

I don't want to sound apocalyptic about these developments. Education is always in ferment and a good thing too. The situation in our universities, I am confident, will soon right

Margin notes:

Is this a partly American phenomenon? *prevents racial and ethnic conflict

Why?—doesn't explain

Note S's use of historical process analysis

Vocab.: infusion stocks zeal Eurocentric apocalyptic ferment Kleagle crucible

Signals a warning—danger

Is this thesis or related to thesis?

Support against multiculturalism

General—where are the specific examples?

Is this an exaggeration? How does he know?

Who are these people? He doesn't mention them specifically.

Reality is stronger than "ideology"? Is this his "solution"?

A sharp conclusion → argument? United States must be example. This is the thesis.

itself. But the impact of separatist pressures on our public schools is more troubling. If a (Kleagle) of the Ku Klux Klan wanted to use the schools to disable and handicap black Americans, he would hardly come up with anything more effective than the "Afrocentric" curriculum. And if separatist tendencies go unchecked, the result can only be the fragmentation, resegregation and tribalization of American life.

I remain optimistic. My impression is that the historic forces driving toward "one people" have not lost their power. The eruption of ethnicity is, I believe, a rather superficial enthusiasm stirred by romantic ideologues on the one hand and by unscrupulous con men on the other: self-appointed spokesmen whose claim to represent their minority groups is carelessly accepted by the media. Most American-born members of minority groups, white or nonwhite, see themselves primarily as Americans rather than primarily as members of one or another ethnic group. A notable indicator today is the rate of intermarriage across ethnic lines, across religious lines, even (increasingly) across racial lines. "We Americans," said Theodore Roosevelt, "are children of the (crucible)."

The growing diversity of the American population makes the quest for unifying ideals and a common culture all the more urgent. In a world savagely rent by ethnic and racial antagonisms, the United States must continue as an example of how a highly differentiated society holds itself together.

What has this annotating accomplished? It has allowed the reader/annotator to consider and think about what she has read, integrate her ideas with the ideas of the author, challenge those she may disagree with, raise issues for further study, find the seeds of ideas that may become the focus of an essay in response to the writing, review what she has read with more facility, and quickly and efficiently return to those parts of the essay she found the most salient.

The aforementioned strategies will assist you in responding intelligently in the classroom, remembering the main points of what you have read, and internalizing the critical reading skill so that it becomes automatic. However, such activities are not as challenging as the ultimate goal of most of your reading assignments, which will be to respond in formal writing to the works you've read. For this, you will need to enhance your study skills a bit further so that they will prepare you to write.

Formal writing assignments require you to demonstrate that you understood what you have read and are able to respond in an informed and intelligent manner to the material. They also require you to use appropriate form, organization, and exposition. Above all, regardless of what you want to express, you will have to communicate your ideas clearly and concisely. To this end, you will need to acquire skills that you can call on when it comes to writing at length about what you have read. To do so, you will find your ability to paraphrase, summarize, and quote directly from the original material particularly helpful.

When you move to this next phase, however, try to avoid a common practice among readers that causes them to waste time and effort put into study. Many students think they have completed a reading assignment when they read the last word of an essay. They utter a sigh of relief, look inside the refrigerator for something to eat, call up friends, or go Web browsing. However, as a critical reader, you need to spend additional time reinforcing what you have read by thinking about the author's views, considering her or his rhetorical methods, and reviewing or adding to your notes and annotations. For example, one culminating activity at this point can be to either mentally or verbally summarize what you have read. You can summarize verbally by enlisting a classmate and simply stating in your own terms the main points of your reading assignment. This oral summarizing will prevent a common problem many readers experience: the natural tendency to forget most of what they read shortly after reading.

BEYOND CONTENT: FOCUSING ON PROCESS

An essayist attempts to communicate a message to his or her audience. This message is the *content*. But "message making" is a process—the exchange of information through a shared system of verbal or visual symbols. Your goal in reading critically is to understand not just the informational content of a text but also how the writer shares meaning and typically tries to influence your beliefs and behavior. A good writer, to paraphrase Plato in *Phadreus*, tries to "enchant" your mind.

From Plato to the present, theorists have stressed this interactional aspect of reading and writing. Someone constructs a message (for our purposes, a written text), transmits it, and we have to receive it, decode it, and respond to it. Thus any "piece" of writing, whether designed to inform, persuade, or entertain, is the product of a complex process of actions and interactions by which we perceive, order, and verify (or make sense of) what we read. Whether we have the capacity to grasp the argument of a text, think logically about a thesis, or understand the cultural background of a writer and how it informs a text depends on how well we *perceive* the ways in which a writer creates meaning in a text.

Defined simply, *perception* is the process by which you create meaning for your world. As a process, it deals with the way you interpret the behavior of others as well as yourself. Thus, understanding perception helps to explain how we process information about self, others, and our world. Our sensory organs—seeing, hearing, touching, tasting, smelling—provide us initial contact with the outside world, enabling us to establish our perceptual field of reference. However, we also perceive what we want to perceive, which we call *psychological selectivity*. Finally, there is a third form of perception known as *cultural selectivity*: from a cultural perspective, we are conditioned by our culture's code of values and modes of understanding. For example, the phenomenon of *binocular rivalry* demonstrates that people of two different cultures exposed to two pictures at the same time will remember elements compatible with their own culture. With critical reading, you can have diverging interpretations of passages or an entire text because you perceive them from different perspectives.

In addition to differences of perception, you should also be mindful of how an author presents information. How an author presents her or his information is as important as what information the author presents. Strategies for writing may include the overall pattern of an essay—for example, is it an argument, an explanation, a definition, an evaluation, a comparison or contrast? While you may not think of essays in terms of genre, as you do literature (which may be presented in the form of poetry, the short story, the play, and so on), such forms can help you understand the motivation behind the writer's work and assist you as you seek out the more significant passages in a piece of writing. For example, if the essay is argumentative, you should focus on the supporting points the author has provided, determining whether they offer adequate support for the author's point of view. In an essay arguing for the return to traditional family values, for instance, the use of one anecdote to prove a point would probably not be enough to persuade most readers.

As you read an essay, you should also consider the author's *purpose* for writing. An essay about a personal experience would probably contain physical description; at the same time, the author's purpose would probably be to communicate an element in his or her life that can provide insight into personal development in general. Among the more common purposes are the following: to inform, to persuade, to disprove, to describe, to narrate, to demonstrate, to compare and contrast, to seek a solution to a problem, to explain a process, to classify, to define, to warn, and to summarize. While most essays contain a variety of purposes, one often will stand out among the others.

PARAPHRASING, SUMMARIZING, QUOTING

As you prepare to respond to the writing of others, you need to develop skills so that your own writing will reflect the hard work that went into the reading process. To this end, you can benefit from learning some shortcuts that will assist you in garnering information about what you have read. These skills include paraphrasing, summarizing, and quoting directly from another author's work.

Paraphrasing

Paraphrasing means taking what you have read and placing it in your own words. Students occasionally complain about this process, using the argument that it is a waste of time to paraphrase when the author's own words are the best way to articulate his or her ideas. However, paraphrasing serves two main purposes. The more obvious one is that it prevents you from plagiarizing, even inadvertently, what you have read. In terms of learning, however, it is particularly helpful because it requires that you digest what you have read and rewrite it. As you do so, you will develop writing patterns that over time will improve your ability to communicate. Paraphrasing forces you to truly think about what you have read and reinforces what you've read, since your mind has now been cognitively stimulated. You may find that paraphrasing often leads you to challenge the text or think more deeply about it simply because the paraphrasing process requires that you fully comprehend what you read.

It is important while paraphrasing to keep in all the essential information of the original while not using any of the author's original vocabulary or style. One rule of thumb is to never use three or more words that appeared together in the original. However, you can keep words such as articles (*a, an, the*) and conjunctions (*and, for, but,* etc.). The following are two examples of paraphrasing that demonstrate unsuccessful and successful application of the technique.

Original
But, pressed too far, the cult of ethnicity has unhealthy consequences. It gives rise, for example, to the conception of the United States as a nation composed not of individuals making their own choices but of inviolable ethnic and racial groups. It rejects the historic American goals of assimilation and integration.

Paraphrase 1
But, pressed too far, the focus on ethnicity has dangerous consequences. It suggests that the United States is a nation made up of separate ethnic and racial groups rather than individuals. It goes against the American ideals of integration and assimilation.

There are several things wrong with paraphrase 1. Rather than change key words, the writer has merely rearranged them. The sentence structure is very similar to that of the original, as is the ordering of ideas. If the student were to incorporate this paraphrase into her or his own essay, the teacher would probably consider it a form of plagiarism. It is simply too close to the original. To truly paraphrase, you must substitute vocabulary, rearrange sentence structure, and change the length and order of sentences. These strategies are more evident in paraphrase 2.

Paraphrase 2
Our country is made up of both individuals and groups. The recent trend to focus on the idea that one's ethnic background should have a major influence on one's perspective as a citizen goes against the moral foundations of the United States. It is the very concept of accepting American culture as one's own that has made our country strong and relatively free from cultural conflict.

Summarizing

A summary is a short, cohesive paragraph or paragraphs that are faithful to the structure and meaning of the original essay you've read, but developed in your own words and including only the most essential elements of the original. Summaries are particularly helpful when you are planning to write lengthy assignments or assignments that require that you compare two or more sources. Because a good summary requires that you use many of the skills of active reading, it helps you to "imprint" the rhetorical features and content of what you have read in your memory, and also provides you with a means of communicating the essence of an essay to another person or group. To summarize successfully, you need to develop the ability to know what to leave out as much as what to include. As you review your source, the annotations and notes you have made previously should help immensely. Since you want to deal with only the essentials of the original, you must delete all unimportant details and redundancies. Unlike paraphrasing, however, most summaries require that you stick to the general order of ideas as they are presented in a text. They also should not be mere retellings of what you have read, but should present the relationships among the ideas in an essay. It may be helpful to think of a summary as analogous to a news story, in which the essential details of what happened are presented in an orderly chronological fashion, because readers can best understand the gist of a story that way. It is simply the way the human mind, at least the Western mind, operates. Another strategy in summarizing is to imagine that the audience you are summarizing for has not read the original. This places a strong responsibility on you to communicate the essentials of the text accurately.

The following six steps should help you in preparing a summary. After you've reviewed them, read the summary that follows and consider whether it seems to have fulfilled these suggestions.

1. Read the entire source at least twice and annotate it at least once before writing.
2. Write an opening sentence that states the author's thesis.
3. Explain the author's main supporting ideas, reviewing your notes to make sure you have included all of them. Be careful not to plagiarize, and use quotations only where appropriate.
4. Restate important concepts, key terms, principles, and so on. Do not include your opinion or judge the essay in any way.
5. Present the ideas in the order in which they originally appeared. Note that in this way summarizing is different from paraphrasing, where staying too close to the original order of words may be detrimental to the process.
6. Review your summary once it has been completed. Consider whether someone who hasn't read the original would find your summary sufficient to understand the essence of the original work. You may also wish to have classmates or friends read the essay and ask them to furnish their verbal understanding of what you've written.

Now, review the following summary of Schlesinger's essay and determine whether it adheres to these points.

Sample Summary

Schlesinger argues that the recent surge of interest in ethnic separatism that is being touted by some whom he considers self-styled spokespersons for various ethnic groups threatens the unifying principle of our country's founders and undermines the strength of our society. This principle is that the American identity that was forged by its creators would be adopted by all peoples arriving here through a process of assimilation to our culture, values, and system of government so that cultural conflict could be avoided. Although he finds some merit in the idea that recognizing the contributions of certain groups who have been kept out of the national focus, for example, "nonwhite minorities," is a positive move, he fears that this can be taken to an extreme. The result could be the development of antagonism between ethnic groups solely on the basis of overemphasizing differences rather than recognizing similarities. He further argues that efforts to fragment American culture into subgroups can have the effect of jeopardizing their own empowerment, the opposite of the movement's intention. He gives the example of "Afrocentric" schooling, which he claims would only harm students enrolled in its curriculum. Despite this new interest in the "cult of ethnicity," the author is optimistic that it is of limited effect. He claims that most Americans still strive toward unity and identify themselves as Americans first, members of ethnic or racial groups second. He buttresses this belief by explaining that intermarriage is growing across racial, religious, and ethnic lines. This striving toward unity and identification with America among groups is particularly important today since their diversity is continuously increasing.

Quoting

Sayings and adages are extremely popular. You find them quoted in everyday speech, printed in calendars, rendered in calligraphy and framed and hung in homes, and spoken by public figures. These are, in effect, direct quotes, although the authors may be anonymous. Direct quotations often have a unique power because they capture the essence of an idea accurately and briefly. Another reason is that they are stylistically powerful. You may find in an essay a sentence or group of sentences that are worded so elegantly that you feel you simply wish to savor them for yourself or plan to use them appropriately for a future writing assignment. Other times, you may wish to use direct quotations to demonstrate to a reader the effectiveness of an original essay or the authoritative voice of the author. And at still other times, it may simply be necessary to quote an author because her or his vocabulary just cannot be changed without injuring the meaning of the original. Review the following quotations taken from the Schlesinger essay, and consider how paraphrasing them would injure their rhetorical power.

Direct Quotations That Reflect the Conciseness of the Original

"The history of the world has been in great part the history of the mixing of peoples."

"On every side today ethnicity is breaking up nations."

"And if separatist tendencies go unchecked, the result can only be the fragmentation, resegregation and tribalization of American life."

Direct Quotations That Have Particular Stylistic Strength

"The pot did not melt everybody."

"The balance is shifting from *unum* to *pluribus*."

Direct Quotations That Establish the Writer's Authority

"The point of America was not to preserve old cultures but to forge a new, American culture. 'By an intermixture with our people,' President George Washington told Vice President John Adams, immigrants will 'get assimilated to our customs, measures and laws: in a word, soon become one people.'"

"A notable indicator today is the rate of intermarriage across ethnic lines, across religious lines, even (increasingly) across racial lines."

Direct Quotation That Demonstrates Conceptual Power

"The eruption of ethnicity, is, I believe, a rather superficial enthusiasm stirred by romantic ideologues on the one hand and by unscrupulous con men on the other."

Avoiding Plagiarism

When you employ summary, paraphrase, and quotation in an essay or a re-search paper, you must avoid *plagiarism*—the attempt to pass off the work of others as your own. The temptation to plagiarize is one of the oldest "crimes" in academe but also an unfortunate by-product of the computer revolution, for there are numerous opportunities for harried, enterprising, or—let's face it—dishonest students to download bits of information or entire texts and appropri-ate them without acknowledgment. At the same time, you should be aware that there are numerous Web sites and software programs that allow your instruc-tors to locate even the most inventive forms of plagiarism—right down to words and phrases—and that when writing research papers, you may be re-quired to attach all downloaded materials. Be warned: College teachers treat plagiarism as academic treason. You can fail a course if you plagiarize, be sus-pended from college, and even expelled.

We will treat plagiarism in greater detail in Chapter 3, which presents infor-mation on writing research papers, but for now you can avoid plagiarism by following these three basic rules:

- Cite (provide a reference for) all quoted, summarized, or paraphrased in-formation in your paper, unless that information is commonly understood. (For example, you would not have to cite the information that two planes flew into the World Trade Center on September 11, 2001, because it is com-mon knowledge.)
- Cite all special phrases or unique stylistic expressions that you derive from another writer's work. You might love a phrase by one of the famous writers in this book—let's say E. B. White or Virginia Woolf—but that writer invented it, it belongs to him or her, and you cannot employ it with-out acknowledging the source.
- Work hard to summarize and paraphrase material in your own words. Constantly check your language and sentence structure against the

language and syntax in the source that you are using. If your words and sentences are too close to the original, change them.

Finally, it is perfectly legitimate to ask your instructor or a tutor in your college's writing center to look at your draft and render a verdict on any information you have summarized, paraphrased, or quoted. Whether this material has been taken intentionally or unintentionally from another source is immaterial. It is your responsibility to present honest written work.

| www.mhhe.com/ **mhreader** | For an interactive tutorial on avoiding plagiarism, go to: **Research > Avoiding Plagiarism** |

READING AND ANALYZING VISUAL TEXTS

In this new era of information technology, we seem to be immersed in a visual culture requiring us to contend with and think critically about the constant flow of images we encounter. From advertising to film to video to the Internet, we must respond with increasing frequency not only to written but also to visual messages—images that typically are reinforced by verbal elements. Consequently, it is important to perceive the powerful linkages that exist in today's culture between visual and verbal experience.

Frequently in courses in engineering, social science, computer science, the humanities, fine arts, and elsewhere, you have to analyze and understand visual elements that are embedded in texts. Textbooks increasingly promote visuals as frames of reference that help readers to comprehend and appreciate information. Some visual elements—charts, tables, and graphs—are integral to an understanding of verbal texts. Other visuals—comic art, drawings, photographs, paintings, advertisements—offer contexts and occasions for enjoyment and deeper understanding of the reading, writing, and thinking processes. Visual images convey messages that often are as powerful as well-composed written texts. When they appear together, image and word are like French doors, both opening to reveal a world of heightened perception and understanding.

When visual elements stand alone, as in painting and photography, they often make profound statements about human experience and frequently reflect certain persuasive purposes that are composed as skillfully as an argumentative essay. Consider, for example, the series that the great Spanish artist Francisco Goya painted, "The Disasters of War," a powerful statement of humankind's penchant for the most grotesque and violent cruelties. In the late 20th century, photographers of the Vietnam War, using a modern visual medium, similarly captured the pain and suffering of armed conflict, as in Eddie Adams's potent stills of the execution of a prisoner by the notorious chief of the Saigon national police, General Nguyen Ngoc Loan. In the framed sequence, the chief of police aims his pistol at the head of the prisoner, presses the trigger, and the viewer, in that captured instant, sees the jolt of the prisoner's head and a sudden spurt of blood. Reproduced widely in the American press in February 1968, this single image did as much as any written editorial to transform the national debate over the Vietnam War. (Both images are reprinted on pages 132–133.)

Although paintings, photographs, advertisements, and other artistic and design forms that rely heavily on visual elements often function as instruments of persuasion, it would be simplistic and simply wrong to embrace uncritically the cliché "A picture is worth a thousand words." For instance, great literary artists from Homer to the present have captured the horrors of war as vividly as artists in other media. Stephen Crane in *The Red Badge of Courage* illustrates the sordidness of America's Civil War in language as graphic as the images of the war's most noted photographer, Mathew Brady. Consider the visual impact of Crane's depiction of battlefield dead:

> The corpse was dressed in a uniform that once had been blue but was now faded to a melancholy shade of green. The eyes, staring at the youth, had changed to the dull hue to be seen on the side of a dead fish. The mouth was opened. Its red had changed to an appalling yellow. Over the grey skin of the face ran little ants. One was trundling some sort of a bundle along the upper lip.

Ultimately the best verbal and visual texts construct meaning in vivid and memorable ways. When used in combination, visual and verbal texts can mix words and images to create uniquely powerful theses and arguments.

Just as you analyze or take apart a verbal text during the process of critical reading, you also have to think critically about visual images or elements. If you encounter charts, graphs, and tables in a text, you have to understand the information these visuals present, the implications of the numbers or statistics, the emphases and highlights that are conveyed, and the way the visual element—the picture, so to speak—shapes your understanding of the material and its relationship to the text. Sometimes the material presented in such visuals is technical, requiring you to carefully analyze, let's say, a bar graph: its structure, the relationship of parts to the whole, the assertions that are advanced, and the validity of the evidence conveyed. In short, critical reading of visual material is as demanding as critical reading of the printed word. Just as you often have to reread a verbal text, you also might have to return to charts, graphs, and tables, perhaps from a fresh perspective, in order to comprehend the content of the visual text.

The following questions can guide your critical analysis of such visual texts as charts, graphs, and tables:

- What is the design or structure of the visual?
- What information do you immediately notice?
- What is the purpose of the visual?
- What thesis or point of view does the information in the visual suggest?
- What is the nature of the evidence and how can it be verified?
- What emphases and relationships do you detect among the visual details?
- How does the visual fit into the context of the verbal text surrounding it?

When responding to charts, tables, and graphs, you must develop the confidence to read such visual texts accurately and critically, taking nothing for granted and trusting your ability to sift through the evidence and the images with a critical eye in order to understand the strategies the author or graphic artist has employed to convey a specific message to the reader.

By and large, informative visuals such as tables and graphs rarely have the striking impact of the sort of graphics found in the best commercial and political advertising or in the illustrations we encounter in slick magazines or cutting-edge cartoon strips. The visual elements used by advertisers, for example, take advantage of our innate capacity to be affected by symbols—McDonald's Golden Arch, the president framed by American flags, a bottle of Coca-Cola beneath the word *America*. Such visual emblems convey unspoken ideas and have enormous power to promote products, personalities, and ideas. For example, the two powerful images on pages 20–21 convey important ideas about the cultures that produced them. Visual symbols achieve even more intense effects when they are reinforced by verbal elements.

When viewing art reproductions, photographs, advertisements, and cartoons from a critical perspective, you often have to detect the explicit and implicit messages being conveyed by certain images and symbols, and the design strategies that condition your response. Because these visuals combine many different elements, you have to consider all critical details: color, light, and shadow; the number and arrangement of objects and the relationships among them; the foregrounding and backgrounding of images within the frame; the impact of typography; the impact of language if it is employed; and the inferences and values that you draw from the overall composition. Learn to treat visuals in any medium as texts that need to be "read" critically. Every visual requires its own form of annotation, in which you analyze the selection and ordering of its parts and interpret the emotional effects and significant ideas and messages it presents. Throughout this text, paired "classic and contemporary" images such as the two on pages 20–21 give you opportunities to read visual texts with a critical eye.

Classic and Contemporary Images
HOW DO WE COMMUNICATE?

Using a Critical Perspective Carefully examine these two illustrations. What is your overall impression of these images? What details and objects in each scene capture your attention? What similarities and differences do you detect? How does each image communicate ideas and values about the culture that has produced it? Does one appeal to you more than the other? Why or why not?

Pulitzer Prize–winning combat photographer Joe Rosenthal captured this scene of U.S. Marines raising the American flag on the Pacific Island of Iwo Jima on February 25, 1945. The campaign to capture the island from Japanese troops cost nearly 7,000 American lives. Rosenthal's photo has been reproduced widely in the media and served as the model for the Marine Corps War Memorial in Washington, D.C.

Photographer Thomas E. Franklin captured a memorable moment
in the wake of the terror attack on the World Trade Center
in New York City in September 2001.

THE WRITING PROCESS

Whether you have been provided an assignment by your instructor or developed your own topic, the various tools for critical reading and analysis that you have mastered should now equip you with the foundation for what is necessary to embark on your own writing assignment. Essays are normally a three-part writing process. The three stages are termed prewriting, drafting, and revising. To illuminate the writing process, we will examine strategies employed by several student and professional writers, including one student, Jamie Taylor, as she read and responded to Schlesinger. But first we require an overview of the writing process, starting with the origins and development of a writer's ideas.

Annie Dillard, one of today's preeminent essayists, stresses the primacy of the creative imagination in the writing process in the selection starting on page 58. Dillard uses the central metaphor of building a house to describe the act of writing, but within her essay one can detect the three stages of the writing process: prewriting ("The line of words is a miner's pick, a woodcarver's gouge, a surgeon's probe. You wield it, and it digs a path you follow"), drafting ("You lay down the words carefully, watching all the angles"), and revising ("The part you must jettison is not only the best-written part; it is also, oddly, the part which was to have been the very point").

Think of the process of writing as a craft involving the planning, transcribing, polishing, and production of a text for an audience. In Old English, the word *craft* signifies strength and power. By treating writing as a craft, you empower yourself to make the most complex compositional tasks manageable. By thinking of the writing process habit of mind involving prewriting, writing, and revision, you can create effective essays and documents.

Prewriting

Prewriting, which you have already been engaged in as you have negotiated the reading-writing connection, is the discovery, exploration, and planning stage of the composing process. It is the stage in which you discover a reason to write, select and narrow a subject, consider audience and purpose, and engage in preliminary writing activities designed to generate textual material. During the prewriting process, you are free to let ideas incubate, to let thoughts and writing strategies ripen. You are free also to get in the mood to write. Ernest Hemingway used to sharpen all his pencils as preparation for a day's writing; the French philosopher Voltaire soaked his feet in cold water to get the creative juices flowing. Professional writers understand the importance of prewriting activities in the composing process, but college writers often undervalue or ignore them completely.

Purpose and Audience   Any writing situation requires you to make choices and decisions about purpose, audience, planning, writing, revision, and transmission of your text. Determining your purpose or goal—the reason why you are writing—at the outset of the composing process is one of the first steps. It prevents you from expending useless energy on thinking that is ultimately unimportant, misdirected, or unrelated to the problem because it forces you to ask, What do I hope to obtain from this text? With a specific purpose in mind, you start to anticipate the type of composing task ahead of you and to identify the problems that might be inherent in this task.

Traditionally, the main forms of writing—narration, description, exposition, and argumentation—help to guide or mold your purpose.

Form	Purpose	Example
Narration	To relate a sequence of events	To tell about an accident
Description	To provide a picture or produce an impression	To describe a moth
Exposition	To explain, inform, analyze	To compare two teachers
Argumentation	To convince or persuade	To oppose abortion

Most writing actually combines more than one of these rhetorical modes or forms, but these basic categories help shape your text to a specific purpose.

Even as you determine your purpose, you must also create common ground between yourself and your audience. In fact, to define your audience is to define part of your problem. Think of your audience as the readers of your text. What do they know about the topic? How do they perceive you—your status, expertise, credibility?

What do you know about their opinions and backgrounds? Are they likely to agree or disagree with you? (This last question is especially important in argumentative writing, which will be discussed in the next chapter.) By defining your audience carefully, you can begin to tailor your text. Only by analyzing your readership will you be able to appeal to an audience.

Freewriting and Brainstorming   Two methods of getting in touch with what you already know or believe are freewriting and brainstorming. Freewriting is quite simple. Merely select a predetermined amount of time, say anywhere from 5 to 15 minutes, and write down everything that you can think of regarding the subject at hand. Don't worry about punctuation or grammar. This activity is mainly to get your cognitive wheels rolling. Brainstorming is a variant of freewriting in which you jot down ideas and questions, often in numbered form. If you find freewriting and brainstorming helpful as techniques, you will probably find the length of time that suits you best. When you have finished, review what you have written. A well-known composition expert, Peter Elbow, explains the value of freewriting in the selection starting on page 61. The freewriting that Elbow describes can help any writer generate ideas, but freewriting and brainstorming can also help writers respond to others' ideas. For example, examine the following freewriting and brainstorming exercises by a student, James Moore, which he wrote after reading an excerpt from Schlesinger's "The Cult of Ethnicity."

Freewriting Sample

This essay shows that the author really knows his history because he cites so many historical figures, places, and can quote word for word authorities that back up his argument. He makes a great argument that America's strength is in its diversity and at the same time its unity. I never thought of these two things as being able to complement one another. I always thought of them as being separate. It opens my mind to a whole new way of thinking. One thing that would have strengthened his argument, though is the fact that although he criticizes people who want to separate themselves into subgroups, he doesn't really mention them by name. He's great when it comes to advancing his own argument but he seems to be a bit too general when he comes to attacking the opposition. I would have liked it if he had mentioned by name people who are undermining America's strengths and listen in their own words.

Brainstorming Sample

1. The author says that ideological conflict isn't such a big problem, but what about the gap between rich and poor? Maybe if there were less of a gap, people wouldn't look for "false idols."
2. Schlesinger seems to be part of the white mainstream. Does this mean he is destined not to understand fully the reasons why people on the margins of society get so tempted to join "cults"?
3. He uses supporting points very well but doesn't exactly explain why "multiculturalism" and "political correctness" are happening now in our society. What is it about today that has opened the door to these ideas?
4. There are so many references to places with ethnic tensions around the world. It would be great to study one of them and see if they have any similarities to the ones that exist in the United States.
5. He seems to be writing for a very educated audience. I wonder really if he can reach the "common person" with this kind of sophisticated writing. I don't know about most of the places he mentions.
6. What's the solution? That could be the start of a topic for my paper. I don't think the author offers any.

Let's consider the benefits these processes can have. First, you can comment on the subject matter of the essay without censoring your thoughts. This prepares you for the second reading by marshaling a more coherent idea of your own perspective. Freewriting or brainstorming can be a tool that helps you understand how you can have something to contribute in the writer-reader "conversation" or helps you see a topic in a new way. For example, in the freewriting example, the student discovered for himself the idea that the strength of American society is a combination of commonality and diversity. Second, during the brainstorming process, you might come up with a potential idea for a response essay, as the student did in the example.

Now let us return to the prewriting process that Jamie Taylor followed.

Brainstorming Notes

1. Schlesinger seems to be saying that multiculturalism poses a danger because it threatens to create ethnic divisiveness rather than healthy identification.

2. This not only undermines us now, but threatens the very democratic principles upon which the United States was founded.
3. He says America must set an example for the rest of the world, which is torn with racial and ethnic strife.
4. He believes that there is a small group of individuals with a "hidden agenda" who are trying to create this divisiveness. These individuals are self-centered and have their own interests at heart, not the interests of the people they represent.
5. One flaw in the essay was that it seemed vague. He didn't mention any names or give specific examples. Only generalities.
6. He suggests the "battle" will be won by ordinary citizens; for example, he cites the many intermarriages occurring today.
7. Although he sees danger, he is optimistic because he thinks democracy is a strong institution.
8. He writes from a position of authority. He cites many historical figures and seems very well read.
9. The major problem I see in his essay is that he seems to lump everyone together in the same boat. He doesn't give enough credit to the average person to see through the hollowness of false idols. You don't need a Ph.D. to see the silliness of so many ideas floating around out there.
10. So many things to consider, how should I focus my essay??? What should be my theme??
11. Hmmm. Idea!!! Since I agree with his basic points, but find he doesn't provide specifics, and doesn't give the average person enough credit to see through the emptiness of cult rhetoric, why not use my personal observations to write a response paper in which I show just how reasonable we are in distinguishing mere rhetoric from substance?

Outlining   In addition to this brainstorming, Ms. Taylor also developed a scratch outline—yet another prewriting strategy—to guide her into the drafting stage of the composing process.

Outline
I. Introduction: Summarize essay and thesis; provide counterthesis.
II. University life as a demonstration of "ethnic" democracies.
III. The emptiness and false promises of self-styled ethnic leaders.
IV. The rejection of "home-grown" cults.
V. Conclusion.

Although Jamie Taylor employed brainstorming and a scratch outline to organize her thoughts prior to writing her essay, not everyone uses these prewriting activities. Some students need to go through a series of prewriting activities, while others can dive into a first draft. Nevertheless, discovering the materials and form for an essay includes a search for ideas, a willingness to discard ideas and strategies that don't work, an ability to look at old ideas in a fresh way, and a talent for moving back and forth across a range of composing activities. Rarely does that flash of insight or first draft produce the ideal flow of words resulting in a well-written and well-ordered essay.

Professional writers have their own unique approaches to the composing process. For example, Annie Dillard is a prolific keeper of journals, from which she extracts ideas for essays and books. She also jots down notes, often in rough outline form. Here are some notes she jotted down, based on journal entries and her essay "Death of a Moth."

Moth in candle:
the poet—materials of world, of bare earth at feet, sucked up, transformed,
    subsumed to spirit, to air, to light
the mystic—not through reason, but through emptiness
the martyr—virgin, sacrifice, death with meaning.

Her "moth essay," as she calls it, evolved from journal entries, doodles, and several drafts, and then fit into a much larger book that she was writing.

 www.mhhe.com/ **mhreader** | For more information on prewriting strategies, go to: **Writing > Prewriting**

Drafting

Everyone approaches the entire composing process differently. There are, however, certain basic principles for the drafting stage that you must consider. These principles are discussed in the following sections.

Developing the Thesis   Every essay requires a main idea or thesis that holds all your information together. What you seek is not just any idea relevant to the bulk of your topic, but the underlying idea that best expresses your purpose in writing the essay. Your thesis is the controlling idea for the entire essay.

The thesis requires you to take a stand on your topic. It is your reason for wanting to inform or persuade an audience. The noted teacher and scholar Sheridan Baker has expressed nicely this need to take a stand or assume an angle of interpretation: "When you have something to say about *cats*, you have found your underlying idea. You have something to fight about: not just 'Cats,' but 'The cat is really man's best friend.'" Not all thesis statements involve arguments or fights. Nevertheless, you cannot have a thesis unless you have something to demonstrate or prove.

The thesis statement, which normally appears as a single sentence near the beginning of your essay, serves five important functions:

- It introduces the topic to the reader.
- It limits the topic to a single idea.

- It expresses your approach to the topic—the opinion, attitude, or outlook that creates your special angle of interpretation for the topic.
- It may provide the reader with hints about the way the essay will develop.
- It should arouse the reader's interest by revealing your originality and your honest commitment to the topic.

Here is a typical thesis statement by a student:

The automobile—America's metallic monster—takes up important public space, pollutes the environment, and makes people lazy, rude, and overweight.

In this thesis, the writer has staked out a position, limited the topic, and given the reader some idea of how the essay actually will develop.

Your thesis cannot always be captured in a single sentence. Indeed, professional writers often offer an implied or unstated thesis or articulate a thesis statement that permeates an entire paragraph. Basically, you should ask if a thesis hooks you. Do you find it provocative? Do you know where the author is coming from? Does the author offer a map for the entire essay? These are some of the issues that you should consider as you compose your own thesis sentences.

✦	www.mhhe.com/ **mhreader**	For more information on developing a thesis, go to: **Writing > Thesis**

Writing Introductory Paragraphs   Your introduction should be like a door opening into the world of your essay. A good introduction entices readers into this world by arousing their curiosity about the topic and thesis with carefully chosen material and through a variety of techniques. The introduction, normally a single, short paragraph composed of a few sentences, serves several important functions:

- It introduces the topic.
- It states the writer's attitude toward the subject, normally in the form of a thesis statement.
- It offers readers a guide to the essay.
- It draws readers into the topic through a variety of techniques.

A solid introduction informs, orients, interests, and engages the audience. "Beginnings," wrote the English novelist George Eliot, "are always troublesome." Getting the introduction just right takes effort, considerable powers of invention, and often several revisions. Fortunately, there are special strategies that make effective introductions possible:

- Use a subject-clarification-thesis format. Present the essay's general subject, clarify and explain the topic briefly, and then present your attitude toward the topic in a thesis statement.

- Offer a brief story or incident that sets the stage for your topic and frames your thesis.
- Start with a shocking, controversial, or intriguing opinion.
- Begin with a comparison or contrast.
- Use a quotation or reference to clarify and illustrate your topic and thesis.
- Ask a question or series of questions directed toward establishing your thesis.
- Offer several relevant examples to support your thesis.
- Begin with a vivid description that supports your main idea.
- Cite a statistic or provide data.
- Correct a false assumption.

All these strategies should introduce your topic and state the thesis of the essay. They should be relatively brief and should direct the reader into the body of the essay. Finally, they should reveal your perspective and your tone or voice. In each introductory paragraph, the reader—your audience—should sense that you are prepared to address your topic in an honest and revealing manner.

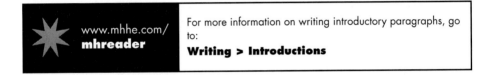

www.mhhe.com/
mhreader

For more information on writing introductory paragraphs, go to:
Writing > Introductions

Writing Body Paragraphs   The body is the middle of the essay. Usually the body consists of a series of paragraphs whose purpose is to satisfy your readers' expectations about the topic and thesis you presented in the introduction. The body of an essay gives substance, stability, and balance to your thesis. It offers facts, details, explanations, and claims supporting your main idea.

Body paragraphs reflect your ability to think critically, logically, and carefully about your topic. They are self-conscious units of expression whose indentations signal a new main point (or topic sentence) or unified and coherent unit of thought. The contour created by the series of body paragraphs that you design grows organically from the rhetorical or composing strategies that you select. As the English critic Herbert Read states in *English Prose Style*, "As thought takes shape in the mind, it takes *a* shape. . . . There is about good writing a visual actuality. It exactly reproduces what we should metaphorically call the contour of our thought. . . . The paragraph is the perception of this contour or shape." In other words, we see in the shape of an essay the shape of our thoughts. The contour created by the series of body paragraphs proceeds naturally from the material you include and the main point that you use to frame this material in each paragraph.

Effective paragraph development depends on your ability to create a unit of thought that is *unified* and *coherent,* and that presents ideas that flesh out the topic sentence or controlling idea for the paragraph, thereby informing or convincing the reader. To achieve a sense of completeness as you develop body paragraphs, be sure to have enough topic sentences and sufficient examples or evidence for each key idea. College writers often have problems writing complete essays with adequately developed body paragraphs. Remember that topic sentences are relatively general ideas. Your primary task is to make readers understand what those ideas mean or why they are important. Your secondary task is to keep readers interested in those central thoughts. The only way to accomplish these two related goals is by explaining the central ideas through various kinds of evidence or support.

Strategies for Development   Different topics and paragraphs lend themselves to different types of development. These types of rhetorical approaches are essentially special writing and reasoning strategies designed to support your critical evaluation of a topic or hypothesis. Among the major rhetorical approaches are description, narration, illustration, process analysis, comparison and contrast, causal analysis, definition, classification, and argumentation. Each strategy might very well serve as your dominant approach to a topic. On the other hand, your essay might reflect a variety of methods. Remember, however, that any blending of rhetorical strategies should not be a random sampling of approaches but should all contribute to your overall point.

Description   Good descriptive writing is often your best tool for explaining your observations about objects, people, scenes, and events. Simply, description is the creation of a picture using words. It is the translation of what the writer sees into what the writer wants the reader to imagine. Description has many applications in academic courses. For example, for a psychology course, you might need to describe the behavior of an autistic child. At an archeological dig or site, you might need to indicate accurately how a section of the excavated area looks. In a botany course, you might need to describe in detail a particular plant.

Effective description depends on several characteristics:

- It conveys ideas through images that appeal to our various senses: sight, hearing, touch, smell, and taste.
- It selects and organizes details carefully in a clearly identifiable spatial ordering—left to right, top to bottom, near to far, and so forth.
- It creates a dominant impression, a special mood or feeling.
- It is objective or subjective depending on the writer's purpose, the demands of an assignment, or the expectations of an audience.

In the following paragraph from her book *Spanish Harlem*, Patricia Cayo Sexton captures the sights, sounds, and rhythms of life in New York's East Harlem:

> Later, when the children return from school, the sidewalks and streets will jump with activity. Clusters of men, sitting on orange crates on the sidewalks, will play checkers or cards. The women will sit on the stoop, arms folded, and

watch the young at play; and the young men, flexing their muscles, will look for some adventure. Vendors, ringing their bells, will hawk hot dogs, orange drinks, ice cream; and the caressing but often jarring noise of honking horns, music, children's games, and casual quarrels, whistles, singing, will go on late into the night. When you are in it you don't notice the noise, but when you stand away and listen to a taped conversation, the sound suddenly appears as a background roar. This loud stimulation of the senses may produce some of the emotionalism of the poor.

Narration   Telling stories—or narration—is a basic pattern of organizing your thoughts. You employ narration on a daily basis—to tell what happened at work, in the cafeteria, or on Saturday night. Narration is also essential to many forms of academic writing ranging from history, to sociology, to science. When planning and writing narration, keep in mind the following guidelines:

- Present the events of your narration in a logical and coherent order. Make certain that you link events through the use of appropriate transitional words.
- Select the narrative details carefully in order to suit the purpose of the essay. Narrate only those aspects of the event that serve to illustrate and support your thesis.
- Choose a point of view and perspective suitable for your topic and audience. Narrative point of view may be either first or third person. A first-person narrative is suitable for stories about yourself. A third-person narrative (*he, she, it, they*) conveys stories about others. The narrative perspective you use depends on your audience and purpose. Obviously, you would use a different perspective and tone in narrating a laboratory experiment than you would narrating a soccer match you participated in.
- Dialogue, if appropriate to your topic, may add realism and interest to your narrative.
- Limit the scope of the event you are narrating and bring it to a suitable conclusion or climax.

When narration is used for informational or expository purposes, the story makes a point, illustrates a principle, or explains something. In other words, in expository narration, the event tends to serve as evidence in support of your thesis.

Here is a sample student paragraph based on narration:

Like most little girls I thought it would be very grown up to get my hair done in a beauty parlor instead of by my mother or older sister. For more than a month I cried and badgered my family. Finally, after hearing enough of my whining, my mother gave in and made an appointment for me. At the beauty parlor, I sat with my mother and a few older women, naively waiting for my transformation into another Shirley Temple. Finally the hairdresser placed me in a chair and began to chop a mass of hair onto the floor and then subject me to a burning sensation as rollers wound my remaining hair tight. The result was a classic example of the overworked permanent. At home later that day, I tried washing and rewashing my hair to remove the tangled mess. It took a week until I would see anyone without a scarf or hat over my head and a month before I could look at someone without feeling that they were making fun of me the minute I turned my back. In a way I feel that such a fruitless journey to the hairdresser actually helped me along the road to adulthood since it was a perfect example of a disappointment that only time and patience, rather than tantrums and senseless worrying, can overcome.

Narration answers the question, What happened? It can be used to tell real or fictional stories, to relate historical events, to present personal experience, to support an analysis of events. It has broad utility in college as a critical writing skill.

Illustration   To make your paragraph or essay complete—without padding, repetition, or digression—be sure to have sufficient examples or illustrations to support key ideas. Different topics and paragraphs lend themselves to different types of examples or supporting evidence. Here are some types of illustration that will help you write well-developed paragraphs and essays:

Fact: The Supreme Court ordered the desegregation of public schools in 1964.

Statistic: A majority of schools in San Diego that were once 90 percent black are now almost 45 percent white.

Example: One example of the success of San Diego's integration effort is its magnet schools.

Personal Experience: I attended the new computer science magnet school from 1996 to 1998. . . .

Quotation: According to the *Phi Delta Kappan,* "On the first day of Los Angeles' mandatory desegregation program, 17,700 out of the total of 40,000 were not on the bus."

Process: With the magnet concept, a school first creates a special theme and emphasis for its curriculum. Then, it . . . .

Comparison and Contrast: By contrast, when Los Angeles announced its forced busing plan, an estimated 15.1 percent of the white population moved out of the system into private schools.

Case Study: Jamie, an eighth-grader, had seen very few black students at the Math-Science Center prior to the implementation of San Diego's desegregation plan. . . .

Illustrations develop your paragraph beyond the topic sentence. Such illustrations or examples may be short or extended. However, to make sure that your paragraphs are complete and properly developed, watch out for weak or poorly presented illustrations. For every main idea or topic sentence in a paragraph, use specific supporting evidence that sufficiently proves or amplifies your point. If you do not have the right evidence in the proper amount, your paragraph and essay will be underdeveloped, as in the following case:

> The concept of choice does seem to appeal to students. On the first day of San Diego's new plan, the only people who were absent from the programs who had volunteered were those who were sick.

This two-sentence paragraph has promise but does not follow through with the main idea adequately. The concept at the heart of the topic sentence is clearer and more complete in the revised version:

> The concept of choice does seem to appeal to San Diego's parents and students. On the first day of San Diego's new plan, the only people who were absent from the program who had volunteered were those who were sick. In contrast, on the first day of Los Angeles's mandatory desegregation program, 17,700 out of the total of 40,000 were not on the bus, according to the *Phi Delta Kappan*. Moreover, when Los Angeles announced its busing plan, an estimated 15.1 percent of the white population moved out of the district or into a private school. In San Diego, there was virtually no "white flight."

In the revision, the student chose to use contrasting evidence, highly specific in nature, to provide adequate support for the topic sentence. Other details and illustrative strategies might have been selected. In selecting illustrative material, you should always ask: Are there other examples that are more lively, specific, concrete, revealing, or interesting? It is not enough to just present examples. Illustration should be as effective as possible.

Process Analysis   When you describe how something works, how something is assembled, how something is done, or how something happens, you are explaining or analyzing a process. The complexity of your explanation will depend on how complex the process itself is, how detailed you want your explanation to be, and what you want your audience to be able to do or understand as a result of reading your explanation. Are you providing relatively simple how-to-do-it instructions for a relatively simple task, or are you attempting to explain a complicated laboratory experiment or computer program? The explanation of a process can make demands on your analytical and problem-solving abilities because you have to break down operations into component parts and actions. Process analysis always involves the systematic presentation of step-by-step or stage-by-stage procedures. You must show *how* the steps or parts in a process lead to its completion or resolution.

The explanation of processes is relevant to many college courses. Such topics as the stages of economic growth, Hobbes's view of the evolution of the state, the origins of the city, the development of the English lyric, the phenomenon of photosynthesis, and the history of abstract art could benefit from process analysis. Often process analysis can be combined with other writing strategies or even be subordinated to a more dominant writing strategy like narration, to which it bears a certain resemblance.

As with all other forms of mature and effective writing, you must assess your audience when writing process papers. You must decide whether you primarily want to inform or to give directions. When you give directions, you normally can assume that your audience wants to learn to do what you tell them about. If your primary purpose is to inform, you must assess the degree of interest of general readers and approach your subject from an objective perspective. Remember that there are natural, physical, mechanical, technical, mental, and historical types of processes. Certain topics might cut across these types, yet in each instance, your purpose is to direct the reader in how to do something or to inform the reader about the nature of the process.

Your analysis of a process can occur at paragraph level or it can control the development of an entire essay. Note how Laurence J. Peter, author of the famous book *The Peter Prescription,* uses process to structure the following paragraph:

> If you are inexperienced in relaxation techniques, begin by sitting in a comfortable chair with your feet on the floor and your hands resting easily in your lap. Close your eyes and breathe evenly, deeply, and gently. As you exhale each breath let your body become more relaxed. Starting with one hand direct your attention to one part of your body at a time. Close your fist and tighten the muscles of your forearm. Feel the sensation of tension in your muscles. Relax your hand and let your forearm and hand become completely limp. Direct all your attention to the sensation of relaxation as you continue to let all tension leave your hand and arm. Continue this practice once or several times each day, relaxing your other hand and arm, your legs, back, abdomen, chest, neck, face, and scalp. When you have this mastered and can relax completely; turn your thoughts to scenes of natural tranquility from your past. Stay with your inner self as long as you wish, whether thinking of nothing or visualizing only the loveliest of images. Often you will become completely unaware of your surroundings. When you open your eyes you will find yourself refreshed in mind and body.

Peter establishes his relationship and his purpose with his audience in the very first sentence, and then offers step-by-step procedures that move readers toward a full understanding of the process. Remember that you are the expert when writing about a process, and that you have to think carefully about the degree of knowledge that your audience shares.

To develop a process paper, follow these guidelines:

- Select an appropriate topic.
- Decide whether your primary purpose is to direct or explain.
- Determine the knowledge gap between you and your audience.
- Explain necessary equipment or define special terms.
- Organize paragraphs in a complete sequence of steps.
- Explain each step clearly and completely.
- State results or outcomes.

Numerous subjects lend themselves to process analysis. You must decide, especially for a particular course, which topic is most appropriate and which topic you know or want to learn about the most.

Comparison and Contrast   Comparison and contrast is an analytical method organizing thought to show similarities and differences between two persons, places, things, or ideas. Comparing and contrasting comes naturally to us. If, for example, you must decide on which candidate to vote for, you might compare the party affiliations, records, and positions on issues of both candidates to find the one that best meets your expectations. Comparison and contrast serves three useful purposes in writing:

1. To evaluate the relative worth or performance of two things by comparing them point-for-point.
2. To increase understanding of two familiar things by exploring them for significant similarities and differences.
3. To explain something unfamiliar by comparing it with something familiar.

The organization of comparison-and-contrast paragraphs and essays is fairly specialized and somewhat more prescribed than other methods of writing. The following are some basic guidelines for preparing comparison-and-contrast papers.

First and foremost, limit your comparison to only two subjects (from here on we'll refer to them as A and B). If you attempt to work with more, you may find your writing becomes confused. Subjects A and B should be from the same category of things. You would do better, for example, to compare two jazz pianists than to compare a jazz pianist and Dixieland jazz as a whole. Moreover, there needs to be a *purpose* for your comparison. Unless you explain your purpose, the comparison, which might otherwise be structurally sound, will ultimately seem meaningless.

The organization of comparison-and-contrast papers generally follows two basic patterns, or methods: the *block method* and the *alternating method*. The block method presents all material on subject A and then all material on subject B. With the block method, each subtopic must be the same for both subjects. The alternating method presents all the material on each subtopic together, analyzing these subtopics in an AB, AB, AB pattern. Although there is no hard-and-fast rule, the alternating method is probably the best choice for most essays in order to avoid the standard pitfalls of the block method. Unless you are an experienced writer, using the block method can lead to an insufficiently developed paper, with some subtopics receiving more attention than others. It can also lead to a paper that seems like two separate essays, with a big chunk about subject A followed by a second disconnected chunk about subject B. Whether you are using the block or alternating method, follow through in an orderly manner, stating clearly the main thesis or reason for establishing the comparison, and providing clear transitions as you move from idea to idea.

Consider the following paragraph, written by a student, John Shin:

The story of Noah and the Great Flood is probably the best known story of a deluge in the Mesopotamian Valley. However, there are several other accounts of a large flood in the valley. Of these, the Akkadian story of Utnapishtim, as told by Gilgamesh, is the most interesting due to its similarities to the biblical story of Noah. Utnapishtim is a king who is forewarned of the coming of a great flood. He is advised to build an ark and does so.

After many days the waters recede and Utnapishtim exits the ark and is turned into a god. The stories of Noah and Utnapishtim bear a striking resemblance in several parts: a god or gods cause a flood to punish men and women; arks, of certain dimensions, are built; animals are taken on board; birds are released to find land; and the arks come to rest on mountains. These parallels are so striking that many think the two to be the same tale.

Given the design of this paragraph, we can assume that the student could develop body paragraphs that deal in detail with each of the key resemblances in the order they are mentioned: the coming of the flood, the building of the ark, the animals taken on board, the release of the birds, and the lodging of both arks on the mountaintop. By employing the alternating method, the student constructs a well-organized comparative framework for his analysis of the story of Noah and the story of Gilgamesh.

Causal Analysis Frequently in college writing you must explain the causes or effects of some event, situation, or phenomenon. This type of investigation is termed *causal analysis*. When you analyze something, you divide it into its logical parts or processes for the purpose of close examination. Thus phenomena as diverse as divorce in America, the Civil War, carcinogens in asbestos, the death of Martin Luther King Jr., or the eruption of Mount St. Helens can be analyzed in terms of their causes and effects.

Cause-and-effect relationships are part of everyday thinking and living. Why did you select the college you now attend? Why did you stop dating Freddy or Barbara, and what effect has this decision had on your life? Why did the football team lose five straight games? You need causal analysis to explain why something occurred, to predict what will occur, and to make informed choices based on your perceptions. With causal analysis, you cannot simply tell a story, summarize an event, or describe an object or phenomenon. Instead, you must explain the *why* and *what* of a topic. The analysis of causes seeks to explain why a particular condition occurred. The analysis of effects seeks to explain what the consequences or results were, are, or will be.

Causal reasoning is common to writing in many disciplines: history, economics, politics, sociology, literature, science, education, and business, to list a few. Some essays and reports focus on causes, others on effects, still others on both causes and effects. Sometimes even the simplest sort of causal reasoning based on personal experience does not admit to the complete separation of causes and effects but depends instead on recognition that causes and effects are interdependent. For example, the following paragraph from a student's sociology paper focuses on a cause-and-effect relationship:

My parents came to New York with the dream of saving enough money to return to Puerto Rico and buy a home with some land and fruit trees. Many Puerto Ricans, troubled by the problem of life on the island, find no relief in migration to New York City. They remain poor, stay in the barrio, are unable to cope with American society and way of life, and experience the destruction of their traditionally close family life. My parents were fortunate. After spending most of their lives working hard, they saved enough to return to the island. Today they tend their orange, lemon, banana, and plantain trees in an area of Puerto Rico called "El Paraíso." It took them most of a lifetime to find their paradise—in their own backyard.

Here the writer blends personal experience with a more objective analysis of causes and effects, presenting the main cause-effect relationship in the first sentence, analyzing typical effects, providing an exception to this conventional effect, and describing the result.

There are times when you will want to focus exclusively on causes or on effects. For example, in a history course the topic might be to analyze why World War II occurred.

It is popularly accepted that Hitler was the major cause of World War II, but the ultimate causes go much deeper than one personality. There were long-standing German grievances against reparations levied on the nation following its defeat in World War I. Moreover, there were severe economic strains that caused resentment among the German people. Compounding these problems was the French and English reluctance to work out a sound disarmament policy and American noninvolvement in the matter. Finally, there was the European fear that Communism was a much greater danger than National Socialism. All these factors contributed to the outbreak of World War II.

Note that in his attempt to explain fully the causes of an event, the writer goes beyond *immediate* causes, that is, the most evident causes that trigger the event being analyzed. He tries to identify the *ultimate* causes, the deep-rooted reasons that completely explain the problem. In order to present a sound analysis of a problem, you need to be able to trace events logically to their underlying origins. Similarly, you have to engage in strategic thinking about immediate and ultimate effects in order to explain fully an event's results.

Writing about cause-and-effect relationships demands sound critical thinking skills with attention to logic and thorough preparation for the demands of the assignment. To write effective and logical essays of causal analysis, follow these guidelines:

- Be honest, objective, and reasonable when establishing your thesis. As a critical thinker, you have to avoid prejudices and logical fallacies, including unsupportable claims, broad generalizations and overstatements, and false relationships. (For a discussion of logical fallacies, see pages 122–124.)
- Distinguish between causes and effects, and decide whether you plan to focus on causes, effects, or both. As a prewriting strategy, draw up a list of causes and a corresponding list of effects. You can then organize your paper around the central causes and effects.
- Distinguish clearly between immediate and ultimate causes and effects. Explore those causes and effects that best serve the purpose of your paper and your audience's expectations.
- Provide evidence. Do not rely on simple assertions. Statistics and testimony from reliable authorities are especially effective types of evidence to support your analysis.

- Try to establish links between causes or effects. Seek a logical sequence of related elements, a chain of causality that helps readers understand the totality of your topic.

Ultimately there are many ways to write about causes and effects, depending on whether you are looking for explanations, reasons, consequences, connections, results, or any combination of these elements.

Definition   Concepts or general ideas often require careful *definition* if readers are to make sense of them or make intelligent decisions. Could you discuss supply and demand in economic theory without knowing the concept of the invisible hand? And isn't it best to know what a political conservative actually believes in before casting your vote? Concepts form the core of any discipline, line of inquiry, or problem. Because concepts are abstract, they may mean different things to different readers. In order to make ourselves understood, we must be able to specify their meaning in a particular context.

There are three types of definition. The simple *lexical* definition, or dictionary definition, is useful when briefly identifying concrete, commonplace, or uncontroversial terms for the reader. Many places, persons, and things can be defined in this manner. The *extended* definition is an explanation that might involve a paragraph or an entire essay. It is frequently used for abstract, complex, or controversial terms. The third form of definition is the *stipulative* definition, in which you offer a special definition of a term or set limitations on your use of the term. A solid definition, whether it is lexical, extended, or stipulative, involves describing the essential nature and characteristics of a concept that distinguish it from related ideas.

Consider the following paragraph by a student, Geeta Berrera:

The degree of loneliness that we feel can range from the mild or temporary case to a severe state which may eventually lead to depression or other psychological disorders. Being able to recognize the signs and signals of loneliness may help you to avoid it in the future. Do you find yourself unable to communicate with others? If so, you might be lonely. Do you find it difficult to put your faith in other human beings? If so, then you are setting up a situation that may be conducive to loneliness because you are preventing yourself from becoming too close to another person. Do you find yourself spending great amounts of time alone on a regular basis? Do you find that you are never invited to parties or other social events? Are you unable to love or care for another human being because you are afraid of permanent responsibilities and commitments? These are all signs and signals of either loneliness or situations that may eventually lead to loneliness. Loneliness is the feeling of sadness or grief experienced by a person at the realization that he or she lacks the companionship of other people.

Notice how the student introduces and emphasizes the central concept—loneliness—that is defined in this paragraph. She adds to the definition through a series of questions and answers—a strategy that permits her to analyze the qualities or manifestations of the concept. These symptoms serve as examples that reveal what is distinctive or representative about the condition of loneliness.

Definition can be used for several purposes. It may explain a difficult concept like phenomenology or a little-known activity like cricket. Definition can be used to identify and illustrate the special nature of a person, object, or abstract idea.

Classification   Classification is a mode of critical thinking and writing based on the division of a concept into groups and subgroups, and the examination of important elements within these groups. We have generalized ideas of classes of objects that help us organize and thereby understand the world. Many of these concepts lend themselves to classification. You think and talk frequently about types of college teachers, types of cars, types of boyfriends or girlfriends, types of movies or music. When registering for courses, you know that English is in the humanities, psychology in the social sciences, geology in the physical sciences; you select these courses on the basis of consistent classification principles, perhaps distribution requirements or the demands of your major. What you are doing is thinking about concepts within a class, sorting out and organizing information, and often evaluating possible alternatives. Classification, in short, is a basic mode of critical thought.

As a pattern of writing, classification enables you to make sense of large and potentially complex concepts. You divide a concept into groups and subgroups, and you classify elements within categories. Assume, for instance, that your politics professor asks for an analysis of the branches of the American federal government. You divide the federal government into the executive, legislative, and judicial branches, and, depending on your purpose, you subdivide even further into departments, agencies, and so forth. Then, according to some consistent principle or thesis—let's say a critical look at the erosion of the division of powers—you develop information for each category reflecting common characteristics. Essentially, if you classify in a rigorous and logical way, you sort out for analysis the parts and ideas within a scheme, progressing from general to specific in your treatment of the topic.

In developing a classification essay, you also have to determine the *system* of classification that works best for the demands of the assignment. The system you select would depend to an extent on your reader's expectations and the nature of the subject. Imagine that you have been asked to write an essay on sports by a physiology teacher, a psychology teacher, or a sociology teacher. Your system might be types of sports injuries for the physiology professor, behavior patterns of tennis players for your psychology professor, or levels of violence and aggression in team sports for your sociology professor. For a broad concept like sports, there are many possible classificatory systems depending on the purpose of your paper.

Although several classification and division strategies might be appropriate for any given concept, the following guidelines should be reviewed and applied for any classification essay:

- Think about the controlling principle for your classification. *Why* are you classifying the concept? *What* is the significance? Create a thesis statement that gives your reader a clear perspective on your classification scheme.
- Divide the subject into major categories and subdivide categories consistently. Make certain that you isolate all important categories and that these categories do not overlap excessively.
- Arrange the classification scheme in an effective, emphatic order—chronological, spatial, in order of importance, or from simple to complex.
- Present and analyze each category in a clear sequence, proceeding through the categories until the classification scheme is complete.
- Define or explain any difficult concepts within each category, providing relevant details and evidence.
- Combine classification with other appropriate writing strategies—comparison and contrast, process analysis, definition, and so forth.

Examine the following student paragraph:

To many people, fishing is finding a "fishy-looking" spot, tossing a hooked worm into the water, and hoping that a hungry fish just happens to be nearby. Anyone who has used this haphazard method can attest to the fact that failures usually outnumber successes. The problem with the "bait and wait" method is that it is very limited. The bait has less chance of encountering a fish than it would if it were presented in different areas of water. A more intelligent approach to fishing is to use the knowledge that at any given moment fish can be in three parts of a lake. Assuming that a lake has fish, anglers will find them on the surface, in the middle, or on the bottom of the lake. Fishing each of these areas involves the use of a separate technique. By fishing the surface, fishing the middle, or fishing the bottom, you greatly increase the chances of catching a fish.

This example is the student's introductory paragraph to a classification essay that blends description, process analysis, comparison and contrast, and the use of evidence to excellent effect. From the outset, however, the reader knows that this will be a classification essay.

Argumentation   Argumentation is a form of critical thinking in which you try to convince an audience to accept your position on a topic or persuade members of this audience to act in a certain way. In a sense, everything is an argument, for much of what you read and write, see and hear, is designed to elicit a desired response. Whether reading texts, viewing various media forms, or listening to the spoken word (especially of politicians), you know that just about anything is potentially debatable.

Argumentation in writing, however, goes beyond ordinary disagreements. With an argumentative essay, your purpose is to convince or persuade readers in a logical, reasonable, and appealing way. In other words, with formal argumentation you must distinguish mere personal opinion from opinions based on reasons derived from solid evidence. An argumentative essay has special features and even step-by-step processes that will be treated in greater detail in the next chapter. For now, it is worth noting that solid argumentative writing can combine many of the forms and purposes that have been discussed in this chapter. Your understanding of such forms and purposes of discourse as narration, illustration, analysis, and comparison and contrast and the ways these strategies can combine in powerful ways will help you compose solid argumentative essays.

Above all, with argumentation you must develop what Virginia Woolf called "some fierce attachment to an idea." Once you commit yourself to a viewpoint on a topic or issue, you will find it easy to bring an argumentative edge to your writing. Consider the following excerpt from a well-known essay by Caroline Bird that begins with the provocative title "College Is a Waste of Time and Money":

> A great majority of our 9 million college students are in school not because they want to be or because they want to learn. They are there because it has become the thing to do or because college is a pleasant place to be; because it's the only way they can get parents or taxpayers to support them without working at a job they don't like; because Mother wanted them to go, or some other reason entirely irrelevant to the course of studies for which college is supposedly organized.

Clearly Bird's claim has that argumentative edge you encounter in essays designed to convince readers of a particular viewpoint or position on an issue. Do you agree or disagree with Bird's claim? How would you respond to her assertions? What evidence would you provide to support your own claim?

Argumentation is a powerful way to tap into the aspirations, values, and conduct of your audience. It makes demands on readers and writers to do something, believe something, or even become somebody different—let's say a more tolerant person or a more active citizen. True, argument can provoke conflict, but it can also resolve it. In fact, many experts today emphasize the value of argument in solving problems and defusing or managing conflicts.

At the outset of any argument process, you must recognize that you have a problem to solve and decisions to make. Problem solving often is at the heart of argumentation; it is a process in which situations, issues, and questions are analyzed and debated or decisions arrived at. The basic steps to problem solving in argumentation are these:

1. Define and analyze the problem. Examine all available information to identify the problem precisely.
2. Interpret the facts and review alternative approaches.
3. Make a claim or a decision—that is, assert the best course of action.
4. Implement the decision in order to persuade or convince your audience that the problem has been addressed and solved.
5. Evaluate the outcome in follow-up documents.

At times, it will be hard to diagnose a problem and find solutions for it. At other times, there is no ideal solution to a problem. Argumentation is not a simple academic exercise but rather an indispensable tool in personal and professional situations. It is indispensable in harnessing increasingly complex political, economic, social, and technological trends on both a domestic and global scale. Moreover, argument can produce ethically constructive and socially responsible results. Argument makes special demands on a writer that will be treated comprehensively in the next chapter.

Writing End Paragraphs   If an essay does not have a strong, appropriate ending, it may leave the reader feeling confused or dissatisfied, a sense that the intention and promises built up in earlier parts of the essay have not been fulfilled. By contrast, an effective closing paragraph leaves the reader with the impression that the essay is complete and satisfying.

The techniques that follow permit you to end your essay emphatically and with grace:

- Use a full-circle pattern. Echo or repeat an opening phrase, idea, or detail that you presented in your introductory paragraph.
- State your conclusions, proofs, or theories based on the facts and supporting ideas of the essay. This strategy works especially well in papers for social science, science, and philosophy courses.
- Show the outcome or effects of the facts and ideas of the essay.
- Suggest a solution as a way to clarify your position on the problem you have discussed.
- Ask a question that sums up the main point of the essay.
- Offer an anecdote, allusion, or lighthearted point that sums up your thesis.
- Use a quotation that supports your main point or illuminates an aspect of the topic.

Other basic ways to end an essay include restating your thesis and main points, calling for action, providing a final summary evaluation, or looking at future consequences based on the essay's analysis or argument. A closing, like your introduction, should be brief. It is your one last attempt at clarity, one last chance to illuminate your topic.

✴ www.mhhe.com/ **mhreader**	For more information on writing endings, go to: **Writing > Conclusions**

Student Essay

Here is the essay that Taylor wrote in response to Schlesinger's "The Cult of Ethnicity." Consider the strategies that she used to make her composing process a success.

Jamie Taylor
Humanities 101, sec. 008
Professor Fred Segal
4 November 2005

Cultist Behavior or Doltish Behavior?

*Introductory
paragraph presents
Schlesinger's main
argument, amplifies
Schlesinger's inferred
claims, and then
presents the writer's
counterargument.*

In Schlesinger's "The Cult of Ethnicity," the author warns that there are forces at work within our nation that undermine our principles of democracy. These forces come in the guise of individuals and groups who claim that they know what's right for the people whom they represent. Although he doesn't mention them all specifically, one can infer he means that certain leaders from the African American community, the Latino community, the Native American community, the Asian community, and so forth are advocating strong identification within groups to keep their identities alive since they claim Eurocentric culture has had a history of stealing and suppressing their own historic roots. But Schlesinger seems to fear that only divisiveness can result. In this regard, he does not give the individual enough credit. Rather than have a paternalistic attitude about what he fears these groups are doing, he should give more credit to the members of these groups to be able to discern which messages regarding ethnicity to accept as being benign and which to reject as being downright silly.

*The first body
paragraph presents
Taylor's first point
supported with
evidence and
examples.*

Take for example, the many clubs in the average college or university. Nearly every ethnic group is represented by one of these organizations. For example, my university has many groups that represent African Americans, Latinos, Asians, Native Americans, even subgroups like the Korean Society, the Chinese Student Association, and so on. Belonging to these groups gives students a healthy place to socialize, discuss common areas of interest and concern, and assist with community outreach. For example, many of these clubs sponsor programs to give demonstrations of cultural traditions such as cooking, dance, clothing, and so on to civic and business groups. They also assist the needy in gaining access to social services, particularly for shut-ins and the elderly who may not speak English. Also, there is strength in numbers, and the fact that these clubs are popular attests to the fact that they tolerate a range of ideas so that no one "ideology" is promoted over another. Besides, if that were to happen, it is the right of the organization to vote a person out of office or membership. To say that these clubs promote divisiveness would be like saying that the Newman Society for Catholic students or Hillel House for Jewish students promotes religious intolerance.

*The second point offers
a unique slant on
divisive ethnic and
racial leaders and the
ability of Americans
to reject their claims.
Again, specific
examples and evidence
buttress Taylor's
argument.*

Second, self-styled leaders of various racial and ethnic groups—in their efforts to be divisive—actually help people to see through their rhetoric, or at least, to apply only that which is reasonable and reject that which is intolerable. Because of to-day's media, such leaders cannot "hide" their views and thus can become their own worst enemies when presenting them in front of a national audience. For example, Louis Farrakhan has not only alienated Jewish individuals owing to his open anti-Semitism, and many among the gay population for his antigay sentiments, but many African Americans as well, particularly women, who often condemn him for his patriarchal views re-garding the family and society. A simple proof of his lack of power is the fact that he has been presenting these antidemo-cratic ideals for decades now, and there is little evidence that anyone is listening to them. Another example is the late Rabbi Meyer Kahane, who advocated the expulsion of all Arabs from Israel. An open opponent of democracy, he was condemned by Jewish leaders in the United States to the point where he was shunned from any discussion regarding religious issues.

Finally, one can feel confident that even within the margins of mainstream white America, cultist groups are their own worst enemy. Take for example, the various groups of survivalists (pri-marily white Americans), white extremists and separatists, anti-gay groups, and radical anti-abortionists. The philosophy and tactics of these organizations are condemned by the vast major-ity of Americans owing to their antidemocratic postures, not to mention their often violent, even murderous activities. They may capture the headlines for a while, but they will never capture the hearts of Americans so long as we stay true to the "measures and laws" that Washington spoke of in his discussion with John Adams.

The writer's third and final point encompasses a variety of "antidemocratic" groups and rejects their "postures."

In conclusion, the open democratic society we have created is just too strong a force to be weakened or undermined by "ro-mantic ideologies" or "unscrupulous con men" as Schlesinger puts it. Mr. Schlesinger has little to worry about. Just look around your school or university cafeteria. There's no white section or Latino section or Asian section: Nowadays, it's just one big American section.

The conclusion returns to Schlesinger, while recapitulating Taylor's main points.

Revising

Revising—the rethinking and rewriting of material—takes place during every stage of the composing process. It is integral to the quest for clarity and mean-ing. "Writing and rewriting are a constant search for what one's saying," de-clares celebrated American author John Updike. Similarly, the famous essayist E. B. White admits, "I rework a lot to make it clear." If these two great prose stylists revise material in order to seek clarity for their ideas, then you also should adopt the professional attitude that you can improve what you first say or think and what you first put down on paper. In fact, one trait that distin-guishes experienced from inexperienced writers is that the professional writers understand fully the need to revise.

Revision is an art. It is the only way to make your writing match the vision of what you want to accomplish. Whether at word, sentence, paragraph, or essay level, you should develop a repertoire of choices that will permit you to solve writing problems and sharpen your ideas. You might also choose to share your draft with another reader who can let you know what is or is not working and give you suggestions for improvement. To make the process of revision worthwhile, you should ask yourself the following questions during your prewriting and drafting activities (you can also give these questions to your reader):

- Is my essay long enough (or too long) to meet the demands of the assignment?
- Is my topic suitable for the assignment?
- Do I have a clear thesis statement?
- Does my writing make sense? Am I communicating with my reader instead of just with myself?
- Have I included everything that is important to the development of my thesis or argument?
- Is there anything I should discard?
- Do I offer enough examples or evidence to support my key ideas?
- Have I ordered and developed paragraphs logically?
- Do I have a clear beginning, middle, and end?

Once you have answered these questions, you will be able to judge the extent to which you have to revise your first draft.

Proofreading   Proofreading is part of the revision process. You do not have final copy until you have carefully checked your essay for mistakes and inconsistencies. It differs from the sort of revision that moves you from an initial draft to subsequent versions of an essay in that it does not offer the opportunity to make major changes in content or organization. It does give you a last chance to correct minor errors that arise from carelessness, haste, or inaccuracy during writing, typing, or word processing.

When you proofread, do so word by word and line by line. Concentrate on spelling, punctuation, grammar, mechanics, and manuscript form. Read each sentence aloud—from the computer screen or your hard copy. If something sounds or looks wrong to you, consult a handbook, dictionary, or other reference. Then make corrections accordingly.

Here are some basic guidelines for proofreading your essay:

1. Check the title. Are words capitalized properly?
2. Check all words in the essay that should be capitalized.
3. Check the spelling of any word you are uncertain about.
4. Check the meaning of any word you think you might have misused.
5. Check to see if you have unintentionally omitted or repeated any words.
6. Check paragraph form. Have you indented each paragraph?

7. Check to make certain you have smooth, grammatically correct sentences. This is your last chance to eliminate awkward and grammatically incorrect sentences.

Responding to Editorial Comments   Even when you submit what you *think* is the final version of your essay, your teacher might not think that the essay has reached its best possible form. Teachers are experienced in detecting essays' strengths and weaknesses, pinpointing mistakes, and suggesting how material can be improved. Their comments are not attacks; they do want you to pay attention to them, to recognize and correct errors, and possibly to revise your essay once again—most likely for a higher grade. If you receive editorial comment in an objective manner and respond to it constructively, you will become a more accurate and effective writer.

When reading your essays, your instructor will use standard correction symbols that appear in English handbooks. He or she will make additional comments in the margins and compose an overall assessment of the paper at the end. Any worthwhile comment on your paper will blend supportive observations with constructive criticism. Often your instructor will offer concrete suggestions for revision. When you receive a graded paper, you typically are expected to make the necessary revisions and either add it to your portfolio or resubmit the essay.

Ultimately, refinement is integral to the entire writing process. From reading materials that you confront at the outset and respond to in various ways, you move through many composing stages to create a finished product. In *The Field of Vision*, the American novelist and critic Wright Morris refers to the important task of refinement that confronts the writer: "By raw material, I mean that comparatively crude ore that has not yet been processed by the imagination—what we refer to as *life*, or as experience, in contrast to art. By technique I mean the way the artist smelts this material down for human consumption." Your best writing is the result of this smelting process, which involves the many strategies covered in this introduction that are designed to help you acquire greater control over the art of critical reading and writing. Donald M. Murray's essay "The Maker's Eye: Revising Your Own Manuscripts," starting on page 65, offers one writer's summation of the stages of the revision process.

A Portfolio on Writing and Communication

✳

How to Mark a Book

Mortimer J. Adler

Mortimer Jerome Adler (1902–2001) was born in New York City and received his PhD from Columbia University in 1928. A staunch advocate for classical philosophy, Adler believed that there are unshakable truths—an idea rejected by most contemporary philosophers. For this reason, Adler has not been taken seriously by the academic estab- lishment. He was a champion of knowledge, believing that philosophy should be a part of everyone's life and that access to the great ideas in philosophy can be of value to everyone. Many of his over 75 books attempt to edify the general reader by explaining basic philosophical concepts in everyday language. He was also chairman of the edito- rial board of the Encyclopedia Britannica. *To make knowledge more accessible to everyone, he also assumed editorship of the Encyclopedia Britannica's Great Books proj- ect, partly sponsored by the University of Chicago. This project, which has put 443 of the world's "classics" into a 54-volume set, graces the bookcases of many dens and stud- ies in middle-class American homes. Despite his advancing years, Adler continued to work on many projects to promote his goal of universal education and enlightenment. "How to Mark a Book" is typical of his didactic, pragmatic approach to education.*

You know you have to read "between the lines" to get the most out of anything. 1 I want to persuade you to do something equally important in the course of your reading. I want to persuade you to "write between the lines." Unless you do, you are not likely to do the most efficient kind of reading.

I contend, quite bluntly, that marking up a book is not an act of mutilation 2 but of love.

You shouldn't mark up a book which isn't yours. Librarians (or your 3 friends) who lend you books expect you to keep them clean, and you should. If you decide that I am right about the usefulness of marking books, you will have to buy them. Most of the world's great books are available today, in reprint edi- tions, at less than a dollar.

There are two ways in which one can own a book. The first is the property 4 right you establish by paying for it, just as you pay for clothes and furniture. But this act of purchase is only the prelude to possession. Full ownership comes only when you have made it a part of yourself, and the best way to make yourself a part of it is by writing in it. An illustration may make the point clear. You buy a beefsteak and transfer it from the butcher's ice-box to your own. But you do not own the beefsteak in the most important sense until you consume it and get it into your bloodstream. I am arguing that books, too, must be absorbed in your bloodstream to do you any good.

Confusion about what it means to own a book leads people to a false reverence for paper, binding, and type—a respect for the physical thing—the craft of the printer rather than the genius of the author. They forget that it is possible for a man to acquire the idea, to possess the beauty, which a great book contains, without staking his claim by pasting his bookplate inside the cover. Having a fine library doesn't prove that its owner has a mind enriched by books; it proves nothing more than that he, his father, or his wife, was rich enough to buy them. 5

There are three kinds of book owners. The first has all the standard sets and best-sellers—unread, untouched. (This deluded individual owns woodpulp and ink, not books.) The second has a great many books—a few of them read through, most of them dipped into, but all of them as clean and shiny as the day they were bought. (This person would probably like to make books his own, but is restrained by a false respect for their physical appearance.) The third has a few books or many—every one of them dog-eared and dilapidated, shaken and loosened by continual use, marked and scribbled in from front to back. (This man owns books.) 6

Is it false respect, you may ask, to preserve intact and unblemished a beautifully printed book, an elegantly bound edition? Of course not. I'd no more scribble all over the first edition of *Paradise Lost* than I'd give my baby a set of crayons and an original Rembrandt! I wouldn't mark up a painting or a statue. Its soul, so to speak, is inseparable from its body. And the beauty of a rare edition or of a richly manufactured volume is like that of a painting or a statue. 7

But the soul of a book *can* be separated from its body. A book is more like the score of a piece of music than it is like a painting. No great musician confuses a symphony with the printed sheets of music. Arturo Toscanini reveres Brahms, but Toscanini's score of the C-minor Symphony is so thoroughly marked up that no one but the maestro himself can read it. The reason why a great conductor makes notations on his musical scores—marks them up again and again each time he returns to study them—is the reason why you should mark up your books. If your respect for magnificent binding or typography gets in the way, buy yourself a cheap edition and pay your respects to the author. 8

Why is marking up a book indispensable to reading it? First, it keeps you awake. (And I don't mean merely conscious; I mean wide awake.) In the second place, reading, if it is active, is thinking, and thinking tends to express itself in words, spoken or written. The marked book is usually the thought-through book. Finally, writing helps you remember the thoughts you had, or the thoughts the author expressed. Let me develop these three points. 9

If reading is to accomplish anything more than passing time, it must be active. You can't let your eyes glide across the lines of a book and come up with an understanding of what you have read. Now an ordinary piece of light fiction, 10 like say, *Gone with the Wind,* doesn't require the most active kind of reading. The books you read for pleasure can be read in a state of relaxation, and nothing is lost. But a great book, rich in ideas and beauty, a book that raises and tries to answer great fundamental questions, demands the most active reading of which you are capable. You don't absorb the ideas of John Dewey the way you absorb the crooning of Mr. Vallee. You have to reach for them. That you cannot do while you're asleep.

If, when you've finished reading a book, the pages are filled with your 11 notes, you know that you read actively. The most famous *active* reader of great books I know is President Hutchins, of the University of Chicago. He also has the hardest schedule of business activities of any man I know. He invariably reads with a pencil, and sometimes, when he picks up a book and pencil in the evening, he finds himself, instead of making intelligent notes, drawing what he calls "caviar factories" on the margins. When that happens, he puts the book down. He knows he's too tired to read, and he's just wasting time.

But, you may ask, why is writing necessary? Well, the physical act of writ- 12 ing, with your own hand, brings words and sentences more sharply before your mind and preserves them better in your memory. To set down your reaction to important words and sentences you have read, and the questions they have raised in your mind, is to preserve those reactions and sharpen those questions.

Even if you wrote on a scratch pad, and threw the paper away when you 13 had finished writing, your grasp of the book would be surer. But you don't have to throw the paper away. The margins (top and bottom, as well as side), the end-papers, the very space between the lines, are all available. They aren't sacred. And, best of all, your marks and notes become an integral part of the book and stay there forever. You can pick up the book the following week or year, and there are all your points of agreement, disagreement, doubt, and inquiry. It's like resuming an interrupted conversation with the advantage of being able to pick up where you left off.

And that is exactly what reading a book should be: a conversation between 14 you and the author. Presumably he knows more about the subject than you do; naturally, you'll have the proper humility as you approach him. But don't let anybody tell you that a reader is supposed to be solely on the receiving end. Understanding is a two-way operation; learning doesn't consist in being an empty receptacle. The learner has to question himself and question the teacher. He even has to argue with the teacher, once he understands what the teacher is saying. And marking a book is literally an expression of your differences, or agreements of opinion, with the author.

There are all kinds of devices for marking a book intelligently and fruitfully. 15 Here's the way I do it:

1. Underlining: Of major points, of important or forceful statements.   16
2. Vertical lines at the margin: To emphasize a statement already underlined.   17
3. Star, asterisk, or other doo-dad at the margin: To be used sparingly, to em- 18 phasize the ten or twenty most important statements in the book. (You may

want to fold the bottom corner of each page on which you use such marks. It won't hurt the sturdy paper on which most modern books are printed, and you will be able to take the book off the shelf at any time and, by opening it at the folded-corner page, refresh your recollection of the book.)

4. Numbers in the margin: To indicate the sequence of points the author [19] makes in developing a single argument.

5. Numbers of other pages in the margin: To indicate where else in the book [20] the author made points relevant to the point marked; to tie up the ideas in a book, which, though they may be separated by many pages, belong together.

6. Circling of key words or phrases. [21]

7. Writing in the margin, or at the top or bottom of the page, for the sake of: [22] Recording questions (and perhaps answers) which a passage raised in your mind; reducing a complicated discussion to a simple statement; recording the sequence of major points right through the book. I use the end-papers at the back of the book to make a personal index of the author's points in the order of their appearance.

The front end-papers are, to me, the most important. Some people reserve [23] them for a fancy bookplate. I reserve them for fancy thinking. After I have finished reading the book and making my personal index on the back end-papers, I turn to the front and try to outline the book, not page by page, or point by point (I've already done that at the back), but as an integrated structure, with a basic unity and an order of parts. This outline is, to me, the measure of my understanding of the work.

If you're a die-hard and anti-book-marker, you may object that the margins, [24] the space between the lines, and the end-papers don't give you room enough. All right. How about using a scratch pad slightly smaller than the page-size of the book—so that the edges of the sheets won't protrude? Make your index, outlines, and even your notes on the pad, and then insert these sheets permanently inside the front and back covers of the book.

Or, you may say that this business of marking books is going to slow up [25] your reading. It probably will. That's one of the reasons for doing it. Most of us have been taken in by the notion that speed of reading is a measure of our intelligence. There is no such thing as the right speed for intelligent reading. Some things should be read quickly and effortlessly, and some should be read slowly and even laboriously. The sign of intelligence in reading is the ability to read different things differently according to their worth. In the case of good books, the point is not to see how many of them you can get through, but rather how many can get through you—how many you can make your own. A few friends are better than a thousand acquaintances. If this be your aim, as it should be, you will not be impatient if it takes more time and effort to read a great book than it does a newspaper.

You may have one final objection to marking books. You can't lend them to [26] your friends because nobody else can read them without being distracted by your notes. Furthermore, you won't want to lend them because a marked copy is a kind of intellectual diary, and lending it is almost like giving your mind away.

If your friend wishes to read your *Plutarch's Lives*, "Shakespeare," or *The* 27 *Federalist Papers*, tell him gently but firmly, to buy a copy. You will lend him your car or your coat—but your books are as much a part of you as your head or your heart.

COMPREHENSION

1. Summarize what Adler means by "marking up a book."
2. In your own words, explain how you believe Adler would define the word *book*.
3. Adler mentions books throughout the essay. What particular type of book is he referring to?

RHETORIC

1. What is the tone of the essay? What can you infer from this tone about Adler's emotional relationship to books?
2. Paragraphs 16 through 22 list devices for marking a book. What is the function of enumerating them in this way? How would the tone of this section have been altered if he had summarized these devices in paragraph form?
3. The author makes reference to various intellectual and artistic figures and works in the essay. How does this help determine for whom the essay has been targeted?
4. Study the relationship between paragraph 9 and paragraphs 10, 11, and 12. What is the rhetorical format of this section? What is the method of argumentation he is employing?
5. Adler uses the analogy that "reading a book should be: a conversation between you and the author." What other analogies can you find in the essay?
6. Adler raises objections to his argument and then refutes the objections. Where does he make use of this rhetorical device? How effective is it in advancing his argument?
7. Adler calls *Gone with the Wind* "light fiction." Is this opinion or fact? Is it a mere observation or a criticism of the book?

WRITING

1. Mark up Adler's essay in the same manner he recommends that you mark up any good piece of writing. Then write an essay using process analysis to summarize the various methods you used.

2. Compare and contrast two books: one that Adler would regard as "light reading" and one that he would regard as worthy of marking up. Indicate the primary differences between these books in terms of their diction, level of discourse, insight, purpose, and scholarship.
3. **WRITING AN ARGUMENT:** Argue for or against the proposition that this essay has lost its relevance owing to the introduction of new forms of educational media.

---

www.mhhe.com/
mhreader

For more information on Mortimer Adler, go to:
More Resources > Ch. 1 Reading & Writing

✳

Writing Matters

Julia Alvarez

Julia Alvarez (b. 1950), a highly regarded novelist and poet, was born in New York City but raised until the age of 10 in the Dominican Republic. She was forced to flee with her family after her father was implicated in a plot to overthrow the dictator Rafael Trujillo, an event alluded to in her semiautobiographical novel, How the Garcia Girls Lost Their Accents *(1991). Another novel,* Yo! *(1997), continues the Garcia/Alvarez family saga. Alvarez attended Middlebury College (BA, 1971) and Syracuse University (MFA, 1975). Since 1978, she has been on the English faculty at Middlebury College. In addition to her fiction, Alvarez has published books of poetry, including* Homecoming: New and Collected Poems *(1995), a collection of essays,* Something to Declare *(1998), and several books for children. In this essay, which appeared in* The Writer, *Alvarez explains her process as a writer.*

One of the questions that always comes up during question-and-answer periods is about the writing life. The more sophisticated, practiced questioners usually ask me, "Can you tell us something about your process as a writer?"

In part, this is the curiosity we all have about each other's "processes," to use the terminology of my experienced questioner. We need to tell, and we also want to know (don't we?) the secret heart of each other's life. Perhaps that is why we love good novels and poems—because we can enter, without shame or without encountering defensiveness or embarrassment, the intimate lives of other people.

But the other part of my questioner's curiosity about the writing life has to do with a sense we all have that if we can only get a hold of the secret ingredients of the writing process, we will become better writers. We will have an easier time of it if we only find that magic pencil or know at which hour to start and at which hour to quit and what to sip that might help us come up with the next word in a sentence.

I always tell my questioners the truth: Listen, there are no magic solutions   4
to the hard work of writing. There is no place to put the writing desk that will draw more words out of you. I had a friend who claimed that an east-west alignment was the best one for writing. The writing would then flow and be more in tune with the positive energies. The north-south alignment would cause blocks as well as bad dreams if your bed was also thus aligned.

See, I tell my questioners, isn't this silly?   5

But even as I say so, I know I am talking out of both sides of my mouth. I   6
admit that after getting my friend's tip, I lined up my writing desk (and my bed) in the east-west configuration. It wasn't that I thought my writing or my dream life would improve, but I am so impressionable that I was afraid that I'd be thinking and worrying about my alignment instead of my line breaks. And such fretting would affect my writing adversely. Even as recently as this very day, I walk into my study first thing in the morning, and I fill up my bowl of clear water and place it on my desk. And though no one told me to do this, I somehow feel this is the right way to start a writing day.

Of course, that fresh bowl of water sits on my desk on good and bad writ-   7
ing days. I know these little ceremonies will not change the kind of day before me. My daily writing rituals are small ways in which I contain my dread and affirm my joy and celebrate the mystery and excitement of the calling to be a writer.

I use the word *calling* in the old religious sense: a commitment to a life con-   8
nected to deeper, more profound forces (or so I hope) than the marketplace, or the academy, or the hectic blur of activity that my daily life is often all about. But precisely because it is a way of life, not just a job, the writing life can be difficult to combine with other lives that require that same kind of passion and commitment—the teaching life, the family life, the parenting life, and so on. And since we writers tend to be intense people, whatever other lives we combine with our writing life, we will want to live them intensely, too. Some of us are better at this kind of juggling than others.

After twenty-five years of clumsy juggling—marriages, friendships, teach-   9
ing, writing, community work, political work, child caring—I think I've finally figured out what the proper balance is for me. Let me emphasize that this is not a prescription for anyone else. But alas, I'm of the Gerald Ford school of writers who can't chew gum and write iambic pentameter at the same time. I can do two, maybe three intense lives at once: writing and being in a family; writing and teaching and being in a family; writing and teaching and doing political work; but if I try to add a fourth or fifth, I fall apart, that is, the writing stops, which for me is the same thing as saying I fall apart.

But still, I keep juggling, picking up one life and another and another, put-   10
ting aside the writing from time to time. We have only one life, after all, and we have to live so many lives with it. (Another reason the writing life appeals so

much is that you can be, at least on paper, all those selves whose lives you can't possibly live out in the one life you've got.)

Living other lives enriches our writing life. The tension between them can 11 sometimes exhaust us, this is true—but the struggle also makes the hard-won hours at the writing desk all the more precious. And if we are committed to our writing, the way we lead our other lives can make them lives-in-waiting to be writing lives.

For me, the writing life doesn't just happen when I sit at the writing desk. It 12 is a life lived with a centering principle, and mine is this: that I will pay close attention to this world I find myself in. "My heart keeps open house," was the way the poet Theodore Roethke put it in a poem. And rendering in language what one sees through the opened windows and doors of that house is a way of bearing witness to the mystery of what it is to be alive in this world.

This is all very high-minded and inspirational, my questioner puts in, but what 13 about when we are alone at our writing desks, feeling wretchedly anxious, wondering if there is anything in us worth putting down?

Let me take you through the trials and tribulations of a typical writing day. 14 It might help as you also set out onto that blank page, encounter one adventure or mishap after another, and wonder—do other writers go through this?

The answer is probably yes. 15

Not much has happened at six-twenty or so in the morning when I enter 16 my writing room above the garage. I like it this way. The mind is free of household details, worries, commitments, voices, problems to solve.

My mood entering the room depends on what happened with my writing the 17 day before. If the previous day was a good one, I look forward to the new writing day. If I was stuck or uninspired, I feel apprehensive. In short, I can't agree more with Hemingway's advice that a writer should always end his writing day knowing where he is headed next. It makes it easier to come back to work.

The first thing I do in my study every morning is read poetry (Jane Kenyon, 18 George Herbert, Rita Dove, Robert Frost, Elizabeth Bishop, Rhina Espaillat, Jane Shore, Emily Dickinson . . .). This is the first music I hear, the most essential. Interestingly, I like to follow the reading of poetry with some prose, as if, having been to the heights I need to come back down to earth.

I consider this early-morning reading a combination of pleasure-reading 19 time when I read the works and authors I most love and finger-exercise reading time, when I am tuning my own voice to the music of the English language as played by its best writers. There's an old Yiddish story about a rabbi who walks out in a rich neighborhood and meets a watchman walking up and down. "For whom are you working?" the rabbi asks. The watchman tells him, and then in his turn, he asks the rabbi, "And whom are you working for, rabbi?" The words strike the rabbi like a shaft. "I am not working for anybody just yet," he barely manages to reply. Then he walks up and down beside the man for a long time and finally asks him, "Will you be my servant?" The watchman says, "I should like to, but what would be my duties?"

"To remind me," the rabbi says.   20

I read my favorite writers to remind me of the quality of writing I am aim-   21
ing for.

Now, it's time to set out: Pencil poised, I read through the hard copy that I ran off   22
at the end of yesterday's writing day. I used to write everything out by longhand,
and when I was reasonably sure I had a final draft, I'd type it up on my old Selectric. But now, I usually write all my prose drafts right out on the computer,
though I need to write out my poems in longhand, to make each word by hand.

This is also true of certain passages of prose and certainly true for times   23
when I am stuck in a novel or story. Writing by hand relieves some of the pressure of seeing something tentative flashed before me on the screen with that authority that print gives to writing. "This is just for me," I tell myself, as I scratch
out a draft in pencil. Often, these scribblings turn into little bridges, tendrils that
take me safely to the other side of silence. When I'm finally on my way, I head
back to the computer.

But even my hard copies look as if they've been written by hand. As I re-   24
vise, I begin to hear the way I want a passage to sound. About the third or
fourth draft, if I'm lucky, I start to see the shape of what I am writing, the way
an essay will go, a character will react, a poem unfold.

Sometimes if Bill and I go on a long car trip, I'll read him what I am work-   25
ing on. This is a wonderful opportunity to "hear" what I've written. The process
of reading my work to someone else does tear apart that beauteous coating of
self-love in which my own creation comes enveloped. I start to hear what I've
written as it would sound to somebody else.

When I'm done with proofing the hard copy of the story or chapter or   26
poem, I take a little break. This is one of the pleasures of working at home. I can
take these refreshing breathers from the intensity of the writing: go iron a shirt
or clean out a drawer or wrap up my sister's birthday present.

After my break, I take a deep breath. What I now do is transcribe all my   27
handwritten revisions on to my computer, before I launch out into the empty
space of the next section of the story or essay or the chapter in a novel. This is
probably the most intense time of the writing day. I am on my way, but I don't
know exactly where it is I am going. But that's why I'm writing: to find out.

On the good days, an excitement builds up as I push off into the language, and   28
sentence seems to follow sentence. I catch myself smiling or laughing out loud
or sometimes even weeping as I move through a scene or a stanza. Certainly
writing seems to integrate parts of me that are usually at odds. As I write, I feel
unaccountably whole; I disappear! That is the irony of this self-absorbed profession: The goal finally is to vanish. On bad days, on the other hand, I don't disappear. Instead, I'm stuck with the blank screen before me. I take more and
more breaks. I wander out on the deck and look longingly south toward the little spire of the Congregational church and wish another life for myself. Oh,
dear, what have I done with my life?

I have chosen it, that's what I've done. So I take several deep breaths and go 29
back upstairs and sit myself down and work over the passage that will not
come. As Flannery O'Connor attested: "Every morning between 9 and 12, I go
to my room and sit before a piece of paper. Many times, I just sit for three hours
with no ideas coming to me. But I know one thing: If an idea does come between 9 and 12, I am there ready for it." The amazing thing for me is that years
later, reading the story or novel or poem, I can't tell the passages that were easy
to write, the ones that came forth like "greased lightning" from those other passages that made me want to give up writing and take up another life.

On occasion, when all else fails, I take the rest of the day "off." I finish read- 30
ing the poet or novelist with whom I began the day or I complain to my journal
or I look through a picture book of shoes one of my characters might wear. But
all the while I am feeling profound self-doubt—as if I were one of those cartoon
characters who runs off a cliff, and suddenly looks down only to discover,
there's no ground beneath her feet!

At the end of the writing day (about two-thirty or three in the afternoon), I 31
leave the room over the garage. I put on my running clothes, and I go for a run.
In part, this exercise does make me feel better. But one of the best perks of running has been that it allows me to follow Hemingway's advice. I don't always
know where I am headed in my writing at the end of the workday, but after I
run, I usually have one or two good ideas. Running helps me work out glitches
in my writing and gives me all kinds of unexpected insights. While I run down
past the Fields's house, through Tucker Development, down to the route that
goes into town, and then back, I've understood what a character is feeling or
how I'm going to organize an essay or what I will title my novel. I've also had a
zillion conversations with dozens of worrisome people, which is much better
than trying to have these conversations with them while I am trying to write.
Also, since I am not near a phone, I am not tempted to call them up and actually
have it out with them. I've saved a lot of friendships and relationships and
spared myself plenty of heartaches this way.

After the run, the rest of the workday is taken up by what I call the writing 32
biz part of being a writer. What this involves, in large part, is responding to the
publicity machine that now seems to be a necessary component of being a published writer. Answering mail, returning phone calls, responding to unsolicited
manuscripts from strangers or to galleys from editors who would so appreciate
my putting in a good word for this young writer or translation or series. Ironically, all this attention can sometimes amount to distraction that keeps me from
doing the work that brought these requests to my door in the first place.

I could just ignore these requests. But all along the way, I found helpers 33
who did read my manuscript, did give me a little of their busy day. These are favors I can never pay back, I can only pass on. And so I do try to answer my own
mail and read as many galleys by new writers as I possibly can and return
phone calls to those who need advice I might be able to give.

When I'm finally finished with my writing biz or I've put it aside in the 34
growing pile for tomorrow, I head to town to run errands or see a friend or

attend a talk at the college. As the fields and farms give way to houses and lawns, I feel as if I'm reentering the world. After having been so intensely a part of a fictional world, I love this daily chance to connect with the small town I live in, to find out how everybody else is doing.

How's it going? everyone asks me, as if they really want to know all about 35 my writing day.

At the end of a good reading, the audience lingers. It's late in Salt Lake City or 36 Portland or Iowa City. Outside the bookstore windows, the sky is dark and star-studded. Then, that last hand goes up, and someone in the back row wants to know, "So, does writing really matter?"

This once really happened to me on a book tour. I felt as if I'd just been hit 37 "upside the head," an expression I like so much because it sounds like the blow was so hard, the preposition got jerked around, too. Does writing matter? I sure hope so, I wanted to say. I've published six books. I've spent most of my think-ing life, which is now over thirty years, writing. Does writing really matter? It was the hardest, and the best, question I've been asked anywhere.

Let's take out the "really," I said. It makes me nervous. I don't really know 38 much of anything, which is why I write, to find things out. Does writing matter?

It matters, of course, it matters. But it matters in such a small, almost invis- 39 ible way that it doesn't seem very important. In fact, that's why I trust it, the tiny rearrangements and insights in our hearts that art accomplishes. It's how I, anyhow, learned to see with vision and perplexity and honesty and continue to learn to see. How I keep the windows and doors open instead of shutting my-self up inside the things I "believe" and have personally experienced. How I move out beyond the safe, small version of my life to live other lives. "Not only to be one self," the poet Robert Desnos wrote about the power of the imagina-tion, "but to become each one."

And this happens not because I'm a writer or, as some questioners put it, "a 40 creative person." I'll bet that even those who aren't writers, those who are con-cerned with making some sense of this ongoing journey would admit this: that it's by what people have written and continue to write, our stories and cre-ations, that we understand who we are. In a world without any books, we would not be the same kind of critter. "Art is not the world," Muriel Rukeyser reminds us, "but a knowing of the world. It prepares us."

Prepares us for what? I have to admit that I don't really know what it pre- 41 pares us for. For our work in the world, I suppose. Prepares us to live our lives more intentionally, ethically, richly. A hand shoots up. "You mean to say that if Hitler had read Tolstoy he would have been a better person?"

Let's say that it would have been worth a try. Let's say that if little Hitler 42 had been caught up in reading Shakespeare or Tolstoy and was moved to the extent that the best books move us, he might not have become who he became. But maybe, Tolstoy or no Tolstoy, Hitler would still have been Hitler. We live, af-ter all, in a flawed world of flawed beings. In fact, some very fine writers who have written some lovely things are not very nice people.

But I still insist that while writing or entering into the writing of another, 43
they were better people. If for no other reason than they were not out there,
causing trouble. Writing is a form of vision, and I agree with that proverb that
says, "Where there is no vision, the people perish." The artist keeps that vision
alive, cleared of the muck and refuse and junk and little dishonesties that always collect and begin to cloud our view of the world around us.

It is the end of the reading. My readers, who for this brief evening have be- 44
come real people with questions about my writing life, come forward to have
their books signed and offer some new insight or ask a further question. That
they care matters. That they are living fuller versions of themselves and of each
one because they have read books matters. This is why writing matters. It clarifies and intensifies; it reduces our sense of isolation and connects us to each
other.

COMPREHENSION

1. How do you interpret the title "Writing Matters"? What does Alvarez say
   about writing in general? How does she describe her own writing process?
   What, for her, is a "typical writing day"?
2. According to Alvarez, *why* does she write? What does she mean when she
   says, "I contain my dread and affirm my joy and celebrate the mystery and
   excitement of the calling to be a writer" (paragraph 7)?
3. Alvarez frames her essay around a public reading. How does she approach
   the reading? Why does she offer herself to a reading public? What does she
   gain from the experience?

RHETORIC

1. From one perspective, this essay involves an extended definition of the
   writing life—or at least Alvarez's experience of this life. What techniques
   and rhetorical strategies does the writer employ to develop this definition?
   Where do they appear?
2. Where does the writer employ process analysis? What are the main steps in
   this process?
3. In addition to process analysis, what other patterns of development does
   Alvarez use? How do these additional patterns help her make her main
   point? Exactly what is this thesis?
4. In paragraph 4, Alvarez writes, "I always tell my questioners the truth."
   What tone is she using here? Where else does she use this tone? What variations does she work on her attitude toward her craft?
5. This essay appeared in *The Writer,* a magazine that offers advice on the craft
   for both professionals and aspiring writers. How does Alvarez compose her
   article for this specialized audience? Why would the essay be of general
   interest—let's say, to college students in a composition class?

6. Alvarez alludes to other writers in this essay. Who are they, and where do the references appear? How many can you identify? What is her purpose in referring to and occasionally quoting from these writers?

WRITING

1. Write a personal essay in which you describe your own writing process. Try to be as detailed and self-reflective as possible in analyzing the importance of this process.
2. Alvarez mentions some of her favorite authors in this essay. Select your own favorite author, and explain why you like her or his writing. What is it, precisely, that you like about this author's style and vision of the world?
3. **WRITING AN ARGUMENT:** Write an argumentative essay criticizing or defending the approach Alvarez takes to the writing process. Do you think that she is "very high-minded and inspirational" (paragraph 13), or is she down-to-earth in her advice? Be certain to evaluate the essay on a simple principle: Did you like or dislike it, and why?

 www.mhhe.com/ **mhreader** | For more information on Julia Alvarez, go to: **More Resources > Ch. 1 Reading & Writing**

The Writing Life

Annie Dillard

Annie Dillard (b. 1945) was born in Pittsburgh and received her BA and MA degrees from Hollins College in Roanoke, Virginia. Her first book, Pilgrim at Tinker Creek, *won the 1975 Pulitzer Prize for general nonfiction. Other published works of nonfiction include* Teaching a Stone to Talk *(1982) and* An American Childhood *(1987). She has received awards from the National Endowment for the Arts and the Guggenheim Foundation as well as many other sources. As an essayist, poet, memoirist, and literary critic, she focuses her themes on the relationships among the self, nature, religion, and faith. Her writing is recognizable by its observations on the minutia of life and its search for meaning in such unlikely places as a stone or an insect. Dillard expanded her range of writing with the publication of her first novel,* The Living *(1992). In this excerpt from* The Writing Life *(1989), the author offers a striking analogy for the creative process.*

When you write, you lay out a line of words. The line of words is a miner's pick, 1
a woodcarver's gouge, a surgeon's probe. You wield it, and it digs a path you
follow. Soon you find yourself deep in new territory. Is it a dead end, or have
you located the real subject? You will know tomorrow, or this time next year.

You make the path boldly and follow it fearfully. You go where the path 2
leads. At the end of the path, you find a box canyon. You hammer out reports,
dispatch bulletins.

The writing has changed, in your hands, and in a twinkling, from an ex- 3
pression of your notions to an epistemological tool. The new place interests you
because it is not clear. You attend. In your humility, you lay down the words
carefully, watching all the angles. Now the earlier writing looks soft and careless. Process is nothing; erase your tracks. The path is not the work. I hope your
tracks have grown over; I hope birds ate the crumbs; I hope you will toss it all
and not look back.

The line of words is a hammer. You hammer against the walls of your 4
house. You tap the walls, lightly, everywhere. After giving many years' attention to these things, you know what to listen for. Some of the walls are bearing
walls; they have to stay, or everything will fall down. Other walls can go with
impunity; you can hear the difference. Unfortunately, it is often a bearing wall
that has to go. It cannot be helped. There is only one solution, which appalls
you, but there it is. Knock it out. Duck.

Courage utterly opposes the bold hope that this is such fine stuff the work 5
needs it, or the world. Courage, exhausted, stands on bare reality: this writing
weakens the work. You must demolish the work and start over. You can save
some of the sentences, like bricks. It will be a miracle if you can save some of the
paragraphs, no matter how excellent in themselves or hard-won. You can waste
a year worrying about it, or you can get it over with now. (Are you a woman, or
a mouse?)

The part you must jettison is not only the best-written part; it is also, oddly, 6
that part which was to have been the very point. It is the original key passage,
the passage on which the rest was to hang, and from which you yourself drew
the courage to begin. Henry James knew it well, and said it best. In his preface
to *The Spoils of Poynton*, he pities the writer, in a comical pair of sentences that
rises to a howl: "Which is the work in which he hasn't surrendered, under dire
difficulty, the best thing he meant to have kept? In which indeed, before the
dreadful *done*, doesn't he ask himself what has become of the thing all for the
sweet sake of which it was to proceed to that extremity?"

So it is that a writer writes many books. In each book, he intended several 7
urgent and vivid points, many of which he sacrificed as the book's form hardened. "The youth gets together his materials to build a bridge to the moon,"
Thoreau noted mournfully, "or perchance a palace or temple on the earth, and
at length the middle-aged man concludes to build a wood-shed with them."
The writer returns to these materials, these passionate subjects, as to unfinished
business, for they are his life's work.

COMPREHENSION

1. What is the central metaphor Dillard uses in describing the act of writing?
2. According to the author, what should a writer's attitude be toward her or his work?
3. Explain the meaning of the quotations taken from Thoreau and James.

RHETORIC

1. How does paragraph 1 prepare you for the tone and diction of what is to follow? How would you describe this tone and diction?
2. The essay is unusual in that it is written in the form of the second-person singular. To whom does *you* refer?
3. What is the intended audience of this essay: writers, would-be writers, the general public, a highly educated public? Explain by providing examples of grammar and sentence structure.
4. Paragraph 4 is distinct in its use of semicolons and short sentences. In fact, the final sentence is only one word. Assuming that good writers try to correlate the meaning of their writing with the style of their writing, how can you find a relationship between meaning and style in this paragraph?
5. What unifying image about the act of writing does the author create in paragraphs 1 and 2? Consider the number of one-syllable words in this section. How does the rhythm of these monosyllabic words complement the overriding image?
6. What is the chief purpose of this essay? Explain your view.

WRITING

1. Imitate Dillard's introductory sentence with one of your own, using the pattern, "When you _____, you _____." Follow the topic sentence with analogies that support your central image, for example, "When you ski, you launch yourself from the Earth" or "When you enter the Internet, you enter distant galaxies."
2. Describe a time in your life when you overcame tremendous obstacles to accomplish a difficult task. Explain how you achieved your goal.
3. Develop a series of interview questions in preparation for interviewing a master crafter, an artist, or a scientist. Interview your subject and write an essay employing process analysis. Explore and explain the way your subject proceeds with his or her work.
4. **WRITING AN ARGUMENT:** Dillard asserts, "You must demolish the work and start over" (paragraph 5). Do you agree or disagree with this proposition? Defend your answer in a brief essay.

www.mhhe.com/
mhreader

For more information on Annie Dillard, go to:
More Resources > Ch. 1 Reading & Writing

✳

Freewriting

Peter Elbow

Peter Elbow (b. 1935) was born in New York and received degrees from Williams College, Exeter College, Oxford, and Brandeis University. He has taught at the University of Massachusetts at Amherst, the State University of New York at Stony Brook, the Massachusetts Institute of Technology, Franconia College, and Evergreen State College. He is considered by some writing teachers to have revolutionized the teaching of writing through his popularization of the concept and practice called "freewriting." He is the author or editor of more than 15 books on writing, including Writing without Teachers, Writing with Power, Embracing Contraries, What Is English? *and most recently* Everyone Can Write: Essays toward a Hopeful Theory of Writing and Teaching Writing *(2000). In "Freewriting," taken from* Writing without Teachers, *Elbow explains an exercise for writing students that he helped popularize in American colleges, universities, and writing workshops.*

The most effective way I know to improve your writing is to do freewriting exercises regularly. At least three times a week. They are sometimes called "automatic writing," "babbling," or "jabbering" exercises. The idea is simply to write for ten minutes (later on, perhaps fifteen or twenty). Don't stop for anything. Go quickly without rushing. Never stop to look back, to cross something out, to wonder how to spell something, to wonder what word or thought to use, or to think about what you are doing. If you can't think of a word or a spelling, just use a squiggle or else write, "I can't think of it." Just put down something. The easiest thing is just to put down whatever is in your mind. If you get stuck it's fine to write "I can't think what to say, I can't think what to say" as many times as you want; or repeat the last word you wrote over and over again; or anything else. The only requirement is that you *never* stop. 1

What happens to a freewriting exercise is important. It must be a piece of writing which, even if someone reads it, doesn't send any ripples back to you. It is like writing something and putting it in a bottle in the sea. The teacherless class helps your writing by providing maximum feedback. Freewritings help you by providing no feedback at all. When I assign one, I invite the writer to let me read it. But also tell him to keep it if he prefers. I read it quickly and make no 2 comments at all and I do not speak with him about it. The main thing is that a freewriting must never be evaluated in any way; in fact there must be no discussion or comment at all.

Here is an example of a fairly coherent exercise (sometimes they are very incoherent, which is fine):   3

> I think I'll write what's on my mind, but the only thing on my mind right now is what to write for ten minutes. I've never done this before and I'm not prepared in any way—the sky is cloudy today, how's that? now I'm afraid I won't be able to think of what to write when I get to the end of the sentence—well, here I am at the end of the sentence—here I am again, again, again, again, at least I'm still writing—Now I ask is there some reason to be happy that I'm still writing—ah yes! Here comes the question again—What am I getting out of this? What point is there in it? It's almost obscene to always ask it but I seem to question everything that way and I was gonna say something else pertaining to that but I got so busy writing down the first part that I forgot what I was leading into. This is kind of fun oh don't stop writing—cars and trucks speeding by somewhere out the window, pens clittering across people's papers. The sky is cloudy—is it symbolic that I should be mentioning it? Huh? I dunno. Maybe I should try colors, blue, red, dirty words—wait a minute—no can't do that, orange, yellow, arm tired, green pink violet magenta lavender red brown black green—now that I can't think of any more colors—just about done—relief? maybe.

Freewriting may seem crazy but actually it makes simple sense. Think of the difference between speaking and writing. Writing has the advantage of permitting more editing. But that's its downfall too. Almost everybody interposes a massive and complicated series of editings between the time words start to be born into consciousness and when they finally come off the end of the pencil or typewriter onto the page. This is partly because schooling makes us obsessed with the "mistakes" we make in writing. Many people are constantly thinking about spelling and grammar as they try to write. I am always thinking about the awkwardness, wordiness, and general mushiness of my natural verbal product as I try to write down words.

But it's not just "mistakes" or "bad writing" we edit as we write. We also   4
edit unacceptable thoughts and feelings, as we do in speaking. In writing there is more time to do it so the editing is heavier: when speaking, there's someone right there waiting for a reply and he'll get bored or think we're crazy if we don't come out with *something*. Most of the time in speaking, we settle for the catch-as-catch-can way in which the words tumble out. In writing, however, there's a chance to try to get them right. But the opportunity to get them right is a terrible burden: you can work for two hours trying to get a paragraph "right" and discover it's not right at all. And then give up.

Editing, *in itself*, is not the problem. Editing is usually necessary if we want   5
to end up with something satisfactory. The problem is that editing goes on *at the same time* as producing. The editor is, as it were, constantly looking ever the shoulder of the producer and constantly fiddling with what he's doing while

he's in the middle of trying to do it. No wonder the producer gets nervous, jumpy, inhibited, and finally can't be coherent. It's an unnecessary burden to try to think of words and also worry at the same time whether they're the right words.

The main thing about freewriting is that it is *nonediting*. It is an exercise in 6 bringing together the process of producing words and putting them down on the page. Practiced regularly, it undoes the ingrained habit of editing at the same time you are trying to produce. It will make writing less blocked because words will come more easily. You will use up more paper, but chew up fewer pencils.

Next time you write, notice how often you stop yourself from writing down 7 something you were going to write down. Or else cross it out after it's written. "Naturally," you say, "it wasn't any good." But think for a moment about the occasions when you spoke well. Seldom was it because you first got the beginning just right. Usually it was a matter of a halting or even garbled beginning, but you kept going and your speech finally became coherent and even powerful. There is a lesson here for writing: trying to get the beginning just right is a formula for failure—and probably a secret tactic to make yourself give up writing. Make some words, whatever they are, and then grab hold of that line and reel in as hard as you can. Afterwards you can throw away lousy beginnings and make new ones. This is the quickest way to get into good writing.

The habit of compulsive, premature editing doesn't just make writing hard. 8 It also makes writing dead. Your voice is damped out by all the interruptions, changes, and hesitations between the consciousness and the page. In your natural way of producing words there is a sound, a texture, a rhythm—a voice— which is the main source of power in your writing. I don't know how it works, but this voice is the force that will make a reader listen to you, the energy that drives the meanings through his thick skull. Maybe you don't *like* your voice; maybe people have made fun of it. But it's the only voice you've got. It's your only source of power. You better get back into it, no matter what you think of it. If you keep writing in it, it may change into something you like better. But if you abandon it, you'll likely never have a voice and never be heard.

Freewritings are vacuums. Gradually you will begin to carry over into your 9 regular writing some of the voice, force, and connectedness that creep into those vacuums.

COMPREHENSION

1. What is the thesis of the essay? Is it implied or stated directly in the text?
2. In paragraph 5, Elbow refers to the "producer" and the "editor." Who are they? Where are they located? How did they develop?
3. In paragraph 8, the author makes a connection between one's personal "voice" and the idea of "power." Why does Elbow focus so strongly on this connection?

RHETORIC

1. Elbow frequently uses the "imperative" (or command) sentence form in the opening paragraph. Why? What would have been the effect had he used the simple declarative form?
2. Writers often use examples to help illustrate their point. Does the example of a freewriting exercise Elbow provides in paragraph 3 help you to understand the method? Why or why not?
3. The author uses colloquial terms such as "squiggle" (paragraph 1); "crazy" and "mushiness" (paragraph 3); and "lousy" (paragraph 7). How does his use of such words affect the tone of the essay?
4. Are there any elements in Elbow's own style that suggest his essay may have started as a freewriting exercise? Consider the reasons he provides for the importance of freewriting, for example, the generating of ideas, discovering one's own voice, expressing oneself succinctly and naturally.
5. Elbow is himself a college writing teacher. Based on your assessment of the tone of the essay, whom do you think is his intended audience? Is it broad or narrow? Specialized or general? Or, could he have in mind more than one type of audience? Explain your answer.
6. Note the number of times Elbow begins his sentences with coordinating conjunctions ("but," "and," "or"). For example, in paragraph 4, he does it three times. Many writing teachers frown on this method of structuring sentences. Why does Elbow employ it?
7. Compare the essay's introduction to its conclusion. Note how the introduction is rather long, and the conclusion is quite short (two sentences in fact). How do these two elements contribute to the overall "pace" of the essay?

WRITING

1. During one week, complete three freewriting exercises. Wait one week, and then review what you have written. Explore any insights your freewriting gives you into your writer's "voice," your concerns, interests, style, and "power."
2. Write an expository paper explaining the difficulties you have when writing a homework assignment that requires submitting an essay or writing an essay-length response during an exam.
3. Write a comparison and contrast essay wherein you examine the similarities and differences of speaking and writing.
4. **WRITING AN ARGUMENT:** Write an essay in which you support or discourage the act of freewriting.

www.mhhe.com/
mhreader   For more information on Peter Elbow, go to:
More Resources > Ch. 1 Reading & Writing

✳

The Maker's Eye: Revising Your Own Manuscripts

Donald M. Murray

Donald M. Murray (b. 1917) has combined a career as teacher, journalist, fiction writer, poet, and author of several important textbooks on writing. He has worked as a teacher, journalist, and editor for Time *magazine. His books include* A Writer Teaches Writing, Write to Learn, Read to Write, *and more recently* Shoptalk: Learning to Write with Writers *(1991),* Crafting a Life in Essay, Story, Poem *(1996), and* The Craft of Revision *(1997). In this essay, originally published in the magazine* The Writer, *Donald Murray argues for the absolute importance of the revision process to the writer. As he presents the stages of the revision process, Murray illustrates their usefulness to any writer—whether beginner or experienced—and offers his personal views and those of other authors.*

When students complete a first draft, they consider the job of writing done— 1
and their teachers too often agree. When professional writers complete a first
draft, they usually feel that they are at the start of the writing process. When a
draft is completed, the job of writing can begin.

That difference in attitude is the difference between amateur and profes- 2
sional, inexperience and experience, journeyman and craftsman. Peter F.
Drucker, the prolific business writer, calls his first draft "the zero draft"—after
that he can start counting. Most writers share the feeling that the first draft, and
all of those which follow, are opportunities to discover what they have to say
and how best they can say it.

To produce a progression of drafts, each of which says more and says it 3
more clearly, the writer has to develop a special kind of reading skill. In school
we are taught to decode what appears on the page as finished writing. Writers,
however, face a different category of possibility and responsibility when they
read their own drafts. To them the words on the page are never finished. Each
can be changed and rearranged, can set off a chain reaction of confusion or clarified meaning. This is a different kind of reading, which is possibly more difficult and certainly more exciting.

Writers must learn to be their own best enemy. They must accept the criti- 4
cism of others and be suspicious of it; they must accept the praise of others and
be even more suspicious of it. Writers cannot depend on others. They must detach themselves from their own pages so that they can apply both their caring
and their craft to their own work.

Such detachment is not easy. Science fiction writer Ray Bradbury suppos- 5
edly puts each manuscript away for a year to the day and then rereads it as a
stranger. Not many writers have the discipline or the time to do this. We must

read when our judgment may be at its worst, when we are close to the euphoric moment of creation.

Then the writer, counsels novelist Nancy Hale, "should be critical of everything that seems to him most delightful in his style. He should excise what he most admires, because he wouldn't thus admire it if he weren't . . . in a sense protecting it from criticism." John Ciardi, the poet, adds, "The last act of the writing must be to become one's own reader. It is, I suppose, a schizophrenic process, to begin passionately and to end critically, to begin hot and to end cold; and, more important, to be passion-hot and critic-cold at the same time." 6

Most people think that the principal problem is that writers are too proud of what they have written. Actually, a greater problem for most professional writers is one shared by the majority of students. They are overly critical, think everything is dreadful, tear up page after page, never complete a draft, see the task as hopeless. 7

The writer must learn to read critically but constructively, to cut what is bad, to reveal what is good. Eleanor Estes, the children's book author, explains: "The writer must survey his work critically, coolly, as though he were a stranger to it. He must be willing to prune, expertly and hard-heartedly. At the end of each revision, a manuscript may look . . . worked over, torn apart, pinned together, added to, deleted from, words changed and words changed back. Yet the book must maintain its original freshness and spontaneity." 8

Most readers underestimate the amount of rewriting it usually takes to produce spontaneous reading. This is a great disadvantage to the student writer, who sees only a finished product and never watches the craftsman who takes the necessary step back, studies the work carefully, returns to the task, steps back, returns, steps back, again and again. Anthony Burgess, one of the most prolific writers in the English-speaking world, admits, "I might revise a page twenty times." Roald Dahl, the popular children's writer, states, "By the time I'm nearing the end of a story, the first part will have been reread and altered and corrected at least 150 times. . . . Good writing is essentially rewriting. I am positive of this." 9

Rewriting isn't virtuous. It isn't something that ought to be done. It is simply something that most writers find they have to do to discover what they have to say and how to say it. It is a condition of the writer's life. 10

There are, however, a few writers who do little formal rewriting, primarily because they have the capacity and experience to create and review a large number of invisible drafts in their minds before they approach the page. And some writers slowly produce finished pages, performing all the tasks of revision simultaneously, page by page, rather than draft by draft. But it is still possible to see the sequence followed by most writers most of the time in rereading their own work. 11

Most writers scan their drafts first, reading as quickly as possible to catch the larger problems of subject and form, then move in closer and closer as they read and write, reread and rewrite. 12

The first thing writers look for in their drafts is *information*. They know that a good piece of writing is built from specific, accurate, and interesting informa- 13 tion. The writer must have an abundance of information from which to construct a readable piece of writing.

Next writers look for *meaning* in the information. The specifics must build a 14 pattern of significance. Each piece of specific information must carry the reader toward meaning.

Writers reading their own drafts are aware of *audience.* They put themselves 15 in the reader's situation and make sure that they deliver information which a reader wants to know or needs to know in a manner which is easily digested. Writers try to be sure that they anticipate and answer the questions a critical reader will ask when reading the piece of writing.

Writers make sure that the *form* is appropriate to the subject and the audi- 16 ence. Form, or genre, is the vehicle which carries meaning to the reader, but form cannot be selected until the writer has adequate information to discover its significance and an audience which needs or wants that meaning.

Once writers are sure the form is appropriate, they must then look at the 17 *structure,* the order of what they have written. Good writing is built on a solid framework of logic, argument, narrative, or motivation which runs through the entire piece of writing and holds it together. This is the time when many writers find it most effective to outline as a way of visualizing the hidden spine by which the piece of writing is supported.

The element on which writers may spend a majority of their time is *develop-* 18 *ment.* Each section of a piece of writing must be adequately developed. It must give readers enough information so that they are satisfied. How much information is enough? That's as difficult as asking how much garlic belongs in a salad. It must be done to taste, but most beginning writers underdevelop, underestimating the reader's hunger for information.

As writers solve development problems, they often have to consider ques- 19 tions of *dimension.* There must be a pleasing and effective proportion among all the parts of the piece of writing. There is a continual process of subtracting and adding to keep the piece of writing in balance.

Finally, writers have to listen to their own voices. *Voice* is the force which 20 drives a piece of writing forward. It is an expression of the writer's authority and concern. It is what is between the words on the page, what glues the piece of writing together. A good piece of writing is always marked by a consistent, individual voice.

As writers read and reread, write and rewrite, they move closer and closer 21 to the page until they are doing line-by-line editing. Writers read their own pages with infinite care. Each sentence, each line, each clause, each phrase, each word, each mark of punctuation, each section of white space between the type has to contribute to the clarification of meaning.

Slowly the writer moves from word to word, looking through language to 22 see the subject. As a word is changed, cut, or added, as a construction is rearranged, all the words used before that moment and all those that follow that moment must be considered and reconsidered.

Writers often read aloud at this stage of the editing process, muttering or 23 whispering to themselves, calling on the ear's experience with language. Does this sound right—or that? Writers edit, shifting back and forth from eye to page to ear to page. I find I must do this careful editing in short runs, no more than fifteen or twenty minutes at a stretch, or I become too kind with myself. I begin to see what I hope is on the page, not what actually is on the page.

This sounds tedious if you haven't done it, but actually it is fun. Making 24 something right is immensely satisfying, for writers begin to learn what they are writing about by writing. Language leads them to meaning, and there is the joy of discovery, of understanding, of making meaning clear as the writer employs the technical skills of language.

Words have double meanings, even triple and quadruple meanings. Each 25 word has its own potential for connotation and denotation. And when writers rub one word against the other, they are often rewarded with a sudden insight, an unexpected clarification.

The maker's eye moves back and forth from word to phrase to sentence to 26 paragraph to sentence to phrase to word. The maker's eye sees the need for variety and balance, for a firmer structure, for a more appropriate form. It peers into the interior of the paragraph, looking for coherence, unity, and emphasis, which make meaning clear.

I learned something about this process when my first bifocals were pre- 27 scribed. I had ordered a larger section of the reading portion of the glass because of my work, but even so, I could not contain my eyes within this new limit of vision. And I still find myself taking off my glasses and bending my nose towards the page, for my eyes unconsciously flick back and forth across the page, back to another page, forward to still another, as I try to see each evolving line in relation to every other line.

When does this process end? Most writers agree with the great Russian 28 writer Tolstoy, who said, "I scarcely ever reread my published writings, if by chance I come across a page, it always strikes me: all this must be rewritten; this is how I should have written it."

The maker's eye is never satisfied, for each word has the potential to ignite 29 new meaning. This article has been twice written all the way through the writing process, and it was published four years ago. Now it is to be republished in a book. The editors make a few small suggestions, and then I read it with my maker's eye. Now it has been re-edited, re-revised, re-read, re-re-edited, for each piece of writing to the writer is full of potential and alternatives.

A piece of writing is never finished. It is delivered to a deadline, torn out of 30 the typewriter on demand, sent off with a sense of accomplishment and shame and pride and frustration. If only there were a couple more days, time for just another run at it, perhaps then . . .

COMPREHENSION

1. In paragraph 1, what does Murray mean by the phrase, "When a draft is completed, the job of writing can begin"? Isn't a draft a form of writing?

2. According to Murray, what are the major differences between student and professional writers? Why do the differences help make the "professional" more accomplished at his or her work?
3. What are the differences between the reading styles of novice and experienced writers? How do the differences affect their own writings?

RHETORIC

1. Compare the introduction of this essay to that of Elbow's "Freewriting." How do they differ in tone and structure?
2. Murray begins to classify various aspects of the writer's concern in paragraph 13. Why does he wait so long to begin this analysis? Why are certain key words in paragraphs 13 through 20 italicized?
3. Murray uses analogy, comparing something with another, very different thing, to make the writing process concrete and familiar. Identify some of these analogies. Why they are models of clarity?
4. Murray refers to a writer as "the maker" several times in the essay. What does he imply by this usage? What other professions might be included in this category?
5. What is the purpose of the essay? Is it to inform? To persuade? To serve as a model? Anything else? Explain your response.
6. Murray ends the essay with ellipses. Why?
7. Notice the sentence in paragraph 29 that has four consecutive words with the prefix "re-." What is the purpose and effect of this rhetorical device?

WRITING

1. Murray focuses on the process, craft, and purpose of the writer, but he does not define "writer." Write an extended definition explaining what he means by this occupation or profession.
2. Write an essay explaining your own writing process. Do not be intimidated if it is not like the one described by Murray. Compare and contrast your method with that of one or more of your classmates.
3. **WRITING AN ARGUMENT:** Murray suggests that revision is actually "fun" (paragraph 24). Do you agree or disagree? Write an essay defending your position.

---

www.mhhe.com/
mhreader   For more information on Donald M. Murray, go to:
More Resources > Ch. 1 Reading & Writing

✳

What Adolescents Miss When We Let Them Grow Up in Cyberspace

Brent Staples

Brent Staples (b. 1951), the oldest of nine children, was born in Chester, Pennsylvania. He received a BA from Widener University (1973) and a PhD in behavioral sciences from the University of Chicago (1977). After a brief career as a professor of psychology at various colleges in Pennsylvania and Illinois, Staples became a newspaper reporter, first for the Chicago Sun-Times *and, since 1983, as an editorial writer for* The New York Times. *Staples is the author of an autobiography,* Parallel Time: Growing Up Black and White *(1994). Although he is of African American descent, Staples warns against being labeled a black writer. "I'm writing about universal themes," he says, "family and leaving home and developing your own identity—which all Americans can enjoy and understand." In the following essay, which appeared as an op-ed piece in* The New York Times *in 2004, Staples turns a skeptical eye on the communication styles of adolescents cruising the Net.*

My 10th-grade heartthrob was the daughter of a fearsome steelworker who   1
struck terror into the hearts of 15-year-old boys. He made it his business to answer the telephone—and so always knew who was calling—and grumbled in
the background when the conversation went on too long. Unable to make time
by phone, the boy either gave up or appeared at the front door. This meant submitting to the intense scrutiny that the girl's father soon became known for.

He greeted me with a crushing handshake, then leaned in close in a trans-   2
parent attempt to find out whether I was one of those bad boys who smoked.
He retired to the den during the visit, but cruised by the living room now and
then to let me know he was watching. He let up after some weeks, but only after getting across what he expected of a boy who spent time with his daughter
and how upset he'd be if I disappointed him.

This was my first sustained encounter with an adult outside my family who   3
needed to be convinced of my worth as a person. This, of course, is a crucial
part of growing up. Faced with same challenge today, however, I would probably pass on meeting the girl's father—and outflank him on the Internet.

Thanks to e-mail, online chat rooms and instant messages—which permit   4
private, real-time conversations—adolescents have at last succeeded in shield-
ing their social lives from adult scrutiny. But this comes at a cost: teenagers
nowadays are both more connected to the world at large than ever, and more
cut off from the social encounters that have historically prepared young people
for the move into adulthood.

The Internet was billed as a revolutionary way to enrich our social lives and   5
expand our civic connections. This seems to have worked well for elderly people and others who were isolated before they got access to the World Wide Web.

But a growing body of research is showing that heavy use of the Net can actually isolate younger socially connected people who unwittingly allow time online to replace face-to-face interactions with their families and friends.

Online shopping, checking e-mail and Web surfing—mainly solitary 6
activities—have turned out to be more isolating than watching television, which friends and family often do in groups. Researchers have found that the time spent in direct contact with family members drops by as much as half for every hour we use the Net at home.

This should come as no surprise to the two-career couples who have seen 7
their domestic lives taken over by e-mail and wireless tethers that keep people working around the clock. But a startling body of research from the Human-Computer Interaction Institute at Carnegie Mellon has shown that heavy Internet use can have a stunting effect outside the home as well.

Studies show that gregarious, well-connected people actually lost friends, 8
and experienced symptoms of loneliness and depression, after joining discussion groups and other activities. People who communicated with disembodied strangers online found the experience empty and emotionally frustrating but were nonetheless seduced by the novelty of the new medium. As Professor Robert Kraut, a Carnegie Mellon researcher, told me recently, such people allowed low-quality relationships developed in virtual reality to replace higher-quality relationships in the real world.

No group has embraced this socially impoverishing trade-off more enthusi- 9
astically than adolescents, many of whom spend most of their free hours cruising the Net in sunless rooms. This hermetic existence has left many of these teenagers with nonexistent social skills—a point widely noted in stories about the computer geeks who rose to prominence in the early days of Silicon Valley.

Adolescents are drawn to cyberspace for different reasons than adults. As 10
the writer Michael Lewis observed in his book *Next: The Future Just Happened,* children see the Net as a transformational device that lets them discard quotidian identities for more glamorous ones. Mr. Lewis illustrated the point with Marcus Arnold, who, as a 15-year-old, adopted a pseudonym a few years ago and posed as a 25-year-old legal expert for an Internet information service. Marcus did not feel the least bit guilty, and wasn't deterred, when real-world lawyers discovered his secret and accused him of being a fraud. When asked whether he had actually read the law, Marcus responded that he found books "boring," leaving us to conclude that he had learned all he needed to know from his family's big-screen TV.

Marcus is a child of the Net, where everyone has a pseudonym, telling a 11
story makes it true, and adolescents create older, cooler, more socially powerful selves any time they wish. The ability to slip easily into a new, false self is tailor-made for emotionally fragile adolescents, who can consider a bout of acne or a few excess pounds an unbearable tragedy.

But teenagers who spend much of their lives hunched over computer 12
screens miss the socializing, the real-world experience that would allow them to leave adolescence behind and grow into adulthood. These vital experiences, like much else, are simply not available in a virtual form.

COMPREHENSION

1. As he explores the impact of communication styles past and present, how does Staples examine the implications of communicating on the Internet? Although the writer's title focuses on adolescents, what is the broader message he wants to convey about communication styles on the Internet?
2. What does Staples say about the creation of multiple personalities on the Internet? Does he approve or disapprove of this phenomenon? Explain.
3. Staples writes that many Internet activities are "solitary activities" (paragraph 6). Would you agree or disagree with this statement, and why?

RHETORIC

1. The writer begins this essay with an anecdote. What is his purpose, and what is the effect in terms of his relationship to his audience?
2. What is Staples's thesis or claim, and where does he state it most clearly?
3. Where does Staples present expert testimony to support his argument? What other forms of illustration does he employ?
4. This essay was written as an op-ed piece (meaning that it appears opposite the editorial page). What aspects of form, length, purpose, and style suggest what a typical op-ed essay attempts to achieve?
5. Examine the writer's final paragraph, and explain why you think (or do not think) it is an effective conclusion.

WRITING

1. Recount a situation in which you felt that you were spending too much time on the Net. Did you experience any of the symptoms described by Staples, or other sensations that you found worrisome? How did you resolve this dilemma?
2. Staples admits that the Internet can be beneficial, especially for older people, but not (especially) for adolescents. Write a comparative essay in which you examine younger and older people's use of the Internet today.
3. **WRITING AN ARGUMENT:** Staples writes, "The Internet was billed as a revolutionary way to enrich our social lives and expand our civic connections." Of course, he then tries to convince his audience that this "revolution" has turned out to be an illusion. Write an argumentative essay in which you claim that the Internet revolution actually has led to richer personal, social, and civic lives. Be certain to provide examples to support your claim.

✳

Sex, Lies and Conversation: Why Is It So Hard for Men and Women to Talk to Each Other?

Deborah Tannen

Deborah Tannen (b. 1945 in Brooklyn, New York) holds a PhD in linguistics from the University of California at Berkeley. She is University Professor and Professor of Linguistics at Georgetown University. Tannen published numerous specialized articles and books on language and linguistics before becoming nationally known as a best-selling author. She publishes regularly in such magazines as Vogue *and* New York, *and her book* That's Not What I Meant: How Conversational Style Makes or Breaks Your Relations with Others *(1986) drew national attention to her work on interpersonal communication. Her other popular books on communication include* You Just Don't Understand: Women and Men in Conversation *(1990),* Talking from 9 to 5: How Women's and Men's Conversational Styles Affect Who Gets Heard, Who Gets Credit, and What Gets Done at Work *(1994), and* I Only Say This Because I Love You: How the Way We Talk Can Make or Break Family Relationships Throughout Our Lives *(2001). The following essay was published in the* Washington Post *in 1990.*

I was addressing a small gathering in a suburban Virginia living room—a women's group that had invited men to join them. Throughout the evening, one man had been particularly talkative, frequently offering ideas and anecdotes, while his wife sat silently beside him on the couch. Toward the end of the evening, I commented that women frequently complain that their husbands don't talk to them. This man quickly concurred. He gestured toward his wife and said, "She's the talker in our family." The room burst into laughter; the man looked puzzled and hurt. "It's true," he explained. "When I come home from work I have nothing to say. If she didn't keep the conversation going, we'd spend the whole evening in silence." 1

This episode crystallizes the irony that although American men tend to talk more than women in public situations, they often talk less at home. And this pattern is wreaking havoc with marriage. 2

The pattern was observed by political scientist Andrew Hacker in the late '70s. Sociologist Catherine Kohler Riessman reports in her new book *Divorce Talk* that most of the women she interviewed—but only a few of the men—gave lack of communication as the reason for their divorces. Given the current divorce rate of nearly 50 percent, that amounts to millions of cases in the United States every year—a virtual epidemic of failed conversation. 3

In my own research, complaints from women about their husbands most often focused not on tangible inequities such as having given up the chance for a career to accompany a husband to his, or doing far more than their share of daily life-support work like cleaning, cooking, social arrangements and errands. 4

Instead, they focused on communication: "He doesn't listen to me," "He doesn't talk to me." I found, as Hacker observed years before, that most wives want their husbands to be, first and foremost, conversational partners, but few husbands share this expectation of their wives.

In short, the image that best represents the current crisis is the stereotypical   5 cartoon scene of a man sitting at the breakfast table with a newspaper held up in front of his face, while a woman glares at the back of it, wanting to talk.

Linguistic Battle of the Sexes

How can women and men have such different impressions of communication   6 in marriage? Why the widespread imbalance in their interests and expectations?

In the April [1990] issue of *American Psychologist,* Stanford University's   7 Eleanor Maccoby reports the results of her own and others' research showing that children's development is most influenced by the social structure of peer interactions. Boys and girls tend to play with children of their own gender, and their sex-separate groups have different organizational structures and interactive norms.

I believe these systematic differences in childhood socialization make talk   8 between women and men like cross-cultural communication, heir to all the attraction and pitfalls of that enticing but difficult enterprise. My research on men's and women's conversations uncovered patterns similar to those described for children's groups.

For women, as for girls, intimacy is the fabric of relationships, and talk is   9 the thread from which it is woven. Little girls create and maintain friendships by exchanging secrets; similarly, women regard conversation as the cornerstone of friendship. So a woman expects her husband to be a new and improved version of a best friend. What is important is not the individual subjects that are discussed but the sense of closeness, of a life shared, that emerges when people tell their thoughts, feelings, and impressions.

Bonds between boys can be as intense as girls', but they are based less on   10 talking, more on doing things together. Since they don't assume talk is the cement that binds a relationship, men don't know what kind of talk women want, and they don't miss it when it isn't there.

Boys' groups are larger, more inclusive, and more hierarchical, so boys must   11 struggle to avoid the subordinate position in the group. This may play a role in women's complaints that men don't listen to them. Some men really don't like to listen, because being the listener makes them feel one-down, like a child listening to adults or an employee to a boss.

But often when women tell men, "You aren't listening," and the men   12 protest, "I am," the men are right. The impression of not listening results from misalignments in the mechanics of conversation. The misalignment begins as soon as a man and a woman take physical positions. This became clear when I studied videotapes made by psychologist Bruce Dorval of children and adults talking to their same-sex best friends. I found that at every age, the girls and women faced each other directly, their eyes anchored on each other's faces. At every age, the boys and men sat at angles to each other and looked elsewhere in the room, periodically glancing at each other. They were obviously attuned to each other, often mirroring each other's movements. But the tendency of men to face away can give women the impression they aren't listening even when they are. A young woman in college was frustrated: Whenever she told her boyfriend she wanted to talk to him, he would lie down on the floor, close his eyes, and put his arm over his face. This signaled to her, "He's taking a nap." But he insisted he was listening extra hard. Normally, he looks around the room, so he is easily distracted. Lying down and covering his eyes helped him concentrate on what she was saying.

Analogous to the physical alignment that women and men take in conver- 13 sation is their topical alignment. The girls in my study tended to talk at length about one topic, but the boys tended to jump from topic to topic. The second-grade girls exchanged stories about people they knew. The second-grade boys teased, told jokes, noticed things in the room and talked about finding games to play. The sixth-grade girls talked about problems with a mutual friend. The sixth-grade boys talked about 55 different topics, none of which extended over more than a few turns.

Listening to Body Language

Switching topics is another habit that gives women the impression men aren't 14 listening, especially if they switch to a topic about themselves. But the evidence of the 10th-grade boys in my study indicates otherwise. The 10th-grade boys sprawled across their chairs with bodies parallel and eyes straight ahead, rarely looking at each other. They looked as if they were riding in a car, staring out the windshield. But they were talking about their feelings. One boy was upset because a girl had told him he had a drinking problem, and the other was feeling alienated from all his friends.

Now, when a girl told a friend about a problem, the friend responded by 15 asking probing questions and expressing agreement and understanding. But the boys dismissed each other's problems. Todd assured Richard that his drinking was "no big problem" because "sometimes you're funny when you're off your butt." And when Todd said he felt left out, Richard responded, "Why should you? You know more people than me."

Women perceived such responses as belittling and unsupportive. But the 16 boys seemed satisfied with them. Whereas women reassure each other by implying, "You shouldn't feel bad because I've had similar experiences," men do so by implying, "You shouldn't feel bad because your problems aren't so bad."

There are even simpler reasons for women's impression that men don't lis- 17 ten. Linguist Lynette Hirschman found that women make more listener-noise, such as "mhm," "uhuh," and "yeah," to show "I'm with you." Men, she found, more often give silent attention. Women who expect a stream of listener-noise interpret silent attention as no attention at all.

Women's conversational habits are as frustrating to men as men's are to   18
women. Men who expect silent attention interpret a stream of listener-noise as
overreaction or impatience. Also, when women talk to each other in a close,
comfortable setting, they often overlap, finish each other's sentences and anticipate what the other is about to say. This practice, which I call "participatory
listenership," is often perceived by men as interruption, intrusion and lack of attention.

A parallel difference caused a man to complain about his wife, "She just   19
wants to talk about her own point of view. If I show her another view, she gets
mad at me." When most women talk to each other, they assume a conversation-
alist's job is to express agreement and support. But many men see their conversational duty as pointing out the other side of an argument. This is heard as
disloyalty by women, and refusal to offer the requisite support. It is not that
women don't want to see other points of view, but that they prefer them
phrased as suggestions and inquiries rather than as direct challenges.

In his book *Fighting for Life,* Walter Ong points out that men use "agonistic"   20
or warlike, oppositional formats to do almost anything; thus discussion becomes debate, and conversation a competitive sport. In contrast, women see
conversation as a ritual means of establishing rapport. If Jane tells a problem
and June says she has a similar one, they walk away feeling closer to each other.
But this attempt at establishing rapport can backfire when used with men. Men
take too literally women's ritual "troubles talk," just as women mistake men's
ritual challenges for real attack.

The Sounds of Silence

These differences begin to clarify why women and men have such different ex-   21
pectations about communication in marriage. For women, talk creates intimacy.
Marriage is an orgy of closeness: you can tell your feelings and thoughts, and
still be loved. Their greatest fear is being pushed away. But men live in a hierarchical world, where talk maintains independence and status. They are on guard
to protect themselves from being put down and pushed around.

This explains the paradox of the talkative man who said of his silent wife,   22
"She's the talker." In the public setting of a guest lecture, he felt challenged to
show his intelligence and display his understanding of the lecture. But at home,
where he has nothing to prove and no one to defend against, he is free to remain
silent. For his wife, being home means she is free from the worry that something
she says might offend someone, or spark disagreement, or appear to be showing off; at home she is free to talk.

The communication problems that endanger marriage can't be fixed by me-   23
chanical engineering. They require a new conceptual framework about the role
of talk in human relationships. Many of the psychological explanations that
have become second nature may not be helpful, because they tend to blame either women (for not being assertive enough) or men (for not being in touch
with their feelings). A sociolinguistic approach by which male-female conversa-

tion is seen as cross-cultural communication allows us to understand the problem and forge solutions without blaming either party.

Once the problem is understood, improvement comes naturally, as it did to 24 the young woman and her boyfriend who seemed to go to sleep when she wanted to talk. Previously, she had accused him of not listening, and he had refused to change his behavior, since that would be admitting fault. But then she learned about and explained to him the differences in women's and men's habitual ways of aligning themselves in conversation. The next time she told him she wanted to talk, he began, as usual, by lying down and covering his eyes. When the familiar negative reaction bubbled up, she reassured herself that he really was listening. But then he sat up and looked at her. Thrilled she asked why. He said, "You like me to look at you when we talk, so I'll try to do it." Once he saw their differences as cross-cultural rather than right and wrong, he independently altered his behavior.

Women who feel abandoned and deprived when their husbands won't lis- 25 ten to or report daily news may be happy to discover their husbands trying to adapt once they understand the place of small talk in women's relationships. But if their husbands don't adapt, the women may still be comforted that for men, this is not a failure of intimacy. Accepting the difference, the wives may look to their friends or family for that kind of talk. And husbands who can't provide it shouldn't feel their wives have made unreasonable demands. Some couples will still decide to divorce, but at least their decisions will be based on realistic expectations.

In these times of resurgent ethnic conflicts, the world desperately needs 26 cross-cultural understanding. Like charity, successful cross-cultural communication should begin at home.

COMPREHENSION

1. What is the thesis or claim of this essay? Where does the author most clearly articulate it?
2. To advance her argument, the author cites political scientists and sociologists, while she, herself, is a linguist. What exactly is the nature of these three professions? What do professionals in the first two fields do? Why does the author use their observations in developing her argument?
3. Why does the author employ a question in her title? What other device does she employ in her title to capture the reader's attention? (*Hint:* It is a reference to the title of a movie.)

RHETORIC

1. The author begins her essay with an anecdote. Is this an effective way of opening this particular essay? Why or why not?

2. Besides anecdotes, the author uses statistics, social science research, appeals to authority, and definition in advancing her argument. Find at least one example of each device. Explain the effectiveness or lack thereof.

3. Where and how does the author imply that she is an authority on the subject? How does this contribute to or detract from her ability to win the reader's confidence?

4. The author divides her essay into four sections: one untitled and three with headings. How does each section relate to the others structurally and thematically?

5. The author dramatically states that "Given the current divorce rate of nearly 50 percent" the United States has a "virtual epidemic of failed conversation" (paragraph 3). Is this fact or opinion? Does it serve to heighten or weaken the import of her thesis?

6. Concerning the lack of proper communication between men and women, the author states, "Once the problem is understood, improvement comes naturally" (paragraph 24). Is this statement substantiated or backed up with evidence? Explain.

7. Explain the analogy the author employs in the final paragraph. Is it a good or poor analogy? Explain.

WRITING

1. Another linguist has written an essay entitled "The Communication Panacea," which argues that much of what is blamed on lack of communication actually has economic and political causes. Argue for or against this proposition in the light of the ideas advanced in Tannen's essay.

2. Using some of the observational methods described in the essay, conduct your own ethnographic research by observing a couple communicating. Write a report discussing your findings.

3. **WRITING AN ARGUMENT:** The author states, "Once the problem is understood, improvement comes naturally." Argue for or against this proposition.

| www.mhhe.com/ **mhreader** | For more information on Deborah Tannen, go to: **More Resources > Ch. 1 Reading & Writing** |

✳

The Language of Discretion

Amy Tan

Amy Tan (b. 1952) was born in California, several years after her mother and father immigrated from China. She was educated at San Jose State and the University of California, Berkeley. Tan has worked as a reporter and as a technical writer; her fiction focuses on the lives of Chinese-American women seeking to reconcile their traditional Chinese heritage with modern American culture. Her books are The Joy Luck Club *(1989),* The Kitchen God's Wife *(1991),* The Hundred Secret Senses *(1996), and* Bonesetter's Daughter *(2001). In this narrative essay from* The State of the Language, *Tan writes with both emotion and clarity about growing up with two languages, and she attacks some linguists who make hasty assumptions.*

At a recent family dinner in San Francisco, my mother whispered to me: "Sau-sau [Brother's Wife] pretends too hard to be polite! Why bother? In the end, she always takes everything." 1

My mother thinks like a *waixiao*, an expatriate, temporarily away from China since 1949, no longer patient with ritual courtesies. As if to prove her point, she reached across the table to offer my elderly aunt from Beijing the last scallop from the Happy Family seafood dish. 2

Sau-sau scowled. *"B'vuo, zhen b'yao!"* (I don't want it, really I don't!) she cried, patting her plump stomach. 3

"Take it! Take it!" scolded my mother in Chinese. 4

"Full, I'm already full," Sau-sau protested weakly, eyeing the beloved scallop. 5

"Ai!" exclaimed my mother, completely exasperated. "Nobody else wants it. If you don't take it, it will only rot!" 6

At this point, Sau-sau sighed, acting as if she were doing my mother a big favor by taking the wretched scrap off her hands. 7

My mother turned to her brother, a high-ranking communist official who was visiting her in California for the first time: "In America a Chinese person could starve to death. If you say you don't want it, they won't ask you again forever." 8

My uncle nodded and said he understood fully: Americans take things quickly because they have no time to be polite. 9

I thought about this misunderstanding again—of social contexts failing in translation—when a friend sent me an article from *The New York Times Magazine* (24 April 1988). The article, on changes in New York's Chinatown, made passing reference to the inherent ambivalence of the Chinese language. 10

Chinese people are so "discreet and modest," the article stated, there aren't even words for "yes" and "no." 11

That's not true. I thought, although I can see why an outsider might think 12
that. I continued reading.

If one is Chinese, the article went on to say. "One compromises, one doesn't 13
hazard a loss of face by an overemphatic response."

My throat seized. Why do people keep saying these things? As if we truly 14
were those little dolls sold in Chinatown tourist shops, heads bobbing up and
down in complacent agreement to anything said!

I worry about the effect of one-dimensional statements on the unwary and 15
guileless. When they read about this so-called vocabulary deficit, do they also
conclude that Chinese people evolved into a mild-mannered lot because the
language only allowed them to hobble forth with minced words?

Something enormous is always lost in translation. Something insidious 16
seeps into the gaps, especially when amateur linguists continue to compare,
one-for-one, language differences and then put forth notions wide open to mis-
interpretation: that Chinese people have no direct linguistic means to make decisions, assert or deny, affirm or negate, just say no to drug dealers, or behave
properly on the witness stand when told, "Please answer yes or no."

Yet one can argue, with the help of renowned linguists, that the Chinese are 17
indeed up a creek without "yes" and "no." Take any number of variations on
the old language-and-reality theory stated years ago by Edward Sapir: "Human
beings . . . are very much at the mercy of the particular language which has become the medium for their society. . . . The fact of the matter is that the 'real
world' is to a large extent built up on the language habits of the group."[1]

This notion was further bolstered by the famous Sapir-Whorf hypothesis, 18
which roughly states that one's perception of the world and how one functions
in it depends a great deal on the language used. As Sapir, Whorf, and new carriers of the banner would have us believe, language shapes our thinking, channels us along certain patterns embedded in words, syntactic structures, and
intonation patterns. Language has become the peg and the shelf that enables us
to sort out and categorize the world. In English, we see "cats" and "dogs"; what
if the language had also specified *glatz*, meaning "animals that leave fur on the
sofa," and *glotz*, meaning "animals that leave fur and drool on the sofa"? How
would language, the enabler, have changed our perceptions with slight vocabulary variations?

And if this were the case—of language being the master of destined 19
thought—think of the opportunities lost from failure to evolve two little words,
yes and *no*, the simplest of opposites! Ghenghis Khan could have been sent back
to Mongolia. Opium wars might have been averted. The Cultural Revolution
could have been sidestepped.

There are still many, from serious linguists to pop psychology cultists, who 20
view language and reality as inextricably tied, one being the consequence of the
other. We have traversed the range from the Sapir-Whorf hypothesis to est and
neurolinguistic programming, which tell us "you are what you say."

[1]Edward Sapir, *Selected Writings*, ed. D. G. Mandelbaum (Berkeley and Los Angeles, 1949).

I too have been intrigued by the theories. I can summarize, albeit badly, 21
ages-old empirical evidence: of Eskimos and their infinite ways to say "snow,"
their ability to *see* the differences in snowflake configurations, thanks to the
richness of their vocabulary, while non-Eskimo speakers like myself founder in
"snow," "more snow," and "lots more where that came from."

I too have experienced dramatic cognitive awakenings via the word. Once I 22
added "mauve" to my vocabulary I began to see it everywhere. When I learned
how to pronounce *prix fixe*, I ate French food at prices better than the easier-to-
say *à la carte* choices.

But just how seriously are we supposed to take this?   23

Sapir said something else about language and reality. It is the part that of- 24
ten gets left behind in the dot-dot-dots of quotes: " . . . No two languages are
ever sufficiently similar to be considered as representing the same social reality.
The worlds in which different societies live are distinct worlds, not merely the
same world with different labels attached."

When I first read this, I thought, Here at last is validity for the dilemmas I 25
felt growing up in a bicultural, bilingual family! As any child of immigrant parents knows, there's a special kind of double bind attached to knowing two languages. My parents, for example, spoke to me in both Chinese and English; I
spoke back to them in English.

"Amy-ah!" they'd call to me.   26

"What?" I'd mumble back.   27

"Do not question us when we call," they scolded me in Chinese. "It is not 28
respectful."

"What do you mean?"   29

"Ai! Didn't we just tell you not to question?"   30

To this day, I wonder which parts of my behavior were shaped by Chinese, 31
which by English. I am tempted to think, for example, that if I am of two minds
on some matter it is due to the richness of my linguistic experiences, not to any
personal tendencies toward wishy-washiness. But which mind says what?

Was it perhaps patience—developed through years of deciphering my 32
mother's fractured English—that had me listening politely while a woman
announced over the phone that I had won one of five valuable prizes? Was it
respect—pounded in by the Chinese imperative to accept convoluted
explanations—that had me agreeing that I might find it worthwhile to drive
seventy-five miles to view a time-share resort? Could I have been at a loss for
words when asked, "Wouldn't you like to win a Hawaiian cruise or perhaps a
fabulous Star of India designed exclusively by Carter and Van Arpels?"

And when this same woman called back a week later, this time complaining 33
that I had missed my appointment, obviously it was my type A language that
kicked into gear and interrupted her. Certainly, my blunt denial—"Frankly I'm
not interested"—was as American as apple pie. And when she said, "But it's in
Morgan Hill," and I shouted, "Read my lips. I don't care if it's Timbuktu," you
can be sure I said it with the precise intonation expressing both cynicism and
disgust.

It's dangerous business, this sorting out of language and behavior. Which 34 one is English? Which is Chinese? The categories manifest themselves: passive and aggressive, tentative and assertive, indirect and direct. And I realize they are just variations of the same theme: that Chinese people are discreet and modest.

Reject them all! 35

If my reaction is overly strident, it is because I cannot come across as too 36 emphatic. I grew up listening to the same lines over and over again, like so many rote expressions repeated in an English phrasebook. And I too almost came to believe them.

Yet if I consider my upbringing more carefully, I find there was nothing dis- 37 creet about the Chinese language I grew up with. My parents made everything abundantly clear. Nothing wishy-washy in their demands, no compromises accepted: "Of course you will become a famous neurosurgeon," they told me. "And yes, a concert pianist on the side."

In fact, now that I remember, it seems that the more emphatic outbursts al- 38 ways spilled over into Chinese: "Not that way! You must wash rice so not a single grain spills out."

I do not believe that my parents—both immigrants from mainland China— 39 are an exception to the modest-and-discreet rule. I have only to look at the number of Chinese engineering students skewing minority ratios at Berkeley, MIT, and Yale. Certainly they were not raised by passive mothers and fathers who said, "It is up to you, my daughter. Writer, welfare recipient, masseuse, or molecular engineer—you decide."

And my American mind says, See, those engineering students weren't able 40 to say no to their parents' demands. But then my Chinese remembers: Ah, but those parents all wanted their sons and daughters to be *pre-med.*

Having listened to both Chinese and English, I also tend to be suspicious of 41 any comparisons between the two languages. Typically, one language—that of the person doing the comparing—is often used as the standard, the benchmark for a logical form of expression. And so the language being compared is always in danger of being judged deficient or superfluous, simplistic or unnecessarily complex, melodious or cacophonous. English speakers point out that Chinese is extremely difficult because it relies on variations in tone barely discernible to the human ear. By the same token, Chinese speakers tell me English is extremely difficult because it is inconsistent, a language of too many broken rules, of Mickey Mice and Donald Ducks.

Even more dangerous to my mind is the temptation to compare both lan- 42 guage and behavior *in translation.* To listen to my mother speak English, one might think she has no concept of past or future tense, that she doesn't see the difference between singular and plural, that she is gender blind because she calls my husband "she." If one were not careful, one might also generalize that, based on the way my mother talks, all Chinese people take a circumlocutory route to get to the point. It is, in fact, my mother's idiosyncratic behavior to ramble a bit.

Sapir was right about differences between two languages and their realities. I 43
can illustrate why word-for-word translation is not enough to translate meaning and intent. I once received a letter from China which I read to non-Chinese
speaking friends. The letter, originally written in Chinese, had been translated
by my brother-in-law in Beijing. One portion described the time when my uncle
at age ten discovered his widowed mother (my grandmother) had remarried—
as a number three concubine, the ultimate disgrace for an honorable family. The
translated version of my uncle's letter read in part:

> In 1925, I met my mother in Shanghai. When she came to me, I didn't have
> greeting to her as if seeing nothing. She pull me to a corner secretly and asked
> me why didn't have greeting to her. I couldn't control myself and cried, "Ma!
> Why did you leave us? People told me: one day you ate a beancake yourself.
> Your sister-in-law found it and sweared at you, called your names. So . . . is it
> true?" She clasped my hand and answered immediately, "It's not true, don't say
> what like this." After this time, there was a few chance to meet her.

"What!" cried my friends. "Was eating a beancake so terrible?"   44

Of course not. The beancake was simply a euphemism; a ten-year-old boy 45
did not dare question his mother on something as shocking as concubinage.
Eating a beancake was his equivalent for committing this selfish act, something
inconsiderate of all family members, hence, my grandmother's despairing response to what seemed like a ludicrous charge of gluttony. And sure enough,
she was banished from the family, and my uncle saw her only a few times before her death.

While the above may fuel people's argument that Chinese is indeed a lan- 46
guage of extreme discretion, it does not mean that Chinese people speak in secrets and riddles. The contexts are fully understood. It is only to those on the
outside that the language seems cryptic, the behavior inscrutable.

I am, evidently, one of the outsiders. My nephew in Shanghai, who recently 47
started taking English lessons, has been writing me letters in English. I had told
him I was a fiction writer, and so in one letter he wrote, "Congratulate to you on
your writing. Perhaps one day I should like to read it." I took it in the same vein
as "Perhaps one day we can get together for lunch." I sent back a cheery note. A
month went by and another letter arrived from Shanghai. "Last one perhaps I
hadn't writing distinctly," he said. "In the future, you'll send a copy of your
works for me."

I try to explain to my English-speaking friends that Chinese language use is 48
more *strategic* in manner, whereas English tends to be more direct; an American
business executive may say, "Let's make a deal," and the Chinese manager may
reply, "Is your son interested in learning about your widget business?" Each to
his or her own purpose, each with his or her own linguistic path. But I hesitate
to add more to the pile of generalizations, because no matter how many examples I provide and explain, I fear that it appears defensive and only reinforces
the image: that Chinese people are "discreet and modest"—and it takes an
American to explain what they really mean.

Why am I complaining? The description seems harmless enough (after all, *The* 49
New York Times Magazine writer did not say "slippery and evasive"). It is precisely the bland, easy acceptability of the phrase that worries me.

I worry that the dominant society may see Chinese people from a limited— 50
and limiting—perspective. I worry that seemingly benign stereotypes may be
part of the reason there are few Chinese in top management positions, in mainstream political roles. I worry about the power of language: that if one says anything enough times—in *any* language—it might become true.

Could this be why Chinese friends of my parents' generation are willing to 51
accept the generalization?

"Why are you complaining?" one of them said to me. "If people think we 52
are modest and polite, let them think that. Wouldn't Americans be pleased to
admit they are thought of as polite?"

And I do believe anyone would take the description as a compliment—at 53
first. But after a while, it annoys, as if the only things that people heard one say
were phatic remarks: "I'm so pleased to meet you. I've heard many wonderful
things about you. For me? You shouldn't have!"

These remarks are not representative of new ideas, honest emotions, or con- 54
sidered thought. They are what is said from the polite distance of social contexts: of greetings, farewells, wedding thank-you notes, convenient excuses, and
the like.

It makes me wonder though. How many anthropologists, how many soci- 55
ologists, how many travel journalists have documented so-called "natural interactions" in foreign lands, all observed with spiral notebook in hand? How
many other cases are there of the long-lost primitive tribe, people who turned
out to be sophisticated enough to put on the stone-age show that ethnologists
had come to see?

And how many tourists fresh off the bus have wandered into Chinatown 56
expecting the self-effacing shopkeeper to admit under duress that the goods are
not worth the price asked? I have witnessed it.

"I don't know," the tourist said to the shopkeeper, a Cantonese woman in 57
her fifties. "It doesn't look genuine to me. I'll give you three dollars."

"You don't like my price, go somewhere else," said the shopkeeper.   58

"You are not a nice person," cried the shocked tourist, "not a nice person at 59
all!"

"Who say I have to be nice," snapped the shopkeeper.   60

"So how does one say 'yes' and 'no' in Chinese?" ask my friends a bit warily.   61

And here I do agree in part with *The New York Times Magazine* article. There 62
is no one word for "yes" or "no"—but not out of necessity to be discreet. If anything, I would say the Chinese equivalent of answering "yes" or "no" is dis*crete,*
that is, specific to what is asked.

Ask a Chinese person if he or she has eaten, and he or she might say *chrle* 63
(eaten already) or perhaps *meiyou* (have not).

Ask, "So you had insurance at the time of the accident?" and the response ₆₄ would be *dwei* (correct) or *meiyou* (did not have).

Ask, "Have you stopped beating your wife?" and the answer refers directly ₆₅ to the proposition being asserted or denied: stopped already, still have not, never beat, have no wife.

What could be clearer? ₆₆

As for those who are still wondering how to translate the language of discre- ₆₇ tion, I offer this personal example.

My aunt and uncle were about to return to Beijing after a three-month visit ₆₈ to the United States. On their last night I announced I wanted to take them out to dinner.

"Are you hungry?" I asked in Chinese. ₆₉

"Not hungry," said my uncle promptly, the same response he once gave me ₇₀ ten minutes before he suffered a low-blood-sugar attack.

"Not too hungry," said my aunt. "Perhaps you're hungry?" ₇₁

"A little," I admitted. ₇₂

"We can eat, we can eat," they both consented. ₇₃

"What kind of food?" I asked. ₇₄

"Oh, doesn't matter. Anything will do. Nothing fancy, just some simple ₇₅ food is fine."

"Do you like Japanese food? We haven't had that yet," I suggested. ₇₆

They looked at each other. ₇₇

"We can eat it," said my uncle bravely, this survivor of the Long March. ₇₈

"We have eaten it before," added my aunt. "Raw fish." ₇₉

"Oh, you don't like it?" I said. "Don't be polite. We can go somewhere else." ₈₀

"We are not being polite. We can eat it," my aunt insisted. ₈₁

So I drove them to Japantown and we walked past several restaurants fea- ₈₂ turing colorful plastic displays of sushi.

"Not this one, not this one either," I continued to say, as if searching for a ₈₃ Japanese restaurant similar to the last. "Here it is," I finally said, turning into a restaurant famous for its Chinese fish dishes from Shandong.

"Oh, Chinese food!" cried my aunt, obviously relieved. ₈₄

My uncle patted my arm. "You think Chinese." ₈₅

"It's your last night here in America," I said. "So don't be polite. Act like an ₈₆ American."

And that night we ate a banquet. ₈₇

COMPREHENSION

1. Why is the writer suspicious of any comparisons made between Chinese and English? What dangerous generalizations may be drawn?
2. What is meant by "the double bind attached to knowing two languages" (paragraph 25)?
3. In your own words, define Sapir's language theory.

RHETORIC

1. What tone does Tan use in her essay? Is her approach objective or subjective? Justify your response.
2. What is Tan's thesis? Is it implied or stated explicitly?
3. How do the anecdotes at the beginning and conclusion of the essay help frame what happens in between? How well do they illustrate or support the essay's body?
4. Cite specific examples of irony or humor in the essay. Is it used consistently throughout the piece? How does its use advance Tan's main ideas?
5. How does Tan employ comparison and contrast to structure this essay?
6. How many sections are there in this essay? What principles of writing and coherence govern each section?

WRITING

1. Tan writes about the generalizations made by "outsiders" about Chinese culture based on the language. Write an essay in which you explore this topic by focusing on misconceptions others may have about you or you may have about others based on language.
2. Tan presents linguistic theories in this essay. Using support from the essay, consider the dangers of linking behavior to language. Can these theories be used to further racist or sexist notions? Are they valid, scientific attempts to study human behavior?
3. **WRITING AN ARGUMENT:** Tan personalizes her essay by referring to family members. Does personal experience strengthen or weaken her thesis? Respond in an argumentative essay.

www.mhhe.com/
mhreader

For more information on Amy Tan, go to:
More Resources > Ch. 1 Reading & Writing

✳

Politics and the English Language

George Orwell

*George Orwell (1903–1950) was the pseudonym of Eric Arthur Blair, an English nov-
elist, essayist, and journalist. Orwell served with the Indian Imperial Police from 1922
to 1927 in Burma, fought in the Spanish Civil War, and acquired from his experience a
disdain of totalitarian and imperialistic systems. This attitude is reflected in the satiric
fable* Animal Farm *(1945) and in the bleak, futuristic novel* 1984 *(1949). This essay,
one of the more famous of the twentieth century, relates sloppy thinking and writing
with political oppression.*

Most people who bother with the matter at all would admit that the English lan-   1
guage is in a bad way, but it is generally assumed that we cannot by conscious
action do anything about it. Our civilization is decadent, and our language—so
the argument runs—must inevitably share in the general collapse. It follows
that any struggle against the abuse of language is a sentimental archaism, like
preferring candles to electric light or hansom cabs to airplanes. Underneath this
lies the half-conscious belief that language is a natural growth and not an instrument which we shape for our own purposes.

Now, it is clear that the decline of a language must ultimately have political   2
and economic causes: it is not due simply to the bad influence of this or that individual writer. But an effect can become a cause, reinforcing the original cause
and producing the same effect in an intensified form, and so on indefinitely. A
man may take to drink because he feels himself to be a failure, and then fail all
the more completely because he drinks. It is rather the same thing that is happening to the English language. It becomes ugly and inaccurate because our
thoughts are foolish, but the slovenliness of our language makes it easier for us
to have foolish thoughts. The point is that the process is reversible. Modern
English, especially written English, is full of bad habits which spread by imitation and which can be avoided if one is willing to take the necessary trouble. If
one gets rid of these habits one can think more clearly, and to think clearly is a
necessary first step towards political regeneration: so that the fight against bad
English is not frivolous and is not the exclusive concern of professional writers.
I will come back to this presently, and I hope that by that time the meaning of
what I have said here will have become clearer. Meanwhile, here are five specimens of the English language as it is now habitually written.

These five passages have not been picked out because they are especially   3
bad—I could have quoted far worse if I had chosen—but because they illustrate
various of the mental vices from which we now suffer. They are a little below
the average, but are fairly representative samples. I number them so that I can
refer back to them when necessary:

1. I am not, indeed, sure whether it is not true to say the Milton who once seemed not unlike a seventeenth-century Shelley had not become, out of an experience even more bitter in each year, more alien (sic) to the founder of that Jesuit sect which nothing could induce him to tolerate.

   —Professor Harold Laski (essay in *Freedom of Expression*)

2. Above all, we cannot play ducks and drakes with a native battery of idioms which prescribes such egregious collocations of vocables as the Basic *put up with* for *tolerate* or *put at a loss* for *bewilder.*

   —Professor Lancelot Hogben (*Interglossa*)

3. On the one side we have the free personality: by definition it is not neurotic, for it has neither conflict nor dream. Its desires, such as they are, are transparent, for they are just what institutional approval keeps in the forefront of consciousness; another institutional pattern would alter their number and intensity; there is little in them that is natural, irreducible, or culturally dangerous. But on the other side, the social bond itself is nothing but the mutual reflection of these self-secure integrities. Recall the definition of love. Is not this the very picture of a small academic? Where is there a place in this hall of mirrors for either personality or fraternity?

   —Essay on psychology in *Politics* (New York)

4. All the "best people" from the gentlemen's clubs, and all the frantic Fascist captains, united in common hatred of Socialism and bestial horror of the rising tide of the mass revolutionary movement, have turned to acts of provocation, to foul incendiarism, to medieval legends of poisoned wells, to legalize their own destruction to proletarian organizations, and rouse the agitated petty-bourgeoisie to chauvinistic fervor on behalf of the fight against the revolutionary way out of the crisis.

   —Communist pamphlet

5. If a new spirit is to be infused into this old country, there is one thorny and contentious reform which must be tackled, and that is the humanization and galvanization of the BBC. Timidity here will bespeak canker and atrophy for the soul. The heart of Britain may be sound and of strong beat, for instance, but the British lion's roar at present is like that of Bottom in Shakespeare's Midsummer Night's Dream—as gentle as any sucking dove. A virile new Britain cannot continue indefinitely to be traduced in the eyes, or rather ears, of the world by the effete languors of Langham Place, brazenly masquerading as "standard English." When the Voice of Britain is heard at nine o'clock, better far and infinitely less ludicrous to hear aitches honestly dropped than the present priggish, inflated, inhibited, schoolma'amish braying of blameless bashful mewing maidens!

   —Letter in *Tribune*

Each of these passages has faults of its own, but, quite apart from avoidable 4 ugliness, two qualities are common to all of them. The first is staleness of imagery; the other is lack of precision. The writer either has a meaning and cannot express it, or he inadvertently says something else, or he is almost indifferent as to whether his words mean anything or not. This mixture of vagueness and sheer incompetence is the most marked characteristic of modern English prose, and especially of any kind of political writing. As soon as certain topics are raised, the concrete melts into the abstract and no one seems able to think of turns of speech that are not hackneyed: prose consists less and less of *words* chosen for the sake of their meaning, and more of *phrases* tacked together like the sections of a prefabricated henhouse. I list below, with notes and examples, various of the tricks by means of which the work of prose construction is habitually dodged:

Dying Metaphors

A newly invented metaphor assists thought by evoking a visual image, while 5 on the other hand a metaphor which is technically "dead" (e.g., *iron resolution*) has in effect reverted to being an ordinary word and can generally be used without loss of vividness. But in between these two classes there is a huge dump of wornout metaphors which have lost all evocative power and are merely used because they save people the trouble of inventing phrases for themselves. Examples are: *Ring the changes on, take up the cudgels for, toe the line, ride roughshod over, stand shoulder to shoulder with, play into the hands of, no axe to grind, grist to the mill, fishing in troubled waters, rift within the lute, on the order of the day, Achilles' heel, swan song, hotbed.* Many of these are used without knowledge of their meaning (what is a "rift," for instance?), and incompatible metaphors are frequently mixed, a sure sign that the writer is not interested in what he is saying. Some metaphors now current have been twisted out of their original meaning without those who use them even being aware of the fact. For example, *toe the line* is sometimes written *tow the line.* Another example is *the hammer and the anvil,* now always used with the implication that the anvil gets the worst of it. In real life it is always the anvil that breaks the hammer, never the other way about: a writer who stopped to think what he was saying would be aware of this, and would avoid perverting the original phrase.

Operators, or Verbal False Limbs

These save the trouble of picking out appropriate verbs and nouns, and at the 6 same time pad each sentence with extra syllables which give it an appearance of symmetry. Characteristic phrases are: *render inoperative, militate against, prove unacceptable, make contact with, be subjected to, give rise to, give grounds for, have the effect of, play a leading part (role) in, make itself felt, take effect, exhibit a tendency to, serve the purpose of,* etc. etc. The keynote is the elimination of simple verbs. Instead of being a single word, such as *break, stop, spoil, mend, kill,* a verb becomes a *phrase,* made up of a noun or adjective tacked on to some general-purposes verb such as *prove, serve, form, play, render.* In addition, the passive voice is wherever possible used in preference to the active, and noun constructions are used instead of gerunds (*by examination of* instead of *by examining*). The range of verbs is further cut down by means of the *-ize* and *de-* formations, and banal statements are given an appearance of profundity by means of the *not un-* formation. Simple conjunctions and prepositions are replaced by such phrases as *with respect to, having regard to, the fact that, by dint of, in view of, in the interests of, on the hypothesis that;* and the ends of sentences are saved from anti-climax by such resounding commonplaces as *greatly to be desired, cannot be left out of account, a development to be expected in the near future, deserving of serious consideration, brought to a satisfactory conclusion,* and so on and so forth.

Pretentious Diction

Words like *phenomenon, element, individual* (as noun), *objective, categorical, effec-*   7
tive, virtual, basic, primary, promote, constitute, exhibit, exploit, utilize, eliminate, liquidate, are used to dress up simple statements and give an air of scientific impartiality to biased judgments. Adjectives like *epoch-making, epic, historic, unforgettable, triumphant, age-old, inevitable, inexorable, veritable,* are used to dignify the sordid processes of international politics, while writing that aims at glorifying war usually takes on an archaic color, its characteristic words being: *realm, throne, chariot, mailed fist, trident, sword, shield, buckler, banner, jackboot, clarion.* Foreign words and expressions such as *cul de sac, ancien régime, deus ex machina, mutatis mutandis, status quo, Gleichschaltung, Weltanschauung,* are used to give an air of culture and elegance. Except for the useful abbreviations *i.e., e.g.,* and *etc.,* there is no real need for any of the hundreds of foreign phrases now current in English. Bad writers, and especially scientific, political and sociological writers, are nearly always haunted by the notion that Latin or Greek words are grander than Saxon ones, and unnecessary words like *expedite, ameliorate, predict, extraneous, deracinated, clandestine, subaqueous* and hundreds of others constantly gain ground from their Anglo-Saxon opposite numbers.[1] The jargon peculiar to Marxist writing (*hyena, hangman, cannibal, petty bourgeois, these gentry, lacquey, flunkey, mad dog, White Guard,* etc.) consists largely of words and phrases translated from Russian, German or French; but the normal way of coining a new word is to use a Latin or Greek root with the appropriate affix and, where necessary, the *-ize* formation. It is often easier to make up words of this kind (*deregionalize, impermissible, extramarital, non-fragmentatory* and so forth) than to think up the English words that will cover one's meaning. The result, in general, is an increase in slovenliness and vagueness.

---

[1]An interesting illustration of this is the way in which the English flower names which were in use till very recently are being ousted by Greek ones, *snapdragon* becoming *antirrhinum, forget-me-not* becoming *myosotis,* etc. It is hard to see any practical reason for this change of fashion: it is probably due to an instinctive turning-away from the more homely word and a vague feeling that the Greek word is scientific.

Meaningless Words

In certain kinds of writing, particularly in art criticism and literary criticism, it   8
is normal to come across long passages which are almost completely lacking in
meaning.[2] Words like *romantic, plastic, values, human, dead, sentimental, natural,
vitality,* as used in art criticism, are strictly meaningless, in the sense that they
not only do not point to any discoverable object, but are hardly even expected
to do so by the reader. When one critic writes, "The outstanding features of Mr.
X's work is its living quality," while another writes, "The immediately striking
thing about Mr. X's work is its peculiar deadness," the reader accepts this as a
simple difference of opinion. If words like *black* and *white* were involved, instead of the jargon words *dead* and *living,* he would see at once that language
was being used in an improper way. Many political words are similarly abused.
The word *Fascism* has now no meaning except in so far as it signifies "something not desirable." The words *democracy, socialism, freedom, patriotic, realistic,
justice,* have each of them several different meanings which cannot be reconciled
with one another. In the case of a word like *democracy,* not only is there no
agreed definition, but the attempt to make one is resisted from all sides. It is almost universally felt that when we call a country democratic we are praising it:
consequently the defenders of every kind of régime claim that it is a democracy,
and fear that they might have to stop using the word if it were tied down to any
one meaning. Words of this kind are often used in a consciously dishonest way.
That is, the person who uses them has his own private definition, but allows his
hearer to think he means something quite different. Statements like *Marshal
Pétain was a true patriot, The Soviet press is the freest in the world, The Catholic
Church is opposed to persecution,* are almost always made with intent to deceive.
Other words used in variable meanings, in most cases more or less dishonestly,
are: *class, totalitarian, science, progressive, reactionary, bourgeois, equality.*

Now that I have made this catalogue of swindles and perversions, let me   9
give another example of the kind of writing that they lead to. This time it must
be of its nature be an imaginary one. I am going to translate a passage of good English into modern English of the worst sort. Here is a well-known verse from
Ecclesiastes:

> I returned, and saw under the sun, that the race is not to the swift, nor the bat-
> tle to the strong, neither yet bread to the wise, nor yet riches to men of under-
> standing, nor yet favor to men of skill; but time and chance happeneth to them
> all.

Here it is in modern English: 10

---

[2]*Example:* "Comfort's catholicity of perception and image, strangely Whitmanesque in range, almost the exact opposite in aesthetic compulsion, continues to evoke that trembling atmospheric
accumulative hinting at a cruel, an inexorably serene timelessness. . . . Wrey Gardiner scores by
aiming at simple bullseyes with precision. Only they are not so simple, and through this contented
sadness runs more than the surface bittersweet of resignation." *(Poetry Quarterly)*

Objective consideration of contemporary phenomena compels the conclusion that success or failure in competitive activities exhibits no tendency to be commensurate with innate capacity, but that a considerable element of the unpredictable must invariably be taken into account.

This is a parody, but not a very gross one. Exhibit (3) above, for instance, 11 contains several patches of the same kind of English. It will be seen that I have not made a full translation. The beginning and ending of the sentence follow the original meaning fairly closely, but in the middle the concrete illustrations—race, battle, bread—dissolve into the vague phrase "success or failure in competitive activities." This had to be so, because no modern writer of the kind I am discussing—no one capable of using phrases like "objective consideration of contemporary phenomena"—would ever tabulate his thoughts in that precise and detailed way. The whole tendency of modern prose is away from concreteness. Now analyze these two sentences a little more closely. The first contains 49 words but only 60 syllables, and all its words are those of everyday life. The second contains 38 words of 90 syllables: 18 of its words are from Latin roots, and one from Greek. The first sentence contains six vivid images, and only one phrase ("time and chance") that could be called vague. The second contains not a single fresh, arresting phrase, and in spite of its 90 syllables it gives only a shortened version of the meaning contained in the first. Yet without a doubt it is the second kind of sentence that is gaining ground in modern English. I do not want to exaggerate. This kind of writing is not yet universal, and outcrops of simplicity will occur here and there in the worst-written page. Still, if you or I were told to write a few lines on the uncertainty of human fortunes, we should probably come much nearer to my imaginary sentence than to the one from *Ecclesiastes*.

As I have tried to show, modern writing at its worst does not consist in 12 picking out words for the sake of their meaning and inventing images in order to make the meaning clearer. It consists in gumming together long strips of words which have already been set in order by someone else, and making the results presentable by sheer humbug. The attraction of this way of writing is that it is easy. It is easier—even quicker, once you have the habit—to say *In my opinion it is a not unjustifiable assumption that* than to say *I think.* If you use ready-made phrases, you not only don't have to hunt about for words; you also don't have to bother with the rhythms of your sentences, since these phrases are generally so arranged as to be more or less euphonious. When you are composing in a hurry—when you are dictating to a stenographer, for instance, or making a public speech—it is natural to fall into a pretentious, latinized style. Tags like *a consideration which we should do well to bear in mind or a conclusion to which all of us would readily assent* will save many a sentence from coming down with a bump. By using stale metaphors, similes and idioms, you save much mental effort, at the cost of leaving your meaning vague, not only for your reader but for yourself. This is the significance of mixed metaphors. The sole aim of a metaphor is to call up a visual image. When these images clash—as in *The Fascist octopus has sung its swan song, the jackboot is thrown into the melting-pot*—can be taken as certain that the writer is not seeing a mental image of the objects he is naming; in other words he is not really thinking. Look again at the examples I gave at the beginning of this essay. Professor Laski (1) uses five negatives in 53 words. One of these is superfluous, making nonsense of the whole passage, and in addition there is the slip *alien* for akin, making further nonsense, and several avoidable pieces of clumsiness which increase the general vagueness. Professor Hogben (2) plays ducks and drakes with a battery which is able to write prescriptions, and, while disapproving of the everyday phrase *put up with*, is unwilling to look *egregious* up in the dictionary and see what it means. In (3), if one takes an uncharitable attitude towards it, [it] is simply meaningless: probably one could work out its intended meaning by reading the whole of the article in which it occurs. In (4) the writer knows more or less what he wants to say, but an accumulation of stale phrases chokes him like tealeaves blocking a sink. In (5) words and meaning have almost parted company. People who write in this manner usually have a general emotional meaning—they dislike one thing and want to express solidarity with another—but they are not interested in the detail of what they are saying. A scrupulous writer, in every sentence that he writes, will ask himself at least four questions, thus: What am I trying to say? What words will express it? What image or idiom will make it clearer? Is this image fresh enough to have an effect? And he will probably ask himself two more: Could I put it more shortly? Have I said anything that is avoidably ugly? But you are not obliged to go to all this trouble. You can shirk it by simply throwing your mind open and letting the ready-made phrases come crowding in. They will construct your sentences for you—even think your thoughts for you, to a certain extent—and at need they will perform the important service of partially concealing your meaning even from yourself. It is at this point that the special connection between politics and the debasement of language becomes clear.

In our time it is broadly true that political writing is bad writing. Where it is 13 not true, it will generally be found that the writer is some kind of rebel, expressing his private opinions, and not a "party line." Orthodoxy, of whatever color, seems to demand a lifeless, imitative style. The political dialects to be found in pamphlets, leading articles, manifestos, White Papers and the speeches of Under-Secretaries do, of course, vary from party to party, but they are all alike in that one almost never finds in them a fresh, vivid, home-made turn of speech. When one watches some tired hack on the platform mechanically repeating the familiar phrases—*bestial atrocities, iron heel, blood-stained tyranny, free peoples of the world, stand shoulder to shoulder*—one often has a curious feeling that one is not watching a live human being but some kind of dummy: a feeling which suddenly becomes stronger at moments when the light catches the speaker's spectacles and turns them into blank discs which seem to have no eyes behind them. And this is not altogether fanciful. A speaker who uses that kind of phraseology has gone some distance towards turning himself into a machine. The appropriate noises are coming out of his larynx, but his brain is not involved as it would be if he were choosing his words for himself. If the speech he is making is one that he is accustomed to make over and over again, he may be almost unconscious of what he is saying, as one is when one utters the responses in church. And this reduced state of consciousness, if not indispensable, is at any rate favorable to political conformity.

In our time, political speech and writing are largely the defense of the inde- 14
fensible. Things like the continuance of British rule in India, the Russian purges and deportations, the dropping of the atom bombs on Japan, can indeed be defended, but only by arguments which are too brutal for most people to face, and which do not square with the professed aims of political parties. Thus political language has to consist largely of euphemism, question-begging and sheer cloudy vagueness. Defenseless villages are bombarded from the air, the inhabitants driven out into the countryside, the cattle machine-gunned, the huts set on fire with incendiary bullets: this is called *pacification*. Millions of peasants are robbed of their farms and sent trudging along the roads with no more than they can carry: this is called *transfer of population or rectification of frontiers*. People are imprisoned for years without trial, or shot in the back of the neck or sent to die of scurvy in Arctic lumber camps: this is called *elimination of unreliable elements*. Such phraseology is needed if one wants to name things without calling up mental pictures of them. Consider for instance some comfortable English professor defending Russian totalitarianism. He cannot say outright, "I believe in killing off your opponents when you can get good results by doing so." Probably, therefore, he will say something like this:

> While freely conceding that the Soviet régime exhibits certain features which the humanitarian may be inclined to deplore, we must, I think, agree that a certain curtailment of the right to political opposition is an unavoidable concomitant of transitional periods, and that the rigors which the Russian people have been called upon to undergo have been amply justified in the sphere of concrete achievement.

The inflated style is itself a kind of euphemism. A mass of Latin words falls 15
upon the facts like soft snow, blurring the outlines and covering up all the details. The great enemy of clear language is insincerity. When there is a gap between one's real and one's declared aims, one turns as it were instinctively to long words and exhausted idioms, like a cuttlefish squirting out ink. In our age there is no such thing as "keeping out of politics." All issues are political issues, and politics itself is a mass of lies, evasions, folly, hatred and schizophrenia. When the general atmosphere is bad, language must suffer. I should expect to find—this is a guess which I have not sufficient knowledge to verify—that the German, Russian and Italian languages have all deteriorated in the last ten or fifteen years, as a result of dictatorship.

But if thought corrupts language, language can also corrupt thought. A bad 16
usage can spread by tradition and imitation, even among people who should and do know better. The debased language that I have been discussing is in some ways very convenient. Phrases like *a not unjustifiable assumption, leaves much to be desired, would serve no good purpose, a consideration which we should do well to bear in mind,* are a continuous temptation, a packet of aspirins always at

one's elbow. Look back through this essay, and for certain you will find that I have again and again committed the very faults I am protesting against. By this morning's post I have received a pamphlet dealing with conditions in Germany. The author tells me that he "felt impelled" to write it. I open it at random, and here is almost the first sentence that I see: "(The Allies) have an opportunity not only of achieving a radical transformation of Germany's social and political structure in such a way as to avoid a nationalistic reaction in Germany itself, but at the same time of laying the foundations of a cooperative and unified Europe." You see, he "feels impelled" to write—feels, presumably, that he has something new to say—and yet his words, like cavalry horses answering the bugle, group themselves automatically into the familiar dreary pattern. This invasion of one's mind by ready-made phrases (*lay the foundations, achieve a radical transformation*) can only be prevented if one is constantly on guard against them, and every such phrase anaesthetizes a portion of one's brain.

I said earlier that the decadence of our language is probably curable. Those 17 who deny this would argue, if they produced an argument at all, that language merely reflects existing social conditions, and that we cannot influence its development by any direct tinkering with words and constructions. So far as the general tone or spirit of a language goes, this may be true, but it is not true in detail. Silly words and expressions have often disappeared, not through any evolutionary process but owing to the conscious action of a minority. Two recent examples were *explore every avenue* and *leave no stone unturned,* which were killed by the jeers of a few journalists. There is a long list of fly-blown metaphors which could similarly be got rid of if enough people would interest themselves in the job; and it should also be possible to laugh the *not un-* formation out of existence,[3] to reduce the amount of Latin and Greek in the average sentence, to drive out foreign phrases and strayed scientific words, and, in general, to make pretentiousness unfashionable. But all these are minor points. The defense of the English language implies more than this, and perhaps it is best to start by saying what it does *not* imply.

To begin with, it has nothing to do with archaism, with the salvaging of ob- 18 solete words and turns of speech, or with the setting up of a "standard English" which must never be departed from. On the contrary, it is especially concerned with the scrapping of every word or idiom which has outworn its usefulness. It has nothing to do with correct grammar and syntax, which are of no importance so long as one makes one's meaning clear, or with the avoidance of Americanisms, or with having what is called a "good prose style." On the other hand it is not concerned with fake simplicity and the attempt to make written English colloquial. Nor does it even imply in every case preferring the Saxon word to the Latin one, though it does imply using the fewest and shortest words that will cover one's meaning. What is above all needed is to let the meaning choose the word, and not the other way about. In prose, the worst thing one can do with

---

[3]One can cure oneself of the *not un-* formation by memorizing this sentence: *A not unblack dog was chasing a not unsmall rabbit across a not ungreen field.*

words is to surrender to them. When you think of a concrete object, you think wordlessly, and then, if you want to describe the thing you have been visualizing, you probably hunt about till you find the exact words that seem to fit it. When you think of something abstract you are more inclined to use words from the start, and unless you make a conscious effort to prevent it, the existing dialect will come rushing in and do the job for you, at the expense of blurring or even changing your meaning. Probably it is better to put off using words as long as possible and get one's meaning as clear as one can through pictures or sensations. Afterwards one can choose—not simply *accept*—the phrases that will best cover the meaning, and then switch around and decide what impression one's words are likely to make on another person. This last effort of the mind cuts out all stale or mixed images, all prefabricated phrases, needless repetitions, and humbug and vagueness generally. But one can often be in doubt about the effect of a word or a phrase, and one needs rules that one can rely on when instinct fails. I think the following rules will cover most cases:

   i. Never use a metaphor, simile or other figure of speech which you are used to seeing in print.

  ii. Never use a long word where a short one will do.

 iii. If it is possible to cut a word out, always cut it out.

  iv. Never use the passive where you can use the active.

   v. Never use a foreign phrase, a scientific word or a jargon word if you can think of an everyday English equivalent.

  vi. Break any of these rules sooner than say anything outright barbarous.

These rules sound elementary, and so they are, but they demand a deep   19 change of attitude in anyone who has grown used to writing in the style now fashionable. One could keep all of them and still write bad English, but one could not write the kind of stuff that I quoted in those five specimens at the beginning of this article.

I have not here been considering the literary use of language, but merely   20 language as an instrument for expressing and not for concealing or preventing thought. Stuart Chase and others have come near to claiming that all abstract words are meaningless, and have used this as a pretext for advocating a kind of political quietism. Since you don't know what Fascism is, how can you struggle against Fascism? One need not swallow such absurdities as this, but one ought to recognize that the present political chaos is connected with the decay of language, and that one can probably bring about some improvement by starting at the verbal end. If you simplify your English, you are freed from the worst follies of orthodoxy. You cannot speak any of the necessary dialects, and when you make a stupid remark its stupidity will be obvious, even to yourself. Political language—and with variations this is true of all political parties, from Conservatives to Anarchists—is designed to make lies sound truthful and murder respectable, and to give an appearance of solidity to pure wind. One cannot change this all in a moment, but one can at least change one's own habits, and from time to time one can even, if one jeers loudly enough, send some worn-out and useless phrase—some *jackboot, Achilles' heel, hotbed, melting pot, acid test, veritable inferno* or other lump of verbal refuse—into the dustbin where it belongs.

COMPREHENSION

1. What is Orwell's purpose? For what type of audience is he writing? Where does he summarize his concerns for readers?
2. According to Orwell, "thought corrupts language" and "language can also corrupt thought" (paragraph 1b). Give examples of these assertions in the essay.
3. In what ways does Orwell believe that politics and language are related?

RHETORIC

1. Orwell himself uses similes and metaphors. Locate five of them, and explain their relationship to the author's analysis.
2. Orwell claims that concrete language is superior to abstract language. Give examples of Orwell's attempt to write concretely.
3. One of the most crucial rhetorical devices in this essay is definition. What important concepts does Orwell define? What methods of definition does he tend to use?
4. Identify an example of hypothetical reasoning in the essay. How does it contribute to the thesis of the essay?
5. After having given five examples of bad English, why does Orwell, in paragraph 10, give another example? How does this example differ from the others? What does it add to the essay?
6. Explain the use of extended analogy in paragraph 14.

WRITING

1. In an analytical essay, assess the state of language in politics today. Cite examples from newspapers and television reports.
2. Prepare an essay analyzing the use and abuse of any word that sparks controversy today—for example, *abortion, AIDS,* or *greed.*
3. **WRITING AN ARGUMENT:** Orwell claims that "the decline of a language must ultimately have political and economic causes" (paragraph 2). Is this claim true? Answer this question in an argumentative essay.

www.mhhe.com/
mhreader

For more information on George Orwell, go to:
More Resources > Ch. 1 Reading & Writing

CONNECTIONS FOR CRITICAL THINKING

1. Examine the "how-to" aspect of the essays by Adler, Elbow, and Murray. What general strategies do they use to develop a comprehensive process analysis of an elusive subject, for example, reading or writing? Write an essay in which you compare the tactics these writers employ to demonstrate their processes.
2. Study the tone of Schlesinger's essay, "The Cult of Ethnicity" (in the chapter introduction). How does he remain "civil" while arguing against a contemporary view he seems to abhor? Next, study Dillard's "The Writer's Life" and examine how she uses imagery, metaphor, and simile in addressing the inscrutable subject of creative writing. Can you make some general observations about how the stylistic elements of an essay contribute to the ability of the author to communicate difficult subjects in a manner that is appealing to the reader?
3. Synthesize the ideas in Elbow's "Freewriting" and those in Murray's "The Maker's Eye: Revising Your Own Manuscripts" so that you can write a coherent essay on writing that takes into account the transition from inspiration to craft.
4. Interview three fellow students to ascertain how they study. Compare and contrast their responses to the suggestions Adler makes in his essay.
5. Create an imaginary dialogue between Mortimer Adler and Julia Alvarez in which the former lauds the joys of reading and the latter celebrates the process of writing.
6. Search a book service Web site such as Amazon, Barnes & Noble, or another source, and review the various readers' remarks concerning the book *The Writing Life* (from which Dillard's excerpt is taken). Write an expository essay commenting on the varied responses that readers have given the book.
7. Alvarez and Tan write personal essays about the impact of language on their lives. Write a comparative essay in which you analyze the similarities and differences in their approach to the topic.
8. Search the Internet for additional information on and essays by Brent Staples. Write a summary of your findings.
9. How do Orwell and Tannen treat the communication process? How are their topics similar and dissimilar? Write a comparative essay on this topic.

chapter **2**

Reading and Writing Effective Arguments

You encounter various forms of argumentation in everyday situations—and most assuredly in many college courses. Consequently, it is important to learn more about this mode of thinking, reading, and writing. As a common form of academic writing, argumentation seeks to explore differences of opinion and attempts to build agreement. As such, argument is not only useful in classroom situations but also in the realm of civic life and discourse, for it provides reasons for people to agree with a particular point of view or at least come to an understanding of an individual's or group's perspective on an issue. Aristotle, who wrote the first major work on argument, thought that the best and most effective argumentative writing blends rational, emotional, and ethical appeals in order to move an audience—whether one person or an entire nation—to desired action.

When you engage in *argumentation,* you offer reasons to support a position, belief, or conclusion. A typical argumentative essay presents a debatable thesis and defends it in logical fashion. Closely allied with argumentation is *persuasion,* in which the writer appeals to readers' intelligence, emotions, and beliefs in order to influence them to adopt a position or act in a certain way. Logic and persuasive appeal often combine when a writer tries to convince an audience that his or her position is valid and other perspectives, while understandable perhaps, require reconsideration.

It is important to distinguish between verbal arguments and written ones. Admittedly both spoken and written arguments have a common purpose in their attempt to convince someone to agree with a particular position, make a certain decision, or take a specific action. In both your verbal and written arguments, you will usually invoke reasons and attempt to manipulate language skillfully. However, with a verbal argument you rarely have access to the types of specific evidence needed to support your reasons, nor do you have the time or ability to martial reasons and evidence in well-organized and coherent ways. Verbal arguments, as you well know, tend to involve excessive emotion; after all, spoken arguments often erupt spontaneously and are rarely thoughtfully constructed and presented.

100

Unlike most verbal arguments, effective written arguments are carefully and logically planned, organized, researched, and revised. The writer analyzes the audience and anticipates objections to the assertions being made. As she or he develops the argument, the writer considers and selects various rhetorical strategies—for example, analysis, definition, or comparison and contrast—to shape the presentation. Moreover, the writer has time to choose the appropriate language and style for the argument, exploring the use of striking diction, figurative language, rhythmic sentence patterns, and various tonalities and shades of meaning during the prewriting, drafting, and revision stages. Finally, especially when composing arguments for college courses, writers must attend to logic and the techniques of valid persuasive appeal.

THE LANGUAGE OF ARGUMENT

Writers of argument often employ various modes of exposition like definition, comparison and contrast, illustration, and analysis, but they incorporate these modes of critical thinking as the means of justifying, or supporting, a logical position. The study of the special language, logic, and structure of argumentation fills volumes. For college writing, there is a core group of critical terms that you should know before you design an argumentative paper:

1. A *claim* is a statement to be justified or upheld. It is the main idea or position that you plan to present in an argument.
2. *Thesis, proposition, assertion,* and *premise* are all similar to a claim in that each is a positive statement or declaration to be supported with reasons and evidence. A premise should be distinguished from the other terms: It is a statement or assumption that is established before an argument is begun and is important to an understanding of logic and various errors or fallacies in reasoning.
3. *Grounds* are the reasons, support, and evidence presented to support your claim.
4. A *warrant* is a stated or unstated belief, rule, or principle that underlies an argument. A *backing* is an even larger principle that serves as the foundation for a warrant.
5. The *major proposition* is the main point of an argument, which is supported by the minor propositions.
6. The *minor propositions* are the reasons you offer in support of the major proposition.
7. *Evidence* is that part of the argument that supports the minor propositions. In argumentation, effective evidence is based either on facts, examples, statistics, and other forms of evidence or on accepted opinions. Without adequate evidence, the audience will not accept your major and minor propositions. Evidence in argument must be accurate and true.

8. A *fact* is a verifiable statement. A valid *opinion* is a judgment based on the facts and careful deductive or inductive reasoning. *Induction* is a process of reasoning by which you develop evidence in order to reach a useful generalization. *Deduction* is a process that proceeds from the general to the particular.
9. A valid *conclusion* of an argument derives logically from the major and minor propositions. The logical conclusion is termed the *inference,* in which you arrive at a decision by reasoning from the previous evidence.
10. A *fallacy* is a line of incorrect reasoning from premises.
11. *Refutation* is the acknowledgment and handling of opposing viewpoints. You must anticipate opposing viewpoints and counter them effectively (what we term *rebuttal*) in order to convince or persuade readers.

Constructing an effective argument depends on the careful arrangement of major and minor propositions, evidence, and refutation. Like a lawyer, you build a position and subject your opponent's position to dissection in an effort to win the case.

THE TEST OF JUSTIFICATION

Whatever its components, whether a writer can construct an argument or not essentially hinges on the concept of *justification*—the recognition that a subject lends itself to legitimate difference of opinion. Justification also involves proving or demonstrating that a claim is in accordance with the reasons and evidence offered to support it.

Not all statements require justification. A statement that is a verifiable fact or a commonly accepted assumption or belief—what we term a *warrant*—generally does not need justification. To test the concept of justification, consider the following four statements.

1. President John F. Kennedy was assassinated on November 22, 1963.
2. Children shouldn't smoke.
3. Abortion is the destruction of a human life.
4. African Americans should receive reparations for the damages caused by slavery.

Which of these statements require justification? The first statement about President Kennedy is a verifiable fact, and the second statement strikes any reasonable audience as common sense. Thus, the first two statements do not require justification and consequently could not be the subject of a useful argument, although the second statement could serve as the warrant for a more specific claim about smoking by young children. By contrast, the third statement concerning abortion makes a critical assumption that would elicit either agreement or disagreement but in either case would demand substantiation. Similarly, the fourth statement about reparations for slavery is an issue that is debatable from a variety of positions. Therefore, the third and fourth statements require justification: They are open to argumentation.

READING AND ANALYZING ARGUMENT

From the time of Aristotle to the present, numerous critical approaches to the study of argument have been devised. One of the most useful recent approaches to argument appears in *An Introduction to Reasoning* and *The Uses of Argument* by British logician and philosopher Stephen Toulmin. In his studies, Toulmin observes that any argument involves a *claim* supported by *reasons* and *evidence*. Whether writing a memo to your instructor contesting a certain grade, or a letter to the editor of your campus newspaper advocating a change in the cafeteria vendor because the food is terrible, or a petition to provide more parking space for commuting students, the argumentative method is the same. Essentially you make a general assertion—a claim—and then offer the smaller propositions or supporting reasons along with the relevant facts, examples, statistics, and expert testimony to justify all claims. And underlying the nature of claims and evidence is recognition of the importance of *warrants*, those unstated beliefs that lead from evidence to claim.

Here is the way that Toulmin presents his model:

Harry is a British subject
CLAIM

Harry was born in Bermuda
REASON

WARRANT
Since a man born in Bermuda is a British subject

In truth, Toulmin's example is basic and perhaps too simple. The claims that you deal with when reading or writing arguments typically are more complex and controversial than Toulmin's diagram suggests, and the need for extensive evidence more demanding. Nevertheless, Toulmin's model offers a useful way to understand the nature of argumentative reasoning.

Understanding Claims and Warrants

When you argue in writing, you make a specific claim, which is an assertion that you plan to prove. You present this claim or proposition as being true, and you support the claim with a series of logically related statements that are true. Think of the claim as the thesis or the main point of the argument that holds all other logically related statements together. The claim is the main idea that you set out to prove, and in a well-reasoned argument, everything makes the claim seem inevitable. Any paper that you write that fails to state a claim—your position in an argument—clearly and emphatically will leave readers shaking their heads and wondering if you actually have an argument to present.

Think of a claim as an arguable point, one that you can build a carefully reasoned paper around. Remember that by applying the test of justification, you need to exclude numerous opinions, nonarguable propositions, and statements of taste and fact that might be common in everyday situations but not legitimate subjects for papers based in sound strategies for written argument. To say "Turn down that rap music" to your roommate is the sort of command (containing perhaps an implied opinion) that doesn't in itself qualify as a claim but could get you involved in a heated conversation. To transform this command into a legitimate claim or an arguable point, you would have to state a proposition that expresses your main idea about rap music.

Suppose, for example, that after reading the essay in this anthology by Henry Louis Gates Jr., "2 Live Crew, Decoded" (beginning on page 534), you *are* asked to write an argumentative paper on hip-hop or rap music. In his brief but provocative essay, Gates raises significant issues about obscenity and censorship, about "mainstream" and ethnic culture, about white perceptions of black males. His claim is that we cannot address the complex, interrelated issues raised by the rap music phenomenon unless we "become literate in the vernacular traditions of African Americans. To do less is to censor through the equivalent of intellectual prior restraint—and censorship is to art what lynching is to justice." Do you readily perceive the argumentative edge in Gates's claim? Do you agree or disagree with his main point that one cannot offer an informed critique of a cultural phenomenon without knowledge of the language of that culture? And what claim would you actually make about rap music and the culture, an increasingly "crossover" youth culture, that supports it? What reasons or grounds would you produce in support of your claim?

A complex, extended argument in essay form often reveals several types of claim that the writer advances. A *claim about meaning* (What is rap music?) is a proposition that defines or interprets a subject as it establishes an arguable point. A *claim about value* (Rap music is good or bad) advances an ideally open-minded view of the subject based on a coherent framework of aesthetic or ethical values. A *claim about policy* (Music stations should be forced to regulate the most offensive forms of rap music) advances propositions concerning laws, regulations, and initiatives designed to produce specific outcomes. Finally, *claims about consequences* (Children who listen to rap musicians begin to mimic their vulgar behavior) are rooted in propositions involving various forms of cause-and-effect relationships. Constructing an argument around one or more of these types of claims is essential in gaining an audience's assent.

Many claims, of course, cannot be presented as absolute propositions—certainly not as absolute as Aristotle's major premise in his famous syllogism (see page 108) that "all human beings are mortal." Writers must seek common ground with readers and foster a degree of trust by anticipating that members of any audience will disagree with their claim, treat it with skepticism, and perhaps even respond with hostility. For this reason, it is important to qualify or clarify the nature of your claim. A *qualifier* restricts the absoluteness of a claim by using such cue words and phrases as *sometimes, probably, usually,* and *in most cases.*

Qualifiers can also explain certain circumstances or conditions under which the claim might not be true. The use of qualifiers enables the writer to anticipate certain audience reactions and handle them in an effective and subtle way.

Even more important than the possible need to qualify a claim is the need to justify it in a new way: by linking the claim with reasons and evidence in such a way that the audience sees the train of thinking that leads from the data to the claim. If you look again at the model that Toulmin provides, you see that the data "Harry was born in Bermuda" does not completely support the claim "Harry is a British subject." What is required is what Toulmin calls a *warrant*, a form of justification—a general belief, principle, or rule—that links the claim and the data or support. Thus, the warrant "Since a man born in Bermuda is a British subject" explains *why* the claim follows from the data.

Another way of understanding this admittedly challenging concept of warrant is to treat it as the process of thinking that leads writers to hold the opinions they present. Thought of from this perspective, we can see that a weak or unclear warrant will undermine an argument and render it invalid. For example, the claim "Sara graduated from an excellent high school and consequently she should do well in college" is based on the warrant or unstated (and untested) belief that all students who graduate from good high schools perform well at the college level. Obviously, this warrant is not satisfactory: To state that college success is based solely on the quality of one's high school education is to base the argument on a warrant that few readers would find acceptable. If, on the other hand, a writer claims that "Sara graduated from an excellent high school with a 3.97 cumulative average, the third highest in her class, and consequently should do well in college," we see that the warrant establishing the link between the claim and conclusion becomes more acceptable. In fact, there is a consensus, or general belief, among experts that a person's grade point average in high school is a sound predictor—perhaps sounder than SAT scores—of a person's potential for success in college.

If you disagree with a writer's assumptions, you basically are questioning the warrants underlying the argument. An effective argument should rest on an acceptable warrant and also on the *backing*—some explanation or support—for it. Remember that even if a warrant, stated or unstated, is clear, understood, and backed with support, readers might still disagree with it. For example, one could argue that Sara might have obtained her lofty GPA in high school by taking easy courses, and that consequently we cannot readily predict her success in college. Not everyone will accept even the most reasonable of warrants.

Reasoning from Evidence

Evidence is the data, or *grounds*, used to make claims or general assertions clear, concrete, and convincing. In argumentation, the presentation of evidence must be examined from the perspective of logic or sound reasoning. Central to logic is the relationship of evidence to a *generalization*, a statement or conclusion that what is applicable in one situation also applies to similar situations. You cannot think and write clearly unless you test evidence to see that it supports your claims, assumptions, or general statements. Evidence in an argumentative essay creates a common ground of understanding that you and your reader can share.

You know that one of the keenest pleasures in reading mystery fiction or viewing whodunits on television or film is the quest for evidence. The great writers of crime and mystery fiction—Edgar Allan Poe, Sir Arthur Conan Doyle, Agatha Christie—were adept at creating a chain of clues, or evidence, leading with the inevitability of logic to the solution to the crime. Whether it is a letter lying on a desk in Poe's "The Purloined Letter" or a misplaced chair in Christie's *The Murder of Roger Ackroyd,* it is evidence that we seek in order to solve the crime.

In argumentative writing, evidence is used more to prove a point than to solve a mystery. College writers must know what constitutes evidence—examples, facts, statistics, quotations and information from authoritative sources, personal experience, careful reasoning—and how to use it to support certain claims. They must also determine if the evidence and assumptions surrounding the evidence are valid.

Here are five basic questions about evidence to consider when reading and writing argumentative essays.

1. *Is the evidence typical and representative?* Examples must fairly represent the condition or situation if your claim is to be valid. If evidence is distorted or unrepresentative, a claim will not be logical or convincing.
2. *Is the evidence relevant?* The evidence should speak directly to the claim. It should not utilize peripheral or irrelevant data.
3. *Is the evidence specific and detailed?* In reading and writing arguments, do not trust broad, catchall statements presented as "evidence." Valid evidence should involve accurate quotations, paraphrases, and presentations of data from authoritative sources.
4. *Is the evidence accurate and reliable?* A claim is only as valid as the data supporting it. Facts should come from reliable sources. (See page 119 for help with evaluating sources.) Current rather than outdated evidence should predominate in a current argument. Sources should be cited accurately for the convenience of the reader. Although personal observation and personal experience are admissible as types of evidence, such testimony rarely serves as conclusive proof for a claim.
5. *Is the evidence sufficient?* There must be enough evidence to support claims and reasons. One extended piece of evidence, no matter how carefully selected, rarely is sufficient to win an argument.

Any argumentative essay should provide a clear, logical link between the writer's claim, assertion, generalization, or conclusion and the evidence. If an argumentative essay reveals false or illogical reasoning—that is, if the step from the evidence to the generalization is wrong, confusing, or deceptive—readers will not accept the truth of the claim or the validity of the evidence.

THINKING CRITICALLY ABOUT ARGUMENTS

Whether you are reading another writer's argument or starting to plan one of your own, you need to consider the purposes of the argument. When you are reading an argument, you should also look for the appeals to reason, emotion, and ethics the writer is using and decide whether or not those appeals are effective for you, the writer's audience. In your own arguments, you will need to decide the types of appeals that will carry the most weight with *your* audience.

The Purpose of Argumentation

As a college writer, your general aim is to communicate or convey messages in essay form to a literate and knowledgeable audience of teachers and scholars. When thinking about the subject for an essay, you also have to consider a more specialized *purpose*—the special nature or aim—behind your composition. You might have to report the result of an experiment in animal behavior, analyze a poem, compare and contrast Mario Puzo's novel *The Godfather* and its film adaptation, or assert the need for capital punishment. In each instance, your essay requires a key rhetorical strategy or set of strategies. These strategies reflect your purpose—your intention—in developing the essay.

An argumentative essay may serve one or more purposes:

1. To present a position, belief, or conclusion in a rational and effective way.
2. To defend a position against critics or detractors.
3. To persuade people to agree with a position or take a certain action.
4. To attack a position without necessarily presenting an alternative or opposing viewpoint.

An effective argumentative essay often combines a variety of forms and purposes. For example, an argumentative essay on legalizing marijuana might have to explain effects, analyze laws, or evaluate experiments, among a broad range of options. When you take time to consider your purpose before you even begin to write, the decisions you make will help you to think more clearly about both the design and intention of your essay.

Appeals to Reason, Emotion, and Ethics

As the definitions of special terms and the discussion of justification presented earlier in the chapter suggest, argumentation places a premium on rational discourse. In fact, the *appeal to reason* is the fundamental purpose of argumentation. However, classical rhetorical theory acknowledges that the *appeal to emotion* and the *appeal to ethics* are also important elements in the construction of argument and the effort to persuade. A mere presentation of reasons is usually not an effective argument. For your argument to be effective, you need to pay attention to the value of strategic emotional and ethical appeal.

Appeals to Reason   The *appeal to reason* or logic is the primary instrument of effective argument. The most common way of developing an argument according to the principles of sound reasoning is *deduction*, which is most readily understood as an ordering of ideas from the general to the particular. With deduction you move from a general assertion through reasons and support focused on the main assertion. Consider the following student paragraph, which uses the deductive method.

Anti-marijuana laws make people contemptuous of the legal system. This contempt is based in part on the key fact that there are too many contradictions and inconsistencies in criminal penalties for marijuana use. Laws vary radically from state to state. In Texas, you can be sentenced to life imprisonment for first-time use of marijuana. By contrast, in the District of Columbia the same "crime" would most likely result in a suspended sentence.

Deduction is a convincing way of arranging ideas and information logically. By stating the proposition or generalization first, you present the most important idea. Then, as in the paragraph above, you move to more specific ideas and details. Examined more rigorously, deductive reasoning involves a process of critical thinking known as *syllogism* in which you move from a major statement or premise, through another minor premise, to a third statement or conclusion. Aristotle's famous syllogism captures this mental process:

Major premise:	All human beings are mortal.
Minor premise:	Socrates is a human being.
Conclusion:	Socrates is mortal.

The soundness of any deductive argument rests on the *truth* of the premises and the *validity* of the syllogism itself. In other words, if you grant the truth of the premises, you must also grant the conclusion. The deductive method can be used effectively in many forms of expository as well as argumentative essays.

Inductive reasoning reverses the process of deduction by moving from the particular ideas to general ones. In the paragraph that follows from an essay by F. M. Esfandiary, "The Mystical West Puzzles the Practical East," the writer presents various ideas and evidence that lead to a major proposition at the end.

Twenty-five hundred years ago, Buddha, like other Eastern philosophers before him, said: "He who sits still, wins." Asia, then immobilized in primitive torpor, had no difficulty responding. It sat still. What it won for sitting still was the perpetuation of famines and terrorizing superstitions, oppression of children, subjugation of women, emasculation of men, fratricidal wars, persecutions, mass killings. The history of Asia, like the history of all mankind, is a horrendous account of human suffering.

By presenting his supporting—and provocative—ideas first, the author is able to interest us before we reach the climactic argument at the end of the paragraph. Of course, whether we accept the Esfandiary's argument—his statement of truth—or are prepared to debate his claim depends on the strength of the reasons and evidence he offers.

Many of the argumentative essays you read and much of the argumentative writing you undertake will reflect the mental processes of deduction and induction. The novelist Robert M. Pirsig offers his version of these critical thinking strategies in his cult classic, *Zen and the Art of Motorcycle Maintenance:*

> If the cycle goes over a bump and the engine misfires, and then goes over another bump and the engine misfires, and then goes over another bump and the engine misfires, and then goes over a long smooth stretch of road and there is no misfiring, and then goes over a fourth bump and the engine misfires again, one can logically conclude that the misfiring is caused by the bumps. That is induction: reasoning from particular experiences to general truths.
>
> Deductive inferences do the reverse. They start with general knowledge and predict a specific observation. For example if, from reading the hierarchy of facts about the machine the mechanic knows the horn of the cycle is powered exclusively by electricity from the battery, then he can logically infer that if the battery is dead the horn will not work. That is deduction.

Constructing an argument through the use of logical reasoning is a powerful way to convince or persuade a particular audience about the validity of your claims.

Appeals to Emotion   In addition to developing your argument logically using the appeal to reason, you should consider the value of incorporating *appeals to emotion* into an argumentative paper. A letter home asking for more money would in all likelihood require a certain carefully modulated emotional appeal. Similarly, Martin Luther King Jr.'s famous "I Have a Dream" speech at the 1963 March on Washington, which begins on page 411 in this textbook, is one of the finest contemporary examples of emotional appeal. King's speech ends with this invocation:

> When we let freedom ring, when we let it ring from every village and every hamlet, from every state and every city, we will be able to speed up that day when all of God's children, black men and white men, Jews and Gentiles, Protestants and Catholics, will be able to join hands and sing in the words of the old Negro spiritual, "Free at last! Free at last! Thank God almighty, we are free at last!"

King's skillful application of balanced biblical cadences, of connotative and figurative language, and a strong, almost prophetic tone demonstrates the value of carefully crafted emotional appeal in the hands of an accomplished writer of argument.

Of course, in constructing an argument you should avoid the sort of cynical manipulation of emotion that is common in the world of spoken discourse and the media in general. (For a list of unfair emotional appeals, see page 122.) But honest emotional appeal provides a human context for the rational ideas and evidence you present in an argumentative essay—ideas that might otherwise be cold and uninteresting to your audience. Assuredly if you want to persuade your audience to undertake a particular course of action, you must draw members of this audience closer to you as a person, perhaps even inspire them by your feelings about the subject or issue. In truth, you *must* establish rapport with your reader in an argumentative essay. If you fail to engage the reader's feelings, the best-constructed rational appeal could fall flat.

Appeals to Ethics   For an emotional appeal to achieve maximum effectiveness, it must not only reinforce the rational strength of your argument but also the ethical basis of your ideas. When you use *ethical appeal,* you present yourself as a well-informed, a fair-minded, and an honest person. Aristotle acknowledged the importance of *ethos,* or the character of the writer in the construction of argument, for if you create a sense that you are trustworthy, your readers or listeners will be inspired or persuaded. The "sound" or "voice" of your essay, which you convey to the reader through your choice of style and tone and which can only be perfected through the process of careful drafting and revision, will help in convincing the audience to share your opinion.

In an appeal to ethics, you try to convince the reader that you are a person of sound character—that you possess good judgment and an acceptable system of values. As a person of goodwill and good sense, you also demonstrate an ability to empathize with your audience, to understand their viewpoints and perspectives. The psychologist Carl Rogers suggests that a willingness to embrace a potentially adversarial audience, to treat this audience more like an ally in an ethical cause, is a highly effective way to establish goodwill and the credibility of your beliefs. In Rogerian argument, your willingness to understand an opposing viewpoint and actually rephrase it reflectively for mutual understanding enables you to further establish your ethical and personal qualities.

You can appreciate the powerful combination of rational, emotional, and ethical appeals in Abraham Lincoln's "Gettysburg Address," which follows for analysis and discussion.

✳

The Gettysburg Address

Abraham Lincoln

Abraham Lincoln (1809–1865) was born the son of a pioneer in 1809 in Hodgesville, Kentucky, and moved to Illinois in 1831. After brief experiences as a clerk, postmaster, and county surveyor, he studied law and was elected to the state legislature in 1834. A prominent member of the newly formed Republican party, Lincoln became president on the eve of the Civil War. In 1862, after Union victory at Antietam, Lincoln issued the Emancipation Proclamation freeing the slaves—the crowning achievement of an illustrious presidency. Although he was an outstanding orator and debater throughout his political career, "The Gettysburg Address" is one of his greatest speeches—and certainly

his most famous one. It was delivered at the dedication of the Gettysburg National Ceme-
tery in 1863. Its form and content reflect the philosophical and moral views of the time
as well as the rhetorical skill of its speaker. Lincoln was assassinated by John Wilkes
Booth in 1865 shortly after Robert E. Lee's surrender and the end of the Civil War.

Four score and seven years ago our fathers brought forth on this continent, a   1
new nation, conceived in Liberty, and dedicated to the proposition that all men
are created equal.

Now we are engaged in a great civil war, testing whether that nation, or any   2
nation so conceived and so dedicated, can long endure. We are met on a great
battlefield of that war. We have come to dedicate a portion of that field as a final
resting-place for those who here gave their lives that that nation might live. It is
altogether fitting and proper that we should do this.

But, in a larger sense, we cannot dedicate—we cannot consecrate—we can-   3
not hallow—this ground. The brave men, living and dead, who struggled here
have consecrated it, far above our poor power to add or detract. The world will
little note, nor long remember, what we say here, but it can never forget what
they did here. It is for us the living, rather, to be dedicated here to the unfinished work which they who fought here have thus far so nobly advanced. It is
rather for us to be here dedicated to the great task remaining before us—that
from these honored dead we take increased devotion to that cause for which
they gave the last full measure of devotion; that we here highly resolve that
these dead shall not have died in vain; this nation, under God, shall have a new
birth of freedom; and that government of the people, by the people, for the people, shall not perish from the earth.

COMPREHENSION

1. Although this speech was supposed to be a "dedication," Lincoln states that
   "we cannot dedicate." What does he mean by this?
2. Lincoln uses abstract words such as "liberty," "freedom," and "nation."
   What does he mean specifically by each of these terms?
3. What exactly happened "four score and seven years ago" in the context of
   the speech? Why is this reference so significant to the purpose of Lincoln's
   address?

RHETORIC

1. Note the progression of imagery from that of "death" to that of "birth."
   How does this structure contribute the theme and coherence of the speech?
2. How do the syntax, punctuation, and choice of the first-person plural form
   of address contribute to our understanding that this message was intended
   to be spoken rather than written?

3. Note how Lincoln refers to the combatants as "brave" and "honored." How does he suggest their struggle was distinguished from that of "us the living"? How does this comparison and contrast create clear similarities and differences between those who fought and those who are present to carry on the work of the soldiers?

4. The American Civil War was a battle between the North and the South as were the opponents at the Battle of Gettysburg. However, Lincoln does not mention this. What is the reason behind this omission? How does it make the speech focus on more comprehensive issues?

5. Besides being president, Lincoln was by definition a politician. In what ways can we determine that this is a political speech as well as a dedication?

6. Speeches are intended to be heard. What are some elements—for example, vocabulary, syntax, length or brevity of the sentences, juxtaposition of sentences, and so on—that appeal to the sense of sound?

7. Does this speech appeal primarily to the intellect, the emotions, or equally between the two? What are two or three sentences that demonstrate one or both of these appeals? What was the rationale behind your selections? Does Lincoln include any ethical appeals?

WRITING

1. Research the actual historical events that occurred during the Battle of Gettysburg. Write an expository essay in which you discuss the significance of this particular speech at this point in the American Civil War. Use a minimum of three secondary source materials.

2. Read the speech three times. Then write a paraphrase of it. Examine your paraphrase to discover what elements you recalled. Then reread the speech and write an expository essay focusing on how the structure of the speech contributed to helping you recall the information you did remember.

3. Engage in an Internet search to find the rhetorical influences on the language and style of Abraham Lincoln. Try such search phrases as "Lincoln, rhetorical influences" and other suitable expressions. Select three authoritative "hits" and write a research paper entitled "Rhetorical Influences on Abraham Lincoln."

4. **WRITING AN ARGUMENT:** Richard Posner in his essay "Security versus Civil Liberties," which appears in the portfolio of argumentative essays at the end of this chapter, asserts that the Emancipation Proclamation "may . . . have been unconstitutional" (paragraph 7). Conduct research on this issue, and then write an argumentative essay in which you agree or disagree with Posner's claim.

WRITING POWERFUL ARGUMENTS

One of the most common writing assignments in college courses, especially courses in the humanities and social sciences, is the argumentative essay. Unlike narrative and descriptive essays and the major forms of expository writing—comparison and contrast, definition, classification, process, and causal analysis—the argumentative paper requires the writer to take a stand and to support a position as effectively as possible. As mentioned earlier, the rhetorical strategies underlying expository or informative writing often appear in argumentative papers. However, given the purpose of the argumentative essay, you must present your ideas as powerfully as possible in order to advance your point of view and convince your readers to accept your position or take a specific course of action. For this reason, you must construct your argumentative paper carefully and effectively.

Argument, as stated in Chapter 1, is not a mere academic enterprise but rather an integral part of our personal, social, and professional lives. Whether you consider the 2004 presidential debates between George W. Bush and John Kerry, recent decisions of the Supreme Court, actions taken by the United States in the Middle East, or tax provisions enacted by Congress—argument determines in many ways the nature of life in the United States. More personally, argument—whether in the form of reviews or commercials—influences the films we watch, the music we download, the clothes we purchase, the food we eat. And argument—the way you sell yourself in an interview—can get you a job or lose it.

In many other ways, argument impacts your professional life. It is most clearly in evidence in newspaper and television journalism and communications, where argument and persuasion express an institution's deepest convictions. Consider the role of editorials and the op-ed page of any newspaper (including your college newspaper) in the nation. Editorial writing offers carefully crafted debates and informed opinions on a variety of topics and issues. Many newspapers also develop guidelines to promote convincing but open-minded positions on the issues of the day.

A deputy editorial page director for the *Chicago Tribune* offers seven questions for essays that, with slight modification, provide an excellent guide for writing an argument:

1. *To whom are you writing?* Are you writing to authorities? Power elites? Professors? Average readers? Yourself?
2. *What's your attitude?* Are you angry? Pleased? Perplexed? What tone will you project?
3. *What, exactly, are you trying to accomplish?* An official response? A public change of attitude? An explanation? Entertainment?
4. *What are you contributing to the debate?* What's the added value here? Just your opinion? New facts? New arguments, contexts, or dimensions to consider?

5. *Do you have something new to say?* Are you advancing the conversation or just rehashing old facts, opinions, and wisdom? Aside from an opinion, do you have a solution?

6. *Have you fiercely attacked your own premise?* Will your position survive scrutiny? How would your opponents answer your most compelling arguments? Are you correct or simply wrong?

7. *Are you stirring up a "three-bowler"?* This borrowed phrase refers to the reader who is so bored with your writing that her face falls in the cereal bowl not once or twice but three times. You must compose your argument in such a way that the reader is hooked by your writing and persuaded by the force of your argument.

If you are further interested in tips for editorial writing, go to www.poynter.org, a Web site for journalists that offers excellent advice in composing articles for the press.

The process for writing powerful arguments that appears in this section is useful, but it is not a formula. Ultimately, you can construct powerful arguments in numerous ways, but you always must consider the relationship between your ideas, your particular purpose, and your audience.

Identify an Issue

Remember that not every subject lends itself to useful or necessary argument. The notion that *"everything* is an argument" probably contains a grain of truth, but in reality some things make for more powerful arguments than others. Certain subjects—for example, stamp collecting—might appeal to you personally and powerfully, but are they worth arguing about? Consequently, your first step in writing an effective argumentative essay is to identify a subject that contains an issue—in other words, a subject that will elicit two or more differing opinions.

Clearly there are certain subjects that touch on current problems and inspire strong opinions. President Lincoln's "Gettysburg Address" dealt with a monumental issue central to the very survival of the nation; for virtually everyone in the United States at that time, the issues raised by Lincoln were debatable. Similarly, the issue of the death penalty produces two diametrically opposed viewpoints in the essays by H. L. Mencken and Coretta Scott King that are reprinted in this chapter. Social and political issues tend to be ripe subjects for debate, fostering pro and con viewpoints. (Remember that there are often more than two sides to a complex issue.) Such issues, by their very nature, often raise powerful and conflicting systems of belief that place heavy burdens on the writer to provide convincing reasons and evidence to support a claim.

Not all issues in argumentative papers have to be of national or global concern, however. Indeed, issues like capital punishment, abortion, or global warming might not be of special interest to you. Of course, if an instructor requires an argumentative essay on one of these broad hot-button topics, you will need to prepare to write the paper by first establishing an argumentative perspective on it—in other words, by choosing your side on the issue. Fortunately, you often have opportunities to select issues of more immediate, personal, or local concern: Should fast-food franchises be permitted in the student cafeteria? Should there be a campus policy on hate language? Should sophomores be required to pass standardized tests in reading, writing, and mathematics before advancing to their junior year? Many powerfully constructed arguments can deal with issues close to home and with subjects that are of considerable personal interest. Whether dealing with an issue mandated by the instructor or selecting your own issue for an argumentative paper, ask yourself at the outset of this critical process what your position on the issue is and how it can be developed through logic and evidence.

Take a Stand and Clarify Your Claim

Once you have identified an issue that lends itself to argumentation—an issue that people might reasonably disagree about—you must take a clear stand on this issue. In other words, your claim will advance your viewpoint over all other viewpoints. The aim is not to defeat an opponent but to persuade readers—your audience—to accept your opinion. Consequently, the first step at this stage is to establish as clearly as possible what your claim is going to be. You might want to experiment with one or more of the following strategies:

- Gather information on the issue from debates on radio, television, or the Internet. Electronic resources on the World Wide Web can be helpful as you begin to research what your position on an issue is going to be. (For help with critically evaluating Web sources, see page 165 in Chapter 3.)
- Brainstorm or write informally about the issue, jotting down your immediate response to it—how it makes you feel or what you think about it. If the issue raises emotional responses, what are the causes of this response? What are your more thoughtful or intellectual responses to the issue?
- List some preliminary reasons you respond to the issue in the way you do. By listing reasons and also listing the types of evidence you will need to support those reasons, you will be able to determine at an early stage if you will have enough material for a solid argumentative paper, and what forms of research you will have to conduct.
- Jot down examples, facts, and ideas that might support your claim.
- Begin to think about possible objections to your position, and list these opposing viewpoints.

As this inventory of strategies suggests, there are numerous ways to think and write critically about your approach to an issue during the prewriting stage of your argument. Essentially, during prewriting you want to begin to articulate and pinpoint your claim, and thereby start to limit, control, and clarify the scope of your argument.

Once you have developed a preliminary approach to an issue, you should be prepared to state your claim in the form of a thesis sentence. From your reading of the information on the thesis statement in the first chapter of this textbook (page 26), you know that you must limit the scope and purpose of your thesis or claim. Too broad a claim will be hard to cover in convincing fashion in a standard argumentative paper. One useful way to limit and clarify your claim is to consider the purpose of your argument. Do you want to argue a position on a particular issue? Do you want to argue that a certain activity, belief, or situation is good or bad, harmful or beneficial, effective or ineffective? Do you want to persuade readers to undertake or avoid a particular course of action? Do you want readers to simply consider an issue in a new light? Do you want readers to endorse your interpretation or evaluation of an artistic or literary work? By sifting through the primary purposes of argument, which involve value judgments, policies, and interpretations, you will arrive at the main point of your argument—your claim.

Analyze Your Audience

All writing can be considered a process of communication, a conversation with an audience of readers. In argumentative writing, it is especially important to establish a common ground of belief with your readers if you expect them to accept your claim or undertake a certain course of action. Of course, you cannot change your ideas and approaches to an issue merely to please a particular audience. However, you do not construct an argumentative paper in order to be misunderstood, disbelieved, or rejected. Within the limits set by who you are, what you believe, and what your purpose is, you can match your argumentative style and approach to audience expectations.

To establish common ground with your audience, it is important to know them well so that you can dispose them favorably to your claim and the reasons and evidence supporting it. Your audience might be a professor, a prospective employer, an admission or financial aid officer, an editor, or a member of your family. If you determine the nature of your audience *before* you compose the first draft of your argumentative paper, you will be able to tailor style, content, and tone to a specific person or group.

Try to imagine and anticipate audience expectations by asking basic questions about your readers:

1. What are the age, gender, professional background, educational level, and political orientation of most of the members of the audience?
2. How much does the audience know about the issue? Is it an audience of experts or a general audience with only limited knowledge of the issue?
3. What does the audience expect from you in terms of the purpose behind your claim? Does the audience expect you to prove your claim or persuade them to accept it, or both?
4. Will the audience be friendly, hostile, or neutral toward your argument? What political, cultural, ethical, or religious factors contribute to the audience's probable position on this issue?
5. What else do you know about the audience's opinions, attitudes, and values? How might these factors shape your approach to the argument?

Suppose, for example, that you are planning to write an argumentative essay on pollution. What common expectations would an English professor, a sociology professor, and a chemistry professor have concerning your argument? What differences in approach and content would be dictated by your decision to write for one of these instructors? Or consider these different audiences for a paper on the topic of pollution: a group of grade school children in your home town, or the Environmental Protection Agency, or the manager of a landfill operation, or a relative in Missouri whose town has been experiencing chemical pollution. In each instance, the type and nature of the audience will influence your approach to the issue and even your purpose. Remember that through *purpose* you find the proper context for your argument. Any writer who wants to communicate effectively with his or her audience will adjust the content and tone of an argument so as not to lose, confuse, or mislead the reader.

Establish Your Tone

By *tone* we mean the attitude you take toward your subject. A word that often is used interchangeably with tone is *voice*. Tone is the personal voice that a reader "hears" in your writing. This voice may vary, depending on the situation, your purpose, and the audience that you are writing for. It may be personal or impersonal and range across a spectrum of attitudes: serious or humorous, subjective or objective, straightforward or ironic, formal or casual, and so forth. You adjust your tone to match your purpose in writing.

In argumentation, an effective tone will be a true and trustworthy reflection of the writing situation. After all, you are writing an argumentative essay in order to convince and persuade, and consequently you need to sound like a reasonable, well-organized, and logical individual. When writing for college instructors, that "community of scholars," you must be especially careful to maintain a reasonable tone. You do not have to sound scholarly, legalistic, or overly technical in presenting your argument, but you do have to employ a personal voice that is appropriate to the writing occasion and audience expectations.

To achieve an appropriate tone in argumentative writing, you will often need to be forceful in presenting your ideas. Remember that you are staking out a position, perhaps on a controversial issue, and you must seem willing to defend it. Try to maintain a consistent voice of authority, but do not be overbearing: Do not move from the podium to the locker room, mixing voices in a way that will confuse or alienate your audience. A tone or voice that exceeds the limits of good taste and commonly accepted norms of argumentative style is likely to be ineffectual. A voice that is too emotional, overblown, or irrational will in all likelihood alienate the reader and erode your claim.

Your tone—your voice—is a revelation of yourself. It derives from your claims and supporting ideas, your language and sentence structure. Even if your audience is one person—typically your professor—you certainly must present yourself to that audience as convincingly as possible. When your tone is adjusted to the issue, the claim, and the supporting evidence, and also to the nature of the opposition, you stand a good chance of writing an effective argumentative essay.

Develop and Organize the Grounds for Your Claim

You establish the validity of your claim by setting out the reasons and evidence—the *grounds*—that support your main point. Whereas the claim presents your general proposition or point of view, as you develop your grounds you organize the argument into minor propositions, evidence, and refutation. By establishing the grounds for your claim, you explain the particular perspective or point of view you take on an issue. The grounds for your claim permit the reader to "see" the strength of your particular position.

There are numerous ways to state the primary reasons or grounds for holding your position. Think of these primary reasons as minor propositions underlying the basis of your claim—reasons that readers would find it difficult to rebut or reject. Three possible models for organizing claims and grounds in an essay can now be considered.

Model 1

Introduction: Statement and clarification of claim.

First minor proposition and evidence.

Second minor proposition and evidence.

Third minor proposition and evidence.

Refutation of opposing viewpoints for minor propositions.

Conclusion.

Model 2

Introduction: Statement and clarification of claim.

First minor proposition and evidence; refutation.

Second minor proposition and evidence; refutation.

Third minor proposition and evidence; refutation.

Conclusion.

Model 3

Statement and clarification of claim.

Summary of opposing viewpoints and refutation.

First minor proposition and evidence.

Second minor proposition and evidence.

Third minor proposition and evidence.

Conclusion.

In practice, arguments rarely adhere slavishly to these models. In fact, you can arrange your argument in numerous ways. However, the models can serve a useful purpose, especially in examinations that require argumentative responses to a question, for they provide a handy template for your answer. In argument, to support your claim, you will need substantial reasons, sometimes more than the three minor propositions illustrated in these models. Remember that one reason generally will not provide sufficient grounds to prove an argument. Moreover, you should keep in mind the need to distinguish between your *opinions*, which in the broadest sense are beliefs that you cannot verify logically, and reasons, which are based on logic, evidence, and direct proof.

Gather and Evaluate Your Evidence

Once you have established your claim and your reasons, you must turn your attention to developing evidence for your claim, a subject already discussed in the first chapter and also here. Collecting evidence is a bit like the strategies for successful fishing presented by the student in his classification paragraph in the first chapter (see page 39): You want to fish the top, the middle, and the bottom of your subject. Phrased somewhat differently, you want to cast a wide net as you seek evidence designed to support your claim and reasons.

At the outset, a carefully designed search of the World Wide Web can yield ample evidence. The Web will permit you to establish links to sites and listservs where you can download or print full or abstracted texts from periodicals, books, documents, and reports. Remember that searching the Web is often like navigating a minefield: Useless "facts," hoaxes, and informational marketing ploys mix with serious research, honest reporting, and critical analysis. To guard against the pitfalls involved in relying exclusively on Web surfing, you should also make a trip to the college library to augment your quest for evidence. Research librarians can help you to evaluate Web sites and direct you to the best sources—both traditional and technological—for the types of evidence you are seeking. (For more on library and Internet research, see Chapter 3.) Depending on your subject, you might consider interviewing individuals who can provide expert testimony designed to support your claim and reasons. Finally, your own personal experience and the experiences of your friends and acquaintances might provide useful evidence, although such kinds of anecdotal or first-hand support should be treated judiciously and not serve as the entire basis for your paper. You and your friends might claim that a current horror movie is great, but such personal evidence must be tempered by a willingness to consult established critics for additional support.

If you cast a wide net and fish the whole lake, you will almost always catch more than you require. Yet the very process of searching comprehensively for evidence can produce exciting unintended consequences. You might, for example, discover that certain evidence suggests a need to revise or qualify your claim. Evidence can also help you to articulate or confirm the warrants that are the foundation of your argument, for experts writing on an issue often state the assumptions, principles, or beliefs that offer connections between a claim and its grounds. The insights gained by considering other evidence might cause you to develop a new reason for your claim that you had not considered initially. You might also discover evidence that helps you to refute the ideas of your anticipated opposition. Having a wealth of evidence at your disposal is an embarrassment of riches that you can exploit skillfully.

After you have collected adequate evidence to bolster your claim and the key reasons supporting that claim, the next necessary step is to evaluate and select the best evidence available to you. Writers who carefully evaluate and select their evidence produce effective arguments. At the outset, the nature of the writing situation—an examination, a term paper, a letter to the editor—will dictate to an extent the type of evidence you need to evaluate. In most instances, however, your evidence should be *credible, comprehensive,* and *current.* Your evidence is credible when the sources of your information are reliable and the evidence itself is representative. Your evidence is comprehensive when you provide a broad range of facts, information, and data designed to cover all aspects of your argument completely. In presenting evidence comprehensively, you also make certain that there is sufficient support for each of your reasons—not too much evidence for one and too little for another, but an even balance between and among the minor propositions. Finally, always try to locate the most current evidence available to support your claim. Data and statistics often do not age well and tend to lose their accuracy. However, in some arguments older evidence can be compared with newer information: For example, a paper arguing that immigration to the United States is out of control could make skillful use of data from the 1960 Census *and* the 2000 Census.

Evidence is the heart of any argument. Without evidence, readers will not be interested in your claim and supporting reasons. Make certain that the evidence —the facts, examples, and details—you present is accurate and skillfully presented so that readers become interested in your more abstract propositions, identify with your position, and come away convinced of the validity of your argument.

Consider Your Warrants

Even as you clarify your claim and assemble your reasons and evidence, you must also consider the assumptions underlying your argument. Think of the assumption or *warrant* as the link between a claim and the supporting evidence—the underlying set of beliefs or principles governing our essential perception of the world and the human condition. Warrants answer the question of *how* the data are connected to the claim. Sometimes these warrants are stated, but often they remain unstated. In either instance, they are not necessarily self-evident or universally accepted. They are significant nevertheless, for as generalizations that are far broader than claims and evidence, warrants serve as the bedrock of an argument.

Warrants help to guarantee that a reader will accept your argument, and consequently it is important to consider them. When you are writing for a friendly or supportive audience, you can usually assume that your readers will accept the warrants supporting your claim, and therefore you might not even need to state them. For example, if you claim in a report for your biology professor that Creationism should not be taught in high school science classrooms, your argument is based on several assumptions or warrants: that the Constitution, for example, requires the separation of church and state, or that there is no scientific basis for Creationism. In fact, when making your claim about Creationism before a scientist, you also are relying on certain *backings,* which are the principles underlying the warrants themselves—for instance, the idea that a scientist is concerned with scientific objectivity rather than literal interpretations of the Bible, or that scientists deal with the empirical reality and not matters of faith. But what if you were make your claim in a letter to a local school board, several of whose members want to revise the ninth-grade earth science curriculum to emphasize Creationism and evolution equally? In this instance, you would be presenting your argument to a potentially skeptical or hostile audience, and you should expect that you will have to state your warrants clearly, bolster them with adequate support, and establish solid causal links between your warrants and your backing.

Whether you are writing an argumentative paper or reading an argumentative essay critically, you need to develop the habit of looking for and evaluating the warrants and the backing behind the argument. If the warrants are stated, it will make this task easier. If the warrants are unstated, you will have to detect and evaluate them if you are reading an argumentative essay. If you are writing an argumentative essay, you should consider whether your audience will probably understand and consent to the warrants that serve as the foundation of your paper. If you have any doubt, then you should include them.

Deal with Opposing Viewpoints

In order to make your argument effectively as possible, you need to acknowledge and deal with opposing viewpoints. Any controversial issue is going to have more than one viewpoint, and you must recognize contending claims and handle them fairly. As suggested in the section on audience analysis, you can enhance your credibility by describing these opposing viewpoints fully and accurately and with a respectful rather than hostile tone, even as you demonstrate that your position in an argument is the most reasonable and valid.

As a prewriting strategy for refutation, you might try dividing a sheet of paper or your computer screen into three columns, labeling them, from left to right, "Supporting Viewpoints," "Opposing Viewpoints," and "Refutation." Then list the main supporting points for your claim, thinking of possible opposing responses and writing them down as you go. Imagining how the opposition will respond to your supporting reasons will help you to develop refutations, or counterarguments. You can use the resulting chart as a guide to organize sections of your argumentative paper.

The listing technique for refutation forces you to acknowledge opposing viewpoints and also refute them in a systematic way. It is perfectly appropriate—

and even necessary—to demonstrate the weakness or insufficiency of opposing arguments, for refutation strengthens your own position. Any complex argument that you present will not be complete unless you skillfully refute all predictable opposing viewpoints, using one of the following four techniques.

- Question the opposition's claim, asking if it is too flimsy, broad, overstated, or improperly grounded in minor propositions.
- Question the evidence. Is it insufficient, outdated, or inaccurate?
- Question the warrants and backing of an opposing argument—those assumptions and beliefs that underpin the opposition's claim.
- Concede some part of the opposition's viewpoint, a subtle but extremely attractive strategy that shows that you are a courteous and unbiased thinker and writer and that therefore constitutes an appeal to ethics.

Avoid Unfair Emotional Appeals and Errors in Reasoning

When you write and revise an argumentative essay, you need to avoid certain temptations and dangers that are unique to this form of discourse. You always have to make certain that your argumentative strategies are fair and appropriate and that you have avoided oversimplifying your argument. You also need to resist the temptation to include persuasive appeals that distort critical reasoning and to avoid errors in logical reasoning.

Emotional appeals are effective when used appropriately in argumentation, but used unfairly they can distort your logical reasoning. Such "loaded" arguments are filled with appeals to the reader's emotions, fears, and prejudices. Here are three of the most common fallacies of emotional distortion to avoid.

1. *Transfer* is the association of a proposition with a famous person. Transfer can be either positive ("In the spirit of President Franklin Delano Roosevelt, we should create a jobs program for the nation's unemployed") or negative ("President George W. Bush is the symbol of unbridled capitalism"). Another term for negative transfer is *name calling*. In both the positive and negative types of transfer, however, there is no logical basis for the connection.
2. *Argumentum ad hominem* ("to the man") is a strategy that discredits a person in an effort to discredit his or her argument. It attacks the person rather than the position: "Richards is a homosexual and consequently cannot understand the sanctity of heterosexual marriage." In this instance, the individual becomes a false issue.
3. *Argumentum ad populum* ("to the people") deliberately arouses an audience's emotions about certain institutions and ideas. Certain words have strong positive or negative connotations. Such words as *patriotism* and *motherhood* are *virtue* words that often prompt the creation of *glittering generalities*. Suggestive words can be used to distort meaning by illogical association and to manipulate an audience to take a stand for or against a proposition: "USC should not take the *totalitarian* step of requiring athletes

to maintain a full course load." A related strategy is the *bandwagon* approach, in which the writer generalizes falsely that the crowd or majority is always right: "Everyone is voting for Erikson and you should too."

These unfair emotional appeals are often found in political speech writing, advertising, and propaganda. When you write argumentative essays, you should use persuasive appeal to reinforce rather than distort the logical presentation of your ideas, blending reasonable claims and valid emotional and ethical appeals to convince rather than trick your audience into agreeing with you.

Equally important is the need to avoid errors in reasoning in the construction of an argument. Here are some types of errors in reasoning, or *logical fallacies,* that are common in argumentative writing.

1. *Hasty generalizations.* A hasty generalization is a conclusion based on insufficient, unrepresentative, or untrue evidence: "The president of the college successfully raised 100 million dollars, so other college presidents should be able to do the same." When you indulge in hasty generalizations you jump to false conclusions. Hasty generalizations are also at the heart of stereotyping—the uncritical application of an oversimplified generalization to a group or to individual members of the group. Make certain that you have adequate and accurate evidence to support any claim or conclusion.

2. *Broad generalizations.* A broad generalization typically employs words like *all, never,* and *always* to state something absolutely or categorically. It is actually a form of overstatement, as in the sentence "Freud always treated sexuality as the basis of human behavior." Usually, readers can easily find exceptions to such sweeping statements, so it is best to qualify them.

3. *Oversimplification.* Oversimplification reduces alternatives. Several forms of oversimplification can be distinguished.

   a. *Either/or.* Don't assume that there are only two sides to an issue, only two possibilities, only yes or no, only right or wrong: "Either we make English a one-year requirement or college students will not be able to write well."

   b. *No choice.* Don't assume that there is only one possibility: "The United States has no other alternative than to build the Star Wars missile defense system." Parents and politicians are prone to no-other-choice propositions.

   c. *No harm or cost.* Don't assume that a potential benefit will not have significant harms, consequences, or costs: "We should sell North Korea as much wheat as it needs." No-harm generalizations or arguments may overlook dangerous implications. Always consider alternative evidence.

   d. *One solution.* Don't assume that a complicated issue has only one solution: "Embryonic stem cell should not be used for research, for using them in this way will lead to the destruction of human life." Always consider evidence for other solutions or alternative approaches to issues and problems.

4. *Begging the question.* Do not assume in your premises or in your evidence what is to be proved in the conclusion. For example, if you argue that vandalism by teenagers is unavoidable because teenagers are young and irresponsible, you are begging the question because you are not proving your premise. Another form of begging the question is to take a conclusion for granted before it is proved.

5. *False cause-and-effect relationships.* Perhaps the most common error in trying to establish causal relationships is known as the *post hoc, ergo propter hoc* fallacy ("after this, therefore because of this"). The fact that one event follows another is not proof that the first caused the second. If you maintain, for instance, that there is an increase in the crime rate every time there is a full moon, you are falsely identifying an unrelated event as a cause. Many superstitions—popular, political, and otherwise—illogically assume that one event somehow causes another.

6. *Disconnected ideas.* Termed in Latin *non sequitur* ("It does not follow"), this fault in reasoning arises when there is no logical connection between two or more ideas. Put differently, an argument's conclusion is not related to its premises: "George W. Bush makes a good president because he was a successful owner of a professional baseball team." Sometimes you think that a connection exists but you fail to state it in writing. For example, you may think that owners of baseball teams and presidents need to have strong people skills and be good judges of character. In other words, *you* may see the logical connection between your ideas about presidents and baseball team owners, but if you don't make it explicit, readers may think there is a non sequitur.

7. *Weak or false analogies.* An *analogy* is a type of comparison that explains a subject by comparing it to the features of another essentially dissimilar subject: "Unless we learn to think critically about the niagara of information that washes over us every day, we will be lost in a flood of rumors and gossip." Analogies can be used to illustrate a point, although they should always be used carefully and with discretion. More significant, an analogy can *never* function as evidence or logical proof of a position.

In conclusion, the hallmark of argumentation is sound critical thinking. If you present your claims, grounds, and evidence carefully, are willing to assemble the best and most objective data, treat the opposition with respect, and are flexible in responding to new ideas, you will be well on your way to constructing a solid argumentative essay. A successful argumentative paper reveals a writer who possesses an inquiring mind—one that is able to judge opinions on the basis of evidence, reason well, and back up ideas and beliefs in a convincing and valid way.

| www.mhhe.com/ **mhreader** | For an interactive tutorial on writing an argument, go to: **Writing > Writing Tutor: Arguments** |

A Portfolio on Argumentation

Classic and Contemporary Essays
How Do We Argue?

We have all been in situations where controversial issues arise. A friendly gathering may evolve into a spirited debate on abortion or cloning. Guests at a family dinner may turn their attention away from the host's expertly prepared cuisine toward a heated exchange over immigration. One issue that seems to inevitably arise when a conversation turns to issues of law and order is the death penalty or capital punishment. Arguments may range from cool statistical analysis of the value of this punishment as a "deterrent" to impassioned pleas regarding the sanctity of all human life. H. L. Mencken, in his classic essay "The Penalty of Death," provides a singular flavor to the argumentative stew by presenting the reasons for maintaining the death penalty; however, it appears evident from his style and tone that he is mocking its proponents by revealing their hypocrisy. He presents no fancy academic studies, nor does he draw on any experts or scholars. His approach is ironic. He contends in his disarming way that deterrence is merely an excuse for the exercise of the ultimate punishment; the true motive is revenge and retribution. Coretta Scott King, in her essay "The Death Penalty Is a Step Back," draws on sociology, law, psychology, morality, and logic to oppose capital punishment. Hers is a multipronged attack against the death penalty, and unlike Mencken, her tone is serious, straightforward, and unadorned. Is there *one* right way to address an issue of such seriousness? Perhaps it is not so much the style and methods one uses, but how well they are used.

✳

The Penalty of Death

H. L. Mencken

H(enry) L(ouis) Mencken (1880–1956) was an American editor, an author, and a critic. Born in Baltimore, he served as an editor for three Baltimore newspapers: the Morning Herald, Evening Herald, *and* The Baltimore Sun. *Noted for his pungent and iconoclastic criticism, he reveled in satirizing the middle classes. He was also a student of philology and published* The American Language, *which went through several editions with added supplements. The topics for his many books ranged from studies of dramatists to the defense of women's rights. He was also a champion for a whole generation of American realist fiction writers, including Theodore Dreiser, Sherwood Anderson, Sinclair Lewis, and Eugene O'Neill. The following well-known essay reveals the hypocrisy behind the rationale many people give for supporting the death penalty and the true reason they support it.*

Of the arguments against capital punishment that issue from uplifters, two are   1
commonly heard most often, to wit:

1. That hanging a man (or frying him or gassing him) is a dreadful business, degrading to those who have to do it and revolting to those who have to witness it.
2. That it is useless, for it does not deter others from the same crime.

The first of these arguments, it seems to me, is plainly too weak to need   2
serious refutation. All it says, in brief, is that the work of the hangman is unpleasant. Granted. But suppose it is? It may be quite necessary to society for all that. There are, indeed, many other jobs that are unpleasant, and yet no one thinks of abolishing them—that of the plumber, that of the soldier, that of the garbage-man, that of the priest hearing confessions, that of the sand-hog, and so on. Moreover, what evidence is there that any actual hangman complains of his work? I have heard none. On the contrary, I have known many who delighted in their ancient art, and practised it proudly.

In the second argument of the abolitionists there is rather more force, but   3
even here, I believe, the ground under them is shaky. Their fundamental error consists in assuming that the whole aim of punishing criminals is to deter other (potential) criminals—that we hang or electrocute A simply in order to so alarm B that he will not kill C. This, I believe, is an assumption which confuses a part with the whole. Deterence, obviously, is *one* of the aims of punishment, but it is surely not the only one. On the contrary, there are at least half a dozen, and some are probably quite as important. At least one of them, practically considered, is *more* important. Commonly, it is described as revenge, but revenge is really not the word for it. I borrow a better term from the late Aristotle: *katharsis*.

Katharsis, so used, means a salubrious discharge of emotions, a healthy letting off of steam. A school-boy, disliking his teacher, deposits a tack upon the pedagogical chair; the teacher jumps and the boy laughs. This is *katharsis.* What I contend is that one of the prime objects of all judicial punishments is to afford the same grateful relief *(a)* to the immediate victims of the criminal punished, and *(b)* to the general body of moral and timorous men.

These persons, and particularly the first group, are concerned only indirectly 4 with deterring other criminals. The thing they crave primarily is the satisfaction of seeing the criminal actually before them suffer as he made them suffer. What they want is the peace of mind that goes with the feeling that accounts are squared. Until they get that satisfaction they are in a state of emotional tension, and hence unhappy. The instant they get it they are comfortable. I do not argue that this yearning is noble; I simply argue that it is almost universal among human beings. In the face of injuries that are unimportant and can be borne without damage it may yield to higher impulses; that is to say, it may yield to what is called Christian charity. But when the injury is serious, Christianity is adjourned, and even saints reach for their sidearms. It is plainly asking too much of human nature to expect it to conquer so natural an impulse. A keeps a store and has a bookkeeper, B. B steals $700, employs it in playing at dice or bingo, and is cleaned out. What is A to do? Let B go? If he does so he will be unable to sleep at night. The sense of injury, of injustice, of frustration will haunt him like pruritus. So he turns B over to the police, and they hustle B to prison. Thereafter A can sleep. More, he has pleasant dreams. He pictures B chained to the wall of a dungeon a hundred feet underground, devoured by rats and scorpions. It is so agreeable that it makes him forget his $700. He has got his *katharsis.*

This same thing precisely takes place on a larger scale when there is a crime 5 which destroys a whole community's sense of security. Every law-abiding citizen feels menaced and frustrated until the criminals have been struck down— until the communal capacity to get even with them, and more than even, has been dramatically demonstrated. Here, manifestly, the business of deterring others is no more than an afterthought. The main thing is to destroy the concrete scoundrels whose act has alarmed everyone, and thus made everyone unhappy. Until they are brought to book that unhappiness continues; when the law has been executed upon them there is a sigh of relief. In other words, there is *katharsis.*

I know of no public demand for the death penalty for ordinary crimes, even 6 for ordinary homicides. Its infliction would shock all men of normal decency of feeling. But for crimes involving the deliberate and inexcusable taking of human life, by men openly defiant of all civilized order—for such crimes it seems, to nine men out of ten, a just and proper punishment. Any lesser penalty leaves them feeling that the criminal has got the better of society—that he is free to add insult to injury by laughing. That feeling can be dissipated only by a recourse to *katharsis,* the invention of the aforesaid Aristotle. It is more effectively and economically achieved, as human nature now is, by wafting the criminal to realms of bliss.

The real objection to capital punishment doesn't lie against the actual exter- 7
mination of the condemned, but against our brutal American habit of putting it
off so long. After all, every one of us must die soon or late, and a murderer, it
must be assumed, is one who makes that sad fact the cornerstone of his meta-
physic. But it is one thing to die, and quite another thing to lie for long months
and even years under the shadow of death. No sane man would choose such a
finish. All of us, despite the Prayer Book, long for a swift and unexpected end.
Unhappily, a murderer, under the irrational American system, is tortured for
what, to him, must seem a whole series of eternities. For months on end he sits
in prison while his lawyers carry on their idiotic buffoonery with writs, injunc-
tions, mandamuses, and appeals. In order to get his money (or that of his
friends) they have to feed him with hope. Now and then, by the imbecility of a
judge or some trick of juridic science, they actually justify it. But let us say that,
his money all gone, they finally throw up their hands. Their client is now ready
for the rope or the chair. But he must still wait for months before it fetches him.

That wait, I believe, is horribly cruel. I have seen more than one man sitting 8
in the death-house, and I don't want to see any more. Worse, it is wholly useless. Why should he wait at all? Why not hang him the day after the last court
dissipates his last hope? Why torture him as not even cannibals would torture
their victims? The common answer is that he must have time to make his peace
with God. But how long does that take? It may be accomplished, I believe, in
two hours quite as comfortably as in two years. There are, indeed, no temporal
limitations upon God. He could forgive a whole herd of murderers in a mil-
lionth of a second. More, it has been done.

COMPREHENSION

1. Based upon your reading of Mencken's essay, is the author for or against
   capital punishment? Explain.
2. Study the last three lines of the essay. Explain what they mean in your own
   words.
3. The author's facility with language is due partly to his impressive vocabulary. Define words such as *salubrious, timorous,* and *manifestly.*

RHETORIC

1. The author uses symbolic logic, classification, and definition as devices in
   paragraphs 1 through 4. Cite examples of each of these rhetorical methods.
   What is each one's function?
2. What is the author's purpose in using a rather droll tone in discussing a
   subject that usually elicits strong emotional responses?
3. In paragraph 3, the author defines *katharsis* as "a healthy letting off of
   steam." In the light of the author's view that carrying out the death penalty
   results in a societal *katharsis,* what is the implicit irony in the definition?

4. What is the author's purpose in using both the placement of a tack on a teacher's seat and the execution of a human being as examples of *katharsis*?

5. What tone does the author use in describing humankind's desire for revenge? Does he support or deride this sentiment? Explain your conclusion by citing particular clues the author provides in his writing.

6. In paragraph 4, the author states, "But when the injury is serious, Christianity is adjourned, and even saints reach for their sidearms." How does this statement relate to the theme of the essay?

WRITING

1. For a creative writing project, pretend you are a legislator. Write an essay wherein you describe a crime and what its proper particular punishment should be. Be sure to fit the punishment to the crime.

2. There is some evidence to suggest that the death penalty may actually *increase* the murder rate. Study this line of inquiry, and write a research paper based upon your findings that either supports or rejects the thesis.

3. **WRITING AN ARGUMENT:** Argue for or against the use of the death penalty in crimes other than murder.

| www.mhhe.com/ **mhreader** | For more information on H. L. Mencken, go to: **More Resources > Ch. 2 Arguments** |

The Death Penalty Is a Step Back

Coretta Scott King

Coretta Scott King (b. 1927) is a civil rights activist, freelance journalist, and, since 1980, writer and commentator for CNN. Born in Alabama, she graduated from Antioch College and the New England Conservatory of Music. She first gained international prominence as the wife of Martin Luther King, Jr., whom she married in 1953. She wrote about her experiences with the revered civil rights leader and orator in a book entitled My Life with Martin Luther King, Jr. *(1969). The following essay states in clear, thoughtful prose her feelings about the death penalty, which she considers both racist and immoral.*

When Steven Judy was executed in Indiana [in 1981] America took another step 1
backwards towards legitimizing murder as a way of dealing with evil in our
society.

Although Judy was convicted of four of the most horrible and brutal mur- 2
ders imaginable, and his case is probably the worst in recent memory for opponents of the death penalty, we still have to face the real issue squarely: Can we
expect a decent society if the state is allowed to kill its own people?

In recent years, an increase of violence in America, both individual and po- 3
litical, has prompted a backlash of public opinion on capital punishment. But
however much we abhor violence, legally sanctioned executions are no deterrent and are, in fact, immoral and unconstitutional.

Although I have suffered the loss of two family members by assassination, 4
I remain firmly and unequivocally opposed to the death penalty for those convicted of capital offenses.

An evil deed is not redeemed by an evil deed of retaliation. Justice is never 5
advanced in the taking of a human life.

Morality is never upheld by legalized murder. Morality apart, there are a 6
number of practical reasons which form a powerful argument against capital
punishment.

First, capital punishment makes irrevocable any possible miscarriage of jus- 7
tice. Time and again we have witnessed the specter of mistakenly convicted
people being put to death in the name of American criminal justice. To those
who say that, after all, this doesn't occur too often, I can only reply that if it happens just once, that is too often. And it has occurred many times.

Second, the death penalty reflects an unwarranted assumption that the 8
wrongdoer is beyond rehabilitation. Perhaps some individuals cannot be rehabil-
itated; but who shall make that determination? Is any amount of academic training sufficient to entitle one person to judge another incapable of rehabilitation?

Third, the death penalty is inequitable. Approximately half of the 711 per- 9
sons now on death row are black. From 1930 through 1968, 53.5 percent of those
executed were black Americans, all too many of whom were represented by
court-appointed attorneys and convicted after hasty trials.

The argument that this may be an accurate reflection of guilt, and homicide 10
trends, instead of a racist application of laws lacks credibility in light of a recent
Florida survey which showed that persons convicted of killing whites were four
times more likely to receive a death sentence than those convicted of killing
blacks.

Proponents of capital punishment often cite a "deterrent effect" as the main 11
benefit of the death penalty. Not only is there no hard evidence that murdering
murderers will deter other potential killers, but even the "logic" of this argument defies comprehension.

Numerous studies show that the majority of homicides committed in this 12
country are the acts of the victim's relatives, friends and acquaintances in the
"heat of passion."

What this strongly suggests is that rational consideration of future conse- 13
quences are seldom a part of the killer's attitude at the time he commits a crime.

The only way to break the chain of violent reaction is to practice nonvio- 14
lence as individuals and collectively through our laws and institutions.

COMPREHENSION

1. On what grounds does King oppose capital punishment?
2. King calls the death penalty "immoral" and "unconstitutional." What does she mean by this?
3. Does King offer any solutions to the problem of crime and violence? What are they?

RHETORIC

1. Where in the essay does King place her main proposition? In your own words, what is this claim?
2. What function do paragraphs 1 to 5 have in the essay?
3. What impact do the words *practical* and *powerful* (in paragraph 6) have on the reader? Who is King's intended audience?
4. Comment on the use of language in King's essay. Is it concrete or abstract? How would you characterize her writing style?
5. Trace King's use of transitions in paragraphs 7, 8, and 9.
6. Where does the writer use refutation in her essay? How does she use it to strengthen her argument? How effective are her responses?
7. Is King's ordering of ideas inductive or deductive? Justify your answer.

WRITING

1. If capital punishment doesn't deter crime, what will? Write an essay in which you offer detailed solutions to the problem of crime and violence. How can society take a step forward in its treatment of criminals?
2. King's essay makes a connection between the death penalty and racism. Develop this theme in an essay. Consider the roles of class, race, legal representation, and political empowerment in determining who goes to prison and who gets executed.
3. **WRITING AN ARGUMENT:** Write an essay for or against capital punishment, using quotes from King's essay either as support or as refutation. Provide examples and your own observations as proof.

Questions comparing H. L. Mencken's "The Penalty of Death" and Coretta Scott King's "The Death Penalty Is a Step Back" appear on page 134.

Classic and Contemporary Images
WHAT IS AN ARGUMENT?

Using a Critical Perspective   What images and strategies do the Spanish artist Goya and the American photographer Adams employ to construct an argument about war? What exactly is their argument? Comment on the nature and effectiveness of the details they use to illustrate their position. Which work do you find more powerful or engaging? Explain.

Horrified by the excesses of the Napoleonic invasion of his homeland and the Spanish war for independence, the Spanish artist Francisco de Goya (1746–1828) painted *The Third of May, 1808*, a vivid rendition of an execution during wartime.

Another wartime execution, this time captured on film by Eddie Adams
in an image that won the Pulitzer Prize for spot news photography
in 1969, brought home to Americans the horrors and
ambiguities of the war in Vietnam.

The questions that follow refer to essays on pages 125–131.

Classic and Contemporary: Questions for Comparison

1. Does Mencken's sarcasm and iconoclastic tone suggest he is writing for a different audience than the more austere and straightforward King? Consider that Mencken was writing at least a half-century before King. To what sorts of audiences would each of the essays appeal? Explain your view.

2. Study the language used in each of the essays. What is similar or different about the style and diction of the two pieces? Does one seem more accessible to the modern reader? Do any of Mencken's references seem dated? Consider such terms as *uplifters* (paragraph 1), *abolitionists* (paragraph 3), and *juridic science* (paragraph 7).

3. Both Mencken and King have had firsthand experience with gruesome events. Mencken mentions that he has observed men in the "death-house" prior to their execution, and Coretta Scott King's husband was assassinated. Does this lend authority to their grievances? Would you be less inclined to trust an opinion from a third arguer who had never had such personal experience?

Debate: Animal Research—Is It Ethical?

Animal Research Saves Human Lives

Heloisa Sabin

Heloisa Sabin is honorary director of Americans for Medical Progress in Alexandria, Virginia. The wife of Albert Sabin, who discovered the oral vaccine for polio, she invokes her husband's name in the following essay to advance her position on animal experimentation. This essay appeared in The Wall Street Journal *on October 18, 1995, shortly after Albert Sabin's death.*

That scene in *Forrest Gump* in which young Forrest runs from his schoolmate tormentors so fast that his leg braces fly apart and his strong legs carry him to safety may be the only image of the polio epidemic of the 1950s etched in the minds of those too young to remember the actual devastation the disease caused. Hollywood created a scene of triumph far removed from the reality of the disease. 1

Some who have benefited directly from polio research, including that of my late husband, Albert, think winning the real war against polio was just as simple. They have embraced a movement that denounces the very process that enables them to look forward to continued good health and promising futures. This "animal rights" ideology—espoused by groups such as People for the Ethical Treatment of Animals, the Humane Society of the United States and the Fund for Animals—rejects the use of laboratory animals in medical research and denies the role such research played in the victory over polio. 2

The leaders of this movement seem to have forgotten that year after year in the early fifties, the very words *infantile paralysis* and *poliomyelitis* struck great fear in young parents that the disease would snatch their children as they slept. Each summer public beaches, playgrounds, and movie theaters were places to be avoided. Polio epidemics condemned millions of children and young adults to lives in which debilitated lungs could no longer breathe on their own and young limbs were left forever wilted and frail. The disease drafted tiny armies of children on crutches and in wheelchairs who were unable to walk, run, or jump. In the United States, polio struck down nearly 58,000 children in 1952 alone. 3

Unlike the braces on Forrest Gump's legs, real ones would be replaced only as the children's misshapen legs grew. Other children and young adults were 4

135

entombed in iron lungs. The only view of the world these patients had was through mirrors over their heads. These memories, however, are no longer part of our collective cultural memory.

Albert was on the front line of polio research. In 1961, thirty years after he 5 began studying polio, his oral vaccine was introduced in the United States and distributed widely. In the nearly forty years since, polio has been eradicated in the Western Hemisphere, the World Health Organization reports, adding that, with a full-scale effort, polio could be eliminated from the rest of the world by the year 2000.

Without animal research, polio would still be claiming thousands of lives 6 each year. "There could have been no oral polio vaccine without the use of innumerable animals, a very large number of animals," Albert told a reporter shortly before his death in 1993. Animals are still needed to test every new batch of vaccine that is produced for today's children.

Animal activists claim that vaccines really didn't end the epidemic—that, 7 with improvements in social hygiene, polio was dying out anyway, before the vaccines were developed. This is untrue. In fact, advanced sanitation was responsible in part for the dramatic *rise* in the number of paralytic polio cases in the fifties. Improvements in sanitation practices reduced the rate of infection, and the average age of those infected by the polio virus went up. Older children and young adults were more likely than infants to develop paralysis from their exposure to the polio virus.

Every child who has tasted the sweet sugar cube or received the drops con 8 taining the Sabin vaccine over the past four decades knows polio only as a word, or an obscure reference in a popular film. Thank heavens it's not part of their reality.

These polio-free generations have grown up to be doctors, teachers, busi 9 ness leaders, government officials, and parents. They have their own concerns and struggles. Cancer, heart disease, strokes, and AIDS are far more lethal realities to them now than polio. Yet, those who support an "animal rights" agenda that would cripple research and halt medical science in its tracks are slamming the door on the possibilities of new treatments and cures.

My husband was a kind man, but he was impatient with those who refused 10 to acknowledge reality or to seek reasoned answers to the questions of life.

The pioneers of polio research included not only the scientists but also the 11 laboratory animals that played a critical role in bringing about the end of polio and a host of other diseases for which we now have vaccines and cures. Animals will continue to be as vital as the scientists who study them in the battle to eliminate pain, suffering, and disease from our lives.

That is the reality of medical progress. 12

COMPREHENSION

1. Summarize the writer's argument. Why does she begin the essay with a reference to the movie *Forrest Gump*, which starred Tom Hanks in an award-

winning role as an American "hero"? Why does she refer repeatedly to "reality"?

2. What disease receives the major part of the writer's attention? Where is it mentioned?
3. List all parts of the essay where the writer refers to her famous husband, Dr. Albert Sabin. What is her purpose? What is she implying?

RHETORIC

1. What is Sabin's major proposition and where does she place it? Is it effective where it is? Justify your answer.
2. What implied or stated warrants affect the writer's argument?
3. Where does the writer acknowledge the opposition? How does she refute those opposed to animal experimentation, and how effective do you think this strategy is? Explain.
4. How does Sabin limit her argument? Does this limitation strengthen or weaken her claim, and why?
5. Describe the logical, ethical, and emotional appeals that Sabin advances in this essay. Explain how each of these types of appeals advances her argument. What other logical appeals can you think of that might have strengthened her argument?
6. Analyze the conclusion. Is it effective? Why or why not?

WRITING

1. Focusing on Sabin's article, write an essay explaining the importance of emotional appeal in argument. Does Sabin use excessive emotion or an appropriate amount? Does her relationship to her recently deceased husband weaken her use of emotion? Does she anticipate this question or ignore it, and why? These are some of the questions you might want to consider.
2. Imagine that you, a family member, or a friend is suffering from an incurable disease. How would you justify not pursuing animal research in an effort to find a cure? Write an essay responding to this question.
3. **WRITING AN ARGUMENT:** Select a disease other than polio—for example, cancer, AIDS, Parkinson's, or Alzheimer's—and argue that animal experimentation is necessary in order to find a cure for it. Conduct Internet research in order to familiarize yourself with the disease and current animal research dealing with it.

✳

A Question of Ethics

Jane Goodall

Jane Goodall (b. 1934) was born in London, England. In 1960, with no university degree or formal training, she began to study chimps in the Gombe Stream Reserve in Tanzania. Living in close proximity to the chimps and gaining their trust over the years, Goodall was the first scientist to discover that chimps are not strictly vegetarians and that the species uses tools—a trait thought previously to belong only to humans. Her books include My Friends the Wild Chimpanzees *(1967) and* The Chimpanzee: The Living Link between Man and Beast *(1992). Goodall today is one of the world's foremost conservationists and animal rights activists. In the following essay, she asks whether it is ethical to use animals in laboratory research.*

David Greybeard first showed me how fuzzy the distinction between animals   1
and humans can be. Forty years ago I befriended David, a chimpanzee, during my first field trip to Gombe in Tanzania. One day I offered him a nut in my open palm. He looked directly into my eyes, took the nut out of my hand and dropped it. At the same moment he very gently squeezed my hand as if to say, I don't want it, but I understand your motives.

Since chimpanzees are thought to be physiologically close to humans, re-   2
searchers use them as test subjects for new drugs and vaccines. In the labs, these very sociable creatures often live isolated from one another in 5-by-5-foot cages, where they grow surly and sometimes violent. Dogs, cats and rats are also kept in poor conditions and subjected to painful procedures. Many people would find it hard to sympathize with rats, but dogs and cats are part of our lives. Ten or 15 years ago, when the use of animals in medical testing was first brought to my attention, I decided to visit the labs myself. Many people working there had forced themselves to believe that animal testing is the only way forward for medical research.

Once we accept that animals are sentient beings, is it ethical to use them in   3
research? From the point of view of the animals, it is quite simply wrong. From our standpoint, it seems ridiculous to equate a rat with a human being. If we clearly and honestly believe that using animals in research will, in the end, reduce massive human suffering, it would be difficult to argue that doing so is unethical. How do we find a way out of this dilemma?

One thing we can do is change our mind-set. We can begin by questioning   4
the assumption that animals are essential to medical research. Scientists have concluded that chimpanzees are not useful for AIDS research because, even though their genetic makeup differs from ours by about 1 percent, their immune systems deal much differently with the AIDS virus. Many scientists test drugs and vaccines on animals simply because they are required to by law rather than out of scientific merit. This is a shame, because our medical technol-

ogy is beginning to provide alternatives. We can perform many tests on cell and tissue cultures without recourse to systemic testing on animals. Computer simulations can also cut down on the number of animal tests we need to run. We aren't exploring these alternatives vigorously enough.

Ten or 15 years ago animal-rights activists resorted to violence against humans in their efforts to break through the public's terrible apathy and lack of imagination on this issue. This extremism is counterproductive. I believe that more and more people are becoming aware that to use animals thoughtlessly, without any anguish or making an effort to find another way, diminishes us as human beings.

COMPREHENSION

1. Why does Goodall begin the essay with the anecdote about the chimpanzee named David Greybeard?
2. How, according to the writer, are laboratory animals treated? What are the physical and behavioral effects on these animals? Why does she visit a laboratory where these experiments are taking place, and what "ethical" conclusions does she draw?
3. According to Goodall, why is it important for people to change their "mindset" or hardened opinions about animal experimentation? What is her attitude toward animal rights extremists? Does she present any beneficial ways to change attitudes? Explain.

RHETORIC

1. Locate Goodall's claim or major proposition. What are her warrants for this essay? Construct an outline of the argument.
2. Identify the forms of evidence that Goodall uses and where each type occurs. Does she present sufficient evidence to support her argument? Why or why not?
3. What assumptions (for example, about David Greybeard) does Goodall make in this essay? Do you find these assumptions reasonable? Explain.
4. What form of reasoning—deduction or induction—does Goodall employ in this essay? Does she strictly use logical appeal, or do ethical and emotional appeals appear? Explain.
5. In what way is this an argumentative essay that presents a problem and offers a solution? Point to specific passages to support your answer.

WRITING

1. Goodall uses the word *ethical* in this essay. Write an essay in which you define *ethics* and analyze the ethical arguments for and against the use of animals in laboratory research.

2. How can you write about animal experimentation without having recourse to emotional appeals? Write an essay responding to this question. Explain how both Sabin and Goodall could have avoided emotional appeals and still constructed effective arguments.
3. **WRITING AN ARGUMENT:** Select either Sabin or Goodall's essay and write a rebuttal to it. Use a combination of logical, ethical, and emotional appeals, as well as a variety of evidence, to support your claim.

| www.mhhe.com/ **mhreader** | For more information on Jane Goodall, go to: **More Resources > Ch. 2 Arguments** |

Debate: The Patriot Act—Should We Sacrifice Civil Liberties for Security?

Security versus Civil Liberties

Richard A. Posner

Richard A. Posner (b. 1939), born in New York City, received a BA from Yale University (1959) and a law degree from Harvard University (1962), where he was editor of the Harvard Law Review *and graduated first in his class. He is a federal appeals court judge for the Seventh Circuit in Chicago and a senior lecturer at the University of Chicago Law School. Termed a "thinking man's conservative," Posner is the author of several influential books, including* Frontiers of Legal Theory *(2001),* Public Intellectuals: A Study of Decline *(2001),* Law, Pragmatism, and Democracy *(2003), and a forthcoming book.* Catastrophe: Risk and Response. *He has also published hundreds of articles in law journals and the popular press. In the following essay, published in the* Atlantic Monthly *shortly after the events of September 11, 2001, Posner offers a considered analysis of the tension between security and civil liberties and a logical argument favoring one over the other.*

In the wake of the September 11 terrorist attacks have come many proposals for tightening security; some measures to that end have already been taken. Civil libertarians are troubled. They fear that concerns about national security will lead to an erosion of civil liberties. They offer historical examples of supposed overreactions to threats to national security. They treat our existing civil liberties—freedom of the press, protections of privacy and of the rights of criminal suspects, and the rest—as sacrosanct, insisting that the battle against international terrorism accommodate itself to them.

I consider this a profoundly mistaken approach to the question of balancing liberty and security. The basic mistake is the prioritizing of liberty. It is a mistake about law and a mistake about history. Let me begin with law. What we take to be our civil liberties—for example, immunity from arrest except upon probable cause to believe we've committed a crime and from prosecution for violating a criminal statute enacted after we committed the act that violates it—were made legal rights by the Constitution and other enactments. The other enactments can be changed relatively easily, by amendatory legislation. Amending the Constitution is much more difficult. In recognition of this the Framers left most of the

141

constitutional provisions that confer rights pretty vague. The courts have made them definite.

Concretely, the scope of these rights has been determined, through an inter-   3
action of constitutional text and subsequent judicial interpretation, by a weigh-ing of competing interests. I'll call them the public-safety interest and the liberty interest. Neither, in my view, has priority. They are both important, and their relative importance changes from time to time and from situation to situation. The safer the nation feels, the more weight judges will be willing to give to the liberty interest. The greater the threat that an activity poses to the nation's safety, the stronger will the grounds seem for seeking to repress that activity, even at some cost to liberty. This fluid approach is only common sense.

Supreme Court Justice Robert Jackson gave it vivid expression many years   4
ago when he said, in dissenting from a free-speech decision he thought doctri-naire, that the Bill of Rights should not be made into a suicide pact. It was not intended to be such, and the present contours of the rights that it confers, hav-ing been shaped far more by judicial interpretation than by the literal text (which doesn't define such critical terms as "due process of law" and "unrea-sonable" arrests and searches) are alterable in response to changing threats to national security.

If it is true, therefore, as it appears to be at this writing, that the events of   5
September 11 have revealed the United States to be in much greater jeopardy from international terrorism than had previously been believed—have revealed it to be threatened by a diffuse, shadowy enemy that must be fought with police measures as well as military force—it stands to reason that our civil liberties will be curtailed. They *should* be curtailed, to the extent that the benefits in greater security outweigh the costs in reduced liberty. All that can reasonably be asked of the responsible legislative and judicial officials is that they weigh the costs as carefully as the benefits.

It will be argued that the lesson of history is that officials habitually exag-   6
gerate dangers to the nation's security. But the lesson of history is the opposite. It is because officials have repeatedly and disastrously underestimated these dangers that our history is as violent as it is. Consider such underestimated dangers as that of secession, which led to the Civil War, of a Japanese attack on the United States, which led to the disaster at Pearl Harbor, of Soviet espionage in the 1940s, which accelerated the Soviet Union's acquisition of nuclear weapons and emboldened Stalin to encourage North Korea's invasion of South Korea; of the installation of Soviet missiles in Cuba, which precipitated the Cuban missile crisis; of political assassinations and outbreaks of urban violence in the 1960s; of the Tet Offensive of 1968; of the Iranian revolution of 1979 and the subsequent taking of American diplomats as hostages; and, for that matter, of the events of September 11.

It is true that when we are surprised and hurt, we tend to overreact—but   7
only with the benefit of hindsight can a reaction be separated into its proper and excess layers. In hindsight we know that interning Japanese Americans did not shorten World War II. But was this known at the time? If not, shouldn't the

Army have erred on the side of caution, as it did? Even today we cannot say with any assurance that Abraham Lincoln was wrong to suspend habeas corpus during the Civil War, as he did on several occasions, even though the Constitution is clear that only Congress can suspend this right. (Another of Lincoln's wartime measures, the Emancipation Proclamation, may also have been unconstitutional.) But Lincoln would have been wrong to cancel the 1864 presidential election, as some urged: by November of 1864 the North was close to victory, and canceling the election would have created a more dangerous precedent than the wartime suspension of habeas corpus. This last example shows that civil liberties remain part of the balance even in the most dangerous of times, and even though their relative weight must then be less.

Lincoln's unconstitutional acts during the Civil War show that even legality   8
must sometimes be sacrificed for other values. We are a nation under law, but first we are a nation. I want to emphasize something else, however: the malleability of law, its pragmatic rather than dogmatic character. The law is not absolute, and the slogan *"Fiat iustitia ruat caelum"* ("Let justice be done though the heavens fall") is dangerous nonsense. The law is a human creation rather than a divine gift, a tool of government rather than a mandarin mystery. It is an instrument for promoting social welfare, and as the conditions essential to that welfare change, so must it change.

Civil libertarians today are missing something else—the opportunity to chal-   9
lenge other public-safety concerns that impair civil liberties. I have particularly in mind the war on drugs. The sale of illegal drugs is a "victimless" crime in the special but important sense that it is a consensual activity. Usually there is no complaining witness, so in order to bring the criminals to justice the police have to rely heavily on paid informants (often highly paid and often highly unsavory), undercover agents, wiretaps and other forms of electronic surveillance, elaborate sting operations, the infiltration of suspect organizations, random searches, and monitoring of airports and highways, the "profiling" of likely suspects on the basis of ethnic or racial identity or national origin, compulsory drug tests, and other intrusive methods that put pressure on civil liberties. The war on drugs has been a big flop; moreover, in light of what September 11 has taught us about the gravity of the terrorist threat to the United States, it becomes hard to take entirely seriously the threat to the nation that drug use is said to pose. Perhaps it is time to redirect law-enforcement resources from the investigation and apprehension of drug dealers to the investigation and apprehension of international terrorists. By doing so we may be able to minimize the net decrease in our civil liberties that the events of September 11 have made inevitable.

COMPREHENSION

1. How, in general, does Judge Posner view the debate over security and civil liberties? Why, as he indicates in paragraph 1, are civil libertarians troubled?

2. Posner's article was written in the wake of the September 11, 2001, attacks. Where does he refer to these events? What other "lessons of history" does he mention?
3. Posner writes of "the malleability of law, its pragmatic rather than dogmatic character" (paragraph 8). What does he mean by these words? How do they influence his analysis of the debate over security and civil liberties?

RHETORIC

1. What is Posner's claim and where does he state it most clearly? What forms of evidence does he provide to support his claim? Would you say that his argumentative method is inductive or deductive, and why?
2. Do you think that Posner's primary purpose is to change his audience's thinking, attack the opposition, justify his position, or perhaps a combination of these possibilities? Explain your conclusion.
3. Posner disagrees with conventional wisdom concerning war, freedom, security, and the Constitution. How does he defend these dissenting positions? What evidence does he provide from history?
4. Why does Posner refer to the drug war in his concluding paragraph? Do you consider this strategy to be effective or a distraction? Justify your response.

WRITING

1. Select one of the historical events mentioned by Posner in his essay, conduct research of the subject, and then write an essay in which you demonstrate how the debate over security and civil liberties was reflected in this episode.
2. Who or what is a "civil libertarian"? Form a group of four or five class members, and discuss this question. Conduct research if necessary, and then write an extended definition of the term.
3. **WRITING AN ARGUMENT:** Judge Posner claims that we often overreact to critical historical events, but overreaction is actually necessary and beneficial to our security, even if civil liberties have to be curtailed. Argue for or against his proposition in an essay. Cite some of Posner's own examples to support your position.

 www.mhhe.com/
mhreader

For more information on Richard A. Posner, go to:
More Resources > Ch. 2 Arguments

✳

Acts of Resistance

Elaine Scarry

Elaine Scarry (b. 1946) was born in Summit, New Jersey. She attended Chatham College (BA, 1968) and received her doctorate from the University of Connecticut in 1974. She is an English professor at Harvard University. Best known for The Body in Pain: The Making and Unmaking of the World *(1985), Scarry has also written* Resisting Representation *(1994) and* On Beauty and Being Just *(1999). In this essay, which appeared originally in the February/March 2004 issue of the* Boston Review *and was republished in* Harper's Magazine, *the noted scholar offers a detailed assessment of the problems raised by the Patriot Act.*

When the U.S.A. Patriot Act arrived in our midst in the fall of 2001, its very title seemed to deliver an injury: "Uniting and Strengthening America by Providing Appropriate Tools Required to Intercept and Obstruct Terrorism." One might have thought that "United States of America" would be a sufficient referent for the letters "U.S.A." and that no one would presume to bestow a new meaning on the word "patriot," with its heavy freight of history (Paul Revere, Patrick Henry, Emma Lazarus) and its always fresh aspiration ("O beautiful for patriot dream").

In the two and a half years since it was passed, the U.S.A. Patriot Act has become the locus of resistance against the unceasing injuries of the Bush-Rumsfeld-Ashcroft triumvirate, as first one community, then two, then eleven, then twenty-seven, and now 272 have passed resolutions against it, as have four state legislatures. The letters "U.S.A." and the word "patriot" are gradually reacquiring their earlier solidity and sufficiency as local and state governments reanimate the practice of self-rule by opposing the Patriot Act's assault on the personal privacy, free flow of information, and freedom of association that lie at the heart of democracy. Each of the resolutions affirms the town's obligation to uphold the constitutional rights of all persons who live there, and many of them explicitly direct police and other residents to refrain from carrying out the provisions of the Act, even when instructed to do so by a federal officer.

When the resistance was first beginning, in the winter of 2001–2002, it took five months for the first five resolutions to come into being; by the winter of 2003–2004, a new resolution was being drafted almost every day. The resolutions come from towns ranging from small villages—Wendell, Massachusetts (986), Riverside, Washington (348), Gaston, Oregon (620)—to huge cities—Philadelphia (1,517,550), Baltimore (651,000), Chicago (2,896,000), Detroit (951,000), Austin (656,300), San Francisco (777,000). Approximately a third of the resolutions come from towns and cities with populations between 20,000 and 200,000.

The fact that the Patriot Act has engendered such resistance may at first   4
seem puzzling. True, its legislative history is sordid: it was rushed through Congress in several days; no hearings were held; it went largely unread; only a few
of its many egregious provisions were modified. But at least it *was* passed by
Congress: many other blows to civil liberties have been delivered as unmodi-
fied executive edicts, such as the formation of military tribunals and the nullification of attorney-client privilege. True, the Patriot Act severed words from
their meanings (beginning with the letters "U.S.A."), but executive statements
associating Iraq with nuclear weapons and with Al Qaeda severed words from
their basis in material fact, at the very great cost of a war that continues to be
materially and mortally destructive. True, the Patriot Act has degraded the legal
stature of the United States by permitting the executive branch to bypass constitutional law, but our legal degradation outside the Patriot Act has gone even
further: Evidence indicates that the Bush Administration has created offshore
torture centers in Bagram, Afghanistan, and on the British island of Diego Garcia, and has sent prisoners to interrogation centers in countries with documented histories of torture such as Egypt, Jordan, Saudi Arabia, and Syria.

The executive edicts, the war against Iraq, and the alleged use of torture   5
have all elicited protest, but what differentiates the opposition to the Patriot Act
is the fact that it has enabled the population to move beyond vocalizing dissent
to retarding, and potentially reversing, the executive's inclination to carry out
actions divorced from the will of the people.

If many members of Congress failed to read the Patriot Act during its swift pas-   6
sage, it is in part because it is almost unreadable. The Patriot Act is written as an
extended sequence of additions to and deletions from previously existing
statutes, instructing the bewildered reader to insert three words into paragraph
X of statute Y without ever providing the altered sentence in either its original
or its amended form. Only someone who had scores of earlier statutes open to
the relevant pages could step painstakingly through the revisions. Reading the
Patriot Act is like standing outside the public library trying to infer the sentences in the books inside by listening to hundreds of mice chewing away on
the pages.

The Act does, however, have a coherent and unitary purpose: to increase   7
the power of the Justice Department and to decrease the rights of individual
persons. The constitutional rights abridged by the Patriot Act are enumerated in
the town resolutions, which most often specify violations of the First Amendment guarantee of free speech and assembly, the Fourth Amendment guarantee
against search and seizure, the Fifth and Fourteenth Amendment guarantees of
due process, and the Sixth and Eighth Amendment guarantees of a speedy and
public trial and of protection against cruel and unusual punishment.

The objective of the Patriot Act becomes even clearer if it is understood con-   8
cretely as making the population *visible* and the Justice Department *invisible*.
The Act inverts the constitutional requirement that people's lives be private and
the work of government officials be public; it instead crafts a set of conditions

that make our inner lives transparent and the workings of the government opaque. Either one of these outcomes would imperil democracy; together they not only injure the country but also cut off the avenues of repair.

When we say democracy requires that the people's privacy be ensured, we ⁹ mean that we ourselves should control the degree to which, and the people to whom, our lives are revealed. Under the Patriot Act, the inner lives of people are made involuntarily transparent by provisions that increase the ability of federal officers to enter and search a person's house, to survey private medical records, business records, library records, and educational records, and to monitor telephone, email, and Internet use. The Fourth Amendment states: "The right of the people to be secure in their persons, houses, papers, and effects, against unreasonable searches and seizures, shall not be violated, and no Warrants shall issue, *but upon probable cause*, supported by Oath or affirmation, and *particularly describing the place to be searched, and the persons or things to be seized*" (emphasis added). The Patriot Act both explicitly lowers the "probable cause" requirement, thereby diminishing judicial review, and eliminates the specificity clause—"particularly describing the place to be searched, and the persons or things to be seized"—which, like "probable cause," puts severe restraints on the scope and duration of the search. The Act is a sweeping license to search and seize, everywhere and anywhere, guided not by court-validated standards of evidence but by Justice Department hunches and racially inflected intuitions.

As necessary to democracy as the nontransparency of persons is the trans- ¹⁰ parency of government actions, and indeed the Constitution pauses again and again to insist upon open records: "Each house [of Congress] shall keep a Journal of its Proceedings, and from time to time publish the same" with "the Yeas and Nays of the Members . . . entered on the Journal"; "a regular Statement and Account of the Receipts and Expenditures of all public Money shall be published from time to time"; presidential objections to a piece of legislation must be forwarded to the house in which the legislation originated and published in its journal; the counting of the Electoral College votes must take place in the presence of the full Congress; treason proceedings will take place in "open Court" and criminal prosecutions in a "public trial," etc.

The obligation of each branch to make its actions public—to make them vis- ¹¹ ible both to the people and to the other branches—is often construed as a right belonging to the populace, the right of "freedom of information." Indeed, it is hard to disagree with the argument that democratic deliberation is impossible without this access to information. Secrecy, the legal theorist Cass Sunstein writes, "is inconsistent with the principle of self-rule." He identifies citizen deliberation as the primary benefit of open government, but there are other benefits, including checks and balances (one branch cannot check the other if it does not know what the other is doing), and "sunlight as a disinfectant" (if deliberations are carried out in secret, "participants may be less careful to ensure that their behavior is unaffected by illegitimate or irrelevant considerations").

Because both the privacy of individual action and the publication of gov- ¹² ernment action are necessary to democratic self-rule, the major complaint of the local resolutions has been the damage done to the liberties of persons and to the integrity of our laws. The most forceful formulation of this worry comes at the conclusion of the Blount County, Tennessee, resolution, which calls upon all residents "to study the Bill of Rights so that they can recognize and resist attempts to undermine our Constitutional Republic . . . and declare null and void all future attempts to establish Martial Law, [or] Declared States of Emergency." Although most of the other resolutions are more measured in their language, they consistently register the view that both the people and the laws of this country are endangered.

The resolutions have a second, closely related focus. Although the Patriot Act   13 enables the federal government to detain and investigate both citizens and non-citizens, and to carry out surveillance of both citizens and non-citizens, its blows fall most heavily on those who are not U.S. citizens.

Consider section 412. As summarized by the city of Ann Arbor, Michigan, it   14 permits the incarceration of non-citizens for seven days without charge and "for six month periods indefinitely, without access to counsel" if the attorney general "determines release would endanger the security of the country or of a specific person." Before it was modified by Congress, the bill authorized the unlimited detention of immigrants, but the revision is less of an improvement than it seems, since various loopholes release the executive branch from the seven-day constraint.

The resolutions collectively work to prevent this imperilment of all resi-   15 dents of the United States. Almost without exception, the 272 resolutions celebrate their commitment to law and liberty for all "persons" or "residents," not only "citizens." This is expressed in part as a matter of constitutional conviction: The very first clause of the very first resolution (Ann Arbor) begins by echoing the 2001 Supreme Court decision *Zadvydas* v. *Davis:* "The due process and equal protection clauses of the Fifth and Fourteenth Amendments to the United States Constitution guarantee certain due process and equal protection rights to all residents of the United States regardless of citizenship or immigration status . . . " Other resolutions remind all residents that discrimination based on "citizenship status" is no more permissible than discrimination based on race or gender. They complain that the Patriot Act tries "to drive a wedge" between citizens and non-citizens, or between police and foreign nationals, a situation held to be intolerable because the town depends on the diversity of its population for its "vitality" and its "economy, culture, and civic character."

Almost the only time when "citizens" are singled out is when the docu-   16 ments place on them the burden of acting to ensure that all "persons" or "residents" enjoy the benefits of due process, protection from unwarranted search and seizure, freedom of speech, freedom of assembly, and privacy. If, in other words, citizens are unique, it is because they are the guardians of rights belonging to citizens and non-citizens alike, not the exclusive holders of those rights.

In addition to aiming blows at our legal framework of self-governance, the   17 Patriot Act licenses the executive branch to harm other institutions—among them, financial markets and universities—and once again its blows appear to be structural.

Take, for example, the provisions that require bankers, broker-dealers, and 18 trading advisers to file "suspicious activity reports" (SARs) when they notice their clients carrying out unusual transfers greater than $5,000. Failure to file is punishable by criminal and civil charges, with fines reaching $10,000. Furthermore, they are prohibited from telling their client about the SAR, which not only taints the client relationship but eliminates at the outset the possibility of determining whether the transfer has some sensible explanation that, if they only knew it, would convince them that the filing was preposterous.

Universities, too, are among the institutions the Patriot Act seeks to change, 19 and the situation may be swiftly assessed by looking at the most widely discussed aspect of the Act, section 215, which applies to both college and public libraries (and, in many cases, bookstores). When approached by an FBI or CIA agent, librarians must turn over a record of the books a specified patron has taken out, and, like the bankers, they are prohibited from telling anyone of the intelligence gathering in which they have just participated.

In his fall 2003 tour of thirty cities to defend the Patriot Act, Attorney Gen- 20 eral John Ashcroft dismissed the idea that the Justice Department could conceivably care about librarians or library records. A University of Illinois study found, however, that by February 2002 (four months after the Patriot Act was passed) 4 percent of all U.S. libraries and 11 percent of libraries in communities of more than 50,000 people had already been visited by FBI agents requesting information about their patrons' reading habits. Ashcroft insisted that not-yet-released FBI records would demonstrate the indifference of the Justice Department to the libraries, but the Justice Department has in fact refused to release these very same records, despite Freedom of Information Act petitions filed by the American Civil Liberties Union and other organizations.

In distilled form, the logic of the Patriot Act and its defense involves four 21 steps: Maximize the power of the Justice Department; erase the public record of Justice Department actions; respond with indignation if anyone protests that the Justice Department might actually be using its newly expanded powers; point out that the protesters are speaking without any hard evidence or facts without mentioning that the executive branch has withheld those very facts from the public.

From the founding of this country the phrase "a government of laws and 22 not of men" has meant that the country cannot pass open-ended laws that will be good if the governors happen to be good and bad if the governors happen to be bad. The goal has always been to pass laws that will protect everyone regardless of the temperament and moral character of the individual governors. The country, as Justice Davis famously observed in the nineteenth century, "has no right to expect that it will always have wise and humane rulers." That's why it is crucial to pass good laws. And crucial, also, to repeal bad ones.

Despite impediments to resistance, 272 towns, cities, and counties have created 23 a firewall against executive trespass in their communities. The resolutions direct residents to decline to assist the federal government in any act that violates the Constitution: local police should abstain from assisting federal officers in house searches that violate the Fourth Amendment, and librarians should abstain from giving out private library records that violate the First and Fourth Amendments.

Here we have the key to why the Patriot Act—rather than the executive 24 edicts—has become the focus of so much resistance. Since military tribunals do not require the assistance of the population, what we think about the military tribunals is a matter of indifference to the executive. Since the country has a standing army rather than a draft, the war against Iraq was neither ours to assist nor ours to decline to assist. If, without the population's assistance, 5,000 foreign nationals can be detained without charges (only three of whom were ever charged with terrorism-related acts), then the population's disapproval of this detention is like smoke rings in the wind. But since the aspirations encoded in the Patriot Act cannot come about without the help of police, bankers, and librarians, the refusal to assist provides a concrete brake on the actions of the federal government.

Although the Justice Department has tried to portray resistance to the Patriot Act as a liberal complaint, the resisters repeatedly assert that they occupy 25 positions across the political spectrum. And, so far, both Congress and the courts appear to be listening. Various congressmen and senators have initiated bills to nullify or limit specific provisions of the Patriot Act. In July 2003 the House passed an amendment to the 2004 Appropriations Bill that withholds all federal funding from section 213—the provision that allows the Justice Department to search a house without notifying the resident. The courts, too, share the concerns of the local resolutions. In January a federal court in Los Angeles ruled one section of the Patriot Act unconstitutional: the judge objected to the provision making it a crime to provide "expert advice or assistance" to terrorists on the grounds that the phrasing is so vague as to license the Justice Department to interfere with First Amendment speech guarantees. In December two federal courts issued rulings declaring acts of detention carried out by the Bush Administration unlawful on grounds similar to those mentioned in the town resolutions.

Sorting out the legal status of the Patriot Act may take some time. The 26 United States Constitution prohibits acts that the Patriot Act licenses, and, although constitutional provisions take legal precedence over contradictory legislation, for the time being the Act appears to empower the federal government not only to call upon the country's residents for assistance but also to impose criminal and civil penalties on those who fail to assist.

Whether the resistance to the Patriot Act gains momentum or is ultimately 27 derailed, the town resolutions remind us that the power of enforcement lies not just with local police but with all those who reside in cities, towns, villages, isolated byways, and country lanes. Law—whether local, state, federal, or constitutional—is only real if, as Patrick Henry said, the rest of us will put our hands to it, put our hearts to it, stand behind it.

COMPREHENSION

1. How does Scarry demonstrate that the Patriot Act "has become the locus of resistance against the unceasing injuries of the Bush-Rumsfeld-Ashcroft triumvirate" (paragraph 2)? How does she characterize the Bush administration? What examples does she provide to support this opinion?
2. According to Scarry, what constitutional rights has the Patriot Act abridged?
3. What does the writer mean by the "nontransparency of persons" and "transparency of government actions"? How do these concepts get to the core of Scarry's understanding of American democracy?

RHETORIC

1. What is the tone of Scarry's introductory paragraph? How does she treat the title of the U.S.A. Patriot Act? Why does she refer to Paul Revere, Patrick Henry, and Emma Lazarus? What assumptions does she seem to be making about her audience?
2. What is Scarry's main proposition? What are her minor propositions? What forms of reasoning and evidence does she provide to support her claim?
3. The writer divides her essay into four sections. What is the purpose of each section? How does the sequence of sections serve to move the argument along?
4. Does Scarry deal effectively with opposing arguments? Why or why not?
5. Scarry's concluding paragraph consists of one lengthy sentence. How does it recapitulate some of the main ideas in the essay? How effective is this last paragraph? Justify your answer.

WRITING

1. Write a 300-word summary of Scarry's essay. Try to capture all of the main features of her argument.
2. Imagine that you are Judge Posner, and compose a response to Scarry's article.
3. **WRITING AN ARGUMENT:** Write an essay in which you take two or three of Scarry's points and refute them, either by posing competing evidence or demonstrating flaws in their logic.

---

www.mhhe.com/
mhreader

For more information on Elaine Scarry, go to:
More Resources > Ch. 2 Arguments

✳

Face Facts:
Patriot Act Aids Security, Not Abuse

Paul Rosenzweig

Paul Rosenzweig (b. 1959), born in New York City, is a lawyer who received a BA from Haverford College (1981), an MA from the University of California at San Diego (1982), and a JD from the University of Chicago (1986). He has served in the United States Department of Justice and the Office of the United States Attorney for Washington, D.C. Rosenzweig is an adjunct professor of law at George Mason University and a senior legal research fellow at the Heritage Foundation. This defense of the Patriot Act appeared in the July 29, 2004, issue of the Christian Science Monitor.

Falsehood, according to Mark Twain's famous dictum, gets halfway around the   1
world before the truth even gets its shoes on. Time and again, outlandish stories
seem to grow legs and find wide distribution before the truth can catch up.

A good example is the U.S.A. Patriot Act. It's so broadly demonized now,   2
you'd never know it passed with overwhelming support in the days immediately after September 11, 2001.

Critics paint the Patriot Act as a caldron of abuse and a threat to civil liber-   3
ties. Advocacy groups run ads depicting anonymous hands tearing up the Constitution and a tearful old man fearful to enter a bookstore. Prominent politicians
who voted for the act call for a complete overhaul, if not outright repeal.

But the truth is catching up. And the first truth is that the Patriot Act was   4
absolutely vital to protect America's security.

Before 9/11, U.S. law enforcement and intelligence agencies were limited by   5
law in what information they could share with each other. The Patriot Act tore
down that wall—and officials have praised the act's value.

As former Attorney General Janet Reno told the 9/11 commission, "Gener-   6
ally, everything that's been done in the Patriot Act has been helpful . . . while at
the same time maintaining the balance with respect to civil liberties."

And as Attorney General John Ashcroft's recent report to Congress makes   7
clear, this change in the law has real, practical consequences. Information-
sharing facilitated by the Patriot Act, for example, was critical to dismantling
terror cells in Portland, OR; Lackawanna, NY; and Virginia. Likewise, the act's
information-sharing provisions assisted the prosecution in San Diego of those
involved with an Al Qaeda drugs-for-weapons plot involving "Stinger" anti-
aircraft missiles.

It also aided in the prosecution of Enaam Arnaout, who had a long-   8
standing relationship with Osama bin Laden and who used his charity organi-

zation to obtain funds illicitly from unsuspecting Americans for terrorist groups and to serve as a channel for people to contribute knowingly to such groups.

These are not trivial successes. They're part of an enormous, ongoing effort   9
to protect America from further terrorist attacks.

We cannot, of course, say that the Patriot Act alone can stop terrorism. But   10
every time we successfully use the new tools at our disposal to thwart a terrorist organization, that's a victory.

Yet remarkably, some of these vital provisions allowing the exchange of in-   11
formation between law enforcement and intelligence agencies will expire at the
end of next year. So here's a second truth: If Congress does nothing, then parts
of the law will return to where they were on the day before 9/11—to a time
when our government couldn't, by law, connect all the dots. Nobody wants a
return to those days, but that is where we are headed if Congress does not set
aside its partisan debates.

But what of the abuses? Time for a third truth: There is no abuse of the Pa-   12
triot Act. None. The Justice Department's inspector general (who is required by
the Patriot Act to examine its use and report any abuse twice a year) reported
that there have been no instances in which the act has been invoked to infringe
on civil rights or civil liberties. Others agree. For example, at a Judiciary Committee hearing on the Patriot Act, Sen. Dianne Feinstein (D) of California said:
"I have never had a single abuse of the Patriot Act reported to me. My staff . . .
asked [the ACLU] for instances of actual abuses. They . . . said they had none."

So the fiction of abuse can be laid to rest. The government is not, to take but   13
one popular myth, invading libraries and scouring your book records. It's a
convenient fiction that calls to mind, as Joseph Bottum, a contributor to *The
Weekly Standard*, has written, the appealing image of "white-haired and apple-
cheeked [librarians] resisting as best they can the terrible forces of McCarthyism, evangelical Christian bookburning, middle-class hypocrisy, and Big
Brother government." But no matter how appealing the image, it has no more
reality than a good Hollywood movie.

Government's obligation is a dual one: to provide security against violence   14
and to preserve civil liberty. This is not a zero-sum game. We can achieve both
goals if we empower government to do sensible things while exercising oversight to prevent any real abuses of authority. The Patriot Act, with its reasonable
extension of authority to allow the government to act effectively with appropriate oversight rules, meets this goal.

And the truth eventually catches up to the fiction.   15

COMPREHENSION

1. Summarize Rosenzweig's position on the Patriot Act. What reasons does he
   give to support his position? What does he mean when he says that government's dual role is not a "zero-sum game" (paragraph 14)?

2. Explain the writer's reference to Mark Twain in your own words. Where else in the essay does Rosenzweig refer to the "truth"?
3. According to Rosenzweig, what are some of the objections to the Patriot Act?

RHETORIC

1. Analyze the first four paragraphs in terms of the stylistic and argumentative strategies the writer employs. How does he introduce his subject? Why does he begin with refutation? What and where is his claim?
2. This article appeared as an op-ed essay in a daily newspaper. What formal features of editorial writing can you identify? How effective are they in conveying an argument?
3. What is the writer's tone? What kind of audience does he seem to be writing for? (It might be helpful to know that the *Christian Science Monitor* is a relatively progressive newspaper.)
4. What evidence does Rosenzweig use to support his claim? Is the evidence real or hypothetical? Does the writer provide sufficient evidence to make his case? Why or why not?
5. How does the writer employ concepts of "truth" and "fiction" to organize his essay?

WRITING

1. Both Posner and Rosenzweig are trained in the law. Write a comparative essay in which you examine the "cases" that they develop in support of laws designed to protect our security, even if these laws provoke questions concerning civil liberties. Do they make the same points? How do they organize their material? What forms of evidence do they employ? Who makes the stronger case? These are some of the questions you might want to consider.
2. Write an op-ed essay telling readers what liberties you would be willing to give up in order to feel secure from the threat of terrorism.
3. **WRITING AN ARGUMENT:** Write a paper in which you refute Rosenzweig point by point, focusing on the three main "truths" that he mentions in his essay. Brainstorm with class members in order to generate ideas, evidence, and other points of view. Refer to information provided by Scarry in her essay opposing the Patriot Act.

www.mhhe.com/
mhreader

For more information on Paul Rosenzweig, go to:
More Resources > Ch. 2 Arguments

✳

The Patriot Act of the 18th Century

Ishmael Reed

Ishmael Reed (b. 1938), born in Chattanooga, Tennessee, is a well-known novelist, poet, essayist, and public commentator who also produces films, writes plays and songs, and edits and publishes other writers. Known for his advocacy of civil rights, especially for African Americans and other people of color, he has been in the forefront of progressive political movements for decades. Reed lives in Oakland, California; he has taught at the University of California at Berkeley and other universities. His impressive body of work includes Mumbo Jumbo *(fiction, 1972),* Secretary to the Spirits *(poetry, 1975), and* Airing Dirty Laundry *(essays, 1993). In this essay, which appeared in the July 5, 2004, issue of* Time *magazine. Reed uses historical examples to develop a subtle commentary on the Patriot Act.*

Nations sometimes lose their bearings when confronted by an enemy. In a state   1
of crisis or even panic, they implement measures that are later viewed as regrettable. From 1798 to 1800, the French were considered terrorists, pirating ships and making things uncomfortable for the fledgling American republic. The Federalist Party led a backlash against the French, and Thomas Jefferson and his Republican Party were seen as Francophiles. The XYZ Affair—a scandal centering on the fact that some French officials demanded bribes from American diplomats—brought relations between France and the U.S. to the breaking point. The Federalist Administration of President John Adams considered such solicitations to be grave insults. There were cultural differences as well. In the view of Abigail Adams, Frenchwomen were risque at best.

The reaction to the threat from France came in the form of the Alien and   2
Sedition Acts, which were championed by the Federalists, passed by Congress and signed by Adams in 1798. The Alien Act required immigrants to reside in the U.S. for 14 years instead of 5 to qualify for citizenship. The act also gave the president the legal right to expel those the government considered "dangerous." The Sedition Act punished "false, scandalous and malicious" writings against the government with fines and imprisonment. Most of those arrested under the Sedition Act were Republican editors, and instead of sending boatloads of aliens back to France, it resulted in no one's deportation. In a foreshadowing of the climate that inspired today's U.S.A. Patriot Act, at the turn of the century 200 years ago, it was common practice to question the patriotism of citizens, immigrants and the political opposition.

Jefferson, who was vice president at the time, drafted his position in secret   3
and wrote it into the Kentucky Resolutions of 1798. James Madison, in collaboration with Jefferson, subsequently authored the Virginia Resolutions. In the second and fourth of the Kentucky Resolutions, Jefferson cited the Tenth

Amendment, which gives the states powers not delegated to the government by the Constitution, to declare the Alien and Sedition Acts unconstitutional. Jefferson feared that a strong central government might put an end to slavery. Jefferson's fight against the Alien and Sedition Acts is often placed in the context of free speech, but it had unintended consequences beyond that. The Kentucky Resolutions were among the first to defend states' rights, and Jefferson had even threatened secession. Similar ideas helped spark the Civil War.

After Jefferson defeated Adams and was elected president in 1800, the Alien   4 and Sedition Acts were allowed to expire. Adams, looking to distance himself from the mess, blamed the whole idea on Alexander Hamilton—who by then had been murdered by Aaron Burr.

The expiration of the acts did not end challenges to the First Amendment or   5 the tendency on the part of some presidents to behave like monarchs, sometimes with the cooperation of Congress. The Espionage Act of 1917 prohibited "false statements" that might "impede military success." During World War II, FBI Director J. Edgar Hoover and President Franklin Roosevelt wanted to use sedition charges to suppress black newspapers, claiming they undermined the war effort with reports of racial dissension and demands for civil rights. It took Chief Justice Earl Warren's Supreme Court on March 9, 1964, in *The New York Times Co.* v. *Sullivan,* to finally declare unconstitutional the Sedition Act of the Adams administration. Though the act had expired under Jefferson's administration, the court's action buried that particular threat to free speech once and for all—or so people hoped. Writing for the majority, Justice William Brennan held that L. B. Sullivan, an Alabama official, had not been libeled in a *New York Times* ad that had been paid for by civil rights proponents. Brennan supported his arguments by citing Jefferson.

COMPREHENSION

1. What parallel does Reed draw between the Alien and Sedition Acts and the Patriot Act? What does he assume about the reader's knowledge of the Patriot Act?
2. Why, according to Reed, were the Alien and Sedition Acts a danger to a democratic society? Why were they declared unconstitutional?
3. What other acts that endanger American society does Reed mention?

RHETORIC

1. Who is Reed's intended audience for this essay? What does he assume about the knowledge and interests of this audience?
2. What is Reed's purpose in writing this article? What is his claim? Is this claim stated, implied, or perhaps a combination of both? Justify your answer.

3. The writer employs process analysis based on a sequence of historical events. Trace this sequence. Where is it complete, and where might it benefit from more evidence or detail?
4. Reed clearly draws parallels—some obvious and others less so—between earlier acts in American history and the Patriot Act. What minor propositions grow from these historical events? Do you feel that they strengthen Reed's argument or detract from it? Do you think that the essay is persuasive? Explain.
5. Evaluate Reed's conclusion. How effective do you find it, and why?

WRITING

1. Go online and find out more about the Alien and Sedition Acts. Then write a paper in which you draw parallels between these earlier acts and the Patriot Act.
2. Write an essay in which you predict the outcome of the Patriot Act and how future historians might view its effects on American democracy.
3. **WRITING AN ARGUMENT:** Write an essay in which you argue for or against the proposition that there is a link, as Reed implies, between the Alien and Sedition Acts and the U.S.A. Patriot Act. Use information presented in the three previous essays in this section, library and online research, and notes from class discussion to construct this argumentative essay.

www.mhhe.com/
mhreader

For more information on Ishmael Reed, go to:
More Resources > Ch. 2 Arguments

 chapter *3*

Writing a Research Paper

A research paper is a report in which you synthesize information on your topic, contributing your own analysis and evaluation to the subject. Research writing is a form of problem solving. You identify a problem, form a hypothesis (an unproven thesis, theory, or argument), gather and organize information from various sources, assess and interpret data, evaluate alternatives, reach conclusions, and provide documentation.

Research writing is both exciting and demanding. American essayist and novelist Joan Didion states, "The element of discovery takes place, in nonfiction, not during the writing but during the research." Nowhere is the interplay of the stages in the composing process more evident than in writing research papers. Prewriting is an especially important stage, for the bulk of your research and bibliographical spadework is done before you actually sit down to draft your report. Moreover, strategic critical thinking skills are required at every step of research writing. Here you sense the active, questioning, reflective activity of the mind as it considers a problem and sifts through the evidence to reach a solution, proof, or conclusion. Developing the ability to do research writing thus represents an integration of problem-solving and composing skills.

Research writing is a skill to be developed rather than a trial to be borne. Contrary to conventional wisdom, research does not simply begin with the library catalog and end with the final bibliographic entry. (In fact, electronic searches and word processing have taken much of the drudgery out of writing research papers.) Nor does research writing exclusively report information in a bland and boring recitation of facts.

Research actually means the careful investigation of a subject in order to discover or revise facts, theories, or applications. Your purpose is to demonstrate how other researchers approach a problem and how you treat that problem. A good research paper subtly blends your ideas and the attitudes or findings of others. In research writing you are dealing with ideas that are already in the public domain, but you are also contributing to knowledge.

158

RESEARCH WRITING:
PRECONCEPTIONS AND PRACTICE

When your ideas—rather than the ideas of others—become the center of the research process, writing a research paper becomes dynamic instead of static. The standard preconception about preparing a research paper is that a researcher simply finds a subject and then assembles information from sources usually found in a library. This strategy does teach disciplined habits of work and thought, and it is a traditional way to conduct research for college courses. Yet, does this conventional preconception match the practices of professional researchers?

Consider the following tasks:

- Evaluate critical responses to a best-selling novel, a book of poetry, a CD, or an award-winning film.
- Analyze the impact of voter turnout on presidential politics during a recent decade.
- Investigate a literary, political, or scientific scandal of the last century.
- Assess the effectiveness of urban, suburban, and rural schools, comparing specific measures of student success.
- Discuss the practical consequences of economic theory, examining work opportunities for men, women, recent immigrants, young people entering the workforce, former welfare recipients, or some other group of workers.
- Define a popular dietary or health-related term, examining how it influences consumer behavior when shopping for food.

How would a professional researcher view these projects? First, the researcher sees a subject as a *problem* rather than a mere topic. Often this problem is authorized or assigned by a collaborator, an editor, or a supervisor in the researcher's workplace. The researcher has the task of developing or testing a hypothesis stemming from the particular problem: for example, whether or not a vegetarian diet effectively wards off cancer. *Hypothesis formation* is at the heart of professional research.

Second, the researcher often conducts primary as well as secondary research. *Primary research* relies on analysis of texts, letters, manuscripts, and other materials, whether written, visual, or aural. *Secondary research* relies on sources that comment on the primary sources. For example, a critic's commentary on *Citizen Kane* or an historian's analysis of the cold war politics of the 1950s would be secondary sources; the film itself or a speech delivered by Senator Joseph McCarthy in 1950 would be primary sources. Because primary sources are not necessarily more reliable than secondary sources, you must always evaluate the reliability of both types of material. Critics can misinterpret, and experts often disagree, forcing you to weigh evidence and reach your own conclusions.

Third, all researchers face deadlines. The solution to a research problem is required to take action, to reach a decision, to influence policy, or to determine a business plan. Confronted with deadlines, professional researchers learn to *telescope* their efforts in order to obtain information quickly. Common strategies include networking (using personal and professional contacts as well as guides to organizations), browsing or searching online, conducting computerized bibliographical searches, and turning to annotated bibliographies (listing articles on the topic with commentaries on each item) and specialized indexes (focusing on a particular field or discipline). Other strategies include consulting review articles, which evaluate other resources, and browsing through current journals and periodicals, which may provide useful background as well as the most current thinking about the topic.

Finally, much professional researching cuts across academic subjects and disciplines, perhaps touching on literature, history, politics, psychology, economics, or more. The interdisciplinary nature of many research projects creates special problems for the researcher, especially in the use of bibliographical materials, which do tend to be subject-oriented. Good researchers know that they cannot confine their search for evidence into one subject area, such as history of physics. Knowledge in the contemporary era tends increasingly toward interdisciplinary concerns, and you must develop the training, discipline, and strong critical thinking skills necessary for any form of college research. Such research is not beyond your talents and abilities. Learn how to use library and electronic sources selectively and efficiently, but also learn how to view the world outside your library as a vast laboratory to be used fruitfully in order to solve your research problems.

THE RESEARCH PROCESS

The research process involves thinking, searching, reading, writing, and rewriting. The final product—the research paper—is the result of your discoveries in and contributions to the realm of ideas about your topic. The process of researching and composing moves back and forth over a series of activities, and the actual act of writing remains unique to the individual researcher.

Writers with little experience in developing research papers do have to be more methodical than experienced researchers who streamline and adjust the composing process to the scope and design of their projects. Despite the idiosyncrasies of individual writers, however, the research process tends to move through several interrelated phases.

Phases in the Research Process

Phase I: Defining Your Objective

Choose a *researchable* topic.

Identify a *problem* inherent in the topic that gives you the reason for writing about the topic.

(*Phases in the Research Process* continued)

Examine the *purpose* of or the benefits to be gained from conducting research on the topic.

Think about the assumptions, interests, and needs of your *audience*.

Decide how you are going to *limit* your topic.

Establish a working *hypothesis* to guide and control the scope and direction of your research.

Phase 2: Locating Your Sources

Decide on your *methodology*—the types or varieties of primary and secondary research you plan to conduct. Determine the method of collecting data.

Go to the library and skim a general article or conduct a computer search to *determine if your topic is researchable* and if your hypothesis is likely to stand up.

Develop a *tentative working bibliography*, a file listing sources that seem relevant to your topic.

Review your bibliography, and *reassess your topic and hypothesis*.

Phase 3: Gathering and Organizing Data

Obtain your sources, taking notes on all information related directly to your thesis.

Analyze and organize your information. Design a *preliminary outline* with a tentative thesis if your findings support your hypothesis.

Revise your thesis if your findings suggest alternative conclusions.

Phase 4: Writing and Submitting the Paper

Write a *rough draft* of the paper, concentrating on the flow of thoughts and integrating research findings into the texture of the report.

Write a *first revision* to tighten organization, improve style, and check on the placement of data. Prepare citations that identify the sources of your information. Assemble a list of the references you have cited in your paper.

Prepare the manuscript using the format called for by the course, the discipline, or the person authorizing the research project.

Phase 1: Defining Your Objective

The first step in research writing is to select a topic that promises an adventure for you in the realm of ideas and that will interest, if not excite, your audience while meeting the expectations and requirements of your assignment.

You reduce wasted time and effort if you approach the research project as a problem to be investigated and solved, a controversy to take a position on, or a question to be answered. As a basis, you need a strong hypothesis or working thesis (which may be little more than a hunch or a calculated guess). The point of your investigation is to identify, illustrate, explain, argue, or prove that thesis. Develop a hypothesis before you actually begin to conduct research; otherwise, you will discover that you are simply reading in or about a topic, instead of reading toward the objective of substantiating your thesis or proposition.

Of course, before you can formulate a hypothesis, you need to start with a general idea of what subject you want to explore, what your purpose is going to be, and how you plan to select and limit a topic from your larger subject area.

FORMULATING A HYPOTHESIS

A topic will lead to a researchable hypothesis if it
- Meets the demands of your assignment.
- Strongly interests you.
- Engages knowledge you already possess.
- Raises questions that will require both primary and secondary research to answer.
- Provokes you toward an opinion or argument.

To help you find and limit a research topic, try the following strategies:

- *Reflect on the assignment.* If your professor gave you a specific written assignment—even if it doesn't include a specific topic—review the assignment with an eye toward key words that indicate the purpose of your research work. Highlight or underline key verbs such as *solve, argue, find, discover,* or *present.* Write out questions for your professor and either ask them in class (other students probably share your questions) or arrange for a conference with your professor.
- *Ask questions.* Ask yourself, in writing, a series of specific questions about your subject. Combine questions that are related. Ask your questions in such a way as to pose problems that demand answers. Then try to determine which topic best fits the demands of the assignment.
- *Prewriting.* Idea generation strategies such as prewriting and brainstorming can help you to determine what you already know, or believe, about

an assigned topic. If your assignment is to research gender roles in popular culture, you might begin by brainstorming on the last two or three movies that you saw and how male characters were depicted. For more information on prewriting strategies, see pages 22–26.

- *Background reading.* Your professor will probably assign a research topic that has something to do with the content of your class. Review the assigned readings for your course as well as your own notes. If your professor has suggested additional readings on the research topic or provided a bibliography, consult a few of those sources as well. Although the purpose of your background reading is to generate ideas, you should still use the note-taking strategies discussed in the following pages to ensure that you give proper credit later for any ideas you use from this preliminary reading.

Your purpose is to solve a *specific* problem, shed light on a *specific* topic, state an opinion on a *specific* controversy, offer *specific* proofs or solutions. Your audience does not want a welter of general information, a bland summary of the known and the obvious, or free associations or meditations on an issue or problem. You know that your audience wants answers; consequently, a way to locate your ideal topic is to ask questions about it.

Phase 2: Locating Your Sources

If you have a sufficiently narrowed topic and a working hypothesis, you should know what type of information will be most useful for your report. Not all information on a topic is relevant, of course; with a hypothesis you can distinguish between useful and irrelevant material.

To use your time efficiently, you have to *streamline* your method for collecting data. Most research writing for college courses relies heavily on secondary research material available in libraries or online. To develop a preliminary list of sources, go directly to general reference works or a list of sources or reserved readings provided by your professor. If you are already knowledgeable about the subject, begin with resources that permit you to find a continuing series of articles and books on a single issue, specifically, periodical indexes, newspaper indexes, and card catalogs. Again, you should be moving as rapidly as possible from the general to the specific.

Should You Begin Your Research Online?   The immense searching capabilities of the Internet make it very tempting to begin your search for information online, via a commercial search engine such as Yahoo! or Google. Although this method can be useful for background reading and idea generation, traditional research—both academic and professional—is generally more productive and efficient if begun in a library. However, if your research topic demands very contemporary and localized knowledge (a current political campaign; a recent medical breakthrough; a trend in popular culture), beginning your search online can be optimal. Research topics that require you to provide deeper contexts and backgrounds, or for which primary and secondary sources are restricted to academic journals and databases, are best begun in a library. Although we begin our discussion of locating sources with guidelines for searching online, only you can determine the most efficient and effective way of beginning your research. (Note, too, that "going to the library" on many campuses often begins at your personal computer with your own access, as a student, to the library's online catalog.)

Finding Online Materials   Your library, your college Web site, or your instructor's home page may list useful sites on the World Wide Web, organized by discipline or interest area. Online clearinghouses and print materials about the Web also identify especially useful sites for researchers. Depending on your topic, there are subject-specific Web pages on the Internet that link you to everything you could want, including both primary and secondary sources. "Findlaw" is a good example for law; most of the sciences and many of the liberal arts have useful pages like this. Once you have located an Internet address—a URL (universal resource locator)—for a site on the World Wide Web, you can go directly to that location. The end of the address can help you assess the kind of location you will reach.

.org = nonprofit organizations, including professional groups

.edu = colleges, universities, and other educational institutions

.com = businesses and commercial enterprises

.gov = government branches and agencies

.mil = branches of the military

.net = major computer networks

If you need to search the Internet for sources, try using one of the search engines supplied by your Internet access program. Search engines such as Google or Yahoo! hunt through vast numbers of pages at Web sites, seeking those that mention keywords that you specify. The search engine then supplies you with a list of those sites. Given the enormous number of Web sites and their component pages, you need to select your search terms carefully so that you locate reasonable numbers of pertinent sources.

A Web page may supply links to other useful sites. If you click on the link, usually highlighted or in color, you can go directly to that related site. For example, the following site *(www.fedworld.gov)*, sponsored by the federal government, includes links to federal databases and a keyword search that can lead to particular resources.

Following a chain of links requires critical thinking to assess whether each link seems reliable and current. This kind of research also can take a great deal of time, especially if you explore each link and then follow it to the next. As you move from link to link, keep your hypothesis in mind so that you are not distracted from your central purpose.

EVALUATING ONLINE SOURCES

1. Is the author identified? Is the site sponsored by a reputable business, agency, or organization? Does the site supply information so that you can contact the author or the sponsor?

2. Does the site provide information comparable to that in other reputable sources, including print sources?

3. Does the site seem accurate and authoritative or quirky and idiosyncratic?

4. Does the site seem unbiased, or is it designed to promote a particular business, industry, organization, political position, or philosophy?

5. Does the site supply appropriate, useful links? Do these links seem current and relevant? Do most of them work? Does the site document sources for the information it supplies directly?

6. Has the site been updated or revised recently?

7. Does the site seem carefully designed? Is it easy and logical to navigate? Are its graphics well integrated and related to the site's overall purpose or topic? Is the text carefully edited?

Using the Library Catalog   The library online or card catalog lists information by author, title, subject, and keyword. Of the four, the subject listings are the best place to look for sources, but they are not necessarily the place to start your research. Begin by determining what your library offers. For instance, the online catalog may include all library materials or only holdings acquired fairly recently. The catalog also may or may not supply up-to-date information because books may take several years to appear in print and some weeks to be cataloged. Thus you may need to turn to separate indexes of articles, primary documents, and online materials for the most current material. Remember also that when you search by subject, you are searching the subject fields that are assigned by the cataloger. This differs from a keyword search in which the researcher—you the writer—selects key terms that describe the research situation and enters them into a search engine that will find these terms anywhere within the item record— whether they happen to be in the title, comments, notes, or subject fields.

On the other hand, if your library has a consolidated online system, you may have immediate access to materials available regionally and to extensive online databases. You may be able to use the same terminal to search for books shelved in your own library, materials available locally through the city or county library, and current periodicals listed in specialized databases. Such access can simplify and consolidate your search.

Subject indexing can be useful when you are researching a topic around which a considerable body of information and analysis has already developed. Identify as many keywords (terms that identify and describe your subject) or relevant subject classifications as possible. Use these same terms as you continue your search for sources, and add additional terms identified in the entries you find. The following example illustrates a keyword search for materials on gender issues and advertising.

Clicking on the Extended Display option for an item supplies full bibliographic information as well as the location of the book in the library and its availability. Following is the information for the fourth item listed on the search screen shown on the next page.

There are two ways to make this search of the keywords *Gender* and *Advertising* more complete: (1) Think of alternate terms that might come into play; for example, *sex* is an synonym for *gender,* and *advertising* is only one form of the verb *to advertise.* A searcher would probably want to include *advertise* or *advertisement.* Using the truncation symbol (in this case, * but it varies in different library catalogs) would help to catch these variations. The best idea is to review the search tips that nearly always accompany any public catalog. (2) Select one of the titles you feel is most closely related to your subject and pull up that record. For example "Sex in advertising" and "Sex role in advertising" could yield fruitful links, directing you to other materials that have been assigned the same headings. This strategy is a much more direct search than thumbing through the red Library of Congress Subject Headings volumes to find out what an appropriate subject heading might be, and it will catch those titles you might have missed when selecting your keywords.

AURARIA LIBRARY

GENDER is in 1561 titles.
ADVERTISING is in 1270 titles.
Both "ADVERTISING" and "GENDER" are in 13 titles.
There are 13 entries with ADVERTISING & GENDER.

| NEXT PAGE | EXTENDED DISPLAY | START OVER | ANOTHER SEARCH | LIMIT THIS SEARCH | (CU-Law) (CSU) (UNC) (DU) (DU-Law) (Jeffco) |

You searched: WORD ⬥ | gender advertising | [Search]

Num	Mark	WORDS (1-12 of 13)	Entries 13 Found
1	☐	Advertising and culture : theoretical perspectives / edited	1
2	☐	Creating Rosie the Riveter : class, gender, and propaganda d	1
3	☐	Education, technology, power : educational computing as a so	1
4	☐	The Electronic grapevine : rumor, reputation, and reporting	1
5	☐	Feminist perspectives on eating disorders / edited by Patric	1
6	☐	Gender advertisements / Erving Goffman.	1
7	☐	Global and multinational advertising / edited by Basil G. En	1
8	☐	Inarticulate longings : The ladies' home journal, gender, an	1
9	☐	Putting on appearances : gender and advertising / Diane Bart	1
10	☐	Russian cultural studies : an introduction / edited by Catri	1
11	☐	Sport business : operational and theoretical aspects / [edit	1
12	☐	Undressing the ad : reading culture in advertising / edited	1

[Save Marked Records]   [JUMP TO] 13

| NEXT PAGE | EXTENDED DISPLAY | START OVER | ANOTHER SEARCH | LIMIT THIS SEARCH | (CU-Law) (CSU) (UNC) (DU) (DU-Law) (Jeffco) |

Search Other Regional Libraries:

 Denver Public Library   Colorado State Publications

Reprinted by permission of Auraria Library

Checking General Reference Sources   General reference sources include encyclopedias, dictionaries, handbooks, atlases, biographies, almanacs, yearbooks, abstracts, and annual reviews of scholarship within a field. Many of these sources are available both in print and in an electronic format, on CD-ROM or online. Begin your search for these sources in your library's reference room. General reference sources can be useful for background reading and for an introduction to your topic. The bibliographies they contain (such as those that end articles in an encyclopedia) are generally limited, however, and frequently out-of-date. Professional researchers do not rely exclusively on general reference sources to solve research problems, and neither should you.

Searching Indexes and Databases   Electronic and print indexes and databases include up-to-date articles in journals, magazines, and newspapers. Indexes usually list materials that you will then need to locate. Some databases, however,

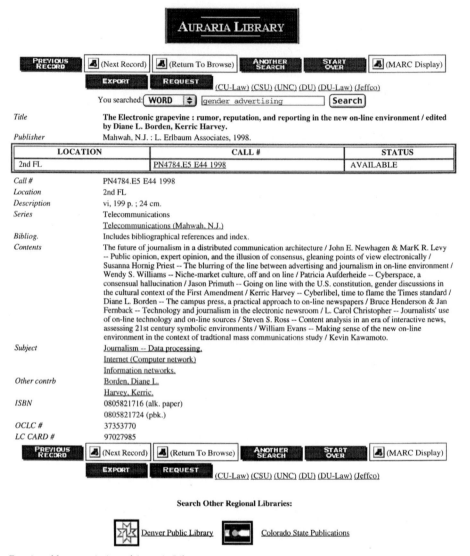

Reprinted by permission of Auraria Library

may include complete texts of articles or even books. Ask a reference librarian how to access materials on CD-ROM or online. If you need historical information or want to trace a topic backwards in time, however, you may need to use print indexes as well because electronic sources may date back only a few years or cover only a certain number of years.

The following indexes and databases are just a few of the many resources that are widely available. Some are general; others are specialized by discipline or field. Such indexes supply ready access to a wide array of useful materials, including articles, books, newspaper stories, statistics, and government docu-

---

ELECTRONIC AND PRINT INDEXES AND DATABASES

General Resources

American Statistics Index

Congressional Information Service Index

Expanded Academic Index

FirstSearch Catalog

Magazine Index

National Newspaper Index

New York Times Index

Specialized Resources

Applied Science and Technology Index

Biological and Agricultural Index

Business Periodicals Index

Education Index

ERIC (Educational Resources Information Center)

General Business File

Humanities Index

Index Medicus or Medline

MLA (Modern Language Association) International Bibliography

PsychLit

Public Affairs Information Service (PAIS)

Social Sciences Index

---

ments. Ask the librarian in the reference area or the catalog area whether these are available in print or online.

Each index or database restricts the sources it lists in specific ways, based on the particular topics covered or the types of sources included. For example, the full title of the *MLA Bibliography* indicates that it lists "Books and Articles on the Modern Languages and Literatures." Besides books and articles, however, it includes essays or chapters collected in a book, conference papers, films, recordings, and other similar sources, but it does not list summaries or encyclopedia articles. Its primary subjects include literary criticism, literary themes and genres, linguistics, and folklore. Thus, you can search for an author's name, a title, a literary period, or subjects as varied as hoaxes, metaphysical poetry, and self-knowledge, all in relationship to studies in language and literature. This bibliography is available in print, on CD-ROM, online, or in other electronic versions. The print version is published every year, but the online version is updated 10 times during the year. A search of the *MLA Bibliography* 1/91–4/01 on CD-ROM for information on gender issues in advertising would turn up items such as the following:

```
TI: Gender Issues in Advertising Language
AU: Artz,-Nancy; Munizer,-Jeanne; Purdy,-Warren
SO: Women-and-Language, Fairfax, VA (W&Lang). 1999 Fall, 22:2, 20-26.
AN: 1999095570


TI: Anglicisms in German Car Advertising: The Problem of Gender
Assignment
AU: Vesterhus,-Sverre-A..
SO: Moderna-Sprak, Goteborg, Sweden (MSpr). 1998, 92:2, 160-70.
AN: 1999091717


TI: Ready or Not: Clothing, Advertising, and Gender in Late Nineteenth-
Century America
AU: Schorman,-Rob
SO: Dissertation-Abstracts-International,-Section-A:-The-Humanities-and-
Social-Sciences, Ann Arbor, MI (DAIA). 1999 Mar, 59:9, 3619 DAI No.:
DA9907365. Degree granting institution: Indiana U, 1998.
AN: 1999079868
```

As your search progresses and your hypothesis evolves, you will find resources even more specifically focused on your interests.

EVALUATING PRINT SOURCES

1. Is the author a credible authority? Does the book jacket, preface, or byline indicate the author's background, education, or other publications? Do other writers refer to this source and accept it as reliable? Is the publisher or publication reputable?

2. Does the source provide information comparable to that in other reputable sources?

3. Does the source seem accurate and authoritative, or does it make claims that are not generally accepted?

4. Does the source seem unbiased, or does it seem to promote a particular business, industry, organization, political position, or philosophy?

(*Evaluating Print Sources* continued)

5. Does the source supply notes, a bibliography, or other information to document its sources?

6. If the source has been published recently, does it include current information? Are its sources current or dated?

7. Does the source seem carefully edited and printed?

Using Nonprint Sources   In the library and online, you have access to potentially useful nonprint materials of all kinds—videos, CD-ROMs, films, slides, works of art, records of performances, or other sources that might relate to your topic. When you search for these sources, you may find them in your library's main catalog or in a separate listing. In the catalog entry, be sure to note the location of the source and its access hours, especially if they are limited. If you need a projector or other equipment to use the material, ask the reference librarian where you go to find such equipment.

Developing Field Resources   You may want to *interview* an expert, *survey* the opinions of other students, *observe* an event or situation, or examine it over a long period of time as a *case study*. Ask your instructor's advice as you design questions for an interview or a survey or procedures for a short- or long-term observation. Also be sure to find out whether you need permission to conduct this kind of research on campus or in the community.

The questions you ask will determine the nature and extent of the responses that you receive; as a result, your questions should be developed after you have established clear objectives for your field research. You also need to plan how you will analyze the answers before, not after, you administer the questionnaire or conduct the interview. Once you have drafted interview or survey questions, test them by asking your friends or classmates to respond. Use these preliminary results to revise any ambiguous questions and to test your method of analysis. If you are an observer, establish in advance what you will observe, how you will record your observations, and how you will analyze them. Get permission, if needed, from the site where you will conduct your observation. Your field sources can help you expand your knowledge of the topic, see its applications or discover real-world surprises, or locate more sources, whether print, electronic, or field.

Using Visuals in Your Paper   Some of the nonprint sources you consult in your research might be useful to include, rather than just cite or refer to, in your own final paper. Technology has made it very easy to cut-and-paste visuals from sources into your own work, as well as to create and incorporate your own visuals. Be sure that when you incorporate other visuals into your own paper (or when you create a visual, such as a graph or chart, that draws on data from another source) you correctly and completely cite the source of the visual data. A caption that briefly describes the visual and gives its source information is usually sufficient.

CONSIDERATIONS FOR USING NONPRINT
SOURCES IN YOUR RESEARCH

- Is a visual/nonprint source the most effective and useful way to present data? For example, if your paper is researching the ways in which photojournalists depict presidential candidates, you will probably want to include sample photographs in your paper. If your topic is trends in voter turnout, you might consult charts and graphs in your research but describe in words the evidence from those visuals rather than reproducing them in your paper.
- Can the visual be easily reproduced? What kinds of technology will you need to capture an image, import it into your own text, and print it legibly (and, if necessary, in color)? What capabilities will your audience need to access a visual? (For example, if you are submitting a paper electronically, remember that large visual files can take a *very* long time to download over a dial-up connection.)
- Have you gathered and noted all of the necessary source information, so you can provide context (and, if necessary, a caption) for the visual as well as an accurate bibliographical citation?

In order to gain expert information, you may wish to contact an informed individual directly by e-mail, following up on contact information supplied at a Web site or through other references. If your topic is of long-term interest to you and you have plenty of time to do your research, you may want to join a *listserv* or *e-mail conference,* a group of people interested in a particular topic, whose messages are sent automatically to all participants. Exchanges among those interested in a topic may also be posted on a *bulletin board server* or a *newsgroup,* where you can read both past and ongoing messages and exchanges. The information you receive from others may be very authoritative and reliable, but it may also represent the biased viewpoint of the individual. Assess it carefully by comparing it with information from other sources, print as well as electronic.

Preparing a Working Bibliography   The purpose of compiling a working bibliography is to keep track of possible sources, to determine the nature and extent of the information available, to provide a complete and accurate list of sources to be presented in the paper, and to make preparing the final bibliography much easier. Include in your working bibliography all sources that you have a hunch are potentially useful. After all, you may not be able to obtain all the items listed, and some material will turn out to be useless, repetitious, or irrelevant to your topic. Such entries can easily be eliminated at a later stage when you prepare your final bibliography.

One way to simplify the task of preparing your final Works Cited or References section is to use a standard form for your working bibliography, whether you use cards or computer entries. The models given later in this chapter are based on two guides, abbreviated as MLA and APA. The *MLA Handbook for Writers of Research Papers* (New York: Modern Language Association of America, 2003; 6th ed.) is generally followed in English, foreign languages, and other fields in the humanities. Instructors in the social sciences, natural sciences, education, and business are likely to favor the style presented in the *Publication Manual of the American Psychological Association* (Washington, DC: American Psychological Association, 2001; 5th ed.). Because the preferred form of citation of sources varies considerably from field to field, check with your instructors to determine which of these two formats they prefer or if they recommend another style. Follow any specific directions from an instructor carefully.

As you locate relevant sources, take down complete information on each item on a 3 × 5 note card or start a bibliographic file on your computer. Complete information, properly recorded, will save you the trouble of having to scurry back to the library or back to a Web page for missing data when compiling your final bibliography. Be sure to list the source's call number and location in the library or its URL; then you can easily find the material once you are ready to begin reading and relocate it if you need to refer to it again. When preparing bibliography cards for entries listed in annotated bibliographies, citation indexes, and abstracts, you might want to jot down notes from any pertinent summaries that are provided. Complete a separate card or file entry for each item that you think is promising.

INFORMATION FOR A WORKING BIBLIOGRAPHY

Record the following information for a book:
1. Name(s) of author(s)
2. Title of book, underlined
3. Place of publication
4. Publisher's name
5. Date of publication
6. Call number or location in library
7. URL and date of access online
(continued)

(Information for a Working Bibliography continued)

Record the following information for an article in a periodical:

1. Name(s) of author(s)
2. Title of article, in quotation marks
3. Title of periodical, underlined
4. Volume number or issue number
5. Date of publication
6. Page numbers on which article appears
7. Call number or location in library
8. URL and date of access online

Author	Dvidhzi, Péter
Title	The Romantic Cult of Shakespeare: Literary Reception in Anthropological Perspective
Place of publication	New York St. Martin's Press
Date of publication	1998
Location	Call Number: PR 2979.H8.D38.1998

If you use your computer to record bibliographic information, you may want to find software designed for this purpose. Your software may provide database categories or options from which you can select the categories required by the style guide you need to use. You also can use the requirements of your style guide to help you develop your own, such as these for a book.

Author's last name:	Pinker
Author's first name:	Steven
Book title:	The Language Instinct: How the Mind Creates Language
Publisher's location:	New York
Publisher (imprint):	HarperCollins Publishers (**HarperPerennial**)
Date published (original date):	1995 (1994)

Once you begin to build a bibliographic database, you can refer to your listings and supplement them each time you are assigned a paper.

Reassessing Your Topic   Once you have compiled your working bibliography, take the time to reassess the entire project before you get more deeply involved in it. Analyze your bibliography cards carefully to determine whether you should proceed to the next stage of information gathering.

Your working bibliography should send out signals that help you shape your thinking about the topic. The dominant signal should indicate that your topic is not too narrow or too broad. Generally, a bibliography of 10 to 15 promising entries for a 1,500-word paper indicates that your topic might be properly limited at this stage. A listing of only three or four entries signals that you must expand the topic or consider discarding it. Conversely, a listing of a hundred entries warns that you might be working yourself into a research swamp.

Another signal from your working bibliography should help you decide whether your hypothesis is on target or could be easily recast to make it more precise. Entry titles, abstracts, and commentaries on articles are excellent sources of confirmation. If established scholarship does not support your hypothesis, it would be best to discard your hypothesis and begin again.

Finally, the working bibliography should provide signals about the categories or parts of your research. Again, titles, abstracts, and commentaries are useful. In other words, as you compile the entries, you can begin to think through the problem and to perceive contours of thought that will dictate the organization of the paper even before you begin to do detailed research. Your working bibliography should be alive with such signals.

Phase 3: Gathering and Organizing Data

If your working bibliography confirms the value, logic, and practicality of your research project, you can then move to the next phase of the research process: taking notes and organizing information. Information shapes and refines your thinking; you move from an overview to a more precise understanding, analysis, and interpretation of the topic. By the end of this third phase, you should be able to transform your hypothesis into a thesis and your assembled notes into an outline.

Plagiarism and Intellectual Property   In this phase of the research process, it is especially critical that you maintain a clear distinction between ideas, opinions, information, words from other sources, and your own interpretation of that information. Plagiarism, or the illicit appropriation of content and ideas, can result from sloppy note taking or poor study habits. Taking care to summarize, paraphrase, and quote from sources with scrupulous care—as well as ensuring that you have given yourself enough time to consider your own argument and write your own paper—will go a long way toward avoiding plagiarism.

The temptation to plagiarize is especially keen in an age when essays can easily be purchased online and when primary and secondary source information can be cut-and-pasted at the click of a button into your own work. Be aware that such behavior in the classroom may result in a failing grade or even suspension. In the professional world, however, plagiarism can result in the loss of a job, the destruction of a reputation, and even criminal charges. Plagiarism is a kind of theft. What a plagiarist steals is called, in legal terms, *intellectual property*—the ideas, opinions, inventions, and discoveries in which another writer or researcher has invested considerable time and resources.

When you are unsure of whether or how to give credit to another source, *always* assume that you should give credit (and ask your professor or a writing center instructor for help with citation guidelines). If you are tempted to buy or "borrow" another person's work because of outside circumstances, remember that it is *always* better to ask for a deadline extension (and additional help with your paper). Just as there is never any excuse for the theft of property, there is never any excuse for the theft of ideas.

Evaluating Sources   As you move into the third phase, begin by skimming your sources. Skimming is a careful examination of the source material to sort out the valuable sources from the not-so-valuable ones. For a book, check the table of contents and index for information on your topic; then determine whether the information is relevant to your problem. For an article, see if the abstract or topic sentences in the body of the essay confirm your research interests. The guidelines below can help you determine if a source will be useful.

CRITERIA FOR ASSESSING THE VALUE OF A SOURCE FOR YOUR PROJECT

1. Is the source directly relevant to your topic?
2. Does it discuss the topic extensively, uniquely, and with good authority?
3. Does it bear on your hypothesis, supporting, qualifying, or contradicting it?
4. Does it present relatively current information, especially for research in the social and natural sciences?
5. Does it meet the criteria for credibility discussed in "Evaluating Online Resources" (page 165) and "Evaluating Print Sources" (page 170)?

---

www.mhhe.com/
mhreader

For more information on evaluating sources, go to:
Research > Source Evaluation Tutor

Taking Notes   Once you have a core of valuable material, you can begin to read these sources closely and take detailed notes. Skillful note taking requires a subtle blend of critical thinking skills. It is not a matter of recording all the information available or simply copying long quotes. You want to select and summarize the general ideas that will form the outline of your paper, to record specific evidence to support your ideas, and to copy exact statements you plan to quote for evidence or interest. You also want to add your own ideas and evaluation of the material. All the notes you take must serve the specific purpose of your paper as it is stated in your hypothesis. It is essential that you record source information for *every* note that you take, whether that note is a summary, a paraphrase, or a direct quotation.

GUIDELINES FOR TAKING NOTES
ABOUT YOUR TOPIC

1.   Write the author's last name, the title of the source, and the page number at the top of each card or entry. (Complete information on the source should already have been recorded on a bibliography card or listed in an entry in a computer file.)

2.   Record only one idea or a group of closely related facts on each card or in each entry.

3.   List a subtopic at the top of the card or entry. This will permit you to arrange your cards or entries from various sources into groups, and these groups can then serve as the basis of your outline.

4.   List three types of information: (*a*) summaries of material; (*b*) paraphrases of material, in which you recast the exact words of the author; and (*c*) direct quotations, accurately transcribed.

5.   Add your own ideas at the bottom of the card or following specific notes.

---

www.mhhe.com/
mhreader

For more information on taking notes, go to:
Research > Taking Notes

Summary, Paraphrase, and Quotation   When you write a *summary* of a source, you focus on the main points of the source and restate them in your own words. Summary notes can be especially helpful to remind you, as you draft, of sources that you might want to revisit and look at more closely. Summaries can also be introduced into the body of your essay, especially in an argument research paper, to provide additional information and support to your thesis. Here is an example of a primary source text and a student's summary:

Primary Source: Carl Elliott, "Humanity 2.0," *The Wilson Quarterly*, Autumn 2003
Even technologies that unambiguously provide enhancements will raise issues of social justice not unlike those we currently face with ordinary medical technologies (wealthy Americans, for example, get liver transplants, while children in the developing world die from diarrhea). We live comfortably with such inequities, in part because we have so enthusiastically embraced an individualistic ethic. But to an outsider, a country's expenditure of billions of dollars on liposuction, face-lifts, and Botox injections while many of its children go without basic health care might well seem obscene.

Student Summary

Topic label	"Transhumanism" and bioethics
Author of article	Elliott
Relevant pages/URL	http://wwics.si.edu
Summary	Bioethicist and philosopher Carl Elliott defines "transhumanism" and describes a conference of "transhumanists" that he attended. As a bioethicist, Elliott argues that we need to pay attention to the ethical implications of the medical "enhancements" currently practiced or being developed that might contribute to the "transhumanist" goal of creating perfect human beings. In particular, our society's emphasis on developing medical technologies to make us more beautiful or intelligent at the expense of those less fortunate is especially disturbing.

A *paraphrase* focuses on a one specific point or piece of information in an article and restates it in your own words. Writing a paraphrase of a source can help you to better understand it. When you paraphrase, follow the original writer's argument but do not mimic the writer's sentence structure or simply replace key words with synonyms.

Topic label	"Transhumanism" and bioethics
Author of article	Elliott
Relevant pages/URL	http://wwics.si.edu
Unacceptable paraphrase	Even procedures designed for cosmetic purposes raise controversies over fairness (rich people get transplants while poor children die from basic diseases). Americans are fine with these inequalities because our culture values the individual. To non-Americans, the money we spend on plastic surgery even though many American children lack health insurance probably appears scandalous.
The sentence structure of the original is imitated and synonyms replace the original terms (<u>scandalous</u> for <u>obscene</u>).	

	(continued)
Acceptable paraphrase	Elliott points out that our culture already seems to overlook the injustice of some individuals spending a great deal of money on cosmetic surgery (or "enhancements") while many lack access to basic health care. Elliott describes this as a uniquely American "individualist ethic" but points out that other cultures might see this inequality as "obscene." By extension, the willingness of the transhumanist movement to explore medical "enhancements" that will only benefit a very few wealthy people is also, ethically, "obscene."
Key terms from the original source are directly quoted. The source argument is rephrased in the student's own terms.	

As these examples demonstrate, an acceptable paraphrase shows that the researcher is genuinely engaged with the source's *argument*—not just the words—and has thought about how this particular component of the argument supports the original author's thesis.

When the language of source material is essential to understanding its argument, *quotation* is the most effective strategy. When you quote directly from a source, put quotation marks around the material that you are selecting.

Topic label	"Transhumanism" and bioethics
Author of article	Elliott
Relevant pages/URL	http://wwics.si.edu
Quotation	"But to an outsider, a country's expenditure of billions of dollars on liposuction, face-lifts, and Botox injections while many of its children go without basic health care might well seem obscene."

When you have completed all your research, organize your notes under the various subtopics or subheadings that you have established. Now is the time to establish your thesis. By reviewing your notes and assessing the data, you should be able to transform the calculated guess that was your hypothesis into a much firmer thesis. Focus your attention on your thesis by stating it at the top of the page where you are working on your outline. If possible or desirable, try to combine some subtopics and eliminate others so that you have between three and five major categories for analysis and development. You are now ready to develop an outline for the research essay.

Designing an Outline   Because you must organize a lot of material in a clear way, an outline is especially valuable in a research essay. Spend as much time as is reasonable drafting an outline. Begin by creating a rough outline that simply lists your general subheadings and their supporting data. Next, work more systematically through your notes and fill in the rough outline with as much detail as possible, developing each point logically and in detail. If you are required to submit an outline with your research paper, you should begin to develop a full, formal outline at this stage. Such an outline would look like this:

I.
   A.
   B.
       1.
       2.
       3.
          a.
          b.
II.

Use roman numerals for your most important points, capital letters for the next most important points, arabic numbers for supporting points, and lowercase letters for pertinent details or minor points. If you are including visuals such as photographs or graphics, include them in the outline as well.

www.mhhe.com/
mhreader

For an interactive tutorial on creating outlines, go to:
Writing > Outline Tutor

Phase 4: Writing and Submitting the Paper

As you enter the fourth and final phase of the research process, keep in mind that a research paper is a formal essay, not a jagged compilation of notes. You should be prepared to take your research effort through several increasingly polished versions, most likely at least a rough draft, a revised draft, and a final manuscript.

Writing the Rough Draft   For your rough draft, concentrate on filling in the shape of your outline. Take the time to rearrange your notes in the topic order that your outline assumes. In this way you will be able to integrate notes and writing more efficiently and effectively.

Even as you adhere to your formal outline in beginning the rough draft, you should also be open to alternate possibilities and prospects for presenting ideas and information. Although your primary task in writing a first draft is to rough out the shape and content of your paper, the flow of your ideas will often be accompanied by self-adjusting operations of your mind, all aimed at making your research effort even better than you thought it could be at the outline stage.

Whether or not you incorporate quotations from your notes into the rough draft is a matter of preference. Some writers prefer to transcribe quotations and paraphrases at this point in order to save time at a later stage. Other writers copy and insert these materials directly from entries in a computer file for notes. Still others feel their thought processes are interrupted by having to write out summarized, quoted, and paraphrased material and to design transitions between their own writing and the transcribed material. They simply write "insert" in the draft with a reference to the appropriate notes. Whatever your strategy, it is essential that you keep track of the sources of summarized, quoted, and paraphrased material so that you can properly cite the sources and avoid plagiarism.

The need to integrate material from several sources tests your reasoning ability during the writing of the rough draft. For any given subtopic in your outline, you will be drawing together information from a variety of sources. To an extent, your outline will tell you how to arrange some of this information. At the same time, you must contribute your own commentary, arrange details in an effective order, and sort out conflicting claims and interpretations. A great deal of thinking as well as writing goes into the design of your first draft. You are not involved in a dull transcription of material when writing the rough draft of a research paper. Instead, you are engaged in a demanding effort to think your way through a problem of considerable magnitude, working in a logical way from the introduction and the statement of your thesis, through the evidence, to the outcome or conclusion that supports everything that has come before.

Revising the Draft   In the rough draft, you thought and wrote your way through the problem. Now you must rethink and rewrite in order to give better form and expression to your ideas. Use the guidelines outlined below to approach your revision. Consider every aspect of your paper, from the most general to the most specific. Look again at the organization of the whole paper, key topics, paragraphs, and sentences; read through for clarity of expression and details of grammar, punctuation, and spelling. A comprehensive revision effort will result in a decidedly more polished version of your paper.

GUIDELINES FOR REVISING
YOUR RESEARCH WRITING

---

1. Does my title illuminate the topic of the essay and capture the reader's interest?

2. Have I created the proper tone to meet the expectations of my audience?

3. Does my opening paragraph hook the reader? Does it clearly establish and limit the topic? Is my thesis statement clear, limited, and interesting?

(continued)

(*Guidelines for Revising Your Research Writing* continued)

4. Do all my body paragraphs support the thesis? Is there a single topic and main idea for each paragraph? Do I achieve unity, coherence, and proper development? Is there sufficient evidence in each paragraph to support the main idea?

5. Are there clear and effective transitions linking my ideas within and between paragraphs?

6. Have I selected the best strategies to meet the demands of the assignment and the expectations of my audience?

7. Are all my assertions clearly stated, defined, and supported? Do I use sound logic and avoid faulty reasoning? Do I acknowledge other peoples' ideas properly?

8. Is my conclusion strong and effective?

9. Are my sentences grammatically correct? Have I avoided errors in the use of verbs, pronouns, adjectives, and prepositions? Have I corrected errors of agreement?

10. Are my sentences complete? Have I corrected all fragments, comma splices, and fused sentences?

11. Have I varied my sentences effectively? Have I employed clear coordination and subordination? Have I avoided awkward constructions?

12. If I include visual information, do I provide adequate context? Is the placement of the visual logical? Is the visual clearly reproduced?

13. Are all words spelled correctly? Do my words mean what I think they mean? Are they specific? Are they concrete? Is my diction appropriate to college writing? Is my language free of clichés, slang, jargon, and euphemism? Do I avoid needless abstractions? Is my usage sound?

14. Have I carefully attended to such mechanical matters as apostrophes, capitals, numbers, and word divisions?

15. Does my manuscript conform to acceptable guidelines for submitting typewritten work?

16. Are all of my summaries, paraphrases, and quotations appropriately cited?

Preparing the Final Manuscript   Leave time in your research effort to prepare a neat, clean, attractively designed manuscript. Store all of your files (notes, drafts, and final version) on a backup CD, and print or duplicate an extra copy of the report. Submit a neat, clear version, and keep the second copy. Consult your instructor for the desired format, and carefully follow the guidelines for manuscript preparation in your final version. Look also at the sample paper near the end of this chapter, which illustrates how to present the final version of a paper in accordance with MLA style (see pages 199–213).

DOCUMENTING SOURCES

Documentation is an essential part of any research paper. Documenting your sources throughout the paper and in a section called Works Cited or References tells your audience just how well you have conducted your research. It offers readers the opportunity to check on authorities, to do further reading, and to assess the originality of your contribution to an established body of opinion. Neglect of proper documentation can lead to charges of plagiarism (see page 175).

Quotations, paraphrases, and summaries obviously require credit, for they are the actual words or the theories or interpretations of others. Paraphrases and summaries also frequently offer statistics or data that are not well known, and this type of information also requires documentation. Facts in a controversy (facts open to dispute or to varying interpretations) also fall within the realm of documentation. Visual information (maps, graphics, and photos) also require documentation, even if they show common knowledge (such as a map of Japan).

MATERIALS THAT REQUIRE DOCUMENTATION

1. Direct quotations
2. Paraphrased material
3. Summarized material
4. Any key idea or opinion adapted and incorporated into your paper
5. Specific data (whether quoted, paraphrased, or tabulated)
6. Disputed facts

Parenthetical documentation—briefly identifying sources within parentheses in the text—is the most common method of indicating sources. The purpose of a parenthetical citation is to identify a source briefly yet clearly enough that it can be located in the list of references at the end of the paper. In MLA style, the author's last name and the page number in the source are included. APA style uses the author's last name and the year of publication; page numbers are included primarily for direct quotations. Then complete information is listed, alphabetically by author or title (if a source has no specific author), in the Works Cited or References section following the text of the paper. The bibliographic information you have collected should provide you with the details needed for the preparation of both parenthetical documentation and a list of sources.

GENERAL GUIDELINES FOR PARENTHETICAL DOCUMENTATION

1. Give enough information so that the reader can readily identify the source in the Works Cited (MLA) or References (APA) section of your paper.
2. Supply the citation information in parentheses placed where the material occurs in your text.
3. Give the specific information required by the documentation system you are using, especially when dealing with multivolume works, editions, newspapers, and legal documents.
4. Make certain that the complete sentence containing the parenthetical documentation is readable and grammatically correct.

With your parenthetical documentation prepared, turn your attention next to a final Works Cited or References section. To prepare this list of sources, simply transcribe those bibliography cards or entries that you actually used to write your paper, following the appropriate format.

GENERAL GUIDELINES FOR PREPARING A LIST OF SOURCES

1. Use the title *Works Cited* (MLA) or *References* (APA).
2. Include only works actually cited in the research paper unless directed otherwise by your instructor.
3. Arrange all works alphabetically according to author's last name or according to the title of a work if there is no author. Ignore *A, An,* or *The.*
4. Begin each entry at the left margin. Indent everything in an entry that comes after the first line by five spaces or ½ inch (MLA style) or by five to seven spaces (following APA style for students, unless your instructor directs otherwise).
5. Double-space every line.
6. Punctuate with periods after the three main divisions in most entries—author, title, and publishing information.

In the following sections, you will find examples of MLA and APA documentation forms. Use these examples to help you cite your sources efficiently and clearly.

MLA (MODERN LANGUAGE ASSOCIATION) DOCUMENTATION

The following examples illustrate how to cite a source in the text and in the list of works cited at the end of a paper.

MLA Parenthetical Documentation   The simplest MLA entry includes the author's last name and the page number, identifying exactly where the quotation or information is located. If the author's name is included in the text, it does not need to be repeated in the citation.

Page Number(s) for a Book
The play offers what many audiences have found a satisfying conclusion (Hansberry 265–76).

Garcia Marquez uses another particularly appealing passage as the opening of the story (105).

Volume and Page Number(s) for One Volume of a Multivolume Work
A strong interest in this literature in the 1960s and 1970s inevitably led to "a significant reassessment of the aesthetic and humanistic achievements of black writers" (Inge, Duke, and Bryer 1: v).

Page Number(s) for an Article in a Journal or Magazine
Barlow's description of the family members includes "their most notable strengths and weaknesses" (18).

Section and Page Number(s) for a Newspaper Article
A report on achievement standards for high school courses found "significant variation among schools" (Mallory B1).

Page Number(s) for a Work without an Author
Computerworld has developed a thoughtful editorial on the issue of government and technology ("Uneasy Silence" 54).

Page Number(s) for a Work by a Group or an Organization
The Commission on the Humanities has concluded that "the humanities are inescapably bound to literacy" (69).

Page Number(s) for Several Works by One Author
In The Coming Fury, Catton identifies the "disquieting omens" (6) which precede the Civil War.

As Catton concludes his history of the Civil War, he notes that "it began with one act of madness and it ended with another" (<u>Never Call Retreat</u> 457).

Page Number(s) for One Work Quoted in Another

Samuel Johnson praises <u>She Stoops to Conquer</u> because Goldsmith's play achieves "the great end of comedy—making an audience merry" (qtd. in Boswell 171).

Online Source

Most online sources do not have page numbers for easy parenthetical citation. When citing from an online source, the author's last name or the title of the article or Web page should be given.

Blackwelder observes that "Depp has [the central conflict of the movie] in his eyes in every scene."

MLA List of Works Cited   Following your paper, list the references you have cited in alphabetical order on a separate page entitled "Works Cited." See the Works Cited page of the sample paper (page 213) for an illustration of how you should prepare this page. Use the following sample entries to help you format your references in MLA style. Pay special attention to abbreviated names of publishers, full names of authors, details of punctuation, and other characteristic features of MLA citations.

Work with One Author

Notice the punctuation and underlining in the basic entry for a book.

Muller, Eddie. <u>Dark City: The Lost World of Film Noir</u>. New York: St. Martin's-Griffin, 1998.

Reynolds, David S. <u>John Brown, Abolitionist</u>. New York: Knopf, 2005.

Several Works by One Author

If you use several books or articles by one author, list the author's name in the initial entry. In the next entry or entries, replace the name with three hyphens.

Said, Edward. <u>Humanism and Democratic Criticism</u>. New York: Columbia UP, 2004.

—. <u>Orientalism</u>. New York: Pantheon, 1978.

Work with Two or Three Authors or Editors

List the names of several authors in the sequence in which they appear in the book or article. Begin with the last name of the author listed first because it is used to determine the alphabetical order for entries. Then identify the other authors by first and last names.

Oakes, Jill, and Rick Riewe. <u>Spirit of Siberia: Traditional Native Life, Clothing, and Footwear</u>. Washington: Smithsonian Inst. P, 1998.

Trueba, Henry T., Grace Pung Guthrie, and Kathryn Hu-Pei Au, eds. <u>Culture and the Bilingual Classroom: Studies in Classroom Ethnography</u>. Rowley: Newbury, 1981.

Work with More than Three Authors or Editors
Name all those involved, or list only the first author or editor with et al., for "and others."

Nordhus, Inger, Gary R. VandenBos, Stig Berg, and Pia Fromholt, eds. <u>Clinical Geropsychology</u>. Washington: APA, 1998.

Nordhus, Inger, et al., eds. <u>Clinical Geropsychology</u>. Washington: APA, 1998.

Work with Group or an Organization as Author
Alphabetize by the name of the group or organization.

National PTA. <u>National Standards for Parent/Family Involvement Programs</u>. Chicago: National PTA, 1997.

Work without an Author
<u>A Visual Dictionary of Art</u>. Greenwich, CT: New York Graphic Society, 1974.

Work in an Anthology of Pieces by the Same Author
Munro, Alice. "Passion." <u>Runaway: Stories</u>. New York: Knopf, 2004. 159–196.

Work in an Anthology of Different Authors
McCorkle, Jill. "Final Vinyl Days." <u>It's Only Rock and Roll: An Anthology of Rock and Roll Short Stories</u>. Ed. Janice Eidus and John Kastan. Boston: Godine, 1998. 19–33.

Anthology Cited as a Whole
Weston-Lews, Aidan, ed. <u>Effigies and Ecstasies: Roman Baroque Sculpture and Design in the Age of Bernini</u>. Edinburgh: Natl. Gallery of Scotland, 1998.

Work in Several Volumes
Smith, Andrew F. <u>The Oxford Encyclopedia of Food and Drink in America</u>. 2 vols. New York: Oxford UP, 2004.

Work Translated from Another Language
The first entry below emphasizes the work of the original author by placing his name first. The next example shifts emphasis to the work of the translators by identifying them first.

Eco, Umberto. <u>On Literature</u>. Trans. Martin McLaughlin. New York: Harcourt, 2004.

Young, David, and Jiann I. Lin, trans. <u>The Clouds Float North: The Complete Poems of Du Xuanji</u>. Bilingual Edition. Hanover: Wesleyan UP, 1998.

Work Appearing as Part of a Series

> Rohn, Suzanne. <u>The Wizard of Oz: Shaping an Imaginary World</u>. Twayne's
> Masterwork Studies 167. New York: Twayne-Simon, 1998.

New Edition of an Older Book

> Wharton, Edith. <u>The Custom of the Country</u>. 1913. NY Public Library Collector's
> Edition. New York: Doubleday, 1998.

Entry from a Reference Volume

Treat less common reference books like other books, including place of publication, publisher, and date. For encyclopedias, dictionaries, and other familiar references, simply note the edition and its date. No page numbers are needed if the
entries appear in alphabetical order in the reference volume.

> "Cretaceous Period." <u>Encyclopedia Americana: International Edition</u>. 2003 ed.

> Minton, John. "Worksong." <u>American Folklore: An Encyclopedia</u>. Ed. Jan Harold
> Brunvand. New York: Garland, 1996.

Work Issued by a Federal, State, or Other Government Agency

Depending on the emphasis you intend, you can start with either the writer or
the government agency responsible for the publication. "GPO" stands for "Government Printing Office," the publisher of most federal documents.

> Brock, Dan W. "An Assessment of the Ethical Issues Pro and Con." <u>Cloning Human
> Beings</u>. National Bioethics Advisory Commission. Vol. 2. Rockville, MD: GPO,
> 1997. E1–E23.

> National Bioethics Advisory Commission. <u>Cloning Human Beings</u>. 2 vols. Rockville,
> MD: GPO, 1997.

> United States. Cong. House. Subcommittee on Oversight and Investigations of the
> Committee on Education and the Workforce. <u>Education at a Crossroads: What
> Works and What's Wasted in Education Today</u>. 105th Cong., 2nd sess.
> Washington: GPO, 1998.

> US Const. Art. 9.

Reference to a Legal Document

When you discuss court cases in your paper, underline their names. In your
Works Cited, do not underline them.

> Aguilar v. Felton. 473 US 402.1985.

Article in a Journal with Continuous Pagination

> Pistol, Todd A. "Unfinished Business: Letters from a Father to His Son, 1922–1928."
> <u>Journal of Men's Studies</u> 7 (1999): 215–31.

Article in a Journal with Pagination Continuing Only through Each Issue

Add the issue number after the volume number.

> Guyer, Jane I. "Traditions of Invention in Equatorial Africa." <u>African Studies Review</u>
> 39.3 (1996): 1–28.

Article in a Weekly or Biweekly Periodical
Lemonick, Michael D. "The Biological Mother Lode." <u>Time</u> 16 Nov. 1998: 96–97.

Zakaria, Fareed. "In Search of the Real Iraq." <u>Newsweek</u> 2 May 2005: 35.

Article in a Monthly or Bimonthly Periodical
If an article in a magazine or a newspaper does not continue on consecutive pages, follow the page number on which it begins with a plus sign.

Waters, Rob. "Medicating Aliah." <u>Mother Jones</u> May–June 2005: 50+.

Article in a Daily Newspaper
Morson, Berny. "Tuft-eared Cats Make Tracks in Colorado." <u>Denver Rocky Mountain News</u> 4 Feb. 1999: 5A+.

Article with No Author
"Machines that Think." <u>New Scientist</u> 25 April 2005: 32.

"Terrorism's Latest Report Card." <u>US News and World Report</u> 9 May 2005: 16.

Editorial in a Periodical
Fogarty, Robert W. "Fictional Families." Editorial. <u>Antioch Review</u> 56 (1998): 388.

Letter Written to the Editor of a Periodical
Posod, Melissa. "That Global Perspective." Letter. <u>Ms.</u> Spring 2005: 6.

Review Article
If a review article has a title, add it after the author's name.

Swain, William N. Rev. of <u>Getting Hits: The Definitive Guide to Promoting Your Website</u>, by Don Sellers. <u>Public Relations Review</u> 24 (1998): 403–09.

Presentation at a Professional Meeting or Conference
Ciardi, John. Address. National Council of Teachers of English Convention. Hilton Hotel, Washington. 19 Nov. 1982.

Film, Slides, Videotape
Start with any actor, producer, director, or other person whose work you wish to emphasize. Otherwise, simply begin with the title of the recording. Note the form cited—DVD, videocassette, and so forth.

<u>America in the Depression Years</u>. Slide program. Laurel: Instructional Resources, 1979.

Olivier, Laurence, prod. and dir. <u>Richard III</u>. By William Shakespeare. Videocassette. London Film Productions, 1955.

<u>Richard III</u>. By William Shakespeare. Prod. and dir. Laurence Olivier. Videocassette. London Film Productions, 1955.

<u>Visions of the Spirit: A Portrait of Alice Walker</u>. By Elena Featherston. Videocassette. Women Make Films, 1989.

Programs on Radio or Television

"Alone on the Ice." The American Experience. PBS. KRMA, Denver. 8 Feb. 1999.

The Life and Adventures of Nicholas Nickleby. By Charles Dickens. Adapt. David
  Edgar. Dir. Trevor Nunn and John Caird. Royal Shakespeare Co. Mobile
  Showcase Network. WNEW, New York. 10–13 Jan. 1983.

CD or Other Recording

Identify the format if the recording is not on a CD.

Basie, Count. "Sunday at the Savoy." 88 Basie Street. Rec. 11–12 May 1983. LP. Pablo
  Records, 1984.

Cherry, Don. "When Will the Blues Leave?" Art Deco. A&M Records, 1989.

Published or Personal Letter

Lasswell, Harold. Letter to the author. 15 July 1976.

Schneider, Alan. "To Sam from Alan." 3 Sept. 1972. No Author Better Served: The
  Correspondence of Samuel Beckett and Alan Schneider. Ed. Maurice Harmon.
  Cambridge: Harvard UP, 1998. 278–82.

Thackeray, William Makepeace. "To George Henry Lewes." 6 Mar. 1848. Letter 452
  of Letters and Private Papers of William Makepeace Thackery. Ed. Gordon N.
  Ray. Cambridge: Harvard UP, 1946. 335–54.

Published or Personal Interview

Freund, Nancy. Telephone interview. 18 June 2004.

Gerard, William. Personal interview. 16 May 2003.

Previn, Andre. Interview. "A Knight at the Keyboard." By Jed Distler. Piano and
  Keyboard. Jan.–Feb. 1999: 24–29.

Computer Software

Biblio-Link II for Windows: Powerful Data Transfer for ProCit. Diskette. Ann Arbor:
  Personal Bibliographic Software, 1993.

Schwartz, Howard F., Robert Hamblen, and Mark S. McMillan, eds. AG Photo CD-1.
  Diskette. Fort Collins: Colorado State U and Advanced Digital Imaging, 1996.

Database Available Online

Bartleby Library. Ed. Steven van Leeuwen. 1999. 27 Apr. 2005
  <http://www.bartleby.com>.

Book, Article, or Other Source Available Online

Besides author and title, add any translator or editor and the date of electronic
publication or last update. Conclude with the date on which you visited the electronic site where the source is located and the site's address. If the URL is too long
or does not lead directly to your source (as with some personal and subscription-
based sites), provide the URL for the site's homepage or search page.

Land-Webber, Ellen. To Save a Life: Stories of Jewish Rescue. 1999. 5 Feb. 1999
  <http://sorrel.humboldt/edu/rescuers>.

Latham, Ernest. "Conducting Research at the National Archives into Art Looting, Recovery, and Restitution." National Archives Library. 4 Dec. 1998. National Archives and Records Administration. 5 Feb. 1999 <http://www.nara.gov/research/assets/sympaper/latham.html>.

Marvell, Andrew. "Last Instructions to a Painter." Poets' Corner. 5 Sept. 2003. 12 Oct 2004 <http://www.theotherpages.org/poems/marvel04.html>.

Wollstonecraft, Mary. "A Vindication of the Rights of Women: With Strictures on Political and Moral Subjects." Project Bartleby Archive. Ed. Steven van Leeuwen. Jan. 1996. 27 Apr. 2005 <http://www.cc.columbia.edu/acis/bartleby/wollstonecraft>.

Magazine Article Available Online

Chatsky, Jean Sherman. "Grow Your Own Employee Benefits." Money.com 30–31 Jan. 1999. 7 Feb. 1999 <http://www.pathfinder.com/money/moneytalk>.

Newspaper Article Available Online

Wolf, Mark. "Finding Art in Albums."@ The Post: World Wide Web Edition of the Cincinnati Post 5 Feb. 1999. 5 Feb. 1999 <http://www.cincypost.com/living/album020599.html>.

Article from an Electronic Journal

Warren, W. L. "Church and State in Angevin Ireland." Chronicon: An Electronic History Journal 1 (1997): 6 pars. 6 Feb. 1999 <http://www.ucc.ie/chronicon/warren.htm>.

Electronic Posting to a Group

Faris, Tommy L. "Tiger Woods." Online posting. 3 Sept. 1996. H-Net: Humanities & Social Sciences Online Posting. 7 Feb. 1999 <http://www.h-net.msu.edu/arete/archives/threads/tiger.html>.

Review Available Online

Holden, Stephen. Rev. of Anne Frank Remembered. 22 Feb. 1996. 5 Feb. 1999 <http://www.english.upenn.edu/afilreis/Holocaust/anne-frank-film.html>.

Public Web Site with Organizational Message

Raab, Jennifer J. "Greeting from Chairman Jennifer J. Raab." Landmarks Preservation Commission New York City. 8 Sept. 1998. 7 Feb. 1999 <http://www.ci.nyc.ny.us/ html/lpc/home.html>.

Database or Other Source Available on CD-ROM

Use "n.p." to indicate either "no place" or "no publisher" if such information is not available. Use "n.d." to indicate "no date."

"Landforms of the Earth: Cause, Course, Effect, Animation." Phenomena of the Earth. CD-ROM. n.p.: Springer Electronic Media/MMCD, 1998.

Life in Tudor Times. CD-ROM. Princeton: Films for the Humanities and Sciences, 1996.

| www.mhhe.com/ **mhreader** | For software that will help you format your citations, go to: **Research > Bibliomaker** |

APA (AMERICAN PSYCHOLOGICAL ASSOCIATION) DOCUMENTATION

The samples below show how to use APA style for citing a source in the text and in the References section at the end of a paper.

APA Parenthetical Documentation   The basic APA parenthetical citation includes the author's last name and the date of publication, information generally sufficient to identify a source in the reference list. Although researchers in the social sciences often cite works as a whole, the page number can be added to identify exactly where a quotation or other specific information is located. If the author's name is included in the text of your paper, it does not need to be repeated in the citation.

Single Author
The city's most current traffic flow analysis (Dunlap, 1998) proposed two alternatives.

Nagle (1998) compared the costs and benefits of both designs.

Two Authors
Use both names each time the source is cited. Use the word *and* to join them in the text; use an ampersand (&) within parentheses and the reference list.

Moll and Greenberg (1990) outline the advantages of a more flexible approach to social context.

An earlier study (Moll & Diaz, 1987) proposed classroom change as one research objective.

Three to Five Authors
Supply all the names the first time the source is cited. If it is cited again, use only the name of the first author and et al., for "and others."

Greene, Rucker, Zauss, and Harris (1998) maintain that anxiety is an important factor in communication.

Greene et al. (1998) address anxiety and communication directly.

More than Five Authors
Use only the name of the first author with *et al.* in the paper. Supply the names of the first six authors in the list of references followed by *et al.* for any additional authors.

Heath et al. (1988) continue to address the problems involved in implementing this methodology.

Group or Organization as Author
The Ford Foundation (1988) outlined several efforts to change decision-making processes.

Work without an Author
"Challenging the Myths" (1995) identifies several traditional beliefs about teacher training.

Page Numbers for a Work
The characteristics of successful charter schools follow an opening definition of the "charter school challenge" (Rowe, 1995, p. 34).

Two or More Works in the Same Citation
If several citations are grouped in one pair of parentheses, arrange them alphabetically.

Recent studies of small groups (Laramie & Nader, 1997; McGrew, 1996; Tiplett, 1999) concentrate on their interactions rather than their context.

Letters, Telephone Calls, E-Mail Messages, and Similar Communications
These communications are personal and thus are cited only in the text, not in the references.

This staffing pattern for nurses is used at four of the six major metropolitan hospitals (G. N. Prescott, personal communication, August 23, 1999).

APA List of References   As you examine the following illustrations, notice how capitalization, italics, punctuation, and other features change with the type of source noted. Note also that authors' names are listed with surnames first, followed by initials only. Although the entries in an APA reference list follow very specific patterns, references in your paper—to titles, for instance—should use standard capitalization (that is, only the first word of titles and subtitles). Similarly, the word and should be spelled out in your paper (except in parenthetical citations) even though the ampersand (&) is used in the references. Note that the names of months (in periodical and online citations) are spelled out, not abbreviated.

For updated examples and further guidance on APA style, see the American Psychological Association Web site at www.apastyle.org.

Book with One Author
Blau, T. H. (1998). *The psychologist as expert witness* (2nd ed.). New York: Wiley.

Nuckalls, C. W. (1998). *Culture: A problem that cannot be solved*. Madison: University of Wisconsin Press.

Several Works by One Author
List the works by year of publication, with the earliest first.

Muller, N. J. (1998). *Mobile telecommunications factbook*. New York: McGraw-Hill.

Muller, N. J. (1999). *Desktop encyclopedia of the Internet*. Boston: Artech House.

Book with Two Authors

Arden, H., & Wall, S. (1998). *Travels in a stone canoe: The return to the wisdomkeepers.* New York: Simon & Schuster.

Book with More than Two Authors or Editors

Greenfield, L. A., Rand, M. R., Craven, D., Klaus, P. A., Perkins, C. A., et al. (1998). *Violence by intimates: Analysis of data on crimes by current or former spouses, boyfriends, and girlfriends* (NCJ-167237). Bureau of Justice Statistics Factbook. Washington: U.S. Department of Justice.

Hair, J. F., Jr., Anderson, R. E., Tatham, R. L., & Black, W. C. (1998). *Multivariate data analysis* (5th ed.). Upper Saddle River, NJ: Prentice Hall, 1998.

Work with a Group or an Organization as Author

When the author is also the publisher, "Author" is used as the publisher's name.

American Public Transit Association. (1986). *The 1986 rail transit report.* Washington, DC: Author.

Amnesty International. (1998). *Children in South Asia: Securing their rights.* New York: Author.

Book without an Author

Ultimate visual dictionary of science. (1998). New York: Dorling Kindersley.

Work in a Collection of Pieces by Different Authors

Ombaka, C. (1998). War and environment in African literature. In P. D. Murphy (Ed.), *Literature of nature: An international sourcebook* (pp. 327–336). Chicago: Fitzroy Dearborn.

Collection of Pieces Cited as a Whole

Young, C. (Ed.). (1998). *Ethnic diversity and public policy.* New York: St. Martin's.

Work in Several Volumes

AFL-CIO. (1960). *American Federation of Labor: History, encyclopedia, and reference book* (Vols. 1–3). Washington, DC: Author.

Work Translated from Another Language

When you cite a translation in your paper, include both its original publication date and the date of the translation you have used, as in (Rousseau, 1762/1954).

Rousseau, J. J. (1954). *The social contract.* (W. Kendall, Trans.) Chicago: Regnery. (Original work published 1762)

Work Appearing as Part of a Series

Frith, K. T. (Vol. Ed.). (1997). *Counterpoints: Vol. 54. Undressing the ad: Reading culture in advertising.* New York: Peter Lang.

New Edition of an Older Book

When you cite an older source in your paper, include the original publication date and the date of the new edition, as in (Packard, 1866/1969).

Packard, F. A. (1969). *The daily public school in the United States.* New York: Arno Press. (Original work published 1866)

Article in a Reference Volume

Breadfruit (1994) In D. Crystal (Ed.), *The Cambridge encyclopedia* (2nd ed., p. 175). Cambridge: Cambridge University Press.

Work Issued by a Federal, State, or Other Government Agency

Nelson, R. E., Ziegler, A. A., Serino, D. F., & Basner, P. J. (1987). Radioactive waste processing apparatus. *Energy research abstracts* Vol. 12, (Abstract No. 34680.) Washington, D.C.: U.S. Department of Energy, Office of Scientific and Technical Information.

Reference to a Legal Document

Individuals with Disabilities Education Act (IDEA), 20 U.S.C. 1400 *et seq.* (1996).

Turner Broadcasting System Inc. v. Federal Communications Commission, 95 U.S. 992 (1997).

Article in a Journal with Pagination Continuing through Each Volume

Dinerman, T. (1998). The case for an American manned Mars mission. *The Journal of Social, Political and Economic Studies, 23,* 369–378.

Greene, J. O., Rucker, M. P., Zauss, E. S., & Harris, A. A. (1998). Communication anxiety and the acquisition of message-production skill. *Communication Education, 47,* 337–347.

Article in a Journal with Pagination Continuing Only through Each Issue

Brune, L. H. (1998). Recent scholarship and findings about the Korean War. *American Studies International, 36*(3), 4–16.

Special Issue of a Periodical

Larsen, C. S. (1994). In the wake of Columbus: Native population biology in the postcontact Americas. In A. T. Steegmann, Jr. (Ed.), *Yearbook of Physical Anthropology: Vol. 37* (pp. 109–154). New York: Wiley-Liss.

Riley, P., & Morse, P. R. (Eds.). (1998). Communication in the global community [Special issue]. *Communication Research, 25*(2).

Article in a Weekly or Biweekly Periodical

Greenwald, J. (1998, November 23). Herbal healing. *Time, 152,* 58–67.

Article in a Monthly or Bimonthly Periodical

Glausiusz, J. (1999, June). Creatures from the bleak lagoon. *Discover, 20,* 76–79.

Gordon, J. S. (1999, May/June). The great crash (of 1792). *American Heritage. 50,* 20–24.

Article in a Daily Newspaper

Levine, S. (1999, January 30). Hearing loss touches a younger generation. *The Washington Post*, pp. A1, A8.

Article with No Author

Fire and lightning. (1998, October 10). *New Scientist*, 25.

Editorial in a Periodical

Zuckerman, M. B. (1999, February 8). Coming to Russia's rescue. *U.S. News and World Report*, p. 68.

Letter Written to the Editor of a Periodical

Triebold, M. (1998, July/August). Digging bones for fun and $$$, [Letter to the editor]. *The Sciences*, 5.

Review Article

Glaeser, E. L. (1997). [Review of the book *Policing space: Territoriality and the Los Angeles Police Department*]. *Contemporary Sociology: A Journal of Reviews, 26,* 750–751.

Presentation at a Professional Meeting or Conference

Achilles, C. M., Keedy, J. L., & Zaharias, J. B. (1996, October). *If we're rebuilding education, let's start with a firm foundation.* Paper presented at the annual meeting of the University Council for Educational Administration, Louisville, KY.

Film, Videotape

If sources do not mention a place of publication, use *n.p.* If no date is mentioned, use *n.d.*

CityTV and Sleeping Giant Productions (Producers). (1994). *Dalai Lama: A portrait in the first person* [Motion picture]. n.p.: Films for the Humanities and Sciences.

Programs on Radio and Television

If appropriate, add the names of contributors or a specific episode before the series title.

The New Detectives: Case Studies in Forensic Science. (1999, February 9). Bethesda, MD: Discovery.

CD or Other Recording

Use *n.d.* and *n.p.* if you need to indicate that a recording or other source does not note the date or place of publication.

Cleveland, J. (1993). Marching to Zion. On *The great gospel men* [CD]. Newton, NJ: Shanachie Records.

Jamal, A. (1961). Night mist blues. On *Ahmad Jamal at the Blackhawk* [Record]. Chicago: Argo.

Letters, Interviews, and Personal Messages

If you have used a communication such as a letter in a print or other medium, follow the form for that type of citation. If the communication is a message or call not available to other researchers, cite it only in your text, not in your list of references. (See page 193.)

Computer Software

Weiss, H. J. (1990). PC-POM: Software for Production and Operations Management (Version 2.10) [Computer software]. Boston: Allyn and Bacon.

Database Available Online

Academic Info: Your Gateway to Quality Internet Resources. (1999, February 4). Retrieved February 5, 1999, from http://www.academicinfo.net

Book, Article, or Other Source Available Online

Hornbeck, D. (1999, January 22). The past in California's landscape. Retrieved February 7, 1999, from California Mission Studies Association, http://www.ca-missions.org/hornbeck.html

1695: Northwestern Indians at Quebec: Huron intrigues. Retrieved February 5, 1999, from State Historical Society of Wisconsin, http://memory.loc.gov/cgibin/query/ r?ammem/lhbum:@field (DOCID+Alit(M7689e42)

Magazine Article Available Online

All hope gone for Hussein, power is passing to Abdullah. (1999, February 7). *Time Daily.* Retrieved February 7, 1999, from http://cgi.pathfinder.com/time/daily/0,2960,19381-101990206,00.html

Spragins, E. E. (1999, February 7). Patient power: How to beat job lock. *Newsweek.* Retrieved February 7, 1999, from http://www.newsweek.com/nw-srv/focus/he/ fohe0224_1.htm

Newspaper Article Available Online

Harden, C., & Long, P. A. (1998, October 28). Grand jury begins work in bid probe. *The Kentucky Post.* Retrieved February 5, 1999, from http://www.kypost.com/news/bids102989.html

Sack, K. On the bipartisan bayou, a brouhaha. (1999, February 5). *New York Times.* Retrieved February 5, 1999, from http://www.nytimes.com/yr/mo/day/news/washpol/la-cooperate.html

Article from an Electronic Journal

Peiss, K. L. (1998, Fall). American women and the making of modern consumer culture. *Journal for MultiMedia History, 1*(1). Retrieved February 5, 1999, from http://www.albany.edu/jmmh/ vol1no1/peiss.html

Abstract Available Online

Gay, H. (1998 August). East end, west end: Science education, culture and class in mid-Victorian London. *Canadian Journal of History, 33*. Retrieved February 5, 1999, from http://www.asask.ca/history/cjh/ABS_897.HTM

Electronic Posting to a Group

French, M. (1996, February 21). Erie Canal? Message posted to http://www.h-net.
msu.edu/aseh/archives/threads/eriecanal.html

Database or Other Source Available on CD-ROM

Real facts about the sun. (2000). *The Dynamic Sun*. Retrieved October 27, 2001, from
NASA database.

| www.mhhe.com/ **mhreader** | For software that will help you format your citations, go to: **Research > Bibliomaker** |

SAMPLE STUDENT PAPER (MLA STYLE)

Lee 1

Clara Lee

Professor Paul Smith

Writing Workshop II

5 May 2004

The Courage of Intimacy:

Movie Masculinity in the Nineties

Mike Newell's 1997 film <u>Donnie Brasco</u> begins and ends with an
extreme close-up of Johnny Depp's eyes. Shot in wide-screen
so that the eyes literally span the entire screen, the image is a
black-and-white snapshot that appears during the opening credits and returns as a full-color close-up at the end of the movie.
Depp's lustrous eyes are large and black and beautiful, and gazing at them up close gives the viewer a surprisingly intimate sensation. Even within the conventional narrative that makes up the
body of the movie, they become noticeably important; Web-site
critic Rob Blackwelder observes that "Depp has [the central conflict of the movie] in his eyes in every scene," and Susan
Wloszczyna of <u>USA Today</u> notes, "It's all in the eyes. Depp's intense orbs focus like surveillance cameras, taking in each crime
and confrontation. He's sucked into the brutal, bullying lifestyle,
and so are we." The close-up image at the beginning and end is
one of the few instances in which the film draws blatant attention

Last name and page number ½ inch below top of page

Heading 1 inch below top of page

All lines double-spaced, including heading and title

Title centered
Title defines topic

Paragraph indented ½ inch or 5 spaces

1-inch side margins

Opening interests reader with detail from film

Quotation from electronic source
Support from print source

1-inch margin at the top

Heading ½ inch below top of page continues last name and page numbering

to its own style, but the device calls attention to the film's central focus, its constant probing into the character at the center of the movie.

Somehow, without restricting the film to a first-person narration by Depp's undercover FBI agent, the audience comes to identify with him and understand the many pressures increasing inside his head simply by watching his eyes. They reflect his watchfulness, his uncertainty, his frustration, and his guilt—all without drawing too much attention to himself from his unsuspecting wise guy companions. He is guarded with his words,

Quotation from film

causing his closest Mafioso friend to remark, "You never say anything without thinking about it first." His quietness invites viewers to read his looks and expressions, to become intimately acquainted with a character who constantly has to hide part of himself from the people around him, until they can virtually feel every twinge of fear or regret that the character feels. Seeing

Thesis stated

this man trapped in situations in which he faces crisis after crisis, unwillingly alienated from his family and eventually his employers, trying only to protect the people he loves, viewers can ultimately recognize him as a more sensitive, struggling, and courageous hero than those celebrated in the past.

Past contrasted with present

Over the decades, Hollywood has glorified the gruff masculinity of actors from Humphrey Bogart to Sylvester Stallone. Joan Mellen notes in her 1977 book <u>Big Bad Wolves: Masculin-</u>

Johnny Depp in the title role of *Donnie Brasco.*

ity in the American Film that in traditional Hollywood films, especially the stoic action films of the 1970s, "physical action unencumbered by effeminate introspection is what characterizes the real man" (5). In the 1990s, it seems that much has changed; introspection has become a central part of leading-male roles. Character-driven films of the past year alone have won accolades for such intimate roles as Robert Duvall's tormented evangelical preacher in The Apostle, Matt Damon's emotionally needy genius and Robin William's mourning therapist in Good Will Hunting, and the unemployed guys struggling over issues like impotence and child custody in The Full Monty. Thoughtful-

Source identified in text

Quotation from book with page number

Other examples noted

Background tied to thesis

Lee 4

John Wayne, the
epitome of mid-20th-
century American
masculinity.

ness, vulnerability, and the ability to handle relationships have become virtual requirements for the male "hero" in the 1990s. The old-fashioned masculinity of characters played by Clint Eastwood or John Wayne in the past has come to be regarded as emotionally repressed and overly macho.

The change is partly cyclical. Mellen cites the 1930s and 1950s as eras in film in which leading men were given greater depth. She says, "despite the limitations imposed by a repressive society [in the fifties], film recovered for men an individual self with a distinctive identity and a flourishing ego" (191–92).

Clarification added in brackets

Quotation with two page numbers

Lee 5

Actors like Marion Brando and James Dean, in particular, played

insecure, emotionally torn rebels who express tenderness in

their relationships with women and with other men. However, in

the sixties, "as the Vietnam War progressed [...] maleness itself   *Ellipses for words omitted added in brackets*

appeared under siege and in need of defense," and "traumatic

events of the sixties induced the Hollywood hero to tighten up,

reveal as little about himself as possible, and to find comfort in

his own recalcitrance" (248–49). Things scarcely got better

when "glorification of the vigilante male [became] the dominant

masculine myth of the seventies" (295) with films like <u>Dirty Harry</u>   *Quotations and summary from source*

and <u>Taxi Driver</u>. "In the seventies film, people are allowed no op-

tion: they must meet force with force" (307). Following two

decades of grim testosterone, there was a definite reaction in

the bubble gum pop culture of the eighties, with flashy cartoon

violence starring Sylvester Stallone or Arnold Schwarzenegger

presenting highly unrealistic images of masculinity, and lighter

portrayals like Marty McFly and Indiana Jones gaining in popu-

larity. By the nineties, American audiences were no longer tak-

ing tough guy masculinity seriously, leading to a trend of ironic

humor in action films from <u>True Lies </u>to <u>Independence Day</u>. It is

doubtful that Will Smith would have been a favorite action hero

in any other era but the 1990s.

However, the crucial underlying shift in American culture is   *Transition back to present day*

the debilitation of the conventional white male hero in a country he once monopolized. Trends in society within the last thirty years have led to greater freedom for women, minorities, and homosexuals, and as pride and power among these groups have increased, there has been a backlash against the white male. Today's hero has to prove that he is sensitive and completely respectful of every group mentioned above in order to remain sympathetic, forcing his previous role of unquestioned dominance to change drastically. In addition, now that women are going to work and less is expected of men in terms of being the provider and protector of the family and society, more is expected of them in their personal relationships. As noted recently by Sylvester Stallone, a fitting symbol of the old macho masculinity who is now trying to change his image to a more sensitive one, "I think the leading man of the future will be one who is beleaguered by the need to constantly define on film the male-female relationship." He also notes, "People want to nurture the underdog. The day of the strongman is over" (94). The themes of inefficacy in society, sensitivity in relationships, and a reaction to the old strongman ideal show up clearly in <u>Donnie Brasco</u>.

Quotation from published interview

Analysis of film

In the movie, FBI agent Joe Pistone, alias Donnie Brasco, goes undercover in the belief that he is on the side of law and order, with the simple goal of booking some major criminals; instead he finds a bunch of endearing but disturbingly violent men who become his closest companions for several years. Particu-

Lee 7

larly perplexing is his relationship with Benjamin "Lefty" Rug-   *Plot summary and interpretation*

giero, the trod-upon hitman whose thirty years of faithful service

are rewarded with dirty-work assignments while younger wise

guys are promoted over him. Lefty is the one who notices Don-

nie and recruits him into the organization, and from the start his

faith in Donnie is clear; as Pistone smugly reports to a contact

early in the movie, "I got my hooks in this guy." However, Pis-

tone's smugness wears off as Lefty repeatedly invites him into

his home, confides in him with his complaints and his dreams,

and says unexpectedly one day waiting in the hospital where his

own son is in the E.R. for a drug overdose, "I love you, Donnie."   *Character analysis*

It is appropriate that the fictional Donnie Brasco is an orphan,

because Lefty essentially becomes a surrogate father to him. Pi-

stone, concerned for Lefty's fate, becomes more and more re-

luctant to "pull out" of his undercover assignment, revealing

Donnie Brasco as a spy and leaving the blame (and death sen-

tence) on Lefty. At one point he stops meeting his FBI contacts

because they are pressuring him to pull out. Instead, he lets

himself take on his mob alter ego more and more, tearing both

his professional and personal lives apart.

In a way, the film is an interesting commentary on how

ideals have changed, because it is set in the 1970s but made

with a 1990s ideology. Because it is based on a book by the real

agent Joe Pistone, who is currently living under the Witness

Lee 8

Protection Program, one might think the portrayal would be strictly fact-based and would not be affected by the recent obsession with the sensitive male; but of course, one must never underestimate the power of filmmakers in any era to interpret their material with their own contemporary vision (note the portrayal of the Three Musketeers as aging and vulnerable in the recent screen adaptation of <u>The Man in the Iron Mask</u>; the seventies version of the same book depicted the Musketeers as brash and irreverent). There is plenty of traditional macho posturing in the Mafia sequences of <u>Donnie Brasco</u>, but director Mike Newell places special emphasis on Pistone's sensitive relationship with Lefty Ruggiero, his mentor in the mob, and on his imperiled relationship with his wife. Newell, a British director most famous for his vastly different romantic comedy <u>Four Weddings and a Funeral</u>, also boasted about <u>Donnie Brasco</u>'s "absolutely novel point of view about the Mob," focusing on "the lowest rung, the have-nots" (Schickel), rather than the rich and powerful men at the top so often depicted in mob movies. The film focuses on the soulful side of a male protagonist in a genre in which sensitivity is rare.

Electronic source cited by author's name only

In fact, <u>Donnie Brasco</u> has been recognized as "a different take on the mob," an evolutionary step in the genre of gangster films. <u>Time</u> calls it a "neo-Scorsesian study of lowlife Mob life" (Schickel), and Blackwelder says that it "rises above the mire of its shopworn genre by showing the cracks in its characters' armor." Chris Grunden sums up the difference when he says, "Newell eschews fancy camera-work and visual flair to remain tightly focused on the human drama—he's made an actors' picture in a genre obsessed with style (GoodFellas, Heat)." Conventional gangster films usually depict the rise and fall of a charismatic criminal. The gangster movies of the thirties and forties featured fast-talking tough guys like James Cagney and Humphrey Bogart; Francis Ford Coppola's 1972 epic The Godfather, which revived the genre, depicted the same glamour, ruthlessness, and power of the Mafia, on an even greater romanticized scale. But after a spate of stylized mob movies in the past twenty years, many reviewers of Donnie Brasco welcomed a new approach in a genre that was growing old and stale. Put another way, Donnie Brasco is the film that finally brings its genre into the nineties by replacing its tough, glamorous hero with a real guy who can't live up to the stereotypes.

Contrasts lead back to thesis

Almost in direct response to the ideal of masculinity presented in the past, Newell shows that although at first Pistone is doing everything right—fitting perfectly into his undercover persona, doing top-rate work for the FBI, and sending checks home regularly to his family—he cannot "be the man in the f—kin' white hat" that he thought he could be, as he puts it late in the movie. He knows how impossible it is to fulfill his male responsibilities in

Analysis of main character

all three of his very different worlds after he has ditched the FBI, almost lost his marriage, and realized that his undercover work, once revealed, will be the cause of Lefty's death. He has failed his own expectations of himself to save the day and make everything right. The contemporary audience recognizes the realism of the situation. As Stallone stated in his interview, "The male is [only] the illusion of the protector and guardian, [… b]ecause in this day and age, there is no security he can offer" (94). By now the audience realizes that a hero cannot always save the day in a conventional sense. In an odd way, viewers even appreciate his failure because it has knocked all of his arrogance out of him and left only an exposed, vulnerable character.

A contemporary audience can especially relate to the issues of family breakdown, recognizing in Donnie's situation the roots of the culture of estrangement and divorce which is so widespread today. Violating the conventional lone male gangster/cop figure, Joe Pistone has not only a wife but three small daughters hidden away in suburbia, and he can't tell them anything about his job without putting them at risk. His visits home are less and less frequent, sometimes months apart, due to the consuming nature of his "job." Although viewers can see from the start the tenderness and love he has for his wife and daughters, his prolonged absences and broken promises (he misses his daughter's first Communion) lead to intensifying arguments

Analysis of relationships with other characters

between him and his wife. As she constantly reminds him, his job is tearing their home apart, and not knowing what he is doing makes it all the more unbearable. Pistone knows, as his identification with the Mafia grows deeper and deeper, that his involvement has serious consequences for his family, and this mounting pressure becomes impossible to resolve when weighed against the life of Lefty Ruggiero.

Regarding the role of women in Mafia movies, Mellen points out that "Well into the seventies the male protagonist of films from <u>The Godfather</u> (I or II) to <u>Serpico</u> uses women solely to discard them" (327). Wives in <u>The Godfather</u> are cheated on, lied to, and in one case, violently beaten. At a pivotal moment at the end of the movie, the wife of Michael Corleone tearfully asks him if he has ordered the death of his sister's husband, and he looks directly into her eyes and lies, saying he did not. She smiles and believes him. Her character is, in fact, constantly under the thumb of her husband who misleads her, ignores her, and coaxes her into marrying him after not contacting her for over a year. She and the other women in the movie are not once consulted or listened to, no matter how much their husbands' actions affect their lives.

<u>Donnie Brasco</u> could have been made in precisely the same way. Pistone's wife Maggie is, after all, left at home for months at a time while her husband is off doing his job for the

Contrasting example

Contrasting example related to film

FBI. However, Newell makes the relationship between them a pivotal storyline in the movie. Repeatedly in the course of the narrative, interrupting the Mafia sequences, the audience sees Pistone call or visit home, reinforcing his identity as a husband and father. Viewers also note the progression as his relationship begins to sour. The lowest point comes when Pistone shows up at his home in the middle of the night to retrieve a bag containing $3 million in cash and confronts Maggie, who has found it and hidden it. When she tells him that he has become "like one of them," he strikes her, and both recoil in surprise, less shocked at the blow than at the realization of what their marriage has become. At this critical moment, he tries to tell her the truth. He awkwardly explains the situation with Lefty and his fear of being responsible for his death. He tells her that he is not sure of what is right anymore. He tells her, "I'm not like them. I am them." It is evident that the troubles of Pistone's marriage hurt himself as much as his wife, and in a sense, dealing with them takes more courage than risking his life as an undercover agent in the Mafia. The film treats this relationship delicately, and the woman here is not merely discarded or lied to, but confronted and confided in, with her concerns presented as clearly as his own.

What makes Pistone's situation so compelling is that he starts out believing that he can be one of the traditional "solitary heroes who solve all problems for themselves" (Mellen 23) and

Incident from film substantiates interpretation

Quotation from film

Source identified in citation

instead comes up against situations that are too difficult to han-

dle. Joe Pistone slaps his wife, not to exert his male dominance,

but because he has lost control. When he tries to make things

right, he doesn't sweep her into his arms (and probably have his

way with her, in the true tradition of male heroes); he is almost

frightened to make a move and instead makes a gesture—

kissing the back of her head—to try and reestablish the emo-

tional (not sexual) intimacy between them. In his early scenes

with Lefty, Pistone is noticeably on his guard and detached from

the affection Lefty is developing for him; later, when he has the

opportunity to be promoted within the ranks of the mob and Lefty

feels betrayed, Pistone tries to express his devotion by visiting

him at the hospital where his son has overdosed. When Lefty or-

ders him to leave, he refuses.

These gestures are some of Pistone's most heroic acts, at

least as Newell presents it. Although he is given a medal and a

check for $500 at the end of the movie for his undercover work

(which is enough to secure scores of convictions), his feelings

about it are clearly mixed; his loyalty to the FBI has been disin-

tegrating as he has lost faith in their good guy/bad guy rhetoric,

and his primary concern—Lefty's safety—is now uncertain. His

success in infiltrating a group of depressed Brooklyn wise guys

is now a cause for guilt. It is at this point at the end of the movie,

as Pistone accepts his reward and his wife tells him it's all over,

Detail from film supports interpretation

Lee 14

Return to detail used in first paragraph

that Newell returns to the extreme close-up of Depp's eyes, and the audience sees how troubled they are. Viewers are left with that image, indicating that Newell intended for them to leave the theater asking themselves what it was all for—whether doing his job was really the right thing or not. True to life, there is no easy, happy ending, in which a man can die in battle or save the day and thus fulfill his "masculine" duties. What matters, however, as viewers return to that close-up, is that they have seen Joe Pistone/Donnie Brasco's vulnerability and his devotion within his relationships. If he feels confused or uncertain at the end, it is because he has faced these emotional issues, which are far

Return to thesis

more subtle than the challenges related to his job. The audience has seen him show more courage in his private struggles than John Wayne ever did out on the frontier and can applaud him for that.

Lee 15

Works Cited

Blackwelder, Rob. "<u>Donnie Brasco</u>." Rev. of <u>Donnie Brasco</u>, dir.

 Mike Newell. <u>The Fairfield [CA] Daily Republic</u> 3 Mar. 1997:

 D5+. <u>Spliced Online</u> Archives Mar. 1977. 23 Sept. 1998

 <http://www.splicedonline.com>.

<u>Donnie Brasco</u>. Dir. Mike Newell. Tristar, 1997.

Grunden, Chris. "<u>Donnie Brasco</u>." Rev. of <u>Donnie Brasco</u>, dir.

 Mike Newell. <u>Film Journal International Online</u> Mar. 1997.

 20 Sept. 1998 <http://www.filmjournal.com/reviews/html/

 mar977.html>.

Mellen, Joan. <u>Big Bad Wolves: Masculinity in American Film</u>.

 New York: Pantheon, 1977.

Schickel, Richard. "Depp Charge." <u>Time Magazine Online</u> 3 Mar.

 1977. 22 Sept. 1998 <http://cgi.pathfinder.com/time/

 magazine/1997/dom/970303/deppcharge.html>.

Stallone, Sylvester. "Masculine Mystique." Interview. <u>Esquire</u>.

 Dec. 1996: 89–96.

Wloszczyna, Susan. "<u>Donnie Brasco</u>: A High Point for Lowlifes."

 Rev. of <u>Donnie Brasco</u>, dir. Mike Newell. <u>USA Today</u> 28

 Feb. 1997: 1D.

Annotations (right margin):

All lines double-spaced, including title and entries

Title 1 inch below top of page and centered

First line at margin

Next lines indented ½ inch or 5 spaces

Entries in alphabetical order

Article title in quotation marks

Movie title underlined

A Portfolio of Professional Research Papers

Your research project will likely serve as your introduction to professional and academic journals. Both in print and online, journals are like a public square or coffeehouse for researchers in a particular academic field or a profession. Unlike mainstream periodicals, which tend to be published for profit and supported by commercial advertising, journals are usually subscription-only and supported by membership dues, university presses, or professional organizations. Periodicals also usually have their own staff—reporters, columnists, cartoonists, and photographers. Journals, on the other hand, publish articles that have been submitted to an editorial board of experts in a field for comment and approval. Contributors to most journals are not paid for their contributions, but instead gain professional credibility and accolades for their journal articles. When you begin your research by delving into professional and academic journals, you are witnessing the creation of knowledge as well as listening to the most current conversation and debate in a field. By citing these journals to support your own hypothesis or claim, your own research is a contribution to this ongoing conversation.

Although individual journals have their own criteria for style, length, and citation guides, you will notice certain conventions reflected across the three disciplines represented in this portfolio. All journal articles draw upon and give credit to contemporary sources and include either a Works Cited or a References list (depending on the citation style). Most journal articles begin with a hypothesis, a problem, or a claim, and end with a resolution, a proposal for action, or a suggestion for further research. Finally, journal articles (unlike articles in general-interest periodicals) make certain assumptions about their audience. For example, articles in medical journals presuppose an audience familiar with human anatomy and physiology. Articles in journals of literary criticism assume that the audience is familiar with canonical literature. And articles in social science journals assume that their audience is comfortable with statistics.

Does this mean that your own research paper needs to conform to the conventions of journal articles in your field? Probably not at this point. The journal articles presented in this portfolio, for example, all describe *original* research. In addition, you are probably not writing your research paper for an audience of specialists or for submission to a professional journal. That said, you can learn a great deal about how professional researchers in a field talk to each other and communicate their ideas and discoveries by closely examining the structure and design, as well as the content, of journal articles.

GUIDELINES FOR READING JOURNAL ARTICLES

- What are the journal's submission guidelines and style expectations? Check the front or back pages of a journal issue, or go to the journal's Web site and look for information on submissions (or solicited/unsolicited manuscripts). Who is eligible to write for this journal? How are articles selected?

- Does the article include an abstract or summary? Begin your reading here. Look up unfamiliar words or concepts, and ask your professor for assistance with any equations, statistics, or other unfamiliar numerical information. Remember that journal articles are written for an audience of experts in the field—you might not yet be an "expert," but you should consider this an opportunity to learn more key concepts and become familiar with your field's jargon.

- What is the hypothesis or claim of this article? Compare it to your own working hypothesis or claim. Does the article support or contradict your main idea? Just because an article contradicts your working hypothesis doesn't mean that you are "wrong," but it is an opportunity to reconsider your working thesis. In addition, if you are writing an argument, you will need to give fair and equal consideration to opposing points of view.

HOW TO CITE AN ARTICLE FROM A JOURNAL OF LITERARY STUDIES OR HUMANITIES JOURNAL

If you are writing a paper about the poet Sylvia Plath, you are probably fulfilling an assignment for a literature course. The citation style used for research papers and journal articles in literature is MLA (see page 185). A reference to the following article by Dahlke on an MLA Works Cited list would use this format:

Dahlke, Laura Johnson. "Plath's Lady Lazarus." The Explicator 60.4 (2002): 234–36.

✳

Plath's "Lady Lazarus"

Laura Johnson Dahlke

Readers know that Plath was a poet, and that "Lady Lazarus" is a poem.

MLA in-text citation style.

Sylvia Plath demonstrated a "long-standing" interest in the biblical story of Lazarus that peaked after her first attempt at suicide in 1953, when she felt she "had been on the other side of life like Lazarus" (qtd. in Sanazaro 55). A. Alvarez states that Plath's "Lady Lazarus," first published in *Ariel* (6–9), is autobiographical: "The deaths of Lady Lazarus correspond to her [Plath's] own crises" (64). This poem, like many others in *Ariel*, features "power not centered in a loving deity" but in a "subject[ion] to dominance by pure power" (McClanahan 168). Unlike the biblical story of Lazarus, in which a loving deity uses power for good, Plath's "Lady Lazarus" reveals a struggle for power with a cruel deity that ends in annihilation.

Thesis statement.

In the story of Lazarus of Bethany in the Gospel of John, Jesus raises Lazarus from the dead so that the "people standing [t]here [...] may believe" (John 11.42). This display of power does not merely advertise God's power, but also benefits Lazarus and his onlookers. Those who witnessed the new life given to Lazarus believed in Jesus and were offered the promise of eternal life in heaven.

In contrast, Lady Lazarus's raising by Herr Doktor produces a struggle for power between them that leads to her eventual destruction. Herr Doktor exploits his power and dominance over Lady Lazarus, and she must fight to control her life. As Theresa Collins points out, "'Lady Lazarus' can be interpreted as a struggle for control [...] a dominion prevented by her torturer, Herr Doktor."

Acknowledgment of other critics.

Like Jesus, Herr Doktor displays his power in front of a crowd, but in contrast, he desires admiration and personal gain from the "peanut-crunching crowd" (28). He does not offer new life to the crowd or Lady Lazarus, but works for his own benefit. She is unwrapped by the male enemy and his assistants to exhibit his power. By calling it a "strip tease" (29) Plath adds "sexual flavour" (Wagner 52) and magnifies the male dominance over Lady Lazarus. As she is unwrapped, she is disembodied—becoming a hand, a knee, skin and bone. She becomes his "pure gold baby," a perfect image of male possession (69).

For this expression of male dominance, however, Lady 5
Lazarus responds. During this strip tease she insists, "I
am the same, identical woman" (34), giving the reader the
sense she is "more than a collection of parts" (Wagner 52).
She also speaks to Herr Doktor, inviting him to "Peel off
the napkin/O my enemy" and smugly questions, "Do I
terrify?—"(10–12). Lady Lazarus's aggressive tone suggests that she wants to see if her appearance startles or
shocks him so that she might gain an advantage over him.
Her language demonstrates her willingness to fight
against this cruel deity.

As he makes a spectacle of her and "betray[s] the per- 6
sona's trust in him" (Wagner 52), she expresses her anger
and yet ironically resigns herself, saying, "I am your
opus/I am your valuable" (67–68). This clear resignation
of power is reminiscent of Plath's own life, as Laura Frost
points out: "Even if she [Plath] rebels against the oppressive patriarchal father [in this poem, Herr Doktor] her
anger is reactive and she does not succeed at freeing herself from him" (52). Being raised by a male figure produces a fight for power that is readily given up and ends
in death, not freedom.

Moreover, Lady Lazarus makes herself vulnerable 7
even as she takes control and charges the crowd for the
"eyeing of [her] scars"(58). These wounds of vulnerability
are symbols of pain caused by her male-dominated life.
They symbolize Plath's own scars from her father's death
and husband's betrayal. Try as she might to overcome
them, she still "turns and burns" (71).

A cyclical pattern appears in the poem, with Lady 8
Lazarus challenging the cruel deity, Herr Doktor, yet
eventually submitting to his power. Lady Lazarus's attempts to excel in her male-dominated world lead her to
a destructive art. She states, "Dying/Is an art, like everything else/I do it exceptionally well" (43–45).

At the end of the poem, Lady Lazarus's final attempt 9
to gain power comes in apparently threatening words:

Beware
Beware.
Out of the ash
I rise with my red hair
And I eat men like air. (80–85)

As Collins states, "The revenge and immortality prom- 10
ised in the last two stanzas are taken out of God's [Herr

Doktor's] hands and attained by the speaker," implying that Plath takes control over her enemy at the end of the poem. This final statement of revenge and immortality, however, is not a true promise. The speaker uses the generic "men" rather than speaking directly to Herr Doktor. In the face of her enemy's power, she becomes incapable of confrontation. The men she promises to eat are abstract people, not the specific man. Had the threat been personalized by using the simple "you" or repeating "brute" or "enemy" she could have redeemed herself, but she chooses not to personalize her warning. This weakens her threat, and again she resigns her power. The poem ends on terms of defeat.

In the biblical story of Lazarus, Jesus' power generates 11 joyous new life. Lazarus lives again on earth with the promise of heaven awaiting. For Lazarus, life is precious both now and in the hereafter. In "Lady Lazarus," the reader senses that life for the speaker is not worth living. Her struggle for power ends in destruction. A cruel deity imposes his power on Lady Lazarus, and she can do nothing but fall prey to his will.

WORKS CITED

MLA Works Cited list.

Alvarez, A. Sylvia Plath." The Art of Sylvia Plath. Ed. Charles Newman. Bloomington: Indiana UP, 1970. 56–68.

Collins, Theresa. "Plath's 'Lady Lazarus.'" Explicator 56 (1998): 156–58.

The Concordia Self-Study Bible: New International Version. Gen. Ed. Robert G. Hoerber. St. Louis: Concordia, 1984.

Frost, Laura. "Woman Adores a Fascist: Feminist Visions of Fascism from Three Guineas to Fear of Flying." Women's Studies 29 (2000): 37–69.

McClanahan, Thomas. "Sylvia Plath." Dictionary of Literary Biography. Ed. Donald J. Greiner. Vol. 5. Detroit: Gale, 1980. 163–68.

Plath, Sylvia. Ariel. New York: Harper, 1966.

Sanazaro, Leonard. "Plath's 'Lady Lazarus.'" Explicator 41.3 (1983): 54–57.

Wagner, Linda W. "Plath's 'Lady Lazarus.'" Explicator 41.1 (1982): 50–52.

Academic and professional affiliations are included in journal articles.

Author Affiliation: Laura Johnson Dahlke, University of Nebraska at Omaha

READING A LITERARY RESEARCH ARTICLE

1. Read Sylvia Plath's poem "Lady Lazarus." Is Dahlke's argument clear and persuasive? What additional information would you need to respond fully to her argument?
2. Summarize Dahlke's article. (See page 177 for guidance on summarizing.)
3. What can Dahlke assume about her audience's knowledge base, and how can you tell? Where does she provide additional context for her argument, and why?

| www.mhhe.com/ **mhreader** | For more information on Laura Johnson Dahlke, go to: **More Resources > Ch. 3 Research** |

HOW TO CITE AN ARTICLE FROM A PUBLIC POLICY OR SOCIAL SCIENCES JOURNAL

A research paper for a political science course is likely to require APA citation style. Note that this article, like many journal articles, follows APA in-text citation style. Because this article was retrieved from an online database (ProQuest), the References list does not include the indentations or italics you would expect in a print journal. A reference to the following article by Gould on an APA References list would use this format:

Gould, Jon B. (2002). Playing with fire: The civil liberties implications of September 11th. *Public Administration Review, 62,* 74–80. Retrieved August 9, 2004, from ProQuest Newspapers database.

✳

Playing with Fire: The Civil Liberties Implications of September 11th

Jon B. Gould

Subjects:	Public policy, Civil rights, Surveillance of citizens, National security, USA PATRIOT Act 2001-US
Classification Codes	9190 United States, 9550 Public sector, 4320 Legislation
Locations:	United States, US
Author(s):	Jon B Gould

Articles accessed through online or CD-ROM databases list all publication data. This information does not appear in this form in a print version.

Article types:	Feature
Publication title:	Public Administration Review. Washington: Sep 2002. Vol. 62 pg. 74, 6 pgs
Supplement:	Special Issue
Source Type:	Periodical
ISSN/ISBN:	00333352
ProQuest document ID:	156494591
Text Word Count	4459
Article URL:	http://gateway.proquest.com/ openurl?url_ver=Z39.88-2004& res_dat=xri:pqd&rft_val_fmt=info:ofi/ fmt:kev:mtx:journal&genre=article& rft_dat=xri:pqd:did=000

Abstract (Article Summary)

APA style requires an abstract.

The aftermath of September 11th has seen a worrisome rise in invasive surveillance measures. Both adopted by statute and initiated by agencies, these provisions provide unprecedented powers for government agents to investigate suspects and search individuals, whether they are directly involved in terrorism or not. The prevailing wisdom has been that the American people will accept these restrictions as the natural cost of heightened security, and initial evidence suggests the public has been willing to tolerate greater limits on civil liberties. However, over time such support will erode, leaving in place permanent restrictions on civil liberties that not only will concern Americans, but also may turn them against government officials and civic participation. Thus, contrary to many interpretations of September 11th, this article argues that the policy response has only sown the seeds for greater detachment from and dissatisfaction with government as the public becomes increasingly separated from the workings and operations of public policy.

Full Text (4,459 words)

[Headnote]

The aftermath of September 11th has seen a worrisome   1
rise in invasive surveillance measures. Both adopted by
statute and initiated by agencies, these provisions pro-

vide unprecedented powers for government agents to investigate suspects and search individuals, whether they are directly involved in terrorism or not. The prevailing wisdom has been that the American people will accept these restrictions as the natural cost of heightened security, and initial evidence suggests the public has been willing to tolerate greater limits on civil liberties. However, over time such support will erode, leaving in place permanent restrictions on civil liberties that not only will concern Americans, but also may turn them against government officials and civic participation. Thus, contrary to many interpretations of September 11th, this article argues that the policy response has only sown the seeds for greater detachment from and dissatisfaction with government as the public becomes increasingly separated from the workings and operations of public policy.

The Legislative Response

The horrors of September 11th have been covered extensively by the popular media, both by same-day reporting of the attacks and lengthier analyses of the long-term effects on victims' families. In response to the terrorist threat—one that, interestingly, was interpreted as rising after the initial attacks—Congress passed and President Bush signed the U.S.A. Patriot Act. Described by Attorney General John Ashcroft as a "package of 'tools' urgently needed to combat terrorism" (McGee 2001), the legislation raises domestic intelligence gathering to an unprecedented level. Among its several provisions, the act stipulates that:

APA author/year citation.

- The standards for wiretapping may be lowered. Whereas previously, the FBI could obtain a court order only if its "primary purpose" was to gather intelligence through wiretapping, the new law permits wiretaps if "a significant purpose" involves intelligence gathering. As a result, people merely suspected of working with terrorists or spies may be wiretapped.
- The FBI may share sensitive grand jury and wiretap information with intelligence agencies without judicial review or any safeguards limiting its future use, so long as the information concerns foreign intelligence or international terrorism.

- Law enforcement may access an individual's internet communications if officials can certify to a court that the information is relevant to an ongoing criminal investigation. This standard is much lower than the showing of probable cause required for most search warrants.
- Financial institutions will be required to closely monitor daily financial transactions and share information with government intelligence services. The law also allows law enforcement agencies secret access to an individual's credit report without judicial review.
- A new crime of domestic terrorism is created, covering conduct that "involves acts dangerous to human life." Presumably, members of Operation Rescue or Greenpeace would be covered under this definition, permitting the FBI to wiretap the homes of individuals who provide lodging or other assistance to activists.
- Non-citizens facing deportation may be held indefinitely on the attorney general's certification that an individual endangers national security.

Even before this act was adopted, the federal government had stepped up security and surveillance, detaining roughly 1,200 people in the weeks following September 11th, proposing military tribunals for captured insurgents, and interviewing nearly 5,000 visa holders. On the home front, security was increased at public buildings and gatherings and, of course, at airports. Most of the public is now aware that a trip through airport security may involve some manner of disrobing. 3

Balancing Civil Liberties

In APA style, book titles are italicized. Database articles do not usually include italics or paragraph indentations, but print journals (and your own papers) should conform to the appropriate style guide.

The first response to these heightened measures has been largely supportive. As Chief Justice William Rehnquist suggests in his book All the Laws but One: Civil Liberties in Wartime, national emergencies shift the balance between freedom and order toward order—in favor of the government's ability to deal with the conditions that threaten the national well-being" (1998, 222). Initial public polling bears out that view. In February 2002, 62 percent of respondents in a Greenberg poll agreed that "Americans will have to accept new restrictions on their 4 civil liberties if we are to win the war on terrorism." During the same period, only 12 percent of respondents in a Newsweek poll feared the Bush administration's response to terrorism was "going too far in restricting civil liberties," a finding virtually unchanged from a similar poll conducted in November 2001. When asked about specific strategies to root out terrorists, 78 percent of respondents in a September 2001 NBC/Wall Street Journal poll said they would be willing to accept surveillance of internet communications, and 63 percent of participants in a similar Harris Poll said they would favor expanded camera surveillance on streets and in public places.

To read these responses as offering the federal govern- 5 ment carte blanche to search and pry, however, misreads the public's calculus of civil liberties. At the same time respondents are expressing support for expanded surveillance measures, they also have reservations about the potential creep of government snooping. When asked whether they believed the "U.S. government might go too far in restricting civil liberties," 62 percent of respondents in a March 2002 Time/CNN poll expressed concerns, a result that is in line with the 58 percent of respondents who, in a November 2001 Investor's Daily poll, said they were concerned about sacrificing "certain civil liberties in light of recently passed anti-terrorism laws."

Still, the issue runs deeper than these potentially con- 6 flicting results. Historically, the American public has expressed generic support for civil liberties principles while at the same time backing restrictions against a clearly identified or understood "other"—particularly a group that is reviled. As Chong explains, the public views civil liberties by balancing on one hand "considerations of [legal] principles and rights" and on the other hand "considerations about the people or groups that are involved in the issue, including considerations about how the issue might affect oneself" (Chong 1993, 870; McClosky and Brill 1983). For this reason, vast majorities in the Harris Poll can simultaneously name individual freedom as "a major contributor to making America great," while at the same time recommending the Ku Klux Klan be placed under electronic surveillance. Respondents balance their attachment to civil liberties against the risk of—or their animosity toward—an "out group."

A similar point is true in the area of criminal proce- 7 dure, where Americans seem willing to countenance surveillance and searches so long as police activity is directed against individuals presumed to be criminals. Over 80 percent of Americans support the "frisking" of individuals who appear "suspicious," and large majorities would allow police officers to search a car for drugs or stolen goods following a stop (Lock 1999). Perhaps the public is balancing the perceived intrusion of the search against the likelihood of uncovering criminal activity, but the more likely answer is that Americans are willing to accept restrictions that do not "directly affect them or the groups to which they belong" (Chong 1993, 887). This is the classic example of the respondent who does not care what the police do to suspected drug dealers—because he is not one—but who opposes home searches because he might have something " embarrassing that would be found (ibid.).

Short for *ibidem*, which means "in the same place." Indicates that the information comes from the same source as the preceding citation (here, Chong 1993, 887).

There is much in the survey data to support this notion. Americans largely accept dogs sniffing their luggage, but they are resistant to police rummaging through their garbage (Lock 1999). Similarly, they oppose warrantless searches of homes (although there are legal grounds to do so), as well as the government's opening of mail (McClosky and Brill 1983). The common denominator is heightened concern when the search or surveillance hits close to home—that is, when individuals fear they may actually be the target of law enforcement. Among other things, this dynamic explains the curious results found in both Canada and the United Kingdom, where elites, who generally are seen as the "carriers of the democratic creed," were much more supportive than the general public of electronic wiretapping (Fletcher 1989, 227; Sullivan and Barnum 1987). Although researchers speculated that the elites' support may be premised on their understanding of the legal safeguards built into wiretapping (Fletcher 1989), the better explanation is that elites, because they have greater social power, need not fear the exercise of government power. By contrast, the general public worries that elites will authorize the surveillance of them (Sullivan and Barnum 1987).

Closer to home, recent surveys identify concerns about the very kinds of surveillance now permitted by Congress. A month before September 11th, over 80 percent of respondents in a Harris Poll said it was extremely important that no one be allowed to watch or listen to them without their permission. Their responses echo previous surveys of internet users, who, by large margins, want to control the information that is collected about them. Although the questions were asked in the context of commercial tracking, the answers paint a consumer—and citizen—base that value its privacy.

Of course, at a time of national emergency, Americans 10 are likely to give government officials increased leeway in surveillance, but in some sense that is the point: Americans' attachment to civil liberties is a balancing test that, if mishandled operationally or politically by government officials, will only backfire. In this respect, I believe there are six factors that help to explain when the public will countenance restrictions—even against themselves— to uncover those individuals who pose a threat. None of these factors is either mutually exclusive or a sufficient condition, but together they provide a checklist of concerns that public administrators ought to consider carefully.

1. When the search or surveillance is not intrusive or the least restrictive method possible.

For several years now, airline personnel have asked trav- 11 elers whether they packed their bags themselves. Presumably this is a personal question, but it is accepted largely because the method is not intrusive. Were the Federal Aviation Administration to order so, a ticket agent could satisfy himself of the answer by prying open a passenger's suitcase and checking the contents against the passenger's memory, but, quite understandably, government officials recognize both the flying public and the airlines are much more likely to accept a simple question. So, too, courthouses and other public buildings use metal detectors to scan for weapons rather than strip searching each individual who enters. Although there are individuals who approach such machines with dread (consider the example of Congressman Dingell, whose artificial hip set off a detector), most of us tolerate the detectors because we recognize they are the least invasive method available to check for weapons.

2. When the perceived threat is great.[1]

There is a long history in this country of restrictions on 12 liberty during times of war or national emergency. Abraham Lincoln suspended habeas corpus during the Civil War; newspapers were censored during World War I;

Japanese Americans were sent to concentration camps during World War II; and the CIA opened mail destined for the USSR during the Cold War. When a national emergency exists, the public is likely to "rally 'round the flag" to support the country or the president and accept such restrictions (Bowen 1989). To reach this point, though, the public must come to see current events as constituting an emergency, a process that relies heavily on news coverage—and with it, the ability of public officials to frame issues as involving national security and not other concerns (Nelson, Clawson, and Oxley 1997). In the post–World War II era, rally effects can be short lived, averaging just under a year (Parker 1995).

3. When those responsible for the search or surveillance are seen as competent.

Interestingly, Attorney General John Ashcroft did not 13 support some of the same measures that are now in the U.S.A. Patriot Act when he served in the U.S. Senate, in part because he did not trust the Clinton administration to exercise the new powers properly.[2] Similarly, in reforming airport security, Congress and the Department of Transportation worried whether airline passengers would accept heightened security measures if those provisions continued to fall under the control of private, low-cost bidders, some of whom employ minimum-wage employees. In addition to providing better oversight, the federalization of airport security was considered necessary to reassure the flying public that screening is being handled competently.

4. When the method employed is considered effective.

People accept metal detectors at courthouses, not only be- 14 cause the intrusion is relatively minor, but also because they believe the systems are capable of identifying—and then stopping—armed individuals set on harm. We will remain content with such measures until the first suspect brings a plastic explosive into court and detonates himself, at which point there undoubtedly will be calls for more sensitive screening to catch explosive materials. At the same time, the public will reject heightened security if its effectiveness does not overcome the level of intrusiveness involved. For example, drivers may tolerate random sobriety checkpoints so long as drunk drivers do not shift their travels to unchecked roads. To accept stops, searches,

Footnotes are found at the end of a database article. In print journals, they are placed either at the bottom of the relevant page or at the end of the article, before the References list.

or surveillance, the public seeks assurance that the invasive methods will be effective.

5. When limiting the search or surveillance to more relevant suspects might smack of illegal discrimination.

Given the demographics of the September 11th hijackers, 15 some might call for intensive screening of young, Middle Eastern men who seek to board an aircraft. Certainly, past experience suggests this profile is more likely to yield a terrorist than, say, an 88-year-old white grandmother from Iowa. But while some criticize current measures that randomly—and thus, equally—search airline passengers at the gate, even more worry that targeted searches would inevitably lead to ethnic or racial profiling. Indeed, one of the surprising findings following September 11th was that 68 percent of respondents in a Newsweek poll said it would be a mistake to "put Arabs and Arab-Americans in this country under special surveillance."

6. When individuals are unaware that the search or surveillance is taking place.

When is a search not intrusive? Potentially when the tar- 16 get is unaware of it. Of course, this is a bit tongue-in-cheek, for liberties are never more at risk than when government agents can intrude without any outside check on their activities. But the public cannot object to surveillance about which it is unaware. This is what makes post–September 11th security so interesting, for the public may object to intrusive searches of which it is aware, but even greater surveillance may take place outside of its purview. On one hand, the U.S.A. Patriot Act has given the FBI and the intelligence community greater latitude to conduct surveillance without the public's knowledge-searches that, even if more intrusive, will likely persist without objection unless agents trip up and their activities are exposed. On the other hand, the public has begun to experience stepped-up security when entering public buildings, traveling by air, or attending notable public events. Such increasingly intrusive searches are probably the closest that members of the general public have come to the types of intrusions or surveillance that they have approved (at least tacitly) in other areas of American life, particularly in the criminal justice arena. As "average citizens" begin to taste the invasiveness of pat-down searches, of airport screeners with dirty plastic gloves "unzipping toiletries bag [and] picking through shoes and dirty laundry" (Hilkevitch 2002), of the newly proposed low-level x-ray scan of passengers (Branom 2002), they may very well rebel against the application of heightened security to "innocent individuals" presumably themselves.

The Aftermath of September 11th—An Increasingly Civil Libertarian Public

The challenge for government officials in the wake of 17 September 11th is that the public will become less supportive of extreme security measures as the perception of a terrorist threat drops. Unless the war in Afghanistan is expanded, or until another terrorist attack is leveled on U.S. soil, the immediate memories of September 11th's horrors will fade, to be replaced by an increasing sense of normalcy. News coverage will shift from a frame of warfare to geopolitics, and, in turn, Americans will rebalance the calculus between heightened surveillance and their own civil liberties. To the extent that major airports continue to grind to a halt from false alarms,[3] the flying public—and the rest of the American public who learn about such mistakes from the media—will begin to doubt the competence of federal agents whose new responsibilities extend to airport security. With these doubts will come an unwillingness to submit to heightened security.

Most important, enterprising reporters undoubtedly 18 will uncover cases in which surveillance measures intended for would-be terrorists extend outward and inadvertently ensnare an innocent, sympathetic individual. Maybe it will be the young mother whose credit dries up after her bank mistakenly turns her name over to intelligence authorities for unusual account activity; perhaps it will be the grandmother, whose interest in Islamic history leads federal agents to track her Internet usage; or maybe it will be the young father on a green card who faces wiretapping, indefinite detention, and eventual deportation because he attended a meeting to plan protests against the International Monetary Fund. There assuredly will be mistakes in the application of new surveillance powers—there almost always are—and the media will be ready to cover the stories. To the extent that the immediate threat of terrorism has begun to recede, these stories will touch an American people tiring of added restrictions on their behavior.

This is not to say that September 11th will turn this 19 country into a land of civil libertarians: Ultimately, Americans are willing to accept restrictions on "others," particularly if the targets are considered threatening. Nonetheless, as government surveillance moves out of the criminal justice arena and Americans begin to see that they, too, may be targeted or searched, we may well experience a renewed debate about the power of government and the wisdom of narrowing civil liberties protection in the name of generic security. In essence, government may actually have created its own backlash in its heightened response to September 11th.[4]

Widening the Distance between Citizens and Government

That the U.S.A. Patriot Act may have raised civil libertar- 20 ian sentiments is only one part of the equation. The stepped-up security following September 11th has widened the distance between citizen and government, potentially dampening citizen participation in government and with it reducing citizens' trust in public institutions and officials. The dynamic here is analogous to the creation of social capital. According to Paxton (1999), social capital is created when individuals share intensive associations and high levels of trust. Given a confluence between interpersonal connections and goodwill, the social capacity for action is increased, in turn facilitating the production of social good. The same is true for political capital. When citizens feel connected to their government or government officials, when they trust these institutions or leaders, citizens are more likely to participate in the governing function, and officials are allowed greater latitude and goodwill to take decisive action to address social problems. By the same token, when citizens feel disconnected from their government, they are far less likely to participate in any type of political activity—including voting—and diminished trust, in turn, strikes a blow at the underlying legitimacy of government institutions and public officials (Lipset and Schneider 1987).

To be sure, the immediate effects of September 11th 21 were to "rally 'round the flag" and the U.S. government.

In the first six months following the attack, Americans reflected overwhelming support for national leaders and government policy, a level not seen in more than 30 years (Moore 2002). But it is worth asking whether such approval reflects support for government in general or for the war on terrorism, particularly since pollsters have not always used the correct question design to estimate the public's trust in government (ibid.).[5]

Apart from these issues of measurement, though, there 22 is a larger concern lurking, for the very security measures installed following the September 11th attacks present the grave risk of further separating Americans from government. Whether the barriers are concrete or merely symbolic, government sends an important message to citizens about their role in democratic self-rule when many of the institutions of government are closed to public access, when individuals must undergo intensive screening to enter public facilities or to interact personally with government officials. The message—that the public should be content to delegate government functions to those inside—is only intensified when ever-increasing security measures limit the number of people who have "passed" and thus are privileged to participate in certain government functions while leaving others to sit outside policy deliberations because they are not deemed sufficiently "secure."

Americans may tolerate these distinctions for a while, 23 seeing them as a necessary price to ensure the continued, safe working of American institutions of government. But ultimately, government, and indeed democratic citizenry, accepts a steep risk in accentuating the differences between those on the inside who run government and those on the outside who are subject to it. For years, scholars have noted that political participation turns partly on an individual's belief that his voice can be heard (Verba and Nie 1987). Indeed, trust in government depends on the citizenry's view that public institutions and government officials are accountable and attentive (Weatherford 1992). Yet, when the public is urged to remain silently supportive of an antiterrorism campaign that may extend indefinitely, when citizens are told they will be ministered to, not participate in the ministering of government, when resources are redirected to defense and surveillance and away from direct government services, those on the outside of government may ultimately extend less goodwill to public officials as they feel increasingly more distanced from government's operations.

This change is not likely in the midst of an immediate 24 military campaign, for public institutions and leaders are viewed more positively in times of crisis. But as time passes and war passions wane and as government returns to its more traditional functions, it will face a citizenry that not only retains reservoirs of doubt about government—in particular, the federal government—but also it has added fuel to the fire by adopting security measures that further distance the public. Having enjoyed popular support during a time of national emergency, public officials may face a sinkhole that few would have predicted from the attack, a public that ultimately will be less supportive of government functions from which it has been kept at arms length.

Acknowledgments

The author thanks Scott Keeter, Ann Springer, and David Rosenbloom for their assistance on this article.

Electronic text clearly indicates what is a footnote.

[Footnote]

Notes

1. Presumably the threat was there all along, just inadequately detected. In times of emergency, the presumption seems to be that further attacks must be coming, if only because we could not predict the ones that just hit.
2. In a 1997 op-ed in the Washington Times, Ashcroft said, "The Clinton administration's paranoid and prurient interest in [monitoring] international e-mail is a wholly unhealthy precedent especially given this administration's track record on FBI files and IRS snooping. Every medium by which people communicate can be subject to exploitation by those with illegal or immoral intentions. Nevertheless, this is no reason to hand Big Brother the keys to unlock our e-mail diaries, open our ATM records or translate our international communications" (A15).
3. The FAA reported that, between February 17 and March 11, 2002, 22 airport terminals had been evacuated nationwide because of "security breaches" (AP

2002). In many of these cases, agents either failed to screen any passengers or were unable to stop an individual whom the x-ray detectors had identified as suspicious.

4. For this reason, Congress may have limited the U.S.A. Patriot Act to 2005 unless reauthorized. Any backlash, however, would likely start before then.

5. Examining an ABC News poll from January of this year, 69 percent of respondents said they trusted the federal government to handle issues of national security and terrorism at least "most of the time." By contrast, only 39 percent of respondents trusted the federal government to handle social issues. The latter numbers are similar to responses from a 2000 National Public Radio poll testing generic trust in government. Then, 5 percent of respondents "just about always" trusted the "federal government to do what is right," with 24 percent saying they agreed "most of the time."

Electronic texts sometimes do not strictly follow style guidelines for Works Cited or Reference lists but still provide all necessary information.

[Reference]

References

Ashcroft, John. 1997. Welcoming Big Brother. Washington Times, August 12, A15.

Associated Press. 2002. Logan Has Had Twice the Number of Evacuations as Similar Airports. April 7. Available at http/www2.bostonherald.com/news/local_regional//ap_logan04072002.htm. Accessed June 10, 2002.

Bowen, Gordon L. 1989. Presidential Action and Public Opinion about U.S. Nicaraguan Policy: Limits to the "Rally Round the Flag" Syndrome. PS: Political Science and Politics 24(4): 793–800.

Branom, Mike. 2002. New Security Devices at Fla. Airport. Associated Press, March 15. Available at http://www_i640.com/handel-newstory.html, Accessed June 10, 2002.

Chong, Dennis. 1993. How People Think, Reason, and Feel about Rights and Liberties. American Journal of Political Science 37(3): 867–99.

Fletcher, Joseph F. 1989. Mass and Elite Attitudes about Wiretapping in Canada: Implications for Democratic

Theory and Politics. Public Opinion Quarterly 53(2): 225–45.

Hilkevitch, Jon. 2002. Where Pawing Dirty Laundry Is Part of the Job. Chicago Tribune (Internet edition), April 1. Available at http://www.chicagotribune.com/classified/automotive/columnists/chi-0204010230 aprOcolumn. Accessed June 10, 2002.

Lipset, Seymour Martin, and William Schneider. 1987. Confidence Gap: Business, Labor and Government in the Public Mind. Baltimore, MD: Johns Hopkins University Press.

Lock, Shmuel. 1999. Crime, Public Opinion, and Civil Liberties. Westport, CT: Praeger.

McClosky, Herbert, and Alida Brill. 1983. Dimensions of Tolerance: What Americans Believe about Civil Liberties. New York: Russell Sage.

McGee, Jim. 2001. An Intelligence Giant in the Making: Antiterrorism Law Likely to Bring Domestic Apparatus of Unprecedented Scope. Washington Post, November 4, A4.

Moore, David W. 2002. Just One Question: The Myth and Mythology of Trust in Government. Public Perspective (January/February): 7–11.

Nelson, Thomas E., Rosalee A. Clawson, and Zoe M. Oxley. 1997. Media Framing of a Civil Liberties Conflict and Its Effect on Tolerance. American Political Science Review 91(3): 567–83.

Parker, Suzanne L. 1995. Toward an Understanding of "Rally" Effects: Public Opinion in the Persian Gulf War. 1995. Public Opinion Quarterly 59(4): 526–46.

Paxton, Pamela. 1999. Is Social Capital Declining in the United States? A Multiple Indicator Assessment. American Journal of Sociology 105(1): 88–127.

Rehnquist, William. 1998. All the Laws but One: Civil Liberties in Wartime. New York: Knopf.

Sullivan, John, and David Barnum. 1987. Attitudinal Tolerance in the United Kingdom: A Comparison of Members of Parliament with the Mass Public. Paper presented at the annual meeting of the American Political Science Association, September 3–5, Chicago, Illinois.

Verba, Sidney, and Norman H. Nie. 1987. Participation in America: Political Democracy and Social Equality. Chicago: University of Chicago Press.

Weatherford, M. Stephen. 1992. Measuring Political Legitimacy. American Political Science Review 86(1):
149–66.

[Author Affiliation]

Jon B. Gould, George Mason University, is an assistant
professor of public and international affairs and a visiting
assistant professor of law at George Mason University
where he is the assistant director of the Administration of
Justice Program. Professor Gould has written on the First
Amendment, hate speech, racial and sexual discrimination, the Fourth Amendment, and justice administration.
Email: jbgould@gmu.edu.

READING A POLITICAL SCIENCE ARTICLE

1. How would you characterize the author's political perspective or bias,
   based on this article? In what ways does he establish and justify his opinion? (For example, what does the article's title tell you about his beliefs?)
2. How many different kinds of sources does Gould consult in his research?
   How recent are his sources? Do you consider his sources to be relevant and
   reliable? Spend some time online and at the library consulting some of these
   sources.
3. Ask your reference librarian how to look up additional articles, books, and
   reviews by and about Jon B. Gould. Although the "author affiliation" at the
   end of this article gives you a sense of a writer's qualifications, searching out
   additional work that a writer has published can give you a better sense of his
   or her place in the academic or professional community.

✦ www.mhhe.com/ **mhreader**	For more information on Jon B. Gould, go to: **More Resources > Ch. 3 Research**

HOW TO CITE AN ARTICLE FROM A SCIENCE OR MEDICINE JOURNAL

Publications in the sciences and medicine follow the CBE/CSE (Council of Science Editors) style guidelines. This is a highly specialized citation style and not
one you will usually be expected to use in most undergraduate writing courses.
However, scientific, medical, and technical journals are valuable sources of information for research in other fields. For example, you might consult the following paper by Ebbeling and colleagues if you were writing an argument for

an education class about providing nutrition education and helpful meals for high-school students. Even though the article is written by and for medical specialists, and its style as well as some of its language is difficult for readers without specialized medical knowledge, the critical reading skills you are developing should help you to work through its argument and determine its key points. Assuming that you would cite this paper for a research project in an education or political science course requiring APA guidelines, you would include the following citation on your References page:

Ebbeling, C. B., Sinclair, K. B., Pereira, M.A., Garcia-Lago, E., Feldman, H.A., & Ludwig, D.S. (2004). Compensation for energy intake from fast food among overweight adolescents. *JAMA, 291,* 2828–2833.

✳

Compensation for Energy Intake from Fast Food among Overweight and Lean Adolescents

Cara B. Ebbeling, PhD
Kelly B. Sinclair, MS, RD
Mark A. Pereira, PhD
Erica Garcia-Lago, BA
Henry A. Feldman, PhD
David S. Ludwig, MD, PhD

A summary like this usually introduces a scientific article. It gives the hypothesis, the parameters of the experiment, and the outcome.

Context Fast food consumption has increased greatly among children in recent years, in tandem with the obesity epidemic. Fast food tends to promote a positive energy balance and, for this reason, may result in weight gain. However, if fast food and obesity are causally related, the question arises of why some children who frequently eat fast food do not become overweight.

Objective To test the hypothesis that overweight adolescents are more susceptible to the adverse effects of fast food than lean adolescents.

Design and Setting In study 1, we fed participants an "extra large" fast food meal in a naturalistic setting (a food

Author Affiliations: Division of Endocrinology. Department of Medicine, Children's Hospital. Boston, Mass (Drs Ebbeling, Feldman, and Ludwig and Mss Sinclair and Garcia-Lago); and Division of Epidemiology, University of Minnesota, Minneapolis (Dr Pereira).
Corresponding Author: David S. Ludwig, MD, PhD, Department of Medicine, Children's Hospital, 300 Longwood Ave. Boston, MA 02115 (david.ludwig@childrens.harvard edu).

court). The participants were instructed to eat as much or little as desired during this 1-hour meal. In study 2, we assessed energy intake under free-living conditions for 2 days when fast food was consumed and 2 days when it was not consumed. Data were collected in Boston, Mass., between July 2002 and March 2003.

Participants   Overweight (n = 26) and lean (n = 28) adolescents aged 13 to 17 years. Overweight was defined as a body mass index exceeding sex- and age-specific 85th percentiles based on the 2000 Centers for Disease Control and Prevention growth charts.

Main Outcome Measures   Energy intake determined by direct observation in study 1 and by unannounced 24-hour dietary recalls, administered by telephone, in study 2.

Results   In study 1, mean (SEM) energy intake from the fast food meal among all participants was extremely large (1652 [87] kcal), accounting for 61.6% (2.2%) of estimated daily energy requirements. Overweight participants ate more than lean participants whether energy was expressed in absolute terms (1860 [129] vs. 1458 [107] kcal, $P = .02$) or relative to estimated daily energy requirements (66.5% [3.1%] vs. 57.0% [2.9%], $P = .03$). In study 2, overweight participants consumed significantly more total energy on fast food days than non–fast food days (2703 [226] vs. 2295 [162] kcal/d; +409 [142] kcal/d; $P = .02$), an effect that was not observed among lean participants (2575 [157] vs. 2622 [191] kcal/d; −47 [173] kcal/d $P = .76$)

Conclusions   In this study, adolescents overconsumed fast food regardless of body weight, although this phenomenon was especially pronounced in overweight participants. Moreover, overweight adolescents were less likely to compensate for the energy in fast food, by adjusting energy intake throughout the day, than their lean counterparts.

Superscript numbers refer to reference list at end of text.

Consumption of fast food has increased rapidly since the 1970s[1] among adolescents from all socioeconomic and racial/ethnic groups across the United States.[2,3] Fast food is ubiquitously available and heavily marketed to adolescents.[4] An estimated 75% of adolescents eat fast food 1 or more times per week.[5]

The increase in fast food consumption parallels the escalating obesity epidemic,[6] raising the possibility that these 2 trends are causally related. Characteristics of fast food previously linked to excess energy intake or adiposity include enormous portion size,[7] high energy density,[8] palatability,[9] excessive amounts of refined starch and added sugars,[10] high fat content,[11] and low levels of dietary fiber.[12] Previous studies, which used between- and within-subject comparisons, consistently demonstrate that consumption of fast food is directly related to total energy intake and inversely related to diet quality.[2,5,13–16] Some studies,[13,14,17] although not all,[5,15] have found a direct association between fast food and body weight.

These studies raise a fundamental question: if most children eat fast food regularly, why do some become overweight, whereas others do not? Perhaps certain individuals are susceptible and others relatively resistant to the adverse effects of fast food. Therefore, we hypothesized that adolescents who eat fast food regularly but are not overweight compensate for the excessive energy in a fast food meal by commensurately decreasing energy intake throughout the day; in contrast, overweight adolescents do not have this tendency.

The purpose of this investigation, which was composed of 2 studies, was to evaluate the effects of fast food on energy intake in overweight vs. lean adolescents. In study 1, we assessed energy intake during a fast food meal consumed in a naturalistic setting. In study 2, we compared energy intake under free-living conditions on days when fast food was consumed and days when it was not consumed.

Methods

Participants

We enrolled 54 adolescents (26 overweight, 28 lean) aged 13 to 17 years who reported eating fast food at least 1 time per week. Fifty-one (24 overweight, 27 lean) of the 54 participants enrolled in study 1 also completed study 2. Newspaper advertisements and fliers, stating that the purpose of the project was to collect information on why and how teenagers eat fast food, were used to recruit participants.

Weight and height were measured using an electronic scale (model 6702, Scale-Tronix, White Plains, NY) and a wall-mounted stadiometer (Holtain Limited, Crymych, Wales), respectively. Body mass index (BMI) was calculated as weight in kilograms divided by the square of height in meters. The Centers for Disease Control and Prevention defines childhood *overweight* as a BMI exceeding sex- and age-specific 95th percentiles and *at risk of overweight* as a BMI between the 85th and 95th percentiles, using the 2000 growth charts.[18] In this investigation, we

Every detail of an experiment is carefully described. Notice the use of chronological arrangement, narrative, and description.

grouped adolescents who were *overweight* and *at risk of overweight* and herein refer to them as *overweight*.[18] Adolescents with a BMI not exceeding the 85th percentiles were considered *lean*. We did not enroll adolescents with a BMI below the 50th percentile or above the 98th percentile and also excluded those diagnosed as having any major medical illness or eating disorder. None of the participants was taking prescription medications or dieting for the purpose of weight loss. As incentive, we offered each participant two $30 gift certificates, one following completion of each study.

The protocol was approved by the institutional review   7
board at Children's Hospital, Boston, Mass. Written informed consent and assent were obtained from parents and participants, respectively. Data were collected between July 2002 and March 2003.

Study 1

Participation involved 1 study visit. We instructed the   8
participants to eat a standard breakfast of cold cereal and milk at 8:30 AM on the day of the visit and then to refrain from eating and drinking (except water) until after the visit. At 1 PM, we fed the participants a fast food meal from a national chain at a food court. All feedings were conducted in groups of 4 participants, on average, to foster socializing that is often part of the fast food experience among adolescents. Participants were grouped by sex and weight status to avoid any self-consciousness about eating that may be associated with these variables (e.g., girls eating less in the presence of boys, overweight adolescents eating less in the presence of their lean peers).

Tables clearly present data.

The same meal, modeled after prevailing "extra large"   9
fast food fare (Table 1), was served to each participant. The following standard instructions were read to the participants before the meal: "In a few minutes, we will bring each of you a meal. Eat as much or as little as you like, until you have had enough. There is more food available, and you may eat as much as you want. Please do not share your food with others in the group. If you need more of anything, just ask." The length of the meal was 1 hour. During this time, a research assistant discreetly monitored food intake to ensure that ample food was always available.

Whenever approximately three fourths of the meal portion of chicken nuggets, fries, or cookies was consumed, a   10

Table 1 Characteristics of Fast Food Meal Fed during Study 1

	"Extra Large" Meal		Refill Portion	
Menu Item	Portion	Energy, kcal*	Portion	Energy, kcal*
Chicken nuggets	9 pieces, 162.45 g	438	4 pieces, 72.29 g	195
French fries	1 "extra large," 199.68 g	584	1 small, 76.04 g	223
Cookies	2 bags, 115.28 g	460	1 bag, 57.64 g	230
Cola†	1 bottle, 20 fl oz	254	1 bottle, 20 fl oz	254
Ketchup	4 packets, 34.40 g	36	Readily available on the table	. . .
Sweet and sour sauce	2 packets, 56.84 g	69	Readily available on the table	. . .

*Energy values represent data derived from the Nutrition Data System for Research Software and are based on the gram weights of the "reference units." The total energy value for the "extra large" meal was 1.841 kcal.
†A refrigerated bottle of cola, rather than a cup of soda, was provided to avoid measurement inaccuracies associated with variable amounts of ice.

Let readers know when and where to consult visual information.

refill portion of the item was added to the tray (Table 1).[5] Empty cola containers were immediately replaced with full containers. Participants could obtain ketchup and sweet and sour sauce from the middle of the table at any time during the meal. This standardized protocol allowed us to provide more of the items that each individual enjoyed the most and, thus, would be likely to order in large portions if given the option. Following the meal, each participant estimated the relative size of the meal consumed during the study compared with the size of fast food meals that he or she typically consumed, using a verbally anchored, 10-cm visual analog scale, ranging from "much smaller than usual" to "much larger than usual."

The difference in weight between the amount of each 11 menu item provided and that remaining on the tray after the meal was used to calculate energy intake. In preparation for data collection, 20 reference units of each menu item were purchased and weighed to evaluate variability in portion sizes. Coefficients of variation, ranging from 0.8% for a packet of sweet and sour sauce to 9.2% for an order of "extra large" french fries, confirmed that portions are highly standardized. Thus, amounts of food provided during the feeding study were estimated based on mean weights of the reference units. Using this method, we were able to serve food immediately after purchasing it, thereby maintaining the temperature, palatability, and visual appearance that are expected by consumers. Leftovers were weighed on an electronic scale (item E1D120,

Ohaus Corporation, Florham Park, NJ). The Nutrition Data System for Research Software (NDS-R; versions 4.04 and 4.05, Nutrition Coordinating Center, University of Minnesota, Minneapolis) was used to convert the gram weight consumed to energy intake (in kilocalories). We relied on the NDS-R, rather than nutrition information available from the restaurant, to allow direct comparison with 24-hour dietary recall data collected for study 2.

Study 2

Four dietary and physical activity recall interviews, 2 for fast food days and 2 for non–fast food days, were administered by telephone to assess energy intake under free-living conditions. We used the NDS-R multiple-pass, 24-hour dietary recall method, which prompted the participant to list in sequence what foods and beverages were consumed during the preceding day, identify gaps in the initial list, and then provide details concerning each reported item. At the end of each recall, participants were asked to confirm the information provided and categorize the amount of food intake for the day as "usual," "more than usual," or "less than usual." Physical activity was quantified using a 24-hour recall protocol modeled after the method of Pate et al.[19] In brief, participants were asked to recall the activity performed most during respective 15-minute time blocks throughout the day and then to rate the relative intensity of each activity as light, moderate, hard, or very hard. A metabolic equivalent (MET level) was assigned to each reported activity to calculate a physical activity factor. As points of reference, resting has a MET level of 1.0, and brisk walking has a level of 5.0.[20] Total energy expenditure (in kilocalories per day) was estimated by multiplying basal metabolic rate, calculated from validated Food and Agriculture Organization, World Health Organization, United Nations University equations that include weight and height as independent variables,[21] by the physical activity factor derived from the four 24-hour recalls.

Two criteria were used to define a fast food day. Criterion 1 specified that the participant eat at 1 of the 5 leading fast food establishments: McDonald's, Burger King, KFC, Wendy's, or Taco Bell.[22] Criterion 2 specified that the participant consume at least 1 menu item containing meat (beef, pork), chicken, fish, beans, or egg plus 1 additional item (e.g., fries, beverage, dessert). A non–fast food day was one that did not meet criterion 1. Days when participants ate at other restaurants, including pizza and sandwich shops, were classified as non–fast food days. Because we were evaluating the effects of fast food meals, as opposed to single menu items, intake was not assessed on days when criterion 1 but not criterion 2 was satisfied. Recalls were unannounced, to avoid reactivity, and conducted on nonconsecutive days. On average, we contacted each participant a mean (SEM) of 6.9 (0.3) times to obtain data for 4 days, including 2 fast food days that satisfied both criteria.

Interstudy Comparison to Evaluate Underreporting

Underreporting of dietary intake is a well-recognized [14] phenomenon, particularly among overweight adolescents, but little is known regarding differential underreporting among foods.[23-25] This phenomenon could bias data in study 2 in either direction: against our primary hypothesis if energy intake from fast food were selectively underreported, or in favor of the hypothesis if energy intake from fast food were reported more completely than energy intake from other foods. To evaluate the potential for bias, we examined underreporting of total energy intake and energy intake from fast food in overweight and lean participants, using data from both studies. Recalled total daily energy intake (study 2), averaged across 2 fast food days and 2 non–fast food days, was expressed as a percentage of estimated total energy expenditure to assess the accuracy of self-report of total energy intake. Recalled energy intake from fast food (study 2), averaged across the 2 fast food days, was expressed as a percentage of observed energy intake during the fast food feeding (study 1) to assess the accuracy of self-report of fast food energy intake.

Statistical Methods

Statistical analyses were conducted using SAS statistical [15] software (release 8.2, SAS Institute Inc., Cary, NC). For study 1, 2-sample t tests were used to compare energy intake during the meal between overweight and lean adolescents. For study 2, analysis of variance was performed using the mixed linear model procedure to evaluate whether the interaction between weight status (overweight vs. lean) and type of day (fast food days vs. non–fast food days) influenced total daily energy intake.

In an additional model, we adjusted for self-reported relative amount of food intake. Preplanned contrasts were estimated from the fitted models for overweight and lean adolescents to determine the effects of fast food on total daily energy intake within groups. For the interstudy comparison, a mixed linear model was used to compare the accuracy of self-report between overweight and lean participants. The model for evaluating self-report of fast food intake was adjusted for the relative meal size rating in study 1. Using a 5% type I error rate, we estimated that a sample of 50 participants (25 overweight, 25 lean) would provide 80% power to detect a between-group difference in energy intake of approximately 150 kcal in study 1 and a difference in effect of approximately 260 kcal between overweight and lean participants in study 2. All results are presented as mean (SEM). Statistical significance was defined as $P < .05$.

Results

Participant characteristics are presented in Table 2. There [16] were no significant differences in demographic variables (sex, race, age) or height between the overweight and lean participants. The overweight adolescents tended to be less physically active than their lean counterparts ($P = .06$) and tended to have a higher total energy expenditure ($P = .07$).

Study 1

When instructed to eat as much or little fast food as de- [17] sired, the participants consumed 1652 (87) kcal, amounting to 61.6% (2.2%) of the estimated total energy expenditure. Overweight participants ate more than lean participants, whether energy intake was expressed in absolute terms or relative to estimated needs (Table 3). Relative meal size ratings did not differ between the overweight and lean adolescents (8.5 [0.4] vs. 7.8 [0.5], $P = .22$).

Study 2

There was a significant interaction between type of day [18] (fast food vs. non–fast food day) and weight status (overweight vs. lean) for total daily energy intake ($P = .05$ unadjusted, $P = .04$ after adjustment for self-reported relative amount of food intake). Overweight participants

Table 2 Characteristics of Overweight and Lean Participants*

Characteristic	Overweight	Lean	P Value†
No. of participants	26	28	
Male/female	14/12	14/14	.78
White/nonwhite	10/16	9/19	.63
Age, y	15.4 (0.3)	15.3 (0.2)	.95
Height, cm	170.0 (1.6)	167.6 (1.6)	.49
Weight, kg	80.5 (2.4)	60.7 (1.4)	<.001
BMI‡	27.8 (0.7)	21.6 (0.3)	<.001
BMI percentile	93.5 (0.8)	65.9 (2.6)	<.001
Physical activity factor, METS§	1.48 (0.03)	1.56 (0.03)	.06
Total energy expenditure, kcal/d§	2767 (113)	2500 (87)	.07

Abbreviations: BMI, body mass index; MET, metabolic equivalent task.
*Values are expressed as frequency for categorical variables or mean (SEM) for continuous variables.
†χ^2 Tests for categorical variables and t tests for continuous variables were used to compare overweight with lean adolescents.
‡A measure of weight in kilograms divided by the square of height in meters.
§Values were calculated based on 4 days of physical activity recall data.

Table 3 Energy Intake From Fast Food Meal, Study 1

	Mean (SEM)		
Variable	Overweight (n = 26)	Lean (n = 28)	P Value
Energy intake, kcal	1860 (129)	1458 (107)	.02
Energy intake, % total energy expenditure	66.5 (3.1)	57.0 (2.9)	.03

consumed 409 (142) kcal/d more on fast food than non–fast food days. In contrast, energy intake was not significantly different on fast food and non–fast food days for the lean participants (Table 4). With regard to physical activity, there was no interaction between type of day and weight status ($P = .46$).

Interstudy Comparison to Evaluate Underreporting

We sought evidence for incomplete reporting of food in-[19] take by examining observed dietary intake data in study 1 and recalled intake data in study 2 (Table 5). Recalled total daily energy intake, expressed as a percentage of estimated total energy expenditure, tended to be lower for the overweight compared with lean participants (−15.3% [8.9%], $P = .09$). Recalled energy intake from fast food in study 2, compared with observed intake in study 1, was also lower for the overweight compared with lean participants (−17.3% [8.7%], $P = .05$). Adjustment for relative

Table 4 Total Daily Energy Intake on Fast Food and Non–Fast Food Days, Study 2

| | Total Energy Intake, Mean (SEM), kcal/d | | | |
| | Fast | Non–Fast | Difference, | |
Group	Food Day*	Food Day	kcal†	P Value‡
Overweight (n = 24)	2703 (226)	2295 (162)	409 (142)	.02
Lean (n = 27)	2575 (157)	2622(191)	−47 (173)	.76

*Recalled energy intake from fast food was a mean (SEM) of 1107(80) kcal for overweight and 1047 (56) kcal for lean adolescents.
†Difference scores are for fast food day − non–fast food day.
‡The P value for the type of day (fast food vs. non–fast food) by obesity status (overweight vs. lean) interaction was .05 unadjusted and .04 after adjustment for self-reported relative amount of food intake.

Table 5 Interstudy Comparison to Evaluate Underreporting*

| | Overweight | Lean | | |
Variable	(n = 24)	(n = 27)	Difference†	P Value
Recalled total energy intake, % of total energy expenditure	91.6 (5.9)	106.9 (6.6)	−15.3 (8.9)	.09
Recalled fast food energy intake, % observed, unadjusted	64.9 (4.9)	82.2 (7.0)	−17.3 (8.7)	.05
Recalled fast food energy intake, % observed, adjusted‡	66.1 (6.2)	81.1 (5.9)	−15.0 (8.6)	.09

*Values are expressed as mean (SEM).
†Difference scores are for overweight − lean.
‡Values are adjusted for relative meal size rating in study 1.

meal size rating in study 1 did not materially affect this difference (-15.0% [8.6%], $P = .09$). Thus, as expected, overweight participants tended to underreport total energy intake compared with lean participants; however, the group difference in reporting accuracy was similar for total and fast food energy intake (-15.3% vs. -17.3%, $P = .84$), providing evidence against the possibility of a false-positive result.

Comment

A "comment" or "discussion" in a scientific article explains and contextualizes the findings.

In 1989, a published statement warned that a lifetime of fast food consumption may place children at increased risk for obesity.[26] However, until recently, the potentially adverse effects of fast food in youth have received limited attention in the medical literature.[2,5] With increasing recognition that excess adiposity confers serious health risks and that environmental factors may be driving the 20 obesity epidemic,[27] the role of fast food in promoting obesity has emerged as a topic of great interest and debate. Some nutrition professionals argue that fast food is contributing to the obesity epidemic,[27,28] whereas others support industry claims that fast food can be part of a healthful diet.[29]

Herein, we present the first investigation, to our [21] knowledge, designed to evaluate the effects of fast food on energy intake in overweight vs. lean adolescents. Assuming a dietary pattern of 3 meals and 1 or 2 snacks per day, average meal size to maintain energy balance should not exceed approximately 30% of daily energy requirements or approximately 790 kcal in our study sample. Compared with this figure, the participants in study 1 massively overate (1652 kcal or 61.6% of estimated total energy expenditure) in the naturalistic setting of a food court. Overeating, observed in both groups of participants, was especially pronounced among the overweight. Moreover, the overweight participants consumed more total energy on days with than without fast food, in contrast to the lean participants, who consumed virtually the same amount on both days. This observation suggests that overweight individuals do not compensate completely for the massive portion sizes characteristic of fast food today.

There are several ways that an individual could main- [22] tain energy balance throughout a day that included large portions of fast food: by decreasing food intake subsequent to a fast food meal, by decreasing food intake in anticipation of a fast food meal, or by adjusting the size of a fast food meal based on how much of other foods have been or will be consumed. Our study does not allow us to determine in which of these ways the lean and overweight participants differed. We also cannot determine whether the failure to compensate fully for energy from large fast food meals is an inherent trait, causing obesity in susceptible individuals, or a secondary event that occurs after development of obesity. Nevertheless, these findings suggest that, at least, fast food consumption serves to maintain or exacerbate obesity in susceptible individuals.

Although excess energy intake in response to large [23] portions is not unique to fast food,[7,30–32] we focused on this dietary pattern because of its dominant position in adolescents' diets and the possibility of a causal link to

the obesity epidemic. Indeed, fast food is designed to promote consumption of a maximum of energy in a minimum of time, a precept of not only the business model
but also the very name. Other dietary scenarios (e.g., a
buffet) might also provoke overeating and incomplete energy compensation if they resembled fast food in critical
respects, including high energy density, low fiber content,
extensive food processing (facilitating rapid swallowing
with minimal chewing), and low satiating value. In those
scenarios, however, the distinction with fast food may be
more one of terminology or marketing than physiology.
By contrast, overeating to the magnitude observed in
study 1 would be virtually impossible with satiating, low-
energy-density, high-fiber foods that require much chewing before swallowing (e.g., fruits, vegetables, legumes,
whole grain products).

Several issues that pertain to study design should be 24
noted. Strengths include evaluation of energy intake in a
naturalistic setting in study 1 and within-subject comparisons in study 2, reducing the possibility of confounding
by demographic and behavioral factors. Limitations include a relatively small sample size, restricted generaliz-
ability, and reliance on self-report for assessment of
energy intake in study 2 (a methodologic issue common
to all studies that aim to assess diet under free-living
conditions).

Consistent with previous studies[23,24] that show that 25
overweight participants have a particularly strong tendency to underreport what they eat, self-reported energy
intake on non–fast food days in study 2 was lower for the
overweight compared with the lean adolescents. However, owing to the within-subject design, underreporting
would lead to a false-positive result only if energy intake
from fast food were reported more completely (i.e., less
underreporting) than total energy intake by the overweight vs. lean adolescents. The interstudy comparison
suggests that this was not the case. Total daily energy intake in study 2, expressed as a percentage of total energy
expenditure, was lower for the overweight than the lean
adolescents. Recalled energy intake from the fast food
meals in study 2, expressed as a percentage of intake observed in study 1, was also lower for the overweight adolescents. However, the magnitude of underreporting of
energy intake from fast food compared with total daily
energy intake by the overweight vs. lean participants was

similar, even after adjustment for meal size rating, suggesting that fast food was not reported more completely than other foods. Moreover, prior studies[33–35] suggest that the opposite is likely to occur: overweight individuals may report high-calorie foods perceived as "fattening" (e.g., fast food) less, rather than more, completely than other foods. This effect, if present, would bias the study toward the null hypothesis.

In conclusion, our investigation suggests that overweight adolescents are less likely to compensate for the energy in large portions of fast food than their lean counterparts. These findings do not imply that fast food is without detrimental effect in lean adolescents. Previous research has shown that fast food consumption among children in a nationally representative sample affects diet quality in ways that would plausibly increase risk for obesity, regardless of baseline body weight.[2] Although the causes of obesity are multifaceted (as emphasized by the fast food industry[22]), public health measures to limit fast food consumption in children may be warranted. Such measures could include nutrition education campaigns, legislation to regulate marketing of fast food to children, and elimination of fast food from schools.

In some academic and professional fields, the contributions of each author are specifically identified in published papers.

Author Contributions

As principal investigator, Dr. Ludwig had full access to all of the data in the study and takes responsibility for the integrity of the data and the accuracy of the data analysis.

Study concept and design: Ebbeling, Sinclair, Pereira, Garcia-Lago, Ludwig.

Acquisition of data: Ebbeling, Sinclair, Garcia-Lago.

Analysis and interpretation of data: Ebbeling, Pereira, Feldman, Ludwig.

Drafting of the manuscript: Ebbeling, Ludwig.

Critical revision of the manuscript for important intellectual content: Sinclair, Pereira, Garcia-Lago, Feldman, Ludwig.

Statistical expertise: Ebbeling, Pereira, Feldman.

Obtained funding: Ludwig.

Administrative, technical, or material support: Sinclair, Garcia-Lago.

Funding/Support

This study was supported by grants R01 DK59240 and K01 DK62237 from the National Institute of Diabetes and Digestive and Kidney Diseases (Bethesda, MD); the Charles H. Hood Foundation (Boston, MA); and grant M01 RR02172 awarded by the National Institutes of Health (Bethesda, MD) to support the General Clinical Research Center at Children's Hospital (Boston, MA).

Role of the Sponsors

The funding organizations played no role in design and conduct of the study; collection, management, analysis, and interpretation of the data; nor preparation, review, or approval of the manuscript.

References

1. Guthrie JF, Lin B-H, Frazao E. Role of food prepared away from home in the American diet, 1977–78 versus 1994–96: changes and consequences. *J Nutr Educ Behav.* 2002;34:140–50.
2. Bowman BA, Gortmaker SL, Ebbeling CB, Pereira MA, Ludwig DS. Effects of fast food consumption on energy intake and diet quality among children in a national household survey. *Pediatrics.* 2004;113:112–118.
3. Nielsen SJ, Siega-Riz AM, Popkin BM. Trends in food locations and sources among adolescents and young adults. *Prev Med.* 2002;35:107–113.
4. Story M, Neumark-Sztainer D, French S. Individual and environmental influences on adolescent eating behaviors. *J Am Diet Assoc.* 2002;102:S40–S51.
5. French SA, Story M, Neumark-Sztainer D, Fulkerson JA, Hannan P. Fast food restaurant use among adolescents: associations with nutrient intake, food choices and behavioral and psychosocial variables. *Int J Obes Relat Metab Disord.* 2001;25:1823–1833.
6. Ogden CL, Flegal KM, Carroll MD, Johnson CL. Prevalence and trends in overweight among US children and adolescents, 1999–2000. *JAMA.* 2002;288:1728–1732.
7. Fisher JO, Rolls BJ, Birch LL. Children's bite size and intake of an entree are greater with large portions than with age-appropriate or self-selected portions. *Am J Clin Nutr.* 2003;77:1164–1170.
8. Rolls BJ, Bell EA, Castellanos VH, Chow M, Pelkman CL, Thorwart ML. Energy density but not fat content of foods affected energy intake in lean and obese women. *Am J Clin Nutr.* 1999;69:863–871.
9. McCrory MA, Suen VM, Roberts SB. Biobehavoral influences on energy intake and adult weight gain. *J Nutr.* 2002;132:3830S–3834S.
10. Ludwig DS. The glycemic index: physiological mechanisms relating to obesity, diabetes, and cardiovascular disease. *JAMA.* 2002;287:2414–2423.
11. Gazzaniga JM, Burns IL. Relationship between diet composition and body fatness, with adjustment for resting energy expenditure and physical activity, in preadolescent children. *Am J Clin Nutr.* 1993;58:21–28.
12. Pereira MA, Ludwig DS. Dietary fiber and bodyweight regulation. *Pediatr Clin North Am.* 2001;48:969–980.
13. Binkley JK, Eales J, Jekanowski M. The relation between dietary change and rising US obesity. *Intl Obes Relat Metab Disord.* 2000;24:1032–1039.
14. French SA, Harnack L, Jeffrey RW. Fast food restaurant use among women in the Pound of Prevention study: dietary, behavioral and demographic correlates. *Intl Obes Relat Metab Disord.* 2000;24:1353–1359.
15. Jeffery RW, French SA. Epidemic obesity in the United States: are fast foods and television viewing contributing? *Am J Public Health.* 1998;88:277–280.

16. McNutt SW, Hu Y, Schreiber GB, Crawford PB, Obarzanek E, Mellin L. A longitudinal study of the dietary practices of black and white girls 9 and 10 years old at enrollment: the NHLBI Growth and Health Study. *J Adolesc Health.* 1997;20:27–37.

17. Pereira MA, Kartashov AI, Ebbeling CB, et al. Fast food meal frequency and the incidence of obesity and abnormal glucose homeostasis in young black and white adults: the CARDIA study [abstract]. *Circulation.* 2003;107:35.

18. Kuczmarski RJ, Ogden CL, Grummer-Strawn LM, et al. CDC growth charts: United States. *Adv Data.* 2000;(314):1–27.

19. Pate RR, Ross R, Dowda M, Trost SG, Sirard JR. Validation of a 3-day physical activity recall instrument in female youth. *Pediatr Exerc Sci.* 2003;15:257–265.

20. Ainsworth BE, Haskell WL, Whitt MC, et al. Compendium of physical activities: an update of activity codes and MET intensities. *Med Sci Sports Exerc.* 2000;32:S498–S504.

21. Dietz WH, Bandini LG, Schoeller DA. Estimates of metabolic rate in obese and nonobese adolescents. *J Pediatr.* 1991;118:144–149.

22. Brownell KD, Horgen KB. *Food Fight: The Inside Story of the Food Industry, America's Obesity Crisis, and What We Can Do About It.* Chicago, Ill: Contemporary Books; 2004.

23. Bandini LG, Schoeller DA, Cyr HN, Dietz WH. Validity of reported energy intake in obese and nonobese adolescents. *Am J Clin Nutr.* 1990;52:421–425.

24. Bandini LG, Vu D, Must A, Cyr H, Goldberg A, Dietz WH. Comparison of high-calorie, low-nutrient-dense food consumption among obese and nonobese adolescents. *Obes Res.* 1999;7:438–443.

25. Krebs-Smith SM, Graubard BI, Kahle LL, Subar AF, Cleveland LE, Ballard-Barbash R. Low energy reporters vs. others: a comparison of reported food intakes. *Eur J Clin Nutr.* 2000;54:281–287.

26. Fast-food fare: consumer guidelines. *N Engl J Med.* 1989; 321:752–756.

27. Ebbeling CB, Pawlak DB, Ludwig DS. Childhood obesity: public-health crisis, common sense cure. *Lancet.* 2002;360:473–482.

28. StOnge M-P, Keller KL, Heymsfield SB. Changes in childhood food consumption patterns: a cause for concern in light of increasing body weights. *Am J Clin Nutr.* 2003;78:1068–1073.

29. Freeland-Graves J, Nitzke S. Position of the American Dietetic Association: total diet approach to communicating food and nutrition information. *J Am Diet Assoc.* 2002;102:100–108.

30. Diliberti N, Bordi PL, Conklin MT, Roe LS, Rolls BJ. Increased portion size leads to increased energy intake in a restaurant meal. *Obes Res.* 2004;12:562–568.

31. Edelman B, Engell D, Bronstein P, Hirsch E. Environmental effects on the intake of overweight and normal-weight men. *Appetite.* 1986;7:71–83.

32. Rolls BJ, Roe LS, Kral TVE, Meengs JS, Wall DE. Increasing the portion size of a packaged snack increases energy intake in men and women. *Appetite.* 2004;42:63–69.

33. Heitmann BL, Lissner L. Dietary underreporting by obese individuals—is it specific or non-specific? *BMJ.* 1995;311:986–989.

34. Johansson L, Solvoll K, Bjorneboe G-EA, Drevon CA. Under- and overreporting of energy intake related to weight status and lifestyle in a nationwide sample. *Am J Clin Nutr.* 1998;68:266–274.

35. Goris AHC, Westerterp-Plantenga MS, Westerterp KR. Undereating and underrecording of habitual food intake in obese men: selective underreporting of fat intake. *Am J Clin Nutr.* 2000; 71:130–134.

READING A SCIENTIFIC, MEDICAL, OR TECHNICAL ARTICLE

1. In your own words, restate the hypothesis that the authors of this article set out to test. In what ways is a hypothesis like an argument? In what ways is it different from other kinds of arguments you may have read or written?
2. Which parts of this article were especially difficult, technical, or specialized? How did you deal with those parts of the article? What kinds of references would you need to consult in order to better understand the methodology and conclusions described here?
3. What is the purpose of the "Comment" section of this article? In what ways does it support, complement, or build on the initial hypothesis?
4. If you were writing a letter to the editor of your local newspaper arguing for better nutrition in school lunchrooms, which portions of this article would you cite and why?

www.mhhe.com/ **mhreader**   For more information on Cara B. Ebbeling et al., go to: **More Resources > Ch. 3 Research**

part 2

Issues Across the Disciplines

 chapter *4*

Education and Society
How, What, and Why Do We Learn?

In "Learning to Read and Write," a chapter from his autobiography, Frederick Douglass offers a spirited affirmation of the rights we all should have to pursue an education. For Douglass, who began his life in slavery, knowledge began not only with experience but also with the need to articulate that experience through literacy. The ability to read and write should be the possession of all human beings, and Douglass was willing to risk punishment—even death—to gain that ability. Today, all over the globe, as ethnic and political conflicts arise, men and women are faced with the same challenge of expressing themselves. For even with a tool like the Internet, if one does not have the tools to express oneself or if the expression of thought is suppressed, the vehicle for conveying ideas, no matter how powerful, is rendered useless.

Perhaps the struggle for an education always involves a certain amount of effort and risk, but the struggle also conveys excitement and the deep, abiding satisfaction that derives from achieving knowledge of oneself and of the world. Time and again in the essays included in this chapter, we discover that there is always a price to be paid for acquiring knowledge, developing intellectual skills, and attaining wisdom. However, numerous task forces and national commissions tell us that students today are not willing to pay this price and that, as a consequence, we have become academically mediocre. Is it true that we no longer delight in educating ourselves through reading, as Richard Rodriguez recounts in "The Lonely Good Company of Books"? Is it true that we take libraries for granted—we expect them to be available but never visit them? A democratic society requires an educated citizenry, people who refuse to commit intellectual suicide or self-neglect. The writers in this chapter, who take many pathways to understanding, remind us that we cannot afford to be passive or compliant when our right to an education is challenged.

Today we are in an era of dynamic change in attitudes toward education. Such issues as sex education, multiculturalism, racism, sexism, and immigration suggest the liveliness of the educational debate on campus. Any debate

252

over contemporary education touches on the themes of politics, economics, religion, or the social agenda, forcing us to recognize that configurations of power are at the heart of virtually all educational issues in society today.

Without education, many of our ideas and opinions can be stereotyped or prejudiced, bearing no relationship to the truth. It is easy to understand how such views can arise if we are merely passive vessels for others' uninformed opinions rather than active learners who seek true knowledge. If we judge the tenor of the essayists in this section, we discover that many of them are subversives, waging war against both ignorance *and* received dogma. These writers treat education as the key to upset the status quo and effect change. Operating from diverse backgrounds, they challenge many assumptions about our educational system and invite us to think critically about its purpose.

Previewing the Chapter

As you read the essays in this chapter and respond to them in discussion and writing, consider the following questions:

- What is the main educational issue that the author deals with?
- What tone does the author establish in treating the subject? Does the author take a positive or a negative value position to the topic?
- Does the author define *education?* If so, how? If not, does the author suggest what he or she means by it?
- What is the impact of society at large on the way the role of education is perceived?
- What forms of evidence do the authors use to support their views on education?
- How do the rhetorical features of the essays that focus on personal experience differ from those of the essays that examine education from a more global perspective?
- What have you learned about the value of education from reading these selections?
- Which essays persuaded you the most? The least? Why?

Classic and Contemporary Images
DOES EDUCATION CHANGE OVER TIME?

Using a Critical Perspective   Consider these two photographs of students in science laboratories, the first from the 19th century and the second from the present. What is the setting of each laboratory like? Who are the people? What does each photographer frame and leave out of the scene? Which educational setting seems more conducive to scientific or educational inquiry? Why?

Founded in 1833, Oberlin College in Ohio was the first U.S. college to grant undergraduate degrees to women. The photograph reprinted here shows both male and female students in a zoology lab at Oberlin sometime during the 1890s.

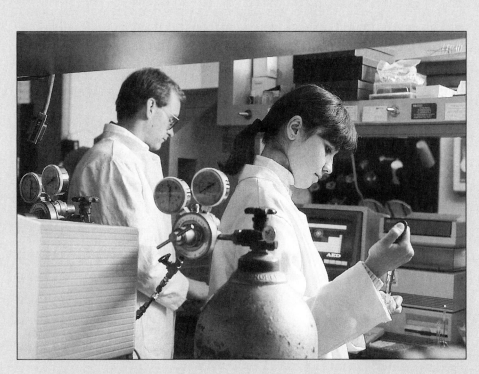

At the beginning of the 21st century, most colleges and universities in the United States are coeducational, and it is no longer unusual to see both male and female students in a laboratory setting, as shown in the contemporary photo of a food science lab at the University of Maine.

Classic and Contemporary Essays

A famous dictum proclaims that "the pen is mightier than the sword." In Frederick Douglass's narrative and the excerpt from the work of Richard Rodriguez, we get two portraits that demonstrate the truth of the adage. Douglass's efforts at becoming fully literate freed him from what would have been a life of slavery. No weapon could have done that for him. It is obvious that Douglass learned his lesson well, for his prose is stately, clear, direct, and precise. His story speaks of a determined youth and man who had a powerful motivation in learning to read and write. Would he have done so without this motivation? Perhaps, because he seems like a very self-directed individual, as is evident from the anecdotes he relates. Rodriguez, too, has a strong motivation to master, even to excel at, reading and writing. Writing nearly 150 years after Douglass, and at a time when he needn't fear slavery looming over him, Rodriguez nevertheless perceived that by emulating his teachers, who promoted book reading, his own reading would make him a better person. He, like Douglass, sensed that there was something about acquiring knowledge and about expanding one's view of the world by learning how others viewed it that would provide him with a certain amount of independence. As you read the following two essays, you may wish to consider whether the quiet, modest tone each author seems to project may have something to do with the subject matter. For reading, although active and mind-opening, is still a private and "lonely" activity.

*

Learning to Read and Write

Frederick Douglass

Frederick Douglass (1817–1895) was an American abolitionist, orator, and journalist. Born of the union between a slave and a white man, Douglass later escaped to Massachusetts. An impassioned antislavery speech brought him recognition as a powerful orator; thereafter he was much in demand for speaking engagements. He described his experience as a black man in America in Narrative of the Life of Frederick Douglass *(1845). After managing to buy his freedom, Douglass founded the* North Star, *a newspaper he published for the next 17 years. In the following excerpt from his stirring autobiography, Douglass recounts the tremendous obstacles he overcame in his efforts to become literate.*

I lived in Master Hugh's family about seven years. During this time, I succeeded in learning to read and write. In accomplishing this, I was compelled to resort to various stratagems. I had no regular teacher. My mistress, who had kindly commenced to instruct me, had, in compliance with the advice and direction of her husband, not only ceased to instruct, but had set her face against my being instructed by any one else. It is due, however, to my mistress to say of her, that she did not adopt this course of treatment immediately. She at first lacked the depravity indispensable to shutting me up in mental darkness. It was at least necessary for her to have some training in the exercise of irresponsible power, to make her equal to the task of treating me as though I were a brute.

My mistress was, as I have said, a kind and tender-hearted woman; and in the simplicity of her soul she commenced, when I first went to live with her, to treat me as she supposed one human being ought to treat another. In entering upon the duties of a slaveholder, she did not seem to perceive that I sustained to her the relation of a mere chattel, and that for her to treat me as a human being was not only wrong, but dangerously so. Slavery proved as injurious to her as it did to me. When I went there, she was a pious, warm, and tender-hearted woman. There was no sorrow or suffering for which she had not a tear. She had bread for the hungry, clothes for the naked, and comfort for every mourner that came within her reach. Slavery soon proved its ability to divest her of these heavenly qualities. Under its influence, the tender heart became stone, and the lamb-like disposition gave way to one of tiger-like fierceness. The first step in her downward course was in her ceasing to instruct me. She now commenced to practise her husband's precepts. She finally became even more violent in her opposition than her husband himself. She was not satisfied with simply doing as well as he had commanded; she seemed anxious to do better. Nothing seemed to make her more angry than to see me with a newspaper. She seemed to think that here lay the danger. I have had her rush at me with a face made all up of fury, and snatch from me a newspaper, in a manner that fully revealed her apprehension. She was an apt woman; and a little experience soon demonstrated, to her satisfaction, that education and slavery were incompatible with each other.

From this time I was most narrowly watched. If I was in a separate room   3
any considerable length of time, I was sure to be suspected of having a book, and was at once called to give an account of myself. All this, however, was too late. The first step had been taken. Mistress, in teaching me the alphabet, had given me the *inch,* and no precaution could prevent me from taking the *ell.*

The plan which I adopted, and the one by which I was most successful, was   4
that of making friends of all the little white boys whom I met in the street. As many of these as I could, I converted into teachers. With their kindly aid, obtained at different times and in different places, I finally succeeded in learning to read. When I was sent on errands, I always took my book with me, and by doing one part of my errand quickly, I found time to get a lesson before my return. I used also to carry bread with me, enough of which was always in the house, and to which I was always welcome; for I was much better off in this regard than many of the poor white children in our neighborhood. This bread I used to bestow upon the hungry little urchins, who, in return, would give me that more valuable bread of knowledge. I am strongly tempted to give the names of two or three of those little boys, a testimonial of the gratitude and affection I bear them; but prudence forbids—not that it would injure me, but it might embarrass them; for it is almost an unpardonable offence to teach slaves to read in this Christian country. It is enough to say of the dear little fellows, that they lived on Philpot Street, very near Durgin and Bailey's shipyard. I used to talk this matter of slavery over with them. I would sometimes say to them, I wished I could be as free as they would be when they got to be men. "You will be free as soon as you are twenty-one, *but I am a slave for life!* Have not I as good a right to be free as you have?" These words used to trouble them; they would express for me the liveliest sympathy, and console me with the hope that something would occur by which I might be free.

I was now about twelve years old, and the thought of being a *slave for life*   5
began to bear heavily upon my heart. Just about this time, I got hold of a book entitled "The Colombian Orator." Every opportunity I got, I used to read this book. Among much of other interesting matter, I found in it a dialogue between a master and his slave. The slave was represented as having run away from his master three times. The dialogue represented the conversation which took place between them, when the slave was retaken the third time. In this dialogue, the whole argument in behalf of slavery was brought forward by the master, all of which was disposed of by the slave. The slave was made to say some very smart as well as impressive things in reply to his master—things which had the desired though unexpected effect; for the conversation resulted in the voluntary emancipation of the slave on the part of the master.

In the same book, I met with one of Sheridan's mighty speeches on and in   6
behalf of Catholic emancipation. These were choice documents to me. I read

them over and over again with unabated interest. They gave tongue to interesting thoughts of my own soul, which had frequently flashed through my mind, and died away for want of utterance. The moral which I gained from the dialogue was the power of truth over the conscience of even a slaveholder. What I got from Sheridan was a bold denunciation of slavery, and a powerful vindication of human rights. The reading of these documents enabled me to utter my thoughts, and to meet the arguments brought forward to sustain slavery; but while they relieved me of one difficulty, they brought on another even more painful than the one of which I was relieved. The more I read, the more I was led to abhor and detest my enslavers. I could regard them in no other light than a band of successful robbers, who had left their homes, and gone to Africa, and stolen us from our homes, and in a strange land reduced us to slavery. I loathed them as being the meanest as well as the most wicked of men. As I read and contemplated the subject, behold! that very discontentment which Master Hugh had predicted would follow my learning to read had already come, to torment and sting my soul to unutterable anguish. As I writhed under it, I would at times feel that learning to read had been a curse rather than a blessing. It had given me a view of my wretched condition, without the remedy. It opened my eyes to the horrible pit, but to no ladder upon which to get out. In moments of agony, I envied my fellow-slaves for their stupidity. I have often wished myself a beast. I preferred the condition of the meanest reptile to my own. Any thing, no matter what, to get rid of thinking! It was this everlasting thinking of my condition that tormented me. There was no getting rid of it. It was pressed upon me by every object within sight or hearing, animate or inanimate. The silver trump of freedom had roused my soul to eternal wakefulness. Freedom now appeared, to disappear no more forever. It was heard in every sound, and seen in every thing. It was ever present to torment me with a sense of my wretched condition. I saw nothing without seeing it, I heard nothing without hearing it, and felt nothing without feeling it. It looked from every star, it smiled in every calm, breathed in every wind, and moved in every storm.

I often found myself regretting my own existence, and wishing myself 7 dead; and but for the hope of being free, I have no doubt but that I should have killed myself, or done something for which I should have been killed. While in this state of mind, I was eager to hear anyone speak of slavery. I was a ready listener. Every little while, I could hear something about the abolitionists. It was some time before I found what the word meant. It was always used in such connections as to make it an interesting word to me. If a slave ran away and succeeded in getting clear, or if a slave killed his master, set fire to a barn, or did any thing very wrong in the mind of a slaveholder, it was spoken of as the fruit of *abolition*. Hearing the word in this connection very often, I set about learning what it meant. The dictionary afforded me little or no help. I found it was "the act of abolishing;" but then I did not know what was to be abolished. Here I was perplexed. I did not dare to ask any one about its meaning, for I was satisfied that it was something they wanted me to know very little about. After a patient waiting, I got one of our city papers, containing an account of the number of petitions from the north, praying for the abolition of slavery in the District of Columbia, and of the slave trade between the States. From this time I understood the words *abolition* and *abolitionist,* and always drew near when that word was spoken, expecting to hear something of importance to myself and fellow-slaves. The light broke in upon me by degrees. I went one day down on the wharf of Mr. Waters; and seeing two Irishmen unloading a scow of stone, I went, unasked, and helped them. When we had finished, one of them came to me and asked me if I were a slave. I told him I was. He asked, "Are ye a slave for life?" I told him that I was. The good Irishman seemed to be deeply affected by the statement. He said to the other that it was a pity so fine a little fellow as myself should be a slave for life. He said it was a shame to hold me. They both advised me to run away to the north; that I should find friends there, and that I should be free. I pretended not to be interested in what they said, and treated them as if I did not understand them; for I feared they might be treacherous. White men have been known to encourage slaves to escape, and then, to get the reward, catch them and return them to their masters. I was afraid that these seemingly good men might use me so; but I nevertheless remembered their advice, and from that time I resolved to run away. I looked forward to a time at which it would be safe for me to escape. I was too young to think of doing so immediately; besides, I wished to learn how to write, as I might have occasion to write my own pass. I consoled myself with the hope that I should one day find a good chance. Meanwhile, I would learn to write.

The idea as to how I might learn to write was suggested to me by being in   8
Durgin and Bailey's ship-yard, and frequently seeing the ship carpenters, after hewing, and getting a piece of timber ready for use, write on the timber the name of that part of the ship for which it was intended. When a piece of timber was intended for the larboard side, it would be marked thus—"L." When a piece was for the starboard side, it would be marked thus—"S." A piece for the larboard side forward, would be marked thus—"L. F." When a piece was for starboard side forward would be marked thus—"S. F." For larboard aft, it would be marked thus—"L. A." For starboard aft, it would be marked thus—"S. A." I soon learned the names of these letters, and for what they were intended when placed upon a piece of timber in the ship-yard. I immediately commenced copying them, and in a short time was able to make the four letters named. After that, when I met with any boy who I knew could write, I would tell him I could write as well as he. The next word would be, "I don't believe you. Let me see you try it." I would then make the letters which I had been so fortunate as to learn, and ask him to beat that. In this way I got a good many lessons in writing, which it is quite possible I should never have gotten in any other way. During this time, my copy-book was the board fence, brick wall, and pavement; my pen and ink was a lump of chalk. With these, I learned mainly how to write. I then commenced and continued copying the Italics in Webster's Spelling Book, until I could make them all without looking on the book. By this time, my little Master Thomas had gone to school, and learned how to write, and had written over a number of copy-books. These had been brought home,

and shown to some of our near neighbors, and then laid aside. My mistress used to go to class meeting at the Wilk Street meetinghouse every Monday afternoon, and leave me to take care of the house. When left thus, I used to spend the time in writing in the spaces left in Master Thomas's copy-book, copying what he had written. I continued to do this until I could write a hand very similar to that of Master Thomas. Thus, after a long, tedious effort for years, I finally succeeded in learning how to write.

COMPREHENSION

1. What strategies does Douglass use to continue his education after his mistress's abandonment?
2. Why did the author's mistress find his reading newspapers particularly threatening?
3. Why does Douglass call learning to read "a curse rather than a blessing" (paragraph 6)?

RHETORIC

1. What is the thesis of Douglass's narration? How well is it supported and developed by the body paragraphs? Explain.
2. The first couple of sentences in the story, though simple, are very powerful. How do they serve to set up the mood of the piece and the reader's expectations?
3. Cite examples of Douglass's use of metaphors, and discuss why they work in those paragraphs.
4. How would you describe Douglass's writing style and level of language? Does it reveal anything about the writer's character? Justify your response.
5. Explain the way in which the author uses comparison and contrast.
6. What is Douglass's definition of *abolition*, and how does Douglass help the reader define it? How does this method contribute to the reader's understanding of the learning process?

WRITING

1. What does Douglass mean when he writes that "education and slavery were incompatible with each other" (paragraph 2)? Write an essay in which you consider the relationship between the two.
2. Both Douglass and his mistress were in inferior positions to Master Hugh. Write an essay in which you compare and contrast their positions in society at the time.
3. Illiteracy is still a major problem in the United States. Write an account of what your day-to-day life would be like if you couldn't write or read. What

impact would this deficiency have on your life? Use concrete examples to
illustrate your narrative.

4. **WRITING AN ARGUMENT:** Write an essay in which you argue for or
against the proposition that American education continues to discriminate
against minority groups.

| www.mhhe.com/ **mhreader** | For more information on Frederick Douglass, go to: **More Resources > Ch. 4 Education** |

*

The Lonely, Good Company of Books

Richard Rodriguez

*Richard Rodriguez (b. 1944) was born in San Francisco and received degrees from
Stanford University and Columbia University. He also did graduate study at the University of California, Berkeley, and at the Warburg Institute, London. Rodriguez became a nationally known writer with the publication of his autobiography,* Hunger of
Memory: The Education of Richard Rodriguez *(1982). In it, he describes the struggles of growing up biculturally—feeling alienated from his Spanish-speaking parents
yet not wholly comfortable in the dominant culture of the United States. He opposes
bilingualism and affirmative action as they are now practiced in the United States, and
his stance has caused much controversy in educational and intellectual circles. Rodriguez continues to write about social issues such as acculturation, education, and language. In the following essay, Rodriguez records his childhood passion for reading.*

From an early age I knew that my mother and father could read and write both   1
Spanish and English. I had observed my father making his way through what, I
now suppose, must have been income tax forms. On other occasions I waited ap-
prehensively while my mother read onion-paper letters air-mailed from Mexico
with news of a relative's illness or death. For both my parents, however, reading
was something done out of necessity and as quickly as possible. Never did I see
either of them read an entire book. Nor did I see them read for pleasure. Their
reading consisted of work manuals, prayer books, newspapers, recipes. . . .

In our house each school year would begin with my mother's careful in-   2
struction: "Don't write in your books so we can sell them at the end of the year."
The remark was echoed in public by my teachers, but only in part: "Boys and
girls, don't write in your books. You must learn to treat them with great care
and respect."

OPEN THE DOORS OF YOUR MIND WITH BOOKS, read the red and white poster   3
over the nun's desk in early September. It soon was apparent to me that reading
was the classroom's central activity. Each course had its own book. And the information gathered from a book was unquestioned. READ TO LEARN, the sign on
the wall advised in December. I privately wondered: What was the connection
between reading and learning? Did one learn something only by reading it?
Was an idea only an idea if it could be written down? In June, CONSIDER BOOKS
YOUR BEST FRIENDS. Friends? Reading was, at best, only a chore. I needed to look
up whole paragraphs of words in a dictionary. Lines of type were dizzying, the
eye having to move slowly across the page, then down, and across. . . . The sentences of the first books I read were coolly impersonal. Toned hard. What most
bothered me, however, was the isolation reading required. To console myself for
the loneliness I'd feel when I read, I tried reading in a very soft voice. Until:
"Who is doing all that talking to his neighbor?" Shortly after, remedial reading
classes were arranged for me with a very old nun.

At the end of each school day, for nearly six months, I would meet with her   4
in the tiny room that served as the school's library but was actually only a store-
room for used textbooks and a vast collection of *National Geographics.* Everything about our sessions pleased me: the smallness of the room; the noise of the
janitor's broom hitting the edge of the long hallway outside the door; the green
of the sun, lighting the wall; and the old woman's face blurred white with a
beard. Most of the time we took turns. I began with my elementary text. Sentences of astonishing simplicity seemed to me lifeless and drab: "The boys ran
from the rain. . . . She wanted to sing. . . . The kite rose in the blue." Then the old
nun would read from her favorite books, usually biographies of early American
presidents. Playfully she ran through complex sentences, calling the words alive
with her voice, making it seem that the author somehow was speaking directly
to me. I smiled just to listen to her. I sat there and sensed for the very first time
some possibility of fellowship between a reader and a writer, a communication,
never *intimate* like that I heard spoken words at home convey, but one nonetheless *personal.*

One day the nun concluded a session by asking me why I was so reluctant   5
to read by myself. I tried to explain; said something about the way written
words made me feel all alone—almost, I wanted to add but didn't, as when I
spoke to myself in a room just emptied of furniture. She studied my face as I
spoke; she seemed to be watching more than listening. In an uneventful voice
she replied that I had nothing to fear. Didn't I realize that reading would open
up whole new worlds? A book could open doors for me. It could introduce me
to people and show me places I never imagined existed. She gestured toward
the bookshelves. (Bare-breasted African women danced, and the shiny hubcaps
of automobiles on the back covers of the *Geographic* gleamed in my mind.) I listened with respect. But her words were not very influential. I was thinking then
of another consequence of literacy, one I was too shy to admit but nonetheless
trusted. Books were going to make me "educated." *That* confidence enabled me,
several months later, to overcome my fear of the silence.

In fourth grade I embarked upon a grandiose reading program. "Give me   6
the names of important books," I would say to startled teachers. They soon
found out that I had in mind "adult books." I ignored their suggestion of anything I suspected was written for children. (Not until I was in college, as a result, did I read *Huckleberry Finn* or *Alice's Adventures in Wonderland*.) Instead, I
read *The Scarlet Letter* and Franklin's *Autobiography*. And whatever I read I read
for extra credit. Each time I finished a book, I reported the achievement to a
teacher and basked in the praise my effort earned. Despite my best efforts, however, there seemed to be more and more books I needed to read. At the library I
would literally tremble as I came upon whole shelves of books I hadn't read. So
I read and I read and I read: *Great Expectations;* all the short stories of Kipling;
The Babe Ruth Story; the entire first volume of the *Encyclopedia Britannica*
(A–ANSTEY); the *Iliad; Moby Dick; Gone with the Wind; The Good Earth; Ramona; Forever Amber; The Lives of the Saints; Crime and Punishment; The Pearl.* . . . Librarians
who initially frowned when I checked out the maximum ten books at a time
started saving books they thought I might like. Teachers would say to the rest of
the class, "I only wish the rest of you took reading as seriously as Richard obviously does."

But at home I would hear my mother wondering, "What do you see in your   7
books?" (Was reading a hobby like her knitting? Was so much reading even
healthy for a boy? Was it the sign of "brains"? Or was it just a convenient excuse
for not helping around the house on Saturday mornings?) Always, "What do
you see . . . ?"

What *did* I see in my books? I had the idea that they were crucial for my ac-   8
ademic success, though I couldn't have said exactly how or why. In the sixth
grade I simply concluded that what gave a book its value was some major idea
or theme it contained. If that core essence could be mined and memorized, I
would become learned like my teachers. I decided to record in a notebook the
themes of the books that I read. After reading *Robinson Crusoe,* I wrote that its
theme was "the value of learning to live by oneself." When I completed *Wuthering Heights,* I noted the danger of "letting emotions get out of control." Rereading these brief moralistic appraisals usually left me disheartened. I couldn't
believe that they were really the source of reading's value. But for many years,
they constituted the only means I had of describing to myself the educational
value of books.

In spite of my earnestness, I found reading a pleasurable activity. I came to   9
enjoy the lonely, good company of books. Early on weekday mornings, I'd read
in my bed. I'd feel a mysterious comfort then, reading in the dawn quiet—the
blue-gray silence interrupted by the occasional churning of the refrigerator motor a few rooms away or the more distant sounds of a city bus beginning its run.
On weekends I'd go to the public library to read, surrounded by old men and
women. Or, if the weather was fine, I would take my books to the park and read
in the shade of a tree. Neighbors would leave for vacation and I would water
their lawns. I would sit through the twilight on the front porches or in back-
yards, reading to the cool, whirling sounds of the sprinklers.

I also had favorite writers. But often those writers I enjoyed most I was least 10
able to value. When I read William Saroyan's *The Human Comedy*, I was immediately pleased by the narrator's warmth and the charm of his story. But as
quickly I became suspicious. A book so enjoyable to read couldn't be very "important." Another summer I determined to read all the novels of Dickens. Reading his fat novels, I loved the feeling I got—after the first hundred pages—of
being at home in a fictional world where I knew the names of the characters and
cared about was going to happen to them. And it bothered me that I was forced
away at the conclusion, when the fiction closed tight, like a fortune-teller's
fist—the futures of all the major characters neatly resolved. I never knew how
to take such feelings seriously, however. Nor did I suspect that these experiences could be part of a novel's meaning. Still, there were pleasures to sustain
me after I'd finish my books. Carrying a volume back to the library, I would be
pleased by its weight. I'd run my fingers along the edge of the pages and marvel at the breadth of my achievement. Around my room, growing stacks of pa-
perback books reinforced my assurance.

I entered high school having read hundreds of books. My habit of reading 11
made me a confident speaker and writer of English. Reading also enabled me to
sense something of the shape, the major concerns, of Western thought. (I was able
to say something about Dante and Descartes and Engels and James Baldwin in
my high school term papers.) In these various ways, books brought me academic
success as I hoped that they would. But I was not a good reader. Merely bookish,
I lacked a point of view when I read. Rather, I read in order to acquire a point of
view. I vacuumed books for epigrams, scraps of information, ideas, themes—
anything to fill the hollow within me and make me feel educated. When one of
my teachers suggested to his drowsy tenth-grade English class that a person
could not have a "complicated idea" until he had read at least two thousand
books, I heard the remark without detecting either its irony or its very complicated truth. I merely determined to compile a list of all the books I had ever read.
Harsh with myself, I included only once a title I might have read several times.
(How, after all, could one read a book more than once?) And I included only
those books over a hundred pages in length. (Could anything shorter be a book?)

There was yet another high school list I compiled. One day I came across a 12
newspaper article about the retirement of an English professor at a nearby state
college. The article was accompanied by a list of the "hundred most important
books of Western Civilization." "More than anything else in my life," the professor told the reporter with finality, "these books have made me all that I am."
That was the kind of remark I couldn't ignore. I clipped out the list and kept it
for the several months it took me to read all of the titles. Most books, of course,
I barely understood. While reading Plato's *Republic*, for instance, I needed to
keep looking at the book jacket comments to remind myself what the text was
about. Nevertheless, with the special patience and superstition of a scholarship
boy, I looked at every word of the text. And by the time I reached the last word,
relieved, I convinced myself that I had read *The Republic*. In a ceremony of great
pride, I solemnly crossed Plato off my list.

COMPREHENSION

1. What was Rodriguez's parents' attitude toward reading? Did it influence his attitude? Cite examples from the essay that support your opinion.
2. What does Rodriguez mean by "the fellowship between a reader and a writer" (paragraph 4)? Why does he differentiate between "intimate" and "personal" forms of communication?
3. Rodriguez hoped that reading would fill "the hollow" inside him. What was the cause of his emptiness? Did he succeed in filling the void? Why did he find reading a lonely experience? Did reading fulfill any of his expectations?

RHETORIC

1. What is the thesis of Rodriguez's essay? Is it stated or implied? Explain.
2. How does the author's use of narrative advance his views on reading and education?
3. What is the writer's tone? How effective is it in conveying his point of view?
4. Rodriguez uses uppercase letters when referring to signs advocating reading. Why does he use this device? How does it support his point of view?
5. The essay ends with an ironic anecdote. Why did Rodriguez choose to conclude this way? Does it satisfactorily illustrate the writer's attitude?
6. What words or phrases imply that there is an ethnic component in Rodriguez's conflict? Is the subtlety effective? Justify your response.

WRITING

1. Rodriguez's parents had a pragmatic attitude toward reading. What was the attitude in your home as you were growing up? Did your parents encourage your interest in reading? Did they read themselves? What is the first book you remember reading by yourself? Write an essay in which you describe your reading history.
2. Is reading still a significant source of information and entertainment, or has it been usurped by television? Is it important (or necessary) to be a reader today?
3. **WRITING AN ARGUMENT:** Rodriguez believed reading would make him "educated." Do you agree or disagree? Is reading vitally important to a person's education? How do you define *education?* Can it be acquired only through reading, or are there other contributing factors? Write an argumentative essay on this topic.

**www.mhhe.com/
mhreader**

For more information on Richard Rodriguez, go to:
More Resources > Ch. 4 Education

Classic and Contemporary:
Questions for Comparison

1. Both Rodriguez and Douglass were motivated to educate themselves in a society inimical to this achievement. Compare and contrast their struggles and attitudes in their quests for knowledge.
2. Pretend you are Richard Rodriguez, and write a letter to Douglass addressing the issues of minorities and education in present-day America. What would Rodriguez say about the progress of minorities in our society?
3. Although Rodriguez and Douglass treat a similar theme, they communicate their messages differently. Which narration do you consider more powerful, and why?
4. Rodriguez explores the theme of isolation in his story. Is there any evidence that this feeling was shared by Douglass in his efforts to learn how to read? Use proof from both narratives to support your view.
5. Slavery was an obvious obstacle to Douglass's attempt to educate himself. What impeded Rodriguez's progress? Were similar forces at work? Cite examples from Rodriguez's narrative to prove your point.

✳

Needed: Affirmative Action
for the Poor

Laura D'Andrea Tyson

Laura D'Andrea Tyson (b. 1947) in 1993 became the first woman to chair the Council of Economic Advisors. Born in Bayonne, New Jersey, Tyson received a BA from Smith College (1969) and a PhD from Massachusetts Institute of Technology (1974). She has taught at the University of California at Berkeley and Princeton University and has been affiliated with the World Bank and the board of Human Genome Sciences, Inc. Tyson has written several books, including Who's Bashing Whom? Trade Conflicts in High-Technology Industries *(1992). In this essay, which appeared in the July 7, 2003, issue of* BusinessWeek, *Tyson analyzes the relationship between income and college graduation.*

In the Information Age, higher education is more and more important as the   1
ticket to economic success. Unfortunately, access to this ticket depends on economic success itself. A young person's chance of getting a college or post-
graduate education depends on family income. Children from low-income
households are much less likely to graduate from college than children raised in
high-income households. As family incomes become more unequal, there are
signs that the relationship between income and college graduation is becoming
stronger.

Children born into high-income households become part of a virtuous cir-   2
cle of success. Parents with university degrees tend to earn more, set higher educational goals for their children, and invest more time in the children's
schooling than parents who have a high-school education or less. In addition,
given local financing of K–12 public education and the segregation of most local communities by income, the children of high-income parents tend to attend
better schools and receive better preparation for college.

A study on socioeconomic status and selective-college admissions by the   3
Century Foundation provides stark evidence of the links between family background and higher education. The recent Supreme Court ruling that it was legal
to give some preferential treatment to disadvantaged minorities speaks to this
point.

Low-income students who graduate from high school with at least minimal   4
qualifications for four-year institutions enroll at half the rate of their high-
income peers. Only 78 percent of students from low-income families who rank
as top achievers on tests of college readiness actually attend college. In contrast,
nearly the same share of students from high-income families who rank at the
bottom of such tests do so. The conventional view is that students from low-
income families don't enroll or complete college because they are not academically qualified. But the New Century evidence paints a different and more
hopeful picture. Despite the considerable obstacles they encounter as they grow
up, many high-school students from low-income, disadvantaged households
are qualified but are choosing not to attend college or to attend colleges that are
less selective than their qualifications justify.

What can be done to help such students get the ticket to college? Clearly,   5
more generous financial aid is part of the answer. The financial barriers to college enrollment among students from low-income families are great—and
growing. Since the early 1970s, the value of federal aid packages for low-income
students has fallen precipitously as a percentage of college costs. In the same period, college costs as a portion of family income have remained unchanged for
the top 40 percent of the family income distribution while increasing substantially for low-income families. Without a big increase in federal and state support for means-tested student aid programs, a growing number of qualified
students from low-income households will find the door to higher education
and upward mobility closed for them and their children.

The nation's colleges and universities should also do more to help children   6
from low-income families. They should mount more aggressive efforts to iden-

tify and recruit students from low-income families with strong academic potential early in their high-school careers, providing them with better information about the course requirements and procedures for college admission. Colleges should also expand their financial aid programs for low-income students. Currently, more four-year colleges offer financial aid to athletes and students with "special nonacademic talents" than to disadvantaged students. Finally, more colleges should follow the lead of the Universities of California, Florida, and Washington and design admissions programs that evaluate the academic accomplishments of applicants in light of such obstacles as family income, parental education, and social environment. Doesn't an SAT total score of 1200 combined with an A average mean something different for an applicant raised in a low-income household and educated in a rundown public school than for an applicant from a high-income home and educated in an outstanding private school?

According to a recent survey, two-thirds of Americans support preferences 7 in college admissions for equally qualified low-income students over high-income students. Century Foundation research demonstrates that such economic affirmative action policies would dramatically increase the share of students from the poorest half of the population in the total number of students admitted to the nation's top 146 colleges. Today, that share is just 10 percent, lower than it would be if admissions decisions were based on grades and test scores alone. Even the most selective colleges have preferential admissions programs for the offspring of their mostly affluent alumni. Surely there is scope for preferential admissions for qualified students who are poor.

COMPREHENSION

1. According to Tyson, what is the relationship between income and college graduation? What conclusions does she draw concerning the need for affirmative action?
2. What are the primary facts mentioned in the New Century report? What conclusions does Tyson draw from this evidence?
3. What recommendations does Tyson make for improving chances of college admission and success for low-income students?

RHETORIC

1. This essay was published in a business magazine. Do you think that Tyson expects her audience to be interested in education? Why or why not? How does she make the subject potentially interesting for this specialized audience?
2. How does Tyson formulate her thesis or claim in paragraph 1? Does any one sentence stand out as the thesis, or does the writer employ a different strategy? Explain.

3. What support does Tyson provide for her claim, and where does it appear? Are there any warrants underlying her argument? If so, what are they?
4. How does Tyson employ comparison and contrast to advance her argument?
5. Although Tyson is a professional economist, her style is relatively informal. Analyze aspects of her style, and explain how they contribute to an essay that is accessible to a broad range of readers.
6. What is the tone of Tyson's concluding paragraph? Why does she cite a "recent survey"? What references to earlier information appear, and how do they reinforce the writer's attitude? How does Tyson's use of data affect the tone of this final paragraph?

WRITING

1. Review each paragraph in Tyson's essay, and in a brief essay analyze the ways in which she develops her argument.
2. Write a personal essay telling how your family's income level has affected the choices you have made about a college education.
3. **WRITING AN ARGUMENT:** Tyson writes, "Surely there is scope for preferential admissions for qualified students who are poor" (paragraph 7). Write an essay in which you agree or disagree with her proposition. Conduct research to obtain data to support your position on affirmative action.

---

www.mhhe.com/
mhreader

For more information on Laura D'Andrea Tyson, go to:
More Resources > Ch. 4 Education

Sex Ed

Anna Quindlen

Anna Quindlen (b. 1953) was born in Philadelphia and educated at Barnard College (BA, 1974). A journalist and novelist, she began her writing career as a reporter for the New York Post *and later moved on to* The New York Times, *where she was a syndicated columnist. Quindlen has written a number of books, including* Living Out Loud *(1986),* Object Lessons *(1991),* One True Thing *(1994), and* Black and Blue *(1998). Quindlen received the Pulitzer Prize for Commentary in 1992. She is currently a columnist for* Newsweek *magazine. In this essay, Quindlen focuses on the problem of teenage pregnancy and suggests children be given more than textbook information to help them cope with their sexuality.*

Several years ago I spent the day at a family planning clinic in one of New York   1
City's poorest neighborhoods. I sat around a Formica table with a half-dozen
sixteen-year-old girls and listened with some amazement as they showed off
their knowledge of human sexuality.

They knew how long sperm lived inside the body, how many women out of   2
a hundred using a diaphragm were statistically likely to get pregnant and the
medical term for the mouth of the cervix. One girl pointed out all the parts of
the female reproductive system on a placard; another recited the stages of the
ovulation cycle from day one to twenty-eight. There was just one problem with
this performance: although the results of their laboratory tests would not be
available for fifteen more minutes, every last one of them was pregnant.

I always think of that day when someone suggests that sex education at   3
school is a big part of the answer to the problem of teenage pregnancy. I happen
to be a proponent of such programs; I think human sexuality is a subject for dispassionate study, like civics and ethics and dozens of other topics that have a
moral component. I'd like my sons to know as much as possible about how
someone gets pregnant, how pregnancy can be avoided, and what it means
when avoidance techniques have failed.

I remember adolescence about as vividly as I remember anything, however,   4
and I am not in the least convinced that that information alone will significantly
alter the rate of teenage pregnancy. It seemed to me that day in the clinic, and on
days I spent at schools and on street corners, that teenage pregnancy has a lot
more to do with what it means to be a teenager than with how someone gets
pregnant. When I was in high school, at the tail end of the sixties, there was a
straightforward line on sex among my friends. Boys could have it; girls
couldn't. A girl who was not a virgin pretended she was. A girl who was sleeping with her boyfriend, no matter how long-playing the relationship, pretended
she was not.

It is the nature of adolescence that there is no past and no future, only the   5
present, burning as fierce, bright, and merciless as a bare light bulb. Girls had
sex with boys because nothing seemed to matter except right now, not pregnancy, not parental disapprobation, nothing but those minutes, this dance, that
face, those words. Most of them knew that pregnancy could result, but they assured themselves that they would be the lucky ones who would not get caught.
Naturally, some of them were wrong, and in my experience they did one of
three things: they went to Puerto Rico for a mysterious weekend trip; visited an
aunt in some faraway state for three months and came back with empty eyes
and a vague reputation, or got married, quickly, in Empire-waist dresses.

What seems to have changed most since then is that there is little philo-   6
sophical counterpoint, hypocritical or not, to the raging hormones of adolescence, and that so many of the once-hidden pregnancies are hidden no more.

Not long after the day at the family planning clinic, I went to a public high   7
school in the suburbs. In the girl's room was this graffito: Jennifer Is a Virgin. I
asked the kids about it and they said it was shorthand for geek, nerd, weirdo,
somebody who was so incredibly out of it that they were in high school and still
hadn't had sex. If you were a virgin, they told me, you just lied about it so that

no one would think you were that immature. The girls in the family planning clinic told me much the same thing—that everyone did it, that the boys wanted it, that not doing it made them seem out of it. The only difference, really, was that the girls in the clinic were poor and would have their babies, and the girls in the high school were well-to-do and would have abortions. Pleasure didn't seem to have very much to do with sex for either group. After she learned she was pregnant, one of the girls at the clinic said, without a trace of irony, that she hoped childbirth didn't hurt as much as sex had. Birth control was easily disposed of in both cases. The pill, the youngsters said, could give you a stroke; the IUD could make you sterile. A diaphragm was disgusting.

    One girl told me the funniest thing her boyfriend—a real original thinker—   8
had told her: they couldn't use condoms because it was like taking a shower with a raincoat on. She was a smart girl and pretty, and I wanted to tell her that it sounded as if she was sleeping with a jerk who didn't deserve her. But that is the kind of basic fact of life that must be taught not in the classroom, not by a stranger, but at home by the family. It is this that, finally, I will try to teach my sons about sex, after I've explained fertile periods and birth control and all the other mechanics that are important to understand but never really go to the heart of the matter: I believe I will say that when you sleep with someone you take off a lot more than your clothes.

COMPREHENSION

1. Does Quindlen approve of sex education? Explain.
2. How does the writer characterize the attitude of adolescents regarding sex and pregnancy?
3. What advice or information about sex will the writer give her sons? Why?

RHETORIC

1. Why did Quindlen choose "Sex Ed" as a title? What is its significance in relation to the thesis?
2. What is Quindlen's thesis? Where is it contained in the essay? Is it directly stated or implied?
3. Is Quindlen's writing an argumentative essay? Support your position with citations from the text.
4. What point is the writer making through use of accumulated details in paragraph 2?
5. What does Quindlen mean by the term "moral component" in paragraph 3? Where else in the essay does she allude to it? How does the author employ definition in this essay?
6. What does Quindlen mean by her final statement that "when you sleep with someone you take off a lot more than your clothes"? How does this ending serve to underscore the thesis of the essay?

WRITING

1. Write a paper analyzing the most common forms of birth control available and listing the advantages and disadvantages of each.
2. In an essay, consider possible solutions to the problem of teenage pregnancy. What role do you think sex education has in ameliorating the problem? Use support from the Quindlen essay if applicable.
3. **WRITING AN ARGUMENT:** Quindlen claims that sex education is like "civics" and "ethics" and should be taught in schools. Do you agree or disagree? Write an essay defending your position.

✶ www.mhhe.com/ **mhreader**	For more information on Anna Quindlen, go to: **More Resources > Ch. 4 Education**

Unplugged: The Myth of Computers in the Classroom

David Gelernter

David Gelernter is a professor of computer science at Yale University. He is a leading figure in the field of human cognition and a seminal thinker in the field known as parallel computing. Gelernter, who was injured by a package sent by the Unabomber in 1993, is the author of Mirror Worlds *(1991),* The Muse in the Machine *(1994),* 1939: The Lost World of the Fair *(1995), and* Machine Beauty *(1998). In the following essay, published in* The New Republic *in 1994, Gelernter offers a cogent analysis of the limits of technology in the classroom.*

Over the last decade an estimated $2 billion has been spent on more than 2 million computers for America's classrooms. That's not surprising. We constantly hear from Washington that the schools are in trouble and that computers are a godsend. Within the education establishment, in poor as well as rich schools, the machines are awaited with nearly religious awe. An inner-city principal bragged to a teacher friend of mine recently that his school "has a computer in every classroom . . . despite being in a bad neighborhood!" 1

Computers should be in the schools. They have the potential to accomplish great things. With the right software, they could help make science tangible or teach neglected topics like art and music. They help students form a concrete idea of society by displaying onscreen a version of the city in which they live— a picture that tracks real life moment by moment. 2

In practice, however, computers make our worst educational nightmares 3 come true. While we bemoan the decline of literacy, computers discount words in favor of pictures and pictures in favor of video. While we fret about the decreasing cogency of public debate, computers dismiss linear argument and promote fast, shallow romps across the information landscape. While we worry about basic skills, we allow into the classroom software that will do a student's arithmetic or correct his spelling.

Take multimedia. The idea of multimedia is to combine text, sound and pictures 4 in a single package that you browse on screen. You don't just *read* Shakespeare; you watch actors performing, listen to songs, view Elizabethan buildings. What's wrong with that? By offering children candy-coated books, multimedia is guaranteed to sour them on unsweetened reading. It makes the printed page look even more boring than it used to look. Sure, books will be available in the classroom, too—but they'll have all the appeal of a dusty piano to a teen who has a Walkman handy.

So what if the little nippers don't read? If they're watching Olivier instead, 5 what do they lose? The text, the written word along with all of its attendant pleasures. Besides, a book is more portable than a computer, has a higher-resolution display, can be written on and dog-eared and is comparatively dirt cheap.

Hypermedia, multimedia's comrade in the struggle for a brave new class- 6 room, is just as troubling. It's a way of presenting documents on screen without imposing a linear start-to-finish order. Disembodied paragraphs are linked by theme; after reading one about the First World War, for example, you might be able to choose another about the technology of battleships, or the life of Woodrow Wilson, or hemlines in the '20s. This is another cute idea that is good in minor ways and terrible in major ones. Teaching children to understand the orderly unfolding of a plot or a logical argument is a crucial part of education. Authors don't merely agglomerate paragraphs; they work hard to make the narrative read a certain way, prove a particular point. To turn a book or a document into hypertext is to invite readers to ignore exactly what counts—the story.

The real problem, again, is the accentuation of already bad habits. Dynamit- 7 ing documents into disjointed paragraphs is one more expression of the sorry fact that sustained argument is not our style. If you're a newspaper or magazine editor and your readership is dwindling, what's the solution? Shorter pieces. If you're a politician and you want to get elected, what do you need? Tasty sound bites. Logical presentation be damned.

Another software species, "allow me" programs, is not much better. These 8 programs correct spelling and, by applying canned grammatical and stylistic rules, fix prose. In terms of promoting basic skills, though, they have all the virtues of a pocket calculator.

In Kentucky, as *The Wall Street Journal* recently reported, students in grades 9 K–3 are mixed together regardless of age in a relaxed environment. It works great, the *Journal* says. Yes, scores on computation tests have dropped 10 percent at one school, but not to worry: "Drilling addition and subtraction in an age of calculators is a waste of time," the principal reassures us. Meanwhile, a Japanese educator informs University of Wisconsin mathematician Richard Akey that in his country, "calculators are not used in elementary or junior high school because the primary emphasis is on helping students develop their mental abilities." No wonder Japanese kids blow the pants off American kids in math. Do we really think "drilling addition and subtraction in an age of calculators is a waste of time"? If we do, then "drilling reading in an age of multimedia is a waste of time" can't be far behind.

Prose-correcting programs are also a little ghoulish, like asking a computer 10 for tips on improving your personality. On the other hand, I ran this article through a spell-checker, so how can I ban the use of such programs in schools? Because to misspell is human; to have no idea of correct spelling is to be semi-literate.

There's no denying that computers have the potential to perform inspiring feats 11 in the classroom. If we are ever to see that potential realized, however, we ought to agree on three conditions. First, there should be a completely new crop of children's software. Most of today's offerings show no imagination. There are hundreds of similar reading and geography and arithmetic programs, but almost nothing on electricity or physics or architecture. Also, they abuse the technical capacities of new media to glitz up old forms instead of creating new ones. Why not build a time-travel program that gives kids a feel for how history is structured by zooming you backward? A spectrum program that lets users twirl a frequency knob to see what happens?

Second, computers should be used only during recess or relaxation periods. 12 Treat them as fillips, not as surrogate teachers. When I was in school in the '60s, we all loved educational films. When we saw a movie in class, everybody won: teachers didn't have to teach, and pupils didn't have to learn. I suspect that classroom computers are popular today for the same reasons.

Most important, educators should learn what parents and most teachers al- 13 ready know: you cannot teach a child anything unless you look him in the face. We should not forget what computers are. Like books—better in some ways, worse in others—they are devices that help children mobilize their own resources and learn for themselves. The computer's potential to do good is modestly greater than a book's in some areas. Its potential to do harm is vastly greater, across the board.

COMPREHENSION

1. State the author's thesis or claim in one sentence.
2. In the final paragraph, Gelernter defines what he believes to be the most important shortcoming of the computer as a teaching tool. Explain the reason why this weakness is so significant.

3. In your own words, explain the author's dislike of hypermedia as a peda-
gogic tool (as expressed in paragraph 6) and why the orderly arrangement
of paragraphs in a book is superior to this newer technological capability.

RHETORIC

1. The introductory paragraph goes from a general fact to a specific quotation.
What is the effect of this method of paragraph patterning?
2. Much of the author's argument hinges on providing evidence that one
medium is superior to another. Explain terms such as *linear argument* (paragraph 3), *agglomerate paragraphs* (paragraph 6), and *allow me programs*
(paragraph 8). How do these terms help Gelernter prove his point?
3. The essay has a three-part structure, each section divided by space. How
would you characterize the purpose of each section? How does the author
use transitions to move from one section to the next?
4. The author states that the overuse of computers in the classroom can hinder
the development of clear thinking and reasoned argument. How clearly
written is *his* essay? How reasoned is his argument? Gather evidence for
your answer by reviewing the essay and determining whether each sentence seems to flow logically from the next and whether each paragraph
seems to move reasonably to the next.
5. The author uses metaphors, similes, and other rhetorical devices. Explain
the effectiveness of expressions such as "have all the appeal of a dusty piano to a teen who has a Walkman handy" (paragraph 4), "dynamiting documents" (paragraph 7), and "software species" (paragraph 8). Locate other
unconventional descriptions.
6. Who is the intended audience for this essay? Educators?, Parents?, Students?, Politicians? What evidence can you show to back up your view?
7. What rhetorical device is the author using in his title? What is the implicit
meaning of the title?

WRITING

1. Visit the writing or reading computer lab in your school. As an objective observer, study the interaction of student and computer. Write a descriptive
essay focusing on the demeanor and behavior of the student and the atmosphere of the classroom. If you wish, compare it to a traditional classroom.
2. Copy a paragraph from the essay, and enter it into a word-processing program that has a grammar-check function. Record any comments that the
program makes in response to its evaluation of the writing. Do the computer's responses to the author's sentence structure make sense?

3. **WRITING AN ARGUMENT:** Select one of the teaching capabilities of modern computers—multimedia, hypertext, or spell- and grammar-check programs. Argue for the benefits of one of these features.

www.mhhe.com/
mhreader

For more information on David Gelernter, go to:
More Resources > Ch. 4 Education

✳

When Bright Girls Decide That Math Is "a Waste of Time"

Susan Jacoby

Susan Jacoby (b. 1945) has worked as an educator and as a reporter for The Washington Post. *As a freelance journalist in the former Soviet Union (from 1969 to 1971), she produced two books about her experiences. Jacoby now contributes to* The Nation *and* McCall's; *her books include* The Possible She *(1979), a collection of autobiographical essays,* Wild Justice: The Evolution of Revenge *(1983), and* Half-Jew: A Daughter's Search for Her Buried Past *(2000). In this essay from* The New York Times, *Jacoby examines the reasons why girls are often deficient in math and science.*

Susannah, a 16-year-old who has always been an A student in every subject   1
from algebra to English, recently informed her parents that she intended to drop physics and calculus in her senior year of high school and replace them with a drama seminar and a work-study program. She expects a major in art or history in college, she explained, and "any more science or math will just be a waste of my time."

Her parents were neither concerned by nor opposed to her decision. "Fine,   2
dear," they said. Their daughter is, after all, an outstanding student. What does it matter if, at age 16, she has taken a step that may limit her understanding of both machines and the natural world for the rest of her life?

This kind of decision, in which girls turn away from studies that would   3
give them a sure footing in the world of science and technology, is a self-inflicted female disability that is, regrettably, almost as common today as it was when I was in high school. If Susannah had announced that she had decided to stop taking English in her senior year, her mother and father would have been horrified. I also think they would have been a good deal less sanguine about her decision if she were a boy.

In saying that scientific and mathematical ignorance is a self-inflicted fe-   4
male wound, I do not, obviously, mean that cultural expectations play no role
in the process. But the world does not conspire to deprive modern women of
access to science as it did in the 1930s, when Rosalyn S. Yalow, the Nobel
Prize–winning physicist, graduated from Hunter College and was advised to go
to work as a secretary because no graduate school would admit her to its
physics department. The current generation of adolescent girls—and their parents, bred on old expectations about women's interests—are active conspirators
in limiting their own intellectual development.

It is true that the proportion of young women in science-related graduate   5
and professional schools, most notably medical schools, has increased significantly in the past decade. It is also true that so few women were studying advanced science and mathematics before the early 1970s that the percentage
increase in female enrollment does not yet translate into large numbers of
women actually working in science.

The real problem is that so many girls eliminate themselves from any seri-   6
ous possibility of studying science as a result of decisions made during the
vulnerable period of midadolescence, when they are most likely to be influenced—on both conscious and subconscious levels—by the traditional belief
that math and science are "masculine" subjects.

During the teen-age years the well-documented phenomenon of "math   7
anxiety" strikes girls who never had any problem handling numbers during
earlier schooling. Some men, too, experience this syndrome—a form of panic,
akin to a phobia, at any task involving numbers—but women constitute the
overwhelming majority of sufferers. The onset of acute math anxiety during the
teen-age years is, as Stalin was fond of saying, "not by accident."

In adolescence girls begin to fear that they will be unattractive to boys if   8
they are typed as "brains." Science and math epitomize unfeminine braininess
in a way that, say, foreign languages do not. High-school girls who pursue an
advanced interest in science and math (unless they are students at special institutions like the Bronx High School of Science where everyone is a brain) usually
find that they are greatly outnumbered by boys in their classes. They are, therefore, intruding on male turf at a time when their sexual confidence, as well as
that of the boys, is most fragile.

A 1981 assessment of female achievement in mathematics, based on re-   9
search conducted under a National Institute for Education grant, found significant differences in the mathematical achievements of 9th and 12th graders. At
age 13 girls were equal to or slightly better than boys in tests involving algebra,
problem solving and spatial ability; four years later the boys had outstripped
the girls.

It is not mysterious that some very bright high-school girls suddenly decide   10
that math is "too hard" and "a waste of time." In my experience, self-sabotage
of mathematical and scientific ability is often a conscious process. I remember
deliberately pretending to be puzzled by geometry problems in my sophomore
year in high school. A male teacher called me in after class and said, in a baffled

tone, "I don't see how you can be having so much trouble when you got straight A's last year in my algebra class."

The decision to avoid advanced biology, chemistry, physics and calculus in 11 high school automatically restricts academic and professional choices that ought to be wide open to anyone beginning college. At all coeducational universities women are overwhelmingly concentrated in the fine arts, social sciences and traditionally female departments like education. Courses leading to degrees in science- and technology-related fields are filled mainly by men.

In my generation, the practical consequences of mathematical and scientific 12 illiteracy are visible in the large number of special programs to help professional women overcome the anxiety they feel when they are promoted into jobs that require them to handle statistics.

The consequences of this syndrome should not, however, be viewed in nar- 13 rowly professional terms. Competence in science and math does not mean one is going to become a scientist or mathematician any more than competence in writing English means one is going to become a professional writer. Scientific and mathematical illiteracy—which has been cited in several recent critiques by panels studying American education from kindergarten through college— produces an incalculably impoverished vision of human experience.

Scientific illiteracy is not, of course, the exclusive province of women. In 14 certain intellectual circles it has become fashionable to proclaim a willed, aggressive ignorance about science and technology. Some female writers specialize in ominous, uninformed diatribes against genetic research as a plot to remove control of childbearing from women, while some well-known men of letters proudly announce that they understand absolutely nothing about computers, or, for that matter, about electricity. This lack of understanding is nothing in which women or men ought to take pride.

Failure to comprehend either computers or chromosomes leads to a terrible 15 sense of helplessness, because the profound impact of science on everyday life is evident even to those who insist they don't, won't, can't understand why the changes are taking place. At this stage of history women are more prone to such feelings of helplessness than men because the culture judges their ignorance less harshly and because women themselves acquiesce in that indulgence.

Since there is ample evidence of such feelings in adolescence, it is up to par- 16 ents to see that their daughters do not accede to the old stereotypes about "masculine" and "feminine" knowledge. Unless we want our daughters to share our intellectual handicaps, we had better tell them no, they can't stop taking mathematics and science at the ripe old age of 16.

COMPREHENSION

1. What reasons does Jacoby give for girls' deficiency in math and science?
2. Why does she call it a "self-inflicted female disability" (paragraph 3)?
3. What are the consequences of being math- and science-illiterate?

RHETORIC

1. Explain the main idea of Jacoby's essay in your own words.
2. Does the writer use abstract or concrete language in her essay? Cite examples to support your response.
3. What technique does Jacoby use in paragraphs 1 and 2? How does it aid in setting up her argument?
4. What rhetorical strategies does the writer use in her essay?
5. How does the use of dialogue aid in developing paragraph 10? What effect does the general use of dialogue have on the writer's point?
6. How is Jacoby's conclusion consistent in tone with the rest of the essay? Does it supply a sense of unity? Why or why not?

WRITING

1. Write an essay describing a school-related phobia you once had, or continue to have (for example, in math, writing, physical education, biology). Explain where you think that fear came from, how it affected your performance in school, and what you did (or are doing) to cope with the problem.
2. Write an essay about the need for math and science literacy in today's world. Use support from Jacoby's essay.
3. **WRITING AN ARGUMENT:** Write an argumentation essay proposing that math and science phobia is not "self-inflicted" but caused primarily by the continued presence of sexism in society.

✦ www.mhhe.com/ **mhreader**	For more information on Susan Jacoby, go to: **More Resources > Ch. 4 Education**

✳

Sure Changes for Colleges in the Future

Arthur E. Levine

Arthur Eliot Levine (b. 1948) is president of Teachers College at Columbia University. Born in New York City, he was educated at Brandeis University (BA, 1970) and the State University of New York at Buffalo (PhD, 1976). One of the most influential voices in public and higher education, Levine has written many books and articles, including

Reform of Undergraduate Education *(1973), which won the Book of the Year Award from the American Council of Education,* Beating the Odds: How the Poor Get to College *(1995), and* When Hope and Fear Collide: A Portrait of Today's College Student *(with Jeannette Cureton, 1998). In this essay, written for the* Education Digest *in 2001, Levine discusses changes colleges will have to make if they are to compete with a changing culture.*

Several major forces today have the power to transform the nation's colleges and universities: shifting demographics, new technologies, the entrance of commercial organizations into higher education, changing relationships between colleges and federal and state governments, and the move from an industrial to an information society.

In addition, the convergence of publishing, broadcasting, telecommunications, and education is blurring the distinction between education and entertainment. A variety of knowledge producers will compete to create courses and other educational services, to develop new ways to distribute knowledge, and to engage larger audiences.

Given such realities, what will happen to higher education as we know it? My answers, based on a quarter-century in colleges, think tanks, and foundations, are speculation. But nine changes seem almost inevitable, each raising thorny questions we dare not ignore, if we are to thrive in the coming years. A number of these changes are, in fact, already well under way:

Higher-education providers will become even more numerous and more diverse. Survival of some institutions, especially less-selective private colleges with small endowments and large programs in adult education, will be increasingly threatened by both domestic and foreign for-profit institutions, as well as nonprofit competitors like libraries and museums that also have entered the educational marketplace. Moreover, technological capabilities are encouraging global universities, which transcend national boundaries. The most successful institutions will be those that can respond quickest and offer a high-quality education to an international student body.

As a result, we should expect new brand names and a new hierarchy of quality in higher-education institutions. Why should a credential from Microsoft University or the British Open University be less prestigious than a regional state college's? Yet, in such an international environment, how can minimum standards be determined and monitored? How should quality-control mechanisms, such as accreditation, be redesigned?

Three basic types of colleges and universities are emerging: "brick universities," or traditional residential institutions; "click universities," or new, usually commercial virtual universities, like Jones International University; and "brick and click" universities, a combination of the two. If current research on e-commerce is correct, the most competitive and attractive higher-education institutions will be "brick and click." While consumers appreciate the convenience, ease, and freedom of services online, they also want a physical space to interact with others and obtain expert advice and assistance face-to-face.

Who will control brick-and-click institutions? Will the for-profit sector buy   7
"bricks"—build physical plants—before traditional colleges develop the capacity to operate in the "click" environment? Or will the opposite occur? How
should each of the nation's colleges determine which of the three categories best
meets its goals?

Higher education is becoming more individualized; students, not institu-   8
tions, will set the educational agenda. Increasingly, students will come from diverse backgrounds and have a widening variety of educational needs. New
technologies will allow education at any time and any place—on campus, in the
office, at home, in the car, on vacation. Each student will be able to choose from
a multitude of knowledge providers the form of instruction and courses most
consistent with how he or she learns.

How can colleges retain and provide services for students with such hetero-   9
geneous backgrounds and individualized educational goals? How can an
institution create a strong sense of identity and community? What can brick-
and-mortar campuses do that online education can't? And, beyond anecdotal
information, can we document what those activities might be?

The focus of higher education is shifting from teaching to learning. Colleges   10
currently emphasize a commonality of process based on "seat time," or the
amount of time each student is taught. Students study for a defined number of
hours, earn credits for each hour of study, and, after earning a specified number
of credits, earn a degree. With more educational providers, the individualiza-
tion of education, and the growing diversity of the student body, however, that
commonality is likely to be lost. The focus will shift to the outcomes students
achieve. Time will become the variable and learning the constant.

Such a development raises very large questions about the meaning of a   11
two-year or a four-year degree. It also shifts the definition of excellence from the
institution's selectivity in admitting students to the value that the institution
demonstrably adds to each student's learning experience.

We also should expect other new forces to gain momentum: 12

Traditional functions of higher education could become unbundled. Col-   13
leges engage in teaching, research, and service—yet teaching is the only function usually thought of as profitable. Research brings in dollars for only a small
number of institutions. Service, by its nature, is not remunerative.

Therefore, for-profit and other new providers in higher education are inter-   14
ested only in teaching—and will compete with traditional colleges solely there.
To the extent colleges lose out to new competitors, financial support from government and private sources for two activities of vital national interest—
research and service—will be lost.

How do we protect them? An institution that engages only in those func-   15
tions is not financially viable, but one that engages only in teaching may be intellectually impoverished. How can we head off the potential unbundling, for
the benefit not only of colleges but also of the nation?

Faculty members will become increasingly independent of colleges and   16
universities. The most renowned faculty members, able to attract tens of thou-

sands of students in an international marketplace, will use the equivalent of an academic William Morris Agency. A talent agent will bring to a professor a book deal with Random House, a weekly PBS program, a consulting contract with IBM, commercial endorsement opportunities, and a distance-learning course with a for-profit company in a total package of $5 million. The names of world-class professors will probably be far more important than their institution's.

Institutions of higher education must ask how they can create communities 17 sufficiently vital to attract and retain faculty in such an environment. Other questions are: Will the rich get richer and the poor get poorer, with only a hand-ful of very prestigious, well-endowed institutions able to afford the most distin-guished professors? What does greater power for the faculty mean in terms of institutional governance? What is the future of tenure if the most sought-after professors leave the academy or become itinerant?

Degrees will wither in importance. Today, the meaning of a degree varies in 18 content and quality, depending on the college. In essence, we offer thousands of different degrees, even if called the same name. A degree now signifies a period of successful college attendance; the class rank indicates the relative success of the student; and the name of the college marks the quality of the degree.

However, with the change in emphasis from institutional process to educa- 19 tional outcomes, degrees will become far less meaningful. A transcript of each student's competencies, including specific information that the student knows or that can be performed, will be far more desirable.

Colleges now have a virtual monopoly on higher-education credentials. If 20 degrees become less important, how will we continue to attract students in a world of limitless educational choices? Why would a student stay at the same college up to five years if degrees give way to specific competencies? What about residential institutions? Will traditional collegiate life be the province of only the most affluent, who have the leisure and money to afford it?

Every person will have an educational passport. In the future, each person's 21 education will occur not only in a cornucopia of different settings and geo-graphic locales, but also via a plethora of different educational providers. As traditional degrees lose importance, the nation will need to establish a central bureau that records each person's educational achievements—however and wherever gained—and that provides documentation. This educational pass-port, or portfolio, will record a lifetime educational history.

We will need common standards for naming and assessing those achieve- 22 ments. In our decentralized system of higher education, how will we do this? Will each state develop its own standards? Or will accreditors, or perhaps foun-dations, take the lead?

Dollars will follow the students more than the educators. With growth in 23 educational providers and emphasis on outcomes, public and private financial supporters will increasingly invest in the educational consumer over the ex-panding grab bag of organizations that offer collegiate instruction. Federal and state aid currently for institutions of higher education quite possibly will be transferred directly to students.

Such a trend will add to questions of how we ensure standards of quality   24
among the increasing number of new providers and require us to ask how academic freedom—demanding institutional autonomy—can be preserved when
colleges are forced to be as market-driven and consumer-oriented as most commercial organizations are today. How can institutions remain economically viable when financial support shifts more to consumers, faculty grow more
independent, and degrees fade in importance?

What I have described is, in some sense, a ghost of Christmas future. While   25
the trends are no more than one individual's halting attempt to predict things to
come, I have no doubt that the forces buffeting higher education today are powerful and will change it considerably.

My fear is that America's colleges will ignore them and the important ques-   26
tions that they demand we confront—or that, simply through complacency or
the glacial speed of our decision-making processes, we will fail to respond in
time to help shape tomorrow.

The Right Question

In the early years of the Industrial Revolution, the Yale Report of 1828 asked   27
whether the needs of a changing society required either major or minor changes
in higher education. The report concluded that it had asked the wrong question.
The right question was, What is the purpose of higher education?

All of the questions I've raised have their deepest roots in that fundamental   28
question. Once more faced with a society in motion, we must not only ask that
question again, but must actively pursue answers, if our colleges and universities are to retain their vitality in a dramatically different world.

COMPREHENSION

1. List the nine changes in higher education that Levine thinks are "almost
   inevitable." According to the writer, which changes already are taking
   place?
2. Describe the three types of colleges that are emerging today.
3. What additional trends in higher education does Levine discuss? What is
   the role of faculty and students in the future? What problems does the
   writer foresee? What, ultimately, is the purpose of higher education and its
   relationship to the trends that Levine analyzes?

RHETORIC

1. What is Levine's thesis? Is it stated or implied? Explain your answer.
2. At whom is this essay aimed, and what is Levine's purpose? On what do
   you base your answer?
3. Explain Levine's use of cause-and-effect analysis and classification to structure his essay.

4. The length of the writer's paragraphs is typically short. What is the effect of this strategy? Do you find it effective or not, and why?
5. Levine poses numerous questions in this essay. List these questions. Does the writer answer all these questions? Does he provide detailed answers or generalities? Do you feel he should have provided more evidence or detail? Justify your answer.
6. At the end of the essay, Levine uses a subheading. What function does this subheading serve? Could it have been omitted? Why or why not?

WRITING

1. Identify your college as one of the three types of institutions of higher education that Levine classifies. Then explain why you think your college fits into this category.
2. In a cause-and-effect essay, analyze some of the ways that the changes and trends Levine examines might have an impact on the college education you will receive.
3. **WRITING AN ARGUMENT:** Argue for or against the proposition that Levine's vision for the future of higher education is correct. For example, do you think that students in the future will have "educational passports" (paragraph 21)? Will the best colleges have "an international student body" (paragraph 4)? Will degrees "whither in importance" (paragraph 18)? These are some of the questions you might want to deal with.

---

www.mhhe.com/
mhreader

For more information on Arthur E. Levine, go to:
More Resources > Ch. 4 Education

✳

Two Cheers for *Brown v. Board of Education*

Clayborne Carson

Clayborne Carson (b. 1944) is professor of history and director of the Martin Luther King Jr. Papers Project at Stanford University. He was born in Buffalo, New York, and educated at the University of California at Los Angeles, where he received his BA (1967), MA (1968), and PhD (1977). A specialist in African American and civil rights history, Carson has written and edited numerous books, including In Struggle: SNCC and the Black Awakening of the 1960s *(1981, revised edition 1995), which received the Frederick Jackson Turner Award;* Eyes on the Prize: America's Civil Rights Years *(1987); and* The Malcolm X File *(1991). Asked by Coretta Scott King to handle her husband's literary estate, Carson is the lead editor of* The Papers of Martin Luther King Jr. *(University of California Press). "It was a job you couldn't say no to," Carson said. In this essay, published in the* Journal of American History *in 2004, Carson offers an evaluation of the impact of a major Supreme Court decision on school segregation.*

My gratuitous opinion of *Brown v. Board of Education* (1954) is somewhat am-   1 bivalent and certainly arrives too late to alter the racial policies of the past fifty years. But for those of us who practice history, hindsight offers a far more reliable kind of wisdom than does foresight. We see clearly now that while the *Brown* decision informed the attitudes that have shaped contemporary American race relations, it did not resolve persistent disputes about the nation's civil rights policies. The Supreme Court's unanimous opinion in *Brown* broke decisively with the racist interpretations of traditional American values set forth in *Scott v. Sandford* (1857) and *Plessy v. Ferguson* (1896), offering instead the optimistic "American Creed" that Gunnar Myrdal saw as the solution to "the Negro problem."[1] Like the two earlier landmark decisions, *Brown* overestimated the extent of ideological consensus among Americans and soon exacerbated racial and regional conflicts instead of resolving them. The Court's ruling against school segregation encouraged African Americans to believe that the entire structure of white supremacy was illegitimate and legally vulnerable. But the civil rights struggles *Brown* inspired sought broader goals than the decision could deliver, and that gap fostered frustration and resentment among many black Americans. In short, the decision's virtues and limitations reflect both the achievements and the failures of the efforts made in the last half century to solve America's racial dilemma and to realize the nation's egalitarian ideals.

That the *Brown* decision spurred subsequent civil rights progress seems ap-   2 parent, but its impact and its significance as a source of inspiration are difficult to measure.[2] Although the Court's initial unwillingness to set firm timetables for school desegregation undercut *Brown*'s immediate impact, African Americans expanded the limited scope of the decision by individual and collective challenges to the Jim Crow system. Small-scale protests escalated during the decade after 1954, becoming a sustained mass movement against all facets of segregation and discrimination in the North as well as the South. Civil rights protests and litigation prompted Congress to pass the Civil Rights Act of 1964 and the Voting Rights Act of 1965, both of which extended the *Brown* decision's egalitar-

---

[1]*Scott v. Sandford,* 19 How. 393 (1857); *Plessy v. Ferguson,* 163 U.S. 537 (1896); *Brown v. Board of Education,* 347 U.S. 483 (1954); Gunnar Myrdal, *An American Dilemma: The Negro Problem and Modern Democracy* (2 vols., New York, 1944). [This and subsequent notes in the selection are the author's.]
[2]On *Brown*'s direct and indirect consequences, see, for example, Michael J. Klarman, "How *Brown* Changed Race Relations: The Backlash Thesis," *Journal of American History,* 81 (June 1994), 81–118. Klarman correctly points out that *Brown* had limited impact on school desegregation, especially in the Deep South, and stimulated southern white resistance to racial reform. He concludes that the contributions of *Brown* to the broader civil rights struggle were mostly indirect.

ian principles well beyond education. The historic mass struggle that followed *Brown* ultimately destroyed the legal foundations of the Jim Crow system, and their destruction prepared the way for a still more far-reaching expansion of prevailing American conceptions of civil rights and of the role of government in protecting those rights. During the past forty years, women and many minority groups, including immigrants and people with disabilities, have gained new legal protections modeled on the civil rights gains of African Americans.[3]

But the *Brown* decision also created racial aspirations that remain unrealized. Although the decision may have been predicated on the notion of a shared American creed, most white Americans were unwilling to risk their own racial privileges to bring about racial equality. The decision was neither universally accepted nor consistently enforced. "Instead, it provoked overwhelming resistance in the South and only tepid interest in the North," the historian John Higham insisted. "In the South the decision released a tidal wave of racial hysteria that swept moderates out of office or turned them into demagogues. State and local officials declined to obstruct a revival of the Ku Klux Klan. Instead, they employed every conceivable device to maintain segregation, including harassment and dissolution of NAACP chapters."[4] By the 1970s, resistance to school desegregation had become national. Northern whites in Boston and elsewhere demonstrated their unwillingness to send their children to predominantly black schools or to allow large-scale desegregation that would drastically alter the racial composition of "their" schools in "their" neighborhoods. Voters in the states of Washington and California passed initiatives to restrict the right of school boards (Washington) and state courts (California) to order busing to achieve school desegregation (the Supreme Court later held the Washington initiative unconstitutional). Nationwide, white racial resentments encouraged an enduring shift of white voters from the Democratic to the Republican party. The 1964 election would be the last presidential contest in which the majority of black voters and of white voters backed the same candidate. Since 1974, when the Supreme Court's *Milliken v. Bradley* decision set limits on busing, the legal meaning of desegregation has been scaled back to conform to American racial and political realities.[5]

[3]Cf. Hugh Davis Graham, *The Civil Rights Era: Origins and Development of National Policy, 1960–1972* (New York, 1990); Hugh Davis Graham, *Collision Course: The Strange Convergence of Affirmative Action and Immigration Policy in America* (New York, 2002); and John D. Skrentny, *The Minority Rights Revolution* (Cambridge, Mass., 2002).

[4]John Higham, "Introduction: A Historical Perspective," in *Civil Rights and Civil Wrongs: Black-White Relations since World War II*, ed. John Higham (University Park, 1997), 4. See also Klarman, "How *Brown* Changed Race Relations"; Numan V. Bartley, *The Rise of Massive Resistance: Race and Politics in the South in the 1950s* (Baton Rouge, 1969); and Neil McMillen, *The Citizens' Council: Organized Resistance to the Second Reconstruction, 1954–1964* (Urbana, 1971).

[5]See Ronald P. Formisano, *Boston against Busing: Race, Class, and Ethnicity in the 1960s and 1970s* (Chapel Hill, 1991); and J. Anthony Lukas, *Common Ground: A Turbulent Decade in the Lives of Three American Families* (New York, 1985). *Washington v. Seattle School District*, 458 U.S. 457 (1982); *Crawford v. Los Angeles Board of Education*, 458 U.S. 527 (1982); *Milliken v. Bradley*, 418 U.S. 717 (1974). See Gary Orfield and Susan E. Eaton, *Dismantling Desegregation: The Quiet Reversal of* Brown v. Board of Education (New York, 1996).

African Americans generally applauded the *Brown* decision when it was an-   4
nounced, but the Court's failure to realize *Brown*'s bold affirmation of egalitarian ideals fueled subsequent black discontent and disillusionment. *Brown* cited
studies that demonstrated the harmful psychological impact of enforced segregation on black students, reporting, "To separate them from others of similar
age and qualifications solely because of their race generates a feeling of inferiority as to their status in the community that may affect their hearts and minds
in a way unlikely ever to be undone." Yet the Court did not offer an effective
means to correct the problem it had identified. During the decades after *Brown*,
most southern black children continued to suffer the psychological consequences of segregation, while a small minority assumed the often considerable
psychological and physical risks of attending newly integrated public schools.
Rather than bringing large numbers of black and white students together in
public schools, the *Brown* decision—and the subsequent years of litigation and
social conflict—enabled a minority of black students to attend predominantly
white schools. Ten years after the *Brown* decision, according to data compiled by
the U.S. Department of Education, almost 98 percent of southern black students
still attended predominantly black schools. Now, at the beginning of the
twenty-first century, the Court's ideal of educational opportunity as "a right
which must be made available to all on equal terms" is still far from being realized. American schools, both public and private, are still highly segregated. According to a recent study, the typical Latino or black student in the United States
still attends a school where members of minority groups are predominant.[6]

Certainly, the *Brown* decision's most significant deficiency is its failure to   5
address the concerns of the majority of African American students who have
been unable or unwilling to seek better educational opportunities by leaving
predominantly black schools for predominantly white ones. While it opened the
door for the Little Rock Nine, who desegregated Central High School in 1957,
the *Brown* decision offered little solace to the hundreds of students who remained at Little Rock's all-black Horace Mann High School. When Arkansas officials reacted to desegregation by closing all of Little Rock's high schools, those
students were denied even segregated educational opportunities.[7] With the encouragement of the lawyers for the National Association for the Advancement
of Colored People's (NAACP) Legal Defense and Education Fund, the Supreme
Court largely abandoned previous efforts to enforce the separate but equal
mandate in order to adopt a narrowly conceived strategy for achieving equal
educational opportunity through desegregation. The pre-*Brown* equalization effort had encouraged social scientists to develop increasingly sophisticated ways

---

[6]*Brown v. Board of Education*, 347 U.S. at 494, 493; Gary Orfield and Chungmei Lee, "*Brown* at Fifty:
King's Dream or Plessy's Nightmare?," Jan. 17, 2004, *The Civil Rights Project, Harvard University*
<http://www.civilrightsproject.harvard.edu/research/reseg04/resegregation04.php> (April 4,
2004). In every region of the nation, at least 30% black students still attend schools with less than
10% white enrollment. *Ibid.*
[7]Cf. Melba Beals, *Warriors Don't Cry: A Searing Memoir of the Battle to Integrate Little Rock's Central
High* (New York, 1995); and Melba Beals, *White Is a State of Mind: A Memoir* (New York, 1995).

of measuring differences in the quality of schools. But during the 1950s, pro–civil rights scholars shifted their focus from the educational environment of black students in black schools to the psychological state of black students experiencing desegregation. The NAACP's initial strategy of forcing southern states to equalize facilities at all-black schools had resulted in tangible improvements, whereas the removal of racial barriers in public schools was advertised as offering intangible psychological gains.

For Thurgood Marshall, who headed the NAACP legal staff, the equalization effort had always been a means of achieving the ultimate goal of desegregation. After the Supreme Court decided in *Sweatt v. Painter* (1950) that a makeshift segregated law school at a black college could not provide educational opportunities equal to those offered by the University of Texas Law School. Marshall exulted, "The complete destruction of *all* enforced segregation is now in sight." Despite having attended predominantly black schools at every stage of his academic career, he saw segregation as a racial stigma that could not be removed by increased state appropriations for Jim Crow schools. In the early 1950s he noted that social scientists were "almost in universal agreement that segregated education produces inequality." He therefore concluded "that segregated schools, perhaps more than any other single factor, are of major concern to the individual of public school age and contribute greatly to the unwholesomeness and unhappy development of the personality of Negroes which the color caste system in the United States has produced."[8]

Few African Americans would wish to return to the pre-*Brown* world of legally enforced segregation, but in the half century since 1954, only a minority of Americans has experienced the promised land of truly integrated public education. By the mid-1960s, with dual school systems still in place in many areas of the Deep South, and with de facto segregation a recognized reality in urban areas, the limitations of *Brown* had become evident to many of those who had spearheaded previous civil rights struggles. The ideological gulf that appeared in African American politics during the period was largely the result of efforts to draw attention to the predominantly black institutions neglected in the drive for racial integration. The black power movement arose in part as an effort by African Americans to control and improve such institutions. Some black power proponents exaggerated the benefits of racial separatism, but their extremism can be best understood as a reaction against the unbalanced post-*Brown* strategy of seeking racial advancement solely through integration. Although James S. Coleman's landmark 1966 study of equality of educational opportunity found that black children attending integrated schools did better than students attending predominantly black schools, it was by no means clear that the gap was the result of interracial interactions rather than of differences in the socioeconomic

[8]*Sweatt v. Painter,* 339 U.S. 629 (1950); *Baltimore Afro-American,* June 17, 1950, quoted in Juan Williams, *Thurgood Marshall: American Revolutionary* (New York, 1998), 195; Thurgood Marshall, "An Evaluation of Recent Efforts to Achieve Racial Integration in Education through Resort to the Courts," *Journal of Negro Education,* 21 (Summer 1952), 316–27, esp. 322.

backgrounds of the students involved. By the late 1960s, growing numbers of black leaders had concluded that improvement of black schools should take priority over school desegregation. In 1967, shortly before the National Advisory Commission on Civil Disorders warned that the United States was "moving toward two societies, one white, one black—separate and unequal," Martin Luther King Jr. acknowledged the need to refocus attention, at least in the short run, on "schools in ghetto areas." He also insisted that "the drive for immediate improvements in segregated schools should not retard progress toward integrated education later." Even veterans of the NAACP's legal campaign had second thoughts. "*Brown* has little practical relevance to central city blacks," Constance Baker Motley commented in 1974. "Its psychological and legal relevance has already had its effect."[9]

Black power advocates sometimes sought to replace the narrow strategy of achieving racial advancement through integration with the equally narrow strategy of achieving it through racial separatism. In both instances, claims of psychological gains often substituted for measurable racial advancements, but the continued popularity of Afrocentric educational experiments indicates that many African Americans now see voluntary segregation as psychologically uplifting. Having personally experienced the burden of desegregating numerous classrooms and having watched my son move with great success from a predominantly black college to a predominantly white law school, I am skeptical of sweeping claims about the impact of racial environment on learning. While believing that debates among African Americans during the last half century about their destiny have been useful, I regret that those debates have often exacerbated ideological conflict rather than encouraging us toward collective action. Rather than having to choose between overcoming racial barriers and improving black community institutions, we should be able to choose both.

In hindsight, the nation would have been better served if the *Brown* decision had evinced a more realistic understanding of the deep historical roots of America's racial problems—perhaps a little more familiarity with the writings of W. E. B. Du Bois and Carter C. Woodson as well as those of Myrdal and his colleagues. Rather than blandly advising that desegregation of public schools be achieved with "all deliberate speed," the Supreme Court—and the NAACP lawyers who argued before it—should have launched a two-pronged attack, not only against racial segregation but also against inferior schools, whatever their racial composition. Such an attack would have heeded the admonition that Du Bois offered in 1935, soon after his forced resignation as editor of the NAACP's journal, the *Crisis:*

---

[9] J. S. Coleman et al., *Equality of Educational Opportunity* (Washington, 1966), *passim; Report of the National Advisory Commission on Civil Disorders* (New York, 1968), 1; Martin Luther King Jr., *Where Do We Go from Here: Chaos or Community?* (New York, 1967), 228. For Constance Baker Motley's statement (quoted from the *New York Times,* May 13, 1974), see James T. Patterson, Brown v. Board of Education: *A Civil Rights Milestone and Its Troubled Legacy* (New York, 2001), 168.

Theoretically, the Negro needs neither segregated schools nor mixed schools. What he needs is Education. . . . Other things being equal, the mixed school is the broader, more natural basis for the education of all youth. It gives wider contacts; it inspires great self-confidence; and suppresses the inferiority complex. But other things seldom are equal, and in that case, Sympathy, Knowledge, and the Truth, outweigh all that the mixed school can offer.[10]

Because the *Brown* decision was a decisive departure from *Plessy's* separate 10 but equal principle, it was an important turning point in African American history. Nevertheless, fifty years later the Court's assumptions about the psychological consequences of legally enforced segregation seem dated. The Jim Crow system no longer exists, but most black American schoolchildren still attend predominantly black public schools that offer fewer opportunities for advancement than typical predominantly white public schools. Moreover, there is no contemporary civil rights movement able to alter that fact. Yet, if *Brown* represents a failed attempt to achieve comprehensive racial advancement, the opinion nonetheless still challenges us by affirming egalitarian ideals that remain relevant: "In these days, it is doubtful that any child may reasonably be expected to succeed in life if he is denied the opportunity of an education. Such an opportunity, where the state has undertaken to provide it, is a right which must be made available to all on equal terms."[11]

[10]*Brown v. Board of Education*, 349 U.S. 294 (1955); W. E. B. Du Bois, "Does the Negro Need Separate Schools?," *Journal of Negro Education*, 4 (July 1935), in *The Oxford W. E. B. Du Bois Reader*, ed. Eric J. Sundquist (New York, 1996), 431.
[11]"*Brown v. Board of Education of Topeka:* Opinion on Segregation Laws," in *Civil Rights and African Americans: A Documentary History*, ed. Albert P. Blaustein and Robert I. Zangrando (Evanston, 1991), 436.

COMPREHENSION

1. According to Carson, what are the benefits and shortcomings of the *Brown v. Board of Education* decision? Why does he give two cheers (instead of the traditional three cheers) for the Supreme Court's 1954 verdict?
2. What are the psychological effects of both segregation and desegregation on African American students?
3. Why does Carson say that the impact of the *Brown* decision is difficult to measure? What evidence does he provide to support this assessment?

RHETORIC

1. Who is Carson's audience for this essay? How does he fit his style to the expectations he holds for this specific audience? Provide examples of vocabulary, syntax, and abstract language to support your response. Why would the article also be of interest to a more general audience?

2. Carson lays out a well-informed argument. What is his claim or major proposition? What are his warrants? What is his support? How does he deal with opposing viewpoints? What conclusions does he draw to convince the reader of his position?
3. Analyze the pattern of cause and effect that Carson presents in this essay.
4. Carson has very strong topic sentences at the start of virtually every paragraph. List these topic sentences, and then show how they control the flow of his thoughts within paragraphs while at the same time advancing his argument.
5. Examine the writer's footnotes. What are his sources? What range and variety of evidence do these notes suggest?

WRITING

1. Using Carson's article as a reference point, write an essay describing the ethnic and racial composition of your former high school or the college you now attend. How does this demographic profile support some of Carson's key insights into *Brown v. Board of Education?*
2. Write an essay in which you offer your own analysis of the ways in which varieties of discrimination you encounter in education can have psychological consequences. Feel free to offer personal experience to support your analysis.
3. **WRITING AN ARGUMENT:** Unlike Carson, who argues both sides of the *Brown* decision in terms of the historical aftermath, write an argumentative essay in which you defend or criticize the results of *Brown v. Board* since 1954. Conduct research and collaborate with class members if you wish.

| www.mhhe.com/ **mhreader** | or more information on Clayborne Carson, go to: **More Resources > Ch. 4 Education** |

CONNECTIONS FOR CRITICAL THINKING

1. Compare and contrast the rhetoric of the personal essay as it is represented in Rodriguez's "The Lonely, Good Company of Books" with the rhetoric of such expository and argumentative essays as Tyson's "Needed: Affirmative Action for the Poor" or Carson's "Two Cheers for *Brown v. Board of Education.*"

2. Analyze an event in your education when you had a disagreement with a teacher or administrator. Explain and explore whether the differences in viewpoint were based on emotional perspective, intellectual perspective, or both.

3. Select the essay in this chapter you find most pertinent to your life as a student. Explain why you have selected the essay, and explore your intellectual and emotional responses to it.

4. Does your college seem to support Jacoby's views regarding the educational lives of women? Explain why or why not.

5. Argue for or against the proposition that the primary function of colleges and universities is to prepare students for the world of work. Use at least two essays from this chapter to support your view.

6. Argue for or against the view that the publicized sexual activities of politicians and other celebrities makes the decision whether or not to keep sex education out of the schools entirely moot.

7. It is 2050. Write an essay in which you explore the physical, technological, and communicative dimensions of a typical college classroom. Your essay should not only describe this environment but also explain why the classroom has the characteristics you have chosen. Refer to the ideas contained in Levine's essay.

8. Open up a computer chat room and discuss the psychological and emotional implications of having children witness the war in Iraq.

9. Write an essay that classifies at least three educational issues that the authors in this chapter examine. Establish a clear thesis to unify the categories that you establish.

10. Analyze the patterns and techniques used by Tyson, Gelernter, and Jacoby to advance their claims about education today.

11. Search the Web for *sex education* and *France* (or another country of your choice). Write an essay describing the policies of your chosen country toward the topic.

12. Look up your college's or university's mission statement on its home Web page. Is the statement relevant? Is it truthful? Explain why or why not. Use personal observation and experience to support your view.

13. Argue for or against the proposition that despite Gelernter's warnings about the purported shortcomings of computers in the classroom, in the future many students will prefer to obtain a bachelor of arts degree completely via computer and the Internet.

chapter **5**

Family Roles and Gender Roles
How Do We Become Who We Are?

Every culture has its own ideas about identity, how it is formed, and where it comes from. What is the influence of family on the creation of identity, of environment, of gender and, as we have seen in Chapter 4, of education? Although it is challenging to reconcile these various cross-cultural ideas, the writers in this chapter attempt to make sense of identity from the perspectives of family and gender, and they invite readers to liberate themselves from the tyranny of stereotyping.

Families nourish us during childhood, and the values our families seek to maintain usually affect our identities in powerful ways, whether we adopt them wholly, modify them, or reject them outright. Writers have always been aware of the importance of the family in human development and behavior, and have written about it from various perspectives, using narration, sociological and psychological analysis, and cultural criticism, among other approaches. Tolstoy wrote that "Happy families are all alike; every unhappy family is unhappy in its own way." But we shall discover that Tolstoy had a limited view of family life and its values—probably circumscribed by the mores of the time he lived in. Some of our finest essayists and observers of social life today demonstrate in this chapter that what constitutes the definition of a family is up for grabs as we begin the new millennium.

The family is one of the few institutions that we find in every society throughout the world, at least every thriving society. Anthropologists, sociologists, and psychologists tell us that family patterns are exceedingly diverse even in the same societies. In the past and even more so today, children grow up in many ways: in nuclear and in nontraditional households; in single-parent and in dual-parent arrangements; in extended families and in the new blended family; and in patriarchal and matriarchal, heterosexual and homosexual, monogamous and polygamous situations. And the dynamics of family life assume added dimension as we move across cultures, studying European families, African American families, Hispanic families, Asian families, and so forth. Even within these groups, we find variables that affect family life and values, such as economic class, social class, and educational levels.

Unlike in previous periods in our history, Americans today seem to be groping for a definition of what constitutes the happy family. With the influences of the media and of peer pressure on children, the rise in the number of latchkey children, and the fact that there is a growing diversity of cultures in America owing to the new wave of immigration, the family appears to be less of a traditional haven than it was even a generation ago. This chapter contains vivid accounts of the long-standing bonds within the family that have been treasured for their capacity to build values of love and sharing. It also contains essays that demonstrate how family life is filled with emotional complexities and conflicts that the child must negotiate as she or he finds meaning and attempts to construct an identity. Each writer, whether writing narration, exposition, or argumentation, shows how significant the family is for the development of our values, personalities, and lifestyles.

As much as our identities are shaped by powerful institutional forces like the family, what we are might be even more powerfully determined by the forces of sexuality and gender. Freud asserted that human behavior is rooted in sexuality, that gender (rather than family or school or any social institution) is destiny. It is clear that notions of what it means to be a man or woman have an impact on the construction of our identities.

The identity issues discussed in this chapter might prove to be controversial, but they will encourage you to confront your own sense of identity. These essays are like a mirror in which you can see and evaluate what you really are.

Previewing the Chapter

As you read the essays in this chapter and respond to them in discussion and writing, consider the following questions:

- What form of rhetoric is the author using: narration, exposition, argumentation? Why is this form appropriate for the author's purpose?
- What perspective does the writer take on the subject of identity formation? Is the writer optimistic, pessimistic, or something else?
- What are the cultural, social, and economic issues addressed in the essay?
- How do you regard the authority of the author? Does she or he seem to be speaking from experience and knowledge? In essays that explain or argue, does the evidence appear substantial or questionable? Explain.
- What stylistic devices does the author employ to recreate a memory, explain a function, or argue a stance regarding an issue of identity?
- Which essays appear alike in purpose and method, and why?
- What have you learned or discovered about your own identity from reading these essays?
- Do you prefer one rhetorical form over another, for example, personal narration over argumentation? If so, why?

Classic and Contemporary Images

HOW DO WE RESPOND TO MARRIAGE?

Using a Critical Perspective   What was your first impression of Brueghel's *Rustic Wedding* and Elise Amendola's *Gay Marriage?* What details do you see? What senses do the artist and the photographer draw on to convey the atmosphere of the wedding? What does each want to say about the institution of marriage? How do you know?

The Flemish artist Pieter Brueghel the Elder (1525–1569), was one of the greatest painters of the 16th century and was renowned for his exuberant depictions of peasant life. His son Pieter Brueghel the Younger (1564–1638) copied many of his father's works and also painted religious subjects. He was responsible for *Rustic Wedding*, shown here.

Hillary, left, and Julie Goodridge, lead plaintiffs in the landmark Massachusetts gay marriage lawsuit, receive their wedding rings from their daughter, Annie, 8, as Unitarian Reverend William Sinkford presides over their marriage ceremony in Boston during the first day of state-sanctioned gay marriage in the United States on May 17, 2004.

Classic and Contemporary Essays
HOW MUCH DO FAMILIES MATTER?

E. B. White and Barbara Kingsolver represent two generations, each raised with different values regarding the function, structure, and role of the family. Both authors are master stylists, but each reflects a style of writing, an intellectual universe, and an external world that views the healthy family differently. White writes in clear, concise, elegiac prose. It marches on in a quiet, evenly patterned rhythm. Perhaps it is a metaphor of his view of life in general and family life in particular. Tradition is to be treasured; continuity is to be celebrated. He attends to the details of a nature outing and suggests that the sights, sounds, and smells that imbue the events he and his son experience are the same as those he experienced years before with his own father. For White, it seems, pleasure is derived from connectivity and permanence.

Kingsolver is passionate about her own perspective on what constitutes a healthy family structure, but it is a family transformed, reconfigured, and rearranged by contemporary events and values. Kingsolver's notion of family is various while White's view is archetypal. Kingsolver seems to believe that change in families creates security, particularly if one is moving from a dysfunctional environment to a more coherent one. Is White conservative in his views? Is Kingsolver a liberal? Perhaps a better way to get a sense of their differences is to inquire whether our amorphous contemporary world requires us to be more flexible and critical. And of course, one must consider that Kingsolver adds a woman's voice to the conversation about family, a voice that was not as frequently heard by the members of White's generation.

✳

Once More to the Lake

E. B. White

Elwyn Brooks White (1899–1985), perhaps the finest American essayist of the twentieth century, was at his most distinctive in his treatments of people and nature. A recipient of the National Medal for Literature, and associated for years with The New Yorker, *White is the author of* One Man's Meat *(1942),* Here Is New York *(1949), and* The Second Tree from the Corner *(1954), among numerous other works. He was also one of the most talented writers of literature for children, the author of* Stuart Little *(1945),* Charlotte's Web *(1952), and* The Trumpet of the Swan *(1970). In this essay, White combines narration and description to make a poignant and vivid statement about past and present, youth and age, life and death.*

One summer, along about 1904, my father rented a camp on a lake in Maine and took us all there for the month of August. We all got ringworm from some kittens and had to rub Pond's Extract on our arms and legs night and morning, and my father rolled over in a canoe with all his clothes on; but outside of that the vacation was a success and from then on none of us ever thought there was any place in the world like that lake in Maine. We returned summer after summer—always on August 1st for one month. I have since become a saltwater man, but sometimes in summer there are days when the restlessness of the tides and the fearful cold of the sea water and the incessant wind which blows across the afternoon and into the evening make me wish for the placidity of a lake in the woods. A few weeks ago this feeling got so strong I bought myself a couple of bass hooks and a spinner and returned to the lake where we used to go, for a week's fishing and to revisit old haunts. 1

I took along my son, who had never had any fresh water up his nose and who had seen lily pads only from train windows. On the journey over to the lake I began to wonder what it would be like. I wondered how time would have marred this unique, this holy spot—the coves and streams, the hills that the sun set behind, the camps and the paths behind the camps. I was sure the tarred road would have found it out and I wondered in what other ways it would be desolated. It is strange how much you can remember about places like that once you allow your mind to return into the grooves which lead back. You remember one thing, and that suddenly reminds you of another thing. I guess I remembered clearest of all the early mornings, when the lake was cool and motionless, remembered how the bedroom smelled of the lumber it was made of and of the wet woods whose scent entered through the screen. The partitions in the camp were thin and did not extend clear to the top of the rooms, and as I was always the first up I would dress softly so as not to wake the others, and sneak out into the sweet outdoors and start out in the canoe, keeping close along the shore in 2 the long shadows of the pines. I remembered being very careful never to rub my paddle against the gunwale for fear of disturbing the stillness of the cathedral.

The lake had never been what you would call a wild lake. There were cot- ₃ tages sprinkled around the shores, and it was in farming country although the shores of the lake were quite heavily wooded. Some of the cottages were owned by nearby farmers, and you would live at the shore and eat your meals at the farmhouse. That's what our family did. But although it wasn't wild, it was a fairly large and undisturbed lake and there were places in it which, to a child at least, seemed infinitely remote and primeval.

I was right about the tar: it led to within half a mile of the shore. But when ₄ I got back there, with my boy, and we settled into a camp near a farmhouse and into the kind of summertime I had known, I could tell that it was going to be pretty much the same as it had been before—I knew it, lying in bed the first morning, smelling the bedroom, and hearing the boy sneak quietly out and go off along the shore in a boat. I began to sustain the illusion that he was I, and therefore, by simple transposition, that I was my father. This sensation per- sisted, kept cropping up all the time we were there. It was not an entirely new feeling, but in this setting it grew much stronger. I seemed to be living a dual ex- istence. I would be in the middle of some simple act, I would be picking up a bait box or laying down a table fork, or I would be saying something, and sud- denly it would be not I but my father who was saying the words or making the gesture. It gave me a creepy sensation.

We went fishing the first morning. I felt the same damp moss covering the ₅ worms in the bait can, and saw the dragonfly alight on the tip of my rod as it hovered a few inches from the surface of the water. It was the arrival of this fly that convinced me beyond any doubt that everything was as it always had been, that the years were a mirage and there had been no years. The small waves were the same, chucking the rowboat under the chin as we fished at an- chor, and the boat was the same boat, the same color green and the ribs broken in the same place, and under the floor-boards the same fresh-water leavings and débris—the dead hellgrammite, the wisps of moss, the rusty discarded fish- hook, the dried blood from yesterday's catch. We stared silently at the tips of our rods, at the dragonflies that came and went. I lowered the tip of mine into the water, tentatively, pensively dislodging the fly, which darted two feet away, poised, darted two feet back, and came to rest again a little farther up the rod. There had been no years between the ducking of this dragonfly and the other one—the one that was part of memory. I looked at the boy, who was silently watching his fly, and it was my hands that held his rod, my eyes watching. I felt dizzy and didn't know which rod I was at the end of.

We caught two bass, hauling them in briskly as though they were mackerel, ₆ pulling them over the side of the boat in a businesslike manner without any land- ing net, and stunning them with a blow on the back of the head. When we got back for a swim before lunch, the lake was exactly where we had left it, the same number of inches from the dock, and there was only the merest suggestion of a breeze. This seemed an utterly enchanted sea, this lake you could leave to its own devices for a few hours and come back to, and find that it had not stirred, this constant and trustworthy body of water. In the shallows, the dark, water-soaked sticks and twigs, smooth and old, were undulating in clusters on the bottom against the clean ribbed sand, and the track of the mussel was plain. A school of minnows swam by, each minnow with its small individual shadow, doubling the attendance, so clear and sharp in the sunlight. Some of the other campers were in swimming, along the shore, one of them with a cake of soap, and the water felt thin and clear and unsubstantial. Over the years there had been this person with the cake of soap, this cultist, and here he was. There had been no years.

Up to the farmhouse to dinner through the teeming, dusty field, the road 7 under our sneakers was only a two-track road. The middle track was missing, the one with the marks of the hooves and the splotches of dried, flaky manure. There had always been three tracks to choose from in choosing which track to walk in; now the choice was narrowed down to two. For a moment I missed terribly the middle alternative. But the way led past the tennis court, and something about the way it lay there in the sun reassured me; the tape had loosened along the backline, the alleys were green with plantains and other weeds, and the net (installed in June and removed in September) sagged in the dry noon, and the whole place steamed with midday heat and hunger and emptiness. There was a choice of pie for dessert, and one was blueberry and one was apple, and the waitresses were the same country girls, there having been no passage of time, only the illusion of it as in a dropped curtain—the waitresses were still fifteen; their hair had been washed, that was the only difference—they had been to the movies and seen the pretty girls with the clean hair.

Summertime, oh summertime, pattern of life indelible, the fade-proof lake, 8 the woods unshatterable, the pasture with the sweetfern and the juniper forever and ever, summer without end; this was the background, and the life along the shore was the design, the cottagers with their innocent and tranquil design, their tiny docks with the flagpole and the American flag floating against the white clouds in the blue sky, the little paths over the roots of the trees leading from camp to camp and the paths leading back to the outhouses and the can of lime for sprinkling, and at the souvenir counters at the store the miniature birch-bark canoes and the post cards that showed things looking a little better than they looked. This was the American family at play, escaping the city heat, wondering whether the newcomers in the camp at the head of the cove were "common" or "nice," wondering whether it was true that the people who drove up for Sunday dinner at the farmhouse were turned away because there wasn't enough chicken.

It seemed to me, as I kept remembering all this, that those times and those 9 summers had been infinitely precious and worth saving. There had been jollity and peace and goodness. The arriving (at the beginning of August) had been so big a business in itself, at the railway station the farm wagon drawn up, the first smell of the pine-laden air, the first glimpse of the smiling farmer, and the great importance of the trunks and your father's enormous authority in such matters, and the feel of the wagon under you for the long ten-mile haul, and at the top of the last long hill catching the first view of the lake after eleven months of not seeing this cherished body of water. The shouts and cries of the other campers when they saw you, and the trunks to be unpacked, to give up their rich burden. (Arriving was less exciting nowadays, when you sneaked up in your car and parked it under a tree near the camp and took out the bags and in five minutes it was all over, no fuss, no loud wonderful fuss about trunks.)

Peace and goodness and jollity. The only thing that was wrong now, really, was the sound of the place, an unfamiliar nervous sound of the outboard motors. This was the note that jarred, the one thing that would sometimes break the illusion and set the years moving. In those other summertimes all motors were inboard; and when they were at a little distance, the noise they made was a sedative, an ingredient of summer sleep. They were one-cylinder and two-cylinder engines, and some were make-and-break and some were jump-spark, but they all made a sleepy sound across the lake. The one-lungers throbbed and fluttered, and the twin-cylinder ones purred and purred, and that was a quiet sound too. But now the campers all had outboards. In the daytime, in the hot mornings, these motors made a petulant, irritable sound; at night, in the still evening when the afterglow lit the water, they whined about one's ears like mosquitoes. My boy loved our rented outboard, and his great desire was to achieve singlehanded mastery over it, and authority, and he soon learned the trick of choking it a little (but not too much), and the adjustment of the needle valve. Watching him I would remember the things you could do with the old one-cylinder engine with the heavy flywheel, how you could have it eating out of your hand if you got really close to it spiritually. Motor boats in those days didn't have clutches, and you would make a landing by shutting off the motor at the proper time and coasting in with a dead rudder. But there was a way of reversing them, if you learned the trick, by cutting the switch and putting it on again exactly on the final dying revolution of the flywheel, so that it would kick back against compression and begin reversing. Approaching a dock in a strong following breeze, it was difficult to slow up sufficiently by the ordinary coasting method, and if a boy felt he had complete mastery over his motor, he was tempted to keep it running beyond its time and then reverse it a few feet from the dock. It took a cool nerve, because if you threw the switch a twentieth of a second too soon you would catch the flywheel when it still had speed enough to go up past center, and the boat would leap ahead, charging bull-fashion at the dock.

We had a good week at the camp. The bass were biting well and the sun shone endlessly, day after day. We would be tired at night and lie down in the accumulated heat of the little bedrooms after the long hot day and the breeze would stir almost imperceptibly outside and the smell of the swamp drift in through the rusty screens. Sleep would come easily and in the morning the red squirrel would be on the roof, tapping out his gay routine. I kept remembering everything, lying in bed in the mornings—the small steamboat that had a long rounded stern like the lip of a Ubangi, and how quietly she ran on the moonlight sails, when the older boys played their mandolins and the girls sang and we ate doughnuts dipped in sugar, and how sweet the music was on the water in the shining night, and what it had felt like to think about girls then. After breakfast we would go up to the store and the things were in the same place—minnows in a bottle, the plugs and spinners disarranged and pawed over by the youngsters from the boys' camp, the fig newtons and the Beeman's gum. Outside, the road was tarred and cars stood in front of the store. Inside, all was just as it had always been, except there was more Coca-Cola and not so much Moxie and root beer and birch beer and sarsaparilla. We would walk out with a bottle of pop apiece and sometimes the pop would backfire up our noses and hurt. We explored the streams, quietly, where the turtles slid off the sunny logs and dug their way into the soft bottom, and we lay on the town wharf and fed worms to the tame bass. Everywhere we went I had trouble making out which was I, the one walking at my side, the one walking in my pants.

One afternoon while we were there at that lake a thunderstorm came up. It 12 was like the revival of an old melodrama that I had seen long ago with childish awe. The second-act climax of the drama of the electrical disturbance over a lake in America had not changed in any important respect. This was the big scene, still the big scene. The whole thing was so familiar, the first feeling of oppression and heat and a general air around camp of not wanting to go very far away. In midafternoon (it was all the same) a curious darkening of the sky, and a lull in everything that had made life tick; and then the way the boats suddenly swung the other way at their moorings with the coming of a breeze out of the new quarter, and the premonitory rumble. Then the kettle drum, then the snare, then the bass drum and cymbals, then crackling light against the dark, and the gods grinning and licking their chops in the hills. Afterward the calm, the rain steadily rustling in the calm lake, the return of light and hope and spirits, and the campers running out in joy and relief to go swimming in the rain, their bright cries perpetuating the deathless joke about how they were getting simply drenched, and the children screaming with delight at the new sensation of bathing in the rain, and the joke about getting drenched linking the generations in a strong indestructible chain. And the comedian who waded in carrying an umbrella.

When the others went swimming my son said he was going in too. He 13 pulled his dripping trunks from the line where they had hung all through the shower, and wrung them out. Languidly, and with no thought of going in, I watched him, his hard little body, skinny and bare, saw him wince slightly as he pulled up around his vitals the small, soggy, icy garment. As he buckled the swollen belt suddenly my groin felt the chill of death.

COMPREHENSION

1. At what point in the essay do you begin to sense White's main purpose? What is his purpose? What type of reader might his purpose appeal to?
2. What motivates White to return to the lake in Maine? Explain the "simple transposition" that he mentions in paragraph 4. List the illustrations that he gives of this phenomenon. What change does he detect in the lake?

3. Explain the significance of White's last sentence. Where are there foreshadowings of this statement?

RHETORIC

1. Describe the author's use of figurative language in paragraphs 2, 10, and 12.
2. Identify those words and phrases that White invokes to establish the sense of mystery about the lake. Why are these words and their connotations important to the nature of the illusion that he describes?
3. Explain the organization of the essay in terms of the following paragraph units: 1 to 4; 5 to 7; 8 to 10; and 11 to 13. Explain the function of paragraphs 8 and 12.
4. There are many vivid and unusual descriptive details in this essay—for example, the dragonfly in paragraph 5 and the two-track road in paragraph 7. How does White create symbolic overtones for these descriptive details and others? Why is the lake itself a complex symbol? Explain with reference to paragraph 6.
5. Describe the persona that White creates for himself in the essay. How does this persona function?
6. What is the relation between the introductory and concluding paragraphs, specifically in terms of irony of statement?

WRITING

1. Explore in an essay the theme of nostalgia in "Once More to the Lake." What are the beauties and the dangers of nostalgia? Can the past ever be recaptured or relived? Justify your answer.
2. White consistently compares an outing with his father to one with his son. How does this structure help to emphasize the continuity of generations? Explain in an analytical and comparative essay.
3. Referring to revisiting a site on the lake that he had visited years before with his father, White remarks in paragraph 4, "I could tell that it was going to be pretty much the same as it had been before." How does this observation reflect the general sentiment White has about the role and function of the family? Respond to the question in an analytical essay.
4. **WRITING AN ARGUMENT:** Argue for or against the proposition that nostalgia can obscure the true nature of family relationships and even suppress painful memories that should be confronted.

www.mhhe.com/
mhreader

For more information on E. B. White, go to:
More Resources > Ch. 5 Family & Gender

✳

Stone Soup

Barbara Kingsolver

Barbara Kingsolver (b. 1955) was born in Annapolis, Maryland, grew up in rural Kentucky, and was educated at DePauw University and the University of Arizona. Her fiction includes The Bean Trees *(1988);* Homeland *(1990);* Animal Dreams *(1991), for which she won a PEN fiction prize and an Edward Abbey Ecofiction Award;* Pigs in Heaven *(1993), which won a Los Angeles Times Book Award for Fiction; and* The Poisonwood Bible *(1998). She has also worked as a biologist, is active in the field of human rights, and plays keyboard with an amateur rock and roll band. The following essay, first published in the January 1995 issue of* Parenting, *eschews the idea of the nuclear family as being the standard by which the healthy family should be judged.*

In the catalog of family values, where do we rank an occasion like this? A curly-haired boy who wanted to run before he walked, age seven now, a soccer player scoring a winning goal. He turns to the bleachers with his fists in the air and a smile wide as a gap-toothed galaxy. His own cheering section of grown-ups and kids all leap to their feet and hug each other, delirious with love for this boy. He's Andy, my best friend's son. The cheering section includes his mother and her friends, his brother, his father and stepmother, a stepbrother and stepsister, and a grandparent. Lucky is the child with this many relatives on hand to hail a proud accomplishment. I'm there too, witnessing a family fortune. But in spite of myself, defensive words take shape in my head. I am thinking: I dare *anybody* to call this a broken home.

Families change, and remain the same. Why are our names for home so slow to catch up to the truth of where we live?

When I was a child, I had two parents who loved me without cease. One of them attended every excuse for attention I ever contrived, and the other made it to the ones with higher production values, like piano recitals and appendicitis. So I was a lucky child too. I played with a set of paper dolls called "The Family of Dolls," four in number, who came with the factory-assigned names of Dad, Mom, Sis, and Junior. I think you know what they looked like, at least before I loved them to death and their heads fell off.

Now I've replaced the dolls with a life. I knit my days around my daughter's survival and happiness, and am proud to say her head is still on. But we aren't the Family of Dolls. Maybe you're not, either. And if not, even though you are statistically no oddity, it's probably been suggested to you in a hundred ways that yours isn't exactly a real family, but an impostor family, a harbinger of cultural ruin, a slapdash substitute—something like counterfeit money. Here at the tail end of our century, most of us are up to our ears in the noisy business of trying to support and love a thing called family. But there's a current in the air with ferocious moral force that finds its way even into political campaigns, claiming there is only one right way to do it, the Way It Has Always Been.

In the face of a thriving, particolored world, this narrow view is so pickled   5 and absurd I'm astonished that it gets airplay. And I'm astonished that it still stings.

Every parent has endured the arrogance of a child-unfriendly grump sitting   6 in judgment, explaining what those kids of ours really need (for example, "a good licking"). If we're polite, we move our crew to another bench in the park. If we're forthright (as I am in my mind, only, for the rest of the day), we fix them with a sweet imperious stare and say, "Come back and let's talk about it after you've changed a thousand diapers."

But it's harder somehow to shrug off the Family-of-Dolls Family Values   7 crew when they judge (from their safe distance) that divorced people, blended families, gay families, and single parents are failures. That our children are at risk, and the whole arrangement is messy and embarrassing. A marriage that ends is not called "finished," it's called *failed*. The children of this family may have been born to a happy union, but now they are called *the children of divorce*.

I had no idea how thoroughly these assumptions overlaid my culture until   8 I went through divorce myself. I wrote to a friend: "This might be worse than being widowed. Overnight I've suffered the same losses—companionship, financial and practical support, my identity as a wife and partner, the future I'd taken for granted. I am lonely, grieving, and hard-pressed to take care of my household alone. But instead of bringing casseroles, people are acting like I had a fit and broke up the family china."

Once upon a time I held these beliefs about divorce: that everyone who   9 does it could have chosen not to do it. That it's a lazy way out of marital problems. That it selfishly puts personal happiness ahead of family integrity. Now I tremble for my ignorance. It's easy, in fortunate times, to forget about the ambush that could leave your head reeling: serious mental or physical illness, death in the family, abandonment, financial calamity, humiliation, violence, despair.

I started out like any child, intent on being the Family of Dolls. I set upon   10 young womanhood believing in most of the doctrines of my generation: I wore my skirts four inches above the knee. I had that Barbie with her zebra-striped swimsuit and a figure unlike anything found in nature. And I understood the Prince Charming Theory of Marriage, a quest for Mr. Right that ends smack dab where you find him. I did not completely understand that another whole story *begins* there, and no fairy tale prepared me for the combination of bad luck and persistent hope that would interrupt my dream and lead me to other arrangements. Like a cancer diagnosis, a dying marriage is a thing to fight, to deny, and finally, when there's no choice left, to dig in and survive. Casseroles would help. Likewise, I imagine it must be a painful reckoning in adolescence (or later on) to realize one's own true love will never look like the soft-focus fragrance ads because Prince Charming (surprise!) is a princess. Or vice versa. Or has skin the color your parents didn't want you messing with, except in the Crayola box.

It's awfully easy to hold in contempt the straw broken home, and that 11
mythical category of persons who toss away nuclear family for the sheer fun of
it. Even the legal terms we use have a suggestion of caprice. I resent the phrase
"irreconcilable differences," which suggests a stubborn refusal to accept a
spouse's little quirks. This is specious. Every happily married couple I know
has loads of irreconcilable differences. Negotiating where to set the thermostat
is not the point. A nonfunctioning marriage is a slow asphyxiation. It is waking
up despised each morning, listening to the pulse of your own loneliness before
the radio begins to blare its raucous gospel that you're nothing if you aren't
loved. It is sharing your airless house with the threat of suicide or other kinds of
violence, while the ghost that whispers, "Leave here and destroy your children," has passed over every door and nailed it shut. Disassembling a marriage
in these circumstances is as much *fun* as amputating your own gangrenous leg.
You do it, if you can, to save a life—or two, or more.

I know of no one who really went looking to hoe the harder row, especially 12
the daunting one of single parenthood. Yet it seems to be the most American of
customs to blame the burdened for their destiny. We'd like so desperately to believe in freedom and justice for all, we can hardly name that rogue bad luck,
even when he's a close enough snake to bite us. In the wake of my divorce,
some friends (even a few close ones) chose to vanish, rather than linger within
striking distance of misfortune.

But most stuck around, bless their hearts, and if I'm any the wiser for my 13
trials, it's from having learned the worth of steadfast friendship. And also, what
not to say. The least helpful question is: "Did you want the divorce, or didn't
you?" Did I want to keep that gangrenous leg, or not? How to explain, in a culture that venerates choice: two terrifying options are much worse than none at
all. Give me any day the quick hand of cruel fate that will leave me scarred but
blameless. As it was, I kept thinking of that wicked third-grade joke in which
some boy comes up behind you and grabs your ear, starts in with a prolonged
tug, and asks, "Do you want this ear any longer?"

Still, the friend who holds your hand and says the wrong thing is made of 14
dearer stuff than the one who stays away. And generally, through all of it, you
live. My favorite fictional character, Kate Vaiden (in the novel by Reynolds
Price), advises: "Strength just comes in one brand—you stand up at sunrise and
meet what they send you and keep your hair combed."

Once you've weathered the straits, you get to cross the tricky juncture from 15
casualty to survivor. If you're on your feet at the end of a year or two, and have
begun putting together a happy new existence, those friends who were kind
enough to feel sorry for you when you needed it must now accept you back to
the ranks of the living. If you're truly blessed, they will dance at your second
wedding. Everybody else, for heavens sake, should stop throwing stones.

Arguing about whether nontraditional families deserve pity or tolerance is a lit- 16
tle like the medieval debate about left-handedness as a mark of the devil. Divorce, remarriage, single parenthood, gay parents, and blended families simply

are. They're facts of our time. Some of the reasons listed by sociologists for these family reconstructions are: the idea of marriage as a romantic partnership rather than a pragmatic one; a shift in women's expectations, from servility to self-respect and independence; and longevity (prior to antibiotics no marriage was expected to last many decades—in Colonial days the average couple lived to be married less than twelve years). Add to all this, our growing sense of entitlement to happiness and safety from abuse. Most would agree these are all good things. Yet their result—a culture in which serial monogamy and the consequent reshaping of families are the norm—gets diagnosed as "failing."

For many of us, once we have put ourselves Humpty-Dumpty–wise back 17 together again, the main problem with our reorganized family is that other people think we have a problem. My daughter tells me the only time she's uncomfortable about being the child of divorced parents is when her friends say they feel sorry for her. It's a bizarre sympathy, given that half the kids in her school and nation are in the same boat, pursuing childish happiness with the same energy as their married-parent peers. When anyone asks how she feels about it, she spontaneously lists the benefits: our house is in the country and we have a dog, but she can go to her dad's neighborhood for the urban thrills of a pool and sidewalks for roller-skating. What's more, she has three sets of grandparents!

Why is it surprising that a child would revel in a widened family and the 18 right to feel at home in more than one house? Isn't it the opposite that should worry us—a child with no home at all, or too few resources to feel safe? The child at risk is the one whose parents are too immature themselves to guide wisely; too diminished by poverty to nurture; too far from opportunity to offer hope. The number of children in the U.S. living in poverty at this moment is almost unfathomably large: twenty percent. There are families among us that need help all right, and by no means are they new on the landscape. The rate at which teenage girls had babies in 1957 (ninety-six per thousand) was twice what it is now. That remarkable statistic is ignored by the religious right—probably because the teen birth rate was cut in half mainly by legalized abortion. In fact, the policy gatekeepers who coined the phrase "family values" have steadfastly ignored the desperation of too-small families, and since 1979 have steadily reduced the amount of financial support available to a single parent. But, this camp's most outspoken attacks seem aimed at the notion of families getting too complex, with add-ons and extras such as a gay parent's partner, or a remarried mother's new husband and his children.

To judge a family's value by its tidy symmetry is to purchase a book for its 19 cover. There's no moral authority there. The famous family comprised by Dad, Mom, Sis, and Junior living as an isolated economic unit is not built on historical bedrock. In *The Way We Never Were*, Stephanie Coontz writes, "Whenever people propose that we go back to the traditional family, I always suggest that they pick a ballpark date for the family they have in mind." Colonial families were tidily disciplined, but their members (meaning everyone but infants) labored incessantly and died young. Then the Victorian family adopted a new division of labor, in which women's role was domestic and children were allowed time for study and play, but this was an upper-class construct supported by myriad slaves. Coontz writes, "For every nineteenth-century middle-class family that protected its wife and child within the family circle, there was an Irish or German girl scrubbing floors . . . a Welsh boy mining coal to keep the home-baked goodies warm, a black girl doing the family laundry, a black mother and child picking cotton to be made into clothes for the family, and a Jewish or an Italian daughter in a sweatshop making 'ladies' dresses or artificial flowers for the family to purchase."

The abolition of slavery brought slightly more democratic arrangements, in which extended families were harnessed together in cottage industries; at the turn of the century came a steep rise in child labor in mines and sweat-shops. Twenty percent of American children lived in orphanages at the time; their parents were not necessarily dead, but couldn't afford to keep them.

During the Depression and up to the end of World War II, many millions of U.S. households were more multigenerational than nuclear. Women my grandmother's age were likely to live with a fluid assortment of elderly relatives, in-laws, siblings, and children. In many cases they spent virtually every waking hour working in the company of other women—a companionable scenario in which it would be easier, I imagine, to tolerate an estranged or difficult spouse. I'm reluctant to idealize a life of so much hard work and so little spousal intimacy, but its advantage may have been resilience. A family so large and varied would not easily be brought down by a single blow: it could absorb a death, long-illness, an abandonment here or there, and any number of irreconcilable differences.

The Family of Dolls came along midcentury as a great American experiment. A booming economy required a mobile labor force and demanded that women surrender jobs to returning soldiers. Families came to be defined by a single breadwinner. They struck out for single-family homes at an earlier age than ever before, and in unprecedented numbers they raised children in suburban isolation. The nuclear family was launched to sink or swim.

More than a few sank. Social historians corroborate that the suburban family of the postwar economic boom, which we have recently selected as our definition of "traditional," was no panacea. Twenty-five percent of Americans were poor in the mid-1950s, and as yet there were no food stamps. Sixty percent of the elderly lived on less than $1,000 a year, and most had no medical insurance. In the sequestered suburbs, alcoholism and sexual abuse of children were far more widespread than anyone imagined.

Expectations soared, and the economy sagged. It's hard to depend on one other adult for everything, come what may. In the last three decades, that amorphous, adaptable structure we call "family" has been reshaped once more by economic tides. Compared with fifties families, mothers are far more likely now to be employed. We are statistically more likely to divorce, and to live in blended families or other extranuclear arrangements. We are also more likely to plan and space our children, and to rate our marriages as "happy." We are less likely to suffer abuse without recourse, or to stare out at our lives through a

glaze of prescription tranquilizers. Our aged parents are less likely to be desti-
tute, and we're half as likely to have a teenage daughter turn up a mother herself. All in all, I would say that if "intact" in modern family-values jargon means
living quietly desperate in the bell jar, then hip-hip-hooray for "broken." A neat
family model constructed to service the Baby Boom economy seems to be returning gradually to a grand, lumpy shape that human families apparently
have tended toward since they first took root in the Olduvai Gorge. We're social
animals, deeply fond of companionship, and children love best to run in packs.
If there is a *normal* for humans, at all, I expect it looks like two or three Families
of Dolls, connected variously by kinship and passion, shuffled like cards and
strewn over several shoeboxes.

The sooner we can let go the fairy tale of families functioning perfectly in   25
isolation, the better we might embrace the relief of community. Even the admirable parents who've stayed married through thick and thin are very likely,
at present, to incorporate other adults into their families—household help and
baby-sitters if they can afford them or neighbors and grandparents if they can't.
For single parents, this support is the rock-bottom definition of family. And
most parents who have split apart, however painfully, still manage to maintain
family continuity for their children, creating in many cases a boisterous phenomenon that Constance Ahrons in her book *The Good Divorce* calls the "binu-
clear family." Call it what you will—when ex-spouses beat swords into
plowshares and jump up and down at a soccer game together, it makes for
happy kids.

Cinderella, look, who needs her? All those evil stepsisters? That story always   26
seemed like too much cotton-picking fuss over clothes. A childhood tale that
fascinated me more was the one called "Stone Soup," and the gist of it is this:
Once upon a time, a pair of beleaguered soldiers straggled home to a village
empty-handed, in a land ruined by war. They were famished, but the villagers
had so little they shouted evil words and slammed their doors. So the soldiers
dragged out a big kettle, filled it with water, and put it on a fire to boil. They
rolled a clean round stone into the pot, while the villagers peered through their
curtains in amazement.

"What kind of soup is that?" they hooted.   27

"Stone soup," the soldiers replied. "Everybody can have some when it's   28
done."

"Well, thanks," one matron grumbled, coming out with a shriveled carrot.   29
"But it'd be better if you threw this in."

And so on, of course, a vegetable at a time, until the whole suspicious vil-   30
lage managed to feed itself grandly.

Any family is a big empty pot, save for what gets thrown in. Each stew   31
turns out different. Generosity, a resolve to turn bad luck into good, and respect
for variety—these things will nourish a nation of children. Name-calling and
suspicion will not. My soup contains a rock or two of hard times, and maybe
yours does too. I expect it's a heck of a bouillabaisse.

COMPREHENSION

1. What is the essay's thesis?
2. According to Kingsolver, why is our society so apt to condemn divorce?
3. What is the author's view of family symmetry (paragraph 19)?

RHETORIC

1. What rhetorical function does the opening anecdote serve in introducing the essay's subject matter?
2. What is the author's purpose in capitalizing, italicizing, and placing quotation marks around certain phrases, for example, "Way It Has Always Been" (paragraph 4); *failed* and *children of divorce* (paragraph 7); "family values" (paragraph 18)?
3. What is the author's purpose in creating a gap between paragraphs 15 and 16? What is the focus of the author's argument after this break?
4. Compare the introductory paragraph with the concluding one. How do they differ? How are they similar? How do they help set the boundaries of the essay?
5. This essay contains personal observation, personal experience, historical data, and anecdote. How would you describe the author's overall method to a person who has not read the essay?
6. Unlike the titles of most essays, the title "Stone Soup" gives no hint at the essay's content. What is the rhetorical purpose in keeping the meaning of the title a mystery until the very end?
7. In paragraph 2, the author asks the question, "Why are our names for home so slow to catch up to the truth of where we live?" Does the author suggest an answer to this question either implicitly or explicitly during the course of the essay? If so, where?

WRITING

1. Interview two individuals at least 25 years apart in age. Compare and contrast their views on divorce.
2. Describe the dynamics of a blended family with which you are familiar. It may be your own or a friend's.
3. **WRITING AN ARGUMENT:** Write an essay arguing that some negative outcomes could occur in the type of family the author celebrates.

www.mhhe.com/
mhreader

For more information on Barbara Kingsolver, go to:
More Resources > Ch. 5 Family & Gender

Classic and Contemporary:
Questions for Comparison

1. Compare and contrast the tone of each writer. How does tone affect purpose? How does it affect mood? Select at least three passages from White and three from Kingsolver that demonstrate how their tones differ. Do they offer any hints as to the "voice" or personality of the writers? Why or why not?
2. What contemporary issues does Kingsolver address that White either ignores or is unaware of? Consider that White was born 58 years before Kingsolver, so his world was quite a different one. Are there other variables that might help us distinguish their concerns and outlooks? For example, gender, class, environment?
3. What central values does each author have regarding the family? How are they similar? How do they differ? How do their values reflect their times?

✳

Marrying Absurd

Joan Didion

Joan Didion (b. 1934) was born in Sacramento, California, and attended the University of California at Berkeley (BA, 1956). A fifth-generation Californian, Didion began her writing career when she won an essay contest sponsored by Vogue *magazine, which then hired her. Today, as essayist, novelist, and screenwriter (with her husband, John Gregory Dunne), she is known as an intensely introspective stylist. Her writing, in its particularity and sharp cultural and political specificity, takes on general significance. Among her many books are the novels* Play It as It Lays *(1970),* The Book of Common Prayer *(1977), and* Democracy *(1984). Didion is best known as a nonfiction writer, whose work includes* Slouching Towards Bethlehem *(1968), from which the following piece is excerpted;* The White Album *(1979);* After Henry *(1992); and* Fixed Ideas: America Since 9.11 *(2003). Here Didion offers a vivid description of marriage Las Vegas style.*

To be married in Las Vegas, Clark County, Nevada, a bride must swear that she   1
is eighteen or has parental permission and a bridegroom that he is twenty-one
or has parental permission. Someone must put up five dollars for the license.
(On Sundays and holidays, fifteen dollars. The Clark County Courthouse issues
marriage licenses at any time of the day or night except between noon and one
in the afternoon, between eight and nine in the evening, and between four and

five in the morning.) Nothing else is required. The State of Nevada, alone among these United States, demands neither a premarital blood test nor a waiting period before or after the issuance of a marriage license. Driving in across the Mojave from Los Angeles, one sees the signs way out on the desert, looming up from that moonscape of rattle-snakes and mesquite, even before the Las Vegas lights appear like a mirage on the horizon: "GETTING MARRIED? Free License Information First Strip Exit." Perhaps the Las Vegas wedding industry achieved its peak operational efficiency between 9:00 P.M. and midnight of August 26, 1965, an otherwise unremarkable Thursday which happened to be, by Presidential order, the last day on which anyone could improve his draft status merely by getting married. One hundred and seventy-one couples were pronounced man and wife in the name of Clark County and the State of Nevada that night, sixty-seven of them by a single justice of the peace, Mr. James A. Brennan. Mr. Brennan did one wedding at the Dunes and the other sixty-six in his office, and charged each couple eight dollars. One bride lent her veil to six others. "I got it down from five to three minutes," Mr. Brennan said later of his feat. "I could've married them *en masse,* but they're people, not cattle. People expect more when they get married."

What people who get married in Las Vegas actually do expect—what, in   2
the largest sense, their "expectations" are—strikes one as a curious and self-contradictory business. Las Vegas is the most extreme and allegorical of American settlements, bizarre and beautiful in its venality and in its devotion to immediate gratification, a place the tone of which is set by mobsters and call girls and ladies' room attendants with amyl nitrite poppers in their uniform pockets. Almost everyone notes that there is no "time" in Las Vegas, no night and no day and no past and no future (no Las Vegas casino, however, has taken the obliteration of the ordinary time sense quite so far as Harold's Club in Reno, which for a while issued, at odd intervals in the day and night, mimeographed "bulletins" carrying news from the world outside); neither is there any logical sense of where one is. One is standing on a highway in the middle of a vast hostile desert looking at an eighty-foot sign which blinks "STARDUST" or "CAESAR'S PALACE." Yes, but what does that explain? This geographical implausibility reinforces the sense that what happens there has no connection with "real" life; Nevada cities like Reno and Carson are ranch towns, Western towns, places behind which there is some historical imperative. But Las Vegas seems to exist only in the eye of the beholder. All of which makes it an extraordinarily stimulating and interesting place, but an odd one in which to want to wear a candle-light satin Priscilla of Boston wedding dress with Chantilly lace insets, tapered sleeves and a detachable modified train.

And yet the Las Vegas wedding business seems to appeal to precisely that   3
impulse. "Sincere and Dignified Since 1954," one wedding chapel advertises. There are nineteen such wedding chapels in Las Vegas, intensely competitive, each offering better, faster, and, by implication, more sincere services than the next: Our Photos Best Anywhere, Your Wedding on a Phonograph Record,

Candlelight with Your Ceremony, Honeymoon Accommodations, Free Transportation from Your Motel to Courthouse to Chapel and Return to Motel, Religious or Civil Ceremonies, Dressing Rooms, Flowers, Rings, Announcements, Witnesses Available, and Ample Parking. All of these services, like most others in Las Vegas (sauna baths, payroll-check cashing, chinchilla coats for sale or rent) are offered twenty-four hours a day, seven days a week, presumably on the premise that marriage, like craps, is a game to be played when the table seems hot.

But what strikes one most about the Strip chapels, with their wishing wells   4
and stained-glass paper windows and their artificial bouvardia, is that so much of their business is by no means a matter of simple convenience, of late-night liaisons between show girls and baby Crosbys. Of course there is some of that. (One night about eleven o'clock in Las Vegas I watched a bride in an orange minidress and masses of flame-colored hair stumble from a Strip chapel on the arm of her bridegroom, who looked the part of the expendable nephew in movies like *Miami Syndicate.* "I gotta get the kids," the bride whimpered. "I gotta pick up the sitter, I gotta get to the midnight show." "What you gotta get," the bridegroom said, opening the door of a Cadillac Coupe de Ville and watching her crumple on the seat, "is sober.") But Las Vegas seems to offer something other than "convenience"; it is merchandising "niceness," the facsimile of proper ritual, to children who do not know how else to find it, how to make the arrangements, how to do it "right." All day and evening long on the Strip, one sees actual wedding parties, waiting under the harsh lights at a crosswalk, standing uneasily in the parking lot of the Frontier while the photographer hired by The Little Church of the West ("Wedding Place of the Stars") certifies the occasion, takes the picture: the bride in a veil and white satin pumps, the bridegroom usually in a white dinner jacket, and even an attendant or two, a sister or a best friend in hot-pink *peau de soie,* a flirtation veil, a carnation nosegay. "When I Fall in Love It Will Be Forever," the organist plays, and then a few bars of Lohengrin. The mother cries; the stepfather, awkward in his role, invites the chapel hostess to join them for a drink at the Sands. The hostess declines with a professional smile; she has already transferred her interest to the group waiting outside. One bride out, another in, and again the sign goes up on the chapel door: "One moment please—Wedding."

I sat next to one such wedding party in a Strip restaurant the last time I was   5
in Las Vegas. The marriage had just taken place; the bride still wore her dress, the mother her corsage. A bored waiter poured out a few swallows of pink champagne ("on the house") for everyone but the bride, who was too young to be served. "You'll need something with more kick than that," the bride's father said with heavy jocularity to his new son-in-law; the ritual jokes about the wedding night had a certain Panglossian character, since the bride was clearly several months pregnant. Another round of pink champagne, this time not on the house, and the bride began to cry. "It was just as nice," she sobbed, "as I hoped and dreamed it would be."

COMPREHENSION

1. What military draft does the author refer to in paragraph 1? How might this fact explain the 171 couples wanting to get married? Why does Didion include this information?
2. Why does Didion think that Las Vegas is "odd" (paragraph 2)? What examples does she provide? According to the author, what sorts of people get married in Las Vegas?
3. According to the author, how does Las Vegas differ from other places?

RHETORIC

1. Why does Didion begin her essay with an overview of the rules governing marriage in Las Vegas? What is her purpose and what tone does she assume here and throughout the essay?
2. What expectations does the author have of her audience? How do you know?
3. What is Didion's claim, where does she place it, and why does it appear where it does? Is the entire argument expressed in one sentence? Why or why not? What warrants are implied?
4. What support does Didion provide for her claim?
5. Why does the author use so much description to advance her argument? Do you find the description useful or excessive, and why? Why does she end the essay with the image of the pregnant bride crying for joy?

WRITING

1. Write an essay about a wedding or other ceremony that was odd or "absurd." Employ narration and description to capture the absurdity of the relationship or event.
2. Analyze and evaluate the tone of Didion's essay. Explain why you think the tone is suitable or perhaps too satirical and opinionated.
3. **WRITING AN ARGUMENT:** Write a rebuttal to Didion's argument. Establish your own claim, provide support, and deal with all of the author's objections.

---

www.mhhe.com/
mhreader

For more information on Joan Didion, go to:
More Resources > Ch. 5 Family & Gender

✳

An American Childhood

Annie Dillard

Annie Dillard (b. 1945 in Pittsburgh) received her BA and MA degrees from Hollins College. Her first book, Pilgrim at Tinker Creek *(1975), won the Pulitzer Prize for general nonfiction. Her other published works of nonfiction include* Teaching a Stone to Talk *(1982) and* An American Childhood *(1987). Dillard expanded her range of writing with the publication of her first novel,* The Living *(1992). She has received awards from the National Endowment for the Arts and the Guggenheim Foundation as well as many other sources. As an essayist, poet, memoirist, and literary critic, she focuses her themes on the relationships among the self, nature, religion, and faith. Her writing is recognizable by its observations of the minutiae of life and its search for meaning in unlikely places, such as a stone or an insect. In this passage from* An American Childhood, *the author gives us a portrait of her mother by focusing on her small idiosyncrasies of speech, gesture, and attitude.*

One Sunday afternoon Mother wandered through our kitchen, where Father   1
was making a sandwich and listening to the ball game. The Pirates were playing the New York Giants at Forbes Field. In those days, the Giants had a utility
infielder named Wayne Terwilliger. Just as Mother passed through, the radio
announcer cried—with undue drama—"Terwilliger bunts one!"

"Terwilliger bunts one?" Mother cried back, stopped short. She turned. "Is   2
that English?"

"The player's name is Terwilliger," Father said. "He bunted."   3

"That's marvelous," Mother said. "'Terwilliger bunts one.' No wonder you   4
listen to baseball. 'Terwilliger bunts one.'"

For the next seven or eight years, Mother made this surprising string of syl-   5
lables her own. Testing a microphone, she repeated, "Terwilliger bunts one";
testing a pen or a typewriter, she wrote it. If, as happened surprisingly often in
the course of various improvised gags, she pretended to whisper something
else in my ear, she actually whispered, "Terwilliger bunts one." Whenever
someone used a French phrase, or a Latin one, she answered solemnly, "Terwilliger bunts one." If Mother had had, like Andrew Carnegie, the opportunity
to cook up a motto for a coat of arms, hers would have read simply and
tellingly, "Terwilliger bunts one." (Carnegie's was "Death to Privilege.")

She served us with other words and phrases. On a Florida trip, she repeated   6
tremulously, "That . . . is a royal poinciana." I don't remember the tree; I remember the thrill in her voice. She pronounced it carefully, and spelled it. She also
liked to say "portulaca."

The drama of the words "Tamiami Trail" stirred her, we learned on the   7
same Florida trip. People built Tampa on one coast, and they built Miami on an-

other. Then—the height of visionary ambition and folly—they piled a slow, tremendous road through the terrible Everglades to connect them. To build the road, men stood sunk in muck to their armpits. They fought off cottonmouth moccasins and six-foot alligators. They slept in boats, wet. They blasted muck with dynamite, cut jungle with machetes; they laid logs, dragged drilling machines, hauled dredges, heaped limestone. The road took fourteen years to build up by the shovelful, a Panama Canal in reverse, and cost hundreds of lives from tropical, mosquito-carried diseases. Then, capping it all, some genius thought of the word Tamiami: they called the road from Tampa to Miami, this very road under our spinning wheels, the Tamiami Trail. Some called it Alligator Alley. Anyone could drive over this road without a thought.

Hearing this, moved, I thought all the suffering of road building was worth   8
it (it wasn't my suffering), now that we had this new thing to hang these new words on—Alligator Alley for those who liked things cute, and, for connoisseurs like Mother, for lovers of the human drama in all its boldness and terror, the Tamiami Trail.

Back home, Mother cut clips from reels of talk, as it were, and played them   9
back at leisure. She noticed that many Pittsburghers confuse "leave" and "let." One kind relative brightened our morning by mentioning why she'd brought her son to visit: "He wanted to come with me, so I left him." Mother filled in Amy and me on locutions we missed. "I can't do it on Friday," her pretty sister told a crowded dinner party, "because Friday's the day I lay in the stores."

(All unconsciously, though, we ourselves used some pure Pittsburghisms.   10
We said "tele pole," pronounced "telly pole," for that splintery sidewalk post I loved to climb. We said "slippy"—the sidewalks are "slippy." We said, "That's all the farther I could go." And we said, as Pittsburghers do say, "This glass needs washed," or "The dog needs walked"—a usage our father eschewed; he knew it was not standard English, nor even comprehensible English, but he never let on.)

"Spell 'poinsettia,'" Mother would throw out at me, smiling with pleasure.   11
"Spell 'sherbet.'" The idea was not to make us whizzes, but, quite the contrary, to remind us—and I, especially, needed reminding—that we didn't know it all just yet.

"There's a deer standing in the front hall," she told me one quiet evening in   12
the country.

"Really?"   13

"No. I just wanted to tell you something once without your saying, 'I   14
know.'"

Supermarkets in the middle 1950s began luring, or bothering, customers by   15
giving out Top Value Stamps or Green Stamps. When, shopping with Mother, we got to the head of the checkout line, the checker, always a young man, asked, "Save stamps?"

"No," Mother replied genially, week after week, "I build model airplanes."   16
I believe she originated this line. It took me years to determine where the joke lay.

Anyone who met her verbal challenges she adored. She had surgery on one 17 of her eyes. On the operating table, just before she conked out, she appealed feelingly to the surgeon, saying, as she had been planning to say for weeks, "Will I be able to play the piano?" "Not on me," the surgeon said. "You won't pull that old one on me."

It was, indeed, an old one. The surgeon was supposed to answer, "Yes, my 18 dear, brave woman, you will be able to play the piano after this operation," to which Mother intended to reply, "Oh, good, I've always wanted to play the piano." This pat scenario bored her; she loved having it interrupted. It must have galled her that usually her acquaintances were so predictably unalert; it must have galled her that, for the length of her life, she could surprise everyone so continually, so easily, when she had been the same all along. At any rate, she loved anyone who, as she put it, saw it coming, and called her on it.

She regarded the instructions on bureaucratic forms as straight lines. "Do 19 you advocate the overthrow of the United States government by force or violence?" After some thought she wrote, "Force." She regarded children, even babies, as straight men. When Molly learned to crawl, Mother delighted in buying her gowns with drawstrings at the bottom, like Swee'pea's, because, as she explained energetically, you could easily step on the drawstring without the baby's noticing, so that she crawled and crawled and crawled and never got anywhere except into a small ball at the gown's top.

When we children were young, she mothered us tenderly and dependably; as 20 we got older, she resumed her career of anarchism. She collared us into her gags. If she answered the phone on a wrong number, she told the caller, "Just a minute," and dragged the receiver to Amy or me, saying, "Here, take this, your name is Cecile," or, worse, just, "It's for you." You had to think on your feet. But did you want to perform well as Cecile, or did you want to take pity on the wretched caller?

During a family trip to the Highland Park Zoo, Mother and I were alone for a 21 minute. She approached a young couple holding hands on a bench by the seals, and addressed the young man in dripping tones: "Where have you been? Still got those baby-blue eyes; always did slay me. And this"—a swift nod at the dumbstruck young woman, who had removed her hand from the man's—"must be the one you were telling me about. She's not so bad, really, as you used to make out. But listen, you know how I miss you, you know where to reach me, same old place. And there's Ann over there—see how she's grown? See the blue eyes?"

And off she sashayed, taking me firmly by the hand, and leading us around 22 briskly past the monkey house and away. She cocked an ear back, and both of us heard the desperate man begin, in a high-pitched wail, "I swear, I never saw her before in my life . . ."

On a long, sloping beach by the ocean, she lay stretched out sunning with Fa- 23 ther and friends, until the conversation gradually grew tedious, when without forethought she gave a little push with her heel and rolled away. People were stunned. She rolled deadpan and apparently effortlessly, arms and legs extended and tidy, down the beach to the distant water's edge, where she lay at ease just as she had been, but half in the surf, and well out of earshot.

She dearly loved to fluster people by throwing out a game's rules at a 24 whim—when she was getting bored, losing in a dull sort of way, and when everybody else was taking it too seriously. If you turned your back, she moved the checkers around on the board. When you got them all straightened out, she denied she'd touched them; the next time you turned your back, she lined them up on the rug or hid them under your chair. In a betting rummy game called Michigan, she routinely played out of turn, or called out a card she didn't hold, or counted backward, simply to amuse herself by causing an uproar and watching the rest of us do double-takes and have fits. (Much later, when serious suitors came to call, Mother subjected them to this fast card game as a trial by ordeal; she used it as an intelligence test and a measure of spirit. If the poor man could stay a round without breaking down or running out, he got to marry one of us, if he still wanted to.)

She excelled at bridge, playing fast and boldly, but when the stakes were 25 low and the hands dull, she bid slams for the devilment of it, or raised her opponents' suit to bug them, or showed her hand, or tossed her cards in a handful behind her back in a characteristic swift motion accompanied by a vibrantly innocent look. It drove our stolid father crazy. The hand was over before it began, and the guests were appalled. How do you score it, who deals now, what do you do with a crazy person who is having so much fun? Or they were down seven, and the guests were appalled. "Pam!" "Dammit, Pam!" He groaned. What ails such people? What on earth possesses them? He rubbed his face.

She was an unstoppable force; she never let go. When we moved across 26 town, she persuaded the U.S. Post Office to let her keep her old address—forever—because she'd had stationery printed. I don't know how she did it. Every new post office worker, over decades, needed to learn that although the Doaks' mail is addressed to here, it is delivered to there.

Mother's energy and intelligence suited her for a greater role in a larger 27 arena—mayor of New York, say—than the one she had. She followed American politics closely; she had been known to vote for Democrats. She saw how things should be run, but she had nothing to run but our household. Even there, small minds bugged her; she was smarter than the people who designed the things she had to use all day for the length of her life.

"Look," she said. "Whoever designed this corkscrew never used one. Why 28 would anyone sell it without trying it out?" So she invented a better one. She showed me a drawing of it. The spirit of American enterprise never faded in Mother. If capitalizing and tooling up had been as interesting as theorizing and thinking up, she would have fired up a new factory every week, and chaired several hundred corporations.

"It grieves me," she would say, "it grieves my heart," that the company that 29 made one superior product packaged it poorly, or took the wrong tack in its advertising. She knew, as she held the thing mournfully in her two hands, that she'd never find another. She was right. We children wholly sympathized, and so did Father; what could she do, what could anyone do, about it? She was Samson in chains. She paced.

She didn't like the taste of stamps so she didn't lick stamps; she licked the 30 corner of the envelope instead. She glued sandpaper to the sides of kitchen drawers, and under kitchen cabinets, so she always had a handy place to strike a match. She designed, and hounded workmen to build against all norms, doubly wide kitchen counters and elevated bathroom sinks. To splint a finger, she stuck it in a lightweight cigar tube. Conversely, to protect a pack of cigarettes, she carried it in a Band-Aid box. She drew plans for an over-the-finger toothbrush for babies, an oven rack that slid up and down, and—the family favorite—Lendalarm. Lendalarm was a beeper you attached to books (or tools) you loaned friends. After ten days, the beeper sounded. Only the rightful owner could silence it.

She repeatedly reminded us of P. T. Barnum's dictum: You could sell any- 31 thing to anybody if you marketed it right. The adman who thought of making Americans believe they needed underarm deodorant was a visionary. So, too, was the hero who made a success of a new product, Ivory soap. The executives were horrified, Mother told me, that a cake of this stuff floated. Soap wasn't supposed to float. Anyone would be able to tell it was mostly whipped-up air. Then some inspired adman made a leap: Advertise that it floats. Flaunt it. The rest is history.

She respected the rare few who broke through to new ways. "Look," she'd 32 say, "here's an intelligent apron." She called upon us to admire intelligent control knobs and intelligent pan handles, intelligent andirons and picture frames and knife sharpeners. She questioned everything, every pair of scissors, every knitting needle, gardening glove, tape dispenser. Hers was a restless mental vigor that just about ignited the dumb household objects with its force.

Torpid conformity was a kind of sin; it was stupidity itself, the mighty stream 33 against which Mother would never cease to struggle. If you held no minority opinions, or if you failed to risk total ostracism for them daily, the world would be a better place without you.

Always I heard Mother's emotional voice asking Amy and me the same few 34 questions: "Is that your own idea? Or somebody else's?" "*Giant* is a good movie," I pronounced to the family at dinner. "Oh, really?" Mother warmed to these occasions. She all but rolled up her sleeves. She knew I hadn't seen it. "Is that your considered opinion?"

She herself held many unpopular, even fantastic, positions. She was 35 scathingly sarcastic about the McCarthy hearings while they took place, right on our living-room television; she frantically opposed Father's wait-and-see calm. "We don't know enough about it," he said. "I do," she said. "I know all I need to know."

She asserted, against all opposition, that people who lived in trailer parks 36 were not bad but simply poor, and had as much right to settle on beautiful land, such as rural Ligonier, Pennsylvania, as did the oldest of families in the finest of hidden houses. Therefore, the people who owned trailer parks, and sought zoning changes to permit trailer parks, needed our help. Her profound belief that the country-club pool sweeper was a person, and that the department-store saleslady, the bus driver, telephone operator, and house-painter were people, and even in groups the steelworkers who carried pickets and the Christmas shoppers who clogged intersections were people—this was a conviction common enough in democratic Pittsburgh, but not altogether common among our friends' parents, or even, perhaps, among our parents' friends.

Opposition emboldened Mother, and she would take on anybody on any 37 issue—the chairman of the board, at a cocktail party, on the current strike; she would fly at him in a flurry of passion, as a songbird selflessly attacks a big hawk.

"Eisenhower's going to win," I announced after school. She lowered her 38 magazine and looked me in the eyes: "How do you know?" I was doomed. It was fatal to say, "Everyone says so." We all knew well what happened. "Do you consult this Everyone before you make your decisions? What if Everyone decided to round up all the Jews?" Mother knew there was no danger of cowing me. She simply tried to keep us all awake. And in fact it was always clear to Amy and me, and to Molly when she grew old enough to listen, that if our classmates came to cruelty, just as much as if the neighborhood or the nation came to madness, we were expected to take, and would be each separately capable of taking, a stand.

COMPREHENSION

1. The writer creates a picture of her mother's personality through a number of anecdotes and explanations. How would you sum up the mother's personality?
2. The mother appears to have a special appreciation for words and language. To what purpose does she apply this appreciation? What effect does it have on her family and acquaintances?
3. What values does the mother hold? What behaviors and attitudes does she abhor and discourage?

RHETORIC

1. In paragraph 7, the author explains that the highway from Tampa to Miami is referred to either as "Tamiami Trail" or "Alligator Alley." What is the connotation of each of these terms? Why does the mother prefer to call it "Tamiami Trail"?
2. The author herself seems to have inherited a special fascination for language. Study her use of dashes and semicolons in paragraphs 26 and 27. How do they help contribute to an energetic use of writing?

3. What are the functions of the spaces between paragraphs 19 and 20; 22 and 23; and 32 and 33? How do these divisions contribute to the structure of the essay as a whole?
4. How does the author use her writing talents to create paragraph 8 out of one long sentence? What other examples can you provide of long sentences in the essay? How do they contribute to the overall style of the writing?
5. What is the overall emotional "tone" of the writer toward her subject—admiring, loving, cautionary? What adjectives does the writer use in describing her mother that provides the reader with clues to the tone?
6. The author quotes her mother directly on several occasions. Can we assume that the author is quoting precisely, given that the essay was written years after the incidents described? Does it matter?
7. The final paragraph not only provides closure to the essay, but transmits a lesson the mother wants her family to learn. How do the style and structure of this paragraph contribute to the ultimate message of the essay? In other words, how does the form help convey the meaning?

WRITING

1. Write a descriptive essay about someone you know, using at least five anecdotes from that person's life, so that by the end of the essay, we have a mental picture of your subject's personality, values, and attitudes.
2. Describe an incident in your life when the unexpected taught you an important lesson.
3. **WRITING AN ARGUMENT:** Argue for or against the proposition that an effective parent should have—at least—a touch of unconventionality.

www.mhhe.com/
mhreader

For more information on Annie Dillard, go to:
More Resources > Ch. 5 Family & Gender

✳

Love, Internet Style

David Brooks

David Brooks (b. 1961) is a columnist for the op-ed page of The New York Times. *Prior to joining the* Times, *he was a senior editor at the* Weekly Standard, *an op-ed page editor at* The Wall Street Journal *and a contributing editor to* Newsweek. *He is also a weekly guest on the* News Hour with Jim Lehrer *on National Public Televi-*

sion. A graduate of the University of Chicago, Brooks writes on a wide range of topics, often from a conservative perspective; he has edited Backward and Upward: The New Conservative Writing. *Brooks's recent books* Bobos in Paradise: The New Upper Class and How They Got There *(2000) and* On Paradise Drive *(2004), which explores the lives of people living in the suburbs, offer fascinating and often amusing insights into contemporary American culture. In this essay, which appeared in* The New York Times *in 2003, Brooks examines the ways in which the Internet facilitates personal relationships.*

The Internet slows things down.   1

If you're dating in the Age of the Hook-Up, sex is this looming possibility   2
from the first moment you meet a prospective partner. But couples who meet
through online dating services tend to exchange e-mail for weeks or months.
Then they'll progress to phone conversations for a few more weeks. Only then
will there be a face-to-face meeting, almost always at some public place early in
the evening, and the first date will often be tentative and Dutch.

Online dating puts structure back into courtship. For generations Americans   3
had certain courtship rituals. The boy would call the girl and ask her to the
movies. He might come in and meet the father. After a few dates he might ask her
to go steady. Sex would progress gradually from kissing to petting and beyond.

But over the past few decades that structure dissolved. And human beings,   4
who are really good at adapting, found that the Internet, of all places, imposes
the restraints they need to let relationships develop gradually. So now 40 million Americans look at online dating sites each month, and we are seeing a revolution in the way people meet and court one another.

The new restraints are not like the old restraints. The online dating scene is   5
like a real estate market where people go to fulfill their most sensitive needs. It
is at once ruthlessly transactional and strangely tender.

It begins with sorting. Online daters can scan through millions of possible   6
partners in an evening and select for age, education, height, politics, religion and
ethnic background. JDate is a popular site for Jews. EHarmony insists that members fill out a long, introspective questionnaire, and thus is one of the few sites
where most members are women. Vanity Date is for the South Beach crowd. "At
Vanity Date," the Web site declares, "we have a vision of creating the largest
database of the world's most good-looking, rich and superficial people."

Most of the sites have programs that link you up with people like yourself.   7
One of the side effects of online dating is that it is bound to accelerate social
stratification, as highly educated people become more efficient at finding and
marrying one another.

Each member at a dating site creates his or her own Web page. The most im-   8
portant feature on the page is the photo; studies show that looks are twice as
powerful as income in attracting mates.

But there are also autobiographical essays. If you judged by these essays,   9
skinny-dipping with intellectuals is the most popular activity in America. All
the writers try to show they are sensual yet smart.

The women on these sites are, or project themselves as being, incredibly self- 10
confident. "I am a vivacious, intelligent, warm-hearted, attractive, cool chick,
with a sharp, witty, and effervescent personality," writes one on Match.com.
Another says: "I am a slender, radiantly beautiful woman on fire with passion
and enthusiasm for life. I am articulate, intelligent and routinely given the acco-
lade of being brilliant."

Still, men almost always make the first contact. Prospective partners begin 11
a long series of e-mail interviews. Internet exchanges encourage both extreme
honesty (the strangers-on-a-train phenomenon) and extreme dishonesty, as people lie about their ages, their jobs, whether they have kids and, most often,
whether they are married. (About a fifth of online daters are married men.)

Whatever else has changed, men are more likely to be predators looking for 12
sex, while women try to hold back. Men will ask women for more photos "from
different angles." A woman, wanting to be reassured that this guy is not some
rapist, will shut off anyone who calls her "hottie" or who mentions sex first.
Women generally control the pace of the relationship.

But despite all the crass competition, all the marketing, all the shopping 13
around, people connect. Studies by Katelyn McKenna at N.Y.U. and others indicate that Internet relationships are at least as powerful as relationships that begin face to face. Many people are better at revealing their true selves through the
keyboard than through conversation. And couples who slow down and prolong
the e-mail phase have a better chance of seeing their relationships last than people who get together more quickly.

The online dating world is superficially cynical. The word "love" will al- 14
most never appear on a member's page, because it is so heavy and intimidating.
But love is what this is all about. And the heart, even in this commercial age,
finds a way.

COMPREHENSION

1. How does the writer describe Internet "love"? What does he mean by his
   opening sentence, "The Internet slows things down"?
2. What features of Internet culture does the writer identify as facilitating human relationships?
3. How do men and women differ in their approach to online relationships?

RHETORIC

1. What is the writer's purpose in beginning his essay with a single-sentence
   paragraph? Is this sentence the thesis? Why or why not?
2. How would you describe the writer's stance? What is his attitude toward
   his subject? Is he an objective observer or does he advance a particular point
   of view? Offer examples to support your answer.

3. What are the writer's main reasons in support of his thesis or claim? What forms of evidence does he offer to support his claim?
4. How does the writer develop an extended definition of Internet "love"?
5. The writer frequently structures his essay by means of comparison and contrast. Why do you think Brooks uses this strategy? Do you find the method effective? Why or why not?
6. Does the final paragraph provide a solid conclusion? Justify your answer.

WRITING

1. Narrate and describe your own experience with Internet relationships or online dating. Explain why the Internet has helped or hindered your relationships with other men or women.
2. Write a definition essay on "Internet love." Be certain to provide examples and utilize other rhetorical strategies like comparison and contrast to develop this extended definition.
3. **WRITING AN ARGUMENT:** Write a persuasive essay arguing that Internet dating is either dangerous or harmless. Provide at least three minor propositions and sufficient evidence to support your position.

✳ www.mhhe.com/ mhreader

For more information on David Brooks, go to:
More Resources > Ch. 5 Family & Gender

✳

Family Values

Richard Rodriguez

Richard Rodriguez (b. 1944) received degrees from Stanford University and Columbia University. He also did graduate study at the University of California, Berkeley, and at the Warburg Institute, London. He is a writer and editor for Pacifica News Service *and a contributing editor and writer for many major American magazines and journals including* Harper's *and the* Los Angeles Times. *His books include* Hunger of Memory: The Education of Richard Rodriguez *(1982) and* Days of Obligation: An Argument with My Mexican Father *(1992). Both books have been profoundly influential in the public discussion on race, bilingualism, affirmative action, and biculturalism. He has also made many appearances as a commentator on the* News Hour with Jim Lehrer. *In the following essay, originally published in the Sunday "Opinion" section of the* Los Angeles Times *in 1992, he addresses the concept of "family values" and focuses on the controversial thesis that homosexuality—rather than being a threat to family values—is actually a buttress against their dissolution.*

I am sitting alone in my car, in front of my parents' house—a middle-aged man 1 with a boy's secret to tell. What words will I use to tell them? I hate the word *gay,* find its little affirming sparkle more pathetic than assertive. I am happier with the less polite *queer.* But to my parents I would say *homosexual,* avoiding the Mexican slang *joto* (I had always heard it said in our house with hints of condescension), though *joto* is less mocking than the sissy-boy *maricon.*

The buzz on everyone's lips now: Family values. The other night on TV, the 2 vice president of the United States, his arm around his wife, smiled into the camera and described homosexuality as "mostly a choice." But how would he know? Homosexuality never felt like a choice to me.

A few minutes ago Rush Limbaugh, the radio guy with a voice that reminds 3 me, for some reason, of a butcher's arms, was banging his console and booming a near-reasonable polemic about family values. Limbaugh was not very clear about which values exactly he considers to be family values. A divorced man who lives alone in New York?

My parents live on a gray, treeless street in San Francisco not far from the 4 ocean. Probably more than half of the neighborhood is immigrant. India lives next door to Greece, who lives next door to Russia. I wonder what the Chinese lady next door to my parents makes of the politicians' phrase *family values.*

What immigrants know, what my parents certainly know, is that when you 5 come to this country, you risk losing your children. The assurance of family— continuity, inevitability—is precisely what America encourages its children to overturn. *Become your own man.* We who are native to this country know this too, of course, though we are likely to deny it. Only a society so guilty about its betrayal of family would tolerate the pieties of politicians regarding family values.

On the same summer day that Republicans were swarming in Houston 6 (buzzing about family values), a friend of mine who escaped family values awhile back and who now wears earrings resembling intrauterine devices, was complaining to me over coffee about the Chinese. The Chinese will never take over San Francisco, my friend said, because the Chinese do not want to take over San Francisco. The Chinese do not even see San Francisco! All they care about is their damn families. All they care about is double-parking smack in front of the restaurant on Clement Street and pulling granny out of the car—and damn anyone who happens to be in the car behind them or the next or the next.

Politicians would be horrified by such an American opinion, of course. But 7 then, what do politicians, Republicans or Democrats, really know of our family life? Or what are they willing to admit? Even in that area where they could reasonably be expected to have something to say—regarding the relationship of family life to our economic system—the politicians say nothing. Republicans celebrate American economic freedom, but Republicans don't seem to connect that economic freedom to the social breakdown they find appalling. Democrats, on the other hand, if more tolerant of the drift from familial tradition, are suspicious of the very capitalism that creates social freedom.

How you become free in America: Consider the immigrant. He gets a job. 8
Soon he is earning more money than his father ever made (his father's authority is thereby subtly undermined). The immigrant begins living a life his father
never knew. The immigrant moves from one job to another, changes houses. His
economic choices determine his home address—not the other way around. The
immigrant is on his way to becoming his own man.

When I was broke a few years ago and trying to finish a book, I lived with 9
my parents. What a thing to do! A major theme of America is leaving home. We
trust the child who forsakes family connections to make it on his own. We call
that the making of a man.

Let's talk about this man stuff for a minute. America's ethos is anti-domestic. 10
We may be intrigued by blood that runs through wealth—the Kennedys or the
Rockefellers—but they seem European to us. Which is to say, they are movies.
They are Corleones. Our real pledge of allegiance: We say in America that nothing about your family—your class, your race, your pedigree—should be as important as what you yourself achieve. We end up in 1992 introducing ourselves
by first names.

What authority can Papa have in a country that formed its identity in an act 11
of Oedipal rebellion against a mad British king? Papa is a joke in America, a
stock sitcom figure—Archie Bunker or Homer Simpson. But my Mexican father
went to work every morning, and he stood in a white smock, making false
teeth, oblivious of the shelves of grinning false teeth mocking his devotion.

The nuns in grammar school—my wonderful Irish nuns—used to push 12
Mark Twain on me. I distrusted Huck Finn, he seemed like a gringo kid I would
steer clear of in the schoolyard. (He was too confident.) I realize now, of course,
that Huck is the closest we have to a national hero. We trust the story of a boy
who has no home and is restless for the river. (Huck's Pap is drunk.) Americans
are more forgiving of Huck's wildness than of the sweetness of the Chinese boy
who walks to school with his mama or grandma. (There is no worse thing in
America than to be a mama's boy, nothing better than to be a real boy—all
boy—like Huck, who eludes Aunt Sally, and is eager for the world of men.)

There's a bent old woman coming up the street. She glances nervously as she 13
passes my car. What would you tell us, old lady, of family values in America?

America is an immigrant country, we say. Motherhood—parenthood—is 14
less our point than adoption. If I had to assign gender to America, I would note
the consensus of the rest of the world. When America is burned in effigy, a male
is burned. Americans themselves speak of Uncle Sam.

Like the Goddess of Liberty, Uncle Sam has no children of his own. He 15
steals children to make men of them, mocks all reticence, all modesty, all memory. Uncle Sam is a hectoring Yankee, a skinflint uncle, gaunt, uncouth, unloved.
He is the American Savonarola—hater of moonshine, destroyer of stills, burner
of cocaine. Sam has no patience with mamas' boys.

You betray Uncle Sam by favoring private over public life, by seeking to ex-   16
empt yourself, by cheating on your income taxes, by avoiding jury duty, by trying to keep your boy on the farm.

Mothers are traditionally the guardians of the family against America—   17
though even Mom may side with America against queers and deserters, at least
when the Old Man is around. Premature gray hair. Arthritis in her shoulders.
Bowlegged with time, red hands. In their fiercely flowered housedresses, mothers are always smarter than fathers in America. But in reality they are betrayed
by their children who leave. In a thousand ways. They end up alone.

We kind of like the daughter who was a tomboy. Remember her? It was al-   18
ways easier to be a tomboy in America than a sissy. Americans admired Annie
Oakley more than they admired Liberace (who, nevertheless, always remembered his mother). But today we do not admire Annie Oakley when we see
Mom becoming Annie Oakley.

The American household now needs two incomes, everyone says. Meaning: 19
Mom is *forced* to leave home out of economic necessity. But lots of us know lots
of moms who are sick and tired of being mom, or only mom. It's like the nuns
getting fed up, teaching kids for all those years and having those kids grow up
telling stories of how awful Catholic school was! Not every woman in America
wants her life's work to be forgiveness. Today there are moms who don't want
their husbands' names. And the most disturbing possibility: What happens
when Mom doesn't want to be Mom at all? Refuses pregnancy?

Mom is only becoming an American like the rest of us. Certainly, people all   20
over the world are going to describe the influence of feminism on women (all
over the world) as their "Americanization." And rightly so.

Nothing of this, of course, will the politician's wife tell you. The politician's   21
wife is careful to follow her husband's sentimental reassurances that nothing
has changed about America except perhaps for the sinister influence of de-
viants. Like myself.

I contain within myself an anomaly at least as interesting as the Republican   22
Party's version of family values. I am a homosexual Catholic, a communicant in
a tradition that rejects even as it upholds me.

I do not count myself among those Christians who proclaim themselves   23
protectors of family values. They regard me as no less an enemy of the family
than the "radical feminists." But the joke about families that all homosexuals
know is that we are the ones who stick around and make families possible. Call
on us. I can think of 20 or 30 examples. A gay son or daughter is the only one
who is "free" (married brothers and sisters are too busy). And, indeed, because
we have admitted the inadmissible about ourselves (that we are queer)—we are
adepts at imagination—we can even imagine those who refuse to imagine us.
We can imagine Mom's loneliness, for example. If Mom needs to be taken to
church or to the doctor or ferried between Christmas dinners, depend on the
gay son or lesbian daughter.

I won't deny that the so-called gay liberation movement, along with femi-   24
nism, undermined the heterosexual household, if that's what politicians mean

when they say family values. Against churchly reminders that sex was for pro- creation, the gay bar as much as the birth-control pill taught Americans not to fear sexual pleasure. In the past two decades—and, not coincidentally, parallel to the feminist movement—the gay liberation movement moved a generation of Americans toward the idea of a childless adulthood. If the women's movement was ultimately more concerned about getting out of the house and into the workplace, the gay movement was in its way more subversive to puritan America because it stressed the importance of play.

Several months ago, the society editor of the morning paper in San Fran- 25 cisco suggested (on a list of "must haves") that every society dame must have at least one gay male friend. A ballet companion. A lunch date. The remark was glib and incorrect enough to beg complaints from homosexual readers, but there was a truth about it as well. Homosexual men have provided women with an alternate model of masculinity. And the truth: The Old Man, God bless him, is a bore. Thus are we seen as preserving marriages? Even Republican marriages?

For myself, homosexuality is a deep brotherhood but does not involve do- 26 mestic life. Which is why, my married sisters will tell you, I can afford the time to be a writer. And why are so many homosexuals such wonderful teachers and priests and favorite aunts, if not because we are freed from the house? On the other hand, I know lots of homosexual couples (male and female) who model their lives on the traditional heterosexual version of domesticity and marriage. Republican politicians mock the notion of a homosexual marriage, but ironically such marriages honor the heterosexual marriage by imitating it.

"The only loving couples I know," a friend of mine recently remarked, "are 27 all gay couples."

This woman was not saying that she does not love her children or that she 28 is planning a divorce. But she was saying something about the sadness of American domestic life: the fact that there is so little joy in family intimacy. Which is perhaps why gossip (public intrusion into the private) has become a national industry. All day long, in forlorn houses, the television lights up a freakish parade of husbands and mothers-in-law and children upon the stage of Sally or Oprah or Phil. They tell on each other. The audience ooohhhs. Then a psychiatrist-shaman appears at the end to dispense prescriptions—the importance of family members granting one another more "space."

The question I desperately need to ask you is whether we Americans have 29 ever truly valued the family. We are famous, or our immigrant ancestors were famous, for the willingness to leave home. And it is ironic that a crusade under the banner of family values has been taken up by those who would otherwise pass themselves off as patriots. For they seem not to understand America, nor do I think they love the freedoms America grants. Do they understand why, in a country that prizes individuality and is suspicious of authority, children are disinclined to submit to their parents? You cannot celebrate American values in the public realm without expecting them to touch our private lives. As Barbara Bush remarked recently, family values are also neighborhood values. It may be harmless enough for Barbara Bush to recall a sweeter America—Midland,

Texas, in the 1950s. But the question left begging is why we chose to leave Midland, Texas. Americans like to say that we can't go home again. The truth is that we don't want to go home again, don't want to be known, recognized. Don't want to respond in the same old ways. (And you know you will if you go back there.)

Little 10-year-old girls know that there are reasons for getting away from the family. They learn to keep their secrets—under lock and key—addressed to Dear Diary. Growing up queer, you learn to keep secrets as well. In no place are those secrets more firmly held than within the family house. You learn to live in closets. I know a Chinese man who arrived in America about 10 years ago. He got a job and made some money. And during that time he came to confront his homosexuality. And then his family arrived. I do not yet know the end of this story.

The genius of America is that it permits children to leave home, it permits us to become different from our parents. But the sadness, the loneliness of America, is clear too.

Listen to the way Americans talk about immigrants. If, on the one hand, there is impatience when today's immigrants do not seem to give up their family, there is also a fascination with this reluctance. In Los Angeles, Hispanics are considered people of family. Hispanic women are hired to be at the center of the American family—to babysit and diaper, to cook and to clean and to ease the dying. Hispanic attachment to family is seen by many Americans, I think, as the reason why Hispanics don't get ahead. But if Asians privately annoy us for being so family oriented, they are also stereotypically celebrated as the new "whiz kids" in school. Don't Asians go to college, after all, to honor their parents?

More important still is the technological and economic ascendancy of Asia, particularly Japan, on the American imagination. Americans are starting to wonder whether perhaps the family values of Asia put the United States at a disadvantage. The old platitude had it that ours is a vibrant, robust society for being a society of individuals. Now we look to Asia and see team effort paying off.

In this time of national homesickness, of nostalgia, for how we imagine America used to be, there are obvious dangers. We are going to start blaming each other for the loss. Since we are inclined, as Americans, to think of ourselves individually, we are disinclined to think of ourselves as creating one another or influencing one another.

But it is not the politician or any political debate about family values that has brought me here on a gray morning to my parents' house. It is some payment I owe to my youth and to my parents' youth. I imagine us sitting in the living room, amid my mother's sentimental doilies and the family photographs, trying to take the measure of the people we have turned out to be in America.

A San Francisco poet, when he was in the hospital and dying, called a priest to his bedside. The old poet wanted to make his peace with Mother Church. He wanted baptism. The priest asked why. "Because the Catholic Church has to accept me," said the poet. "Because I am a sinner."

Isn't willy-nilly inclusiveness the point, the only possible point to be de- 37 rived from the concept of family? Curiously, both President Bush and Vice President Quayle got in trouble with their constituents recently for expressing a real family value. Both men said that they would try to dissuade a daughter or granddaughter from having an abortion. But, finally, they said they would support her decision, continue to love her, never abandon her.

There are families that do not accept. There are children who are forced to 38 leave home because of abortions or homosexuality. There are family secrets that Papa never hears. Which is to say there are families that never learn the point of families.

But there she is at the window. My mother has seen me and she waves me 39 in. Her face asks: Why am I sitting outside? (Have they, after all, known my secret for years and kept it, out of embarrassment, not knowing what to say?) Families accept, often by silence. My father opens the door to welcome me in.

COMPREHENSION

1. The title of this essay is "Family Values." What does Rodriguez mean by family values? According to the author, do Americans respect family values as they claim? Why or why not?
2. According to the author, do immigrants newly arrived to the United States possess a traditional allegiance to family values? Explain your answer.
3. In the conclusion, the author reflects—regarding his own homosexuality—that "Families accept, often in silence." Is this an aspect of traditional family values? Why or why not?
4. Why does Rodriguez think that gay men and women are often the primary upholders of family values within their own families?
5. What is the thesis of the essay? Is it implicit or explicit? Explain.

RHETORIC

1. Although much of this essay is expositional, Rodriguez begins and ends his essay with an event, that is, visiting his family to announce his homosexuality. Why has he shaped his essay in this way? What is problematic about his own relationship to the "gay" community? Why does Rodriguez feel uncomfortable with the term "gay" to denote homosexual?
2. Rodriguez employs considerable irony in his essay. For example, in paragraph 15, he notes that two icons of American democracy, the Statue of Liberty and "Uncle Sam," are childless. Select two other ironic statements the author makes in order to point out the contradiction between the "idea" of family values in America and the actual state of family values.
3. In paragraph 20, Rodriguez uses the term *Americanization*. What does he mean by this term? How is it central to his thesis?

4. Does Rodriguez suggest that much of what is said in public regarding "family values" in America is hypocritical? If so, what group or groups does he focus on? How does he support his argument?
5. Explain the meaning of the following stylistic flourishes: "the word *gay* . . . [is a] little affirming sparkle more pathetic than assertive" (paragraph 1); "America's ethos is anti-domestic" (paragraph 10); "Oedipal rebellion" (paragraph 11); "American Savonarola" (paragraph 15); "psychiatrist-shaman" (paragraph 28); "national homesickness" (paragraph 34); and "willy-nilly inclusiveness" (paragraph 37).
6. In paragraph 10, Rodriguez states that the Kennedys and Rockefellers "are movies." What does he mean?
7. Describe the emotional tone of this essay, considering that it is written by a man who is openly gay and understands that he is considered suspect and outside the mainstream of the American "value" system. Is it angry, thoughtful, defiant, sympathetic? Select three or four passages that led you to your conclusion regarding tone.

WRITING

1. Interview a member of your grandparents' generation; a member of your parents' generation; and a member of your own generation regarding their views on family values. Write an essay summarizing the similarities and differences among the three views.
2. Interview a counselor at your college or university. Ask the counselor to explain the various issues surrounding family conflict he or she comes across in the course of his or her job. Write an essay exploring your interview findings. Be sure to obtain permission from your interviewee and follow appropriate guidelines for protecting his or her anonymity.
3. For a creative writing project, imagine yourself 25 years from today. Write a letter to an old classmate, describing your family life.
4. **WRITING AN ARGUMENT:** In an essay, argue for or against the proposition that choosing to follow a tradition of "family values" is entirely the choice of the individual. Be sure to use at least three supporting points to advance your argument.

www.mhhe.com/ **mhreader**   For more information on Richard Rodriguez, go to: **More Resources > Ch. 5 Family & Gender**

✳

The Female Body

Margaret Atwood

Margaret Atwood (b. 1939) is a Canadian poet, novelist, short story writer, and critic whose work explores the role of personal consciousness in a troubled world. Her second collection of poems, The Circle Game *(1966), brought her recognition; she is also well known for her novels, including* Surfacing *(1973),* Life before Man *(1979),* The Handmaid's Tale *(1986),* Cat's Eye *(1988),* Alias Grace *(1996), and* The Blind Assassin *(2000). Atwood is interested in the complexities of language, and her subjects are wide-ranging, from the personal to the global. In the following essay from* Good Bones *(1992), Atwood uses a lively, unconventional style to address a serious theme.*

. . . entirely devoted to the subject of "The Female Body." Knowing how well you have written on this topic . . . this capacious topic . . .
       —*Letter from the* Michigan Quarterly Review

1.   I agree, it's a hot topic. But only one? Look around, there's a wide   1
range. Take my own, for instance. I get up in the morning. My topic feels like
hell. I sprinkle it with water, brush parts of it, rub it with towels, powder it, add
lubricant. I dump in the fuel and away goes my topic, my topical topic, my controversial topic, my capacious topic, my limping topic, my nearsighted topic,
my topic with back problems, my badly behaved topic, my vulgar topic, my
outrageous topic, my aging topic, my topic that is out of the question and anyway still can't spell, in its oversized coat and worn winter boots, scuttling along
the sidewalk as if it were flesh and blood, hunting for what's out there, an avo-
cado, an alderman, an adjective, hungry as ever.

2.   The basic Female Body comes with the following accessories: garter   2
belt, panti-girdle, crinoline, camisole, bustle, brassiere, stomacher, chemise, virgin zone, spike heels, nose ring, veil, kid gloves, fishnet stockings, fichu, bandeau, Merry Widow, weepers, chokers, barrettes, bangles, beads, lorgnette,
feather boa, basic black, compact, Lycra stretch one-piece with modesty panel,
designer peignoir, flannel nightie, lace teddy, bed, head.

3.   The Female Body is made of transparent plastic and lights up when you   3
plug it in. You press a button to illuminate the different systems. The circulatory
system is red, for the heart and arteries, purple for the veins; the respiratory system is blue; the lymphatic system is yellow; the digestive system is green, with
liver and kidneys in aqua. The nerves are done in orange and the brain is pink.
The skeleton, as you might expect, is white.

The reproductive system is optional, and can be removed. It comes with or without a miniature embryo. Parental judgment can thereby be exercised. We do not wish to frighten or offend. [4]

4.   He said, I won't have one of those things in the house. It gives a young girl a false notion of beauty, not to mention anatomy. If a real woman was built like that she'd fall on her face. [5]

She said, If we don't let her have one like all the other girls she'll feel singled out. It'll become an issue. She'll long for one and she'll long to turn into one. Repression breeds sublimation. You know that. [6]

He said, It's not just the pointy plastic tits, it's the wardrobes. The wardrobes and that stupid male doll, what's his name, the one with the underwear glued on. [7]

She said, Better to get it over with when she's young. He said, All right, but don't let me see it. [8]

She came whizzing down the stairs, thrown like a dart. She was stark naked. Her hair had been chopped off, her head was turned back to front, she was missing some toes and she'd been tattooed all over her body with purple ink in a scrollwork design. She hit the potted azalea, trembled there for a moment like a botched angel, and fell. [9]

He said, I guess we're safe. [10]

5.   The Female Body has many uses. It's been used as a door knocker, a bottle opener, as a clock with a ticking belly, as something to hold up lampshades, as a nutcracker, just squeeze the brass legs together and out comes your nut. It bears torches, lifts victorious wreaths, grows copper wings and raises aloft a ring of neon stars; whole buildings rest on its marble heads. [11]

It sells cars, beer, shaving lotion, cigarettes, hard liquor; it sells diet plans and diamonds, and desire in tiny crystal bottles. Is this the face that launched a thousand products? You bet it is, but don't get any funny big ideas, honey, that smile is a dime a dozen. [12]

It does not merely sell, it is sold. Money flows into this country or that country, flies in, practically crawls in, suitful after suitful, lured by all those hairless pre-teen legs. Listen, you want to reduce the national debt, don't you? Aren't you patriotic? That's the spirit. That's my girl. [13]

She's a natural resource, a renewable one luckily, because those things wear out so quickly. They don't make 'em like they used to. Shoddy goods. [14]

6.   One and one equals another one. Pleasure in the female is not a requirement. Pair-bonding is stronger in geese. We're not talking about love, we're talking about biology. That's how we all got here, daughter. [15]

Snails do it differently. They're hermaphrodites, and work in threes. [16]

7.   Each Female Body contains a female brain. Handy. Makes things work. Stick pins in it and you get amazing results. Old popular songs. Short circuits. Bad dreams. [17]

Anyway: each of these brains has two halves. They're joined together by a 18
thick cord; neural pathways flow from one to the other, sparkles of electric information washing to and fro. Like light on waves. Like a conversation. How
does a woman know? She listens. She listens in.

The male brain, now, that's a different matter. Only a thin connection. Space 19
over here, time over there, music and arithmetic in their own sealed compart-
ments. The right brain doesn't know what the left brain is doing. Good for aiming though, for hitting the target when you pull the trigger. What's the target?
Who's the target? Who cares? What matters is hitting it. That's the male brain
for you. Objective.

This is why men are so sad, why they feel so cut off, why they think of 20
themselves as orphans cast adrift, footloose and stringless in the deep void.
What void? she asks. What are you talking about? The void of the universe, he
says, and she says Oh and looks out the window and tries to get a handle on it,
but it's no use, there's too much going on, too many rustlings in the leaves, too
many voices, so she says, Would you like a cheese sandwich, a piece of cake, a
cup of tea? And he grinds his teeth because she doesn't understand, and wanders off, not just alone but Alone, lost in the dark, lost in the skull, searching for
the other half, the twin who could complete him.

Then it comes to him: he's lost the Female Body! Look, it shines in the 21
gloom, far ahead, a vision of wholeness, ripeness, like a giant melon, like an apple, like a metaphor for "breast" in a bad sex novel; it shines like a balloon, like
a foggy noon, a watery moon, shimmering in its egg of light.

Catch it. Put it in a pumpkin, in a high tower, in a compound, in a chamber, 22
in a house, in a room. Quick, stick a leash on it, a lock, a chain, some pain, settle
it down, so it can never get away from you again.

COMPREHENSION

1. Why do you think this essay was written? Justify your response.
2. List the different ways in which Atwood views the female body.
3. What distinction does Atwood make between male and female brains?

RHETORIC

1. What is the tone of Atwood's essay? Supply concrete evidence from her
   writing.
2. Does the essay contain a thesis? Is it stated or implied?
3. Define the following words in section 2: *fichu, bandeau, Merry Widow, weepers.* Why do the words *bed* and *head* also appear in this list?
4. How does Atwood's use of details and metaphors strengthen her points in
   the essay? Cite specific examples.
5. What is the object being described in section 4? How does its inclusion help
   underscore Atwood's point?

6. Why did Atwood choose this particular way to organize her essay? What does it tell the reader about her attitude toward the subject?
7. Is the tone of the final paragraph similar to that of the rest of the essay? Provide evidence from the writing and explain.

WRITING

1. Using a style similar to Atwood's, write a brief essay in which you describe the female brain, the male brain, or the male body.
2. Analyze the ways in which sex and the female body have been used in sales and advertising.
3. **WRITING AN ARGUMENT:** In an argumentative essay, consider the role played by sex-specific toys in reinforcing sexual stereotyping in children. Use Atwood's essay as well as your personal experience as support.

| www.mhhe.com/ **mhreader** | For more information on Margaret Atwood, go to: **More Resources > Ch. 5 Family & Gender** |

*

Why Men Don't Last: Self-Destruction as a Way of Life

Natalie Angier

Natalie Angier (b. 1958) grew up in New York City and graduated from Barnard College in 1978. She has worked as a magazine staff writer for Discover *and* Time *and became a reporter for* The New York Times *in 1990. Her work as a* Times *science correspondent led to a Pulitzer Prize in 1991. She is also a recipient of the Lewis Thomas Award and was one of only seven journalists to receive four stars in the* Forbes Media Guide *that rated 500 reporters. She has also published in the* Atlantic, Parade, Washington Monthly, *and* Reader's Digest. *Her books include* The Beauty of the Beastly: New Views on the Nature of Life *(1995) and* Women: An Intimate Geography *(1999). In the following essay, first published in* The New York Times *in 1999, Angier examines the biological, social, and psychological differences between men and women in order to explain the reason why there is a marked difference in life expectancy between the genders.*

My father had great habits. Long before ficus trees met weight machines, he 1
was a dogged exerciser. He did pushups and isometrics. He climbed rocks. He
went for long, vigorous walks. He ate sparingly and avoided sweets and grease.
He took such good care of his teeth that they looked fake.

My father had terrible habits. He was chronically angry. He threw things 2
around the house and broke them. He didn't drink often, but when he did, he
turned more violent than usual. He didn't go to doctors, even when we begged
him to. He let a big, ugly mole on his back grow bigger and bigger, and so he
died of malignant melanoma, a curable cancer, at 51.

My father was a real man—so good and so bad. He was also Everyman.   3

Men by some measures take better care of themselves than women do and 4
are in better health. They are less likely to be fat, for example; they exercise
more, and suffer from fewer chronic diseases like diabetes, osteoporosis and
arthritis.

By standard measures, men have less than half the rate of depression seen 5
in women. When men do feel depressed, they tend to seek distraction in an activity, which, many psychologists say, can be a more effective technique for dispelling the mood than is a depressed woman's tendency to turn inward and
ruminate. In the United States and many other industrialized nations, women
are about three times more likely than men to express suicidal thoughts or to attempt to kill themselves.

And yet . . . men don't last. They die off in greater numbers than women do 6
at every stage of life, and thus their average life span is seven years shorter.
Women may attempt suicide relatively more often, but in the United States, four
times more men than women die from the act each year.

Men are also far more likely than women to die behind the wheel or to kill 7
others as a result of their driving. From 1977 to 1995, three and a half times
more male drivers than female drivers were involved in fatal car crashes. Death
by homicide also favors men; among those under 30, the male-to-female ratio is
8 to 1.

Yes, men can be impressive in their tendency to self-destruct, explosively or 8
gradually. They are at least twice as likely as women to be alcoholics and three
times more likely to be drug addicts. They have an eightfold greater chance
than women do of ending up in prison. Boys are much more likely than girls to
be thrown out of school for a conduct or antisocial personality disorder, or to
drop out on their own surly initiative. Men gamble themselves into a devastating economic and emotional pit two to three times more often than women do.

"Between boys' suicide rates, dropout rates and homicide rates, and men's 9
self-destructive behaviors generally, we have a real crisis in America," said
William S. Pollack, a psychologist at Harvard Medical School and co-director of
the Center for Men at McLean Hospital in Belmont, Mass. "Until recently, the
crisis has gone unheralded."

It is one thing to herald a presumed crisis, though, and to cite a ream of 10
gloomy statistics. It is quite another to understand the crisis, or to figure out
where it comes from or what to do about it. As those who study the various

forms of men's self-destructive behaviors realize, there is not a single, glib, over-arching explanation for the sex-specific patterns they see.

A crude evolutionary hypothesis would have it that men are natural risk- 11
takers, given to showy displays of bravado, aggression and daring all for the sake of attracting a harem of mates. By this premise, most of men's self-destructive, violent tendencies are a manifestation of their need to take big chances for the sake of passing their genes into the river of tomorrow.

Some of the data on men's bad habits fit the risk-taker model. For example, 12
those who study compulsive gambling have observed that men and women tend to display very different methods and preferences for throwing away big sums of money.

"Men get enamored of the action in gambling," said Linda Chamberlain, a 13
psychologist at Regis University in Denver who specializes in treating gambling disorders. "They describe an overwhelming rush of feelings and excitement as-sociated with the process of gambling. They like the feeling of being a player, and taking on a struggle with the house to show that they can overcome the odds and beat the system. They tend to prefer the table games, where they can feel powerful and omnipotent while everybody watches them."

Dr. Chamberlain noted that many male gamblers engage in other risk- 14
taking behaviors, like auto racing or hang gliding. By contrast, she said, "Women tend to use gambling more as a sedative, to numb themselves and es-cape from daily responsibilities, or feelings of depression or alienation. Women tend to prefer the solitary forms of gambling, the slot machines or video poker, where there isn't as much social scrutiny."

Yet the risk-taking theory does not account for why men outnumber 15
women in the consumption of licit and illicit anodynes. Alcohol, heroin and marijuana can be at least as numbing and sedating as repetitively pulling the arm of a slot machine. And some studies have found that men use drugs and alcohol for the same reasons that women often overeat: as an attempt to self-medicate when they are feeling anxious or in despair.

"We can speculate all we want, but we really don't know why men drink 16
more than women," said Enoch Gordis, the head of the National Institute on Al-cohol Abuse and Alcoholism. Nor does men's comparatively higher rate of sui-cide appear linked to the risk-taking profile. To the contrary, Paul Duberstein, as assistant professor of psychiatry and oncology at the University of Rochester School of Medicine, has found that people who complete a suicidal act are often low in a personality trait referred to as "openness to experience," tending to be rigid and inflexible in their behaviors. By comparison, those who express suici-dal thoughts tend to score relatively high on the openness-to-experience scale.

Given that men commit suicide more often that women, and women talk 17
about it more, his research suggests that, in a sense, women are the greater risk-takers and novelty seekers, while the men are likelier to feel trapped and help-less in the face of changing circumstances.

Silvia Cara Canetto, an associate professor of psychology at Colorado State 18
University in Fort Collins, has extensively studied the role of gender in suicidal

behaviors. Dr. Canetto has found that cultural narratives may determine why women attempt suicide more often while men kill themselves more often. She proposes that in Western countries, to talk about suicide or to survive a suicidal act is often considered "feminine," hysterical, irrational, and weak. To actually die by one's own hand may be viewed as "masculine," decisive, strong. Even the language conveys the polarized, weak-strong imagery: a "failed" suicide attempt as opposed to a "successful" one.

"There is indirect evidence that there is negative stigma toward men who 19 survive suicide," Dr. Canetto said. "Men don't want to 'fail,' even though failing in this case means surviving." If the "suicidal script" that identifies completing the acts as "rational, courageous and masculine" can be "undermined and torn to pieces," she said, we might have a new approach to prevention.

Dr. Pollak of the Center for Men also blames many of men's self-destructive 20 ways on the persistent image of the dispassionate, resilient, action-oriented male—the Marlboro Man who never even gasps for breath. For all the talk of the sensitive "new man," he argues, men have yet to catch up with women in expanding their range of acceptable emotions and behaviors. Men in our culture, Dr. Pollack says, are pretty much limited to a menu of three strong feelings: rage, triumph, lust. "Anything else and you risk being seen as a sissy," he said.

In a number of books, most recently, *Real Boys: Rescuing Our Sons from the* 21 *Myths of Boyhood,* he proposes that boys "lose their voice, a whole half of their emotional selves," beginning at age 4 or 5. "Their vulnerable, sad feelings and sense of need are suppressed or shamed out of them," he said—by their peers, parents, the great wide televised fist in their face.

He added: "If you keep hammering it into a kid that he has to look tough 22 and stop being a crybaby and a mama's boy, the boy will start creating a mask of bravado."

That boys and young men continue to feel confused over the proper har- 23 monics of modern masculinity was revealed in a study that Dr. Pollack conducted of 200 eighth-grade boys. Through questionnaires, he determined their scores on two scales, one measuring their "egalitarianism"—the degree to which they think men and women are equal, that men should change a baby's diapers, that mothers should work and the like—and the other gauging their "traditionalism" as determined by their responses to conventional notions, like the premise that men must "stand on their own two feet" and must "always be willing to have sex if someone asks."

On average, the boys scored high on both scales. "They are split on what it 24 means to be a man," said Dr. Pollack.

The cult of masculinity can beckon like a siren song in baritone. Dr. Franklin 25 L. Nelson, a clinical psychologist at the Fairbanks Community Mental Health Center in Alaska, sees many men who get into trouble by adhering to sentimental notions of manhood. "A lot of men come up here hoping to get away from a wimpy world and live like pioneers by old-fashioned masculine principles of individualism, strength and ruggedness," he said. They learn that nothing is

simple; even Alaska is part of a wider, interdependent world and they really do
need friends, warmth and electricity.

"Right now, it's 35 degrees below zero outside," he said during a January 26
interview. "If you're not prepared, it doesn't take long at that temperature to
freeze to death."

COMPREHENSION

1. What does the second sentence in the essay mean? What comment is the author suggesting about modern life by including it?
2. Angier makes a number of comparisons between the lifestyles of men and women. Does the author suggest one overriding principle regarding why "men don't last," or is it really a compilation of many factors? Explain.
3. Does Angier suggest that ideas of "masculinity" are hereditary or environmental or both? Explain by using examples from the text.
4. Angier begins her essay with a personal anecdote about her father. She claims that he was "Everyman." Does he appear to you to have acted like the "typical male"?

RHETORIC

1. Angier often mixes facts with theory. For example, in paragraph 11, she refers to a hypothesis that "men are natural risk-takers," while in other sections, she provides hard data about men's mortality rates. Does this combination make her argument more robust or does it make it less convincing?
2. Describe Angier's tone. Is she sympathetic that men die younger, or does she seem to castigate them? Select three examples from her essay that support your view.
3. Writers will often mention opposing viewpoints to buttress their own arguments. Angier doesn't. Does this strengthen or weaken her main premise? Explain.
4. Angier is noted for her lively prose style. Examine the following phrases and discuss their meaning and why they add "color" to her writing: "the great wide televised fist in their face" (paragraph 21); "the proper harmonics of modern masculinity" (paragraph 23); and "beckon like a siren song in baritone" (paragraph 25).
5. Compare and contrast the introduction and conclusion of the essay. How do they differ in imagery and tone? How does the subject matter of the conclusion help Angier achieve closure in her argument?
6. Angier often employs the "vocabulary of masculinity" in her argument. For example, she states, "My father was a real man" (paragraph 3); other masculine terms include "action-oriented" and "Marlboro man" (paragraph 20) and "away from a wimpy world" (paragraph 25). How does the use of this vocabulary contribute to her portrayal of the "masculine image"?

7. In paragraphs 5 through 14, Angier lists a number of statistics and behaviors that she attributes to male self-destructiveness. Is there a rationale behind the order in which she lists them, or does it seem more like a compendium of facts? Regardless of your answer, what is the rhetorical effect?

WRITING

1. Write an essay in the form of a process analysis with the title "How Men Can Live Longer." Be sure to provide examples and illustrations to support your report.
2. Write a narrative essay in which you describe an event in which you took or witnessed an unnecessary risk. Reflect on why the risk was taken and what its consequences were—whether positive or negative.
3. For a research paper, compare and contrast the "masculine traits" of American males with those of males from a different country or culture. Include at least four secondary sources.
4. **WRITING AN ARGUMENT:** Respond to Angier, arguing for or against the idea that women are more self-destructive than men.

✴ www.mhhe.com/ mhreader

For more information on Natalie Angier, go to:
More Resources > Ch. 5 Family & Gender

✴

Parenting as an Industry

Amitai Etzioni

Amitai Etzioni (b. 1929) is a noted sociologist and social policy adviser and a leader of the Communitarian movement to improve society. He graduated from Hebrew University in Jerusalem and received a PhD in sociology from the University of California at Berkeley. In addition, he has received numerous honorary doctorates. Among his many elected offices is president of the American Sociological Association from 1994 to 1995. He has taught at the University of Cologne, Columbia, George Washington University, and Harvard. He is the author of numerous books, including The Moral Dimension *(1988);* A Responsive Society *(1991);* The Spirit of Community *(1993), from which the following work is excerpted; and* The New Golden Rule: Community and Morality in a Democratic Society *(1997). In this essay he discusses what he perceives to be the woeful state of parenting in America.*

Consider for a moment parenting as an industry. As farming declined, most fa-   1
thers left to work away from home. Over the past twenty years millions of
American mothers have sharply curtailed their work in the "parenting industry" by moving to work outside the home. By 1991 two-thirds (66.7 percent) of
all mothers with children under eighteen were in the labor force and more than
half (55.4 percent) of women with children under the age of three. At the same
time a much smaller number of child care personnel moved into the parenting
industry.[1]

If this were any other business, say, shoemaking, and more than half of the   2
labor force had been lost and replaced with fewer, less-qualified hands and still
we asked the shoemakers to produce the same number of shoes of the same
quality (with basically no changes in technology), we would be considered
crazy. But this is what happened to parenting. As first men and then women left
to work outside the home, they were replaced by some child care services, a relatively small increase in baby-sitters and nannies, and some additional service
by grandparents—leaving parenting woefully shorthanded. The millions of
latchkey children, who are left alone for long stretches of time, are but the most
visible result of the parenting deficit.

Is this the "fault" of the women's movement, feminism, or mothers per se?   3
Obviously not. All women did was demand for themselves what men had long
possessed, working outside the home not only for their own personal satisfaction, but because of what they often perceived as the economic necessity. Whatever the cause, the result is an empty nest. Only it isn't the small fry who grew
up and took off: it is the parents who flew the coop. Those who did not leave altogether increased their investment of time, energy, involvement, and commitment outside the home.

Although parenting is the responsibility of both parents—and may well be   4
discharged most effectively in two-parent families immersed in a community
context of kin and neighbors—*most important is the scope of commitment.* Single
parents may do better than two-career absentee parents. Children require attention, as Robert Beliah and the other authors of *The Good Society* declared. Kids
also require a commitment of time, energy, and, above all, of self.

The prevalent situation is well captured by a public service commercial in   5
which a mother calls her child and reassures him that she has left money for
him next to the phone. "Honey, have some dinner," she mutters as the child
takes the twenty-dollar bill she left behind, rolls it up, and snorts cocaine. One
might add that the father didn't even call.

The fact is that parenting cannot be carried out over the phone, however   6
well meaning and loving the calls may be. It requires physical presence. The notion of "quality time" (not to mention "quality phone calls") is a lame excuse for
parental absence; it presupposes that bonding and education can take place in
brief time bursts, on the run. *Quality time occurs within quantity time.* As you

---

[1]*Current Population Survey,* Bureau of Labor Statistics, unpublished tabulations, 1991. [This and
subsequent notes in the selection are the author's.]

spend time with one's children—fishing, gardening, camping, or "just" eating a meal—there are unpredictable moments when an opening occurs and education takes hold. As family expert Barbara Dafoe Whitehead puts it: "Maybe there is indeed such a thing as a one-minute manager, but there is no such thing as a one-minute parent."[2]

Is the answer to the parenting deficit building more child care centers? After all, other societies have delegated the upbringing of their children, from black nannies in the antebellum South to Greek slaves in ancient Rome. True enough. But in these historical situations the person who attended to the children was an adjunct to the parents rather than a replacement for them and an accessory reserved mostly for upper-class families with leisure. A caregiver remained with the family throughout the children's formative years and often beyond; she was, to varying degrees, integrated into the family. The caregiver, in turn, reflected, at least in part, the family's values and educational posture. Some children may have been isolated from their parents, but as a rule there was a warm, committed figure dedicated to them, one who bonded and stayed with them.

Today most child care centers are woefully understaffed with poorly paid and underqualified personnel. Child care workers are in the lowest tenth of all wage earners (with an average salary of $5.35 per hour in 1988), well below janitors.[3] They frequently receive no health insurance or other benefits, which makes child care an even less attractive job. As Edward Zigler, a professor of child development at Yale, put it: "We pay these people less than we do zoo keepers—and then we expect them to do wonders."[4] The personnel come and go, at a rate of 41 percent per year at an average day care center.

Bonding between children and caregivers under these circumstances is very difficult to achieve. Moreover, children suffer a loss every time their surrogate parents leave. It would be far from inaccurate to call the worst of these facilities "kennels for kids." Sure, there are exceptions. There are a few fine, high-quality child care centers, but they are as rare and almost as expensive as the nannies that some truly affluent households can command. These exceptions should not distract us from the basically dismal picture: substandard care and all-too-frequent warehousing of children, with overworked parents trying frantically to make up the deficit in their free time.

Government or social supervision of the numerous small institutions and home facilities in which child care takes place to ensure proper sanitation and care, even to screen out child abusers, is difficult and is often completely

[2]Barbara Whitehead, "The New Politics in Action—Fortifying the Family," presentation at the conference "Left and Right: The Emergence of a New Politics in the 1990s?" sponsored by the Heritage Foundation and the Progressive Foundation, October 30, 1991, Washington, D.C. (see transcript, 25).

[3]Richard T. Gill, Nathan Glazer, Stephen A. Thernstrom, *Our Changing Population* (Englewood Cliffs, NJ: Prentice Hall, 1992), 278. Child care workers' average salary: *Who Cares? Child Care and the Quality of Care in America* (Oakland, CA: Child Care Employee Project, 1989), 49.

[4]Kenneth Labich, "Can Your Career Hurt Your Kids?" *Fortune*, May 20, 1991, 49.

neglected or only nominally carried out. We should not be surprised to encounter abuses such as the case of the child care home in which fifty-four children were left in the care of a sixteen-year-old and were found strapped into
child car seats for the entire day.[5]

Certainly many low-income couples and single parents have little or no 11
choice except to use the minimum that such centers provide. All we can offer
here is to urge that before parents put their children in such institutions, they
should check them out as extensively as possible (including surprise visits in
the middle of the day). Moreover, we should all support these parents' quest for
additional support from corporations and government if they cannot themselves spend more on child care.

Particularly effective are cooperative arrangements that require each parent 12
to contribute some time—four hours a week—to serve at his or her child's center. Not only do such arrangements reduce the center's costs, they also allow
parents to see firsthand what actually goes on, ensuring some measure of *built-
in accountability*. It provides for continuity—while staff come and go, parents
stay. (Even if they divorce, they may still participate in their child care center.)
And as parents get to know other parents of children in the same stage of development, they form social bonds, which can be drawn upon to work together to
make these centers more responsive to children's needs.

Above all, age matters. Infants under two years old are particularly vulner- 13
able to separation anxiety. Several bodies of data strongly indicate that infants
who are institutionalized at a young age will not mature into well-adjusted
adults.[6] As Edward Zigler puts it: "We are cannibalizing children. Children are
dying in the system, never mind achieving optimum development."[7] A study of
third-graders by two University of Texas researchers compared children who

---

[5]Ibid., 49.

[6]N. Baydar and Jeanne Brooks-Gunn, "Effects of Maternal Employment and Child Care Arrangements on Preschoolers' Cognitive and Behavioral Outcomes: Evidence from the Children of the
National Longitudinal Survey of Youth," *Developmental Psychology* 27 (November 1991): 932–46;
J. Belsky and Michael J. Rovine, "Nonmaternal Care in the First Year of Life and the Security of
Infant-Parent Attachment," *Child Development* 59 (February 1988): 157–67; T. B. Brazelton, "Issues
for Working Parents," *American Journal of Orthopsychiatry* 56 (1986): 14–25; J. Belsky and D. Egge-
been, "Early and Extensive Maternal Employment in Young Children's Socioemotional Development: Children of the National Longitudinal Survey of Youth," *Journal of Marriage and Family* 53
(November 1991): 1083–1110; B. E. Vaughn, K. E. Deane, and E. Waters, "The Impact of Out-of-
Home Care on Child-Mother Attachment Quality: Another Look at Some Enduring Questions,"
1–2, in I. Bretherton and E. Water, eds., *Growing Points of Attachment Theory and Research Mono-
graphs for the Society for Research in Child Development* 50 (1985): 1–2, serial no. 209.

Some studies have found that the effects of child care are not different from parental care. For
example, see K. A. Clarke-Stewart and G. G. Fein, "Early Childhood Programs," 917–99, in P. H.
Mussen, ed., *Handbook of Child Psychology*, Vol. 2 (New York: Wiley, 1983). And a few studies show
that child care rather than parental care is more effective for the intellectual development of poor
children. For example, see Jay Belsky, "Two Waves of Day Care Research: Development Effects
and Conditions of Quality," 1–34, in R. C. Ainslie, ed., *The Child and the Day Care Setting: Qualitative
Variations and Development* (New York: Praeger, 1984).

[7]Kenneth Labich, "Can Your Career Hurt Your Kids?" *Fortune*, May 20, 1991, 38.

returned home after school to their mothers with children who remained in day care centers:

> children who stayed at the day care centers after school were having problems. They received more negative peer nominations, and their negative nominations outweighed their positive nominations. In addition, the day care third-graders made lower academic grades on their report card and scored lower on standardized tests. There was some evidence of poor conduct grades.[8]

Unless the parents are absent or abusive, infants are better off at home. 14 Older children, between two and four, may be able to handle some measure of institutionalization in child care centers, but their personalities often seem too unformed to be able to cope well with a nine-to-five separation from a parent.

As a person who grew up in Israel, I am sometimes asked whether it is true 15 that kibbutzim succeed in bringing up toddlers in child care centers. I need to note first that unlike the personnel in most American child care centers, the people who care for children in kibbutzim are some of the most dedicated members of the work force because these communities consider child care to be a very high priority. As a result, child care positions are highly sought after and there is little turnover, which allows for essential bonding to take place. In addition, both parents are intimately involved in bringing up their children, and they frequently visit the child care centers, which are placed very close to where they live and work. Even so, Israeli kibbutzim are rapidly dismantling their collective child care centers and returning children to live with their families—because both the families and the community established that even a limited disassociation of children from their parents at a tender age is unacceptable.

There is no sense looking back and beating our breasts over how we got 16 ourselves into the present situation. But we must acknowledge that as a matter of social policy (as distinct from some individual situations) we have made a mistake in assuming that strangers can be entrusted with the effective personality formation of infants and toddlers. Over the last twenty-five years we have seen the future, and it is not a wholesome one. With poor and ineffective community child care, and with ever more harried parents, it will not suffice to tell their graduates to "just say no" and expect them to resist all temptations, to forgo illegal drugs and alcohol, and to postpone sexual activity. If we fervently wish them to grow up in a civilized society, and if we seek to live in one, let's face facts: it will not happen unless we dedicate more of ourselves to our children and their care and education. . . .

Nobody likes to admit it, but between 1960 and 1990 American society al- 17 lowed children to be devalued, while the golden call of "making it" was put on a high pedestal. Recently, college freshmen listed "being well off financially" as more important than "raising a family." (In 1990 the figures were 74 percent

---

[8]Deborah Lowe Vandell and Mary Anne Corasaniti, "The Relationship between Third-Graders' After-School Care and Social, Academic, and Emotional Functioning," *Child Development* 59 (August 1988): 874.

versus 70 percent, respectively, and in 1991 they were 74 percent versus 68 percent.)[9] . . .

Some blame this development on the women's rights movement, others on 18 the elevation of materialism and greed to new historical heights. These and other factors may have all combined to devalue children. However, women are obviously entitled to all the same rights men are, including the pursuit of greed.

But few people who advocated equal rights for women favored a society in 19 which sexual equality would mean a society in which all adults would act like men, who in the past were relatively inattentive to children. The new gender-equalized world was supposed to be a combination of all that was sound and ennobling in the traditional roles of women and men. Women were to be free to work any place they wanted, and men would be free to show emotion, care, and domestic commitment. For children this was not supposed to mean, as it too often has, that they would be bereft of dedicated parenting. Now that we have seen the result of decades of widespread neglect of children, the time has come for both parents to revalue children and for the community to support and recognize their efforts. . . .

We return then to the value we as a community put on having and bringing 20 up children. In a society that places more value on Armani suits, winter skiing, and summer houses than on education, parents are under pressure to earn more, whatever their income. They feel that it is important to work overtime and to dedicate themselves to enhancing their incomes and advancing their careers. We must recognize now, after two decades of celebrating greed and in the face of a generation of neglected children, the importance of educating one's children.

[9]American Enterprise Institute, 1990.

COMPREHENSION

1. Does Etzioni consider parenting a true industry, or is he using the term as a metaphor? Explain.
2. What does the author mean by "quality time occurs within quantity time" (paragraph 6)?
3. What sort of community does Etzioni suggest America has become? Does he state this explicitly or suggest it through examples? Explain.

RHETORIC

1. Etzioni states that if we applied the same criteria to any other business as we do to child care, we would be "considered crazy" (paragraph 2) and calls child care centers "kennels for kids" (paragraph 9). Why does he judge the current child care situation so severely?

2. What purpose do the citations to studies and surveys serve in advancing Etzioni's argument? Do they contribute significantly to advancing his thesis? Why or why not?
3. What audience does the author seem to be addressing? What is the implied educational level of the readership to which the essay is geared?
4. Is the author's concluding paragraph based on fact or opinion? Is his depiction of American society a caricature, or is it based on reality? Or is it both?
5. The author addresses the plight of poorer parents in paragraphs 11 and 12. How does his tone regarding their dilemma differ from the one he uses when he criticizes more affluent members of society?
6. The author has chosen to address all aspects of the "parenting industry" in a relatively short essay. Does the author sacrifice depth of understanding for breadth of inclusiveness, or does he do a good job of presenting all major facets of the problem? Explain your view.
7. In paragraph 15, Etzioni veers somewhat from his objective analysis by citing personal experience. Does this added dimension aid his argument? Why or why not?

WRITING

1. Write a 250- to 300-word summary of the essay. Follow the suggestions for summary writing discussed in Chapter 1.
2. Enlist three child care workers and have them read the article. Ask them to respond to Etzioni's complaints; then write a response essay, using their views as its focus.
3. Research the child care policy of the United States and that of another modern technological country. Compare and contrast the two.
4. **WRITING AN ARGUMENT:** Argue for or against the proposition that by not including a comprehensive child care plan, the author is not contributing to solving the crisis he addresses.

www.mhhe.com/ **mhreader** | For more information on Amitai Etzioni, go to: **More Resources > Ch. 5 Family & Gender**

CONNECTIONS FOR CRITICAL THINKING

1. Both Annie Dillard's "An American Childhood" and E. B. White's "Once More to the Lake" explore the experience of childhood from a different perspective. Do they share a common voice or mood? What is distinctive about each essay? Which essay do you prefer? Why? Consider the style and emotional impact of the writing.

2. Both Barbara Kingsolver's "Stone Soup" and Richard Rodriguez's "Family Values" attempt to alter stereotypes commonly held about contemporary families. What type of family does each author address? How do the authors differ in their rhetorical strategies and their use of supporting points to buttress their arguments? Who is the implied audience for each of the essays? How did you reach your conclusion?

3. In his essay "Parenting as an Industry," Amitai Etzioni describes the gross shortcomings of current child care practice with regard to its clients. Using Dillard's "An American Childhood," discuss how the communal nature of childhood experience could not be achieved in the typical "industrial" setting of contemporary child care.

4. Argue for or against the idea that Didion's portrayal of marriage and Angier's analysis of men are biased.

5. Argue for or against the idea that descriptions of the relatively new types of family relationships described by Kingsolver in "Stone Soup" or in Rodriguez's "Family Values" are presented in a biased, romanticized manner.

6. Argue for or against the view that changes in society and its norms, specifically, increased geographical mobility, an evolving workplace, ideas about economic class, and individual liberties, and sexual preference, have resulted in new forms of identity. Use examples from the work of Atwood, Etzioni, Brooks, and Rodriguez.

7. Select the two more substantially argued essays in this chapter, Etzioni's "Parenting as an Industry" and Rodriguez's "Family Values." Compare and contrast their methods of argumentation.

8. Angier criticizes many of the traits associated with manhood. Analyze her critique and argue for or against the proposition that if men were to abandon many of the behaviors and attitudes she describes, they would "last" longer, thus solving the issue of men's lower life expectancy as discussed in "Why Men Don't Last."

9. Establish your own definition of what it means to be a male or a female. Refer to the essays of Atwood, Angier, and Rodriguez.

10. Search the Web for *family values*. Write an extended analysis of the results of your research.

11. Join several newsgroups or chat rooms that focus on Internet dating. Compare and contrast the ideological focus of the conversations among members.

12. Create your own home page, and develop an interface that includes a selected quote regarding the family taken from one of your essays. Create a link to the home page with a response page where fellow students can make comments regarding the quotation. At the end of the semester, write a report and summary of the responses you receive. You may ask students to include their country of origin or their ethnicity to help you find possible connections between these factors and the responses.

chapter **6**

History, Culture, and Civilization

Are We Citizens of the World?

At the start of a new century, the paroxysms caused by conflicts among peoples, nations, ethnic groups, and cultures continue to shake continents. The United States might have emerged from the cold war as the dominant superpower, but numerous local and global threats remain. We seem to be at a crossroads in history, culture, and civilization, but does the future hold great promise or equally great danger?

The future assuredly holds significant peril as well as promise. History tells us that while there has never been complete absence of barbarism and nonrational behavior in human affairs, there have been societies, cultures, and nations committed to harmonious, or civil, conduct within various social realms. While it is clear that we have not attained an ideal state of cultural or world development, at the same time, we have advanced beyond the point in primitive civilization where someone chipped at a stone in order to make a better tool.

As we consider the course of contemporary civilization, we must contend with our own personal histories and cultures as well as with the interplay of contradictory global forces. We have become increasingly concerned with finding a purpose beyond the parameters of our very limiting personal and nationalistic identities, something that the Czech writer and statesman Václav Havel calls the "divine revolution." Indeed, we have entered an era of renewed ethnic strife, where a preoccupation with cultural difference seems stronger than the desire for universal civilization. The writers assembled here grapple with these contradictions; they search for those constituents of history and culture that might hasten the advent of a civilized world.

The idea of civilization suggests a pluralistic ethos whereby people of diverse histories and backgrounds can maintain cultural identities but also coexist with other cultural representatives in a spirit of tolerance and mutual respect. The wars, upheavals, and catastrophes of the 20th century were spawned by a narrow consciousness. Hopefully, as we enter a new century, all of us can advance the goal of a universal civilization based on the best that we have been able to create for humankind.

Previewing the Chapter

As you read the essays in this chapter and respond to them in discussion and writing, consider the following questions:

- How does the author define *culture, history,* or *civilization?* Is this definition stated or implied? Is the definition broad or narrow? Explain.
- Is the writer hopeful or pessimistic about the state of culture and history?
- What values does the author seem to think are necessary to advance the idea of history and culture?
- Do you find the author's tone to be objective or subjective? What is the author's purpose? Does the author have a personal motive in addressing the topic in the way he or she does?
- Which areas of knowledge—for example, history, philosophy, political science—does the author bring to bear on the subject?
- Do you agree or disagree with the author's view of the contemporary state of civilization?
- What cultural problems and historical conflicts are raised by the author in his or her treatment of the subject?
- Does the author have a narrow or broad focus on the relationship of history and culture to the larger society?
- Which authors altered your perspective on a topic, and why?
- Based on your reading of these essays, how would you define *civilization?* Are you hopeful about the current state of civilization?

Classic and Contemporary Images
HOW DO WE BECOME AMERICANS?

Using a Critical Perspective   Compare the scene of early 20th-century immigrants at New York City's Ellis Island with the March 1999 X-ray photo taken by Mexican authorities of human forms and cargo in a truck. What mood is conveyed by each representation? Does each photograph have a thesis or argument? Explain. Which photo do you find more engaging and provocative, and why?

From the time of the first European settlers, the North American continent has experienced wave after wave of immigration from every part of the world. One period of heavy immigration occurred in the late 19th and early 20th centuries, when millions of people from eastern and southern Europe entered the United States through Ellis Island in New York City, as shown in the classic photograph.

More recently, immigrants continue to come from all over the world, often entering the country illegally. The X-ray photo shows a wide shot and a close-up image of people being smuggled across Mexico's border with Guatemala.

Classic and Contemporary Essays
ARE WE HEADING TOWARD A WORLD CULTURE?

Both of these essays address the issues of prejudice and national identification. As you read them, consider not only the differing styles and methods of discourse of their authors (to be expected of essays written more than 200 years apart) but also their themes and import. In addition, consider that Reed is writing from the perspective of a racial (and to his mind) cultural minority and may therefore be particularly aware of those who would claim America's values, beliefs, doctrines, and cultural influences are homogeneous. Additionally, it may be ironic to note that Reed is writing within the context of a modern democratic state, while in 1762, the publication year of Goldsmith's essay, the United States did not even exist. Goldsmith's tone and style are immediately identifiable as being from another era. However, you should not confuse the formality of his rhetoric with the casual and intimate relationship to his reader. He addresses the reader quite directly, and the inspiration for his essay comes, he claims, from an informal conversation at a local tavern or coffeehouse. So, while his writing may appear abstruse, recall that Shakespeare wrote for an audience that had little formal education but understood his work with ease. Contemporary readers, on the other hand, often have trouble deciphering him because temporal distance often makes phrasing, vocabulary, and specific references obscure. Goldsmith uses no statistics, no historical record, no ideological analysis in his testimony. Rather, he assumes the tone of the "gentleman observer," a common form of address in his era. He may sound formal, but his tone is friendly. He disputes his colleagues' views that the English character represents the height of human development, but disarms his opponents through civility and deference. Reed is a contemporary African American writer of both fiction and nonfiction. From the perspective of a creative artist and keen observer of the modern intellectual and political scene, he combines personal experience and the historical record to demonstrate that much of "Western Civilization" draws its roots from a multiplicity of sources, not just European. Ironically, Reed, whose style is more colloquial than that of his distinguished predecessor, presents his argument using more of the traditional modes of support, for example, cultural analysis, appeals to authority, and historical evidence. Reed, however, unlike the courteous Goldsmith, is more forthright and direct in his disagreements. Can this be attributed to the reserved style of the English character versus the more direct discourse of the modern American? Or are we examining the reflections of two different personalities with different agendas?

＊

National Prejudices

Oliver Goldsmith

Oliver Goldsmith (1730–1774), the son of an Anglican curate, was an Anglo-Irish es-
sayist, poet, novelist, dramatist, and journalist. His reputation as an enduring figure in
English literature is based on his novel The Vicar of Wakefield *(1766); his play* She
Stoops to Conquer *(1773); his poem* The Deserted Village *(1770); and the essays*
and satiric letters collected in The Bee *(1759) and* The Citizen of the World *(1762).*
In this essay from the latter, Goldsmith argues quietly for a new type of citizen, one who
can transcend the xenophobia governing national behavior.

As I am one of that sauntering tribe of mortals, who spend the greatest part of   1
their time in taverns, coffee houses, and other places of public resort, I have
thereby an opportunity of observing an infinite variety of characters, which, to
a person of a contemplative turn, is a much higher entertainment than a view of
all the curiosities of art or nature. In one of these, my late rambles, I accidentally
fell into the company of half a dozen gentlemen, who were engaged in a warm
dispute about some political affair; the decision of which, as they were equally
divided in their sentiments, they thought proper to refer to me, which naturally
drew me in for a share of the conversation.

Amongst a multiplicity of other topics, we took occasion to talk of the dif-   2
ferent characters of the several nations of Europe; when one of the gentlemen,
cocking his hat, and assuming such an air of importance as if he had possessed
all the merit of the English nation in his own person, declared that the Dutch
were a parcel of avaricious wretches; the French a set of flattering sycophants;
that the Germans were drunken sots, and beastly gluttons; and the Spaniards
proud, haughty, and surly tyrants; but that in bravery, generosity, clemency, and
in every other virtue, the English excelled all the rest of the world.

This very learned and judicious remark was received with a general smile of   3
approbation by all the company—all, I mean, but your humble servant; who, endeavoring to keep my gravity as well as I could, and reclining my head upon my
arm, continued for some time in a posture of affected thoughtfulness, as if I had
been musing on something else, and did not seem to attend to the subject of conversation; hoping by these means to avoid the disagreeable necessity of explaining myself, and thereby depriving the gentleman of his imaginary happiness.

But my pseudo-patriot had no mind to let me escape so easily. Not satisfied   4
that his opinion should pass without contradiction, he was determined to have
it ratified by the suffrage of every one in the company; for which purpose addressing himself to me with an air of inexpressible confidence, he asked me if
I was not of the same way of thinking. As I am never forward in giving my

opinion, especially when I have reason to believe that it will not be agreeable; so, when I am obliged to give it, I always hold it for a maxim to speak my real sentiments. I therefore told him that, for my own part, I should not have ventured to talk in such a peremptory strain, unless I had made the tour of Europe, and examined the manners of these several nations with great care and accuracy: that, perhaps, a more impartial judge would not scruple to affirm that the Dutch were more frugal and industrious, the French more temperate and polite, the Germans more hardy and patient of labor and fatigue, and the Spaniards more staid and sedate, than the English; who, though undoubtedly brave and generous, were at the same time rash, headstrong, and impetuous; too apt to be elated with prosperity, and to despond in adversity.

I could easily perceive that all the company began to regard me with a jeal-   5
ous eye before I had finished my answer, which I had no sooner done, than the patriotic gentleman observed, with a contemptuous sneer, that he was greatly surprised how some people could have the conscience to live in a country which they did not love, and to enjoy the protection of a government, to which in their hearts they were inveterate enemies. Finding that by this modest declaration of my sentiments I had forfeited the good opinion of my companions, and given them occasion to call my political principles in question, and well knowing that it was in vain to argue with men who were so very full of themselves, I threw down my reckoning and retired to my own lodgings, reflecting on the absurd and ridiculous nature of national prejudice and prepossession.

Among all the famous sayings of antiquity, there is none that does greater   6
honor to the author, or affords greater pleasure to the reader (at least if he be a person of a generous and benevolent heart), than that of the philosopher, who, being asked what "countryman he was," replied, that he was, "a citizen of the world."—How few are there to be found in modern times who can say the same, or whose conduct is consistent with such a profession!—We are now become so much Englishmen, Frenchmen, Dutchmen, Spaniards or Germans, that we are no longer citizens of the world; so much the natives of one particular spot, or members of one petty society, that we no longer consider ourselves as the general inhabitants of the globe, or members of that grand society which comprehends the whole human kind.

Did these prejudices prevail only among the meanest and lowest of the peo-   7
ple, perhaps they might be excused, as they have few, if any, opportunities of correcting them by reading, travelling, or conversing with foreigners; but the misfortune is, that they infect the minds, and influence the conduct, even of our gentlemen; of those, I mean, who have every title to this appellation but an exemption from prejudice, which however, in my opinion, ought to be regarded as the characteristical mark of a gentleman; for let a man's birth be ever so high, his station ever so exalted, or his fortune ever so large, yet if he is not free from national and other prejudices, I should make bold to tell him, that he had a low and vulgar mind, and had no just claim to the character of a gentleman. And in fact, you will always find that those are most apt to boast of national merit, who

have little or no merit of their own to depend on; than which, to be sure, nothing is more natural: the slender vine twists around the sturdy oak, for no other reason in the world but because it has not strength sufficient to support itself.

Should it be alleged in defense of national prejudice, that it is the natural 8 and necessary growth of love to our country, and that therefore the former cannot be destroyed without hurting the latter, I answer, that this is a gross fallacy and delusion. That it is the growth of love to our country, I will allow; but that it is the natural and necessary growth of it, I absolutely deny. Superstition and enthusiasm too are the growth of religion; but who ever took it in his head to affirm that they are the necessary growth of this noble principle? They are, if you will, the bastard sprouts of this heavenly plant, but not its natural and genuine branches, and may safely enough be lopped off, without doing any harm to the parent stock; nay, perhaps, till once they are lopped off, this goodly tree can never flourish in perfect health and vigor.

Is it not very possible that I may love my own country, without hating the 9 natives of other countries? that I may exert the most heroic bravery, the most undaunted resolution, in defending its laws and liberty, without despising all the rest of the world as cowards and poltroons? Most certainly it is; and if it were not—But why need I suppose what is absolutely impossible?—But if it were not, I must own, I should prefer the title of the ancient philosopher, viz. a citizen of the world, to that of an Englishman, a Frenchman, a European, or to any other appellation whatever.

COMPREHENSION

1. Why does Goldsmith maintain that he is "a citizen of the world"? According to the author, could such an individual also be a patriot? Explain.
2. What connection does Goldsmith make between national prejudices and the conduct of gentlemen? Why does he allude to the manners of gentlemen?
3. Compare and contrast Goldsmith's observations with those of Schlesinger in "The Cult of Ethnicity" (page 8).

RHETORIC

1. Locate in the essay examples of the familiar style in writing. What is the relationship between this style and the tone and substance of the essay?
2. Explain the metaphors at the end of paragraphs 7 and 8.
3. What is the relevance of the introductory narrative, with its description of characters, to the author's declaration of thesis? Where does the author state his proposition concerning national prejudices?
4. Analyze the function of classification and contrast in paragraphs 2 to 5. How does the entire essay serve as a pattern of definition?

5. Examine the pattern of reasoning involved in the author's presentation of his argument in the essay, notably in paragraphs 6 to 8. What appeals to emotion and to reason does he make?
6. Assess the rhetorical effectiveness of Goldsmith's concluding paragraph.

WRITING

1. Why has it been difficult to eliminate the problem that Goldsmith posed in 1762? Are we better able today to function as citizens of the world? In what ways? What role does the United Nations play in this issue? What factors contribute to a new world citizenry? Explore these questions in an essay.
2. Write a paper on contemporary national prejudices—from the viewpoint of an ingenious foreigner.
3. **WRITING AN ARGUMENT:** Write an argumentative essay on the desirability of world government or on the need to be a citizen of the world.

www.mhhe.com/
mhreader

For more information on Oliver Goldsmith, go to:
More Resources > Ch. 6 History & Culture

※

America: The Multinational Society

Ishmael Reed

Ishmael Reed (b. 1938), an American novelist, poet, and essayist, is the founder and editor (along with Al Young) of Quilt *magazine, begun in 1981. In his writing, Reed uses a combination of standard English, black dialect, and slang to satirize American society. He believes that African Americans must move away from identification with Europe in order to rediscover their African qualities. Reed's books include* Flight to Canada *(1976),* The Terrible Twos *(1982),* The Terrible Threes *(1989), and* Japanese by Spring *(1993). In addition, he has written volumes of verse, including* Secretary to the Spirits *(1975), and has published collections of his essays, including* Airing Dirty Laundry *(1993). In the following essay from* Writin' Is Fightin' *(1990), Reed seeks to debunk the myth of the European ideal and argues for a universal definition of culture.*

At the annual Lower East Side Jewish Festival yesterday, a Chinese woman ate a pizza slice in front of Ty Thuan Duc's Vietnamese grocery store. Beside her a Spanish-speaking family patronized a cart with two signs: "Italian Ices" and "Kosher by Rabbi Alper." And after the pastrami ran out, everybody ate knishes.

—The New York Times, June 23, 1983

On the day before Memorial Day, 1983, a poet called me to describe a city he   1
had just visited. He said that one section included mosques, built by the Islamic
people who dwelled there. Attending his reading, he said, were large numbers
of Hispanic people, forty thousand of whom lived in the same city. He was not
talking about a fabled city located in some mysterious region of the world. The
city he'd visited was Detroit.

A few months before, as I was leaving Houston, Texas, I heard it announced   2
on the radio that Texas's largest minority was Mexican American, and though a
foundation recently issued a report critical of bilingual education, the taped
voice used to guide the passengers on the air trams connecting terminals in Dallas Airport is in both Spanish and English. If the trend continues, a day will
come when it will be difficult to travel through some sections of the country
without hearing commands in both English and Spanish; after all, for some
western states, Spanish was the first written language and the Spanish style
lives on in the western way of life.

Shortly after my Texas trip, I sat in an auditorium located on the campus of   3
the University of Wisconsin at Milwaukee as a Yale professor—whose original
work on the influence of African cultures upon those of the Americas has led to
his ostracism from some monocultural intellectual circles—walked up and
down the aisle, like an old-time southern evangelist, dancing and drumming
the top of the lectern, illustrating his points before some serious Afro-American
intellectuals and artists who cheered and applauded his performance and his
mastery of information. The professor was "white." After his lecture, he joined
a group of Milwaukeeans in a conversation. All of the participants spoke
Yoruban, though only the professor had ever traveled to Africa.

One of the artists told me that his paintings, which included African and   4
Afro-American mythological symbols and imagery, were hanging in the local
McDonald's restaurant. The next day I went to McDonald's and snapped pictures of smiling youngsters eating hamburgers below paintings that could grace
the walls of any of the country's leading museums. The manager of the local
McDonald's said, "I don't know what you boys are doing, but I like it," as he
commissioned the local painters to exhibit in his restaurant.

Such blurring of cultural styles occurs in everyday life in the United States to   5
a greater extent than anyone can imagine and is probably more prevalent than
the sensational conflict between people of different backgrounds that is played
up and often encouraged by the media. The result is what the Yale professor,
Robert Thompson, referred to as a cultural bouillabaisse, yet members of the nation's present educational and cultural Elect still cling to the notion that the
United States belongs to some vaguely defined entity they refer to as "Western
civilization," by which they mean, presumably, a civilization created by the people of Europe, as if Europe can be viewed in monolithic terms. Is Beethoven's
Ninth Symphony, which includes Turkish marches, a part of Western civilization, or the late nineteenth- and twentieth-century French paintings, whose creators were influenced by Japanese art? And what of the cubists, through whom
the influence of African art changed modern painting, or the surrealists, who

were so impressed with the art of the Pacific Northwest Indians that, in their map of North America, Alaska dwarfs the lower forty-eight in size?

Are the Russians, who are often criticized for their adoption of "Western" 6 ways by Tsarist dissidents in exile, members of Western civilization? And what of the millions of Europeans who have black African and Asian ancestry, black Africans having occupied several countries for hundreds of years? Are these "Europeans" members of Western civilization, or the Hungarians, who originated across the Urals in a place called Greater Hungary, or the Irish, who came from the Iberian Peninsula?

Even the notion that North America is part of Western civilization because 7 our "system of government" is derived from Europe is being challenged by Native American historians who say that the founding fathers, Benjamin Franklin especially, were actually influenced by the system of government that had been adopted by the Iroquois hundreds of years prior to the arrival of large numbers of Europeans.

Western civilization, then, becomes another confusing category like Third 8 World, or Judeo-Christian culture, as man attempts to impose his small-screen view of political and cultural reality upon a complex world. Our most publicized novelist recently said that Western civilization was the greatest achievement of mankind, an attitude that flourishes on the street level as scribbles in public restrooms: "White Power," "Niggers and Spics Suck," or "Hitler was a prophet," the latter being the most telling, for wasn't Adolph Hitler the archetypal monoculturalist who, in his pigheaded arrogance, believed that one way and one blood was so pure that it had to be protected from alien strains at all costs? Where did such an attitude, which has caused so much misery and depression in our national life, which has tainted even our noblest achievements, begin? An attitude that caused the incarceration of Japanese-American citizens during World War II, the persecution of Chicanos and Chinese Americans, the near-extermination of the Indians, and the murder and lynchings of thousands of Afro-Americans.

Virtuous, hardworking, pious, even though they occasionally would wan- 9 der off after some fancy clothes, or rendezvous in the woods with the town prostitute, the Puritans are idealized in our schoolbooks as "a hardy band" of no-nonsense patriarchs whose discipline razed the forest and brought order to the New World (a term that annoys Native American historians). Industrious, responsible, it was their "Yankee ingenuity" and practicality that created the work ethic. They were simple folk who produced a number of good poets, and they set the tone for the American writing style, of lean and spare lines, long before Hemingway. They worshiped in churches whose colors blended in with the New England snow, churches with simple structures and ornate lecterns.

The Puritans were a daring lot, but they had a mean streak. They hated the 10 theater and banned Christmas. They punished people in a cruel and inhuman manner. They killed children who disobeyed their parents. When they came in contact with those whom they considered heathens or aliens, they behaved in such a bizarre and irrational manner that this chapter in the American history

comes down to us as a late-movie horror film. They exterminated the Indians, who taught them how to survive in a world unknown to them, and their encounter with the calypso culture of Barbados resulted in what the tourist guide in Salem's Witches' House refers to as the Witchcraft Hysteria.

The Puritan legacy of hard work and meticulous accounting led to the establishment of a great industrial society; it is no wonder that the American industrial revolution began in Lowell, Massachusetts, but there was the other side, the strange and paranoid attitudes toward those different from the Elect. 11

The cultural attitudes of that early Elect continue to be voiced in everyday life in the United States: the president of a distinguished university, writing a letter to the *Times*, belittling the study of African civilizations; the television network that promoted its show on the Vatican art with the boast that this art represented "the finest achievements of the human spirit." A modern up-tempo state of complex rhythms that depends upon contacts with an international community can no longer behave as if it dwelled in a "Zion Wilderness" surrounded by beasts and pagans. 12

When I heard a schoolteacher warn the other night about the invasion of the American educational system by foreign curriculums, I wanted to yell at the television set, "Lady, they're already here." It has already begun because the world is here. The world has been arriving at these shores for at least ten thousand years from Europe, Africa, and Asia. In the late nineteenth and early twentieth centuries, large numbers of Europeans arrived, adding their cultures to those of the European, African, and Asian settlers who were already here, and recently millions have been entering the country from South America and the Caribbean, making Yale Professor Bob Thompson's bouillabaisse richer and thicker. 13

One of our most visionary politicians said that he envisioned a time when the United States could become the brain of the world, by which he meant the repository of all of the latest advanced information systems. I thought of that remark when an enterprising poet friend of mine called to say that he had just sold a poem to a computer magazine and that the editors were delighted to get it because they didn't carry fiction or poetry. Is that the kind of world we desire? A humdrum homogeneous world of all brains but no heart, no fiction, no poetry; a world of robots with human attendants bereft of imagination, of culture? Or does North America deserve a more exciting destiny? To become a place where the cultures of the world crisscross. This is possible because the United States is unique in the world: The world is here. 14

COMPREHENSION

1. Why does Reed believe that the notion of Western or European civilization is fallacious?
2. According to Reed, what are the origins of our monoculturalist view?
3. What are the dangers of such a narrow view? What historical examples does Reed allude to?

RHETORIC

1. How do paragraphs 1 to 4 help set the stage for Reed's discourse? Does this section contain Reed's thesis?
2. Does the computer analogy in Reed's conclusion work? Do his rhetorical questions underscore the thesis?
3. Comment on the author's extensive use of details and examples. How do they serve to support his point? Which examples are especially illuminating? Why?
4. What kind of humor does Reed use in his essay? Does its use contribute to the force of his essay? Why or why not?
5. Is Reed's reasoning inductive or deductive? Justify your answer.
6. How does Reed employ definitions to structure his essay?

WRITING

1. How does America's insistence that it is a European country affect its dealings with other nations? How does it influence the way it treats its own citizens? Explore these questions in a causal-analysis essay, using support from Reed.
2. Write an essay in which you consider how a multinational United States affects you on a day-to-day basis. How does it enrich your life or the life of the country? Use specific examples and details to support your opinion.
3. **WRITING AN ARGUMENT:** Write an essay arguing that a multinational society is often riddled with complex problems. What are some of the drawbacks or disadvantages of such a society? What causes these conflicts? Explore these issues in your writing.

> www.mhhe.com/
> **mhreader**
>
> For more information on Ishmael Reed, go to:
> **More Resources > Ch. 6 History & Culture**

Classic and Contemporary: Questions for Comparison

1. How do the respective tones of Goldsmith's and Reed's essays differ? What clues are contained in the texts that make this difference evident? Use examples from both.
2. In his essay, Goldsmith argues against nationalism and professes to be a citizen of the world. Compare Goldsmith's view with Reed's argument that to be an "American" means to accept the variety of influences that have converged into a "multinational society." How do these arguments differ? How are they similar?

3. Discuss both Goldsmith's and Reed's essays in terms of formality of voice. Does one author speak with more authority than the other? Or are they equally authoritative, but employing the stylistic modes of their times? Use examples from both essays.

✳

The Multicultural Mistake

K. Anthony Appiah

K. Anthony Appiah (b. 1954) was born in London, the son of a British mother and Ghanian father. Educated at Cambridge University, where he received his BA (1975), MA (1980), and PhD (1982), Appiah has taught in Ghana as well as at Harvard, Yale, Cornell, and Duke universities. He currently is professor of African American studies at Princeton University. His many books include In My Father's House: Africa in the Philosophy of Culture *(1992),* Globalization and Its Discontents *(with Saskia Sasser, 1999),* Africana: The Encyclopedia of the African and African-American Experience *(1999), and* Thinking It Through: An Introduction to Contemporary Philosophy *(2003). In the following selection, Appiah offers a critique of America's "much-vaunted variety."*

Have you noticed that *culture*—the word—has been getting a heavy workout recently? Anthropologists, of course, have used it zealously for over a century, though the term's active life in literature and politics began long before that. But some current ways in which the concept of culture has been put to use would have surprised even midcentury readers, especially the idea that everything from anorexia to zydeco is illuminated by being displayed as the product of some group's culture. 1

Culture's main competitor in its kudzu-like proliferation is *diversity,* a favorite now of corporate and educational CEOs, politicians, and pundits. And *cultural diversity* brings the two together. Is it not, indeed, one of the most pious of the pieties of our age that the United States is a society of enormous cultural diversity? And isn't Nathan Glazer right to say, in his new book, *We Are All Multiculturalists Now* (Harvard, 1997), that "multiculturalism is just the latest in [a] sequence of terms describing how American society, particularly American education, should respond to its diversity"? 2

Well, yes, American diversity is easily granted, and so is the need for a response to that diversity. But what isn't so clear is that it's our *cultural* diversity that deserves attention. 3

When Jews from the *shtetl* and Italians from the *villaggio* arrived at Ellis Island, they brought with them a rich mixture of what we call culture. That is, they brought a language and stories and songs and sayings; they transplanted a 4

religion with specific rituals, beliefs, and traditions, a cuisine of a certain hearty peasant quality, and distinctive modes of dress; and they came with particular ideas about family life. It was often reasonable for their new neighbors to ask what these first-generation immigrants were doing, and why; and a sensible answer would frequently have been, "It's an Italian thing" or "a Jewish thing," or, simply, "It's their culture."

It's striking how much of this form of difference has disappeared. There are still seders and nuptial masses, still gefilte fish and spaghetti, but how much does an Italian name tell you these days about church attendance, or knowledge of Italian, or taste in food or spouses? Even Jews, whose status as a small non-Christian group in an overwhelmingly Christian society might have been expected to keep their "difference" in focus, are getting harder to identify as a cultural group. (At the seder I go to every Passover, nearly half of those in attendance are gentiles.)

One way—the old way—of describing what has happened would be to say that the families that arrived during the turn-of-the-century wave of immigration have assimilated, become American. But, from another perspective, we might say that they became white. When the Italians and the Jews of Eastern Europe arrived, they were thought of as racially different both from African Americans and from the white Protestant majority. Now hardly anybody thinks of their descendants this way. They are Americans, but unless their ancestors include people from Africa or Asia, they are also white. And nobody, except perhaps a few oddballs in the Aryan Nation, thinks white people share a culture different from everybody else's.

The contrast between blacks and whites seems very evident, of course. White people rarely think of anything in their culture as white: normal, no doubt, middle-class, maybe, and even, sometimes, American; but not white. Black Americans, by contrast, do think of much in their lives in racial terms; they may speak black English (which some respectfully call Ebonics), go to black churches, listen and dance to black music. (And this isn't just how black people think; other people think that way about them.)

Yet to contrast black and white stories is to neglect much that they have in common. There are, indeed, forms of English speech that are black, even if there are also large regional and class variations in black, as in white, speech. But these are all forms of English we're talking about. Indeed, despite the vast waves of immigrants of the past few decades, something like 97 percent of adult Americans, whatever their color, speak English "like a native"; and, with the occasional adjustment for an accent here and there, those 97 percent can all understand one another. Leave out recent immigrants and the number gets close to 100 percent.

Language is only one of many things most Americans share. This is, for example, a country where almost every citizen knows something about baseball and basketball. Americans also share a familiarity with the consumer culture. They shop American style and know a good deal about the same consumer goods: Coca-Cola, Nike, Levi's, Ford, Nissan, GE. They have seen Hollywood

movies and know the names of some stars; and even the few who watch little or no television can probably tell you the names of some of its personalities. Even the supposedly persisting differences of religion turn out to be shallower than you might think. American Judaism is, as is often observed, extraordinarily American. Catholics in this country are a nuisance for Rome just because they are . . . well, so Protestant.

Coming as I do from Ghana, I find the broad cultural homogeneity of 10 America more striking than its much-vaunted variety. So why, in this society, which has less diversity of culture than most others, are we so preoccupied with diversity and so inclined to conceive of it as cultural?

Let me offer a name—not an explanation, just a piece of terminology for our 11 much-vaunted diversity. Let's say that we are creatures of *diverse social identities.* The cozy truism that we are a diverse society reflects the fact that many people now insist that they are profoundly shaped by the groups to which they belong, that their social identity—their membership in these groups—is central to who they are. Moreover, they go on to pursue what the Canadian philosopher Charles Taylor calls a "politics of recognition"; they ask the rest of us to ac- knowledge publicly their "authentic" identities.

The identities that demand recognition are multifarious. Some groups have 12 the names of the earlier ethnic cultures: Italian, Jewish, Polish. Some correspond to the old races (black, Asian, Indian) or to religions (Baptist, Catholic, Jewish). Some are basically regional (Southern, Western, Puerto Rican). Yet others are new groups modeled on the old ethnicities (Hispanic, Asian American) or are social categories (woman, gay, bisexual, disabled, deaf) that are none of these.

Nowadays, we are not the slightest bit surprised when someone remarks 13 upon a feature of the "culture" of groups like these. Gay culture, deaf culture, Chicano culture, Jewish culture—see how these phrases trip off the tongue. But if you ask what distinctively marks off gay people or deaf people or Jews from others, it is not obviously the fact that to each identity there corresponds, a dis- tinct culture. *Hispanic* sounds like the name of a cultural group defined by shar- ing the cultural trait of speaking Spanish, but half the second-generation Hispanics in California don't speak Spanish fluently, and in the next generation the proportion will fall even further.

You may wonder, in fact, whether there isn't a connection between the thin- 14 ning of the cultural content of identities and the rising stridency of their claims. Those European immigrants who lived in their rich ethnic cultures were busy demanding the linguistic Americanization of their children, making sure they learned America's official culture. One suspects that they didn't need to insist on the public recognition of their culture, because—whether or not they were happy with it—they simply took it for granted. Their middle-class descendants, whose domestic lives are conducted in English and extend eclectically from *Seinfeld* to Chinese takeout, are discomfited by a sense that their identities are shallow by comparison with those of their grandparents; some of them fear that unless the rest of us acknowledge the importance of their difference, there soon won't be anything worth acknowledging.

For many middle-class Americans, families have changed. Grandparents 15
have moved into retirement communities, cousins no longer live down the
street, parents have separated. In sum, many of the social preconditions of that
extended intergenerational family life have disappeared, and, for many Americans, the will to live that way has vanished too. Given the connection between
the old family life and the old cultural identities, it is not surprising that the loss
of the former has produced nostalgia for the latter.

The trouble with appealing to cultural difference is that it obscures rather 16
than illuminates this situation. It's not black culture that the racist disdains, but
blacks. No amount of knowledge of the architectural achievements of Nubia or
Kush guarantees respect for African Americans. No African American is entitled
to greater concern because he is descended from a people who created jazz or
produced Toni Morrison. Culture is not the problem, and it is not the solution.

So maybe we should conduct our discussions of education and citizenship, 17
toleration and social peace, without the talk of cultures. Long ago, in the mists of
prehistory, our ancestors learned that it is sometimes good to let a field lie fallow.

COMPREHENSION

1. How does Appiah define the following concepts: *culture, cultural diversity,
   cultural homogeneity, diverse social identities?* What connections does he draw
   among these terms? What, specifically, is the difference *between cultural diversity* and *diverse social identities?*
2. According to Appiah, why is "the broad cultural homogeneity of America
   more striking than its much-vaunted variety" (paragraph 10)?
3. What problems does the writer see in appealing to cultural difference? Why
   does Appiah suggest that we "do away" with talk of culture? Do you think
   he is serious? Why or why not?

RHETORIC

1. How does the writer introduce the subject? What is his tone? How does he
   manage to appeal to his audience even though the subject—*culture*—is
   somewhat abstract? How does Appiah establish himself as an authority on
   the subject of culture?
2. Trace the writer's use of examples. What are the major extended examples?
   What are the minor examples? How do they serve to support his key ideas
   about culture and related topics?
3. Where does the writer employ the comparative method? What is his purpose in drawing comparisons and contrasts?
4. Appiah's style ranges from the colloquial to the abstract. Cite examples of
   each. What is the effect?
5. Trace the pattern of definition that the writer develops. What are the key elements and related terms in this definition?

6. What thesis or claim emerges from Appiah's definition of culture? How does the writer use the end paragraphs to persuade his audience to accept his conclusions about culture in America?

WRITING

1. Write your own definition of culture, using it to inquire into the related ideas of cultural diversity and homogeneity.
2. In an essay, describe your college's "culture"? Does student life suggest "cultural diversity" or "cultural homogeneity"? What has caused this campus culture, as you see it, and what are the effects?
3. **WRITING AN ARGUMENT:** Appiah finds America's homogeneity to be more striking than its diversity. Do you agree or disagree with him? Using your own reasons and evidence, respond to this question in an argumentative essay.

| www.mhhe.com/ **mhreader** | For more information on K. Anthony Appiah, go to: **More Resources > Ch. 6 History & Culture** |

✳

The Myth of the Latin Woman: I Just Met a Girl Named Maria

Judith Ortiz Cofer

Judith Ortiz Cofer (b. 1952) was born in Puerto Rico and immigrated to the United States in 1956. Once a bilingual teacher in Florida public schools, Cofer has written two books of poetry; several plays; a novel, The Line of the Sun *(1989); an award-winning collection of essays and poems,* Silent Dancing: A Partial Remembrance of a Puerto Rican Childhood *(1990); and a collection of short stories,* An Island Like You: Stories of the Barrio *(1995). She is a professor of English and creative writing at the University of Georgia. In the following essay, she offers both personal insight and philosophical reflection on the theme of ethnic stereotyping.*

On a bus trip to London from Oxford University where I was earning some   1
graduate credits one summer, a young man, obviously fresh from a pub, spotted me and as if struck by inspiration went down on his knees in the aisle. With both hands over his heart he broke into an Irish tenor's rendition of "Maria" from *West Side Story.* My politely amused fellow passengers gave his lovely voice the round of gentle applause it deserved. Though I was not quite as

amused, I managed my version of an English smile: no show of teeth, no extreme contortions of the facial muscles—I was at this time of my life practicing reserve and cool. Oh, that British control, how I coveted it. But "Maria" had followed me to London, reminding me of a prime fact of my life: you can leave the island, master the English language, and travel as far as you can, but if you are a Latina, especially one like me who so obviously belongs to Rita Moreno's gene pool, the island travels with you.

This is sometimes a very good thing. it may win you that extra minute of someone's attention. But with some people, the same things can make *you* an island—not a tropical paradise but an Alcatraz, a place nobody wants to visit. As a Puerto Rican girl living in the United States and wanting like most children to "belong," I resented the stereotype that my Hispanic appearance called forth from many people I met.

Growing up in a large urban center in New Jersey during the 1960s, I suffered from what I think of as "cultural schizophrenia." Our life was designed by my parents as a microcosm of their *casas* on the island. We spoke in Spanish, ate Puerto Rican food bought at the *bodega*, and practiced strict Catholicism at a church that allotted us a one-hour slot each week for mass, performed in Spanish by a Chinese priest trained as a missionary for Latin America.

As a girl I was kept under strict surveillance by my parents, since my virtue and modesty were, by their cultural equation, the same as their honor. As a teenager I was lectured constantly on how to behave as a proper *senorita*. But it was a conflicting message I received, since the Puerto Rican mothers also encouraged their daughters to look and act like women and to dress in clothes our Anglo friends and their mothers found too "mature" and flashy. The difference was, and is, cultural; yet I often felt humiliated when I appeared at an American friend's party wearing a dress more suitable to a semiformal than to a playroom birthday celebration. At Puerto Rican festivities, neither the music nor the colors we wore could be too loud.

I remember Career Day in our high school, when teachers told us to come dressed as if for a job interview. It quickly became obvious that to the Puerto Rican girls "dressing up" meant wearing their mother's ornate jewelry and clothing, more appropriate (by mainstream standards) for the company Christmas party than as daily office attire. That morning I had agonized in front of my closet, trying to figure out what a "career girl" would wear. I knew how to dress for school (at the Catholic school I attended, we all wore uniforms), I knew how to dress for Sunday mass, and I knew what dresses to wear for parties at my relatives' homes. Though I do not recall the precise details of my Career Day outfit, it must have been a composite of these choices. But I remember a comment my friend (an Italian American) made in later years that coalesced my impressions of that day. She said that at the business school she was attending, the Puerto Rican girls always stood out for wearing "everything at once." She meant, of course, too much jewelry, too many accessories. On that day at school we were simply made the negative models by the nuns, who were themselves not credible fashion experts to any of us. But it was painfully obvious to me that

to the others, in their tailored skirts and silk blouses, we must have seemed "hopeless" and "vulgar." Though I now know that most adolescents feel out of step much of the time, I also know that for the Puerto Rican girls of my generation that sense was intensified. The way our teachers and classmates looked at us that day in school was just a taste of the cultural clash that awaited us in the real world, where prospective employers and men on the street would often misinterpret our tight skirts and jingling bracelets as a "come-on."

Mixed cultural signals have perpetuated certain stereotypes—for example, 6 that of the Hispanic woman as the "hot tamale" or sexual firebrand. It is a one-dimensional view that the media have found easy to promote. In their special vocabulary, advertisers have designated "sizzling" and "smoldering" as the adjectives of choice for describing not only the foods but also the women of Latin America. From conversations in my house I recall hearing about the harassment that Puerto Rican women endured in factories where the "boss-men" talked to them as if sexual innuendo was all they understood, and worse, often gave them the choice of submitting to their advances or being fired.

It is custom, however, not chromosomes, that leads us to choose scarlet over 7 pale pink. As young girls it was our mothers who influenced our decisions about clothes and colors—mothers who had grown up on a tropical island where the natural environment was a riot of primary colors, where showing your skin was one way to keep cool as well as to look sexy. Most important of all, on the island, women perhaps felt freer to dress and move more provocatively since, in most cases, they were protected by the traditions, mores, and laws of a Spanish/ Catholic system of morality and machismo whose main rule was: *You may look at my sister, but if you touch her I will kill you.* The extended family and church structure could provide a young woman with a circle of safety in her small pueblo on the island; if a man "wronged" a girl, everyone would close in to save her family honor.

My mother has told me about dressing in her best party clothes on Saturday 8 nights and going to the town's plaza to promenade with her girlfriends in front of the boys they liked. The males were thus given an opportunity to admire the women and to express their admiration in the form of *piropos:* erotically charged street poems they composed on the spot. (I have myself been subjected to a few *piropos* while visiting the island, and they can be outrageous, although custom dictates that they must never cross into obscenity.) This ritual, as I understand it, also entails a show of studied indifference on the woman's part; if she is "decent," she must not acknowledge the man's impassioned words. So I do understand how things can be lost in translation. When a Puerto Rican girl dressed in her idea of what is attractive meets a man from the mainstream culture who has been trained to react to certain types of clothing as a sexual signal, a clash is likely to take place. I remember the boy who took me to my first formal dance leaning over to plant a sloppy, overeager kiss painfully on my mouth; when I didn't respond with sufficient passion, he remarked resentfully: "I thought you Latin girls were supposed to mature early," as if I were expected to *ripen* like a fruit or vegetable, not just grow into womanhood like other girls.

It is surprising to my professional friends that even today some people, in- 9
cluding those who should know better, still put others "in their place." It happened to me most recently during a stay at a classy metropolitan hotel favored
by young professional couples for weddings. Late one evening after the theater,
as I walked toward my room with a colleague (a woman with whom I was coordinating an arts program), a middle-aged man in a tuxedo, with a young girl
in satin and lace on his arm, stepped directly into our path. With his champagne
glass extended toward me, he exclaimed "Evita!"

Our way blocked, my companion and I listened as the man half-recited, 10
half-bellowed "Don't Cry for Me, Argentina." When he finished, the young girl
said: "How about a round of applause for my daddy?" We complied, hoping
this would bring the silly spectacle to a close. I was becoming aware that our little group was attracting the attention of the other guests. "Daddy" must have
perceived this too, and he once more barred the way as we tried to walk past
him. He began to shout-sing a ditty to the tune of "La Bamba"—except the
lyrics were about a girl named Maria whose exploits rhymed with her name
and gonorrhea. The girl kept saying "Oh, Daddy" and looking at me with
pleading eyes. She wanted me to laugh along with the others. My companion
and I stood silently waiting for the man to end his offensive song. When he finished, I looked not at him but at his daughter. I advised her calmly never to ask
her father what he had done in the army. Then I walked between them and to
my room. My friend complimented me on my cool handling of the situation,
but I confessed that I had really wanted to push the jerk into the swimming
pool. This same man—probably a corporate executive, well-educated, even
worldly by most standards—would not have been likely to regale an Anglo
woman with a dirty song in public. He might have checked his impulse by assuming that she could be somebody's wife or mother, or at least *somebody* who
might take offense. But, to him, I was just an Evita or a Maria: merely a character in his cartoon-populated universe.

Another facet of the myth of the Latin woman in the United States is the me- 11
nial, the domestic—Maria the housemaid or countergirl. It's true that work as
domestics, as waitresses, and in factories is all that's available to women with little English and few skills. But the myth of the Hispanic menial—the funny maid,
mispronouncing words and cooking up a spicy storm in a shiny California
kitchen—has been perpetuated by the media in the same way that "Mammy"
from *Gone with the Wind* became America's idea of the black woman for generations. Since I do not wear my diplomas around my neck for all to see, I have on
occasion been sent to that "kitchen" where some think I obviously belong.

One incident has stayed with me, though I recognize it as a minor offense. 12
My first public poetry reading took place in Miami, at a restaurant where a
luncheon was being held before the event. I was nervous and excited as I
walked in with notebook in hand. An older woman motioned me to her table,
and thinking (foolish me) that she wanted me to autograph a copy of my newly
published slender volume of verse, I went over. She ordered a cup of coffee
from me, assuming that I was the waitress. (Easy enough to mistake my poems

for menus, I suppose.) I know it wasn't an intentional act of cruelty. Yet of all the good things that happened later, I remember that scene most clearly, because it reminded me of what I had to overcome before anyone would take me seriously. In retrospect I understand that my anger gave my reading fire. In fact, I have almost always taken any doubt in my abilities as a challenge, the result most often being the satisfaction of winning a convert, of seeing the cold, appraising eyes warm to my words, the body language change, the smile that indicates I have opened some avenue for communication. So that day as I read, I looked directly at that woman. Her lowered eyes told me she was embarrassed at her faux pas, and when I willed her to look up at me, she graciously allowed me to punish her with my full attention. We shook hands at the end of the reading and I never saw her again. She has probably forgotten the entire incident, but maybe not.

Yet I am one of the lucky ones. There are thousands of Latinas without the 13 privilege of an education or the entrees into society that I have. For them life is a constant struggle against the misconceptions perpetuated by the myth of the Latina. My goal is to try to replace the old stereotypes with a much more interesting set of realities. Every time I give a reading, I hope the stories I tell, the dreams and fears I examine in my work, can achieve some universal truth that will get my audience past the particulars of my skin color, my accent, or my clothes.

I once wrote a poem in which I called all Latinas "God's brown daughters." 14 This poem is really a prayer of sorts, offered upward, but also, through the human-to-human channel of art, outward. It is a prayer for communication and for respect. In it, Latin women pray "in Spanish to an Anglo God/with a Jewish heritage," and they are "fervently hoping/that if not omnipotent,/ at least He be bilingual."

COMPREHENSION

1. What is the theme of the essay?
2. What does Cofer mean by the expression "cultural schizophrenia" (paragraph 3)?
3. Define the following words: *coveted* (paragraph 1), *Anglo* (paragraph 4), *coalesced* (paragraph 5), *machismo* (paragraph 7), and *entrees* (paragraph 13).

RHETORIC

1. Cofer uses many anecdotes in her discussion of stereotyping. How does this affect the tone of the essay?
2. What is the implied audience for this essay? What aspects of the writing led you to your conclusion?
3. This essay is written in the first person, which tends to reveal a lot about the writer's personality. What adjectives come to mind when you think of the writer's singular voice?

4. Although this essay has a sociological theme, Cofer demonstrates that she has a poet's sensitivity toward language. What in the following sentence from paragraph 7 demonstrates this poetic style: "It is custom, however, not chromosomes, that leads us to choose scarlet over pale pink"? Select two other sentences from the essay that demonstrate Cofer's stylistic talent, and explain why they, too, are poetic.

5. In paragraph 8, Cofer demonstrates differing cultural perceptions between Hispanic and Anglo behavior. How is the paragraph structured so that this difference is demonstrated dramatically?

6. Cofer uses quotation marks to emphasize the connotation of certain words. Explain the significance of the following words: *mature* (paragraph 4); *hopeless* (paragraph 5); *hot tamale* (paragraph 6); *wronged* (paragraph 7); and *decent* (paragraph 8).

WRITING

1. Write a problem-solution essay in which you (*a*) discuss the reasons behind cultural stereotyping and (*b*) provide suggestions on how to overcome stereotyped thinking.

2. Select an ethnic, racial, or cultural group, and explain how group members undergo stereotyping through their depiction in the media.

3. **WRITING AN ARGUMENT:** In an essay, argue for or against the proposition that stereotyping is excusable because it often is based on learned assumptions about which an individual cannot be expected to have knowledge.

---

www.mhhe.com/ **mhreader**	For more information on Judith Ortiz Cofer, go to: **More Resources > Ch. 6 History & Culture**

✴

Yellow Woman and a Beauty of the Spirit

Leslie Marmon Silko

Leslie Marmon Silko (b.1948) was born in Albuquerque, New Mexico, and grew up on the Laguna Pueblo Reservation on the Rio Grande plateau. Of mixed Laguna, Mexican, and European American ancestry, Silko attended the University of New Mexico (BA, 1969) and briefly enrolled in law school before deciding to pursue a career as a writer. Associated with the Native American Renaissance, Silko has written stories, novels, essays, and poetry exploring Native American myths and traditions as well as the relationship of the tribes to contemporary culture. Silko has taught at the University of New Mexico and the University of Arizona and has received numerous awards, including a prestigious five-year MacArthur Foundation grant. Her best-known work includes the novel Ceremony *(1977), a collection of poetry,* Laguna Woman *(1974), a collection of short stories,* Storyteller *(1981), and an autobiography,* Sacred Water *(1993). Silko has also published a collection of essays,* Yellow Woman and a Beauty of the Spirit *(1996); in the title essay from this collection, Silko examines her mixed ancestry and explains traditional Pueblo culture.*

From the time I was a small child, I was aware that I was different. I looked different from my playmates. My two sisters looked different too. We didn't look quite like the other Laguna Pueblo children, but we didn't look quite white either. In the 1880s, my great-grandfather had followed his older brother west from Ohio to the New Mexico Territory to survey the land for the U.S. government. The two Marmon brothers came to the Laguna Pueblo reservation because they had an Ohio cousin who already lived there. The Ohio cousin was involved in sending Indian children thousands of miles away from their families to the War Department's big Indian boarding school in Carlisle, Pennsylvania. Both brothers married full-blood Laguna Pueblo women. My great-grandfather had first married my great-grandmother's older sister, but she died in childbirth and left two small children. My great-grandmother was fifteen or twenty years younger than my great-grandfather. She had attended Carlisle Indian School and spoke and wrote English beautifully.

I called her Grandma A'mooh because that's what I heard her say whenever she saw me. *A'mooh* means "granddaughter" in the Laguna language. I remember this word because her love and her acceptance of me as a small child were so important. I had sensed immediately that something about my appearance was not acceptable to some people, white and Indian. But I did not see any signs of that strain or anxiety in the face of my beloved Grandma A'mooh.

Younger people, people my parents' age, seemed to look at the world in a more modern way. The modern way included racism. My physical appearance seemed not to matter to the old-time people. They looked at the world very differently; a person's appearance and possessions did not matter nearly as much as a person's behavior. For them, a person's value lies in how that person interacts with other people, how that person behaves toward the animals and the earth. That is what matters most to the old-time people. The Pueblo people believed this long before the Puritans arrived with their notions of sin and damnation, and racism. The old-time beliefs persist today; thus I will refer to the old-time people in the present tense as well as the past. Many worlds may coexist here.

I spent a great deal of time with my great-grandmother. Her house was next to our house, and I used to wake up at dawn, hours before my parents or younger sisters, and I'd go wait on the porch swing or on the back steps by her kitchen door. She got up at dawn, but she was more than eighty years old, so she needed a little while to get dressed and to get the fire going in the cookstove. I had been carefully instructed by my parents not to bother her and to behave, and to try to help her any way I could. I always loved the early mornings when the air was so cool with a hint of rain smell in the breeze. In the dry New Mexico air, the least hint of dampness smells sweet.

My great-grandmother's yard was planted with lilac bushes and iris; there were four o'clocks, cosmos, morning glories, and hollyhocks, and old-fashioned rosebushes that I helped her water. If the garden hose got stuck on one of the big rocks that lined the path in the yard, I ran and pulled it free. That's what I came to do early every morning: to help Grandma water the plants before the heat of the day arrived.

Grandma A'mooh would tell about the old days, family stories about relatives who had been killed by Apache raiders who stole the sheep our relatives had been herding near Swahnee. Sometimes she read Bible stories that we kids liked because of the illustrations of Jonah in the mouth of a whale and Daniel surrounded by lions. Grandma A'mooh would send me home when she took her nap, but when the sun got low and the afternoon began to cool off, I would be back on the porch swing, waiting for her to come out to water the plants and to haul in firewood for the evening. When Grandma was eighty-five, she still chopped her own kindling. She used to let me carry in the coal bucket for her, but she would not allow me to use the ax. I carried armloads of kindling too, and I learned to be proud of my strength.

I was allowed to listen quietly when Aunt Susie or Aunt Alice came to visit Grandma. When I got old enough to cross the road alone, I went and visited them almost daily. They were vigorous women who valued books and writing. They were usually busy chopping wood or cooking but never hesitated to take time to answer my questions. Best of all they told me the *hummah-hah* stories, about an earlier time when animals and humans shared a common language. In the old days, the Pueblo people had educated their children in this manner; adults took time out to talk to and teach young people. Everyone was a teacher, and every activity had the potential to teach the child.

But as soon as I started kindergarten at the Bureau of Indian Affairs day school, I began to learn more about the differences between the Laguna Pueblo world and the outside world. It was at school that I learned just how different I looked from my classmates. Sometimes tourists driving past on Route 66 would stop by Laguna Day School at recess time to take photographs of us kids. One day, when I was in the first grade, we all crowded around the smiling white tourists, who peered at our faces. We all wanted to be in the picture because afterward the tourists sometimes gave us each a penny. Just as we were all posed and ready to have our picture taken, the tourist man looked at me. "Not you," he said and motioned for me to step away from my classmates. I felt so embarrassed that I wanted to disappear. My classmates were puzzled by the tourists' behavior, but I knew the tourists didn't want me in their snapshot because I looked different, because I was part white.

In the view of the old-time people, we are all sisters and brothers because the Mother Creator made all of us—all colors and all sizes. We are sisters and brothers, clanspeople of all the living beings around us. The plants, the birds, fish, clouds, water, even the clay—they are all related to us. The old-time people believe that all things, even rocks and water, have spirit and being. They understood that all things want only to continue being as they are; they need only to be left as they are. Thus the old folks used to tell us kids not to disturb the earth unnecessarily. All things as they were created exist already in harmony with one another as long as we do not disturb them.

As the old story tells us, Tse'itsi'nako, Thought Woman, the Spider, thought 10 of her three sisters, and as she thought of them, they came into being. Together with Thought Woman, they thought of the sun and the stars and the moon. The Mother Creators imagined the earth and the oceans, the animals and the people, and the *ka'tsina* spirits that reside in the mountains. The Mother Creators imagined all the plants that flower and the trees that bear fruit. As Thought Woman and her sisters thought of it, the whole universe came into being. In this universe, there is no absolute good or absolute bad; they are only balances and harmonies that ebb and flow. Some years the desert receives abundant rain, other years there is too little rain, and sometimes there is so much rain that floods cause destruction. But rain itself is neither innocent nor guilty. The rain is simply itself.

My great-grandmother was dark and handsome. Her expression in photo- 11 graphs is one of confidence and strength. I do not know if white people then or now would consider her beautiful. I do not know if the old-time Laguna Pueblo people considered her beautiful or if the old-time people even thought in those terms. To the Pueblo way of thinking, the act of comparing one living being with another was silly, because each being or thing is unique and therefore incomparably valuable because it is the only one of its kind. The old-time people thought it was crazy to attach such importance to a person's appearance. I understood very early that there were two distinct ways of interpreting the world. There was the white people's way and there was the Laguna way. In the Laguna way, it was bad manners to make comparisons that might hurt another person's feelings.

In everyday Pueblo life, not much attention was paid to one's physical ap- 12 pearance or clothing. Ceremonial clothing was quite elaborate but was used only for the sacred dances. The traditional Pueblo societies were communal and strictly egalitarian, which means that no matter how well or how poorly one might have dressed, there was no social ladder to fall from. All food and other resources were strictly shared so that no one person or group had more than another. I mention social status because it seems to me that most of the definitions of beauty in contemporary Western culture are really codes for determining social status. People no longer hide their face-lifts and they discuss their liposuctions because the point of the procedures isn't just cosmetic, it is social. It says to the world, "I have enough spare cash that I can afford surgery for cosmetic purposes."

In the old-time Pueblo world, beauty was manifested in behavior and in 13 one's relationships with other living beings. Beauty was as much a feeling of harmony as it was a visual, aural, or sensual effect. The whole person had to be beautiful, not just the face or the body; faces and bodies could not be separated from hearts and souls. Health was foremost in achieving this sense of well-being and harmony; in the old-time Pueblo world, a person who did not look healthy inspired feelings of worry and anxiety, not feelings of well-being. A healthy person, of course, is in harmony with the world around her; she is at peace with herself too. Thus an unhappy person or spiteful person would not be considered beautiful.

In the old days, strong, sturdy women were most admired. One of my most 14 vivid preschool memories is of the crew of Laguna women, in their forties and fifties, who came to cover our house with adobe plaster. They handled the ladders with great ease, and while two women ground the adobe mud on stones and added straw, another woman loaded the hod with mud and passed it up to the two women on ladders, who were smoothing the plaster on the wall with their hands. Since women owned the houses, they did the plastering. At Laguna, men did the basket making and the weaving of fine textiles; men helped a great deal with the child care too. Because the Creator is female, there is no stigma on being female; gender is not used to control behavior. No job was a man's job or a woman's job; the most able person did the work.

My Grandma Lily had been a Ford Model A mechanic when she was a 15 teenager. I remember when I was young, she was always fixing broken lamps and appliances. She was small and wiry, but she could lift her weight in rolled roofing or boxes of nails. When she was seventy-five, she was still repairing washing machines in my uncle's coin-operated laundry.

The old-time people paid no attention to birthdays. When a person was 16 ready to do something, she did it. When she no longer was able, she stopped. Thus the traditional Pueblo people did not worry about aging or about looking old because there were no social boundaries drawn by the passage of years. It was not remarkable for young men to marry women as old as their mothers. I never heard anyone talk about "women's work" until after I left Laguna for college. Work was there to be done by any able-bodied person who wanted to do it. At the same time, in the old-time Pueblo world, identity was acknowledged to be always in a flux; in the old stories, one minute Spider Woman is a little spider under a yucca plant, and the next instant she is a sprightly grandmother walking down the road.

When I was growing up, there was a young man from a nearby village who 17 wore nail polish and women's blouses and permed his hair. People paid little attention to his appearance; he was always part of a group of other young men from his village. No one ever made fun of him. Pueblo communities were and still are very independent, but they also have to be tolerant of individual eccentricities because survival of the group means everyone has to cooperate.

In the old Pueblo world, differences were celebrated as signs of the Mother 18 Creator's grace. Persons born with exceptional physical or sexual differences were highly respected and honored because their physical differences gave them special positions as mediators between this world and the spirit world. The great Navajo medicine man of the 1920s, the Crawler, had a hunchback and could not walk upright, but he was able to heal even the most difficult cases.

Before the arrival of Christian missionaries, a man could dress as a woman 19 and work with the women and even marry a man without any fanfare. Likewise, a woman was free to dress like a man, to hunt and go to war with the men, and to marry a woman. In the old Pueblo worldview, we are all a mixture of male and female, and this sexual identity is changing constantly. Sexual inhibition did not begin until the Christian missionaries arrived. For the old-time people, marriage was about teamwork and social relationships, not about sexual excitement. In the days before the Puritans came, marriage did not mean an end to sex with people other than your spouse. Women were just as likely as men to have a *si'ash,* or lover.

New life was so precious that pregnancy was always appropriate, and preg- 20 nancy before marriage was celebrated as a good sign. Since the children belonged to the mother and her clan, and women owned and bequeathed the houses and farmland, the exact determination of paternity wasn't critical. Although fertility was prized, infertility was no problem because mothers with unplanned pregnancies gave their babies to childless couples within the clan in open adoption arrangements. Children called their mother's sisters "mother" as well, and a child became attached to a number of parent figures.

In the sacred kiva ceremonies, men mask and dress as women to pay hom- 21 age and to be possessed by the female energies of the spirit beings. Because differences in physical appearance were so highly valued, surgery to change one's face and body to resemble a model's face and body would be unimaginable. To be different, to be unique was blessed and was best of all.

The traditional clothing of Pueblo women emphasized a woman's sturdiness. 22 Buckskin leggings wrapped around the legs protected her from scratches and injuries while she worked. The more layers of buckskin, the better. All those layers gave her legs the appearance of strength, like sturdy tree trunks. To demonstrate sisterhood and brotherhood with the plants and animals, the old-time people make masks and costumes that transform the human figures of the dancers into the animal beings they portray. Dancers paint their exposed skin; their postures and motions are adapted from their observations. But the motions are stylized. The observer sees not an actual eagle or actual deer dancing, but witnesses a human being, a dancer, gradually changing into a woman/buffalo or a man/deer. Every impulse is to reaffirm the urgent relationships that human beings have with the plant and animal world.

In the high desert plateau country, all vegetation, even weeds and thorns, 23 becomes special, and all life is precious and beautiful because without the plants, the insects, and the animals, human beings living here cannot survive. Perhaps human beings long ago noticed the devastating impact human activity can have on the plants and animals; maybe this is why tribal cultures devised the stories about humans and animals intermarrying, and the clans that bind humans to animals and plants through a whole complex of duties.

We children were always warned not to harm frogs or toads, the beloved 24 children of the rain clouds, because terrible floods would occur. I remember in the summer the old folks used to stick bog bolls of cotton on the outside of their screen doors as bait to keep the flies from going in the house when the door was opened. The old folks staunchly resisted the killing of flies because once, long, long ago, when human beings were in a great deal of trouble, a Green Bottle Fly carried the desperate messages of human beings to the Mother Creator in the Fourth World, below this one. Human beings had outraged the Mother Creator by neglecting the Mother Corn altar while they dabbled with sorcery and magic. The Mother Creator disappeared, and with her disappeared the rain clouds, and the plants and the animals too. The people began to starve, and they had no way of reaching the Mother Creator down below. Green Bottle Fly took the message to the Mother Creator, and the people were saved. To show their gratitude, the old folks refused to kill any flies.

The old stories demonstrate the interrelationships that the Pueblo people have   25 maintained with their plant and animal clanspeople. Kochininako, Yellow Woman, represents all women in the old stories. Her deeds span the spectrum of human behavior and are mostly heroic acts, though in at least one story, she chooses to join the secret Destroyer Clan, which worships destruction and death. Because Laguna Pueblo cosmology features a female Creator, the status of women is equal with the status of men, and women appear as often as men in the old stories as hero figures. Yellow Woman is my favorite because she dares to cross traditional boundaries of ordinary behavior during times of crisis in order to save the Pueblo; her power lies in her courage and in her uninhibited sexuality, which the old-time Pueblo stories celebrate again and again because fertility was so highly valued.

The old stories always say that Yellow Woman was beautiful, but remember   26 that the old-time people were not so much thinking about physical appearances. In each story, the beauty that Yellow Woman possesses is the beauty of her passion, her daring, and her sheer strength to act when catastrophe is imminent.

In one story, the people are suffering during a great drought and accompa-   27 nying famine. Each day, Kochininako has to walk farther and farther from the village to find fresh water for her husband and children. One day she travels far, far to the east, to the plains, and she finally locates a freshwater spring. But when she reaches the pool, the water is churning violently as if something large had just gotten out of the pool, Kochininako does not want to see what huge creature had been at the pool, but just as she fills her water jar and turns to hurry away, a strong, sexy man in buffalo skin leggings appears by the pool. Little drops of water glisten on his chest. She cannot help but look at him because he is so strong and so good to look at. Able to transform himself from human to buffalo in the wink of an eye, Buffalo Man gallops away with her on his back. Kochininako falls in love with Buffalo Man, and because of this liaison, the Buffalo People agree to give their bodies to the hunters to feed the starving Pueblo. Thus Kochininako's fearless sensuality results in the salvation of the people of her village, who are saved by the meat the Buffalo People "give" to them.

My father taught me and my sisters to shoot .22 rifles when we were seven;   28 I went hunting with my father when I was eight, and I killed my first mule deer buck when I was thirteen. The Kochininako stories were always my favorite because Yellow Woman had so many adventures. In one story, as she hunts rabbits to feed her family, a giant monster pursues her, but she has the courage and presence of mind to outwit it.

In another story, Kochininako has a fling with Whirlwind Man and returns 29 to her husband ten months later with twin baby boys. The twin boys grow up to be great heroes of the people. Once again, Kochininako's vibrant sexuality benefits her people.

The stories about Kochininako made me aware that sometimes an individ- 30 ual must act despite disapproval, or concern for appearances or what others may say. From Yellow Woman's adventures, I learned to be comfortable with my differences. I even imagined that Yellow Woman had yellow skin, brown hair, and green eyes like mine, although her name does not refer to her color, but rather to the ritual color of the east.

There have been many other moments like the one with the camera-toting 31 tourist in the schoolyard. But the old-time people always say, remember the stories, the stories will help you be strong. So all these years I have depended on Kochininako and the stories of her adventures.

Kochininako is beautiful because she has the courage to act in times of great 32 peril, and her triumph is achieved by her sensuality, not through violence and destruction. For these qualities of the spirit, Yellow Woman and all women are beautiful.

COMPREHENSION

1. Silko devotes part of this essay to recollections of her great-grandmother, Grandma A'mooh. What is her great-grandmother like? What does the writer learn from Grandma A'mooh? Why is the essay more about Pueblo women than men?
2. Explain what you learn about traditional Pueblo culture from this essay. What values does the writer associate with this "old-time" culture? According to Silko, how does this culture contrast both explicitly and implicitly with modern Anglo culture? How does this traditional culture sustain Silko as a young girl?
3. The writer summarizes several Pueblo stories. What are the main ones? Why does Silko especially like the story of Kochininako or Yellow Woman?

RHETORIC

1. What is the writer's thesis? Does this thesis appear in a single sentence? If so, what is it? If not, what is the implied thesis?
2. What strategy does the writer use to both start and conclude this essay? Is this strategy effective? Justify your response.
3. Silko provides an extended definition of Pueblo culture in this selection. Explain how she uses description, narration, comparison and contrast, and analysis to develop this definition.

4. While Silko's primary purpose is to define or explain Pueblo culture, she also provides several supporting definitions. Identify them, and explain how they contribute to the broader definition.
5. Is the diction in this essay concrete or abstract, specific or general? Identify specific passages to support your answer.
6. Consider the essay as an argument. What is the claim? What is the supporting evidence? What warrants underpin the argument? How effective is the argument, and why?

WRITING

1. Working in a group, create a list of all the features of traditional Pueblo culture that Silko discusses. Choose two or three and write brief summaries of these features.
2. Using Silko's essay as a frame of reference, write a comparative essay in which you discuss contemporary American cultural values in relationship to "old-time" Pueblo values and traditions.
3. **WRITING AN ARGUMENT:** Argue for or against the proposition that traditional Pueblo culture is superior to contemporary American culture. Use at least three topics drawn from Silko's essay—for example, approach to diversity and difference, treatment of women, or respect for the environment—to develop your argumentative essay.

 www.mhhe.com/ **mhreader** | For more information on Leslie Marmon Silko, go to: **More Resources > Ch. 6 History & Culture**

A World Not Neatly Divided

Amartya Sen

Amartya Sen (b. 1933), born in Santiniketan, India, was awarded the Nobel Prize in Economics in 1988 for his groundbreaking work on welfare economics. Educated at Presidency College in Calcutta and Cambridge University (PhD, 1959), Sen has taught at Harvard University, the London School of Economics, and Oxford University; currently he is a professor at Trinity College, Cambridge University. His major works, all of which investigate the role of poverty and inequality in the world, include Collective Choice and Social Welfare *(1970),* On Economic Inequality *(1973),* Poverty and Famines: An Essay on Entitlement and Deprivation *(1981),* Commodities and Capabilities *(1985), and* Development as Freedom *(1999). In the following essay, which appeared in* The New York Times *in 2001, Sen suggests that generalizations about "civilization" tend to blur the realities of complex cultures.*

When people talk about clashing civilizations, as so many politicians and aca-   1
demics do now, they can sometimes miss the central issue. The inadequacy of
this thesis begins well before we get to the question of whether civilizations
must clash. The basic weakness of the theory lies in its program of categorizing
people of the world according to a unique, allegedly commanding system of
classification. This is problematic because civilizational categories are crude and
inconsistent and also because there are other ways of seeing people (linked to
politics, language, literature, class, occupation, or other affiliations).

The befuddling influence of a singular classification also traps those who dis-   2
pute the thesis of a clash: To talk about "the Islamic world" or "the Western
world" is already to adopt an impoverished vision of humanity as unalterably divided. In fact, civilizations are hard to partition in this way, given the diversities
within each society as well as the linkages among different countries and cultures.
For example, describing India as a "Hindu civilization" misses the fact that India
has more Muslims than any other country except Indonesia and possibly Pakistan. It is futile to try to understand Indian art, literature, music, food, or politics
without seeing the extensive interactions across barriers of religious communities.
These include Hindus and Muslims, Buddhists, Jains, Sikhs, Parsees, Christians
(who have been in India since at least the fourth century, well before England's
conversion to Christianity), Jews (present since the fall of Jerusalem), and even
atheists and agnostics. Sanskrit has a larger atheistic literature than exists in any
other classical language. Speaking of India as a Hindu civilization may be comforting to the Hindu fundamentalist, but it is an odd reading of India.

A similar coarseness can be seen in the other categories invoked, like "the Is-   3
lamic world." Consider Akbar and Aurangzeb, two Muslim emperors of the
Mogul dynasty in India. Aurangzeb tried hard to convert Hindus into Muslims
and instituted various policies in that direction, of which taxing the non-Muslims
was only one example. In contrast, Akbar reveled in his multiethnic court and
pluralist laws, and issued official proclamations insisting that no one "should be
interfered with on account of religion" and that "anyone is to be allowed to go
over to a religion that pleases him."

If a homogeneous view of Islam were to be taken, then only one of these   4
emperors could count as a true Muslim. The Islamic fundamentalist would
have no time for Akbar; Prime Minister Tony Blair, given his insistence that tolerance is a defining characteristic of Islam, would have to consider excommuni-
cating Aurangzeb. I expect both Akbar and Aurangzeb would protest, and so
would I. A similar crudity is present in the characterization of what is called
"Western civilization." Tolerance and individual freedom have certainly been
present in European history. But there is no dearth of diversity here, either.
When Akbar was making his pronouncements on religious tolerance in Agra, in
the 1590s, the Inquisitions were still going on; in 1600, Giordano Bruno was
burned at the stake, for heresy, in Campo dei Fiori in Rome.

Dividing the world into discrete civilizations is not just crude. It propels us   5
into the absurd belief that this partitioning is natural and necessary and must
overwhelm all other ways of identifying people. That imperious view goes not
only against the sentiment that "we human beings are all much the same," but
also against the more plausible understanding that we are diversely different.
For example, Bangladesh's split from Pakistan was not connected with religion,
but with language and politics.

Each of us has many features in our self-conception. Our religion, impor-   6
tant as it may be, cannot be an all-engulfing identity. Even a shared poverty can
be a source of solidarity across the borders. The kind of division highlighted by,
say, the so-called "antiglobalization" protesters—whose movement is, inciden-
tally, one of the most globalized in the world—tries to unite the underdogs of
the world economy and goes firmly against religious, national, or "civilizational" lines of division.

The main hope of harmony lies not in any imagined uniformity, but in the   7
plurality of our identities, which cut across each other and work against sharp
divisions into impenetrable civilizational camps. Political leaders who think
and act in terms of sectioning off humanity into various "worlds" stand to make
the world more flammable—even when their intentions are very different. They
also end up, in the case of civilizations defined by religion, lending authority to
religious leaders seen as spokesmen for their "worlds." In the process, other
voices are muffled and other concerns silenced. The robbing of our plural identities not only reduces us; it impoverishes the world.

COMPREHENSION

1. According to Sen, what is the "basic weakness" underlying the idea that the
   world is composed of "clashing civilizations" (paragraph 1)?
2. What does the writer mean by "singular classification" (paragraph 2)? Why
   is classifying people in terms of their civilization "crude and inconsistent"?
   Why is applying singular classification to religions and other features of society wrong?
3. What, according to Sen, is "the main hope for harmony" (paragraph 7) in
   the world?

RHETORIC

1. What argumentative strategy does Sen employ in the introductory paragraph? What point of view is he arguing against?
2. While arguing against a certain type of classification, Sen actually uses classification as a rhetorical strategy in this essay. How, precisely, does Sen employ classification to organize his argument?
3. What examples does the writer use to support his argument? Why does he
   use them? Why does he decide not to provide illustrations near the end of
   the selection?

4. What transitional devices serve to unify the essay?
5. How effective is Sen's concluding paragraph? Does it serve to confirm his claim? Why or why not?

WRITING

1. Write an essay about the problems you see in your community or on campus. Explain how singular classification might explain some of these problems.
2. In an analytical essay, explain how singular classification might help explain the events of September 11, 2001.
3. **WRITING AN ARGUMENT:** Write an essay in which you demonstrate that singular classification actually can be helpful in framing public discourse about nations, groups, or civilizations.

www.mhhe.com/
mhreader

For more information on Amartya Sen, go to:
More Resources > Ch. 6 History & Culture

✳

The Arab World

Edward T. Hall

Edward T. Hall (b. 1914) was born in Missouri and earned a master's degree at the University of Arkansas and a PhD in anthropology at Columbia University. He was a professor of anthropology at the Illinois Institute of Technology and at Northwestern University. Hall is also the author of many books on anthropology and culture, among the most famous of which are The Silent Language *(1959),* The Hidden Dimension *(1966),* The Dance of Life *(1983),* Hidden Differences: Doing Business with the Japanese *(1987), and* Understanding Cultural Differences: Germans, French and Americans *(1990). In this selection from* The Hidden Dimension, *Hall demonstrates how such basic concepts as public and private space are perceived far differently depending upon one's culture of origin.*

In spite of over two thousand years of contact, Westerners and Arabs still do not   1
understand each other. Proxemic research reveals some insights into this difficulty. Americans in the Middle East are immediately struck by two conflicting sensations. In public they are compressed and overwhelmed by smells, crowding, and high noise levels; in Arab homes Americans are apt to rattle around, feeling exposed and often somewhat inadequate because of too much space!

(The Arab houses and apartments of the middle and upper classes which Americans stationed abroad commonly occupy are much larger than the dwellings such Americans usually inhabit.) Both the high sensory stimulation which is experienced in public places and the basic insecurity which comes from being in a dwelling that is too large provide Americans with an introduction to the sensory world of the Arab.

Behavior in Public

Pushing and shoving in public places is characteristic of Middle Eastern culture.   2
Yet it is not entirely what Americans think it is (being pushy and rude) but stems from a different set of assumptions concerning not only the relations between people but how one experiences the body as well. Paradoxically, Arabs consider northern Europeans and Americans pushy, too. This was very puzzling to me when I started investigating these two views. How could Americans who stand aside and avoid touching be considered pushy? I used to ask Arabs to explain this paradox. None of my subjects was able to tell me specifically what particulars of American behavior were responsible, yet they all agreed that the impression was widespread among Arabs. After repeated unsuccessful attempts to gain insight into the cognitive world of the Arab on this particular point, I filed it away as a question that only time would answer. When the answer came, it was because of a seemingly inconsequential annoyance.

While waiting for a friend in a Washington, D.C., hotel lobby and wanting   3
to be both visible and alone, I had seated myself in a solitary chair outside the normal stream of traffic. In such a setting most Americans follow a rule, which is all the more binding because we seldom think about it, that can be stated as follows: as soon as a person stops or is seated in a public place, there balloons around him a small sphere of privacy which is considered inviolate. The size of the sphere varies with the degree of crowding, the age, sex, and the importance of the person, as well as the general surroundings. Anyone who enters this zone and stays there is intruding. In fact, a stranger who intrudes, even for a specific purpose, acknowledges the fact that he has intruded by beginning his request with "Pardon me, but can you tell me . . . ?"

To continue, as I waited in the deserted lobby, a stranger walked up to   4
where I was sitting and stood close enough so that not only could I easily touch him but I could even hear him breathing. In addition, the dark mass of his body filled the peripheral field of vision on my left side. If the lobby had been crowded with people, I would have understood his behavior, but in an empty lobby his presence made me exceedingly uncomfortable. Feeling annoyed by this intrusion, I moved my body in such a way as to communicate annoyance. Strangely enough, instead of moving away, my actions seemed only to encourage him, because he moved even closer. In spite of the temptation to escape the annoyance, I put aside thoughts of abandoning my post, thinking, "To hell with it. Why should I move? I was here first and I'm not going to let this fellow drive me out even if he is a boor." Fortunately, a group of people soon arrived whom my tormentor immediately joined. Their mannerisms explained his behavior,

for I knew from both speech and gestures that they were Arabs. I had not been able to make this crucial identification by looking at my subject when he was alone because he wasn't talking and he was wearing American clothes.

In describing the scene later to an Arab colleague, two contrasting patterns 5 emerged. My concept and my feelings about my own circle of privacy in a "public" place immediately struck my Arab friend as strange and puzzling. He said, "After all, it's a public place, isn't it?" Pursuing this line of inquiry, I found that in Arab thought I had no rights whatsoever by virtue of occupying a given spot; neither my place nor my body was inviolate! For the Arab, there is no such thing as an intrusion in public. Public means public. With this insight, a great range of Arab behavior that had been puzzling, annoying, and sometimes even frightening began to make sense. I learned, for example, that if *A* is standing on a street corner and *B* wants his spot, *B* is within his rights if he does what he can to make *A* uncomfortable enough to move. In Beirut only the hardy sit in the last row in a movie theater, because there are usually standees who want seats and who push and shove and make such a nuisance that most people give up and leave. Seen in this light, the Arab who "intruded" on my space in the hotel lobby had apparently selected it for the very reason I had: it was a good place to watch two doors and the elevator. My show of annoyance, instead of driving him away, had only encouraged him. He thought he was about to get me to move.

Another silent source of friction between Americans and Arabs is in an area 6 that Americans treat very informally—the manners and rights of the road. In general, in the United States we tend to defer to the vehicle that is bigger, more powerful, faster, and heavily laden. While a pedestrian walking along a road may feel annoyed he will not think it unusual to step aside for a fast-moving automobile. He knows that because he is moving he does not have the right to the space around him that he has when he is standing still (as I was in the hotel lobby). It appears that the reverse is true with the Arabs who apparently *take on rights to space as they move.* For someone else to move into a space an Arab is also moving into is a violation of his rights. It is infuriating to an Arab to have someone else cut in front of him on the highway. It is the American's cavalier treatment of moving space that makes the Arab call him aggressive and pushy.

Concepts of Privacy

The experience described above and many others suggested to me that Arabs 7 might actually have a wholly contrasting set of assumptions concerning the body and the rights associated with it. Certainly the Arab tendency to shove and push each other in public and to feel and pinch women in public conveyances would not be tolerated by Westerners. It appeared to me that they must not have any concept of a private zone outside the body. This proved to be precisely the case.

In the Western world, the person is synonymous with an individual inside 8 a skin. And in northern Europe generally, the skin and even the clothes may be inviolate. You need permission to touch either if you are a stranger. This rule applies in some parts of France, where the mere touching of another person during an argument used to be legally defined as assault. For the Arab the location of the person in relation to the body is quite different. The person exists somewhere down inside the body. The ego is not completely hidden, however, because it can be reached very easily with an insult. It is protected from touch but not from words. The dissociation of the body and the ego may explain why the public amputation of a thief's hand is tolerated as standard punishment in Saudi Arabia. It also sheds light on why an Arab employer living in a modern apartment can provide his servant with a room that is a boxlike cubicle approximately 5 by 10 by 4 feet in size that is not only hung from the ceiling to conserve floor space but has an opening so that the servant can be spied on.

As one might suspect, deep orientations toward the self such as the one just 9 described are also reflected in the language. This was brought to my attention one afternoon when an Arab colleague who is the author of an Arab English dictionary arrived in my office and threw himself into a chair in a state of obvious exhaustion. When I asked him what had been going on, he said: "I have spent the entire afternoon trying to find the Arab equivalent of the English word 'rape.' There is no such word in Arabic. All my sources, both written and spoken, can come up with no more than an approximation, such as 'He took her against her will.' There is nothing in Arabic approaching your meaning as it is expressed in that one word."

Differing concepts of the placement of the ego in relation to the body are not 10 easily grasped. Once an idea like this is accepted, however, it is possible to understand many other facets of Arab life that would otherwise be difficult to explain. One of these is the high population density of Arab cities like Cairo, Beirut, and Damascus. According to the animal studies described in the earlier chapters [of *The Hidden Dimension*], the Arabs should be living in a perpetual behavioral sink. While it is probable that Arabs are suffering from population pressures, it is also just as possible that continued pressure from the desert has resulted in a cultural adaptation to high density which takes the form described above. Tucking the ego down inside the body shell not only would permit higher population densities but would explain why it is that Arab communications are stepped up as much as they are when compared to northern European communication patterns. Not only is the sheer noise level much higher, but the piercing look of the eyes, the touch of the hands, and the mutual bathing in the warm moist breath during conversation represent stepped up sensory inputs to a level which many Europeans find unbearably intense.

The Arab dream is for lots of space in the home, which unfortunately many 11 Arabs cannot afford. Yet when he has space, it is very different from what one finds in most American homes. Arab spaces inside their upper middle-class homes are tremendous by our standards. They avoid partitions because Arabs *do not like to be alone*. The form of the home is such as to hold the family together inside a single protective shell, because Arabs are deeply involved with each other. Their personalities are intermingled and take nourishment from each other like the roots and soil. If one is not with people and actively involved in some way, one is deprived of life. An old Arab saying reflects this value: "Paradise without people should not be entered because it is Hell." Therefore, Arabs in the United States often feel socially and sensorially deprived and long to be back where there is human warmth and contact.

Since there is no physical privacy as we know it in the Arab family, not even 12 a word for privacy, one could expect that the Arabs might use some other means to be alone. Their way to be alone is to stop talking. Like the English, an Arab who shuts himself off in this way is not indicating that anything is wrong or that he is withdrawing, only that he wants to be alone with his own thoughts or does not want to be intruded upon. One subject said that her father would come and go for days at a time without saying a word, and no one in the family thought anything of it. Yet for this very reason, an Arab exchange student visiting a Kansas farm failed to pick up the cue that his American hosts were mad at him when they gave him the "silent treatment." He only discovered something was wrong when they took him to town and tried forcibly to put him on a bus to Washington, D.C., the headquarters of the exchange program responsible for his presence in the U.S.

Arab Personal Distances

Like everyone else in the world, Arabs are unable to formulate specific rules for 13 their informal behavior patterns. In fact, they often deny that there are any rules, and they are made anxious by suggestions that such is the case. Therefore, in order to determine how the Arab sets distances, I investigated the use of each sense separately. Gradually, definite and distinctive behavioral patterns began to emerge.

Olfaction occupies a prominent place in the Arab life. Not only is it one of 14 the distance-setting mechanisms, but it is a vital part of a complex system of behavior. Arabs consistently breathe on people when they talk. However, this habit is more than a matter of different manners. To the Arab good smells are pleasing and a way of being involved with each other. To smell one's friend is not only nice but desirable, for to deny him your breath is to act ashamed. Americans, on the other hand, trained as they are not to breathe in people's faces, automatically communicate shame in trying to be polite. Who would expect that when our highest diplomats are putting on their best manners they are also communicating shame? Yet this is what occurs constantly, because diplomacy is not only "eyeball to eyeball" but breath to breath.

By stressing olfaction, Arabs do not try to eliminate all the body's odors, 15 only to enhance them and use them in building human relationships. Nor are they self-conscious about telling others when they don't like the way they smell. A man leaving his house in the morning may be told by his uncle, "Habib, your stomach is sour and your breath doesn't smell too good. Better not talk too close to people today." Smell is even considered in the choice of a mate. When couples are being matched for marriage, the man's go-between will sometimes ask to smell the girl, who may be turned down if she doesn't "smell nice." Arabs recognize that smell and disposition may be linked.

In a word, the olfactory boundary performs two roles in Arab life. It enfolds   16
those who want to relate and separates those who don't. The Arab finds it essential to stay inside the olfactory zone as a means of keeping tab on changes in
emotion. What is more, he may feel crowded as soon as he smells something
unpleasant. While not much is known about "olfactory crowding," this may
prove to be as significant as any other variable in the crowding complex because it is tied directly to the body chemistry and hence to the state of health
and emotions. . . . It is not surprising, therefore, that the olfactory boundary constitutes for the Arabs an informal distance-setting mechanism in contrast to the
visual mechanisms of the Westerner.

Facing and Not Facing

One of my earliest discoveries in the field of intercultural communication was   17
that the position of the bodies of people in conversation varies with the culture.
Even so, it used to puzzle me that a special Arab friend seemed unable to walk
and talk at the same time. After years in the United States, he could not bring
himself to stroll along, facing forward while talking. Our progress would be arrested while he edged ahead, cutting slightly in front of me and turning side-
ways so we could see each other. Once in this position, he would stop. His
behavior was explained when I learned that for the Arabs to view the other person peripherally is regarded as impolite, and to sit or stand back-to-back is considered very rude. You must be involved when interacting with Arabs who are
friends.

One mistaken American notion is that Arabs conduct all conversations at   18
close distances. This is not the case at all. On social occasions, they may sit on
opposite sides of the room and talk across the room to each other. They are,
however, apt to take offense when Americans use what are to them ambiguous
distances, such as the four- to seven-foot social-consultative distance. They frequently complain that Americans are cold or aloof or "don't care." This was
what an elderly Arab diplomat in an American hospital thought when the
American nurses used "professional" distance. He had the feeling that he was
being ignored, that they might not take good care of him. Another Arab subject
remarked, referring to American behavior, "What's the matter? Do I smell bad?
Or are they afraid of me?"

Arabs who interact with Americans report experiencing a certain flatness   19
traceable in part to a very different use of the eyes in private and in public as
well as between friends and strangers. Even though it is rude for a guest to walk
around the Arab home eyeing things, Arabs look at each other in ways which
seem hostile or challenging to the American. One Arab informant said that he
was in constant hot water with Americans because of the way he looked at them
without the slightest intention of offending. In fact, he had on several occasions
barely avoided fights with American men who apparently thought their masculinity was being challenged because of the way he was looking at them. As
noted earlier, Arabs look each other in the eye when talking with an intensity
that makes most Americans highly uncomfortable.

Involvement

As the reader must gather by now, Arabs are involved with each other on many 20 different levels simultaneously. Privacy in a public place is foreign to them. Business transactions in the bazaar, for example, are not just between buyer and seller, but are participated in by everyone. Anyone who is standing around may join in. If a grownup sees a boy breaking a window, he must stop him even if he doesn't know him. Involvement and participation are expressed in other ways as well. If two men are fighting, the crowd must intervene. On the political level, *to fail to intervene* when trouble is brewing is to take sides, which is what our State Department always seems to be doing. Given the fact that few people in the world today are even remotely aware of the cultural mold that forms their thoughts, it is normal for Arabs to view *our* behavior as though it stemmed from *their* own hidden set of assumptions.

Feelings about Enclosed Spaces

In the course of my interviews with Arabs the term "tomb" kept cropping up in 21 conjunction with enclosed space. In a word, Arabs don't mind being crowded by people but hate to be hemmed in by walls. They show a much greater overt sensitivity to architectural crowding than we do. Enclosed space must meet at least three requirements that I know of if it is to satisfy the Arabs: there must be plenty of unobstructed space in which to move around (possibly as much as a thousand square feet); very high ceilings—so high in fact that they do not normally impinge on the visual field; and, in addition, there must be an unobstructed view. It was spaces such as these in which the Americans referred to earlier felt so uncomfortable. One sees the Arab's need for a view expressed in many ways, even negatively, for to cut off a neighbor's view is one of the most effective ways of spiting him. In Beirut one can see what is known locally as the "spite house." It is nothing more than a thick, four-story wall, built at the end of a long fight between neighbors, on a narrow strip of land for the express purpose of denying a view of the Mediterranean to any house built on the land behind. According to one of my informants, there is also a house on a small plot of land between Beirut and Damascus which is completely surrounded by a neighbor's wall built high enough to cut off the view from all windows!

Boundaries

Proxemic patterns tell us other things about Arab culture. For example, the 22 whole concept of the boundary as an abstraction is almost impossible to pin down. In one sense, there are no boundaries. "Edges" of towns, yes, but permanent boundaries out in the country (hidden lines), no. In the course of my work with Arab subjects I had a difficult time translating our concept of a boundary into terms which could be equated with theirs. In order to clarify the distinctions between the two very different definitions, I thought it might be helpful to pinpoint acts which constituted trespass. To date, I have been unable to discover anything even remotely resembling our own legal concept of trespass.

Arab behavior in regard to their own real estate is apparently an extension 23
of, and therefore consistent with, their approach to the body. My subjects simply failed to respond whenever trespass was mentioned. They didn't seem to
understand what I meant by this term. This may be explained by the fact that
they organize relationships with each other according to closed social systems
rather than spatially. For thousands of years Moslems, Marinites, Druses, and
Jews have lived in their own villages, each with strong kin affiliations. Their hierarchy of loyalties is: first to one's self, then to kinsman, townsman, or
tribesman, co-religionist and/or countryman. Anyone not in these categories is
a stranger. Strangers and enemies are very closely linked, if not synonymous, in
Arab thought. Trespass in this context is a matter of who you are, rather than a
piece of land or a space with a boundary that can be denied to anyone and
everyone, friend and foe alike.

In summary, proxemic patterns differ. By examining them it is possible to 24
reveal hidden cultural frames that determine the structure of a given people's
perceptual world. Perceiving the world differently leads to differential definitions of what constitutes crowded living, different interpersonal relations, and
a different approach to both local and international politics.

COMPREHENSION

1. This excerpt is from Hall's book entitled *The Hidden Dimension*. What is the
   hidden dimension, according to the author?
2. In paragraph 10, Hall explains that "differing concepts of the placement of
   the ego in relation to the body are not easily grasped." What does he mean
   by this statement? How is it relevant to the theme of his essay?
3. The title of this essay is "The Arab World." What does the term *world* mean
   within the context of the writing?
4. Define the following words: *proxemic* (paragraph 1), *paradox* (paragraph 2),
   inviolate (paragraph 3), *defer* (paragraph 6), *olfaction* (paragraph 14), and *peripherally* (paragraph 17).

RHETORIC

1. Anthropology is often thought of as an intellectual pursuit. How would
   you characterize Hall's voice, considering his style of language and method
   of analysis?
2. How does Hall develop his comparison and contrast of the American versus the Arab perception of manners and driving?
3. People often favor their own perspective of life over a foreign perspective.
   Is Hall's comparison value-free, or does he seem to prefer one cultural system to another? Explain by making reference to his tone.
4. Who is the implied audience for this essay? Explain your view.

5. Hall makes use of personal anecdote in explaining his theme. What other forms of support does he offer the reader? Cite at least two others and provide an example of each.
6. Writers often have various purposes for writing, for example, to entertain, to inform, to effect change, to advise, or to persuade. What is Hall's purpose or purposes in writing this essay? Explain your view.

WRITING

1. Write an expository essay in which you explain the use and interpretation of personal space by observing students in social situations at your college or university.
2. Write a personal anecdote about a time in your life in which cultural perception caused a conflict between yourself and another person.
3. **WRITING AN ARGUMENT:** In a persuasive essay, argue for or against the proposition that some cultures are better than others.

| www.mhhe.com/ **mhreader** | For more information on Edward T. Hall, go to: **More Resources > Ch. 6 History & Culture** |

Strangers from a Distant Shore

Ronald Takaki

Ronald Takaki (b. 1939) was born in Honolulu, Hawaii. He earned his BA from the college of Wooster in 1961 and his MA and PhD from the University of California, Berkeley, where he is currently a professor of ethnic studies. He also taught American history at the College of San Mateo from 1965 to 1967. Takaki feels that American history is still viewed from a predominantly white perspective and is devoted to more accurately representing all Americans in our society. He has written a number of books: A Pro-Slavery Crusade: The Agitation to Reopen the African Slave Trade *(1971),* Strangers from a Distant Shore: A History of Asian Americans *(1989),* Hiroshima: Why America Dropped the Bomb *(1996),* Iron Cages: Race and Culture in 19th Century America *(1999), and* Double Victory: A Multicultural History of America in World War II *(2000). He is a frequent lecturer and a frequent contributor to history journals as well as other publications. In the following essay, he reflects on the internment of Japanese citizens during World War II and describes his attempts to establish a connection with his Japanese relatives.*

To confront the current problems of racism, Asian Americans know they must   1
remember the past and break its silence. This need was felt deeply by Japanese
Americans during the hearings before the commission reviewing the issue of redress and reparations for Japanese Americans interned during World War II.
Memories of the internment nightmare have haunted the older generation like
ghosts. But the former prisoners have been unable to exorcise them by speaking
out and ventilating their anger.

> When we were children,
> you spoke Japanese
> in lowered voices
> between yourselves.
> Once you uttered secrets
> which we should not know,
> were not to be heard by us.
> When you spoke
> of some dark secret
> you would admonish us,
> "Don't tell it to anyone else."
> It was a suffocated vow of silence.[1]

"Stigmatized," the ex-internees have been carrying the "burden of shame"   2
for over forty painful years. "They felt like a rape victim," explained Congressman Norman Mineta, a former internee of the Heart Mountain internment camp.
"They were accused of being disloyal. They were the victims but they were on
trial and they did not want to talk about it." But Sansei, or third-generation
Japanese Americans, want their elders to tell their story. Warren Furutani, for example, told the commissioners that young people like himself had been asking
their parents to tell them about the concentration camps and to join them in pil-
grimages to the internment camp at Manzanar. "Why? Why!" their parents
would reply defensively. "Why would you want to know about it? It's not important, we don't need to talk about it." But, Furutani continued, they need to
tell the world what happened during those years of infamy.[2]

Suddenly, during the commission hearings, scores of Issei and Nisei came   3
forward and told their stories. "For over thirty-five years I have been the stereotype Japanese American," Alice Tanabe Nehira told the commission. "I've kept
quiet, hoping in due time we will be justly compensated and recognized for our
years of patient effort. By my passive attitude, I can reflect on my past years to
conclude that it doesn't pay to remain silent." The act of speaking out has en-

---

[1]Richard Oyama, poem published in *Transfer 38* (San Francisco, 1979), p. 43, reprinted in Elaine
Kim, *Asian American Literature: An Introduction to the Writings and Their Social Context* (Philadelphia, 1982). [This and subsequent notes in the selection are the author's.]
[2]Congressman Robert Matsui, speech in the House of Representatives on bill 442 for redress and
reparations, September 17, 1987, *Congressional Record* (Washington, 1987), p. 7584; Congressman
Norman Mineta, interview with author, March 26, 1988; Warren Furutani, testimony, reprinted in
Amerasia, vol. 8, no. 2 (1981), p. 104.

abled the Japanese-American community to unburden itself of years of anger and anguish. Sometimes their testimonies before the commission were long and the chair urged them to conclude. But they insisted the time was theirs. "Mr. Commissioner," protested poet Janice Mirikitani,

> So when you tell me my time is
> up I tell you this.
> Pride has kept my lips
> pinned by nails,
> my rage coffined.
> But I exhume my past
> to claim this time.[3]

The former internees finally had spoken, and their voices compelled the na- 4 tion to redress the injustice of internment. In August 1988, Congress passed a bill giving an apology and a payment of $20,000 to each of the survivors of the internment camps. When President Ronald Reagan signed the bill into law, he admitted that the United States had committed "a grave wrong," for during World War II, Japanese Americans had remained "utterly loyal" to this country. "Indeed, scores of Japanese Americans volunteered for our Armed Forces— many stepping forward in the internment camps themselves. The 442nd Regimental Combat Team, made up entirely of Japanese Americans, served with immense distinction to defend this nation, their nation. Yet, back at home, the soldiers' families were being denied the very freedom for which so many of the soldiers themselves were laying down their lives." Then the president recalled an incident that happened forty-three years ago. At a ceremony to award the Distinguished Service Cross to Kazuo Masuda, who had been killed in action and whose family had been interned, a young actor paid tribute to the slain Nisei soldier. "The name of that young actor," remarked the president, who had been having trouble saying the Japanese names"—I hope I pronounce this right—was Ronald Reagan." The time had come, the president acknowledged, to end "a sad chapter in American history."[4]

Asian Americans have begun to claim their time not only before the com- 5 mission on redress and reparations but elsewhere as well, in the novels of Maxine Hong Kingston and Milton Murayama, the plays of Frank Chin and Philip Gotanda, the scholarly writings of Sucheng Chan and Elaine Kim, the films of Steve Okazaki and Wa Wang, and the music of Hiroshima and Fred Houn. Others, too, have been breaking silences. Seventy-five-year-old Tomo Shoji, for example, had led a private life, but in 1981 she enrolled in an acting course because she wanted to try something frivolous and to take her mind off her husband's illness. In the beginning, Tomo was hesitant, awkward on the stage. "Be

---

[3]Alice Tanabe Nehira, testimony, reprinted in *Amerasia,* vol. 8, no. 2 (1981), p. 93; Janice Mirikitani, "Breaking Silences," reprinted ibid., p. 109.
[4]Text of Reagan's Remarks: reprinted in *Pacific Citizen,* August 19–26, 1988, p. 5; *San Francisco Chronicle,* August 5 and 11, 1988.

yourself," her teacher urged. Then suddenly she felt something surge through her, springing from deep within, and she began to tell funny and also sad stories about her life. Now Tomo tours the West Coast, a wonderful wordsmith giving one-woman shows to packed audiences of young Asian Americans. "Have we really told our children all we have gone through?" she asks. Telling one of her stories, Tomo recounts: "My parents came from Japan and I was born in a lumber camp. One day, at school, my class was going on a day trip to a show, and I was pulled aside and told I would have to stay behind. All the white kids went." Tomo shares stories about her husband: "When I first met him, I thought, wow. Oh, he was so macho! And he wanted his wife to be a good, submissive wife. But then he married me." Theirs had been at times a stormy marriage. "Culturally we were different because he was Issei and I was American, and we used to argue a lot. Well, one day in 1942 right after World War II had started he came home and told me we had to go to an internment camp. 'I'm not going to one because I'm an American citizen,' I said to him. 'You have to go to camp, but not me.' Well, you know what, that was one time my husband was right!" Tomo remembers the camp: "We were housed in barracks, and we had no privacy. My husband and I had to share a room with another couple. So we hanged a blanket in the middle of the room as a partition. But you could hear everything from the other side. Well, one night, while we were in bed, my husband and I got into an argument, and I dumped him out of the bed. The other couple thought we were making violent love." As she stands on the stage and talks stories excitedly, Tomo cannot be contained: "We got such good, fantastic stories to tell. All our stories are different."[5]

Today, young Asian Americans want to listen to these stories—to shatter images of themselves and their ancestors as "strangers" and to understand who they are as Asian Americans. "What don't you know?" their elders ask. Their question seems to have a peculiar frame: it points to the blank areas of collective memory. And the young people reply that they want "to figure out how the invisible world the emigrants built around [their] childhoods fit in solid America." They want to know more about their "no name" Asian ancestors. They want to decipher the signs of the Asian presence here and there across the landscape of America—railroad tracks over high mountains, fields of cane virtually carpeting entire islands, and verdant agricultural lands.

Deserts to farmlands
Japanese-American
Page in history.[6]

---

[5]Tomo Shoji, "Born Too Soon . . . It's Never Too Late: Growing Up Nisei in Early Washington," presentations at the University of California, Berkeley, September 19, 1987, and the Ohana Cultural Center, Oakland, California, March 4, 1988.
[6]Kingston, Maxine Hong, *The Woman Warrior*, p. 6; poem in Kazuo Ito, *Issei: A History of Japanese Immigrants in North America* (Seattle, 1973), p. 493.

They want to know what is their history and "what is the movies." They ⁷
want to trace the origins of terms applied to them. "Why are we called 'Oriental'?" they question, resenting the appellation that has identified Asians as exotic, mysterious, strange, and foreign. "The word 'orient' simply means 'east.'
So why are Europeans 'West' and why are Asians 'East'? Why did empire-
minded Englishmen in the sixteenth century determine that Asia was 'east' of
London? Who decided what names would be given to the different regions and
peoples of the world? Why does 'American' usually mean 'white'?" Weary of
Eurocentric history, young Asian Americans want their Asian ancestral lives in
America chronicled, "given the name of a place." They have earned the right to
belong to specific places like Washington, California, Hawaii, Puunene,
Promontory Point, North Adams, Manzanar, Doyers Street. "And today, after
125 years of our life here," one of them insists, "I do not want just a home that
time allowed me to have." Seeking to lay claim to America, they realize they can
no longer be indifferent to what happened in history no longer embarrassed by
the hardships and humiliations experienced by their grandparents and parents.

> My heart, once bent and cracked, once
> ashamed of your China ways.
> Ma, hear me now, tell me your story
> again and again.⁷

As they listen to the stories and become members of a "community of mem- ⁸
ory," they are recovering roots deep within this country and the homelands of
their ancestors. Sometimes the journey leads them to discover rich and interesting things about themselves. Alfred Wong, for example, had been told repeatedly for years by his father, "Remember your Chinese name. Remember your
village in Toisha. Remember you are Chinese. Remember all this and you will
have a home." One reason why it was so important for the Chinese immigrants
to remember was that they never felt sure of their status in America. "Unlike
German and Scottish immigrants, the Chinese immigrants never felt comfortable here," Wong explained. "So they had a special need to know there was a
place, a home for them somewhere."⁸

But Wong had a particular reason to remember. His father had married by ⁹
mutual agreement two women on the same day in China and had come to
America as a merchant in the 1920s. Later he brought over one of his wives. But
she had to enter as a "paper wife," for he had given the immigration authorities
the name of the wife he had left behind. Born here in 1938, Wong grew up
knowing about his father's other wife and the other half of the family in China;

---

⁷Kingston, *The Woman Warrior,* p. 6; Robert Kwan, "Asian v. Oriental: A difference that Counts,"
Pacific Citizen, April 25, 1980; Sir James Augustus Henry Murry (ed.), *The Oxford English Dictionary*
(Oxford, 1933), vol. 7, p. 200; Aminur Rahim, "Is Oriental an Occident?" in *The Asiandian,* vol. 5,
no. 1, April 1983, p. 20; Shawn Wong, *Homebase* (New York, 1979), p. 111; Nellie Wong, "From a
Heart of Rice Straw," in Nellie Wong, *Dreams in Harrison Railroad Park* (Berkeley, 1977), p. 41.
⁸Robert Bellah, et al., *Habits of the Heart: Individualism and Commitment in American Life* (Berkeley,
1985), p. 153; Alfred Wong, interviewed by Carol Takaki, April 6 and 13, 1988.

his parents constantly talked about them and regularly sent money home to Quangdong. For years the "family plan" had been for him to see China someday. In 1984 he traveled to his father's homeland, and there in the family home—the very house his father had left decades earlier—Alfred Wong was welcomed by his *Chunk Gwok Ma* ("China Mama"). "You look just like I had imagined you would look," she remarked. On the walls of the house, he saw hundreds of photographs—of himself as well as sisters, nieces, nephews, and his own daughter—that had been placed there over the years. He suddenly realized how much he had always belonged there, and had a warm connectedness. "It's like you were told there was this box and there was a beautiful diamond in it," Wong said. "But for years and years you couldn't open the box. Then finally you got a chance to open the box and it was as wonderful as you had imagined it would be."[9]

Mine is a different yet similar story. My father, Toshio Takaki, died in 1945, [10] when I was only five years old; my mother married Koon Keu Young about a year later, and I grew up knowing very little about my father. Many years later, in 1968, after my parents had moved to Los Angeles, my mother passed away and I had to clear out her room after the funeral. In one of her dresser drawers, I found an old photograph of my father as a teenager: it was his immigration photograph. I noticed some Japanese writing on the back. Later a friend translated: "This is Toshio Takaki, registered as an emigrant in Mifune, Kumamoto Prefecture, 1918." I wondered how young Toshio managed to come to the United States. Why did he go to Hawaii? Did he go alone? What dreams burned within the young boy? But a huge silence stood before me, and I could only speculate that he must have come alone and entered as a student, since the 1908 Gentlemen's Agreement had prohibited the immigration of Japanese laborers. In Hawaii, he met and married my mother, Catherine Okawa, a Nisei. I had no Takaki relatives in Hawaii, I thought.

Ten years later, while on a sabbatical in Hawaii, I was "talking story" with [11] my uncle Richard Okawa. I was telling him about the book I was then writing—*Iron Cages*, a study of race and culture in America. Suddenly his eyes lit up as he exclaimed: "Hey, why you no go write a book about us, huh? About the Japanese in Hawaii. After all, your grandparents came here as plantation workers and your mother and all your aunts and uncles were born on the plantation." Smiling, I replied: "Why not?" I went on to write a history of the plantation laborers. The book was published in 1983, and I was featured on television news and educational programs in Hawaii. One of the programs was aired in January 1985; a plantation laborer on the Puunene Plantation, Maui, was watching the discussion on television when he exclaimed to his wife: "Hey, that's my cousin, Ronald!" "No joke with me," she said, and he replied: "No, for real, for real."

A few months later, in July, I happened to visit Maui to give a lecture on the [12] plantation experience. While standing in the auditorium shortly before my

[9]Ibid.

presentation, I noticed two Japanese men approaching me. One of them draped a red carnation lei around my shoulders and smiled: "You remember me, don't you?" I had never seen this man before and was confused. Then he said again. "You remember me?" After he asked for the third time, he pulled a family photograph from a plastic shopping bag. I saw among the people in the picture my father as a young man, and burst out excitedly: "Oh, you're a Takaki!" He replied: "I'm your cousin, Minoru. I saw you on television last January and when I found out you were going to come here I wanted to see you again. You were five years old when I last saw you. I was in the army on my way to Japan and I came by your house in Palolo Valley. But I guess you don't remember. I've been wondering what happened to you for forty years." Our families had lost contact with each other because of the war, the isolation of the plantation located on another island, my father's death, and my mother's remarriage. Minoru introduced me to his brother Susumu and his son, Leighton, who works on the Puunene Plantation and represents the fourth generation of Takaki plantation workers. Afterward they took me to the Puunene Plantation, showing me McGerrow Camp, where my branch of the Takaki family had lived, and filling me with stories about the old days. "You also have two cousins, Jeanette and Lillian in Honolulu," Minoru said, "and a big Takaki family in Japan."

A year later I visited my Takaki family in Japan. On the day I arrived, my 13 cousin Nobuo showed me a box of old photographs that had been kept for decades in an upstairs closet. "We don't know who this baby is," he said, pointing to a picture of a baby boy. "That's me!" I exclaimed in disbelief. The box contained many photographs of my father, mother, sister. and me. My father had been sending pictures to the family in Kumamoto. I felt a part of me had been there all along and I had in a sense come home. Nobuo's wife Keiko told me that I was *Kumamoto kenjin*—"one of the people of Kumamoto." During my visit, I was taken to the farm where my father was born. We drove up a narrow winding road past waterfalls and streams, tea farms, and rice paddies, to a village nestled high in the mountains. The scene reminded me of old Zen paintings of Japanese landscapes and evoked memories of my mother telling me the story of Momotaro. Toshino Watanabe, an old woman in her eighties, gave me a family portrait that her sister had sent in 1915, there they were in fading sepia—my uncle Teizo, grandfather Santaro, aunt Yukino, cousin Tsutako, uncle Nobuyoshi, and father Toshio, just fourteen years old—in McGerrow Camp on the Puunene Plantation.

The stories of Alfred Wong and myself branch from the late history of Asian 14 Americans and America itself—from William Hooper and Aaron Palmer, westward expansion, the economic development of California and Hawaii, the Chinese Exclusion Act, the Gentlemen's Agreement. The history of America is essentially the story of immigrants, and many of them, coming from a "different shore" than their European brethren, had sailed east to this new world. After she had traveled across the vast Pacific Ocean and settled here, a woman captured the vision of the immigrants from Asia in haiku's seventeen syllables:

All the dreams of youth
Shipped in emigration boats
To reach this far shore.[10]

In America, Asian immigrants and their offspring have been actors in   15
history—the first Chinese working on the plantations of Hawaii and in the gold
fields of California, the early Japanese immigrants transforming the brown San
Joaquin Valley into verdant farmlands, the Korean immigrants struggling to
free their homeland from Japanese colonialism, the Filipino farm workers and
busboys seeking the America in their hearts, the Asian-Indian immigrants picking fruit and erecting Sikh temples in the West, the American-born Asians like
Jean Park and Jade Snow Wong and Monica Sone trying to find an identity for
themselves as Asian Americans, the second-wave Asian immigrants bringing
their skills and creating new communities as well as revitalizing old communities with culture and enterprise, and the refugees from the war-torn countries of
Southeast Asia trying to put their shattered lives together and becoming our
newest Asian Americans. Their dreams and hopes unfurled here before the
wind, all of them—from the first Chinese miners sailing through the Golden
Gate to the last Vietnamese boat people flying into Los Angeles International
Airport—have been making history in America. And they have been telling us
about it all along.

[10]Poem by Shigeko, in Kazuo Ito, *Issei,* p. 40.

COMPREHENSION

1. Why does Takaki think it important for younger Asian Americans to know
   about their culture?
2. Why were the adults reluctant to talk about their experiences in the internment camps?
3. What does the writer mean when he says, "Asian Americans have begun to
   claim their time" (paragraph 5)?

RHETORIC

1. Takaki uses poetry and dialogue in his essay. How do these devices help to
   advance his point of view?
2. What is the main idea of the essay? Where in the writing does it appear?
3. What use does Takaki make of historical facts and details? Cite some
   examples of these in the essay, and discuss what effect they have on the
   theme?
4. How does Takaki organize his essay? Trace his ideas through the first five
   paragraphs. What transitions does he use to shift focus? What is the reasoning behind this strategy?

5. Much of the information Takaki provides comes from individuals. What makes this a powerful technique? Cite examples from the essay to support your response.
6. Examine the conclusion. Why does Takaki crowd so much information into it? How does it work to reinforce Takaki's thesis?

WRITING

1. Photographs play an important role in Takaki's essay as reminders of the past. Find an old family photograph (preferably one taken before your birth), and describe it in detail, including the people in it, their relation to you, the setting, the year it was taken, and its significance to you and your family. Use details and sensory images.
2. Write a research paper about one of the major historical events mentioned in Takaki's essay (the internment of Asian Americans during WWII or the building of railroads across the United States).
3. **WRITING AN AGRUMENT:** Takaki claims that ethnic groups in the United States, especially Asian Americans, "must remember the past and break its silence." In an essay, argue that this effort has already been accomplished— or argue the opposite, that the effort to explore one's ethnic or racial identity must continue.

 www.mhhe.com/ **mhreader** | For more information on Ronald Takaki, go to: **More Resources > Ch. 6 History & Culture**

CONNECTIONS FOR CRITICAL THINKING

1. Cofer writes about Latino culture and Silko about Native American culture in their respective essays. What connections do they make between their subjects and cultural affiliation and alienation? How do they present their ideas? How are their tones similar? How are they different?

2. Write an essay exploring the topic of culture and civilization in the essays by Appiah, Sen, and Silko.

3. Consider the current position of women in culture. Refer to any three essays in this chapter to support your main observations.

4. How does one's experience of being an outsider or stranger to a culture affect one's understanding of that culture? Use essays from this chapter to support your key points.

5. Write an essay exploring the shape of contemporary American culture as it is reflected in the essays in this chapter. Cite specific support from at least three of the selections you have read.

6. How does a nation maintain a strong sense of self and still remain open to outside influences? Is a national identity crucial to a nation's survival? Use the opinions of representative authors in this chapter to address the question.

7. Is there such a thing as ethnic character, something that distinguishes Native Americans from African Americans, or Latinos from Asian immigrants? What factors contribute to identification with culture and with nation? Cite at least three essays in this chapter.

8. Argue for or against the proposition that Americans are ignorant of both the contributions and values of non-Western cultures in our country. Refer specifically to Reed and Takaki.

9. Design a chat room with five class members and discuss the differences in cultural perspectives among the writers in this chapter.

10. Search the Web for information on Judith Ortiz Cofer and Ishmael Reed. Download appropriate material, and then write a brief research paper on their perceptions of ethnicity and the American experience.

 chapter **7**

Government, Politics, and Social Justice
How Do We Decide What Is Fair?

Recent studies indicate that American students have an extremely weak understanding of government and politics. In fact, one-third of all high school juniors cannot identify the main purpose of the Declaration of Independence or say in which century it was signed. This document is one of the selections in this chapter. If we are ignorant of such a basic instrument in the making of our history and society, what can that say about our concepts of citizenship? Do we now see ourselves purely in terms of economic units, that is, our ability or potential to make money, or as consumers, the roles we play in spending it? Other notable essays on government, politics, and social justice in this chapter will help us understand our cultural legacies, and what has traditionally been thought of as the impetus in developing America as a country.

Major writers can bring politics and issues of social justice to life, enabling us to develop a sense of the various processes that have influenced the development of cultures over time. By studying the course of history and politics, we develop causal notions of how events are interrelated and how traditions have evolved. The study of history and politics can be an antidote to the continuous "present tense" of the media, which often have the power to make us believe we live from moment to moment, discouraging reflection on serious issues, such as why we live the way we do and how we came to be the people we are. Essays, speeches, documents, biographies, narratives, and many other literary forms capture events and illuminate the past while holding up a mirror to the present. Our political story can be brought to life out of the plain but painfully eloquent artifacts of oral culture. On the other hand, Thomas Jefferson employs classical rhetorical structures—notably argumentation—in outlining democratic vistas in the Declaration of Independence.

Even the briefest reflection will remind us of how important political processes and institutions are. Put simply, a knowledge of government and politics, and of our quest for social justice, validates our memory, a remembrance of how important the past is to our current existence. When, for example, Martin Luther King Jr. approaches the subject of oppression from a theological perspective, we are reminded of how important the concept of freedom is to our heritage and the various ways it can be addressed. Indeed, had we been more familiar with chapters in human history, we might have avoided some of the commensurate responses to the crises in our own era. The essays in this chapter help remind us—as the philosopher Santayana warned—"Those who forget the lessons of history are doomed to repeat them."

Only through a knowledge of government and politics can we make informed choices. Through a study of government and politics, we learn about challenges and opportunities, conflicts and their resolutions, and the use and abuse of power across time in numerous cultures and civilizations. It is through the study of historical processes and political institutions that we seek to define ourselves and to learn how we have evolved.

Previewing the Chapter

As you read the selections in this chapter and respond to them in discussion and writing, consider the following questions:

- On what specific events does the author concentrate? What is the time frame?
- What larger historical and political issues concern the author?
- From what perspective does the author treat the subject, from that of participant, observer, commentator, or some other perspective?
- What is the author's purpose in treating events and personalities: to explain, to instruct, to amuse, to criticize, or to celebrate?
- What does the author learn about history and politics from his or her inquiry into events?
- What sorts of conflicts—historical, political, economic, social, religious—emerge in the essay?
- Are there any correspondences among the essays? What analogies do the authors themselves draw?
- What is the relationship of people and personalities to the events under consideration?
- Which biases and ideological positions do you detect in the authors' works?
- How has your understanding of history and politics been challenged by the essays in this chapter?

Classic and Contemporary Images

HAVE WE MADE ADVANCES IN CIVIL RIGHTS?

Using a Critical Perspective   Are you optimistic or skeptical about the lofty words in the Declaration of Independence that announce "everyone is created equal"? How do these two visual texts, one advertising a slave auction and the other presenting Dr. Martin Luther King Jr. and a young friend, affect your response? What aspects of these visual texts stand out? What is your emotional and ethical response to the images? What do the two illustrations tell us about the evolution of equal rights and justice in the United States?

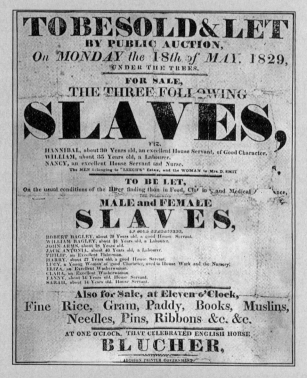

Advertisement of slaves for sale by the company Hewlett & Bright,
May 13, 1835.

Six-year-old Robin Arrington, daughter of a Miami Southern Christian
Leadership Conference attorney, leans on Dr. Martin Luther King's
shoulder as Dr. King holds a press conference, April 11, 1966,
in Miami, Florida. Dr. King arrived in Miami for a meeting
to establish a local chapter of his SCLC.

Classic and Contemporary Essays
WHAT IS THE AMERICAN DREAM?

Both Thomas Jefferson and Martin Luther King Jr. are now safely placed within the pantheon of American historical figures. The following two writing samples help indicate why. Both are concerned with perhaps the most significant issue that concerns contemporary humankind: freedom. Jefferson creates a doctrine that is powerful owing to his use of concise and powerful language, which he employs both to enumerate British offenses as well as to call upon his fellow Americans to revolt if need be. While his list of grievances may seem unquestionably correct to the contemporary mind, one must consider that Jefferson was a product of the Enlightenment, when philosophers had finally turned their attention to the primacy of individual rights after millennia of living under monarchic rule. King also provides us with the powerful theme of freedom in his famous speech; while his reflections address the peculiarly American racial divide, his style contains many biblical references and his rhetoric is that of the sermon. The reader should consider why these two documents, regardless of their historical context, seem to be milestones in our nation's history.

✳

The Declaration of Independence

In Congress, July 4, 1776

Thomas Jefferson

Thomas Jefferson (1743–1826) was governor of Virginia during the American Revolution, America's first secretary of state, and the third president of the United States. He had a varied and monumental career as politician, public servant, scientist, architect, educator (he founded the University of Virginia), and man of letters. Jefferson attended the Continental Congress in 1775, where he wrote the rough draft of the Declaration of Independence and revised it. Other hands made contributions to the document that was signed on July 4, 1776, but the wording, style, structure, and spirit of the final version are distinctly Jefferson's. Like Thomas Paine, Benjamin Franklin, James Madison, and other major figures of the Revolutionary era, Jefferson was notable for his use of prose as an instrument for social and political change. In the Declaration of Independence, we see the direct, precise, logical, and persuasive statement of revolutionary principles that makes the document one of the best known and best written texts in world history. Jefferson died in his home at Monticello on July 4, fifty years to the day from the signing of the Declaration of Independence.

When in the Course of human events it becomes necessary for one people to dissolve the political bands which have connected them with another, and to assume among the powers of the earth, the separate and equal station to which the Laws of Nature and of Nature's God entitle them, a decent respect to the opinions of mankind requires that they should declare the causes which impel them to the separation. 1

We hold these truths to be self-evident, that all men are created equal, that they are endowed by their Creator with certain unalienable Rights, that among these are Life, Liberty and the pursuit of Happiness.—That to secure these rights, Governments are instituted among Men, deriving their just powers from the consent of the governed.—That whenever any Form of Government becomes destructive of these ends, it is the Right of the People to alter or to abolish it, and to institute new Government, laying its foundation on such principles and organizing its powers in such form, as to them shall seem most likely to effect their Safety and Happiness. Prudence, indeed, will dictate that Governments long established should not be changed for light and transient causes; and accordingly all experience hath shewn that mankind are more disposed to suffer, while evils are sufferable, than to right themselves by abolishing the forms to which they are accustomed. But when a long train of abuses and usurpations, pursuing invariably the same Object evinces a design to reduce them under absolute Despotism, it is their right, it is their duty, to throw off 2 such Government, and to provide new Guards for their future security.—Such has been the patient sufferance of these Colonies; and such is now the necessity which constrains them to alter their former Systems of Government. The history of the present King of Great Britain is a history of repeated injuries and usurpations, all having in direct object the establishment of an absolute Tyranny over these States. To prove this, let Facts be submitted to a candid world.

He has refused his Assent to Laws, the most wholesome and necessary for the public good. 3

He has forbidden his Governors to pass Laws of immediate and pressing importance, unless suspended in their operation till his Assent should be obtained; and when so suspended, he has utterly neglected to attend to them. 4

He has refused to pass other Laws for the accommodation of large districts of people, unless those people would relinquish the right of Representation in the Legislature, a right inestimable to them and formidable to tyrants only. 5

He has called together legislative bodies at places unusual, uncomfortable, and distant from the depository of their public Records, for the sole purpose of fatiguing them into compliance with his measures. 6

He has dissolved Representative Houses repeatedly, for opposing with manly firmness his invasions on the rights of the people. 7

He has refused for a long time, after such dissolutions, to cause others to be elected; whereby the Legislative powers, incapable of Annihilation, have returned to the People at large for their exercise; the State remaining in the mean time exposed to all the dangers of invasion from without, and convulsions within. 8

He has endeavored to prevent the population of these States; for that purpose obstructing the Laws for Naturalization of Foreigners; refusing to pass others to encourage their migrations hither, and raising the conditions of new Appropriations of Lands. 9

He has obstructed the Administration of Justice, by refusing his Assent to Laws for establishing Judiciary powers. 10

He has made Judges dependent on his Will alone, for the tenure of their offices, and the amount and payment of their salaries. 11

He has erected a multitude of New Offices, and sent hither swarms of Officers to harass our people, and eat out their substance. 12

He has kept among us, in times of peace, Standing Armies without the Consent of our legislatures. 13

He has affected to render the Military independent of and superior to the Civil power. 14

He has combined with others to subject us to a jurisdiction foreign to our constitution, and unacknowledged by our laws; giving his Assent to their Acts of pretended Legislation: 15

For quartering large bodies of armed troops among us:

For protecting them, by a mock Trial, from punishment for any Murders which they should commit on the Inhabitants of these States:

For cutting off our Trade with all parts of the world:

For imposing Taxes on us without our Consent:

For depriving us in many cases, of the benefits of Trial by jury:

For transporting us beyond Seas to be tried for pretended offences:

For abolishing the free System of English Laws in a neighboring Province, establishing therein an Arbitrary government, and enlarging its Boundaries so as to render it at once an example and fit instrument for introducing the same absolute rule into these Colonies:

For taking away our Charters, abolishing our most valuable Laws and altering fundamentally the Forms of our Governments:

For suspending our own Legislatures, and declaring themselves invested with power to legislate for us in all cases whatsoever.

He has abdicated Government here, by declaring us out of his Protection 16 and waging War against us.

He has plundered our seas, ravaged our Coasts, burnt our towns, and de- 17 stroyed the lives of our people.

He is at this time transporting large Armies of foreign Mercenaries to com- 18 plete the works of death, desolation and tyranny, already begun with circumstances of Cruelty & Perfidy scarcely paralleled in the most barbarous ages, and totally unworthy the Head of a civilized nation.

He has constrained our fellow Citizens taken Captive on the high Seas to 19 bear Arms against their Country, to become the executioners of their friends and Brethren, or to fall themselves by their Hands.

He has excited domestic insurrections amongst us, and has endeavored to 20 bring on the inhabitants of our frontiers, the merciless Indian Savages, whose known rule of warfare, is an undistinguished destruction of all ages, sexes and conditions.

In every stage of these Oppressions We have Petitioned for Redress in the 21 most humble terms: Our repeated Petitions have been answered only by repeated injury. A Prince, whose character is thus marked by every act which may define a Tyrant, is unfit to be the ruler of a free people.

Nor have We been wanting in attentions to our British brethren. We have 22 warned them from time to time of attempts by their legislature to extend an unwarrantable jurisdiction over us. We have reminded them of the circumstances of our emigration and settlement here. We have appealed to their native justice and magnanimity, and we have conjured them by the ties of our common kindred to disavow these usurpations, which would inevitably interrupt our connections and correspondence. They too have been deaf to the voice of justice and of consanguinity. We must, therefore, acquiesce in the necessity, which denounces our Separation, and hold them, as we hold the rest of mankind, Enemies in War, in Peace Friends.

We, therefore, the Representatives of the United States of America, in Gen- 23 eral Congress, Assembled, appealing to the Supreme Judge of the world for the rectitude of our intentions, do, in the Name, and by Authority of the good People of these Colonies, solemnly publish and declare, That these United Colonies are, and of Right ought to be Free and Independent States; that they are Absolved from all Allegiance to the British Crown, and that all political connection between them and the State of Great Britain, is and ought to be totally dissolved; and that as Free and Independent States, they have full Power to levy War, conclude Peace, contract Alliances, establish Commerce, and to do all other Acts and Things which Independent States may of right do. And for the support of this Declaration, with a firm reliance on the protection of divine Providence, we mutually pledge to each other our Lives, our Fortunes and our sacred Honor.

COMPREHENSION

1. Explain Jefferson's main and subordinate purposes in this document.
2. What is Jefferson's key assertion, or argument? Mention several reasons that he gives to support his argument.
3. Summarize Jefferson's definition of human nature and government.

RHETORIC

1. There are many striking words and phrases in the Declaration of Independence, notably in the beginning. Locate three such examples, and explain their connotative power and effectiveness.
2. Jefferson and his colleagues had to draft a document designed for several audiences. What audiences did they have in mind? How do their language and style reflect their awareness of multiple audiences?
3. The Declaration of Independence is a classic model of syllogistic reasoning and deductive argument (see the Glossary). What is its major premise, and where is this premise stated? The minor premise? The conclusion?
4. What sort of inductive evidence does Jefferson offer?
5. Why is the middle portion, or body, of the Declaration of Independence considerably longer than the introduction or conclusion? What holds the body together?
6. Explain the function and effect of parallel structure in this document.

WRITING

1. Discuss the relevance of the Declaration of Independence to politics today.
2. Explain why the Declaration of Independence is a model of effective prose.

3. Write your own declaration of independence—from family, employer, required courses, or the like. Develop this declaration as an op-ed piece for a newspaper.
4. **WRITING AN ARGUMENT:** Do you believe that "all men are created equal"? Justify your answer in an argumentative essay.

www.mhhe.com/ **mhreader** | For more information on Thomas Jefferson, go to: **More Resources > Ch. 7 Government & Politics**

✳

I Have a Dream

Martin Luther King Jr.

Martin Luther King Jr. (1929–1968) was born in Atlanta, Georgia, and received degrees from Morehouse College, Crozer Theological Seminary, Boston University, and Chicago Theological Seminary. As Baptist clergyman, civil rights leader, founder and president of the Southern Christian Leadership Conference, and, in 1964, Nobel Peace Prize winner, King was a celebrated advocate of nonviolent resistance to achieve equality and racial integration in the world. King was a gifted orator and a highly persuasive writer. His books include Stride toward Freedom *(1958);* Letter from Birmingham City Jail *(1963);* Strength to Love *(1963);* Why We Can't Wait *(1964); and* Where Do We Go from Here: Chaos or Community? *(1967), a book published shortly before Reverend King was assassinated on April 4, 1968, in Memphis, Tennessee. This selection, a milestone of American oratory, was the keynote address at the March on Washington, August 28, 1963.*

I am happy to join with you today in what will go down in history as the greatest demonstration for freedom in the history of our nation.　1

Fivescore years ago, a great American, in whose symbolic shadow we stand today, signed the Emancipation Proclamation. This momentous decree came as a great beacon light of hope to millions of Negro slaves who had been seared in the flames of withering injustice. It came as a joyous daybreak to end the long night of their captivity.　2

But one hundred years later, the Negro still is not free; one hundred years later, the life of the Negro is still sadly crippled by the manacles of segregation and the chains of discrimination; one hundred years later, the Negro lives on a lonely island of poverty in the midst of a vast ocean of material prosperity; one hundred years later, the Negro is still languishing in the corners of American society and finds himself in exile in his own land.　3

So we've come here today to dramatize a shameful condition. In a sense ₄ we've come to our nation's capital to cash a check. When the architects of our republic wrote the magnificent words of the Constitution and the Declaration of Independence, they were signing a promissory note to which every American was to fall heir. This note was the promise that all men, yes, black men as well as white men, would be guaranteed the unalienable rights of life, liberty, and the pursuit of happiness.

It is obvious today that America has defaulted on this promissory note in so ₅ far as her citizens of color are concerned. Instead of honoring this sacred obligation, America has given the Negro people a bad check; a check which has come back marked "insufficient funds." We refuse to believe that there are insufficient funds in the great vaults of opportunity of this nation. And so we've come to cash this check, a check that will give us upon demand the riches of freedom and the security of justice.

We have also come to this hallowed spot to remind America of the fierce ur- ₆ gency of now. This is no time to engage in the luxury of cooling off or to take the tranquilizing drug of gradualism. Now is the time to make real the promises of democracy; now is the time to rise from the dark and desolate valley of segregation to the sunlit path of racial justice; now is the time to lift our nation from the quicksands of racial injustice to the solid rock of brotherhood; now is the time to make justice a reality for all God's children. It would be fatal for the nation to overlook the urgency of the moment. This sweltering summer of the Negro's legitimate discontent will not pass until there is an invigorating autumn of freedom and equality.

Nineteen sixty-three is not an end, but a beginning. And those who hope ₇ that the Negro needed to blow off steam and will now be content, will have a rude awakening if the nation returns to business as usual.

There will be neither rest nor tranquility in America until the Negro is ₈ granted his citizenship rights. The whirlwinds of revolt will continue to shake the foundations of our nation until the bright day of justice emerges.

But there is something that I must say to my people who stand on the warm ₉ threshold which leads into the palace of justice. In the process of gaining our rightful place we must not be guilty of wrongful deeds.

Let us not seek to satisfy our thirst for freedom by drinking from the cup of ₁₀ bitterness and hatred. We must forever conduct our struggle on the high plane of dignity and discipline. We must not allow our creative protest to degenerate into physical violence. Again and again we must rise to the majestic heights of meeting physical force with soul force.

The marvelous new militancy which has engulfed the Negro community ₁₁ must not lead us to a distrust of all white people, for many of our white brothers, as evidenced by their presence here today, have come to realize that their destiny is tied up with our destiny and they have come to realize that their freedom is inextricably bound to our freedom. This offense we share mounted to storm the battlements of injustice must be carried forth by a biracial army. We cannot walk alone.

And as we walk, we must make the pledge that we shall always march 12 ahead. We cannot turn back. There are those who are asking the devotees of civil rights, "When will you be satisfied?" We can never be satisfied as long as the Negro is the victim of the unspeakable horrors of police brutality.

We can never be satisfied as long as our bodies, heavy with fatigue of travel, 13 cannot gain lodging in the motels of the highways and the hotels of the cities. We cannot be satisfied as long as the Negro's basic mobility is from a smaller ghetto to a larger one.

We can never be satisfied as long as our children are stripped of their self- 14 hood and robbed of their dignity by signs stating "for whites only." We cannot be satisfied as long as a Negro in Mississippi cannot vote and a Negro in New York believes he has nothing for which to vote. No, we are not satisfied, and we will not be satisfied until justice rolls down like waters and righteousness like a mighty stream.

I am not unmindful that some of you have come here out of excessive trials 15 and tribulation. Some of you have come fresh from narrow jail cells. Some of you have come from areas where your quest for freedom left you battered by the storms of persecution and staggered by the winds of police brutality. You have been the veterans of creative suffering. Continue to work with the faith that unearned suffering is redemptive.

Go back to Mississippi; go back to Alabama; go back to South Carolina; go 16 back to Georgia; go back to Louisiana; go back to the slums and ghettos of the northern cities, knowing that somehow this situation can, and will be changed. Let us not wallow in the valley of despair.

So I say to you, my friends, that even though we must face the difficulties of 17 today and tomorrow, I still have a dream. It is a dream deeply rooted in the American dream that one day this nation will rise up and live out the true meaning of its creed—we hold these truths to be self-evident, that all men are created equal.

I have a dream that one day on the red hills of Georgia, sons of former 18 slaves and sons of former slave-owners will be able to sit down together at the table of brotherhood.

I have a dream that one day, even the state of Mississippi, a state sweltering 19 with the heat of injustice, sweltering with the heat of oppression, will be transformed into an oasis of freedom and justice.

I have a dream my four little children will one day live in a nation where 20 they will not be judged by the color of their skin but by the content of their character. I have a dream today!

I have a dream that one day, down in Alabama, with its vicious racists, with 21 its governor having his lips dripping with the words of interposition and nullification, that one day, right there in Alabama, little black boys and black girls will be able to join hands with little white boys and white girls as sisters and brothers. I have a dream today!

I have a dream that one day every valley shall be exalted, every hill and 22 mountain shall be made low, the rough places shall be made plain, and the crooked places shall be made straight and the glory of the Lord will be revealed and all flesh shall see it together.

This is our hope. This is the faith that I go back to the South with.   23

With this faith we will be able to hear out of the mountain of despair a stone   24
of hope. With this faith we will be able to transform the jangling discords of our nation into a beautiful symphony of brotherhood.

With this faith we will be able to work together, to pray together, to strug-   25
gle together, to go to jail together, to stand up for freedom together, knowing that we will be free one day. This will be the day when all of God's children will be able to sing with new meaning—"my country 'tis of thee; sweet land of liberty; of thee I sing; land where my fathers died, land of the pilgrims' pride; from every mountain side, let freedom ring"—and if America is to be a great nation, this must become true.

So let freedom ring from the prodigious hilltops of New Hampshire.   26
Let freedom ring from the mighty mountains of New York.   27
Let freedom ring from the heightening Alleghenies of Pennsylvania.   28
Let freedom ring from the snow-capped Rockies of Colorado.   29
Let freedom ring from the curvaceous slopes of California.   30
But not only that.   31
Let freedom ring from Stone Mountain of Georgia.   32
Let freedom ring from Lookout Mountain of Tennessee.   33
Let freedom ring from every hill and molehill of Mississippi, from every   34
mountainside, let freedom ring.

And when we allow freedom to ring, when we let it ring from every village   35
and hamlet, from every state and city, we will be able to speed up that day when all of God's children—black men and white men, Jews and Gentiles, Catholics and Protestants—will be able to join hands and to sing in the words of the old Negro spiritual, "Free at last, free at last; thank God Almighty, we are free at last."

COMPREHENSION

1. What is the main purpose behind this speech? Where does King state this purpose most clearly?
2. Why does King make use of "fivescore years ago" (paragraph 2)? How is this more appropriate than simply saying, "a hundred years ago"?
3. Who is King's audience? Where does he acknowledge the special historic circumstances influencing his speech?

RHETORIC

1. From what sources does King adapt phrases to give his work allusive richness?
2. What do the terms *interposition* and *nullification* (paragraph 21) mean? What is their historical significance?

3. Why does King make use of repetition? Does this technique work well in print? Explain.
4. What is the purpose of the extended metaphor in paragraphs 4 and 5? Which point in paragraph 3 does it refer to?
5. In which paragraphs does King address the problems of African Americans?
6. Why is this selection entitled "I Have a Dream"? How do dreams serve as a motif for this speech?

WRITING

1. "I Have a Dream" is considered by many people to be among the greatest speeches delivered by an American. Do you think that it deserves to be? Explain in an essay.
2. Write a comparative essay analyzing King's assessment of black Americans' condition in 1963 and their condition today. What do you think King would say if he knew of contemporary conditions?
3. Write your own "I Have a Dream" essay, basing it on your vision of America or of a special people.
4. **WRITING AN ARGUMENT:** Prepare a newspaper editorial advocating a solution to one aspect of racial, ethnic, or sexual injustice.

 www.mhhe.com/ **mhreader** | For more information on Martin Luther King Jr., go to: **More Resources > Ch. 7 Government & Politics**

Classic and Contemporary: Questions for Comparison

1. Compare the Declaration of Independence with King's speech in terms of the level of language, style, and content. Are they equally powerful and resonant? Cite specific passages from the essays to illustrate your responses.
2. Rewrite the Declaration of Independence in modern English as you believe Dr. King might, reflecting his concerns about the African American and other minorities in this country. Include a list of grievances similar to the one concerning British rule.
3. Write a research paper about the lives and times of King and Jefferson. Compare and contrast any significant events or pertinent biographical data in their backgrounds.

✳

To See or Not To See

Patricia J. Williams

Patricia Joyce Williams (b. 1951) was born in Boston, Massachusetts. Following her education at Wellesley College (BA, 1972) and Harvard University Law School (JD 1975), Williams worked in the Office of the City Attorney, Los Angeles, and at the Western Center on Law and Poverty. Subsequently, she held a number of academic positions as a professor of law at the City University of New York, the University of Wisconsin, and Columbia University, where, since 1992, she has taught. "Rights are islands of empowerment," Williams contends; in her writing she often deals with the ways in which those who are disempowered by race, class, or gender have difficulty retaining their rights. Williams's books include The Alchemy of Race and Rights *(1991),* The Rooster's Egg: On the Persistence of Prejudice *(1995), and* Seeing a Color-Blind Future: The Paradox of Race *(1998). In the following essay, published in* The Nation *in 2004, Williams examines the policy, increasingly common in American cities, of stopping people "randomly" to request identification.*

Boston's Massachusetts Bay Transportation Authority recently announced a 1
new policy of stopping people "randomly" to request identification from those
whom police believe to be acting "suspiciously"; they will also be asking to
check packages and bags "at random." In New York City, meanwhile, the MTA
has instituted a new policy forbidding cameras anywhere in the subway. One
angry photographer protesting the ban cited a friend who was questioned by
police after taking a picture of his wife near the Whitestone Bridge. Apparently
in response to recent scandals, Donald Rumsfeld has issued a ban on all digital
cameras, cell phones with digital cameras and camcorders from all military
compounds. And on the planetary line of defense, the Pentagon, having had the
foresight to purchase the right to publish satellite photographs of Earth a few
years ago, has never disseminated the same—thus removing aerial photos of
hotspots like Afghanistan from the public realm. Although the contracts are
said to have expired, John Pike of globalsecurity.org says that imaging companies don't want to "gratuitously annoy their biggest customer, the U.S. military.
These companies are run by businessmen, not crusading journalists."

Down on the ground, a Pakistani immigrant was detained after he asked a 2
stranger to snap a picture of him amid the fall foliage of upstate New York. The
lovely colors were reflected in the lapping pool of a water-treatment plant; the
stranger thought he might be casing the joint and called the police. At the borders, it is not only citizens of designated countries who are scrutinized and detained but also artwork, music and books. Curators, conductors and academics
are frustrated in their ability to plan for conferences or shows involving work
shipped from places like Cuba, Africa, the Middle East.

The flip side to all this banning and blindfolding is that the police have ₃
cameras trained on the public all over New York City. Private security firms
have cameras guarding every inch of work and shopping space. Antiterrorism
measures allow law enforcement to "sneak and peek" into private homes and
personal computers based on the suspicions of individual officers, without judicial oversight or accountability. Not that oversight will help in a time of panic:
As of this writing, an art professor at the University of Buffalo named Stephen
Kurtz awaits the outcome of a grand jury investigation into his series of gallery
installations protesting the genetic modification of food. When Kurtz's wife
died recently of a heart condition, the paramedic who came to his home saw
petri dishes and a DNA extractor used to analyze food for possible genetic alter-
ation. The paramedic reported him to the FBI, who confiscated the extractor, his
computer and papers, as well as his wife's body. Although nothing hazardous
was found, a grand jury has been convened to consider whether he should be
prosecuted under a provision of the "U.S. Biological Weapons Anti-Terrorism
Act of 1989," a law recently expanded by the U.S.A. Patriot Act to prohibit the
possession of "any biological agent, toxin, or delivery system" that has no "pro-
phylactic, protective, bona fide research, or other peaceful purpose."

Perhaps the emergence of this lumbering Panopticon wouldn't be quite so ₄
worrisome if we could be sure that there was equality of sneaking and peeking.
Given specific events like the upcoming conventions and election, most of us
might not mind if guards searched everyone who entered the subway, regardless of race, religion or rank. Since that is obviously impossible—we depend on
a transit system that transports rather than grinds to a standstill—most of us
also probably would accept random, truly random, searches as a more efficient
deterrent. For example, we wouldn't be so bothered if officials were stopping
not just brown people "randomly" but, say, every third person through the
turnstile, or everyone with a backpack—if, in other words, the winnowing were
consistently neutral. But "random" is often employed as though it were synonymous with the idle suspicions of individual officers–despite history attesting to
the manner in which free-floating "suspicion" is too frequently a cipher for ethnic stereotyping, racial voyeurism, unconstitutional animus.

Conducting the searches and manning the cameras are so many security ₅
guards, private contractors, prison wardens, housing police, sheriffs and regular police. A random sampling of law-enforcement personnel would no doubt
reveal ordinary Americans: They pray to God, beat their spouses, pay their
taxes, molest their children and love their dogs at approximately the same frequency as everyone else. Some of them are well-trained professionals, some
barely more than neglected, out-of-control kids, like Lynndie England. For better or worse, they carry within them the likes, prejudices, violence and ideals of
our very complex society. Some of them live next door. Some want to make sure
you never get within twenty miles of their neighborhood. Some believe that you
are the Antichrist. Some want to marry your daughter.

But all that personal preference and human idiosyncrasy is beside the point ₆
if they conform to reasonably clear guidelines informed by as broad a range of

public input and oversight as possible. If, on the other hand, the network of information gatherers and secret surveillers is allowed to become an ever more closed society and to indulge their own prejudices and paranoia, we will see a very different, if no less human, reaction: the increased conforming of our society to the standards of sober bureaucrats, information analysts so narrowly focused as to be narrow-minded, media rumor-mills and traumatized military men. Our homes, backpacks, offices, pockets and cars will have to be suitably sanitized of books they haven't read, of science they don't understand, or art that unsettles them, of looks that trigger flashbacks and of ideas that are so creative as to seem foreign.

We must not cede the power to witness what is happening to us, to know how we are seen, to oversee our own representation. Without that freedom, we must recognize ourselves in the awful words spoken by the despairing family of one Iraqi man who has disappeared into U.S. custody: "It's because they have absolute force. No one sees what they do." 7

COMPREHENSION

1. This article appeared in Williams's column for *The Nation*, "Diary of a Mad Law Professor." What is she angry about in this essay?
2. List the examples of government intrusion into personal lives that Williams cites in this essay.
3. What does the writer mean by "equality of sneaking and peeking" (paragraph 4)? What is her point about this phenomenon?

RHETORIC

1. For whom was this article written? Lawyers? Liberals? Conservatives? A general audience? How do you know?
2. As a specialist trained in the law, Williams adopts an argumentative or adversarial approach to her subject. What aspects of legal training do you find in this essay? How does she make her "case" or her claim? What does she imply about her adversaries—the people and institutions aligned against her in the court of inquiry?
3. This essay was written on the eve of the 2004 Democratic and Republican National Conventions, held in Boston and New York City, respectively. Where does Williams make reference to these events? How does the writer make certain that this topical fact does not "date" the essay or make it irrelevant to the present?
4. What varieties of evidence does Williams present in this essay? Do you find that there is too little or too much to support her claim? Explain.
5. How does Williams adjust her style to her message? Is this style legalistic or plainspoken? Where does the writer use figurative language? Where is she colloquial? Concrete? Abstract? What is the overall effect?

6. How does the writer's end paragraph serve as a concise reformulation of the main ideas in her essay?

WRITING

1. Imagine that you have been stopped by a person in authority who asks for your identification. (Perhaps you have actually experienced this situation.) What would be your response? How would you handle the situation and its aftermath? What would it tell you about the state of social and political conditions in the United States today?
2. Compare and contrast Williams's views in this essay with at least one of the articles on the Patriot Act appearing in the "Debate" at the end of Chapter 2.
3. **WRITING AN ARGUMENT:** Use the material in Williams's essay to write an argumentative essay in which you defend or criticize the "sneaking and peeking" that has become commonplace in the United States in the aftermath of 9/11.

---

www.mhhe.com/ mhreader | For more information on Patricia J. Williams, go to: **More Resources > Ch. 7 Government & Politics**

Cyberspace: If You Don't Love It, Leave It

Esther Dyson

Esther Dyson (b. 1951) was born in Zurich, Switzerland, grew up in Princeton, New Jersey, and received a BA in economics from Harvard University. She is the daughter of Freeman Dyson, a physicist prominent in arms control. She is the editor and publisher of the widely respected computer newsletter Release 1.0, *which is circulated to many computer industry leaders. She is also chairperson of the Electronic Frontier Foundation and on the board of the Santa Fe Institute, the Global Business Network, and the Institute for East/West Studies. She served as a reporter for* Forbes *magazine for four years. The following essay appeared in* The New York Times Magazine *in July 1995. In it, Dyson defends the free-market approach to cyberspace content, arguing that regulation of the Internet is simply impossible and counterproductive.*

Something in the American psyche loves new frontiers. We hanker after wide-   1
open spaces; we like to explore; we like to make rules instead of follow them.
But in this age of political correctness and other intrusions on our national cult
of independence, it's hard to find a place where you can go and be yourself
without worrying about the neighbors.

There is such a place: cyberspace. Lost in the furor over porn on the Net is   2
the exhilarating sense of freedom that this new frontier once promised—and
still does in some quarters. Formerly a playground for computer nerds and
techies, cyberspace now embraces every conceivable constituency: schoolchildren, flirtatious singles, Hungarian-Americans, accountants—along with ped-
erasts and porn fans. Can they all get along? Or will our fear of kids surfing for
cyberporn behind their bedroom doors provoke a crackdown?

The first order of business is to grasp what cyberspace *is*. It might help to   3
leave behind metaphors of highways and frontiers and to think instead of real
estate. Real estate, remember, is an intellectual, legal, artificial environment constructed *on top of* land. Real estate recognizes the difference between parkland
and shopping mall, between red-light zone and school district, between church,
state and drugstore.

In the same way, you could think of cyberspace as a giant and unbounded   4
world of virtual real estate. Some property is privately owned and rented out;
other property is common land; some places are suitable for children, and others are best avoided by all but the kinkiest citizens. Unfortunately, it's those
places that are now capturing the popular imagination: places that offer bomb-
making instructions, pornography, advice on how to procure stolen credit
cards. They make cyberspace sound like a nasty place. Good citizens jump to a
conclusion: Better regulate it.

The most recent manifestation of this impulse is the Exon-Coats Amend-   5
ment, a well-meaning but misguided bill drafted by Senators Jim Exon, Demo-
crat of Nebraska, and Daniel R. Coats, Republican of Indiana, to make
cyberspace "safer" for children. Part of the telecommunications reform bill
passed by the Senate and awaiting consideration by the House, the amendment
would outlaw making "indecent communication" available to anyone under
18.[1] Then there's the Amateur Action bulletin board case, in which the owners
of a porn service in Milpitas, Calif., were convicted in a Tennessee court of violating "community standards" after a local postal inspector requested that the
material be transmitted to him.

Regardless of how many laws or lawsuits are launched, regulation won't   6
work.

Aside from being unconstitutional, using censorship to counter indecency   7
and other troubling "speech" fundamentally misinterprets the nature of cyberspace. Cyberspace isn't a frontier where wicked people can grab unsuspecting
children, nor is it a giant television system that can beam offensive messages at

---

[1]The Communications Decency Act (CDA) was passed by Congress, but the Supreme Court ruled
that it was unconstitutional in 1996. [This note is the author's.]

unwilling viewers. In this kind of real estate, users have to *choose* where they visit, what they see, what they do. It's optional, and it's much easier to bypass a place on the Net than it is to avoid walking past an unsavory block of stores on the way to your local 7-Eleven.

Put plainly, cyberspace is a voluntary destination—in reality, many destina- 8 tions. You don't just get "onto the net"; you have to go someplace in particular. That means that people can choose where to go and what to see. Yes, community standards should be enforced, but those standards should be set by cyberspace communities themselves, not by the courts or by politicians in Washington. What we need isn't Government control over all these electronic communities: We need self-rule.

What makes cyberspace so alluring is precisely the way in which it's *different* 9 from shopping malls, television, highways and other terrestrial jurisdictions. But let's define the territory:

First, there are private e-mail conversations, akin to the conversations you 10 have over the telephone or voice mail. These are private and consensual and require no regulation at all.

Second, there are information and entertainment services, where people can 11 download anything from legal texts and lists of "great new restaurants" to game software or dirty pictures. These places are like bookstores, malls and movie houses—places where you go to buy something. The customer needs to request an item or sign up for a subscription; stuff (especially pornography) is not sent out to people who don't ask for it. Some of these services are free or included as part of a broader service like Compuserve or America Online; others charge and may bill their customers directly.

Third, there are "real" communities—groups of people who communicate 12 among themselves. In real-estate terms, they're like bars or restaurants or bathhouses. Each active participant contributes to a general conversation, generally through posted messages. Other participants may simply listen or watch. Some are supervised by a moderator; others are more like bulletin boards—anyone is free to post anything. Many of these services started out unmoderated but are now imposing rules to keep out unwanted advertising, extraneous discussions or increasingly rude participants. Without a moderator, the decibel level often gets too high.

Ultimately, it's the rules that determine the success of such places. Some of 13 the rules are determined by the supplier of content; some of the rules concern prices and membership fees. The rules may be simple: "Only high-quality content about oil-industry liability and pollution legislation: $120 an hour." Or: "This forum is unmoderated, and restricted to information about copyright issues. People who insist on posting advertising or unrelated material will be asked to desist (and may eventually be barred)." Or: "Only children 8 to 12, on school-related topics and only clean words. The moderator will decide what's acceptable."

Cyberspace communities evolve just the way terrestrial communities do: 14 people with like-minded interests band together. Every cyberspace community has its own character. Overall, the communities on Compuserve tend to be more techy or professional; those on America Online, affluent young singles; Prodigy, family oriented. Then there are independents like Echo, a hip, downtown New York service, or Women's Wire, targeted to women who want to avoid the male culture prevalent elsewhere on the Net. There's SurfWatch, a new program allowing access only to locations deemed suitable for children. On the Internet itself, there are lots of passionate noncommercial discussion groups on topics ranging from Hungarian politics (Hungary-Online) to copyright law.

And yes, there are also porn-oriented services, where people share dirty 15 pictures and communicate with one another about all kinds of practices, often anonymously. Whether these services encourage the fantasies they depict is subject to debate—the same debate that has raged about pornography in other media. But the point is that no one is forcing this stuff on anybody.

What's unique about cyberspace is that it liberates us from the tyranny of 16 government, where everyone lives by the rule of the majority. In a democracy, minority groups and minority preferences tend to get squeezed out, whether they are minorities of race and culture or minorities of individual taste. Cyberspace allows communities of any size and kind to flourish; in cyberspace, communities are chosen by the users, not forced on them by accidents of geography. This freedom gives the rules that preside in cyberspace a moral authority that rules in terrestrial environments don't have. Most people are stuck in the country of their birth, but if you don't like the rules of a cyberspace community, you can just sign off. Love it or leave it. Likewise, if parents don't like the rules of a given cyberspace community, they can restrict their children's access to it.

What's likely to happen in cyberspace is the formation of new communities, 17 free of the constraints that cause conflict on earth. Instead of a global village, which is a nice dream but impossible to manage, we'll have invented another world of self-contained communities that cater to their own members' inclinations without interfering with anyone else's. The possibility of a real market-style evolution of governance is at hand. In cyberspace, we'll be able to test and evolve rules governing what needs to be governed—intellectual property, content and access control, rules about privacy and free speech. Some communities will allow anyone in; others will restrict access to members who qualify on one basis or another. Those communities that prove self-sustaining will prosper (and perhaps grow and split into subsets with ever-more-particular interests and identities). Those that can't survive—either because people lose interest or get scared off—will simply wither away.

In the near future, explorers in cyberspace will need to get better at defining 18 and identifying their communities. They will need to put in place—and accept—their own local governments, just as the owners of expensive real estate often prefer to have their own security guards rather than call in the police. But they will rarely need help from any terrestrial government.

Of course, terrestrial governments may not agree. What to do, for instance, 19 about pornography? The answer is labeling—not banning—questionable material. In order to avoid censorship and lower the political temperature, it makes sense for cyberspace participants themselves to agree on a scheme for questionable items, so that people or automatic filters can avoid them. In other words, posting pornography in "alt.sex.bestiality" would be OK; it's easy enough for software manufacturers to build an automatic filter that would prevent you—or your child—from ever seeing that item on a menu. (It's as if all the items were wrapped with labels on the wrapper.) Someone who posted the same material under the title "Kid-Fun" could be sued for mislabeling.

Without a lot of fanfare, private enterprises and local groups are already 20 producing a variety of labeling and ranking services, along with kid-oriented sites like Kidlink, EdWeb and Kids' Space. People differ in their tastes and values and can find services or reviewers on the Net that suit them in the same way they select books and magazines. Or they can wander freely if they prefer, making up their own itinerary.

In the end, our society needs to grow up. Growing up means understand- 21 ing that there are no perfect answers, no all-purpose solutions, no government-sanctioned safe havens. We haven't created a perfect society on earth and we won't have one in cyberspace either. But at least we can have individual choice—and individual responsibility.

COMPREHENSION

1. The title of the essay is a variation of a phrase popularized in the 1960s. What is the original expression and what was its significance? What is its relevance to this essay?
2. What is Dyson's thesis? Is it stated explicitly? If so, where in the essay does it occur? If it is merely suggested, how is it suggested and where?
3. There are many forms of new media that are not considered communities. Why does Dyson refer to cyberspace as a community?
4. According to Dyson, what distinguishes cyberspace from physical space?
5. What does Dyson mean when she states that cyberspace needs "self-rule" (paragraph 8)?

RHETORIC

1. How does Dyson use her introduction to foreshadow her main concerns about censorship in cyberspace?
2. How does Dyson use metaphor in paragraphs 11 through 13 to help us understand the structure of cyberspace? Why is metaphor a particularly good literary device to use when explaining a new concept?
3. Key to Dyson's views on cyberspace is that it is a "voluntary destination" (paragraph 8). What evidence does Dyson present that it is voluntary? What argument can be made that it is not always "voluntary"?
4. Who is the implied audience for this essay? What level of education does one need to have and how sophisticated about the world of cyberspace

does one need to be in order to comprehend and process the author's views? Explain your answer.

5. Dyson refers to laws, rules, and regulations as strategies that various interest groups may use to determine access to content in cyberspace. How does Dyson distinguish these three related tactics? What significance does differentiating these methods have in the author's presentation of her argument?

6. Dyson concludes her essay with an analogy between human society and cyberspace culture. Why has she saved this final support for last? How does it extend her argument rather than merely restate it?

WRITING

1. In a comparison-and-contrast essay, select three cyberspace communities and describe each one's character (refer to Dyson's reference to cyberspace character in paragraph 14).

2. In paragraph 17, Dyson refers to the "global village," a term coined by the media critic Marshall McLuhan. For a research project, study McLuhan's views on the nature of the global village and compare and contrast them to Dyson's views of the nature of cyberspace.

3. **WRITING AN ARGUMENT:** Dyson argues that technology can create filters, labeling and ranking services to prevent children from viewing inappropriate material. In an essay, argue for or against the proposition that there can be a nontechnological solution to this issue, for example, instilling values in children or developing a society that does not create a mystique about taboo subject matter.

---

www.mhhe.com/ **mhreader**	For more information on Esther Dyson, go to: **More Resources > Ch. 7 Government & Politics**

✳

The Circle of Governments

Niccolò Machiavelli

Niccolò Machiavelli (1469–1527), Italian patriot, statesman, and writer, is one of the seminal figures in the history of Western political thought. His inquiries into the nature of the state, the amoral quality of political life, and the primacy of power are distinctly modernist in outlook. He began his studies of political and historical issues after being forced to retire from Florentine politics in 1512. Exiled outside the city, Machiavelli wrote The Prince *(1513),* The Discourses *(1519),* The Art of War *(1519–1520), and* The Florentine History *(1525). The following selection from* The Discourses *(conceived by the author as commentaries on the first 10 books of Livy's* History of Rome*) analyzes the varieties of government and their political implications in history.*

Having proposed to myself to treat of the kind of government established at   1
Rome, and of the events that led to its perfection, I must at the beginning observe that some of the writers on politics distinguished three kinds of government, vis. the monarchical, the aristocratic, and the democratic; and maintain that the legislators of a people must choose from these three the one that seems to them most suitable. Other authors, wiser according to the opinion of many, count six kinds of governments, three of which are very bad, and three good in themselves, but so liable to be corrupted that they become absolutely bad. The three good ones are those which we have just named; the three bad ones result from the degradation of the other three, and each of them resembles its corresponding original, so that the transition from the one to the other is very easy. Thus monarchy becomes tyranny; aristocracy degenerates into oligarchy; and the popular government lapses readily into licentiousness. So that a legislator who gives to a state which he founds either of these three forms of government, constitutes it but for a brief time; for no precautions can prevent either one of the three that are reputed good from degenerating into its opposite kind; so great are in these the attractions and resemblances between the good and the evil.

Chance has given birth to these different kinds of governments amongst   2
men; for at the beginning of the world the inhabitants were few in number and lived for a time dispersed, like beasts. As the human race increased, the necessity for uniting themselves for defense made itself felt; the better to attain this object they chose the strongest and most courageous from amongst themselves and placed him at their head promising to obey him. Thence they began to know the good and the honest, and to distinguish them from the bad and vicious; for seeing a man injure his benefactor aroused at once two sentiments in every heart, hatred against the ingrate and love for the benefactor. They blamed the first, and on the contrary honored those the more who showed themselves grateful, for each felt that he in turn might be subject to a like wrong; and to prevent similar evils, they set to work to make laws, and to institute punishments for those who contravened them. Such was the origin of justice. This caused them, when they had afterwards to choose a prince, neither to look to the strongest nor bravest, but to the wisest and most just. But when they began to make sovereignty hereditary and non-elective, the children quickly degenerated from their fathers; and, so far from trying to equal their virtues, they considered that a prince had nothing else to do than to excel all the rest in luxury, indulgence, and every other variety of pleasure. The prince consequently soon drew upon himself the general hatred. An object of hatred, he naturally felt fear; fear in turn dictated to him precautions and wrongs, and thus tyranny quickly developed itself. Such were the beginning and causes of disorders, conspiracies, and plots against the sovereigns, set on foot, not by the feeble and timid, but by

those citizens who, surpassing the others in grandeur of soul, in wealth, and in courage, could not submit to the outrages and excesses of their princes.

Under such powerful leaders the masses armed themselves against the tyrant, and after having rid themselves of him, submitted to these chiefs as their liberators. These, abhorring the very name of prince, constituted themselves a new government; and at first bearing in mind the past tyranny, they governed in strict accordance with the laws which they had established themselves; preferring public interests to their own, and to administer and protect with greatest care both public and private affairs. The children succeeded their fathers, and ignorant of the changes of fortune, having never experienced its reverses, and indisposed to remain content with this civil equality, they in turn gave themselves up to cupidity, ambition, libertinage, and violence, and soon caused the aristocratic government to degenerate into an oligarchic tyranny, regardless of all civil rights. They soon, however, experienced the same fate as the first tyrant; the people, disgusted with their government, placed themselves at the command of whoever was willing to attack them, and this disposition soon produced an avenger, who was sufficiently well seconded to destroy them. The memory of the prince and the wrongs committed by him being still fresh in their minds, and having overthrown the oligarchy, the people were not willing to return to the government of a prince. A popular government was therefore resolved upon, and it was so organized that the authority would not again fall into the hands of a prince or a small number of nobles. And as all governments are at first looked up to with some degree of reverence, the popular state also maintained itself for a time, but which was never of long duration, and lasted generally only about as long as the generation that had established it; for it soon ran into that kind of licence which inflicts injury upon public as well as private interests. Each individual only consulted his own passions, and a thousand acts of injustice were daily committed, so that, constrained by necessity, or directed by the counsels of some good man, or for the purpose of escaping from this anarchy, they returned anew to the government of a prince, and from this they generally lapsed again into anarchy, step-by-step, in the same manner and from the same causes as we have indicated.

Such is the circle which all republics are destined to run through. Seldom, however, do they come back to the original form of government, which results from the fact that their duration is not sufficiently long to be able to undergo these repeated changes and preserve their existence. But it may well happen that a republic lacking strength and good counsel in its difficulties becomes subject after a while to some neighboring state, that is better organized than itself; and if such is not the case, then they will be apt to revolve indefinitely in the circle of revolutions. I say, then, that all kinds of government are defective; those three which we have qualified as good because they are too short-lived, and the three bad ones because of their inherent viciousness. Thus sagacious legislators, knowing the vices of each of these systems of government by themselves, have chosen one that should partake of all of them, judging that to be the most stable and solid. In fact, when there is combined under the same constitution a prince, a nobility, and the power of the people, then these three powers will watch and keep each other reciprocally in check.

COMPREHENSION

1. Where in the essay does Machiavelli state his thesis? What is his thesis?
2. Explain in your own words the three types of government Machiavelli describes, their origins, and their pitfalls.
3. Ultimately, who determines what system of government a country will have—the governed or the legislators? Explain your view.

RHETORIC

1. In paragraph 1, Machiavelli states the motivation for writing his essay. How does he create a transition from explaining this motivation to addressing his subject directly?
2. Machiavelli explains the three forms of government in a particular order. What is the unifying rhetoric behind the order in which he describes them? How does it relate to the theme of the essay?
3. Both paragraphs 2 and 3 describe the process by which governments are formed. What methods does the author use to create coherent paragraphs in providing a step-by-step description of these formations?
4. Would you consider this essay descriptive, narrative, expository, or a combination of two or more of these methods? Explain your answer.
5. Define *oligarchy, benefactor, licentiousness, cupidity, libertinage,* and *sagacious.* What does the use of these words in the essay suggest about the author and his intended audience?
6. From what vantage point does the author appear to view his subject matter: participant, reporter, critic, or teacher? Explain your view.

WRITING

1. Using the terms *monarchy, oligarchy,* and *democracy,* describe the various governing bodies of your school, their functions, and where they fit into Machiavelli's taxonomy.
2. **WRITING AN ARGUMENT:** Argue for or against the proposition that the United States has an ideal form of government, according to Machiavelli's view of what a government should be.
3. **WRITING AN ARGUMENT:** Argue for or against *one* of the forms of government that Machiavelli describes in his essay.

 www.mhhe.com/ **mhreader**

For more information on Niccolò Machiavelli, go to:
More Resources > Ch. 7 Government & Politics

✳

Grant and Lee:
A Study in Contrasts

Bruce Catton

Bruce Catton (1899–1978) was born in Petosky, Michigan. After serving in the Navy during World War I, he attended Oberlin College but left in his junior year to pursue a career in journalism. From 1942 to 1952, Catton served in the government, first on the War Production Board and later in the departments of Commerce and the Interior. He left government to devote himself to literary work as a columnist for The Nation *and a historian of the Civil War. His many works include* A Stillness at Appomattox *(1953), which won the 1954 Pulitzer Prize;* Mr. Lincoln's Army *(1951);* The Centennial History of the Civil War *(1961–1965); and* Prefaces to History *(1970). In the following selection, Catton presents vivid portraits of two well-known but little understood figures from American history.*

When Ulysses S. Grant and Robert E. Lee met in the parlor of a modest house at   1
Appomattox Court House, Virginia, on April 9, 1865, to work out the terms for
the surrender of Lee's Army of Northern Virginia, a great chapter in American
life came to a close, and a great new chapter began.

These men were bringing the Civil War to its virtual finish. To be sure, other   2
armies had yet to surrender, and for a few days the fugitive Confederate government would struggle desperately and vainly, trying to find some way to go
on living now that its chief support was gone. But in effect it was all over when
Grant and Lee signed the papers. And the little room where they wrote out
the terms was the scene of one of the poignant, dramatic contrasts in American
history.

They were two strong men, these oddly different generals, and they repre-   3
sented the strengths of two conflicting currents that, through them, had come
into final collision.

Back of Robert E. Lee was the notion that the old aristocratic concept might   4
somehow survive and be dominant in American life.

Lee was tidewater Virginia, and in his background were family, culture, and   5
tradition . . . the age of chivalry transplanted to a New World which was making its own legends and its own myths. He embodied a way of life that had
come down through the age of knighthood and the English country squire.
America was a land that was beginning all over again, dedicated to nothing
much more complicated than the rather hazy belief that all men had equal
rights and should have an equal chance in the world. In such a land Lee stood
for the feeling that it was somehow of advantage to human society to have a
pronounced inequality in the social structure. There should be a leisure class,

backed by ownership of land; in turn, society itself should be keyed to the land as the chief source of wealth and influence. It would bring forth (according to this ideal) a class of men with a strong sense of obligation to the community; men who lived not to gain advantage for themselves, but to meet the solemn obligations which had been laid on them by the very fact that they were privileged. From them the country would get its leadership; to them it could look for the higher values—of thought, of conduct, of personal deportment—to give it strength and virtue.

Lee embodied the noblest elements of this aristocratic ideal. Through him, 6 the landed nobility justified itself. For four years, the Southern states had fought a desperate war to uphold the ideals for which Lee stood. In the end, it almost seemed as if the Confederacy fought for Lee; as if he himself was the Confederacy . . . the best thing that the way of life for which the Confederacy stood could ever have to offer. He had passed into legend before Appomattox. Thousands of tired, underfed, poorly clothed Confederate soldiers, long since past the simple enthusiasm of the early days of the struggle, somehow considered Lee the symbol of everything for which they had been willing to die. But they could not quite put this feeling into words. If the Lost Cause, sanctified by so much heroism and so many deaths, had a living justification, its justification was General Lee.

Grant, the son of a tanner on the Western frontier, was everything Lee was 7 not. He had come up the hard way and embodied nothing in particular except the eternal toughness and sinewy fiber of the men who grew up beyond the mountains. He was one of a body of men who owed reverence and obeisance to no one, who were self-reliant to a fault, who cared hardly anything for the past but who had a sharp eye for the future.

These frontier men were the precise opposites of the tidewater aristocrats. 8 Back of them, in the great surge that had taken people over the Alleghenies and into the opening Western country, there was a deep, implicit dissatisfaction with a past that had settled into grooves. They stood for democracy, not from any reasoned conclusion about the proper ordering of human society, but simply because they had grown up in the middle of democracy and knew how it worked. Their society might have privileges, but they would be privileges each man had won for himself. Forms and patterns meant nothing. No man was born to anything, except perhaps to a chance to show how far he could rise. Life was competition.

Yet along with this feeling had come a deep sense of belonging to a national 9 community. The Westerner who developed a farm, opened a shop, or set up in business as a trader, could hope to prosper only as his own community prospered—and his community ran from the Atlantic to the Pacific and from Canada down to Mexico. If the land was settled, with towns and highways and accessible markets, he could better himself. He saw his fate in terms of the nation's own destiny. As its horizons expanded, so did his. He had, in other words, an acute dollars-and-cents stake in the continued growth and development of his country.

And that, perhaps, is where the contrast between Grant and Lee becomes 10 most striking. The Virginia aristocrat, inevitably, saw himself in relation to his own region. He lived in a static society which could endure almost anything except change. Instinctively, his first loyalty would go to the locality in which that society existed. He would fight to the limit of endurance to defend it, because in defending it he was defending everything that gave his own life its deepest meaning.

The Westerner, on the other hand, would fight with an equal tenacity for the 11 broader concept of society. He fought so because everything he lived by was tied to growth, expansion, and a constantly widening horizon. What he lived by would survive or fall with the nation itself. He could not possibly stand by unmoved in the face of an attempt to destroy the Union. He would combat it with everything he had, because he could only see it as an effort to cut the ground out from under his feet.

So Grant and Lee were in complete contrast, representing two diametrically 12 opposed elements in American life. Grant was the modern man emerging; beyond him, ready to come on the stage, was the great age of steel and machinery, of crowded cities and a restless burgeoning vitality. Lee might have ridden down from the old age of chivalry, lance in hand, silken banner fluttering over his head. Each man was the perfect champion of his cause, drawing both his strengths and his weaknesses from the people he led.

Yet it was not all contrast, after all. Different as they were—in background, 13 in personality, in underlying aspiration—these two great soldiers had much in common. Under everything else, they were marvelous fighters. Furthermore, their fighting qualities were really very much alike.

Each man had, to begin with, the great virtue of utter tenacity and fidelity. 14 Grant fought his way down the Mississippi Valley in spite of acute personal discouragement and profound military handicaps. Lee hung on in the trenches at Petersburg after hope itself had died. In each man there was an indomitable quality . . . the born fighter's refusal to give up as long as he can still remain on his feet and lift his two fists.

Daring and resourcefulness they had, too; the ability to think faster and 15 move faster than the enemy. These were the qualities which gave Lee the dazzling campaigns of Second Manassas and Chancellorsville and won Vicksburg for Grant.

Lastly, and perhaps greatest of all, there was the ability, at the end, to turn 16 quickly from war to peace once the fighting was over. Out of the way these two men behaved at Appomattox came the possibility of a peace of reconciliation. It was a possibility not wholly realized, in the years to come, but which did, in the end, help the two sections to become one nation again . . . after a war whose bitterness might have seemed to make such a reunion wholly impossible. No part of either man's life became him more than the part he played in their brief meeting in the McLean house at Appomattox. Their behavior there put all succeeding generations of Americans in their debt. Two great Americans, Grant and

Lee—very different, yet under everything very much alike. Their encounter at Appomattox was one of the great moments of American history.

COMPREHENSION

1. What is the central purpose of Catton's study? Cite evidence to support your view. Who is his audience?
2. What is the primary appeal to readers of describing history through the study of individuals rather than through the recording of events? How does Catton's essay reflect this appeal?
3. According to Catton, what special qualities did Grant and Lee share, and what qualities set them apart?

RHETORIC

1. What role does the opening paragraph have in setting the tone for the essay? Is the tone typical of what you would expect of an essay describing military generals? Explain your view. How does the conclusion echo the introductory paragraph?
2. Note that the sentence, "Two great Americans, Grant and Lee—very different, yet under everything very much alike" (paragraph 16), has no verb. What does this indicate about Catton's style? What other sentences contain atypical syntax? What is their contribution to the unique quality of the writing?
3. Although this essay is about a historical era, there is a notable lack of specific facts—dates, statistics, and events. What has Catton focused on instead?
4. What is the function of the one-sentence paragraph 3?
5. Paragraphs 9, 10, 12, and 13 begin with coordinating conjunctions. How do these transitional words give the paragraphs their special coherence? How would more typical introductory expressions, such as *in addition, furthermore,* or *moreover,* have altered this coherence?
6. What strategy does Catton use in comparing and contrasting the two generals? Study paragraphs 5 through 16. Which are devoted to describing each man separately, and which include aspects of each man? What is the overall development of the comparisons?

WRITING

1. Does Lee's vision of society exist in the United States today? If not, why not? If so, where do you find this vision? Write a brief essay on this topic.
2. Select two well-known individuals in the same profession—for example, politics, entertainment, or sports. Make a list for each, enumerating the dif-

ferent aspects of their character, behavior, beliefs, and background. Using this as an outline, devise an essay comparing and contrasting the two.

3. **WRITING AN ARGUMENT:** Apply, in an argumentative essay, Catton's observation about "two diametrically opposed elements in American life" (paragraph 12) to the current national scene.

www.mhhe.com/ mhreader   For more information on Bruce Catton, go to:
More Resources > Ch. 7 Government & Politics

✳

American Dreamer

Bharati Mukherjee

Bharati Mukherjee (b. 1940) was born in Calcutta, India, and learned to read and write by the age of three. In 1947, she moved to Britain with her family. After receiving her BA from the University of Calcutta and her MA in English and Ancient Indian Culture from the University of Boroda, she came to the United States, where she received an MFA in Creative Writing and a PhD in English and Comparative Literature at the University of Iowa. Mukherjee is the author of Jasmine *(1989) and* The Middleman and Other Stories, *which won the 1988 National Book Critic's Circle Award for Fiction. Her more recent work includes the novels* The Holder of the World *(1993) and* Leave It to Me *(1997). She is currently professor at the University of California, Berkeley. Mukherjee is often interested in and writing about issues of cultural identity. In the following essay, which first appeared in the magazine* Mother Jones *in 1997, she examines why "hyphenated Americans" always seem to be members of nonwhite groups.*

The United States exists as a sovereign nation. "America," in contrast, exists as   1
a myth of democracy and equal opportunity to live by, or as an ideal goal to reach.

I am a naturalized U.S. citizen, which means that, unlike native-born citi-   2
zens, I had to prove to the U.S. government that I merited citizenship. What I didn't have to disclose was that I desired "America," which to me is the stage for the drama of self-transformation.

I was born in Calcutta and first came to the United States—to Iowa City, to   3
be precise—on a summer evening in 1961. I flew into a small airport surrounded by cornfields and pastures, ready to carry out the two commands my father had written out for me the night before I left Calcutta: Spend two years studying creative writing at the Iowa Writers' Workshop, then come back home and marry the bridegroom he selected for me from our caste and class.

In traditional Hindu families like ours, men provided and women were ⁴ provided for. My father was a patriarch and I a pliant daughter. The neighborhood I'd grown up in was homogeneously Hindu, Bengali-speaking, and middle-class. I didn't expect myself to ever disobey or disappoint my father by setting my own goals and taking charge of my future.

When I landed in Iowa 35 years ago, I found myself in a society in which almost everyone was Christian, white, and moderately well-off. In the women's ⁵ dormitory I lived in my first year, apart from six international graduate students (all of us were from Asia and considered "exotic"), the only non-Christian was Jewish, and the only nonwhite an African-American from Georgia. I didn't anticipate then, that over the next 35 years, the Iowa population would become so diverse that it would have 6,931 children from non-English-speaking homes registered as students in its schools, nor that Iowans would be in the grip of a cultural crisis in which resentment against immigrants, particularly refugees from Vietnam, Sudan, and Bosnia, as well as unskilled Spanish-speaking workers, would become politicized enough to cause the Immigration and Naturalization Service to open an "enforcement" office in Cedar Rapids in October for the tracking and deporting of undocumented aliens.

In Calcutta in the '50s, I heard no talk of "identity crisis"—communal or individual. The concept itself—a person not knowing who he or she is—was ⁶ unimaginable in our hierarchical, classification-obsessed society. One's identity was fixed, derived from religion, caste, patrimony, and mother tongue. A Hindu Indian's last name announced his or her forefathers' caste and place of origin. A Mukherjee could only be a Brahmin from Bengal. Hindu tradition forbade inter-caste, interlanguage, interethnic marriages. Bengali tradition even discouraged emigration: To remove oneself from Bengal was to dilute true culture.

Until the age of 8, I lived in a house crowded with 40 or 50 relatives. My ⁷ identity was viscerally connected with ancestral soil and genealogy. I was who I was because I was Dr. Sudhir Lal Mukherjee's daughter, because I was a Hindu Brahmin, because I was Bengali-speaking, and because my *desh*—the Bengali word for homeland—was an East Bengal village called Faridpur.

The University of Iowa classroom was my first experience of coeducation. And ⁸ after not too long, I fell in love with a fellow student named Clark Blaise, an American of Canadian origin, and impulsively married him during a lunch break in a lawyer's office above a coffee shop.

That act cut me off forever from the rules and ways of upper-middle-class ⁹ life in Bengal, and hurled me into a New World life of scary improvisations and heady explorations. Until my lunch-break wedding, I had seen myself as an Indian foreign student who intended to return to India to live. The five-minute ceremony in the lawyer's office suddenly changed me into a transient with conflicting loyalties to two very different cultures.

The first 10 years into marriage, years spent mostly in my husband's native ¹⁰ Canada, I thought of myself as an expatriate Bengali permanently stranded in North America because of destiny or desire. My first novel, *The Tiger's Daughter,* embodies the loneliness I felt but could not acknowledge, even to myself, as I negotiated the no man's land between the country of my past and the continent of my present. Shaped by memory, textured with nostalgia for a class and culture I had abandoned, this novel quite naturally became an expression of the expatriate consciousness.

It took me a decade of painful introspection to put nostalgia in perspective and to make the transition from expatriate to immigrant. After a 14-year stay in Canada, I forced my husband and our two sons to relocate to the United States. But the transition from foreign student to U.S. citizen, from detached onlooker to committed immigrant, has not been easy.

The years in Canada were particularly harsh. Canada is a country that officially, and proudly, resists cultural fusion. For all its rhetoric about a cultural "mosaic," Canada refuses to renovate its national self-image to include its changing complexion. It is a New World country with Old World concepts of a fixed, exclusivist national identity. Canadian official rhetoric designated me as one of the "visible minority" who, even though I spoke the Canadian languages of English and French, was straining "the absorptive capacity" of Canada. Canadians of color were routinely treated as "not real" Canadians. One example: In 1985 a terrorist bomb, planted in an Air-India jet on Canadian soil, blew up after leaving Montreal, killing 329 passengers, most of whom were Canadians of Indian origin. The prime minister of Canada at the time, Brian Mulroney, phoned the prime minister of India to offer Canada's condolences for India's loss.

Those years of race-related harassments in Canada politicized me and deepened my love of the ideals embedded in the American Bill of Rights. I don't forget that the architects of the Constitution and the Bill of Rights were white males and slaveholders. But through their declaration, they provided us with the enthusiasm for human rights, and the initial framework from which other empowerments could be conceived and enfranchised communities expanded.

I am a naturalized U.S. citizen and I take my American citizenship very seriously. I am not an economic refugee, nor am I a seeker of political asylum. I am a voluntary immigrant. I became a citizen by choice, not by simple accident of birth.

Yet these days, questions such as who is an American and what is American culture are being posed with belligerence, and being answered with violence. Scapegoating of immigrants has once again become the politicians' easy remedy for all that ails the nation. Hate speeches fill auditoriums for demagogues willing to profit from stirring up racial animosity. An April [1996] Gallup poll indicated that half of Americans would like to bar almost all legal immigration for the next five years.

The United States, like every sovereign nation, has a right to formulate its immigration policies. But in this decade of continual, large-scale diasporas, it is imperative that we come to some agreement about who "we" are, and what our goals are for the nation, now that our community includes people of many races, ethnicities, languages, and religions.

The debate about American culture and American identity has to date 17
been monopolized largely by Eurocentrists and ethnocentrists whose rhetoric
has been flamboyantly divisive, pitting a phantom "us" against a demonized
"them."

All countries view themselves by their ideals. Indians idealize the cultural con- 18
tinuum, the inherent value system of India, and are properly incensed when
foreigners see nothing but poverty, intolerance, strife, and injustice. Americans
see themselves as the embodiments of liberty, openness, and individualism,
even as the world judges them for drugs, crime, violence, bigotry, militarism,
and homelessness. I was in Singapore in 1994 when the American teenager
Michael Fay was sentenced to caning for having spraypainted some cars.
While I saw Fay's actions as those of an individual, and his sentence as too
harsh, the overwhelming local sentiment was that vandalism was an "American" crime, and that flogging Fay would deter Singapore youths from becoming "Americanized."

Conversely, in 1994, in Tavares, Florida, the Lake County School Board an- 19
nounced its policy (since overturned) requiring middle school teachers to instruct their students that American culture, by which the board meant
European-American culture, is inherently "superior to other foreign or historic
cultures." The policy's misguided implication was that culture in the United
States has not been affected by the American Indian, African-American, Latin-
American, and Asian-American segments of the population. The sinister implication was that our national identity is so fragile that it can absorb diverse and
immigrant cultures only by recontextualizing them as deficient.

Our nation is unique in human history in that the founding idea of "Amer- 20
ica" was in opposition to the tenet that a nation is a collection of like-looking,
like-speaking, like-worshipping people. The primary criterion for nationhood
in Europe is homogeneity of culture, race, and religion—which has contributed
to blood-soaked balkanization in the former Yugoslavia and the former Soviet
Union.

America's pioneering European ancestors gave up the easy homogeneity of 21
their native countries for a new version of Utopia. Now, in the 1990s, we have
the exciting chance to follow that tradition and assist in the making of a new
American culture that differs from both the enforced assimilation of a "melting
pot" and the Canadian model of a multicultural "mosaic."

The multicultural mosaic implies a contiguity of fixed, self-sufficient, ut- 22
terly distinct cultures. Multiculturalism, as it has been practiced in the United
States in the past 10 years, implies the existence of a central culture, ringed by
peripheral cultures. The fallout of official multiculturalism is the establishment
of one culture as the norm and the rest as aberrations. At the same time, the
multiculturalist emphasis on race- and ethnicity-based group identity leads to a
lack of respect for individual differences within each group, and to vilification
of those individuals who place the good of the nation above the interests of
their particular racial or ethnic communities.

We must be alert to the dangers of an "us" vs. "them" mentality. In Califor-   23
nia, this mentality is manifesting itself as increased violence between minority,
ethnic communities. The attack on Korean American merchants in South Central Los Angeles in the wake of the Rodney King beating trial is only one recent
example of the tragic side effects of this mentality. On the national level, the
politicization of ethnic identities has encouraged the scapegoating of legal immigrants, who are blamed for economic and social problems brought about by
flawed domestic and foreign policies.

We need to discourage the retention of cultural memory if the aim of that   24
retention is cultural balkanization. We must think of American culture and nationhood as a constantly reforming, transmogrifying "we."

In this age of diasporas, one's biological identity may not be one's only   25
identity. Erosions and accretions come with the act of emigration. The experience of cutting myself off from a biological homeland and settling in an adopted
homeland that is not always welcoming to its dark-complexioned citizens has
tested me as a person, and made me the writer I am today.

I choose to describe myself on my own terms, as an American, rather than as an   26
Asian-American. Why is it that hyphenation is imposed only on nonwhite
Americans? Rejecting hyphenation is my refusal to categorize the cultural landscape into a center and its peripheries; it is to demand that the American nation
deliver the promises of its dream and its Constitution to all its citizens equally.

My rejection of hyphenation has been misrepresented as race treachery by   27
some India-born academics on U.S. campuses who have appointed themselves
guardians of the "purity" of ethnic cultures. Many of them, though they reside
permanently in the United States and participate in its economy, consistently
denounce American ideals and institutions. They direct their rage at me because, by becoming a U.S. citizen and exercising my voting rights, I have invested in the present and not the past; because I have committed myself to help
shape the future of my adopted homeland; and because I celebrate racial and
cultural mongrelization.

What excites me is that as a nation we have not only the chance to retain   28
those values we treasure from our original cultures but also the chance to acknowledge that the outer forms of those values are likely to change. Among Indian immigrants, I see a great deal of guilt about the inability to hang on to
what they commonly term "pure culture." Parents express rage or despair at
their U.S.-born children's forgetting of, or indifference to, some aspects of Indian culture. Of those parents I would ask: What is it we have lost if our children are acculturating into the culture in which we are living? Is it so terrible
that our children are discovering or are inventing homelands for themselves?

Some first-generation Indo-Americans, embittered by racism and by unof-   29
ficial "glass ceilings," construct a phantom identity, more-Indian-than-Indians-
in-India, as a defense against marginalization. I ask: Why don't you get actively
involved in fighting discrimination? Make your voice heard. Choose the forum
most appropriate for you. If you are a citizen, let your vote count. Reinvest your

energy and resources into revitalizing your city's disadvantaged residents and neighborhoods. Know your constitutional rights, and when they are violated, use the agencies of redress the Constitution makes available to you. Expect change, and when it comes, deal with it!

As a writer, my literary agenda begins by acknowledging that America has 30 transformed me. It does not end until I show that I (along with the hundreds of thousands of immigrants like me) am minute by minute transforming America. The transformation is a two-way process: It affects both the individual and the national-cultural identity.

Others who write stories of migration often talk of arrival at a new place as 31 a loss, the loss of communal memory and the erosion of an original culture. I want to talk of arrival as a gain.

COMPREHENSION

1. What is the significance of the title? In what way is Mukherjee a "dreamer?" In what way does the United States inspire "dreaming?"
2. In paragraph 6, Mukherjee states that in India, she had a strong sense of identity. Why was it difficult for her to feel at ease with her "American identity"?
3. A country is a geographical area with national boundaries as well as a concept and an ideal. Does Mukherjee focus on these aspects of the United States and Canada equally, or does she emphasize one more than the other? Explain.

RHETORIC

1. The essay is divided into four parts. Why has the author contrived this structure? What is the focus of each? How does each section function rhetorically in relation to the other three?
2. Mukherjee introduces her essay with her own explanations of the terms "America" and "the United States." What is her purpose, considering that this is an autobiographical essay?
3. Mukherjee explores her transition from "expatriate" to "immigrant" to "U.S. citizen" in paragraphs 10 and 11. Explain the significance of each term in general and each term's particular role in the author's cultural metamorphosis.
4. Mukherjee rejects and condemns the belligerence toward and scapegoating of immigrants. How would you characterize the effect of these attacks on Mukherjee, an immigrant herself? Note, in particular, her statements in paragraphs 13 and 26.
5. How does Mukherjee employ irony in paragraph 12 to demonstrate the double standard imposed on individuals who do not fit the stereotypical mold of what it means to be a "citizen?"

6. In paragraph 14, the author states, "I take my American citizenship very seriously." Is the tone of the essay serious? Explain your view.
7. As you define the following words, identify the intended audience for this essay: *exclusivist* (paragraph 12), *demagogues* (paragraph 15), *diasporas* (paragraph 16), *ethnocentrists* and *demonized* (paragraph 17), and *balkanization* (paragraph 24).

WRITING

1. In a personal essay, write about a time in your life when your allegiance, honesty, or integrity were unfairly questioned. Be sure to use specifics such as the circumstances of who, what, when, where, and why. Also describe how you felt at the time and the emotional outcome.
2. Write an essay based on personal experience or observation, explaining whether Mukherjee is correct in stating that "hyphenation is imposed only on nonwhite Americans" (paragraph 26). A variation on this theme might be an exploration why "hyphenated" terms used to describe certain white American groups have a different tone and purpose than terms used for nonwhites.
3. **WRITING AN ARGUMENT:** In an essay, argue for or against the proposition that a course on cultural diversity should be taught at your college or university. Consider whether other ways of approaching the subject would be more profitable, or whether the subject needs to be addressed at all.
4. **WRITING AN ARGUMENT:** In her conclusion, Mukherjee criticizes "guardians of the 'purity' of ethnic cultures." Is there such a thing as a "pure" ethnic culture? Write an essay arguing your viewpoint.

www.mhhe.com/
mhreader

For more information on Bharati Mukherjee, go to:
More Resources > Ch. 7 Government & Politics

✳

Stranger in the Village

James Baldwin

James Baldwin (1924–1988), a major American essayist, novelist, short story writer, and playwright, was born and grew up in Harlem. He won a Eugene Saxon Fellowship and lived in Europe from 1948 to 1956. Always an activist in civil rights causes, Baldwin focused in his essays and fiction on the black search for identity in modern Amer-

ica and on the myth of white superiority. Among his principal works are Go Tell It on the Mountain *(1953),* Notes of a Native Son *(1955),* Giovanni's Room *(1956),* Nobody Knows My Name *(1961),* Another Country *(1962), and* If Beale Street Could Talk *(1974). One of the finest contemporary essayists, Baldwin had a rare talent for portraying the deepest concerns about civilization in an intensely personal style, as the following essay indicates.*

From all available evidence no black man had ever set foot in this tiny Swiss village before I came. I was told before arriving that I would probably be a "sight" for the village; I took this to mean that people of my complexion were rarely seen in Switzerland, and also that city people are always something of a "sight" outside of the city. It did not occur to me—possibly because I am an American— that there could be people anywhere who had never seen a Negro.

It is a fact that cannot be explained on the basis of the inaccessibility of the village. The village is very high, but it is only four hours from Milan and three hours from Lausanne. It is true that it is virtually unknown. Few people making plans for a holiday would elect to come here. On the other hand, the villagers are able, presumably, to come and go as they please—which they do: to another town at the foot of the mountain, with a population of approximately five thousand, the nearest place to see a movie or go to the bank. In the village there is no movie house, no bank, no library, no theater; very few radios, one jeep, one station wagon; and, at the moment, one typewriter, mine, an invention which the woman next door to me here had never seen. There are about six hundred people living here, all Catholic—I conclude this from the fact that the Catholic church is open all year round, whereas the Protestant chapel, set off on a hill a little removed from the village, is open only in the summertime when the tourists arrive. There are four or five hotels, all closed now, and four or five *bistros,* of which, however, only two do any business during the winter. These two do not do a great deal, for life in the village seems to end around nine or ten o'clock. There are a few stores, butcher, baker, *épicerie,* a hardware store, and a money-changer—who cannot change travelers' checks, but must send them down to the bank, an operation which takes two or three days. There is something called the *Ballet Haus,* closed in the winter and used for God knows what, certainly not ballet, during the summer. There seems to be only one schoolhouse in the village, and this for the quite young children; I suppose this to mean that their older brothers and sisters at some point descend from these mountains in order to complete their education—possibly, again, to the town just below. The landscape is absolutely forbidding, mountains towering on all four sides, ice and snow as far as the eye can reach. In this white wilderness, men and women and children move all day, carrying washing, wood, buckets of milk or water, sometimes skiing on Sunday afternoons. All week long boys and young men are to be seen shoveling snow off the rooftops, or dragging wood down from the forest in sleds.

The village's only real attraction, which explains the tourist season, is the hot spring water. A disquietingly high proportion of these tourists are cripples, or semi-cripples, who come year after year—from other parts of Switzerland, usually—to take the waters. This lends the village, at the height of the season, a rather terrifying air of sanctity, as though it were a lesser Lourdes. There is often something beautiful, there is always something awful, in the spectacle of a person who has lost one of his faculties, a faculty he never questioned until it was gone, and who struggles to recover it. Yet people remain people, on crutches or indeed on deathbeds; and wherever I passed, the first summer I was here, among the native villagers or among the lame, a wind passed with me—of astonishment, curiosity, amusement, and outrage. The first summer I stayed two weeks and never intended to return. But I did return in the winter, to work; the village offers, obviously, no distractions whatever and has the further advantage of being extremely cheap. Now it is winter again, a year later, and I am here again. Everyone in the village knows my name, though they scarcely ever use it, knows that I come from America—though this, apparently, they will never really believe: black men come from Africa—and everyone knows that I am the friend of the son of a woman who was born here, and that I am staying in their chalet. But I remain as much a stranger today as I was the first day I arrived, and the children shout *Neger! Neger!* as I walk along the streets.

It must be admitted that in the beginning I was far too shocked to have any 4
real reaction. In so far as I reacted at all, I reacted by trying to be pleasant—it being a great part of the American Negro's education (long before he goes to school) that he must make people "like" him. This smile-and-the-world-smiles-with-you routine worked about as well in this situation as it had in the situation for which it was designed, which is to say that it did not work at all. No one, after all, can be liked whose human weight and complexity cannot be, or has not been, admitted. My smile was simply another unheard-of phenomenon which allowed them to see my teeth—they did not, really, see my smile and I began to think that, should I take to snarling, no one would notice any difference. All of the physical characteristics of the Negro which had caused me, in America, a very different and almost forgotten pain were nothing less than miraculous—or infernal—in the eyes of the village people. Some thought my hair was the color of tar, that it had the texture of wire, or the texture of cotton. It was jocularly suggested that I might let it all grow long and make myself a winter coat. If I sat in the sun for more than five minutes some daring creature was certain to come along and gingerly put his fingers on my hair, as though he were afraid of an electric shock, or put his hand on my hand, astonished that the color did not rub off. In all of this, in which it must be conceded there was the charm of genuine wonder and in which there was certainly no element of intentional unkindness, there was yet no suggestion that I was human: I was simply a living wonder.

I knew that they did not mean to be unkind, and I know it now; it is necessary, nevertheless, for me to repeat this to myself each time I walk out of the chalet. The children who shout *Neger!* have no way of knowing the echoes this sound raises in me. They are brimming with good humor and the more daring swell with pride when I stop to speak with them. Just the same, there are days when I cannot pause and smile, when I have no heart to play with them; when, indeed, I mutter sourly to myself, exactly as I muttered on the streets of a city these children have never seen, when I was no bigger than these children are now: *Your* mother *was a nigger.* Joyce is right about history being a nightmare—but it may be the nightmare from which no one *can* awaken. People are trapped in history and history is trapped in them.

There is a custom in the village—I am told it is repeated in many villages— 6 of "buying" African natives for the purpose of converting them to Christianity. There stands in the church all year round a small box with a slot for money, decorated with a black figurine, and into this box the villagers drop their francs. During the *carnaval* which precedes Lent, two village children have their faces blackened—out of which bloodless darkness their blue eyes shine like ice—and fantastic horsehair wigs are placed on their blond heads; thus disguised, they solicit among the villagers for money for the missionaries in Africa. Between the box in the church and the blackened children, the village "bought" last year six or eight African natives. This was reported to me with pride by the wife of one of the *bistro* owners and I was careful to express astonishment and pleasure at the solicitude shown by the village for the souls of black folk. The *bistro* owner's wife beamed with a pleasure far more genuine than my own and seemed to feel that I might now breathe more easily concerning the souls of at least six of my kinsmen.

I tried not to think of these so lately baptized kinsmen, of the price paid for 7 them, or the peculiar price they themselves would pay, and said nothing about my father, who having taken his own conversion too literally never, at bottom, forgave the white world (which he described as heathen) for having saddled him with a Christ in whom, to judge at least from their treatment of him, they themselves no longer believed. I thought of white men arriving for the first time in an African village, strangers there, as I am a stranger here, and tried to imagine the astounded populace touching their hair and marveling at the color of their skin. But there is a great difference between being the first white man to be seen by Africans and being the first black man to be seen by whites. The white man takes the astonishment as tribute, for he arrives to conquer and to convert the natives, whose inferiority in relation to himself is not even to be questioned; whereas I, without a thought of conquest, find myself among a people whose culture controls me, has even, in a sense, created me, people who have cost me more in anguish and rage than they will ever know, who yet do not even know of my existence. The astonishment with which I might have greeted them, should they have stumbled into my African village a few hundred years ago, might have rejoiced their hearts. But the astonishment with which they greet me today can only poison mine.

And this is so despite everything I may do to feel differently, despite my 8 friendly conversations with the *bistro* owner's wife, despite their three-year-old son who has at last become my friend, despite the *saluts* and *bonsoirs* which I exchange with people as I walk, despite the fact that I know that no individual can be taken to task for what history is doing, or has done. I say that the culture of these people controls me—but they can scarcely be held responsible for

European culture. America comes out of Europe, but these people have never seen America nor have most of them seen more of Europe than the hamlet at the foot of their mountain. Yet they move with an authority which I shall never have; and they regard me, quite rightly, not only as a stranger in their village but as a suspect latecomer, bearing no credentials, to everything they have—however unconsciously—inherited.

For this village, even were it incomparably more remote and incredibly more primitive, is the West, the West onto which I have been so strangely grafted. These people cannot be, from the point of view of power, strangers anywhere in the world; they have made the modern world, in effect, even if they do not know it. The most illiterate among them is related, in a way that I am not, to Dante, Shakespeare, Michelangelo, Aeschylus, Da Vinci, Rembrandt, and Racine; the cathedral at Chartres says something to them which it cannot say to me, as indeed would New York's Empire State Building, should anyone here ever see it. Out of their hymns and dances come Beethoven and Bach. Go back a few centuries and they are in their full glory—but I am in Africa, watching the conquerors arrive.

The rage of the disesteemed is personally fruitless, but it is also absolutely inevitable; this rage, so generally discounted, so little understood even among the people whose daily bread it is, is one of the things that makes history. Rage can only with difficulty, and never entirely, be brought under the domination of the intelligence and is therefore not susceptible to any arguments whatever. This is a fact which ordinary representatives of the *Herrenvolk*, having never felt this rage and being unable to imagine it, quite fail to understand. Also, rage cannot be hidden, it can only be dissembled. This dissembling deludes the thoughtless, and strengthens rage and adds, to rage, contempt. There are, no doubt, as many ways of coping with the resulting complex of tensions as there are black men in the world, but no black man can hope ever to be entirely liberated from this internal warfare—rage, dissembling, and contempt having inevitably accompanied his first realization of the power of white men. What is crucial here is that, since white men represent in the black man's world so heavy a weight, white men have for black men a reality which is far from being reciprocal; and hence all black men have toward all white men an attitude which is designed, really, either to rob the white man of the jewel of his naïveté, or else to make it cost him dear.

The black man insists, by whatever means he finds at his disposal, that the white man cease to regard him as an exotic rarity and recognize him as a human being. This is a very charged and difficult moment, for there is a great deal of will power involved in the white man's naïveté. Most people are not naturally reflective any more than they are naturally malicious, and the white man prefers to keep the black man at a certain human remove because it is easier for him thus to preserve his simplicity and avoid being called to account for crimes committed by his forefathers, or his neighbors. He is inescapably aware, nevertheless, that he is in a better position in the world than black men are, nor can he quite put to death the suspicion that he is hated by black men therefore. He does not wish to be hated, neither does he wish to change places, and at this point in his uneasiness he can scarcely avoid having recourse to those legends which white men have created about black men, the most usual effect of which is that the white man finds himself enmeshed, so to speak, in his own language which describes hell, as well as the attributes which lead one to hell, as being as black as night.

Every legend, moreover, contains its residuum of truth, and the root function of language is to control the universe by describing it. It is of quite considerable significance that black men remain, in the imagination, and in overwhelming numbers in fact, beyond the disciplines of salvation; and this despite the fact the West has been "buying" African natives for centuries. There is, I should hazard, an instantaneous necessity to be divorced from this so visibly unsaved stranger, in whose heart, moreover, one cannot guess what dreams of vengeance are being nourished; and, at the same time, there are few things on earth more attractive than the idea of the unspeakable liberty which is allowed the unredeemed. When, beneath the black mask, a human being begins to make himself felt one cannot escape a certain awful wonder as to what kind of human being it is. What one's imagination makes of other people is dictated, of course, by the laws of one's own personality and it is one of the ironies of black-white relations that, by means of what the white man imagines the black man to be, the black man is enabled to know who the white man is. 12

I have said, for example, that I am as much a stranger in this village today as I was the first summer I arrived, but this is not quite true. The villagers wonder less about the texture of my hair than they did then, and wonder rather more about me. And the fact that their wonder now exists on another level is reflected in their attitudes and in their eyes. There are the children who make those delightful, hilarious, sometimes astonishingly grave overtures of friendship in the unpredictable fashion of children; other children, having been taught that the devil is a black man, scream in genuine anguish as I approach. Some of the older women never pass without a friendly greeting, never pass, indeed, if it seems that they will be able to engage me in conversation; other women look down or look away or rather contemptuously smirk. Some of the men drink with me and suggest that I learn how to ski—partly, I gather, because they cannot imagine what I would look like on skis—and want to know if I am married, and ask questions about my *métier*. But some of the men have accused *le sale négre*—behind my back—of stealing wood and there is already in the eyes of some of them that peculiar, intent, paranoiac malevolence which one sometimes surprises in the eyes of American white men when, out walking with their Sunday girl, they see a Negro male approach. 13

There is a dreadful abyss between the streets of this village and the streets of the city in which I was born, between the children who shout *Neger!* today and those who shouted *Nigger!* Yesterday—the abyss is experience, the American experience. The syllable hurled behind me today expresses, above all, wonder: I am a stranger here. But I am not a stranger in America and the same syllable riding on the American air expresses the war my presence has occasioned in the American soul. 14

For this village brings home to me this fact: that there was a day, and not re-   15
ally a very distant day, when Americans were scarcely Americans at all but dis-
contented Europeans, facing a great unconquered continent and strolling, say,
into a marketplace and seeing black men for the first time. The shock this spectacle afforded is suggested, surely, by the promptness with which they decided
that these black men were not really men but cattle. It is true that the necessity
on the part of the settlers of the New World of reconciling their moral assumptions with the fact—and the necessity—of slavery enhanced immensely the
charm of this idea, and it is also true that this idea expresses, with a truly American bluntness, the attitude which to varying extents all masters have had toward all slaves.

But between all former slaves and slave owners and the drama which be-   16
gins for Americans over three hundred years ago at Jamestown, there are at
least two differences to be observed. The American Negro slave could not suppose, for one thing, as slaves in past epochs had supposed and often done, that
he would ever be able to wrest the power from his master's hands. This was a
supposition which the modern era, which was to bring about such vast changes
in the aims and dimensions of power, put to death; it only begins, in unprecedented fashion, and with dreadful implications, to be resurrected today. But
even had this supposition persisted with undiminished force, the American Negro slave could not have used it to lend his condition dignity, for the reason that
this supposition rests on another: that the slave in exile yet remains related to
his past, has some means—if only in memory—of revering and sustaining the
forms of his former life, is able, in short, to maintain his identity.

This was not the case with the American Negro slave. He is unique among   17
the black men of the world in that his past was taken from him, almost literally,
at one blow. One wonders what on earth the first slave found to say to the first
dark child he bore. I am told that there are Haitians able to trace their ancestry
back to African kings, but any American Negro wishing to go back so far will
find his journey through time abruptly arrested by the signature on the bill of
sale which served as the entrance paper for his ancestor. At the time—to say
nothing of the circumstances—of the enslavement of the captive black man who
was to become the American Negro, there was not the remotest possibility that
he would ever take power from his master's hands. There was no reason to suppose that his situation would ever change, nor was there, shortly, anything to
indicate that his situation had ever been different. It was his necessity, in the
words of E. Franklin Frazier, to find a "motive for living under American culture or die." The identity of the American Negro comes out of this extreme situation, and the evolution of this identity was a source of the most intolerable
anxiety in the minds and the lives of his masters.

For the history of the American Negro is unique also in this: that the ques-   18
tion of his humanity, and of his rights therefore as a human being, became a
burning one for several generations of Americans, so burning a question that it
ultimately became one of those used to divide the nation. It is out of this argument that the venom of the epithet *Nigger!* is derived. It is an argument which

Europe has never had, and hence Europe quite sincerely fails to understand how or why the argument arose in the first place, why its effects are so frequently disastrous and always so unpredictable, why it refuses until today to be entirely settled. Europe's black possessions remained—and do remain—in Europe's colonies, at which remove they represented no threat whatever to European identity. If they posed any problem at all for the European conscience, it was a problem which remained comfortingly abstract: in effect, the black man, *as a man,* did not exist for Europe. But in America, even as a slave, he was an inescapable part of the general social fabric and no American could escape having an attitude toward him. Americans attempt until today to make an abstraction of the Negro, but the very nature of these abstractions reveals the tremendous effects the presence of the Negro has had on the American character.

When one considers the history of the Negro in America it is of the greatest 19 importance to recognize that the moral beliefs of a person, or a people, are never really as tenuous as life—which is not moral—very often causes them to appear; these create for them a frame of reference and a necessary hope, the hope being that when life has done its worst they will be enabled to rise above themselves and to triumph over life. Life would scarcely be bearable if this hope did not exist. Again, even when the worst has been said, to betray a belief is not by any means to have put oneself beyond its power; the betrayal of a belief is not the same thing as ceasing to believe. If this were not so there would be no moral standards in the world at all. Yet one must also recognize that morality is based on ideas and that all ideas are dangerous—dangerous because ideas can only lead to action and where the action leads no man can say. And dangerous in this respect: that confronted with the impossibility of becoming free of them, one can be driven to the most inhuman excesses. The ideas on which American beliefs are based are not, though Americans often seem to think so, ideas which originated in America. They came out of Europe. And the establishment of democracy on the American continent was scarcely as radical a break with the past as was the necessity, which Americans faced, of broadening this concept to include black men.

This was, literally, a hard necessity. It was impossible, for one thing, for 20 Americans to abandon their beliefs, not only because these beliefs alone seemed able to justify the sacrifices they had endured and the blood that they had spilled, but also because these beliefs afforded them their only bulwark against a moral chaos as absolute as the physical chaos of the continent it was their destiny to conquer. But in the situation in which Americans found themselves, these beliefs threatened an idea which, whether or not one likes to think so, is the very warp and woof of the heritage of the West, the idea of white supremacy.

Americans have made themselves notorious by the shrillness and the bru- 21 tality with which they have insisted on this idea, but they did not invent it; and it has escaped the world's notice that those very excesses of which Americans have been guilty imply a certain, unprecedented uneasiness over the idea's life and power, if not, indeed, the idea's validity. The idea of white supremacy rests simply on the fact that white men are the creators of civilization (the present civilization, which is the only one that matters; all previous civilizations are simply "contributions" to our own) and are therefore civilization's guardians and defenders. Thus it was impossible for Americans to accept the black man as one of themselves, for to do so was to jeopardize their status as white men. But not so to accept him was to deny his human reality, his human weight and complexity, and the strain of denying the overwhelmingly undeniable forced Americans into rationalizations so fantastic that they approached the pathological.

At the root of the American Negro problem is the necessity of the American 22 white man to find a way of living with the Negro in order to be able to live with himself. And the history of this problem can be reduced to the means used by Americans—lynch law and law, segregation and legal acceptance, terrorization and concession—either to come to terms with this necessity, or to find a way around it, or (most usually) to find a way of doing both these things at once. The resulting spectacle, at once foolish and dreadful, led someone to make the quite accurate observation that "the Negro-in-America is a form of insanity which overtakes white men."

In this long battle, a battle by no means finished, the unforeseeable effects 23 of which will be felt by many future generations, the white man's motive was the protection of his identity; the black man was motivated by the need to establish an identity. And despite the terrorization which the Negro in America endured and endures sporadically until today, despite the cruel and totally inescapable ambivalence of his status in his country, the battle for his identity has long ago been won. He is not a visitor to the West, but a citizen there, an American; as American as the Americans who despise him, the Americans who fear him, the Americans who love him—the Americans who became less than themselves, or rose to be greater than themselves by virtue of the fact that the challenge he represented was inescapable. He is perhaps the only black man in the world whose relationship to white men is more terrible, more subtle, and more meaningful than the relationship of bitter possessed to uncertain possessor. His survival depended, and his development depends, on his ability to turn his peculiar status in the Western world to his own advantage and, it may be, to the very great advantage of that world. It remains for him to fashion out of his experience that which will give him sustenance, and a voice.

The cathedral at Chartres, I have said, says something to the people of this 24 village which it cannot say to me; but it is important to understand that this cathedral says something to me which it cannot say to them. Perhaps they are struck by the power of the spires, the glory of the windows; but they have known God, after all, longer than I have known him, and in a different way, and I am terrified by the slippery bottomless well to be found in the crypt, down which heretics were hurled to death, and by the obscene, inescapable gargoyles jutting out of the stone and seeming to say that God and the devil can never be divorced. I doubt that the villagers think of the devil when they face a cathedral because they have never been identified with the devil. But I must accept the status which myth, if nothing else, gives me in the West before I can hope to change the myth.

Yet, if the American Negro has arrived at his identity by virtue of the absoluteness of his estrangement from his past, American white men still nourish the illusion that there is some means of recovering the European innocence, of returning to a state in which black men do not exist. This is one of the greatest errors Americans can make. The identity they fought so hard to protect has, by virtue of that battle, undergone a change: Americans are as unlike any other white people in the world as it is possible to be. I do not think, for example, that it is too much to suggest that the American vision of the world—which allows so little reality, generally speaking, for any of the darker forces in human life, which tends until today to paint moral issues in glaring black and white—owes a great deal to the battle waged by Americans to maintain between themselves and black men a human separation which could not be bridged. It is only now beginning to be borne in on us—very faintly, it must be admitted, very slowly, and very much against our will—that this vision of the world is dangerously inaccurate, and perfectly useless. For it protects our moral high-mindedness at the terrible expense of weakening our grasp of reality. People who shut their eyes to reality simply invite their own destruction, and anyone who insists on remaining in a state of innocence long after that innocence is dead turns himself into a monster. 25

The time has come to realize that the interracial drama acted out on the American continent has not only created a new black man, it has created a new white man, too. No road whatever will lead Americans back to the simplicity of this European village where white men still have the luxury of looking on me as a stranger. I am not, really, a stranger any longer for any American alive. One of the things that distinguishes Americans from other people is that no other people has ever been so deeply involved in the lives of black men, and vice versa. This fact faced, with all its implications, it can be seen that the history of the American Negro problem is not merely shameful, it is also something of an achievement. For even when the worst has been said, it must also be added that the perpetual challenge posed by this problem was always, somehow, perpetually met. It is precisely this black-white experience which may prove of indispensable value to us in the world we face today. This world is white no longer, and it will never be white again. 26

COMPREHENSION

1. According to Baldwin, what distinguishes Americans from other people? What is his purpose in highlighting these differences?
2. What connections between Europe, Africa, and America emerge from this essay? What is the relevance of the Swiss village to this frame of reference?
3. In the context of the essay, explain what Baldwin means by his statement, "People are trapped in history and history is trapped in them" (paragraph 5).

RHETORIC

1. Analyze the effect of Baldwin's repetition of "there is" and "there are" constructions in paragraph 2. What does the parallelism at the start of paragraph 8 accomplish? Locate other examples of parallelism in the essay.
2. Analyze the image of winter in paragraph 3 and its relation to the rest of the essay.
3. Where in the essay is Baldwin's complex thesis condensed for the reader? What does this placement of thesis reveal about the logical method of development in the essay?
4. How does Baldwin create his introduction? What is the focus? What key motifs does the author present that. will inform the rest of the essay? What is the relationship of paragraph 5 to paragraph 6?
5. What paragraphs constitute the second section of the essay? What example serves to unify this section? What major shift in emphasis occurs in the third part of the essay? Explain the cathedral of Chartres as a controlling motif between these two sections.
6. What comparisons and contrasts help structure and unify the essay?

WRITING

1. Examine the paradox implicit in Baldwin's statement in the last paragraph that the history of the American Negro problem is "something of an achievement."
2. Describe a time when you felt yourself a "stranger" in a certain culture.
3. **WRITING AN ARGUMENT:** Write an argumentative essay on civilization based on the last sentence in Baldwin's essay: "This world is white no longer, and it will never be white again."

 www.mhhe.com/ **mhreader** | For more information on James Baldwin, go to: **More Resources > Ch. 7 Government & Politics**

*

Some Reflections on American Manners

Alexis de Tocqueville

*Alexis Charles Henri Clerél de Tocqueville (1805–1859), descended from an aristo-
cratic Norman family, was a French lawyer, politician, statesman, and historian. Sent
to the United States in 1831 to study the American penal system, he wrote instead one
of the most penetrating inquiries into the nature of the American system,* Democracy
in America *(1835). In this chapter from his study, Tocqueville compares and contrasts
manners as manifested in the political and social contexts of democracy and aristocracy.*

Nothing, at first sight, seems less important than the external formalities of hu- 1
man behavior, yet there is nothing to which men attach more importance. They
can get used to anything except living in a society which does not share their
manners. The influence of the social and political system on manners is therefore worth serious examination.

Manners, speaking generally, have their roots in mores; they are also some- 2
times the result of an arbitrary convention agreed between certain men. They
are both natural and acquired.

When some see that, without dispute or effort of their own, they stand first 3
in society; when they daily have great aims in view which keep them occupied,
leaving details to others; and when they live surrounded by wealth they have
not acquired and do not fear to lose, one can see that they will feel a proud disdain for all the petty interests and material cares of life and that there will be a
natural grandeur in their thoughts that will show in their words and manners.

In democracies there is generally little dignity of manner, as private life is 4
very petty. Manners are often vulgar, as thoughts have small occasion to rise
above preoccupation with domestic interests.

True dignity in manners consists in always taking one's proper place, not 5
too high and not too low; that is as much within the reach of a peasant as of a
prince. In democracies everybody's status seems doubtful; as a result, there is
often pride but seldom dignity of manners. Moreover, manners are never well
regulated or well thought out.

There is too much mobility in the population of a democracy for any defi- 6
nite group to be able to establish a code of behavior and see that it is observed.
So everyone behaves more or less after his own fashion, and a certain incoher-
ence of manners always prevails, because they conform to the feelings and ideas
of each individual rather than to an ideal example provided for everyone to
imitate.

In any case, this is much more noticeable when an aristocracy has just fallen ₇ than when it has long been destroyed.

New political institutions and new mores then bring together in the same ₈ places men still vastly different in education and habits and compel them to a life in common; this constantly leads to the most ill-assorted juxtapositions. There is still some memory of the former strict code of politeness, but no one knows quite what it said or where to find it. Men have lost the common standard of manners but have not yet resolved to do without it, so each individual tries to shape, out of the ruins of former customs, some rule, however arbitrary and variable. Hence manners have neither the regularity and dignity frequent in aristocracies nor the qualities of simplicity and freedom which one sometimes finds in democracies; they are both constrained and casual.

But this is not a normal state of things. ₉

When equality is complete and old-established, all men, having roughly the ₁₀ same ideas and doing roughly the same things, do not need to come to an understanding or to copy each other in order to behave and talk in the same way; one sees a lot of petty variations in their manners but no great differences. They are never exactly alike, since they do not copy one pattern; they are never very unlike, because they have the same social condition. At first sight one might be inclined to say that the manners of all Americans are exactly alike, and it is only on close inspection that one sees all the variations among them.

The English make game of American manners, but it is odd that most of ₁₁ those responsible for those comic descriptions belong themselves to the English middle classes, and the cap fits them very well too. So these ruthless critics generally themselves illustrate just what they criticize in America; they do not notice that they are abusing themselves, to the great delight of their own aristocracy.

Nothing does democracy more harm than its outward forms of behavior; ₁₂ many who could tolerate its vices cannot put up with its manners.

But I will not admit that there is nothing to praise in democratic manners. ₁₃

In aristocracies, all within reach of the ruling class are at pains to imitate it, ₁₄ and very absurd and insipid imitations result. Democracies, with no models of high breeding before them, at least escape the necessity of daily looking at bad copies thereof.

In democracies manners are never so refined as among aristocracies, but ₁₅ they are also never so coarse. One misses both the crude words of the mob and the elegant and choice phrases of the high nobility. There is much triviality of manner, but nothing brutal or degraded.

I have already said that a precise code of behavior cannot take shape in ₁₆ democracies. That has its inconveniences and its advantages. In aristocracies rules of propriety impose the same demeanor on all, making every member of the same class seem alike in spite of personal characteristics; they bedizen and conceal nature. Democratic manners are neither so well thought out nor so regular, but they often are more sincere. They form, as it were, a thin, transparent veil through which the real feelings and personal thoughts of each man can be easily seen. Hence there is frequently an intimate connection between the form and the substance of behavior; we see a less decorative picture, but one truer to life. One may put the point this way: democracy imposes no particular manners, but in a sense prevents them from having manners at all.

Sometimes the feelings, passions, virtues, and vices of an aristocracy may 17 reappear in a democracy, but its manners never. They are lost and vanish past return when the democratic revolution is completed. It would seem that nothing is more lasting than the manners of an aristocratic class, for it preserves them for some time after losing property and power, nor more fragile, for as soon as they have gone, no trace of them is left, and it is even difficult to discover what they once were when they have ceased to exist. A change in the state of society works this marvel, and a few generations are enough to bring it about.

The principal characteristics of the aristocracy remain engraved in history 18 after its destruction, but the slight and delicate forms of its manners are lost to memory almost immediately after its fall. No one can imagine them when they are no longer seen. Their disappearance is unnoted and unfelt. For the heart needs an apprenticeship of custom and education to appreciate the refined pleasure derived from distinguished and fastidious manners; once the habit is lost, the taste for them easily goes too.

Thus, not only are democratic peoples unable to have aristocratic manners, 19 but they cannot even conceive or desire them. As they cannot imagine them, from their point of view it is as if they had never existed.

One should not attach too much importance to this loss, but it is permissi- 20 ble to regret it.

I know it has happened that the same men have had very distinguished 21 manners and very vulgar feelings; the inner life of courts has shown well enough what grand appearances may conceal the meanest hearts. But though the manners of an aristocracy by no means create virtue, they may add grace to virtue itself. It was no ordinary sight to see a numerous and powerful class whose every gesture seemed to show a constant and natural dignity of feeling and thought, an ordered refinement of taste and urbanity of manners.

The manners of the aristocracy created a fine illusion about human nature; 22 though the picture was often deceptive, it was yet a noble satisfaction to look on it.

COMPREHENSION

1. Summarize Tocqueville's observations about American manners, and explain why he believes they got that way.
2. Explain the positive and negative aspects that Tocqueville finds in both aristocratic manners and democratic ones.
3. Why are manners the one element in the transition from an aristocracy to a democracy that cannot be transmitted?

RHETORIC

1. The author makes a number of points concerning the nature of manners. What method, if any, does he use to reach his conclusions?
2. The author seems quite concerned about the concepts of formality and informality. Would you rate his writing as formal or informal? What educational level does the author assume his intended audience has attained? Explain your answer.
3. Paragraph 3 is one long sentence. What punctuation devices does the author use to achieve this? How does his use of the word *when* help give the paragraph a logical structure?
4. We ordinarily think of rhythm as a component of music, yet, by mixing long and short sentences, the author is able to establish a rhythm to his prose. How do the short sentences help keep the prose moving? How do they function as transitional devices?
5. One commonly learns in school not to begin a sentence with the word *but*. Tocqueville breaks this convention three times in his essay—in paragraphs 9, 13, and 21. Explain why this is or is not effective.
6. The author uses comparison and contrast in many of his sentences. For example, in paragraph 15, he lists three distinctions between democratic manners and aristocratic ones. How often does he use this device in the essay? What is the total effect of using it so consistently?
7. Paragraph 13 offers the reader a rare example of the double negative in English. How does this reflect upon the style of the writing? How would the tone be different if the sentence were, "But I will admit there is something to praise in democratic manners"?

WRITING

1. Select an aspect of American behavior or perspective—such as language, attire, or taste—and write a brief essay explaining your subject, using Tocqueville's writing style.
2. For a research project, use anthropological, cultural, and historical source material in your library to write an essay about daily life in one American city during the early nineteenth century.
3. All cultures have rituals concerning things such as conversation, comfort zones, greeting and leave-taking signals. Browse through a book featuring photographs of a range of people from another era or culture, and write a brief descriptive essay describing their gestures or expressions.
4. **WRITING AN ARGUMENT:** Argue for or against the proposition that manners have nothing to do with democracy.

| ✦ | www.mhhe.com/ **mhreader** | For more information on Alexis de Tocqueville, go to: **More Resources > Ch. 7 Government & Politics** |

CONNECTIONS FOR CRITICAL THINKING

1. Discuss whether the attributes of cyberspace as noted by Dyson can promote or discourage the ideas in the Declaration of Independence.
2. Discuss the views that Mukherjee and Baldwin have in common regarding the refusal of American culture to accept the "otherness" of those it perceives as not behaving like or looking like the conventional "American." Expand your discussion to present your own views about the similarities and differences in the ways "white" America views immigrants and African Americans.
3. Compare and contrast the diction, level of discourse, style, and vocabulary of Dyson and Baldwin.
4. Both Thomas Jefferson and Martin Luther King Jr. made powerful appeals to the government in power on behalf of their people. Write a comparison and contrast essay that examines the language, style, and content of both essays.
5. Select the essays you find the most and the least appealing or compelling in this chapter. Discuss why you selected them, and explore the way you developed your viewpoint.
6. Compare and contrast the narrative style of Catton with the more analytical style of Machiavelli.
7. Compare and contrast the difficulties Baldwin had in attempting to "fit in" to an alien European culture with the experiences Mukherjee describes of a nonwhite American trying to assimilate into the dominant culture.
8. Create a "group" chat room with three students from different sections of your course, and discuss both Dyson's views on cyberspace regulation and Williams's opinions on surveillance. Provide a summary of your discussion to your classmates.
9. Interview five parents who have children under the age of 10, and ask them if and how they control the Internet content their children view. Report your findings to your class.
10. Discuss with a nonnative student Tocqueville's "Reflections on American Manners." Explore to what degree your interviewee agrees with his classic assessment, and to what degree his views pertain today.

chapter **8**

Business and Economics
How Do We Earn Our Keep?

Work is central to the human experience; in fact, it is work and its economic and social outcomes that provide us with the keys to an understanding of culture and civilization. Work tells us much about scarcity and abundance, poverty and affluence, the haves and have-nots in any society, as well as a nation's economic imperatives. Whether it is the rise and fall of cities, the conduct of business and corporations, or the economic policies of government, we see in the culture of work an attempt to impose order on nature. Work is our handprint upon the world.

The work we perform and the careers we pursue also define us in very personal ways. "I'm a professor at Harvard" or "I work for IBM" serve as identity badges. (Robert Reich, a contributor to this chapter, did work at Harvard.) For what we do explains, at least in part, what and who we are. The very act of looking for work illuminates one's status in society, one's background, one's aspirations. Jonathan Swift, in his classic essay "A Modest Proposal," written in 1729, demonstrates how labor reveals economic and political configurations of power. Over 250 years later, Robert Reich tells us the same thing in his analysis of the changing nature of work and how these changes create an even broader gap between the rich and the poor.

Work is not merely an important human activity but an essential one for social and psychological health. You might like your work, or you might loathe it; be employed or unemployed; enjoy the reputation of a workaholic or a person who lives for leisure time; view work as a curse or as a duty. Regardless, it is work that occupies a central position in your relationship to society. In fact, Sigmund Freud spoke of work as the basis of one's social reality.

Regardless of your perspective on the issue, it is important to understand the multiple dimensions of work. In both traditional and modern societies, work prepares us for economic and social roles. It affects families, school curricula, public policy. Ultimately, as many authors here suggest, it determines our self-esteem. Through work we come to terms with ourselves and our environment. The nature and purpose of the work we do provide us with a powerful measure of our worth.

Previewing the Chapter

As you read the essays in this chapter and respond to them in discussion and writing, consider the following questions:

- What are the significant forms of support the author uses in viewing the world of work: observation, statistics, personal experience, history, and so on?
- What assumptions does the author make about the value of work?
- Does the author discuss work in general or focus on one particular aspect of work?
- How does the writer define *work*? In what ways, if any, does the author expand on the simple definition of *work* as "paid employment"?
- What issues of race, class, and gender does the author raise?
- What is the relationship of work to the changing social, political, and economic systems depicted in the author's essay?
- What tone does the writer take in his or her presentation of the work experience?
- What psychological insights does the author offer into the culture of work?
- What does the writer's style reveal about his or her attitude toward work?

Classic and Contemporary Images
WILL WORKERS BE DISPLACED BY MACHINES?

Using a Critical Perspective Diego Rivera's mural and George Haling's photograph present industrial scenes that reveal the impact of technology on workers. What details are emphasized in each illustration? How are these two images similar and dissimilar? What, for example, is the relation of human beings to the machines that are the centerpiece of each photograph? Are the artist and photographer objective or subjective in the presentation of each scene? Explain.

In the era known as the "Machine Age," 1918–1941, many artists, industrial
designers, and architects in the United States and Europe evoked the
mechanisms and images of industry in their works. During this time,
the Mexican painter Diego Rivera (1886–1957) created a mural
for the Detroit Institute of Arts (1932–33),
a portion of which is reprinted here.

Today, computers are used to help control assembly lines, as shown in this recent photo of a Chrysler assembly line.

Classic and Contemporary Essays
DOES EQUAL OPPORTUNITY EXIST?

Virginia Woolf's "Professions for Women" is ironic from the start as she readily admits she can speak expertly of only one profession, her own, which is writing. But her message is clear regarding the effect of living in a male-dominated society. Simply put, it is very difficult to break the shackles of conditioning that one acquires from being told over and over again by one's culture that gender is destiny, regardless of what one aspires to. The author—through personal experience—demonstrates how this discrimination has a profound effect on the ability to see with one's own eyes and to think with one's own head. Henry Louis Gates Jr. presents an interesting variation on this theme. Although the outcome is the same, the premise is reversed. He demonstrates how correlating supposedly positive attributes to a group—that is, superior athletic performance and race—results in the same deadening of the sense of personal ambition and a limiting of the scope of what one can aspire to. The thoughtful reader should be able to learn valuable lessons from comparing and contrasting these essays—one of which is that misguided perception all too often can be a self-fulfilling prophecy.

＊

Professions for Women

Virginia Woolf

Virginia Woolf (1882–1941), novelist and essayist, was the daughter of Sir Leslie Stephen, a famous critic and writer on economics. An experimental novelist, Woolf at-tempted to portray consciousness through a poetic, symbolic, and concrete style. Her novels include Jacob's Room *(1922),* Mrs. Dalloway *(1925),* To the Lighthouse *(1927), and* The Waves *(1931). She was also a perceptive reader and critic; her criti-cism appears in* The Common Reader *(1925) and* The Second Common Reader *(1933). In the following essay, which was delivered originally as a speech to the Women's Service League in 1931, Woolf argues that women must overcome several "angels," or phantoms, in order to succeed in professional careers.*

When your secretary invited me to come here, she told me that your Society is 1
concerned with the employment of women and she suggested that I might tell you something about my own professional experiences. It is true I am a woman; it is true I am employed; but what professional experiences have I had? It is dif-ficult to say. My profession is literature; and in that profession there are fewer experiences for women than in any other, with the exception of the stage—fewer, I mean, that are peculiar to women. For the road was cut many years ago—by Fanny Burney, by Aphra Behn, by Harriet Martineau, by Jane Austen, by George Eliot—many famous women, and many more unknown and forgot-ten, have been before me, making the path smooth, and regulating my steps. Thus, when I came to write, there were very few material obstacles in my way. Writing was a reputable and harmless occupation. The family peace was not broken by the scratching of a pen. No demand was made upon the family purse. For ten and sixpence one can buy paper enough to write all the plays of Shakespeare—if one has a mind that way. Pianos and models, Paris, Vienna and Berlin, masters and mistresses, are not needed by a writer. The cheapness of writing paper is, of course, the reason why women have succeeded as writers before they have succeeded in the other professions.

But to tell you my story—it is a simple one. You have only got to figure to 2
yourselves a girl in a bedroom with a pen in her hand. She had only to move that pen from left to right—from ten o'clock to one. Then it occurred to her to do what is simple and cheap enough after all—to slip a few of those pages into an envelope, fix a penny stamp in the corner, and drop the envelope into the red box at the corner. It was thus that I became a journalist; and my effort was re-warded on the first day of the following month—a very glorious day it was for me—by a letter from an editor containing a check for one pound ten shillings and sixpence. But to show you how little I deserve to be called a professional woman, how little I know of the struggles and difficulties of such lives, I have

to admit that instead of spending that sum upon bread and butter, rent, shoes and stockings, or butcher's bills, I went out and bought a cat—a beautiful cat, a Persian cat, which very soon involved me in bitter disputes with my neighbors.

What could be easier than to write articles and to buy Persian cats with the profits? But wait a moment. Articles have to be about something. Mine, I seem to remember, was about a novel by a famous man. And while I was writing this review, I discovered that if I were going to review books I should need to do battle with a certain phantom. And the phantom was a woman, and when I came to know her better I called her after the heroine of a famous poem, "The Angel in the House." It was she who used to come between me and my paper when I was writing reviews. It was she who bothered me and wasted my time and so tormented me that at last I killed her. You who come of a younger and happier generation may not have heard of her—you may not know what I mean by the Angel in the House. I will describe her as shortly as I can. She was intensely sympathetic. She was immensely charming. She was utterly unselfish. She excelled in the difficult arts of family life. She sacrificed herself daily. If there was a chicken, she took the leg; if there was a draught she sat in it—in short she was so constituted that she never had a mind or a wish of her own, but preferred to sympathize always with the minds and wishes of others. Above all—I need not say it—she was pure. Her purity was supposed to be her chief beauty—her blushes, her great grace. In those days—the last of Queen Victoria—every house had its Angel. And when I came to write I encountered her with the very first words. The shadow of her wings fell on my page; I heard the rustling of her skirts in the room. Directly, that is to say, I took my pen in hand to review that novel by a famous man, she slipped behind me and whispered: "My dear, you are a young woman. You are writing about a book that has been written by a man. Be sympathetic; be tender; flatter; deceive; use all the arts and wiles of our sex. Never let anybody guess that you have a mind of your own. Above all, be pure." And she made as if to guide my pen. I now record the one act for which I take some credit to myself, though the credit rightly belongs to some excellent ancestors of mine who left me a certain sum of money—shall we say five hundred pounds a year—so that it was not necessary for me to depend solely on charm for my living. I turned upon her and caught her by the throat. I did my best to kill her. My excuse, if I were to be had up in a court of law, would be that I acted in self-defense. Had I not killed her she would have killed me. She would have plucked the heart out of my writing. For, as I found, directly I put pen to paper, you cannot review even a novel without having a mind of your own, without expressing what you think to be the truth about human relations, morality, sex. And all these questions, according to the Angel in the House, cannot be dealt with freely and openly by women; they must charm, they must conciliate, they must—to put it bluntly—tell lies if they are to succeed. Thus, whenever I felt the shadow of her wing or the radiance of her halo upon my page, I took up the inkpot and flung it at her. She died hard. Her fictitious nature was of great assistance to her. It is far harder to kill a phantom than a reality. She was always creeping back when I thought I had dispatched her. Though I flatter myself that I killed her in the end, the struggle was severe; it took much time that had better have been spent upon learning Greek grammar; or in roaming the world in search of adventures. But it was a real experience; it was an experience that was bound to befall all women writers at that time. Killing the Angel in the House was part of the occupation of a woman writer.

But to continue my story. The Angel was dead; what then remained? You 4 may say that what remained was a simple and common object—a young woman in a bedroom with an inkpot. In other words, now that she had rid herself of falsehood, that young woman had only to be herself. Ah, but what is "herself"? I mean, what is a woman? I assure you, I do not know. I do not believe that you know. I do not believe that anybody can know until she has expressed herself in all the arts and professions open to human skill. That indeed is one of the reasons why I have come here—out of respect for you, who are in process of showing us by your experiments what a woman is, who are in process of providing us, by your failures and successes, with that extremely important piece of information.

But to continue the story of my professional experiences. I made one pound 5 ten and six by my first review; and I bought a Persian cat with the proceeds. Then I grew ambitious. A Persian cat is all very well, I said; but a Persian cat is not enough. I must have a motor car. And it was thus that I became a novelist— for it is a very strange thing that people will give you a motor car if you will tell them a story. It is a still stranger thing that there is nothing so delightful in the world as telling stories. It is far pleasanter than writing reviews of famous novels. And yet, if I am to obey your secretary and tell you my professional experiences as a novelist, I must tell you about a very strange experience that befell me as a novelist. And to understand it you must try first to imagine a novelist's state of mind. I hope I am not giving away professional secrets if I say that a novelist's chief desire is to be as unconscious as possible. He has to induce in himself a state of perpetual lethargy. He wants life to proceed with the utmost quiet and regularity. He wants to see the same faces, to read the same books, to do the same things day after day, month after month, while he is writing, so that nothing may break the illusion in which he is living—so that nothing may disturb or disquiet the mysterious nosings about, feelings round, darts, dashes and sudden discoveries of that very shy and illusive spirit, the imagination. I suspect that this state is the same both for men and women. Be that as it may, I want you to imagine me writing a novel in a state of trance. I want you to figure to yourselves a girl sitting with a pen in her hand, which for minutes, and indeed for hours, she never dips into the inkpot. The image that comes to my mind when I think of this girl is the image of a fisherman lying sunk in dreams on the verge of a deep lake with a rod held out over the water. She was letting her imagination sweep unchecked round every rock and cranny of the world that lies submerged in the depths of our unconscious being. Now came the experience, the experience that I believe to be far commoner with women writers than with men. The line raced through the girl's fingers. Her imagination had rushed away. It had sought the pools, the depths, the dark places where the largest fish slumber. And then there was a smash. There was an explosion. There was foam and confusion. The imagination had dashed itself against something hard. The girl was roused from her dream. She was indeed in a state of the most acute and difficult distress. To speak without figure she had thought of something, something about the body, about the passions which it was unfitting for her as a woman to say. Men, her reason told her, would be shocked. The consciousness of what men will say of a woman who speaks the truth about her passions had roused her from her artist's state of unconsciousness. She could write no more. The trance was over. Her imagination could work no longer. This I believe to be a very common experience with women writers—they are impeded by the extreme conventionality of the other sex. For though men sensibly allow themselves great freedom in these respects, I doubt that they realize or can control the extreme severity with which they condemn such freedom in women.

These then were two very genuine experiences of my own. These were two 6 of the adventures of my professional life. The first—killing the Angel in the House—I think I solved. She died. But the second, telling the truth about my own experiences as a body, I do not think I solved. I doubt that any woman has solved it yet. The obstacles against her are still immensely powerful—and yet they are very difficult to define. Outwardly, what is simpler than to write books? Outwardly, what obstacles are there for a woman rather than for a man? Inwardly, I think, the case is very different; she has still many ghosts to fight, many prejudices to overcome. Indeed it will be a long time still, I think, before a woman can sit down to write a book without finding a phantom to be slain, a rock to be dashed against. And if this is so in literature, the freest of all professions for women, how is it in the new professions which you are now for the first time entering?

Those are the questions that I should like, had I time, to ask you. And in- 7 deed, if I have laid stress upon these professional experiences of mine, it is because I believe that they are, though in different forms, yours also. Even when the path is nominally open—when there is nothing to prevent a woman from being a doctor, a lawyer, a civil servant—there are many phantoms and obstacles, as I believe, looming in her way. To discuss and define them is I think of great value and importance; for thus only can the labor be shared, the difficulties be solved. But besides this, it is necessary also to discuss the ends and the aims for which we are fighting, for which we are doing battle with these formidable obstacles. Those aims cannot be taken for granted; they must be perpetually questioned and examined. The whole position, as I see it—here in this hall surrounded by women practicing for the first time in history I know not how many different professions—is one of extraordinary interest and importance. You have won rooms of your own in the house hitherto exclusively owned by men. You are able, though not without great labor and effort, to pay the rent. You are earning your five hundred pounds a year. But this freedom is only a beginning; the room is your own, but it is still bare. It has to be furnished; it has to be decorated; it has to be shared. How are you going to furnish it, how are you going to decorate it? With whom are you going to share it, and upon what terms? These, I think, are questions of the utmost importance and interest. For the first time in history you are able to ask for them; for the first time you are able to decide for yourselves what the answers should be. Willingly would I stay and discuss those questions and answers—but not tonight. My time is up; and I must cease.

COMPREHENSION

1. This essay was presented originally as a speech. What internal evidence indicates that it was intended as a talk? How do you respond to it today as a reader?
2. Who or what is the "angel" that Woolf describes in this essay? Why must she kill it? What other obstacles does a professional woman encounter?
3. Paraphrase the last two paragraphs of this essay. What is the essence of Woolf's argument?

RHETORIC

1. There is a significant amount of figurative language in the essay. Locate and explain examples. What does the figurative language contribute to the tone of the essay? Compare and contrast the figurative language in this essay and in Woolf's "The Death of the Moth" in Chapter 11.
2. How do we know that Woolf is addressing an audience of women? Why does she pose so many questions, and what does this strategy contribute to the rapport that she wants to establish? Explain the effect of the last two sentences.
3. How does Woolf use analogy to structure part of her argument?
4. Why does Woolf rely on personal narration? How does it affect the logic of her argument?
5. Evaluate Woolf's use of contrast to advance her argument.
6. Where does Woolf place her main proposition? How emphatic is it, and why?

WRITING

1. How effectively does Woolf use her own example as a professional writer to advance a broader proposition concerning all women entering professional life? Answer this question in a brief essay.
2. Discuss the problems and obstacles that you anticipate when you enter your chosen career.
3. **WRITING AN ARGUMENT:** Argue for or against the proposition that Woolf's essay has very little relevance for women planning careers today.

www.mhhe.com/
mhreader

For more information on Virginia Woolf, go to:
More Resources > Ch. 8 Business & Economics

✳

Delusions of Grandeur

Henry Louis Gates Jr.

Henry Louis Gates Jr. (b. 1950) is an educator, writer, and editor. He was born in West Virginia and educated at Yale and at Clare College in Cambridge. Gates has had a varied career, working as a general anesthetist in Tanzania and as a staff correspondent for Time *magazine in London. His essays have appeared in such diverse publications as* Black American Literature Forum, Yale Review, The New York Times Book Review, *and* Sports Illustrated. *He is also the author of* Figures in Black: Words, Signs and the Racial Self *(1987) and* The Signifying Monkey: A Theory of Afro-American Literary Criticism *(1988) and is the editor, with Nellie Y. McKey, of* The Norton Anthology of African American Literature *(1996). In this article from* Sports Illustrated, *Gates turns his attention to the limited career choices presented as viable to African American youth and to public misconceptions about blacks in sports.*

Standing at the bar of an all-black VFW post in my hometown of Piedmont,   1
W.Va., I offered five dollars to anyone who could tell me how many African-American professional athletes were at work today. There are 35 million African-Americans, I said.

"Ten million!" yelled one intrepid soul, too far into his cups.   2

"No way . . . more like 500,000," said another.   3

"You mean *all* professional sports," someone interjected, "including golf   4
and tennis, but not counting the brothers from Puerto Rico?" Everyone laughed.

"Fifty thousand, minimum," was another guess.   5

Here are the facts: 6

There are 1,200 black professional athletes in the U.S.

There are 12 times more black lawyers than black athletes.

There are 2½ times more black dentists than black athletes.

There are 15 times more black doctors than black athletes.

Nobody in my local VFW believed these statistics; in fact, few people   7
would believe them if they weren't reading them in the pages of *Sports Illustrated.* In spite of these statistics, too many African-American youngsters still believe that they have a much better chance of becoming another Magic Johnson or Michael Jordan than they do of matching the achievements of Baltimore Mayor Kurt Schmoke or neurosurgeon Dr. Benjamin Carson, both of whom, like Johnson and Jordan, are black.

In reality, an African-American youngster has about as much chance of be-   8
coming a professional athlete as he or she does of winning the lottery. The tragedy for our people, however, is that few of us accept that truth.

Let me confess that I love sports. Like most black people of my generation—   9
I'm 40—I was raised to revere the great black athletic heroes, and I never tired of

listening to the stories of triumph and defeat that, for blacks, amount to a collective epic much like those of the ancient Greeks: Joe Louis's demolition of Max Schmeling; Satchel Paige's dazzling repertoire of pitches; Jesse Owens's in-your-face performance in Hitler's 1936 Olympics; Willie Mays's over-the-shoulder basket catch; Jackie Robinson's quiet strength when assaulted by racist taunts; and a thousand other grand tales.

Nevertheless, the blind pursuit of attainment in sports is having a devastat- 10 ing effect on our people. Imbued with a belief that our principal avenue to fame and profit is through sport, and seduced by a win-at-any-cost system that corrupts even elementary school students, far too many black kids treat basketball courts and football fields as if they were classrooms in an alternative school system. "O.K., I flunked English," a young athlete will say. "But I got an A plus in slamdunking."

The failure of our public schools to educate athletes is part and parcel of the 11 schools' failure to educate almost everyone. A recent survey of the Philadelphia school system, for example, stated that "more than half of all students in the third, fifth and eighth grades cannot perform minimum math and language tasks." One in four middle school students in that city fails to pass to the next grade each year. It is a sad truth that such statistics are repeated in cities throughout the nation. Young athletes—particularly young black athletes—are especially ill-served. Many of them are functionally illiterate, yet they are passed along from year to year for the greater glory of good old Hometown High. We should not be surprised to learn, then, that only 26.6 percent of black athletes at the collegiate level earn their degrees. For every successful educated black professional athlete, there are thousands of dead and wounded. Yet young blacks continue to aspire to careers as athletes, and it's no wonder why; when the University of North Carolina recently commissioned a sculptor to create archetypes of its student body, guess which ethnic group was selected to represent athletes?

Those relatively few black athletes who do make it in the professional ranks 12 must be prevailed upon to play a significant role in the education of all of our young people, athlete and nonathlete alike. While some have done so, many others have shirked their social obligations: to earmark small percentages of their incomes for the United Negro College Fund; to appear on television for educational purposes rather than merely to sell sneakers; to let children know the message that becoming a lawyer, a teacher or a doctor does more good for our people than winning the Super Bowl; and to form productive liaisons with educators to help forge solutions to the many ills that beset the black community. These are merely a few modest proposals.

A similar burden falls upon successful blacks in all walks of life. Each of us 13 must strive to make our young people understand the realities. Tell them to cheer Bo Jackson but to emulate novelist Toni Morrison or businessman Reginald Lewis or historian John Hope Franklin or Spelman College president Johnetta Cole—the list is long.

Of course, society as a whole bears responsibility as well. Until colleges stop 14 using young blacks as cannon fodder in the big-business wars of so-called nonprofessional sports, until training a young black's mind becomes as important as training his or her body, we will continue to perpetuate a system akin to that of the Roman gladiators, sacrificing a class of people for the entertainment of the mob.

COMPREHENSION

1. What is the general assumption made about African Americans in sports?
2. Why do American schools continue to perpetuate the myth that Gates is writing about?
3. According to Gates, what should successful African American athletes do to help guide the career choices of young black males?

RHETORIC

1. What is Gates's thesis? Where does it appear?
2. How does the introductory paragraph work to set up the writer's focus?
3. State Gates's purpose in using statistics in his essay.
4. What is the tone of Gates's essay? Cite specific sections where this tone seems strongest.
5. Examine the accumulation of facts in paragraph 11. How does this technique underscore Gates's point?
6. Explain Gates's allusion to Roman gladiators in his conclusion. How does it aid in emphasizing his main point?

WRITING

1. Write a brief essay in which you analyze your personal reaction to Gates's statistics. Were you surprised by them? What assumptions did you have about the number of black athletes? Why do you think most Americans share these assumptions?
2. Write a biographical research paper on the life and career of an African American athlete.
3. **WRITING AN ARGUMENT:** Pretend you are addressing a group of young African Americans at an elementary school. Argue that sports and entertainment should (or should not) be their career choices.

Classic and Contemporary: Questions for Comparison

1. Examine the argumentative styles of Woolf and Gates. What are their main propositions? What are their minor propositions? What evidence do they provide?
2. Woolf first presented her paper as a speech before an audience of women. Gates wrote his essay as an opinion piece for *Sports Illustrated*. Write a comparative audience analysis of the two selections. Analyze purpose, tone, style, and any other relevant aspects of the two essays.
3. Argue for or against the proposition that white women and African American men face the same barriers to employment in today's professions. Refer to the essays by Woolf and Gates to support your position.

✳

Men at Work

Anna Quindlen

Anna Quindlen (b. 1958) was born in Philadelphia. She has worked as a reporter and columnist for the New York Post *and* The New York Times *and is currently a columnist for* Newsweek *magazine. She published a novel entitled* Object Lessons *in 1991. In 1992, she received the Pulitzer Prize for commentary. An outspoken feminist, she stated in an interview in* Commonweal *that "I write for me . . . I tend to write about what we have come, unfortunately, to call women's issues. Those are issues that directly affect my life and those are issues that are historically underreported." The following is an essay about the contemporary father, published in a collection of her work entitled* Thinking Out Loud *(1993).*

> *Overheard in a Manhattan restaurant, one woman to another: "He's a terrific father, but he's never home."*

The five o'clock dads can be seen on cable television these days, just after that time in the evening the stay-at-home moms call the arsenic hours. They are sixties sitcom reruns, Ward and Steve and Alex, and fifties guys. They eat dinner with their television families and provide counsel afterward in the den. Someday soon, if things keep going the way they are, their likenesses will be enshrined in a diorama in the Museum of Natural History, frozen in their recliner chairs. The sign will say, "Here sit lifelike representations of family men who worked only eight hours a day."

The five o'clock dad has become an endangered species. A corporate culture that believes presence is productivity, in which people of ambition are afraid to be seen leaving the office, has lengthened his workday and shortened his home-life. So has an economy that makes it difficult for families to break even at the end of the month. For the man who is paid by the hour, that means never saying no to overtime. For the man whose loyalty to the organization is measured in time at his desk, it means goodbye to nine to five.

To lots of small children it means a visiting father. The standard joke in one   3
large corporate office is that the dads always say their children look like angels when they're sleeping because that's the only way they ever see them. A Gallup survey taken several years ago showed that roughly 12 percent of the men surveyed with children under the age of six worked more than sixty hours a week, and an additional 25 percent worked between fifty and sixty hours. (Less than 8 percent of the working women surveyed who had children of that age worked those hours.)

No matter how you divide it up, those are twelve-hour days. When the   4
talk-show host Jane Wallace adopted a baby recently, she said one reason she was not troubled by becoming a mother without becoming a wife was that many of her married female friends were "functionally single," given the hours their husbands worked. The evening commuter rush is getting longer. The 7:45 to West Backofbeyond is more crowded than ever before. The eight o'clock dad. The nine o'clock dad.

There's a horribly sad irony to this, and it is that the quality of fathering is   5
better than it was when the dads left work at five o'clock and came home to cafe curtains and tuna casserole. The five o'clock dad was remote, a "Wait till your father gets home" kind of dad with a newspaper for a face. The roles he and his wife had were clear: she did nurture and home, he did discipline and money.

The role fathers have carved out for themselves today is a vast improve-   6
ment, a muddling of those old boundaries. Those of us obliged to convert be-havior into trends have probably been a little heavy-handed on the shared childbirth and egalitarian diaper-changing. But fathers today do seem to be more emotional with their children, more nurturing, more open. Many say, "My father never told me he loved me," and so they tell their own children all the time that they love them.

When they're home.   7

There are people who think that this is changing even as we speak, that   8
there is a kind of perestroika of home and work that we will look back on as be-ginning at the beginning of the 1990s. A nonprofit organization called the Fam-ilies and Work Institute advises corporations on how to balance personal and professional obligations and concerns, and Ellen Galinsky, its cofounder, says she has noticed a change in the last year.

"When we first started doing this the groups of men and of women   9
sounded very different," she said. "If the men complained at all about long hours, they complained about their wives' complaints. Now if the timbre of the voice was disguised I couldn't tell which is which. The men are saying: 'I don't want to live this way anymore. I want to be with my kids.' I think the corporate culture will have to begin to respond to that."

This change can only be to the good, not only for women but especially for 10 men, and for kids, too. The stereotypical five o'clock dad belongs in a diorama, with his "Ask your mother" and his "Don't be a crybaby." The father who believes hugs and kisses are sex-blind and a dirty diaper requires a change, not a woman, is infinitely preferable. What a joy it would be if he were around more.

"This is the man's half of having it all," said Don Conway-Long, who 11 teaches a course at Washington University in St. Louis about men's relationships that drew 135 students this year for thirty-five places. "We're trying to do what women want of us, what children want of us, but we're not willing to transform the workplace." In other words, the hearts and minds of today's fathers are definitely in the right place. If only their bodies could be there, too.

COMPREHENSION

1. According to the author, contemporary fathers and traditional fathers both have faults. Describe the specific problems in each group.
2. What is the thesis of this essay? Where in the essay is this thesis most succinctly articulated?
3. Is this a regional essay? Does it address a particular class or geographic area of America? Would the examples need to be expanded if the author were to include *all* types of American fathers? Explain.

RHETORIC

1. The author uses some unique phrasing and vocabulary in her essay. What is the effect on the tone of the essay of expressions such as "arsenic hours" (paragraph 1), "visiting father" (paragraph 3), "functionally single" (paragraph 4), "West Backofbeyond" (paragraph 4), and "perestroika" (paragraph 8)?
2. The opening paragraph of an essay often sets the tone for the rest. How does the tone of paragraph 1 help direct the tone of the essay's argument?
3. In paragraph 5, the author states that there is a "horribly sad irony" in the fact that fathers are better nurturers now but have less time to nurture. What other ironies does the author use to advance her argument?
4. What is the purpose of the rhetorical strategy of using the three-word paragraph "When they're home"? Does it add or detract from the coherence of the essay?
5. Paragraph 5 contains the rather oddly structured sentence. "The roles he and his wife had were clear: she did nurture and home, he did discipline and money." Conduct a grammatical analysis of the sentence. Does it make sense? Does it transgress any rules of grammar? Explain.
6. In paragraph 6, the author states, "fathers today do seem to be more emotional with their children, more nurturing, more open." Where in the essay

does she provide documentation of this? Is this assertion argued sufficiently, or is it merely presented as an assumption without evidence?

7. Does the author provide a solution for or a recommendation on how to
solve the problems she raises? Does its presence or absence strengthen or
weaken the argument? Explain.

WRITING

1. The author cites television portraits of fifties fathers as her evidence for the
family behavior of the traditional father. Explore whether her comparison
between the contemporary father and the traditional one is accurate by
comparing two television sitcoms: one written some time ago, the other
more recently.

2. For a creative writing project, write an imaginary letter to a supervisor, stating your view that your work hours should be reduced so that you can
spend more time with your family. Be sure to include appropriate supporting material.

3. **WRITING AN ARGUMENT:** Argue for or against the proposition that the
author's description of the modern father is a narrow one, based on biases
of class and culture.

www.mhhe.com/ **mhreader**

For more information on Anna Quindlen, go to:
More Resources > Ch. 8 Business & Economics

※

Globalization: The Super-Story

Thomas L. Friedman

*Thomas L. Friedman (b. 1953) was born in Minneapolis, Minnesota. He majored in
Mediterranean Studies at Brandeis University (BA 1975) and received an MA in Modern Middle Eastern Studies from Oxford University in 1978. As journalist, author, television commentator, and op-ed contributor to* The New York Times, *Friedman tries to
provide unbiased viewpoints on cultural, political, and economic issues. From 1979 to
1984 he was the* Times *correspondent in Beirut, Lebanon, and subsequently until 1988
served as bureau chief in Jerusalem. His book recounting his 10 years in the Middle East,*
From Beirut to Jerusalem *(1983), received the National Book Award for nonfiction.
Friedman also has published* The Lexus and the Olive Tree: Understanding Globalization *(2000) and a collection of essays,* Longitudes and Attitudes: Explaining the
World after September 11 *(2002), which contains the following selection.*

I am a big believer in the idea of the super-story, the notion that we all carry 1
around with us a big lens, a big framework, through which we look at the
world, order events, and decide what is important and what is not. The events
of 9/11 did not happen in a vacuum. They happened in the context of a new international system—a system that cannot explain everything but *can* explain
and connect more things in more places on more days than anything else. That
new international system is called globalization. It came together in the late
1980s and replaced the previous international system, the cold war system,
which had reigned since the end of World War II. This new system is the lens,
the super-story, through which I viewed the events of 9/11.

I define globalization as the inexorable integration of markets, transporta- 2
tion systems, and communication systems to a degree never witnessed before—
in a way that is enabling corporations, countries, and individuals to reach
around the world farther, faster, deeper, and cheaper than ever before, and in a
way that is enabling the world to reach into corporations, countries, and individuals farther, faster, deeper, and cheaper than ever before.

Several important features of this globalization system differ from those of 3
the cold war system in ways that are quite relevant for understanding the
events of 9/11. I examined them in detail in my previous book, *The Lexus and the
Olive Tree*, and want to simply highlight them here.

The cold war system was characterized by one overarching feature—and 4
that was *division*. That world was a divided-up, chopped-up place, and whether
you were a country or a company, your threats and opportunities in the cold
war system tended to grow out of who you were divided from. Appropriately,
this cold war system was symbolized by a single word—*wall*, the Berlin Wall.

The globalization system is different. It also has one overarching feature— 5
and that is *integration*. The world has become an increasingly interwoven place,
and today, whether you are a company or a country, your threats and opportunities increasingly derive from who you are connected to. This globalization
system is also characterized by a single word—*web*, the World Wide Web. So in
the broadest sense we have gone from an international system built around division and walls to a system increasingly built around integration and webs. In
the cold war we reached for the hotline, which was a symbol that we were divided but at least two people were in charge—the leaders of the United States
and the Soviet Union. In the globalization system we reach for the Internet,
which is a symbol that we are all connected and nobody is quite in charge.

Everyone in the world is directly or indirectly affected by this new system, 6
but not everyone benefits from it, not by a long shot, which is why the more it
becomes diffused, the more it also produces a backlash by people who feel overwhelmed by it, homogenized by it, or unable to keep pace with its demands.

The other key difference between the cold war system and the globalization 7
system is how power is structured within them. The cold war system was built
primarily around nation-states. You acted on the world in that system through
your state. The cold war was a drama of states confronting states, balancing
states, and aligning with states. And, as a system, the cold war was balanced at

the center by two superstates, two superpowers: The United States and the Soviet Union.

The globalization system, by contrast, is built around three balances, which 8 overlap and affect one another. The first is the traditional balance of power between nation-states. In the globalization system, the United States is now the sole and dominant superpower and all other nations are subordinate to it to one degree or another. The shifting balance of power between the United States and other states, or simply between other states, still very much matters for the stability of this system. And it can still explain a lot of the news you read on the front page of the paper, whether it is the news of China balancing Russia, Iran balancing Iraq, or India confronting Pakistan.

The second important power balance in the globalization system is between 9 nation-states and global markets. These global markets are made up of millions of investors moving money around the world with the click of a mouse. I call them the Electronic Herd, and this herd gathers in key global financial centers— such as Wall Street, Hong Kong, London, and Frankfurt—which I call the Supermarkets. The attitudes and actions of the Electronic Herd and the Supermarkets can have a huge impact on nation-states today, even to the point of triggering the downfall of governments. Who ousted Suharto in Indonesia in 1998? It wasn't another state, it was the Supermarkets, by withdrawing their support for, and confidence in, the Indonesian economy. You also will not understand the front page of the newspaper today unless you bring the Supermarkets into your analysis. Because the United States can destroy you by dropping bombs, but the Supermarkets can destroy you by downgrading your bonds. In other words, the United States is the dominant player in maintaining the globalization game board, but it is hardly alone in influencing the moves on that game board.

The third balance that you have to pay attention to—the one that is really 10 the newest of all and the most relevant to the events of 9/11—is the balance between individuals and nation-states. Because globalization has brought down many of the walls that limited the movement and reach of people, and because it has simultaneously wired the world into networks, it gives more power to *individuals* to influence both markets and nation-states than at any other time in history. Whether by enabling people to use the Internet to communicate instantly at almost no cost over vast distances, or by enabling them to use the Web to transfer money or obtain weapons designs that normally would have been controlled by states, or by enabling them to go into a hardware store now and buy a five-hundred-dollar global positioning device, connected to a satellite, that can direct a hijacked airplane—globalization can be an incredible force-multiplier for individuals. Individuals can increasingly act on the world stage directly, unmediated by a state.

So you have today not only a superpower, not only Supermarkets, but also 11 what I call "super-empowered individuals." Some of these super-empowered individuals are quite angry, some of them quite wonderful—but all of them are now able to act much more directly and much more powerfully on the world stage.

Osama bin Laden declared war on the United States in the late 1990s. After 12
he organized the bombing of two American embassies in Africa, the U.S. Air
Force retaliated with a cruise missile attack on his bases in Afghanistan as
though he were another nation-state. Think about that: on one day in 1998, the
United States fired 75 cruise missiles at bin Laden. The United States fired 75
cruise missiles, at $1 million apiece, at a person! That was the first battle in his-
tory between a superpower and a super-empowered angry man. September 11
was just the second such battle.

Jody Williams won the Nobel Peace Prize in 1997 for helping to build an in- 13
ternational coalition to bring about a treaty outlawing land mines. Although
nearly 120 governments endorsed the treaty, it was opposed by Russia, China,
and the United States. When Jody Williams was asked, "How did you do that?
How did you organize one thousand different citizens' groups and nongovernmental organizations on five continents to forge a treaty that was opposed by
the major powers?" she had a very brief answer: "E-mail." Jody Williams used
e-mail and the networked world to super-empower herself.

Nation-states, and the American superpower in particular, are still hugely 14
important today, but so too now are Supermarkets and super-empowered individuals. You will never understand the globalization system, or the front page
of the morning paper—or 9/11—unless you see each as a complex interaction
between all three of these actors: states bumping up against states, states bumping up against Supermarkets, and Supermarkets and states bumping up against
super-empowered individuals—many of whom, unfortunately, are super-
empowered angry men.

COMPREHENSION

1. What is the writer's "super-story"? How does he define it?
2. What are the main features of globalization? How does globalization differ
   from the system characterized by the cold war? Explain the "three balances" (paragraph 8) that Friedman writes about.
3. What does Friedman mean by "super-empowered" individuals (paragraph
   11)?

RHETORIC

1. What is the writer's thesis or claim in this essay? Where does it appear?
2. How and why does Friedman create a personal voice as well as a colloquial
   style in this selection? What is the effect?
3. What definitions does the writer establish? Are the definitions too abstract,
   or does Friedman provide sufficient explanations and evidence? Explain.
4. Locate instances of classification and comparison and contrast. Why does
   Friedman use these rhetorical strategies? How do the two methods complement each other?

5. Friedman uses several metaphors in this essay. What are they, and how do they function to enhance meaning?
6. Why does the writer introduce 9/11 in the final three paragraphs? What is the effect on the overall message and purpose of the essay?

WRITING

1. In groups of three or four, use Friedman's essay to brainstorm about globalization. Construct a list of ideas and attributes. Using this list, write a definition essay exploring the subject of globalization. Include comparison and contrast or classification, or both, to help organize this essay of definition.
2. Write a personal essay on how you think globalization is affecting your life.
3. **WRITING AN ARGUMENT:** Write a letter to Friedman, either agreeing or disagreeing with his opinions concerning globalization, supporting or refuting his ideas, or offering alternative views.

www.mhhe.com/ **mhreader**

For more information on Thomas L. Friedman, go to:
More Resources > Ch. 8 Business & Economics

✳

Nickel and Dimed

Barbara Ehrenreich

Barbara Ehrenreich (b. 1941) was born in Butte, Montana. The daughter of working-class parents, she attended Reed College (BA 1963) and Rockefeller University, where she received a doctorate in biology in 1968. After deciding not to pursue a career in science, Ehrenreich turned to political causes, using her scientific training to investigate a broad range of social issues. A prolific writer, Ehrenreich has contributed to Time, The New Republic, The Progressive, *and other magazines. She also has written several books, including* The American Health Empire *(1970),* Complaints and Disorders: The Sexual Politics of Sickness *(1978), and, most recently,* Nickel and Dimed: On (Not) Getting By in America *(2001). In the following excerpt from* Nickel and Dimed, *Ehrenreich recounts her experience working for a large cleaning agency.*

I am rested and ready for anything when I arrive at The Maids' office suite   1
Monday at 7:30 A.M. I know nothing about cleaning services like this one, which, according to the brochure I am given, has over three hundred franchises nationwide, and most of what I know about domestics in general comes from nineteenth-century British novels and *Upstairs, Downstairs.* Prophetically

enough, I caught a rerun of that very show on PBS over the weekend and was struck by how terribly correct the servants looked in their black-and-white uniforms and how much wiser they were than their callow, egotistical masters. We too have uniforms, though they are more oafish than dignified—ill-fitting and in an overloud combination of kelly-green pants and a blinding sunflower-yellow polo shirt. And, as is explained in writing and over the next day and a half of training, we too have a special code of decorum. No smoking anywhere, or at least not within fifteen minutes of arrival at a house. No drinking, eating, or gum chewing in a house. No cursing in a house, even if the owner is not present, and—perhaps to keep us in practice—no obscenities even in the office. So this is Downstairs, is my chirpy first thought. But I have no idea, of course, just how far down these stairs will take me.

Forty minutes go by before anyone acknowledges my presence with more 2 than a harried nod. During this time the other employees arrive, about twenty of them, already glowing in their uniforms, and breakfast on the free coffee, bagels, and doughnuts The Maids kindly provides for us. All but one of the others are female, with an average age I would guess in the late twenties, though the range seems to go from prom-fresh to well into the Medicare years. There is a pleasant sort of bustle as people get their breakfasts and fill plastic buckets with rags and bottles of cleaning fluids, but surprisingly little conversation outside of a few references to what people ate (pizza) and drank (Jell-O shots are mentioned) over the weekend. Since the room in which we gather contains only two folding chairs, both of them occupied, the other new girl and I sit cross-legged on the floor, silent and alert, while the regulars get sorted into teams of three or four and dispatched to the day's list of houses. One of the women explains to me that teams do not necessarily return to the same houses week after week, nor do you have any guarantee of being on the same team from one day to the next. This, I suppose, is one of the advantages of a corporate cleaning service to its customers: there are no sticky and possibly guilt-ridden relationships involved, because the customers communicate almost entirely with Tammy, the office manager, or with Ted, the franchise owner and our boss. The advantage to the cleaning person is harder to determine, since the pay compares so poorly to what an independent cleaner is likely to earn—up to $15 an hour, I've heard. While I wait in the inner room, where the phone is and Tammy has her desk, to be issued a uniform, I hear her tell a potential customer on the phone that The Maids charges $25 per person-hour. The company gets $25 and we get $6.65 for each hour we work? I think I must have misheard, but a few minutes later I hear her say the same thing to another inquirer. So the only advantage of working here as opposed to freelancing is that you don't need a clientele or even a car. You can arrive straight from welfare or, in my case, the bus station—fresh off the boat.

At last, after all the other employees have sped off in the company's eye- 3 catching green-and-yellow cars, I am led into a tiny closet-sized room off the inner office to learn my trade via videotape. The manager at another maid service where I'd applied had told me she didn't like to hire people who had done cleaning before because they were resistant to learning the company's system, so I prepare to empty my mind of all prior house-cleaning experience. There are four tapes—dusting, bathrooms, kitchen, and vacuuming—each starring an attractive, possibly Hispanic young woman who moves about serenely in obedience to the male voiceover: For vacuuming, begin in the master bedroom; when dusting, begin with the room directly off the kitchen. When you enter a room, mentally divide it into sections no wider than your reach. Begin in the section to your left and, within each section, move from left to right and top to bottom. This way nothing is ever overlooked.

I like *Dusting* best, for its undeniable logic and a certain kind of austere   4 beauty. When you enter a house, you spray a white rag with Windex and place it in the left pocket of your green apron. Another rag, sprayed with disinfectant, goes in the middle pocket, and a yellow rag bearing wood polish in the right-hand pocket. A dry rag, for buffing surfaces, occupies the right-hand pocket of your slacks. Shiny surfaces get Windexed, wood gets wood polish, and everything else is wiped dust-free with disinfectant. Every now and then Ted pops in to watch with me, pausing the video to underscore a particularly dramatic moment: "See how she's working around the vase? That's an accident waiting to happen." If Ted himself were in a video, it would have to be a cartoon, because the only features sketched onto his pudgy face are brown buttonlike eyes and a tiny pug nose; his belly, encased in a polo shirt, overhangs the waistline of his shorts. "You know, all this was figured out with a stopwatch," he tells me with something like pride. When the video warns against oversoaking our rags with cleaning fluids, he pauses it to tell me there's a danger in undersoaking too, especially if it's going to slow me down. "Cleaning fluids are less expensive than your time." It's good to know that *something* is cheaper than my time, or that in the hierarchy of the company's values I rank above Windex.

Vacuuming is the most disturbing video, actually a double feature beginning   5 with an introduction to the special backpack vacuum we are to use. Yes, the vacuum cleaner actually straps onto your back, a chubby fellow who introduces himself as its inventor explains. He suits up, pulling the straps tight across and under his chest and then says proudly into the camera: "See, I *am* the vacuum cleaner." It weighs only ten pounds, he claims, although, as I soon find out, with the attachments dangling from the strap around your waist, the total is probably more like fourteen. What about my petulant and much-pampered lower back? The inventor returns to the theme of human/machine merger: when properly strapped in, we too will be vacuum cleaners, constrained only by the cord that attaches us to an electrical outlet, and vacuum cleaners don't have backaches. Somehow all this information exhausts me, and I watch the second video, which explains the actual procedures for vacuuming, with the detached interest of a cineast. Could the model maid be an actual maid and the model home someone's actual dwelling? And who are these people whose idea of decorating is matched pictures of mallard ducks in flight and whose house is perfectly characterless and pristine even before the model maid sets to work?

At first I find the videos on kitchens and bathrooms baffling, and it takes 6
me several minutes to realize why: there is no *water*, or almost no water, involved. I was taught to clean by my mother, a compulsive housekeeper who
employed water so hot you needed rubber gloves to get into it and in such Ni-
agara-like quantities that most microbes were probably crushed by the force of
it before the soap suds had a chance to rupture their cell walls. But germs are
never mentioned in the videos provided by The Maids. Our antagonists exist
entirely in the visible world—soap scum, dust, counter crud, dog hair, stains,
and smears—and are to be attacked by damp rag or, in hard-core cases, by Dobie (the brand of plastic scouring pad we use). We scrub only to remove impu-
rities that might be detectable to a customer by hand or by eye; otherwise our
only job is to wipe. Nothing is said about the possibility of transporting bacteria, by rag or by hand, from bathroom to kitchen or even from one house to the
next. It is the "cosmetic touches" that the videos emphasize and that Ted, when
he wanders back into the room, continually directs my eye to. Fluff up all throw
pillows and arrange them symmetrically. Brighten up stainless steel sinks with
baby oil. Leave all spice jars, shampoos, etc., with their labels facing outward.
Comb out the fringes of Persian carpets with a pick. Use the vacuum cleaner to
create a special, fernlike pattern in the carpets. The loose ends of toilet paper
and paper towel rolls have to be given a special fold (the same one you'll find in
hotel bathrooms). "Messes" of loose paper, clothing, or toys are to be stacked
into "neat messes." Finally, the house is to be sprayed with the cleaning service's signature floral-scented air freshener, which will signal to the owners, the
moment they return home, that, yes, their house has been "cleaned."

After a day's training, I am judged fit to go out with a team, where I soon 7
discover that life is nothing like the movies, at least not if the movie is *Dusting*.
For one thing, compared with our actual pace, the training videos were all in
slow motion. We do not walk to the cars with our buckets full of cleaning fluids
and utensils in the morning, we run, and when we pull up to a house, we run
with our buckets to the door. Liza, a good-natured woman in her thirties who is
my first team leader, explains that we are given only so many minutes per house,
ranging from under sixty for a 1½-bathroom apartment to two hundred or more
for a multibathroom "first timer." I'd like to know why anybody worries about
Ted's time limits if we're being paid by the hour but hesitate to display anything
that might be interpreted as attitude. As we get to each house, Liza assigns our
tasks, and I cross my fingers to ward off bathrooms and vacuuming. Even dusting, though, gets aerobic under pressure, and after about an hour of it—reaching
to get door tops, crawling along floors to wipe baseboards, standing on my
bucket to attack the higher shelves—I wouldn't mind sitting down with a tall
glass of water. But as soon as you complete your assigned task, you report to the
team leader to be assigned to help someone else. Once or twice, when the normal
process of evaporation is deemed too slow, I am assigned to dry a scrubbed floor
by putting rags under my feet and skating around on it. Usually, by the time I get
out to the car and am dumping the dirty water used on floors and wringing out

rags, the rest of the team is already in the car with the motor running. Liza assures me that they've never left anyone behind at a house, not even, presumably, a very new person whom nobody knows.

In my interview, I had been promised a thirty-minute lunch break, but this turns out to be a five-minute pit stop at a convenience store, if that. I bring my own sandwich—the same turkey breast and cheese every day—as do a couple of the others; the rest eat convenience store fare, a bagel or doughnut salvaged from our free breakfast, or nothing at all. The two older married women I'm teamed up with eat best—sandwiches and fruit. Among the younger women, lunch consists of a slice of pizza, a "pizza pocket" (a roll of dough surrounding some pizza sauce), or a small bag of chips. Bear in mind we are not office workers, sitting around idling at the basal metabolic rate. A poster on the wall in the office cheerily displays the number of calories burned per minute at our various tasks, ranging from about 3.5 for dusting to 7 for vacuuming. If you assume an average of 5 calories per minute in a seven-hour day (eight hours minus time for travel between houses), you need to be taking in 2,100 calories in addition to the resting minimum of, say, 900 or so. I get pushy with Rosalie, who is new like me and fresh from high school in a rural northern part of the state, about the meagerness of her lunches, which consist solely of Doritos—a half-bag from the day before or a freshly purchased small-sized bag. She just didn't have anything in the house, she says (though she lives with her boyfriend and his mother), and she certainly doesn't have any money to buy lunch, as I find out when I offer to fetch her a soda from a Quik Mart and she has to admit she doesn't have eighty-nine cents. I treat her to the soda, wishing I could force her, mommylike, to take milk instead. So how does she hold up for an eight- or even nine-hour day? "Well," she concedes, "I get dizzy sometimes."

How poor are they, my coworkers? The fact that anyone is working this job at all can be taken as prima facie evidence of some kind of desperation or at least a history of mistakes and disappointments, but it's not for me to ask. In the prison movies that provide me with a mental guide to comportment, the new guy doesn't go around shaking hands and asking, "Hi there, what are you in for?" So I listen, in the cars and when we're assembled in the office, and learn, first, that no one seems to be homeless. Almost everyone is embedded in extended families or families artificially extended with housemates. People talk about visiting grandparents in the hospital or sending birthday cards to a niece's husband; single mothers live with their own mothers or share apartments with a coworker or boyfriend. Pauline, the oldest of us, owns her own home, but she sleeps on the living room sofa, while her four grown children and three grandchildren fill up the bedrooms.

But although no one, apparently, is sleeping in a car, there are signs, even at the beginning, of real difficulty if not actual misery. Half-smoked cigarettes are returned to the pack. There are discussions about who will come up with fifty cents for a toll and whether Ted can be counted on for prompt reimbursement. One of my teammates gets frantic about a painfully impacted wisdom tooth and keeps making calls from our houses to try to locate a source of free dental care. When my—or, I should say, Liza's—team discovers there is not a single Dobie in our buckets, I suggest that we stop at a convenience store and buy one rather than drive all the way back to the office. But it turns out I haven't brought any money with me and we cannot put together $2 between the four of us.

The Friday of my first week at The Maids is unnaturally hot for Maine in 11 early September—95 degrees, according to the digital time-and-temperature displays offered by banks that we pass. I'm teamed up with the sad-faced Rosalie and our leader, Maddy, whose sullenness, under the circumstances, is almost a relief after Liza's relentless good cheer. Liza, I've learned, is the highest-ranking cleaner, a sort of supervisor really, and said to be something of a snitch, but Maddy, a single mom of maybe twenty-seven or so, has worked for only three months and broods about her child care problems. Her boyfriend's sister, she tells me on the drive to our first house, watches her eighteen-month-old for $50 a week, which is a stretch on The Maids' pay, plus she doesn't entirely trust the sister, but a real day care center could be as much as $90 a week. After polishing off the first house, no problem, we grab "lunch"—Doritos for Rosalie and a bag of Pepperidge Farm Goldfish for Maddy—and head out into the exurbs for what our instruction sheet warns is a five-bathroom spread and a first-timer to boot. Still, the size of the place makes us pause for a moment, buckets in hand, before searching out an appropriately humble entrance. It sits there like a beached ocean liner, the prow cutting through swells of green turf, windows without number. "Well, well," Maddy says, reading the owner's name from our instruction sheet. "Mrs. W. and her big-ass house. I hope she's going to give us lunch."

Mrs. W. is not in fact happy to see us, grimacing with exasperation when 12 the black nanny ushers us into the family room or sunroom or den or whatever kind of specialized space she is sitting in. After all, she already has the nanny, a cooklike person, and a crew of men doing some sort of finishing touches on the construction to supervise. No, she doesn't want to take us around the house, because she already explained everything to the office on the phone, but Maddy stands there, with Rosalie and me behind her, until she relents. We are to move everything on all surfaces, she instructs during the tour, and get underneath and be sure to do every bit of the several miles, I calculate, of baseboards. And be mindful of the baby, who's napping and can't have cleaning fluids of any kind near her.

Then I am let loose to dust. In a situation like this, where I don't even know 13 how to name the various kinds of rooms, The Maids' special system turns out to be a lifesaver. All I have to do is keep moving from left to right, within rooms and between rooms, trying to identify landmarks so I don't accidentally do a room or a hallway twice. Dusters get the most complete biographical overview, due to the necessity of lifting each object and tchotchke individually, and I learn that Mrs. W. is an alumna of an important women's college, now occupying herself by monitoring her investments and the baby's bowel movements. I find special charts for this latter purpose, with spaces for time of day, most recent fluid intake, consistency, and color. In the master bedroom, I dust a whole shelf of books on pregnancy, breastfeeding, the first six months, the first year, the first two years—and I wonder what the child care–deprived Maddy makes of all this. Maybe there's been some secret division of the world's women into breeders and drones, and those at the maid level are no longer supposed to be reproducing at all. Maybe this is why our office manager, Tammy, who was once a maid herself, wears inch-long fake nails and tarty little outfits—to show she's advanced to the breeder caste and can't be sent out to clean anymore.

It is hotter inside than out, un-air-conditioned for the benefit of the baby, I 14 suppose, but I do all right until I encounter the banks of glass doors that line the side and back of the ground floor. Each one has to be Windexed, wiped, and buffed—inside and out, top to bottom, left to right, until it's as streakless and invisible as a material substance can be. Outside, I can see the construction guys knocking back Gatorade, but the rule is that no fluid or food item can touch a maid's lips when she's inside a house. Now, sweat, even in unseemly quantities, is nothing new to me. I live in a subtropical area where even the inactive can expect to be moist nine months out of the year. I work out, too, in my normal life and take a certain macho pride in the Vs of sweat that form on my T-shirt after ten minutes or more on the StairMaster. But in normal life fluids lost are immediately replaced. Everyone in yuppie-land—airports, for example—looks like a nursing baby these days, inseparable from their plastic bottles of water. Here, however, I sweat without replacement or pause, not in individual drops but in continuous sheets of fluid soaking through my polo shirt, pouring down the backs of my legs. The eyeliner I put on in the morning—vain twit that I am— has long since streaked down onto my cheeks, and I could wring my braid out if I wanted to. Working my way through the living room(s), I wonder if Mrs. W. will ever have occasion to realize that every single doodad and *objet* through which she expresses her unique, individual self is, from another vantage point, only an obstacle between some thirsty person and a glass of water.

When I can find no more surfaces to wipe and have finally exhausted the 15 supply of rooms, Maddy assigns me to do the kitchen floor. OK, except that Mrs. W. is *in* the kitchen, so I have to go down on my hands and knees practically at her feet. No, we don't have sponge mops like the one I use in my own house; the hands-and-knees approach is a definite selling point for corporate cleaning services like The Maids. "We clean floors the old-fashioned way—*on our hands and knees*" (emphasis added), the brochure for a competing firm boasts. In fact, whatever advantages there may be to the hands-and-knees approach—you're closer to your work, of course, and less likely to miss a grimy patch—are undermined by the artificial drought imposed by The Maids' cleaning system. We are instructed to use less than half a small bucket of lukewarm water for a kitchen and all adjacent scrubbable floors (breakfast nooks and other dining areas), meaning that within a few minutes we are doing nothing more than redistributing the dirt evenly around the floor. There are occasional customer complaints about the cleanliness of our floors—for example, from a man who wiped up a spill on his freshly "cleaned" floor only to find the paper towel he employed for this purpose had turned gray. A mop and a full bucket of hot soapy water would not only get a floor cleaner but would be a lot more dignified for the person who does the cleaning. But it is this primal posture of submission—and of what is ultimately anal accessibility—that seems to gratify the consumers of maid services.

I don't know, but Mrs. W.'s floor is hard—stone, I think, or at least a stone- 16 like substance—and we have no knee pads with us today. I had thought in my middle-class innocence that knee pads were one of Monica Lewinsky's prurient fantasies, but no, they actually exist, and they're usually a standard part of our equipment. So here I am on my knees, working my way around the room like some fanatical penitent crawling through the stations of the cross, when I realize that Mrs. W. is staring at me fixedly—so fixedly that I am gripped for a moment by the wild possibility that I may have once given a lecture at her alma mater and she's trying to figure out where she's seen me before. If I were recognized, would I be fired? Would she at least be inspired to offer me a drink of water? Because I have decided that if water is actually offered, I'm taking it, rules or no rules, and if word of this infraction gets back to Ted, I'll just say I thought it would be rude to refuse. Not to worry, though. She's just watching that I don't leave out some stray square inch, and when I rise painfully to my feet again, blinking through the sweat, she says, "Could you just scrub the floor in the entryway while you're at it?"

I rush home to the Blue Haven at the end of the day, pull down the blinds 17 for privacy, strip off my uniform in the kitchen—the bathroom being too small for both a person and her discarded clothes—and stand in the shower for a good ten minutes, thinking all this water is *mine*. I have paid for it, in fact, I have earned it. I have gotten through a week at The Maids without mishap, injury, or insurrection. My back feels fine, meaning I'm not feeling it at all; even my wrists, damaged by carpal tunnel syndrome years ago, are issuing no complaints. Coworkers warned me that the first time they donned the backpack vacuum they felt faint, but not me. I am strong and I am, more than that, good. Did I toss my bucket of filthy water onto Mrs. W.'s casual white summer outfit? No. Did I take the wand of my vacuum cleaner and smash someone's Chinese porcelain statues or Hummel figurines? Not once. I was at all times cheerful, energetic, helpful, and as competent as a new hire can be expected to be. If I can do one week, I can do another, and might as well, since there's never been a moment for job-hunting. The 3:30 quitting time turns out to be a myth; often we don't return to the office until 4:30 or 5:00. And what did I think? That I was going to go out to interviews in my soaked and stinky postwork condition? I decide to reward myself with a sunset walk on Old Orchard Beach.

On account of the heat, there are still a few actual bathers on the beach, but 18 I am content to sit in shorts and T-shirt and watch the ocean pummel the sand. When the sun goes down I walk back into the town to find my car and am amazed to hear a sound I associate with cities like New York and Berlin. There's a couple of Peruvian musicians playing in the little grassy island in the street near the pier, and maybe fifty people—locals and vacationers—have gathered around, offering their bland end-of-summer faces to the sound. I edge my way through the crowd and find a seat where I can see the musicians up close—the beautiful young guitarist and the taller man playing the flute. What are they doing in this rinky-dink blue-collar resort, and what does the audience make of this surprise visit from the dark-skinned South? The melody the flute lays out over the percussion is both utterly strange and completely familiar, as if it had been imprinted in the minds of my own peasant ancestors centuries ago and forgotten until this very moment. Everyone else seems to be as transfixed as I am. The musicians wink and smile at each other as they play, and I see then that they are the secret emissaries of a worldwide lower-class conspiracy to snatch joy out of degradation and filth. When the song ends, I give them a dollar, the equivalent of about ten minutes of sweat.

COMPREHENSION

1. Why do women work for The Maids when they could earn more money as independent cleaners? How does Ehrenreich distinguish her cleaning practices from her coworkers'? Why do the maids emphasize "cosmetic touches" (paragraph 6)?
2. Describe the plight of the writer's coworkers. What "signs . . . of real difficulty if not actual misery" (paragraph 10) does she detect? What, if anything, does Ehrenreich do to help them?
3. Who is Mrs. W? What is her lifestyle like, and what does she expect of the maids? How does she treat Ehrenreich?

RHETORIC

1. How does the writer structure her narrative? How much time elapses? What elements of conflict develop? What transitional devices does the writer employ to unify the action?
2. Where does the writer employ description, and for what purpose? What descriptive details seem most striking to you? How, for example, does Ehrenreich bring her coworkers and Mrs. W. to life?
3. Identify those instances where the writer uses process analysis and comparison and contrast to organize her essay. Why does she select these strategies?
4. Explain the tone of this selection. What elements of irony and sarcasm do you detect?
5. Do you think this essay provides a straightforward account of the writer's experience working for The Maids, or does she have an argumentative point? Justify your response.
6. How does the writer conclude this selection? What elements in the last paragraph capture the main purpose behind her account?

WRITING

1. Write a narrative and descriptive essay of a job you have held that involved menial labor. Establish a time frame. Describe any colleagues who worked with you. Have a thesis or an argument that you either state explicitly or permit to emerge from the account.
2. Compare and contrast a bad job that you have held and a job that provided you with a degree of satisfaction.
3. **WRITING AN ARGUMENT:** In *Nickel and Dimed,* the writer set out to find minimum-wage jobs in several parts of the United States, including a Wal-Mart in Minnesota and a restaurant in Florida. However, Ehrenreich knew at the outset that these jobs were temporary and that she had the luxury of going back to her comfortable life and her career as a writer and activist. Argue for or against the proposition that Ehrenreich was being unethical and exploitative in her behavior. Refer to this selection to support your position.

www.mhhe.com/
mhreader

For more information on Barbara Ehrenreich, go to:
More Resources > Ch. 8 Business & Economics

✳

Los Pobres

Richard Rodriguez

Richard Rodriguez, (b. 1944), born in San Francisco, received degrees from both Stanford University and Columbia University. He also did graduate study at the University of California, Berkeley, and at the Warburg Institute, London. He is a writer and editor for Pacifica News Service. *Rodriguez became a nationally known writer with the publication of his autobiography,* Hunger of Memory: The Education of Richard Rodriguez *(1982). In it, he describes the struggles of growing up biculturally—feeling alienated from his Spanish-speaking parents yet not wholly comfortable in the dominant culture of the United States. He opposes bilingualism and affirmative action as they are now practiced in the United States, and his stance has caused much controversy in educational and intellectual circles. Rodriguez continues to write about social issues such as acculturation, education, and language. In "Los Pobres," Rodriguez shows us how what starts off as a summer job ends with a personal revelation about social and personal identity.*

It was at Stanford, one day near the end of my senior year, that a friend told me   1
about a summer construction job he knew was available. I was quickly alert.
Desire uncoiled within me. My friend said that he knew I had been looking for
summer employment. He knew I needed some money. Almost apologetically
he explained: It was something I probably wouldn't be interested in, but a
friend of his, a contractor, needed someone for the summer to do menial jobs.
There would be lots of shoveling and raking and sweeping. Nothing too hard.
But nothing more interesting either. Still, the pay would be good. Did I want it?
Or did I know someone who did?

I did. Yes, I said, surprised to hear myself say it.   2

In the weeks following, friends cautioned that I had no idea how hard   3
physical labor really is. ("You only *think* you know what it is like to shovel for
eight hours straight.") Their objections seemed to me challenges. They resolved
the issue. I became happy with my plan. I decided, however, not to tell my parents. I wouldn't tell my mother because I could guess her worried reaction. I
would tell my father only after the summer was over, when I could announce
that, after all, I did know what "real work" is like.

The day I met the contractor (a Princeton graduate, it turned out), he asked   4
me whether I had done any physical labor before. "In high school, during the
summer," I lied. And although he seemed to regard me with skepticism, he decided to give me a try. Several days later, expectant, I arrived at my first construction site. I would take off my shirt to the sun. And at last grasp desired
sensation. No longer afraid. At last become like a *bracero*. "We need those tree
stumps out of here by tomorrow," the contractor said. I started to work.

I labored with excitement that first morning—and all the days after. The   5
work was harder than I could have expected. But it was never as tedious as my
friends had warned me it would be. There was too much physical pleasure in
the labor. Especially early in the day, I would be most alert to the sensations of
movement and straining. Beginning around seven each morning (when the air
was still damp but the scent of weeds and dry earth anticipated the heat of the
sun), I would feel my body resist the first thrusts of the shovel. My arms, tight-
ened by sleep, would gradually loosen; after only several minutes, sweat would
gather in beads on my forehead and then—a short while later—I would feel my
chest silky with sweat in the breeze. I would return to my work. A nervous
spark of pain would fly up my arm and settle to burn like an ember in the thick
of my shoulder. An hour, two passed. Three. My whole body would assume
regular movements. Even later in the day, my enthusiasm for primitive sensation would survive the heat and the dust and the insects pricking my back. I
would strain wildly for sensation as the day came to a close. At three thirty,
quitting time, I would stand upright and slowly let my head fall back, luxuriat-
ing in the feeling of tightness relieved.

Some of the men working nearby would watch me and laugh. Two or three   6
of the older men took the trouble to teach me the right way to use a pick, the
correct way to shovel. "You're doing it wrong, too fucking hard," one man
scolded. Then proceeded to show me—what persons who work with their bod-

ies all their lives quickly learn—the most economical way to use one's body in labor.

"Don't make your back do so much work," he instructed. I stood impatiently   7
listening, half listening, vaguely watching, then noticed his work-thickened fingers clutching the shovel. I was annoyed. I wanted to tell him that I enjoyed shoveling the wrong way. And I didn't want to learn the right way. I wasn't afraid of back pain. I liked the way my body felt sore at the end of the day.

I was about to, but, as it turned out, I didn't say a thing. Rather it was at that   8
moment I realized that I was fooling myself if I expected a few weeks of labor to gain me admission to the world of the laborer. I would not learn in three months what my father had meant by "real work." I was not bound to this job; I could imagine its rapid conclusion. For me the sensations of exertion and fatigue could be savored. For my father or uncle, working at comparable jobs when they were my age, such sensations were to be feared. Fatigue took a different toll on their bodies—and minds.

It was, I know, a simple insight. But it was with this realization that I took   9
my first step that summer toward realizing something even more important about the "worker." In the company of carpenters, electricians, plumbers, and painters at lunch, I would often sit quietly, observant. I was not shy in such company. I felt easy, pleased by the knowledge that I was casually accepted, my presence taken for granted by men (exotics) who worked with their hands. Some days the younger men would talk and talk about sex, and they would howl at women who drove by in cars. Other days the talk at lunchtime was subdued; men gathered in separate groups. It depended on who was around. There were rough, good-natured workers. Others were quiet. The more I remember that summer, the more I realize that there was no single *type* of worker. I am embarrassed to say I had not expected such diversity. I certainly had not expected to meet, for example, a plumber who was an abstract painter in his off hours and admired the work of Mark Rothko. Nor did I expect to meet so many workers with college diplomas. (They were the ones who were not surprised that I intended to enter graduate school in the fall.) I suppose what I really want to say here is painfully obvious, but I must say it nevertheless: The men of that summer were middle-class Americans. They certainly didn't constitute an oppressed society. Carefully completing their work sheets; talking about the fortunes of local football teams; planning Las Vegas vacations; comparing the gas mileage of various makes of campers—they were not *los pobres* my mother had spoken about.

On two occasions, the contractor hired a group of Mexican aliens. They   10
were employed to cut down some trees and haul off debris. In all, there were six men of varying age. The youngest in his late twenties; the oldest (his father?) perhaps sixty years old. They came and they left in a single old truck. Anonymous men. They were never introduced to the other men at the site. Immediately upon their arrival, they would follow the contractor's directions, start working—rarely resting—seemingly driven by a fatalistic sense that work which had to be done was best done as quickly as possible.

I watched them sometimes. Perhaps they watched me. The only time I saw 11
them pay me much notice was one day at lunchtime when I was laughing with
the other men. The Mexicans sat apart when they ate, just as they worked by
themselves. Quiet. I rarely heard them say much to each other. All I could hear
were their voices calling out sharply to one another, giving directions. Otherwise, when they stood briefly resting, they talked among themselves in voices
too hard to overhear.

The contractor knew enough Spanish, and the Mexicans—or at least the 12
oldest of them, their spokesman—seemed to know enough English to communicate. But because I was around, the contractor decided one day to make me
his translator. (He assumed I could speak Spanish.) I did what I was told. Shyly
I went over to tell the Mexicans that the patrón wanted them to do something
else before they left for the day. As I started to speak, I was afraid with my old
fear that I would be unable to pronounce the Spanish words. But it was a simple instruction I had to convey. I could say it in phrases.

The dark sweating faces turned toward me as I spoke. They stopped their 13
work to hear me. Each nodded in response. I stood there. I wanted to say something more. But what could I say in Spanish, even if I could have pronounced
the words right? Perhaps I just wanted to engage in small talk, to be assured of
their confidence, our familiarity. I thought for a moment to ask them where in
Mexico they were from. Something like that. And maybe I wanted to tell them
(a lie, if need be) that my parents were from the same part of Mexico.

I stood there.   14

Their faces watched me. The eyes of the man directly in front of me moved 15
slowly over my shoulder, and I turned to follow his glance toward *el patrón*
some distance away. For a moment I felt swept up by that glance into the Mexicans' company. But then I heard one of them returning to work. And then the
others went back to work. I left them without saying anything more.

When they had finished, the contractor went over to pay them in cash. (He 16
later told me that he paid them collectively—"for the job," though he wouldn't
tell me their wages. He said something quickly about the good rate of exchange
"in their own country.") I can still hear the loudly confident voice he used with
the Mexicans. It was the sound of the *gringo* I had heard as a very young boy.
And I can still hear the quiet, indistinct sounds of the Mexican, the oldest who
replied. At hearing that voice I was sad for the Mexicans. Depressed by their
vulnerability. Angry at myself. The adventure of the summer seemed suddenly
ludicrous. I would not shorten the distance I felt from *los pobres* with a few
weeks of physical labor. I would not become like them. They were different
from me. . . .

In the end my father was right—though perhaps he did not know how right 17
or why—to say that I would never know what real work is. I will never know
what he felt at his last factory job. If tomorrow I worked at some kind of factory,
it would go differently for me. My long education would favor me. I could act
as a public person—able to defend my interests, to unionize, to petition, to
speak up—to challenge and demand. (I will never know what real work is.) I

will never know what the Mexicans knew, gathering their shovels and ladders and saws.

Their silence stays with me now. The wages those Mexicans received for their labor were only a measure of their disadvantaged condition. Their silence is more telling. They lack a public identity. They remain profoundly alien. Persons apart. People lacking a union obviously, people without grounds. They depend upon the relative good will or fairness of their employers each day. For such people, lacking a better alternative, it is not such an unreasonable risk. 18

Their silence stays with me. I have taken these many words to describe its impact. Only: the quiet. Something uncanny about it. Its compliance. Vulnerability. Pathos. As I heard their truck rumbling away, I shuddered, my face mirrored with sweat. I had finally come face to face with *los pobres*. 19

COMPREHENSION

1. How does Rodriguez set the scene for his narrative? What contrasts does he develop in the course of the essay?
2. What are the chief revelations Rodriguez receives from his work experience?
3. Why does Rodriguez focus on the silence of the Mexicans in the final two paragraphs? What is the relationship between this silence and the "real work" his father knows?

RHETORIC

1. In paragraph 9, Rodriguez puts quotation marks around worker and parentheses around *exotics*, and he italicizes *type*. What is the purpose of each choice of punctuation or style?
2. There are several fragments in each of the final two paragraphs. What is the effect of using this sentence structure? Where else are fragments employed in the essay?
3. Why are paragraphs 2 and 14 so short? How does the length of these paragraphs help delineate Rodriguez's mood?
4. What sensations does Rodriguez focus on in paragraph 5? Which words contribute most to evoking them?
5. In what way do the first three paragraphs prepare or fail to prepare you for the narrative that follows?
6. The opening sentence of paragraph 19 repeats that of paragraph 18. What is the purpose of this repetition?

WRITING

1. Imagine yourself in the same situation as Rodriguez. Would your presumptions about hard work and your coworkers have been the same? Would you be more or less naive than Rodriguez? Explain in a brief essay.

2. What are the major differences between the Mexican workers and the American workers in the essay? Write an essay focusing on these differences.
3. **WRITING AN ARGUMENT:** In an essay, argue for or against the proposition that illegal aliens are needed in the United States because they accept work that Americans are unwilling to consider.

| www.mhhe.com/ **mhreader** | For more information on Richard Rodriguez, go to: **More Resources > Ch. 8 Business & Economics** |

✳

Why the Rich Are Getting Richer and the Poor, Poorer

Robert Reich

Robert Reich (b. 1946) is a University Professor in the Heller Graduate School at Brandeis University. He served as secretary of labor in the first Clinton administration and, before that, as a professor of economics at Harvard University. He has written numerous books on economics and has been a prominent lecturer for a dozen years. His books include The Next American Frontier *(1983) and* The Work of Nations *(1991), which takes its title from Adam Smith's classic work on economics* The Wealth of Nations, *written in 1776. Reich is known for his ability to "think outside the box," in other words, to see things from a unique and original perspective. Here he warns of what exists—perhaps in front of our very noses—but that we are too caught up in the moment to consider.*

The division of labor is limited by the extent of the market.
—*Adam Smith,* An Inquiry into the Nature and Causes of the Wealth of Nations
(1776)

Regardless of how your job is officially classified (manufacturing, service, managerial, technical, secretarial, and so on), or the industry in which you work (automotive, steel, computer, advertising, finance, food processing), your real competitive position in the world economy is coming to depend on the function you perform in it. Herein lies the basic reason why incomes are diverging. The fortunes of routine producers are declining. In-person servers are also becoming poorer, although their fates are less clear-cut. But symbolic analysts—who solve, identify, and broker new problems—are, by and large, succeeding in the world economy.

All Americans used to be in roughly the same economic boat. Most rose or   2
fell together as the corporations in which they were employed, the industries
comprising such corporations, and the national economy as a whole became
more productive—or languished. But national borders no longer define our economic fates. We are now in different boats, one sinking rapidly, one sinking
more slowly, and the third rising steadily.

The boat containing routine producers is sinking rapidly. Recall that by mid-   3
century routine production workers in the United States were paid relatively
well. The giant pyramidlike organizations at the core of each major industry
coordinated their prices and investments—avoiding the harsh winds of competition and thus maintaining healthy earnings. Some of these earnings, in
turn, were reinvested in new plant and equipment (yielding ever-larger-scale
economies); another portion went to top managers and investors. But a large
and increasing portion went to middle managers and production workers.
Work stoppages posed such a threat to high-volume production that organized
labor was able to exact an ever-larger premium for its cooperation. And the pattern of wages established within the core corporations influenced the pattern
throughout the national economy. Thus the growth of a relatively affluent middle class, able to purchase all the wondrous things produced in high volume by
the core corporations.

But, as has been observed, the core is rapidly breaking down into global   4
webs which earn their largest profits from clever problem-solving, -identifying,
and brokering. As the costs of transporting standard things and of communicating information about them continue to drop, profit margins on high-volume,
standardized production are thinning, because there are few barriers to entry.
Modern factories and state-of-the-art machinery can be installed almost anywhere on the globe. Routine producers in the United States, then, are in direct
competition with millions of routine producers in other nations. Twelve thousand people are added to the world's population every hour, most of whom,
eventually, will happily work for a small fraction of the wages of routine producers in America.[1]

The consequence is clearest in older, heavy industries, where high-volume,   5
standardized production continues its ineluctable move to where labor is
cheapest and most accessible around the world. Thus, for example, the
Maquiladora factories cluttered along the Mexican side of the U.S. border in the
sprawling shanty towns of Tijuana, Mexicali, Nogales, Agua Prieta, and Ciudad
Juárez—factories owned mostly by Americans, but increasingly by Japanese—

---

[1]The reader should note, of course, that lower wages in other areas of the world are of no particular attraction to global capital unless workers there are sufficiently productive to make the labor
cost of producing *each unit* lower there than in higher-wage regions. Productivity in many low-
wage areas of the world has improved due to the ease with which state-of-the-art factories and
equipment can be installed there. [This and subsequent notes in the selection are the author's.]

in which more than a half million routine producers assemble parts into finished goods to be shipped into the United States.

The same story is unfolding worldwide. Until the late 1970s, AT&T had depended on routine producers in Shreveport, Louisiana, to assemble standard telephones. It then discovered that routine producers in Singapore would perform the same tasks at a far lower cost. Facing intense competition from other global webs, AT&T's strategic brokers felt compelled to switch. So in the early 1980s they stopped hiring routine producers in Shreveport and began hiring cheaper routine producers in Singapore. But under this kind of pressure for ever lower high-volume production costs, today's Singaporean can easily end up as yesterday's Louisianan. By the late 1980s, AT&T's strategic brokers found that routine producers in Thailand were eager to assemble telephones for a small fraction of the wages of routine producers in Singapore. Thus, in 1989, AT&T stopped hiring Singaporeans to make telephones and began hiring even cheaper routine producers in Thailand. 6

The search for ever lower wages has not been confined to heavy industry. Routine data processing is equally footloose. Keypunch operators located anywhere around the world can enter data into computers, linked by satellite or transoceanic fiber-optic cable, and take it out again. As the rates charged by satellite networks continue to drop, and as more satellites and fiber-optic cables become available (reducing communication costs still further), routine data processors in the United States find themselves in ever more direct competition with their counterparts abroad, who are often eager to work for far less. 7

By 1990, keypunch operators in the United States were earning, at most, $6.50 per hour. But keypunch operators throughout the rest of the world were willing to work for a fraction of this. Thus, many potential American data-processing jobs were disappearing, and the wages and benefits of the remaining ones were in decline. Typical was Saztec International, a $20-million-a-year data-processing firm headquartered in Kansas City, whose American strategic brokers contracted with routine data processors in Manila and with American-owned firms that needed such data-processing services. Compared with the average Philippine income of $1,700 per year, data-entry operators working for Saztec earn the princely sum of $2,650. The remainder of Saztec's employees were American problem-solvers and -identifiers, searching for ways to improve the worldwide system and find new uses to which it could be put.[2] 8

By 1990, American Airlines was employing over 1,000 data processors in Barbados and the Dominican Republic to enter names and flight numbers from used airline tickets (flown daily to Barbados from airports around the United States) into a giant computer bank located in Dallas. Chicago publisher R. R. Donnelley was sending entire manuscripts to Barbados for entry into computers in preparation for printing. The New York Life Insurance Company was dispatching insurance claims to Castleisland, Ireland, where routine producers, 9

[2]John Maxwell Hamilton, "A Bit Player Buys into the Computer Age," *New York Times Business World*, December 3, 1989, p. 14.

guided by simple directions, entered the claims and determined the amounts due, then instantly transmitted the computations back to the United States. (When the firm advertised in Ireland for twenty-five data-processing jobs, it received six hundred applications.) And McGraw-Hill was processing subscription renewal and marketing information for its magazines in nearby Galway. Indeed, literally millions of routine workers around the world were receiving information, converting it into computer-readable form, and then sending it back—at the speed of electronic impulses—whence it came.

The simple coding of computer software has also entered into world commerce. India, with a large English-speaking population of technicians happy to do routine programming cheaply, is proving to be particularly attractive to global webs in need of this service. By 1990, Texas Instruments maintained a software development facility in Bangalore, linking fifty Indian programmers by satellite to TI's Dallas headquarters. Spurred by this and similar ventures, the Indian government was building a teleport in Poona, intended to make it easier and less expensive for many other firms to send their routine software design specifications for coding.[3]

This shift of routine production jobs from advanced to developing nations is a great boon to many workers in such nations who otherwise would be jobless or working for much lower wages. These workers, in turn, now have more money with which to purchase symbolic-analytic services from advanced nations (often embedded within all sorts of complex products). The trend is also beneficial to everyone around the world who can now obtain high-volume, standardized products (including information and software) more cheaply than before.

But these benefits do not come without certain costs. In particular the burden is borne by those who no longer have good-paying routine production jobs within advanced economies like the United States. Many of these people used to belong to unions or at least benefited from prevailing wage rates established in collective bargaining agreements. But as the old corporate bureaucracies have flattened into global webs, bargaining leverage has been lost. Indeed, the tacit national bargain is no more.

Despite the growth in the number of new jobs in the United States, union membership has withered. In 1960, 35 percent of all nonagricultural workers in America belonged to a union. But by 1980 that portion had fallen to just under a quarter, and by 1989 to about 17 percent. Excluding government employees, union membership was down to 13.4 percent.[4] This was a smaller proportion even than in the early 1930s, before the National Labor Relations Act created a legally protected right to labor representation. The drop in membership has been accompanied by a growing number of collective bargaining agreements to

[3]Udayan Gupta, "U.S.-Indian Satellite Link Stands to Cut Software Costs," *Wall Street Journal*, March 6, 1989, p. B2.
[4]*Statistical Abstract of the United States* (Washington, D.C.: U.S. Government Printing Office, 1989), p. 416, table 684.

freeze wages at current levels, reduce wage levels of entering workers, or reduce wages overall. This is an important reason why the long economic recovery that began in 1982 produced a smaller rise in unit labor costs than any of the
eight recoveries since World War II—the low rate of unemployment during its
course notwithstanding.

Routine production jobs have vanished fastest in traditional unionized in-   14
dustries (autos, steel, and rubber, for example), where average wages have kept
up with inflation. This is because the jobs of older workers in such industries
are protected by seniority; the youngest workers are the first to be laid off.
Faced with a choice of cutting wages or cutting the number of jobs, a majority
of union members (secure in the knowledge that there are many who are junior
to them who will be laid off first) often have voted for the latter.

Thus the decline in union membership has been most striking among   15
young men entering the work force without a college education. In the early
1950s, more than 40 percent of this group joined unions; by the late 1980s, less
than 20 percent (if public employees are excluded, less than 10 percent).[5] In
steelmaking, for example, although many older workers remained employed,
almost half of all routine steelmaking jobs in America vanished between 1974
and 1988 (from 480,000 to 260,000). Similarly with automobiles: During the
1980s, the United Auto Workers lost 500,000 members—one-third of their total
at the start of the decade. General Motors alone cut 150,000 American production jobs during the 1980s (even as it added employment abroad). Another consequence of the same phenomenon: the gap between the average wages of
unionized and nonunionized workers widened dramatically—from 14.6 percent in 1973 to 20.4 percent by end of the 1980s.[6] The lesson is clear. If you drop
out of high school or have no more than a high school diploma, do not expect a
good routine production job to be awaiting you.

Also vanishing are lower- and middle-level management jobs involving   16
routine production. Between 1981 and 1986, more than 780,000 foremen, super-
visors, and section chiefs lost their jobs through plant closings and layoffs.[7]
Large numbers of assistant division heads, assistant directors, assistant managers, and vice presidents also found themselves jobless. GM shed more than
40,000 white-collar employees and planned to eliminate another 25,000 by the
mid-1990s.[8] As America's core pyramids metamorphosed into global webs,
many middle-level routine producers were as obsolete as routine workers on
the line.

As has been noted, foreign-owned webs are hiring some Americans to do   17
routine production in the United States. Philips, Sony, and Toyota factories are

[5]Calculations from Current Population Surveys by L. Katz and A. Revenga, "Changes in the Structure of Wages: U.S. and Japan," National Bureau of Economic Research, September 1989.
[6]U.S. Department of Commerce, Bureau of Labor Statistics, "Wages of Unionized and Nonunionized Workers," various issues.
[7]U.S. Department of Labor, Bureau of Labor Statistics, "Reemployment Increases among Displaced
Workers," BLS News, USDL 86-414, October 14, 1986, table 6.
[8]Wall Street Journal, February 16, 1990, p. A5.

popping up all over—to the self-congratulatory applause of the nation's governors and mayors, who have lured them with promises of tax abatements and new sewers, among other amenities. But as these ebullient politicians will soon discover, the foreign-owned factories are highly automated and will become far more so in years to come. Routine production jobs account for a small fraction of the cost of producing most items in the United States and other advanced nations, and this fraction will continue to decline sharply as computer-integrated robots take over. In 1977, it took routine producers thirty-five hours to assemble an automobile in the United States; it is estimated that by the mid-1990s, Japanese-owned factories in America will be producing finished automobiles using only eight hours of a routine producer's time.[9]

The productivity and resulting wages of American workers who run such 18 robotic machinery may be relatively high, but there may not be many such jobs to go around. A case in point: in the late 1980s, Nippon Steel joined with America's ailing Inland Steel to build a new $400 million cold-rolling mill fifty miles west of Gary, Indiana. The mill was celebrated for its state-of-the-art technology, which cut the time to produce a coil of steel from twelve days to about one hour. In fact, the entire plant could be run by a small team of technicians, which became clear when Inland subsequently closed two of its old cold-rolling mills, laying off hundreds of routine workers. Governors and mayors take note: your much-bally-hooed foreign factories may end up employing distressingly few of your constituents.

Overall, the decline in routine jobs has hurt men more than women. This is 19 because the routine production jobs held by men in high-volume metal bending manufacturing industries had paid higher wages than the routine production jobs held by women in textiles and data processing. As both sets of jobs have been lost, American women in routine production have gained more equal footing with American men—equally poor footing, that is. This is a major reason why the gender gap between male and female wages began to close during the 1980s.

The second of the three boats, carrying in-person servers, is sinking as well, but 20 somewhat more slowly and unevenly. Most in-person servers are paid at or just slightly above the minimum wage and many work only part-time, with the result that their take-home pay is modest, to say the least. Nor do they typically receive all the benefits (health care, life insurance, disability, and so forth) garnered by routine producers in large manufacturing corporations or by symbolic analysts affiliated with the more affluent threads of global webs.[10] In-person servers are sheltered from the direct effects of global competition and, like

[9]Figures from the International Motor Vehicles Program, Massachusetts Institute of Technology, 1989.
[10]The growing portion of the American labor force engaged in in-person services, relative to routine production, thus helps explain why the number of Americans lacking health insurance increased by at least 6 million during the 1980s.

everyone else, benefit from access to lower-cost products from around the world. But they are not immune to its indirect effects.

For one thing, in-person servers increasingly compete with former routine 21 production workers, who, no longer able to find well-paying routine production jobs, have few alternatives but to seek in-person service jobs. The Bureau of Labor Statistics estimates that of the 2.8 million manufacturing workers who lost their jobs during the early 1980s, fully one-third were rehired in service jobs paying at least 20 percent less.[11] In-person servers must also compete with high school graduates and dropouts who years before had moved easily into routine production jobs but no longer can. And if demographic predictions about the American work force in the first decades of the twenty-first century are correct (and they are likely to be, since most of the people who will comprise the work force are already identifiable), most new entrants into the job market will be black or Hispanic men, or women—groups that in years past have possessed relatively weak technical skills. This will result in an even larger number of people crowding into in-person services. Finally, in-person servers will be competing with growing numbers of immigrants, both legal and illegal, for whom in-person services will comprise the most accessible jobs. (It is estimated that between the mid-1980s and the end of the century, about a quarter of all workers entering the American labor force will be immigrants.[12])

Perhaps the fiercest competition that in-person servers face comes from 22 labor-saving machinery (much of it invented, designed, fabricated, or assembled in other nations, of course). Automated tellers, computerized cashiers, automatic car washes, robotized vending machines, self-service gasoline pumps, and all similar gadgets substitute for the human beings that customers once encountered. Even telephone operators are fast disappearing, as electronic sensors and voice simulators become capable of carrying on conversations that are reasonably intelligent and always polite. Retail sales workers—among the largest groups of in-person servers—are similarly imperiled. Through personal computers linked to television screens, tomorrow's consumers will be able to buy furniture, appliances, and all sorts of electronic toys from their living rooms—examining the merchandise from all angles, selecting whatever color, size, special features, and price seem most appealing, and then transmitting the order instantly to warehouses from which the selections will be shipped directly to their homes. So, too, with financial transactions, airline and hotel reservations, rental car agreements, and similar contracts, which will be executed between consumers in their homes and computer banks somewhere else on the globe.[13]

Advanced economies like the United States will continue to generate siz- 23 able numbers of new in-person service jobs, of course, the automation of older

[11]U.S. Department of Labor, Bureau of Labor Statistics, "Reemployment Increases among Disabled Workers," October 14, 1986.

[12]Federal Immigration and Naturalization Service, *Statistical Yearbook* (Washington, D.C.: U.S. Government Printing Office, 1986, 1987).

[13]See Claudia H. Deutsch, "The Powerful Push for Self-Service," *New York Times,* April 9, 1989, section 3, p. 1.

ones notwithstanding. For every bank teller who loses her job to an automated teller, three new jobs open for aerobics instructors. Human beings, it seems, have an almost insatiable desire for personal attention. But the intense competition nevertheless ensures that the wages of in-person servers will remain relatively low. In-person servers—working on their own, or else dispersed widely amid many small establishments, filling all sorts of personal-care niches—cannot readily organize themselves into labor unions or create powerful lobbies to limit the impact of such competition.

In two respects, demographics will work in favor of in-person servers, 24 buoying their collective boat slightly. First, as has been noted, the rate of growth of the American work force is slowing. In particular, the number of young workers is shrinking. Between 1985 and 1995, the number of the eighteen- to twenty-four-year-olds will have declined by 17.5 percent. Thus, employers will have more incentive to hire and train in-person servers whom they might previously have avoided. But this demographic relief from the competitive pressures will be only temporary. The cumulative procreative energies of the postwar baby-boomers (born between 1946 and 1964) will result in a new surge of workers by 2010 or thereabouts.[14] And immigration—both legal and illegal—shows every sign of increasing in years to come.

Next, by the second decade of the twenty-first century, the number of 25 Americans aged sixty-five and over will be rising precipitously, as the baby-boomers reach retirement age and live longer. Their life expectancies will lengthen not just because fewer of them will have smoked their way to their graves and more will have eaten better than their parents, but also because they will receive all sorts of expensive drugs and therapies designed to keep them alive—barely. By 2035, twice as many Americans will be elderly as in 1988, and the number of octogenarians is expected to triple. As these decaying baby-boomers ingest all the chemicals and receive all the treatments, they will need a great deal of personal attention. Millions of deteriorating bodies will require nurses, nursing-home operators, hospital administrators, orderlies, home-care providers, hospice aides, and technicians to operate and maintain all the expensive machinery that will monitor and temporarily stave off final disintegration. There might even be a booming market for euthanasia specialists. In-person servers catering to the old and ailing will be in strong demand.[15]

One small problem: the decaying baby-boomers will not have enough 26 money to pay for these services. They will have used up their personal savings years before. Their Social Security payments will, of course, have been used by the government to pay for the previous generation's retirement and to finance much of the budget deficits of the 1980s. Moreover, with relatively fewer young

[14]U.S. Bureau of the Census, Current Population Reports, Series P-23, no. 138, tables 2-1, 4-6. See W. Johnson, A. Packer, et al., *Workforce 2000: Work and Workers for the 21st Century* (Indianapolis: Hudson Institute, 1987).

[15]The Census Bureau estimates that by the year 2000, at least 12 million Americans will work in health services—well over 6 percent of the total work force.

Americans in the population, the supply of housing will likely exceed the demand, with the result that the boomers' major investments—their homes—will be worth less (in inflation-adjusted dollars) when they retire than they planned for. In consequence, the huge cost of caring for the graying boomers will fall on many of the same people who will be paid to care for them. It will be like a great sump pump: in-person servers of the twenty-first century will have an abundance of health-care jobs, but a large portion of their earnings will be devoted to Social Security payments and income taxes, which will in turn be used to pay their salaries. The net result: no real improvement in their standard of living.

The standard of living of in-person servers also depends, indirectly, on the   27
standard of living of the Americans they serve who are engaged in world commerce. To the extent that *these* Americans are richly rewarded by the rest of the world for what they contribute, they will have more money to lavish upon in-person services. Here we find the only form of "trickle-down" economics that has a basis in reality. A waitress in a town whose major factory has just been closed is unlikely to earn a high wage or enjoy much job security; in a swank resort populated by film producers and banking moguls, she is apt to do reasonably well. So, too, with nations. In-person servers in Bangladesh may spend their days performing roughly the same tasks as in-person servers in the United States, but have a far lower standard of living for their efforts. The difference comes in the value that their customers add to the world economy.

Unlike the boats of routine producers and in-person servers, however, the   28
vessel containing America's symbolic analysts is rising. Worldwide demand for their insights is growing as the ease and speed of communicating them steadily increases. Not every symbolic analyst is rising as quickly or as dramatically as every other, of course; symbolic analysts at the low end are barely holding their own in the world economy. But symbolic analysts at the top are in such great demand worldwide that they have difficulty keeping track of all their earnings. Never before in history has opulence on such a scale been gained by people who have earned it, and done so legally.

Among symbolic analysts in the middle range are American scientists and   29
researchers who are busily selling their discoveries to global enterprise webs. They are not limited to American customers. If the strategic brokers in General Motors' headquarters refuse to pay a high price for a new means of making high-strength ceramic engines dreamed up by a team of engineers affiliated with Carnegie Mellon University in Pittsburgh, the strategic brokers of Honda or Mercedes-Benz are likely to be more than willing.

So, too, with the insights of America's ubiquitous management consultants,   30
which are being sold for large sums to eager entrepreneurs in Europe and Latin America. Also, the insights of America's energy consultants, sold for even larger sums to Arab sheikhs. American design engineers are providing insights to Olivetti, Mazda, Siemens, and other global webs; American marketers, techniques for learning what worldwide consumers will buy; American advertisers, ploys for ensuring that they actually do. American architects are issuing designs and blueprints for opera houses, art galleries, museums, luxury hotels, and

residential complexes in the world's major cities; American commercial property developers, marketing these properties to worldwide investors and purchasers.

Americans who specialize in the gentle art of public relations are in demand 31
by corporations, governments, and politicians in virtually every nation. So, too, are American political consultants, some of whom, at this writing, are advising the Hungarian Socialist Party, the remnant of Hungary's ruling Communists, on how to salvage a few parliamentary seats in the nation's first free election in more than forty years. Also at this writing, a team of American agricultural consultants is advising the managers of a Soviet farm collective employing 1,700 Russians eighty miles outside Moscow. As noted, American investment bankers and lawyers specializing in financial circumnavigations are selling their insights to Asians and Europeans who are eager to discover how to make large amounts of money by moving large amounts of money.

Developing nations, meanwhile, are hiring American civil engineers to ad- 32
vise on building roads and dams. The present thaw in the Cold War will no doubt expand these opportunities. American engineers from Bechtel (a global firm notable for having employed both Caspar Weinberger and George Shultz for much larger sums than either earned in the Reagan administration) have begun helping the Soviets design and install a new generation of nuclear reactors. Nations also are hiring American bankers and lawyers to help them renegotiate the terms of their loans with global banks, and Washington lobbyists to help them with Congress, the Treasury, the World Bank, the IMF, and other politically sensitive institutions. In fits of obvious desperation, several nations emerging from communism have even hired American economists to teach them about capitalism.

Almost everyone around the world is buying the skills and insights of 33
Americans who manipulate oral and visual symbols—musicians, sound engineers, film producers, makeup artists, directors, cinematographers, actors and actresses, boxers, scriptwriters, songwriters, and set designers. Among the wealthiest of symbolic analysts are Steven Spielberg, Bill Cosby, Charles Schulz, Eddie Murphy, Sylvester Stallone, Madonna, and other star directors and performers—who are almost as well known on the streets of Dresden and Tokyo as in the Back Bay of Boston. Less well rewarded but no less renowned are the unctuous anchors on Turner Broadcasting's Cable News, who appear daily, via satellite, in places ranging from Vietnam to Nigeria. Vanna White is the world's most-watched game-show hostess. Behind each of these familiar faces is a collection of American problem-solvers, -identifiers, and brokers who train, coach, advise, promote, amplify, direct, groom, represent, and otherwise add value to their talents.[16]

---

[16]In 1989, the entertainment business summoned to the United States $5.5 billion in foreign earnings—making it among the nation's largest export industries, just behind aerospace. U.S. Department of Commerce, International Trade Commission, "Composition of U.S. Exports," various issues.

There are also the insights of senior American executives who occupy the 34
world headquarters of global "American" corporations and the national or regional headquarters of global "foreign" corporations. Their insights are duly exported to the rest of the world through the webs of global enterprise. IBM does
not export many machines from the United States, for example. Big Blue makes
machines all over the globe and services them on the spot. Its prime American
exports are symbolic and analytic. From IBM's world headquarters in Armonk,
New York, emanate strategic brokerage and related management services
bound for the rest of the world. In return, IBM's top executives are generously
rewarded.

The most important reason for this expanding world market and increasing 35
global demand for the symbolic and analytic insights of Americans has been the
dramatic improvement in worldwide communication and transportation technologies. Designs, instructions, advice, and visual and audio symbols can be
communicated more and more rapidly around the globe, with ever greater precision and at ever-lower cost. Madonna's voice can be transported to billions of
listeners, with perfect clarity, on digital compact discs. A new invention emanating from engineers in Battelle's laboratory in Columbus, Ohio, can be sent almost anywhere via modem, in a form that will allow others to examine it in
three dimensions through enhanced computer graphics. When face-to-face
meetings are still required—and videoconferencing will not suffice—it is relatively easy for designers, consultants, advisers, artists, and executives to board
supersonic jets and, in a matter of hours, meet directly with their worldwide
clients, customers, audiences, and employees.

With rising demand comes rising compensation. Whether in the form of 36
licensing fees, fees for service, salaries, or shares in final profits, the economic
result is much the same. There are also nonpecuniary rewards. One of the best-
kept secrets among symbolic analysts is that so many of them enjoy their work.
In fact, much of it does not count as work at all, in the traditional sense. The
work of routine producers and in-person servers is typically monotonous; it
causes muscles to tire or weaken and involves little independence or discretion.
The "work" of symbolic analysts, by contrast, often involves puzzles, experiments, games, a significant amount of chatter, and substantial discretion over
what to do next. Few routine producers or in-person servers would "work" if
they did not need to earn the money. Many symbolic analysts would "work"
even if money were no object.

At mid-century, when America was a national market dominated by core 37
pyramid-shaped corporations, there were constraints on the earnings of people
at the highest rungs. First and most obviously, the market for their services was
largely limited to the borders of the nation. In addition, whatever conceptual
value they might contribute was small relative to the value gleaned from large
scale—and it was dependent on large scale for whatever income it was to summon. Most of the problems to be identified and solved had to do with enhanc-

ing the efficiency of production and improving the flow of materials, parts, assembly, and distribution. Inventors searched for the rare breakthrough revealing an entirely new product to be made in high volume; management consultants, executives, and engineers thereafter tried to speed and synchronize its manufacture, to better achieve scale efficiencies; advertisers and marketers sought then to whet the public's appetite for the standard item that emerged. Since white-collar earnings increased with larger scale, there was considerable incentive to expand the firm; indeed, many of America's core corporations grew far larger than scale economies would appear to have justified.

By the 1990s, in contrast, the earnings of symbolic analysts were limited nei- 38 ther by the size of the national market nor by the volume of production of the firms with which they were affiliated. The marketplace was worldwide, and conceptual value was high relative to value added from scale efficiencies.

There had been another constraint on high earnings, which also gave way 39 by the 1990s. At mid-century, the compensation awarded to top executives and advisers of the largest of America's core corporations could not be grossly out of proportion to that of low-level production workers. It would be unseemly for executives who engaged in highly visible rounds of bargaining with labor unions, and who routinely responded to government requests to moderate prices, to take home wages and benefits wildly in excess of what other Americans earned. Unless white-collar executives restrained themselves, moreover, blue-collar production workers could not be expected to restrain their own demands for higher wages. Unless both groups exercised restraint, the government could not be expected to forbear from imposing direct controls and regulations.

At the same time, the wages of production workers could not be allowed to 40 sink too low, lest there be insufficient purchasing power in the economy. After all, who would buy all the goods flowing out of American factories if not American workers? This, too, was part of the tacit bargain struck between American managers and their workers.

Recall the oft-repeated corporate platitude of the era about the chief ex- 41 ecutive's responsibility to carefully weigh and balance the interests of the corporation's disparate stakeholders. Under the stewardship of the corporate statesman, no set of stakeholders—least of all white-collar executives—was to gain a disproportionately large share of the benefits of corporate activity; nor was any stakeholder—especially the average worker—to be left with a share that was disproportionately small. Banal though it was, this idea helped to maintain the legitimacy of the core American corporation in the eyes of most Americans, and to ensure continued economic growth.

But by the 1990s, these informal norms were evaporating, just as (and 42 largely because) the core American corporation was vanishing. The links between top executives and the American production worker were fading: an ever-increasing number of subordinates and contractees were foreign, and a steadily growing number of American routine producers were working for foreign-owned firms. An entire cohort of middle-level managers, who had once been deemed "white collar," had disappeared; and, increasingly, American executives were exporting their insights to global enterprise webs.

As the American corporation itself became a global web almost indistin- 43 guishable from any other, its stakeholders were turning into a large and diffuse group, spread over the world. Such global stakeholders were less visible, and far less noisy, than national stakeholders. And as the American corporation sold its goods and services all over the world, the purchasing power of American workers became far less relevant to its economic survival.

Thus have the inhibitions been removed. The salaries and benefits of Amer- 44 ica's top executives, and many of their advisers and consultants, have soared to what years before would have been unimaginable heights, even as those of other Americans have declined.

COMPREHENSION

1. To what does the title allude? Why is this allusion significant to the meaning of the title?
2. To whom does Reich refer when he mentions "symbolic analysts"? Regardless of their occupation, what do all symbolic analysts have in common regarding the nature of their work?
3. What has traditionally been the image of and the nature of work among the white-collar workers to which Reich alludes? Why are they now one of the groups in danger of losing employment opportunities?

RHETORIC

1. Reich uses the central metaphor of the "boat" in describing the state of economics and employment. Why? What connotations are associated with this image in regard to financial security?
2. How does Reich's introduction prepare you for the major themes he addresses in the body of his essay?
3. Examine the section breaks at the start of paragraphs 3, 11, 20, 35, and 37. How does each section relate to the theme of the essay as a whole? What transitional devices does Reich use to bridge one section to the next?
4. Paragraphs 5, 6, 9, and 16 cite specific and detailed examples of the effects of the changing global economy. How does this contribute to conveying Reich's authority regarding the subject he is discussing?
5. Reich describes a dire situation for the American worker. How would you characterize the tone of this description? Is it angry, resigned, impartial, accusatory? You may use these or any other adjectives as long as you explain your view.
6. Why has Reich opened his essay with an epigraph from Adam Smith? What is the relationship of the quotation to the overall theme of the essay? How does the tone of the epigraph contrast with the tone of the title?

7. What is the author's purpose? Is it to inform, to explain, to warn, to enlighten, to offer solutions, or a combination of any of these? Explain your view.

WRITING

1. In a classification essay, describe three areas of academic concentration at your college or university that can help prepare one for a job as a symbolic analyst.
2. In an expository essay, explain whether you believe the discrepancy between high-wage and low-wage workers will increase, decrease, or remain the same.
3. **WRITING AN ARGUMENT:** In an essay, argue for or against the proposition that as long as one knows which careers command the highest salaries, it is up to the individual to decide whether he or she should pursue a job in those fields.

www.mhhe.com/
mhreader

For more information on Robert Reich, go to:
More Resources > Ch. 8 Business & Economics

✳

A Modest Proposal
Preventing the Children of Poor People in Ireland from Being a Burden to Their Parents or Country, and for Making Them Beneficial to the Public

Jonathan Swift

Jonathan Swift (1667–1745) is best known as the author of three satires: A Tale of a Tub *(1704),* Gulliver's Travels *(1726), and* A Modest Proposal *(1729). In these satires, Swift pricks the balloon of many of his contemporaries' and our own most cherished prejudices, pomposities, and delusions. He was also a famous churchman, an eloquent spokesman for Irish rights, and a political journalist. The following selection, perhaps the most famous satiric essay in the English language, offers modest advice to a nation suffering from poverty, overpopulation, and political injustice.*

It is a melancholy object to those who walk through this great town or travel in the country, when they see the streets, the roads, and cabin doors, crowded with   1 beggars of the female-sex, followed by three, four, or six children, all in rags and importuning every passenger for an alms. These mothers, instead of being able to work for their honest livelihood, are forced to employ all their time in strolling to beg sustenance for their helpless infants, who, as they grow up, either turn thieves for want of work, or leave their dear native country to fight for the Pretender in Spain, or sell themselves to the Barbadoes.

I think it is agreed by all parties that this prodigious number of children in the arms, or on the backs, or at the heels of their mothers, and frequently of their fathers, is in the present deplorable state of the kingdom a very great additional grievance; and therefore whoever could find out a fair, cheap, and easy method of making these children sound, useful members of the commonwealth would deserve so well of the public as to have his statue set up for a preserver of the nation.

But my intention is very far from being confined to provide only for the children of professed beggars; it is of a much greater extent, and shall take in the whole number of infants at a certain age who are born of parents in effect as little able to support them as those who demand our charity in the streets.

As to my own part, having turned my thoughts for many years upon this important subject, and maturely weighted the several schemes of other projectors, I have always found them grossly mistaken in their computation. It is true, a child just dropped from its dam may be supported by her milk for a solar year, with little other nourishment; at most not above the value of two shillings, which the mother may certainly get, or the value in scraps, by her lawful occupation of begging; and it is exactly at one year old that I propose to provide for them in such a manner as instead of being a charge upon their parents or the parish, or wanting food and raiment for the rest of their lives, they shall on the contrary contribute to the feeding, and partly to the clothing, of many thousands.

There is likewise another great advantage in my scheme, that it will prevent those voluntary abortions, and that horrid practice of women murdering their bastard children, alas, too frequent among us, sacrificing the poor innocent babes, I doubt, more to avoid the expense than the shame, which would move tears and pity in the most savage and inhuman breast.

The number of souls in this kingdom being usually reckoned one million and a half, of these I calculate there may be about two hundred thousand couples whose wives are breeders; from which number I subtract thirty thousand couples who are able to maintain their own children, although I apprehend there cannot be so many under the present distresses of the kingdom; but this being granted, there will remain an hundred and seventy thousand breeders. I again subtract fifty thousand for those women who miscarry, or whose children die by accident or disease within the year. There only remain an hundred and twenty thousand children of poor parents annually born. The question therefore is, how this number shall be reared and provided for, which, as I have already said, under the present situation of affairs, is utterly impossible by all the methods hitherto proposed. For we can neither employ them in handicraft or agriculture; we neither build houses (I mean in the country) nor cultivate land. They can very seldom pick up a livelihood by stealing till they arrive at six years old, except where they are of towardly parts; although I confess they learn the rudiments much earlier, during which time they can however be looked upon only as probationers, as I have been informed by a principal gentleman in the county of Cavan, who protested to me that he never knew above one or two instances under the age of six, even in a part of the kingdom so renowned for the quickest proficiency in that art.

I am assured by our merchants that a boy or girl before twelve years old is 7 no salable commodity; and even when they come to this age they will not yield above three pounds, or three pounds and half a crown at most on the Exchange; which cannot turn to account either to the parents or the kingdom, the charge of nutriment and rags having been at least four times that value.

I shall now therefore humbly propose my own thoughts, which I hope will 8 not be liable to the least objection.

I have been assured by a very knowing American of my acquaintance in 9 London, that a young healthy child well nursed is at a year old a most delicious, nourishing, and wholesome food, whether stewed, roasted, baked or boiled; and I make no doubt that it will equally serve in a fricassee or a ragout.

I do therefore humbly offer it to public consideration that of the hundred 10 and twenty thousand children, already computed, twenty thousand may be reserved for breed, whereof only one fourth part to be males, which is more than we allow to sheep, black cattle, or swine; and my reason is that these children are seldom the fruits of marriage, a circumstance not much regarded by our savages, therefore one male will be sufficient to serve four females. That the remaining hundred thousand may at a year old be offered in sale to the persons of quality and fortune through the kingdom, always advising the mother to let them suck plentifully in the last month, so as to render them plump and fat for a good table. A child will make two dishes at an entertainment for friends; and when the family dines alone, the fore or hind quarter will make a reasonable dish, and seasoned with a little pepper or salt will be very good boiled on the fourth day, especially in winter.

I have reckoned upon a medium that a child just born will weigh twelve 11 pounds, and in a solar year if tolerably nursed increaseth to twenty-eight pounds.

I grant this food will be somewhat dear, and therefore very proper for land- 12 lords, who, as they have already devoured most of the parents, seem to have the best title to the children.

Infant's flesh will be in season throughout the year, but more plentiful in 13 March, and a little before and after. For we are told by a grave author, an eminent French physician, that fish being a prolific diet, there are more children born in Roman Catholic countries about nine months after Lent than at any other season: therefore, reckoning a year after Lent, the markets will be more glutted than usual, because the number of popish infants is at least three to one in this kingdom; and therefore it will have one other collateral advantage, by lessening the number of Papists among us.

I have already computed the charge of nursing a beggar's child (in which 14 list I reckon all cottagers, laborers, and four fifths of the farmers) to be about two shillings per annum, rags included: and I believe no gentleman would repine to give ten shillings for the carcass of a good fat child, which, as I have said, will make four dishes of excellent nutritive meat, when he hath only some particular friend or his own family to dine with him. Thus the squire will learn to be a good landlord, and grow popular among the tenants; the mother will have eight shillings net profit, and be fit for work till she produces another child.

Those who are more thrifty (as I must confess the times require) may flay 15 the carcass; the skin of which artificially dressed will make admirable gloves for ladies, and summer boots for fine gentlemen.

As to our city of Dublin, shambles may be appointed for this purpose in the 16 most convenient parts of it, and butchers we may be assured will not be wanting; although I rather recommend buying the children alive, and dressing them hot from the knife as we do roasting pigs.

A very worthy person, a true lover of his country, and whose virtues I 17 highly esteem, was lately pleased in discoursing on this matter to offer a refinement upon my scheme. He said that many gentlemen of this kingdom, having of late destroyed their deer, he conceived that the want of venison might be well supplied by the bodies of young lads and maidens, not exceeding fourteen years of age nor under twelve, so great a number of both sexes in every county being now ready to starve for want of work and service; and these to be disposed of by their parents, if alive, or otherwise by their nearest relations. But with due deference to so excellent a friend and so deserving a patriot, I cannot be altogether in his sentiments; for as to the males, my American acquaintance assured me from frequent experience that their flesh was generally tough and lean, like that of our schoolboys, by continual exercise, and their taste disagreeable; and to fatten them would not answer the charge. Then as to the females, it would, I think with humble submission, be a loss to the public, because they soon would become breeders themselves: and besides, it is not improbable that some scrupulous people might be apt to censure such a practice (although indeed very unjustly) as a little bordering upon cruelty; which, I confess, hath always been with me the strongest objection against any project, how well so ever intended.

But in order to justify my friend, he confessed that this expedient was put 18 into his head by the famous Psalmanazar, a native of the island Formosa, who came from thence to London above twenty years ago, and in conversation told my friend that in his country when any young person happened to be put to death, the executioner sold the carcass to persons of quality as a prime dainty; and that in his time the body of a plump girl of fifteen, who was crucified for an attempt to poison the emperor, was sold to his Imperial Majesty's prime minister of state, and other great mandarins of the court, in joints from the gibbet, at four hundred crowns. Neither indeed can I deny that if the same use were made of several plump young girls in this town, who without one single groat to their fortunes cannot stir abroad without a chair, and appear at the playhouse and assemblies in foreign fineries which they never will pay for, the kingdom would not be the worse.

Some persons of a desponding spirit are in great concern about that vast 19 number of poor people who are aged, diseased, or maimed, and I have been desired to employ my thoughts what course may be taken to ease the nation of so grievous an encumbrance. But I am not in the least pain upon that matter, because it is very well known that they are every day dying and rotting by cold and famine, and filth and vermin, as fast as can be reasonably expected. And as to the younger laborers, they are now in almost as hopeful a condition. They cannot get work, and consequently pine away for want of nourishment to a degree that if at any time they are accidentally hired to common labor, they have not strength to perform it; and thus the country and themselves are happily delivered from the evils to come.

I have too long digressed, and therefore shall return to my subject. I think 20 the advantages by the proposal which I have made are obvious and many, as well as of the highest importance.

For first, as I have already observed, it would greatly lessen the number of 21 Papists, with whom we are yearly overrun, being the principal breeders of the nation as well as our most dangerous enemies; and who stay at home on purpose to deliver the kingdom to the Pretender, hoping to take their advantage by the absence of so many good Protestants, who have chosen rather to leave their country than to stay at home and pay tithes against their conscience to an Episcopal curate.

Secondly, the poorer tenants will have something valuable of their own, 22 which by law may be made liable to distress, and help to pay their landlord's rent, their corn and cattle being already seized and money a thing unknown.

Thirdly, whereas the maintenance of an hundred thousand children, from 23 two years old and upwards, cannot be computed at less than ten shillings a piece per annum, the nation's stock will be thereby increased fifty thousand pounds per annum, besides the profit of a new dish introduced to the tables of all gentlemen of fortune in the kingdom who have any refinement in taste. And the money will circulate among ourselves, the goods being entirely of our own growth and manufacture.

Fourthly, the constant breeders, besides the gain of eight shillings sterling 24 per annum by the sale of their children, will be rid of the charge of maintaining them after the first year.

Fifthly, this food would likewise bring great custom to taverns, where the 25 vintners will certainly be so prudent as to procure the best receipts for dressing it to perfection, and consequently have their houses frequented by all the fine gentlemen, who justly value themselves upon their knowledge in good eating; and a skillful cook, who understands how to oblige his guests, will contrive to make it as expensive as they please.

Sixthly, this would be a great inducement to marriage, which all wise na- 26 tions have either encouraged by rewards or enforced by laws and penalties. It would increase the care and tenderness of mothers toward their children, when they were sure of a settlement for life to the poor babes, provided in some sort by the public, to their annual profit instead of expense. We should see an honest emulation among the married women, which of them could bring the fattest child to the market. Men would become as fond of their wives during the time of their pregnancy as they are now of their mares in foal, their cows in calf, or sows when they are ready to farrow; nor offer to beat or kick them (as is too frequent a practice) for fear of a miscarriage.

Many other advantages might be enumerated. For instance, the addition of some thousand carcasses in our exportation of barreled beef, the propagation of swine's flesh, and improvement in the art of making good bacon, so much wanted among us by the great destruction of pigs, too frequent at our tables, which are no way comparable in taste or magnificence to a well-grown, fat yearling child, which roasted whole will make a considerable figure at a lord mayor's feast or any other public entertainment. But this and many others I omit, being studious of brevity. ²⁷

Supposing that one thousand families in this city would be constant customers for infants' flesh, besides others who might have it at merry meetings, particularly weddings and christenings, I compute that Dublin would take off annually about twenty thousand carcasses, and the rest of the kingdom (where probably they will be sold somewhat cheaper) the remaining eighty thousand. ²⁸

I can think of no one objection that will possibly be raised against this proposal, unless it should be urged that the number of people will be thereby much lessened in the kingdom. This I freely own, and it was indeed one principal design in offering it to the world. I desire the reader will observe, that I calculate my remedy for this one individual kingdom of Ireland and for no other that ever was, is, or I think ever can be upon earth. Therefore let no man talk to me of other expedients: of taxing our absentees at five shillings a pound: of using neither clothes nor household furniture except what is of our own growth and manufacture: of utterly rejecting the materials and instruments that promote foreign luxury: of curing the expensiveness of pride, vanity, idleness, and gaming in our women: of introducing a vein of parsimony, prudence, and temperance: of learning to love our country, in the want of which we differ even from Laplanders and the inhabitants of Topinamboo: of quitting our animosities and factions, nor acting any longer like the Jews, who were murdering one another at the very moment their city was taken: of being a little cautious not to sell our country and conscience for nothing: of teaching landlords to have at least one degree of mercy toward their tenants: lastly, of putting a spirit of honesty, industry, and skill into our shopkeepers; who, if a resolution could be now taken to buy only our native goods, would immediately unite to cheat and exact upon us in the price, the measure and the goodness, nor could ever yet be brought to make one fair proposal of just dealing, though often and earnestly invited to it. ²⁹

Therefore I repeat, let no man talk to me of these and the like expedients, till he hath at least some glimpse of hope that there will ever be some hearty and sincere attempt to put them in practice. ³⁰

But as to myself, having been wearied out for many years with offering 31 vain, idle, visionary thoughts, and at length utterly despairing of success, I fortunately fell upon this proposal, which, as it is wholly new, so it hath something solid and real, of no expense and little trouble, full in our own power, and whereby we can incur no danger in disobliging England. For this kind of commodity will not bear exportation, the flesh being of too tender a consistence to admit a long continuance in salt, although perhaps I could name a country which would be glad to eat up our whole nation without it.

After all, I am not so violently bent upon my own opinion as to reject any 32 offer proposed by wise men, which shall be found equally innocent, cheap, easy, and effectual. But before something of that kind shall be advanced in contradiction to my scheme, and offering a better, I desire the author or authors will be pleased maturely to consider two points. First, as things now stand, how they will be able to find food and raiment for an hundred thousand useless mouths and backs. And secondly, there being a round million of creatures in human figure throughout this kingdom, whose sole subsistence put into a common stock would leave them in debt two millions of pounds sterling, adding those who are beggars by profession to the bulk of farmers, cottagers, and laborers, with their wives and children who are beggars in effect; I desire those politicians who dislike my overture, and may perhaps be so bold to attempt an answer, that they will first ask the parents of these mortals whether they would not at this day think it a great happiness to have been sold for food at a year old in the manner I prescribe, and thereby have avoided such a perpetual scene of misfortunes as they have since gone through by the oppression of landlords, the impossibility of paying rent without money or trade, the want of common sustenance, with neither house nor clothes to cover them from the inclemencies of the weather, and the most inevitable prospect of entailing the like or greater miseries upon their breed forever.

I profess, in the sincerity of my heart, that I have not the least personal in- 33 terest in endeavoring to promote this necessary work, having no other motive than the public good of my country, by advancing our trade, providing for infants, relieving the poor, and giving some pleasure to the rich. I have no children by which I can propose to get a single penny; the youngest being nine years old, and my wife past childbearing.

COMPREHENSION

1. Who is Swift's audience for this essay? Defend your answer.
2. Describe the persona in this essay. How is the unusual narrative personality (as distinguished from Swift's personality) revealed by the author in degrees? How can we tell that the speaker's opinions are not shared by Swift?
3. What are the major propositions behind Swift's modest proposal? What are the minor propositions?

RHETORIC

1. Explain the importance of the word *modest* in the title. What stylistic devices does this "modesty" contrast with?
2. What is the effect of Swift's persistent reference to people as "breeders," "dams," "carcass," and the like? Why does he define *children* in economic terms? Find other words that contribute to this motif.
3. Analyze the purpose of the relatively long introduction, consisting of paragraphs 1 to 7. How does Swift establish his ironic-satiric tone in this initial section?
4. What contrasts and discrepancies are at the heart of Swift's ironic statement in paragraphs 9 and 10? Explain both the subtlety and savagery of the satire in paragraph 12.
5. Paragraphs 13 to 20 develop six advantages of Swift's proposal, while paragraphs 21 to 26 list them in enumerative manner. Analyze the progression of these propositions. What is the effect of the listing? Why is Swift parodying argumentative techniques?
6. How does the author both sustain and suspend the irony in paragraph 29? How is the strategy repeated in paragraph 32? How does the concluding paragraph cap his satiric commentary on human nature?

WRITING

1. Discuss Swift's social, political, religious, and economic views as they are revealed in the essay.
2. Write a comprehensive critique of America's failure to address the needs of its poor.
3. **WRITING AN ARGUMENT:** Write a modest proposal—on, for example, how to end the drug problem—advancing an absurd proposition through various argumentative techniques.

www.mhhe.com/
mhreader

For more information on Jonathan Swift, go to:
More Resources > Ch. 8 Business & Economics

CONNECTIONS FOR CRITICAL THINKING

1. Using the essays of Quindlen, Ehrenreich, and Rodriguez, compare the effects of work on human relationships.
2. Compare and contrast the social context and opportunities of the workers in the essay by Rodriguez with the workers in the essay by Ehrenreich.
3. Write a definition essay entitled "What Is Work?" Refer to any of the selections in this chapter to substantiate your opinions.
4. Describe the potential effect of the global marketplace as described by Reich and Friedman on the "visiting father" mentioned by Quindlen.
5. Compare the writings of Swift, Rodriguez, Ehrenreich, and Reich in terms of the options of those on the lowest rungs of the economic system in Western society.
6. Locate the Web sites of three major newspapers in three different large cities in America. Review their classified sections, and compare and contrast the types of jobs advertised in the three cities. Do a similar search and comparison for federal and state government jobs.
7. Examine your own college or university's Web site, and review its philosophy regarding the relationship of college studies to the world of work.
8. Join a newsgroup with a special interest in the global economy. Post a general question to its members, asking whether they agree with Reich's analysis of the changing job market. Collect the responses.
9. Woolf and Swift are considered to be "classic" writers. What makes their essays "classics"?
10. Monitor and videotape three business news television shows that focus on analysis and commentary. Analyze the discourse of the moderators, hosts, and guests. As an alternative method, review selected business news Web sites, and subject them to the same process.
11. Locate several Web sites for job seekers, for example, Monster.com. Enter the job classification you are interested in, and compare and contrast the number and types of jobs advertised for three cities. Do a similar search and comparison for federal and state government jobs advertised on the Web.
12. To what extent is American society guided by a business ethic? For example, is there a historical continuum in which Americans have been so preoccupied with wealth that the quest for money actually becomes a distinguishing mark of the national character? Or do you think that the merging of business and religion (what social scientists term the Protestant work ethic) captures a uniquely American trait? Discuss these issues in an analytical and argumentative essay.

chapter *9*

Media and Popular Culture
What Is the Message?

We are surrounded today as never before by images, sounds, and texts—by what Todd Gitlin, who has an essay in this chapter, terms a "media torrent." Radio and television programs, newspapers and Internet sites, MTV and video games, iPods and cell-phone gadgetry increasingly mold our place in culture and society. Indeed, the power of media in our waking and subliminal lives might very well condition our understanding (or misunderstanding) of reality. Today, we can download "reality."

Today's media universe, fueled by new technology, is transforming our sense of the world. Consider the ways in which computers permit us to enter the media stream, making us willing, even compulsive participants and consumers of popular culture. Video games, streaming advertisements, wraparound music, newsgroups, chat rooms, and more—all provide data and sensation at warp computer speed. Some slow down the torrent: Bloggers (a word that didn't exist until recently) interrogate their lives and the "facts," even holding newspapers and television news channels accountable for information. But if, as Marshall McLuhan declared, the medium is the message, then any medium, whether old or new, has the power to reflect or construct versions of reality.

Perhaps Americans have moved from a print-based culture to an aural/visual one, preferring electronic media for information, distraction, and entertainment. For centuries, books were the molders of popular taste and culture. Tocqueville in *Democracy in America* was amazed by the fact that in the rudest pioneer's hut there was a copy of Shakespeare—and probably, we might add, a copy of the best seller of all time, the Bible. Today's typical household might have more media for DVD players, VCRs, MP3 players, and PlayStations than books in a library. Of course, print media—let's say a book on the ways in which major political parties manipulate the media—offer us the opportunity to scrutinize facts and sources in ways that shock jocks on radio and television or participants in chat rooms cannot. The *medium* or *source* from which we receive sounds, images, and text—the place from which we enter the media torrent—determines the version of reality we carry with us. We can even drown in this torrent, as people who have been captured in "virtual reality" can attest.

The writers in this chapter invite us to enter the media torrent from a variety of places. They ask broad cultural questions about the ways we conduct our everyday lives and the choices we make. These questions have both local and worldwide implications, because media and their technological helpmates have created a global village permitting the instant transmission of ideas and images, as well as a subtle transfer of culture—typically American—to the remotest parts of the planet. Whether we navigate the torrent intelligently or succumb passively to the images and sounds washing over us, it is clear that the media in this century will have a significant impact on human experience.

Previewing the Chapter

As you read the essays in this chapter and respond to them in discussion and writing, consider the following questions:

- What is the main media form and issue on which the author focuses?
- What tone or attitude does the author take toward the subject?
- How does the author fit his or her analysis into the context of popular culture? What social, economic, psychological, or political problems or controversies are treated?
- Which areas of expertise do the author bring to bear on the subject?
- What form of rhetoric—narration and description, exposition, argument—does the writer use? If argument, what is the author's claim? What evidence does the writer provide to support this claim?
- What essays are similar in subject, thesis, style, purpose, or method, and why?
- What have you learned about the media and popular culture from reading these essays?
- Which essays did you find the most compelling or persuasive? The least? Why?

Classic and Contemporary Images

WHAT DO GANGSTER FILMS REVEAL ABOUT US?

Using a Critical Perspective   The best gangster films—like *Little Caesar, The Godfather,* and the *Sopranos* series—challenge viewers to form ethical opinions about and interpretations of the tale of crime that unfolds. Even if you haven't seen these two films, what do you think is happening in each frame? What details do you focus on? Do the two characters capture the essence of the gangster life, and why? What aspects of film art—framing, the use of close-up or distance shots, the handling of light, shadow, and color—convey an ethical statement? More broadly, in what ways can film art serve as a commentary on American life?

During the 1930s, the first decade of sound films, actors such as
Edward G. Robinson, James Cagney, and Paul Muni created the
classic portrait of the gangster as a tough-talking, violent outlaw
in films such as *Little Caesar* (1930), *The Public Enemy* (1931),
and *Scarface* (1932).

HBO's award-winning program *The Sopranos* provides a more nuanced
portrait of the gangster as a member of a corrupt, and corrupting,
organization with shifting loyalties. For example, Tony Soprano,
the character played by James Gandolfini in the series,
is a family man beset by so many problems that
he must seek psychiatric help.

Classic and Contemporary Essays
WHY ARE WE FASCINATED BY GANGSTERS?

Both Robert Warshow and Ellen Willis examine the portrayal of the gangster in modern culture, although each has a somewhat different interpretive slant. Warshow sees the gangster as a "tragic hero" at odds with—although at times admired by—our value system. Willis, on the other hand, sees him as a representation of a flawed mythology, a world doomed from the start owing to its own internal inadequacies. Warshow demonstrates that the gangster is ill-fated because he is compelled to operate in an "outlawed" system that ultimately must be squashed by the dominant one. Willis demonstrates—through her analysis of the HBO series *The Sopranos*—that the gangster is destined to failure because at heart his system is intrinsically corrupt, and it is the very system to which he has pledged his allegiance that is responsible for his anxiety. Both critics acknowledge that the mythological gangster exists outside of "real time," and although he may be surrounded by family and acquaintances—in fact, embroiled with them in an alternative social structure—he is ultimately alone. Warshow's gangster is doomed because he must die, because as the author states, "The gangster's whole life is an effort to assert himself as an individual, to draw himself out of the crowd, and he always dies *because* he is an individual." For Willis, Tony Soprano, while always conscious of the reality of death and in fact perpetrating it against his adversaries, wills himself to survive in a split culture. Tony's dilemma, which causes him to seek psychiatric help, is that he exists simultaneously in a "dying tribal culture" and in mainstream American popular culture. If gangsters are doomed, the reason is that in the end, corruption, by its very nature, cannot renew itself. Each of these highly articulate essays demonstrates that an observant writer can glean issues of major moral significance from what the general public views as mere "entertainment."

✳

The Gangster as Tragic Hero

Robert Warshow

Robert Warshow (1917–1955) attended the University of Michigan and worked for the U.S. Army Security Agency from 1942 to 1946. After the war, he served as an editor of Commentary, *writing film criticism for this magazine and also for* Partisan Review. *Before his untimely death from a heart attack, Warshow had written several brilliant essays on film and on popular culture. Writing of Warshow, Lionel Trilling observed, "I believe that certain of his pieces establish themselves in the line of Hazlitt, a tradition in which I would place only one other writer of our time, George Orwell." One of these brilliant essays, focusing on the interrelation of film and society, is "The Gangster as Tragic Hero,"* which appeared in The Immediate Experience: Movies, Games, Theatre and Other Aspects of Popular Culture *(1962).*

America, as a social and political organization, is committed to a cheerful view   1
of life. It could not be otherwise. The sense of tragedy is a luxury of aristocratic societies, where the fate of the individual is not conceived of as having a direct and legitimate political importance, being determined by a fixed and supra-political—that is, non-controversial—moral order or fate. Modern equalitarian societies, however, whether democratic or authoritarian in their political forms, always base themselves on the claim that they are making life happier; the avowed function of the modern state, at least in its ultimate terms, is not only to regulate social relations, but also to determine the quality and possibilities of human life in general. Happiness thus becomes the chief political issue—in a sense, the only political issue—and for that reason it can never be treated as an issue at all. If an American or a Russian is unhappy, it implies a certain reprobation of his society, and therefore, by a logic of which we can all recognize the necessity, it becomes an obligation of citizenship to be cheerful; if the authorities find it necessary, the citizen may even be compelled to make a public display of his cheerfulness on important occasions, just as he may be conscripted into the army in time of war.

Naturally, this civic responsibility rests most strongly upon the organs of   2
mass culture. The individual citizen may still be permitted his private unhappiness so long as it does not take on political significance, the extent of this tolerance being determined by how large an area of private life the society can accommodate. But every production of mass culture is a public act and must conform with accepted notions of the public good. Nobody seriously questions the principle that it is the function of mass culture to maintain public morale, and certainly nobody in the mass audience objects to having his morale

maintained.[1] At a time when the normal condition of the citizen is a state of anxiety, euphoria spreads over our culture like the broad smile of an idiot. In terms
of attitudes towards life, there is very little difference between a "happy" movie
like *Good News,* which ignores death and suffering, and a "sad" movie like *A
Tree Grows in Brooklyn,* which uses death and suffering as incidents in the service of a higher optimism.

But, whatever its effectiveness as a source of consolation and a means of 3
pressure for maintaining "positive" social attitudes, this optimism is fundamentally satisfying to no one, not even to those who would be most disoriented
without its support. Even within the area of mass culture, there always exists a
current of opposition, seeking to express by whatever means are available to it
that sense of desperation and inevitable failure which optimism itself helps to
create. Most often, this opposition is confined to rudimentary or semi-literate
forms: in mob politics and journalism, for example, or in certain kinds of religious enthusiasm. When it does enter the field of art, it is likely to be disguised
or attenuated: in an unspecific form of expression like jazz, in the basically
harmless nihilism of the Marx Brothers, in the continually reasserted strain of
hopelessness that often seems to be the real meaning of the soap opera. The
gangster film is remarkable in that it fills the need for disguise (though not sufficiently to avoid arousing uneasiness) without requiring any serious distortion.
From its beginnings, it has been a consistent and astonishingly complete presentation of the modern sense of tragedy.[2]

In its initial character, the gangster film is simply one example of the 4
movies' constant tendency to create fixed dramatic patterns that can be repeated
indefinitely with a reasonable expectation of profit. One gangster film follows
another as one musical or one Western follows another. But this rigidity is not
necessarily opposed to the requirements of art. There have been very successful
types of art in the past which developed such specific and detailed conventions
as almost to make individual examples of the type interchangeable. This is true,
for example, of Elizabethan revenge tragedy and Restoration comedy.

For such a type to be successful means that its conventions have imposed 5
themselves upon the general consciousness and become the accepted vehicles
of a particular set of attitudes and a particular aesthetic effect. One goes to any
individual example of the type with very definite expectations, and originality
is to be welcomed only in the degree that it intensifies the expected experience

---

[1]In her testimony before the House Committee on Un-American Activities, Mrs. Leila Rogers said
that the movie *None But the Lonely Heart* was un-American because it was gloomy. Like so much
else that was said during the unhappy investigation of Hollywood, this statement was at once stupid and illuminating. One knew immediately what Mrs. Rogers was talking about; she had simply
been insensitive enough to carry her philistinism to its conclusion. [This and subsequent notes in
the selection are the author's.]

[2]Efforts have been made from time to time to bring the gangster film into line with the prevailing
optimism and social constructiveness of our culture; *Kiss of Death* is a recent example. These efforts
are usually unsuccessful; the reasons for their lack of success are interesting in themselves, but I
shall not be able to discuss them here.

without fundamentally altering it. Moreover, the relationship between the conventions which go to make up such a type and the real experience of its audience or the real facts of whatever situation it pretends to describe is of only secondary importance and does not determine its aesthetic force. It is only in an ultimate sense that the type appeals to its audience's experience of reality; much more immediately, it appeals to previous experience of the type itself: it creates its own field of reference.

Thus the importance of the gangster film, and the nature and intensity of its 6
emotional and aesthetic impact, cannot be measured in terms of the place of the gangster himself or the importance of the problem of crime in American life. Those European moviegoers who think there is a gangster on every corner in New York are certainly deceived, but defenders of the "positive" side of American culture are equally deceived if they think it relevant to point out that most Americans have never seen a gangster. What matters is that the experience of the gangster *as an experience of art* is universal to Americans. There is almost nothing we understand better or react to more readily or with quicker intelligence. The Western film, though it seems never to diminish in popularity, is for most of us no more than the folklore of the past, familiar and understandable only because it has been repeated so often. The gangster film comes much closer. In ways that we do not easily or willingly define, the gangster speaks for us, expressing that part of the American psyche which rejects the qualities and the demands of modern life, which rejects "Americanism" itself.

The gangster is the man of the city, with the city's language and knowledge, 7
with its queer and dishonest skills and its terrible daring, carrying his life in his hands like a placard, like a club. For everyone else, there is at least the theoretical possibility of another world—in that happier American culture which the gangster denies, the city does not really exist; it is only a more crowded and more brightly lit country—but for the gangster there is only the city; he must inhabit it in order to personify it: not the real city, but that dangerous and sad city of the imagination which is so much more important, which is the modern world. And the gangster—though there are real gangsters—is also, and primarily, a creature of the imagination. The real city, one might say, produces only criminals; the imaginary city produces the gangster: he is what we want to be and what we are afraid we may become.

Thrown into the crowd without background or advantages, with only those 8
ambiguous skills which the rest of us—the real people of the real city—can only pretend to have, the gangster is required to make his way, to make his life and impose it on others. Usually, when we come upon him, he has already made his choice or the choice has already been made for him, it doesn't matter which: we are not permitted to ask whether at some point he could have chosen to be something else than what he is.

The gangster's activity is actually a form of rational enterprise, involving, 9
fairly definite goals and various techniques for achieving them. But thus rationality is usually no more than a vague background; we know, perhaps, that the gangster sells liquor or that he operates a numbers racket; often we are not

given even that much information. So his activity becomes a kind of pure criminality: he hurts people. Certainly our response to the gangster film is most consistently and most universally a response to sadism; we gain the double
satisfaction of participating vicariously in the gangster's sadism and then seeing it turned against the gangster himself.

But on another level the quality of irrational brutality and the quality of 10
rational enterprise become one. Since we do not see the rational and routine
aspects of the gangster's behavior, the practice of brutality—the quality of
unmixed criminality—becomes the totality of his career. At the same time, we
are always conscious that the whole meaning of this career is a drive for success:
the typical gangster film presents a steady upward progress followed by a very
precipitate fall. Thus brutality itself becomes at once the means to success and
the content of success—a success that is defined in its most general terms, not as
accomplishment or specific gain, but simply as the unlimited possibility of aggression. (In the same way, film presentations of businessmen tend to make
it appear that they achieve their success by talking on the telephone and holding conferences and that success *is* talking on the telephone and holding
conferences.)

From this point of view, the initial contact between the film and its audience 11
is an agreed conception of human life: that man is a being with the possibilities
of success or failure. This principle, too, belongs to the city; one must emerge
from the crowd or else one is nothing. On that basis the necessity of the action
is established, and it progresses, by inalterable paths to the point where the
gangster lies dead and the principle has been modified: there is really only one
possibility—failure. The final meaning of the city is anonymity and death.

In the opening scene of *Scarface*, we are shown a successful man; we know 12
he is successful because he has just given a party of opulent proportions and
because he is called Big Louie. Through some monstrous lack of caution, he
permits himself to be alone for a few moments. We understand from this immediately that he is about to be killed. No convention of the gangster film is
more strongly established than this: it is dangerous to be alone. And yet the
very conditions of success make it impossible not to be alone, for success is always the establishment of an *individual* pre-eminence that must be imposed on
others, in whom it automatically arouses hatred; the successful man is an outlaw. The gangster's whole life is an effort to assert himself as an individual, to
draw himself out of the crowd, and he always dies *because* he is an individual;
the final bullet thrusts him back, makes him, after all, a failure. "Mother of
God," says the dying Little Caesar, "is this the end of Rico?"—speaking of himself thus in the third person because what has been brought low is not the
undifferentiated *man,* but the individual with a name, the gangster, the success;
even to himself he is a creature of the imagination. (T. S. Eliot has pointed out
that a number of Shakespeare's tragic heroes have this trick of looking at themselves dramatically; their true identity, the thing that is destroyed when they
die, is something outside themselves—not a man, but a style of life, a kind of
meaning.)

At bottom, the gangster is doomed because he is under the obligation to 13 succeed, not because the means he employs are unlawful. In the deeper layers of the modern consciousness, *all* means are unlawful, every attempt to succeed is an act of aggression, leaving one alone and guilty and defenseless among enemies: one is *punished* for success. This is our intolerable dilemma: that failure is a kind of death and success is evil and dangerous, is—ultimately—impossible. The effect of the gangster film is to embody this dilemma in the person of the gangster and resolve it by his death. The dilemma is resolved because it is *his* death, not ours. We are safe; for the moment, we can acquiesce in our failure, we can choose to fail.

COMPREHENSION

1. What are the "organs of mass culture" (paragraph 2)? What properties do they all have in common?
2. Define the term *tragic hero* as Warshow uses it in his title.
3. Compare and contrast Warshow's concepts of the "real city" with those of the "imaginary city" as they relate to modern life and mass culture.

RHETORIC

1. Although the ultimate focus of the essay is on the "gangster," the subject is not referred to until paragraph 4. Why does the author need so much exposition before focusing on his main topic?
2. What do terms such as *supra-political* (paragraph 1), *harmless nihilism* (paragraph 3), and *general consciousness* (paragraph 5) suggest about the tone of the essay? What do they imply concerning the target audience for the essay?
3. Study the topic sentence of each paragraph. Are the topic sentences successful in setting up the material that follows? How does this strategy enhance or detract from the coherence of the author's argument?
4. Essayists usually provide their thesis at the beginning of their essays. Where does Warshow provide the thesis in his essay? What is the purpose and effect of placing it where it is?
5. The author explains the nature of the gangster film genre—its function, characters, themes, plots, meanings, and so on—*before* he cites specific films. Is this a rhetorical weakness in the essay, or does it give the essay particular potency? Explain.
6. Does the conclusion summarize the main points of the essay, bolster them, provide new insights into them, or does it do a combination of these things? Explain.
7. Study the introductory paragraph and the conclusion. What themes are reiterated or complemented? How do these two paragraphs serve to provide both thematic and structural coherence?

WRITING

1. Select a genre of television show or movie. Analyze its conventions and to what degree these conventions transgress the implicit values of our society.
2. Select a contemporary gangster movie or television program. Using Warshow's criteria, demonstrate—via reference to its characters, plot, and theme—how your selection reinforces the author's thesis.
3. **WRITING AN ARGUMENT:** Argue for or against the proposition that genre movies are a form of escapism that distorts the individual's concept of the actual society he or she lives in and its citizens.

---

www.mhhe.com/
mhreader

For more information on Robert Warshow, go to:
More Resources > Ch. 9 Media & Pop Culture

✴

Our Mobsters, Ourselves:
Why *The Sopranos* Is Therapeutic TV

Ellen Willis

Ellen Willis (b. 1941) was born in New York City; she received a BA from Barnard College in 1962 and was enrolled for graduate study at the University of California at Berkeley in 1962–63. Willis has been a freelance writer and editor for four decades, contributing articles on the media and popular culture to the New Yorker, Ms., Rolling Stone, Village Voice, *and other periodicals. She also has been a fellow at the Nation Institute and has directed the cultural reporting and criticism program at New York University. Her books include* Beginning to See the Light: Pieces of a Decade *(1981),* Beginning to See the Light: Sex, Hope, and Rock-and-Roll *(1992),* No More Nice Girls: Countercultural Essays *(1992), and* Don't Think, Smile! Notes on a Decade of Denial *(1999). In the following essay, published in* The Nation *in 2001, Willis places the popular HBO series* The Sopranos *in a broad cultural context.*

Midway through the first season of *The Sopranos*, the protagonist's psychother-   1
apist, Jennifer Melfi, has a not-exactly-traditional family dinner with her middle-class Italian parents, son and ex-husband Richard. She lets slip (hmm!) that one of her patients is a mobster, much to Richard's consternation. An activist in Italian anti-defamation politics, he is incensed at the opprobrium the Mafia has brought on all Italians. What is the point, he protests, of trying to help such a

person? In a subsequent scene he contemptuously dismisses Jennifer and her profession for purveying "cheesy moral relativism" in the face of evil. His challenge boldly proclaims what until then has been implicit: The richest and most compelling piece of television—no, of popular culture—that I've encountered in the past twenty years is a meditation on the nature of morality, the possibility of redemption and the legacy of Freud.

To be sure, *The Sopranos* is much else as well. For two years (the third season 2 began March 4) David Chase's HBO series has served up a hybrid genre of post-Godfather decline-of-the-mob movie and soap opera, with plenty of sex, violence, domestic melodrama and comic irony; a portrait of a suburban landscape that does for northern New Jersey what film noir did for Los Angeles, with soundtrack to match; a deft depiction of class and cultural relations among various subgroups and generations of Italian-Americans; a gloss on the manners and mores of the fin-de-siècle American middle-class family; and perfect-pitch acting, especially by James Gandolfini as Tony Soprano; Edie Falco as his complicated wife, Carmela; Lorraine Bracco as Dr. Melfi; and the late Nancy Marchand as the Sopranos' terrifying matriarch, Livia.

Cumulatively, these episodes have the feel of an as yet unfinished nineteenth- 3 century novel. While the sheer entertainment and suspense of the plot twists are reminiscent of Dickens and his early serials, the underlying themes evoke George Eliot: The world of Tony Soprano is a kind of postmodern *Middlemarch*, whose inhabitants' moral and spiritual development (or devolution) unfolds within and against the norms of a parochial social milieu. This era being what it is, however, the Sopranos' milieu has porous boundaries, and the norms that govern it are a moving target. In one scene, the family is in mid-breakfast when Tony and Carmela's teenage daughter, Meadow, apropos a recent scandal brought on by a high school classmate's affair with her soccer coach, declaims about the importance of talking openly about sex. Yes, Tony agrees, but not during breakfast. "Dad, this is the 1990s," Meadow protests. "Outside it may be the 1990s," Tony retorts, "but in this house it's 1954." It's wishful thinking, and Tony knows it. What 1950s gangster would take Prozac and make weekly visits to a shrink or, for that matter, have a daughter named Meadow?

In fact, contemporary reality pervades the Sopranos' suburban manse. A 4 school counselor tries to persuade them that their son, Anthony Jr., has attention deficit disorder. Meadow hosts a clandestine party in her grandmother's empty house that gets busted for drugs and alcohol. Tony's sister Janice, who years ago decamped to Seattle, became a Buddhist and changed her name to Parvati, shows up at his door flaunting her postcounterculture reinvented self. And while Tony displays some of the trappings of the stereotypical Italian patriarch—he is proud of supporting his family in style, comes and goes as he pleases, leaves the running of the household to Carmela and cheats on her with the obligatory goomah—his persona as fear-inspiring gangster does not translate to his home life. Carmela is his emotional equal; she does what she likes, tells him off without hesitation and, unlike old-style mob wives, knows plenty about the business. Nor, despite periodic outbursts of temper, is Tony an intimidating father. Caught between empathy for their children and the urge to whip them into line, the Sopranos share the dirty little secret of nineties middle-class parenthood: You can't control teenagers' behavior without becoming full-time prison guards. "Let's not overplay our hand," Tony cautions after Meadow's party caper, "'cause if she knows we're powerless, we're fucked."

In Tony's other "house"—represented by his office in the Bada Bing strip   5 club—1954 is also under siege. Under pressure of the RICO laws, longtime associates turn government witness. Neophytes chafe at their lowly status in the hierarchy, disobey their bosses, take drugs, commit gratuitous freelance crimes and in general fail to understand that organized crime is a business, not a vehicle for self-expression or self-promotion. The line between reality and media image has become as tenuous here as elsewhere: Tony and his men love *GoodFellas* and the first two *Godfathers* (by general agreement III sucks) and at the same time are objects of fantasy for civilians steeped in the same movies. Tony accepts an invitation to play golf with his neighbor Dr. Cusamano, who referred him to Melfi, and finds that his function is to titillate the doctor's friends; during a falling out with Jennifer he tries to connect with another therapist, who demurs, explaining that he has seen *Analyze This* ("It's a fucking comedy," Tony protests). Tony's fractious nephew Christopher, pissed because press coverage of impending mob indictments doesn't mention him, reprises *GoodFellas* by shooting an insufficiently servile clerk in the foot. He aspires to write screenplays about mob life, and in pursuit of this dream is used for material and kicks by a Hollywood film director and his classy female assistant. Meanwhile Jennifer's family debates whether wiseguy movies defame Italians or rather should be embraced as American mythology, like westerns. *The Sopranos*, of course, has provoked the same argument, and its continual reflection of its characters in their media mirrors is also a running commentary on the show itself.

Self-consciousness, then, is a conspicuous feature of Tony Soprano's world   6 even aside from therapy; in fact, it's clear that self-consciousness has provoked the anxiety attack that sends him to Jennifer Melfi. It's not just a matter of stressful circumstances. Tony's identity is fractured, part outlaw rooted in a dying tribal culture, part suburbanite enmeshed in another kind of culture altogether—a split graphically exemplified by the famous episode in which Tony, while taking Meadow on a tour of colleges in Maine, spots a mobster-turned-informer hiding in the witness protection program and manages to juggle his fatherly duties with murder. Despite his efforts at concealment, his criminal life is all too evident to his children (after all, they too have seen *The Godfather*), a source of pain and confusion on both sides. Tony's decision to seek therapy also involves an identity crisis. In his first session, which frames the first episode, he riffs on the sad fate of the strong and silent Gary Cooper: Once they got him in touch with his feelings, he wouldn't shut up. "I have a semester and a half of college," he tells Dr. Melfi, "so I understand Freud. I understand therapy as a concept, but in my world it does not go down." In his wiseguy world, that is: Carmela thinks it's a great idea.

Richard Melfi's charge of moral relativism is highly ironic, for Jennifer finds 7
that her task is precisely to confront the tribal relativism and cognitive disso-
nance that keep Tony Soprano from making sense of his life. He sees his business as the Sicilians' opportunity to get in on the American Dream, the violence
that attends it as enforcement of rules known to all who choose to play the
game: Gangsters are soldiers, whose killing, far from being immoral, is impelled
by positive virtues—loyalty, respect, friendship, willingness to put one's own
life on the line. It does not strike Tony as inconsistent to expect his kids to behave or to send them to Catholic school, any more than he considers that nights
with his Russian girlfriend belie his reverence for the institution of the family.
Nor does he see a contradiction in his moral outrage at a sadistic, pathologically
insecure associate who crushes a man with his car in fury over an inconsequential slight.

In its original literal sense, "moral relativism" is simply moral complexity. 8
That is, anyone who agrees that stealing a loaf of bread to feed one's children is
not the moral equivalent of, say, shoplifting a dress for the fun of it, is a relativist
of sorts. But in recent years, conservatives bent on reinstating an essentially religious vocabulary of absolute good and evil as the only legitimate framework
for discussing social values have redefined "relative" as "arbitrary." That con-
flation has been reinforced by social theorists and advocates of identity politics
who argue that there is no universal morality, only the value systems of particular cultures and power structures. From this perspective, the psychoanalytic—
and by extension the psychotherapeutic—worldview is not relativist at all.

Its values are honesty, self-knowledge, assumption of responsibility for the 9
whole of what one does, freedom from inherited codes of family, church, tribe
in favor of a universal humanism: in other words, the values of the Enlightenment, as revised and expanded by Freud's critique of scientific rationalism for
ignoring the power of unconscious desire. What eludes the Richard Melfis is
that the neutral, unjudging stance of the therapist is not an end in itself but a
strategy for pursuing this moral agenda by eliciting hidden knowledge.

Predictably, the cultural relativists have no more use for Freud than the reli- 10
gious conservatives. Nor are the devotees of "rational choice" economics and of
a scientism that reduces all human behavior to genes or brain chemistry eager to
look below the surface of things, or even admit there's such a thing as "below the
surface." Which is why, in recent years, psychoanalysis has been all but banished
from the public conversation as a serious means of discussing our moral and
cultural and political lives. And as the Zeitgeist goes, so goes popular culture:
Though a continuing appetite for the subject might be inferred from the popularity of memoirs, in which psychotherapy is a recurring theme, it has lately been
notably absent from movies and television. So it's more than a little interesting
that *The Sopranos* and *Analyze This* plucked the gangster-sees-therapist plot from
the cultural unconscious at more or less the same time and apparently by coincidence. In *The Sopranos,* however, therapy is no fucking comedy, nor does it recycle old Hollywood cliches about shamanlike shrinks and sudden cathartic cures.

It's a serious battle for a man's soul, carried on in sessions that look and sound a lot like the real thing (at least as I've experienced it)—full of silence, evasive chatter, lies, boredom and hostility, punctuated by outbursts of painful emotion, moments of clarity and insights that almost never sink in right away. Nor is it only the patient's drama; the therapist is right down there in the muck, sorting out her own confusions, missteps, fantasies and fears, attraction and repulsion, as she struggles to understand.

The parallels between psychotherapy and religion are reinforced by the ad- 11 ventures of the other *Sopranos* characters, who are all defined by their spiritual state. Some are damned, like Livia, whose nihilism is summed up in her penchant for smiling at other people's misfortunes and in her bitter remark to her grandson, "It's all a big nothing. What makes you think you're so special?" Some are complacent, like the respectable bourgeois Italian-Americans, or the self-regarding but fatally unself-aware Father Phil, Carmela's young spiritual adviser, who feeds (literally as well as metaphorically) on the neediness of the mob wives. The older, middle-level mobsters see themselves as working stiffs who expect little from life and for whom self-questioning is a luxury that's out of their class. (One of them is temporarily jolted when Tony's nephew Christopher is shot and has a vision of himself in hell; but the crisis passes quickly.) Charmaine Bucco, a neighborhood girl and old friend of Carmela's who with her husband, Artie, owns an Italian restaurant, is the embodiment of passionate faith in the virtues of honesty, integrity and hard work; she despises the mobsters, wishes they would stop patronizing the restaurant and does her best to pull the ambivalent Artie away from his longtime friendship with Tony. And then there are the strugglers, like Christopher, who inchoately wants something more out of life but also wants to rise in the mob, and Big Pussy, Tony's close friend as well as crew member, who rats to the Feds to ward off a thirty-year prison term, agonizes over his betrayal and ultimately takes refuge in identifying with his FBI handlers.

Carmela Soprano is a struggler, an ardent Catholic who feels the full weight 12 of her sins and Tony's and lets no one off the hook. She keeps hoping Tony will change but knows he probably will not; and despite the many discontents of her marriage, anger at Tony's infidelity and misgivings about her complicity in his crimes, she will not leave him. Though she rationalizes her choice on religious grounds ("The family is a sacred institution"), she never really deceives herself: She still loves Tony, and furthermore she likes the life his money provides. Nor does she hesitate to trade on his power in order to do what she feels is a mother's duty: She intimidates Cusamano's lawyer sister-in-law into writing Meadow a college recommendation. Guilt and frustration drive her to Father Phil, who gives her books on Buddhism, foreign movies and mixed sexual signals, but after a while she catches on to his bullshit, and in a scene beloved of *Sopranos* fans coolly nails him: "He's a sinner, Father. You come up here and you eat his steaks and use his home entertainment center . . . I think you have this MO where you manipulate spiritually thirsty women, and I think a lot of it's tied up with food somehow, as well as the sexual tension game." Compromised as she is, Carmela is a moral touchstone because of her clear eye.

But Tony's encounters with Melfi are the spiritual center of the show. The 13
short version of Tony's psychic story is this: His gangster persona provides him with constant excitement and action, a sense of power and control, a definition of masculinity. Through violence rationalized as business or impersonal soldiering he also gets to express his considerable unacknowledged rage without encroaching on his alter ego as benevolent husband and father. But when the center fails to hold, the result is panic, then—as Melfi probes the cracks—depression, self-hatred, sexual collapse and engulfing, ungovernable anger. There are glimmers along the way, as when Tony sees the pointlessness of killing the sexually wayward soccer coach, calls off the hit and lets the cops do their job (after which he feels impelled to get so drunk he passes out). But the abyss always looms.

Tony's heart of darkness is personified by Livia Soprano, who at first seems 14
peggable as a better-done-than-usual caricature of the overbearing ethnic mother but is gradually revealed as a monstrous Medea. Furious at Tony for consigning her to a fancy "retirement community," Livia passes on some well-chosen pieces of information—including the fact that he's seeing a shrink—to Tony's malleable Uncle Junior, who orders him killed. When the hit is botched, she suddenly begins to show symptoms of Alzheimer's. Jennifer Melfi puts it together; worried that Tony's life is in danger, she breaks the therapeutic rule that patients must make their own discoveries and confronts him with her knowledge. He reacts with a frightening, hate-filled paroxysm of denial—for the first time coming close to attacking Jennifer physically—but is forced to admit the truth when he hears a damning conversation between Livia and Junior, caught on tape by the FBI.

This is a turning point in the story, but not, as the standard psychiatric 15
melodrama would have it, because the truth has made Tony free. The truth has knocked him flat. "What kind of person can I be," he blurts to Carmela, "where his own mother wants me dead?" Afraid that Junior will go after Jennifer, he orders her to leave town; when she comes back she is angry and fearful and tells him to get out of her life. He is lost, his face a silent Munchian scream. Later Jennifer has a change of heart, but things are not the same: The trust is gone. And yet, paradoxically, her rejection has freed him to be more honest, throwing the details of his gang's brutality in her face, railing at her for making him feel like a victim, at himself for becoming the failed Gary Cooper he once mocked, at the "happy wanderers" who still seem in control.

Jennifer encourages him to feel the sadness under the rage, but what comes 16
through is hard and bleak. He tells anyone who mentions his mother, "She's dead to me," but it's really he who feels dead. During this time, Anthony Jr. shocks his mother by announcing that God is dead; "Nitch" says so. (At its most serious, the show never stops being funny.) Tony mentions this to Jennifer, who gives him a minilecture on existential angst: When some people realize they're

solely responsible for their lives, and all roads lead to death, they feel "intense dread" and conclude that "the only absolute truth is death." "I think the kid's onto something," Tony says.

As if to validate Richard Melfi's contempt, he uses what he's learned in therapy—that you can't compartmentalize your life—to more fully accept his worst impulses. Against his more compassionate instincts, he allows an old friend who is the father of a classmate of Meadow's and a compulsive gambler to join his high-stakes card game. When David inevitably piles up a debt he can't pay, Tony moves in on his business, sucking it dry and draining his son's college fund. Amid a torrent of self-pity, David asks why Tony let him in the game. Tony answers jocularly that it's his nature—you know, as in the tale of the frog and the scorpion. In the last episode of season two Tony whacks Pussy, whose perfidy has been revealed, choosing his mob code over his love and sorrow for the man. He then walks out on Jennifer, as if to say, this is who I am and will be.

Jennifer's trip is also a rocky one. In her person, the values of Freud and the Enlightenment are filtered through the cultural radical legacy of the 1960s: She is a woman challenging a man whose relationship to both legitimate and outlaw patriarchal hierarchies is in crisis. It's a shaky and vulnerable role, the danger of physical violence an undercurrent from the beginning, but there are also bonds that make the relationship possible. Tony chooses her over a Jewish male therapist because "you're a paisan, like me," and she is drawn to the outlaw, no doubt in rebellion against the safe smugness of her own social milieu. Predictably, Tony loses all sexual interest in his wife and girlfriend and falls in love with his doctor (if there is any answering spark, it stays under the professional surface), but after the initial "honeymoon" of therapy, trouble, as always, begins. Tony gives Jennifer "gifts" like stealing her car and getting it fixed; it's his way of assuring her, and himself, that his power is benevolent, but of course she only feels violated. Wanting to find out about her life, he has her followed by a corrupt con who harasses her boyfriend, thinking he's doing Tony a favor; she can't help but be suspicious. By inviting her family to object to her criminal patient, she gives voice to her own doubts: Perhaps she is not only endangering herself but abetting evil.

Her conflict intensifies when she tells Tony she must charge for a missed session, and he throws the money at her, calling her a whore. It explodes in the aftermath of the attempt on his life. But then the other side of her ambivalence reasserts itself; she feels she has irresponsibly abandoned a patient and takes him back against the advice of her own (Jewish male) therapist. Now it is Jennifer who is in crisis, treating her anxiety with heavy drinking. She is frightened and morally repulsed by Tony's graphic revelations, yet also feels an erotically tinged fascination (it's like watching a train wreck, she tells her shrink). She still cares about Tony but seems to have lost faith in her ability to exorcise the demonic by making contact with the suffering human being. In the last episode, with Tony closed as a clam, she admits that she blew it, that she stopped pushing him because she was afraid. But he can't hear her.

No false optimism here. Yet it's no surprise that by the second hour of the 20 third season premiere Tony is back in Jennifer Melfi's office. The requirements of the show's premise aside, his untenable situation has not changed. Having glimpsed the possibility of an exit from despair, it would be out of character for him simply to close that door and walk away. For the same reason, I suspect our culture's flight from psychoanalysis is not permanent. It's grandiose, perhaps, to see in one television series, however popular, a cultural trend; and after all *The Sopranos* is on HBO, not CBS or NBC. But ultimately the show is so gripping because, in the words of Elaine Showalter, it's a "cultural Rorschach test." It has been called a parable of corruption and hypocrisy in the postmodern middle class, and it is that; a critique of sexuality, the family and male-female relations in the wake of feminism, and it's that too. But at the primal level, the inkblot is the unconscious. The murderous mobster is the predatory lust and aggression in all of us; his lies and cover-ups are ours; the therapist's fear is our own collective terror of peeling away those lies. The problem is that we can't live with the lies, either. So facing down the terror, a little at a time, becomes the only route to sanity, if not salvation.

In the tumultuous last episode of *The Sopranos*' first season, another in- 21 former is killed. Tony finds out about his mother and sends Jennifer into hiding. Uncle Junior and two of his underlings are arrested, arousing fears that one of them will flip. Artie Bucco nearly kills Tony after being told—by Livia—that Tony is responsible for the fire that destroyed his restaurant (the idea was to help the Buccos by heading off a planned mob hit in the restaurant, which would have ruined the business—this way they could get the insurance and rebuild), but Tony swears "on my mother" it isn't true. Carmela tells off Father Phil. At the end, Tony, Carmela and the kids are caught in a violent storm in their SUV; they can't see a thing but suddenly realize they're in front of the Buccos' (rebuilt) restaurant. There's no power, but Artie graciously ushers them in, lights a candle and cooks them a meal. Tony proposes a toast: "To my family. Someday soon you're gonna have families of your own. And if you're lucky, you'll remember the little moments. Like this. That were good." The moment feels something like sanity. The storm, our storm, goes on.

COMPREHENSION

1. In her introduction, the writer states that HBO's *Sopranos* series is "a meditation of the nature of morality, the possibility of redemption and the legacy of Freud." In the context of the overall essay, what does she mean by this statement?
2. What observations does Willis make about "self-consciousness" or "self-knowledge" as a crucial aspect of our understanding of *The Sopranos*?
3. According to Willis, what does *The Sopranos* tell us about contemporary American culture? Why is the program, in the words of a critic whom the writer quotes, "a cultural Rorschach test"?

RHETORIC

1. Throughout the essay, Willis summarizes various plot lines in *The Sopranos* and describes certain main characters. Why does she devote so much space to this type of overview? What assumptions is she making about her audience? What information does she disclose that the average viewer might not be likely to know?
2. How does the author make a case for considering *The Sopranos* "a gloss on the manners and mores of the fin-de-siècle American middle-class family" (paragraph 2)? What evidence does she provide to support this premise?
3. What comparative elements appear in this essay? How do they serve to bolster the author's main critical points and advance her argument?
4. Is the writer's reference to Freud consistent with your understanding of him? If not, how would you clarify her references to Freud?
5. Identify these allusions and explain what they contribute to the writer's analysis: Dickens and Eliot (paragraph 3); *GoodFellas* and *Analyze This* (paragraph 5); Gary Cooper (paragraph 6); the American Dream (paragraph 7); "heart of darkness" (paragraph 14); the Enlightenment (paragraph 18).
6. Willis makes a number of assertions regarding moral and cultural relativism, psychotherapy, and religion. Are these facts or opinions? Explain your view.
7. How does the writer bring this essay to a conclusion? Do you find the end paragraph effective? Why or why not?

WRITING

1. Write a review of any gangster film or television series that you are familiar with, referring to aspects of Willis's essay to support your commentary and analysis.
2. Why is *The Sopranos* so popular? Write an essay in response to this question.
3. **WRITING AN ARGUMENT:** It could be argued that movies and television programs like *The Sopranos* engage in stereotyping of Italian Americans. Which side of the argument do you take? Write an argumentative essay on this topic. Evaluate core premises, offer evidence, and structure the argument carefully.

Classic and Contemporary: Questions for Comparison

1. Warshow critiques the function and role of the gangster in the popular media, whereas Willis focuses on an analysis of one television series in which the life of the gangster is articulated. How do these different focuses determine the themes of each essay? What are the positive and negative consequences of addressing the general issue of "The Gangster as Tragic Hero" without an in-depth explication of an example, as in the essay by Warshow, as opposed to the detailed analysis of one cable series without a discussion of the gangster genre, as in Willis's essay?

2. Warshow explores the significance of the gangster film within the greater context of American culture. If you were to ascribe a genre to his essay, what would you call it: cultural criticism? media analysis? genre definition? There need not necessarily be one correct response, but be sure to support your answer with examples from the essay. On the other hand, is Willis's essay a review, a critique, or an analysis of a television series, or is it more than one of these? Explain your view.

3. Willis introduces her essay with a summary of a scene from *The Sopranos*. Warshow, on the other hand, begins by making general observations on the nature of society, the modern state, and politics. How do these different rhetorical strategies set up the mode of analysis, tone, and purpose of the two essays?

4. Is it fair to say that Willis's essay is about "television" while Warshow's is about society at large? Explain your view.

✳

Loose Ends

Rita Dove

Rita Dove (b. 1953) grew up in Akron, Ohio, and graduated from Miami University of Ohio and the University of Iowa. She later studied at Tübingen University in West Germany on a Fulbright scholarship and received fellowships from the Guggenheim Foundation and the National Endowment for the Arts. In 1987, she received the Pulitzer Prize for her third book of poetry Thomas and Beulah. *In 1993, Dove was named Poet Laureate, the first African American and, at 40 years of age, the youngest person ever to hold that post. Rita Dove continues to write and to teach at the University of Virginia, where she is Commonwealth Professor of English. About her writing, she has said: "I am concerned with race but certainly not every poem of mine mentions the fact of being black. They are poems about humanity and sometimes*

humanity happens to be black. I cannot run from, I won't run from, any kind of truth."
Her most recent book of poems is American Smooth *(2004). The following essay is*
from The Poet's World *(1995).*

For years the following scene would play daily at our house: Home from   1
school, my daughter would heave her backpack off her shoulder and let it thud
to the hall floor, then dump her jacket on top of the pile. My husband would tell
her to pick it up—as he did every day—and hang it in the closet. Begrudgingly
with a snort and a hrrumph, she would comply. The ritual interrogation began:
    "Hi, Aviva. How was school?"   2
    "Fine."   3
    "What did you do today?"   4
    "Nothing."   5
    And so it went, every day. We cajoled, we pleaded, we threatened with ra- 6
tioned ice cream sandwiches and new healthy vegetable casseroles, we attempted subterfuges such as: "What was Ms. Boyers wearing today?" or: "Any
new pets in science class?" but her answer remained the same: I dunno.
    Asked, however, about that week's episodes of "MathNet," her favorite se- 7
ries on Public Television's "Square One," or asked for a quick gloss of a segment
of "Lois and Clark" that we happened to miss, and she'd spew out the details of
a complicated story, complete with character development, gestures, every
twist and backflip of the plot.
    Is TV greater than reality? Are we to take as damning evidence the soap 8
opera stars attacked in public by viewers who obstinately believe in the on-
screen villainy of Erica or Jeannie's evil twin? Is an estrangement from real life
the catalyst behind the escalating violence in our schools, where children imitate the gun-'em-down pyrotechnics of cop-and-robber shows?
    Such a conclusion is too easy. Yes, the influence of public media on our per- 9
ceptions is enormous, but the relationship of projected reality—i.e., TV—to
imagined reality—i.e., an existential moment—is much more complex. It is not
that we confuse TV with reality, but that we prefer it to reality—the manageable
struggle resolved in twenty-six minutes, the witty repartee within the family
circle instead of the grunts and silence common to most real families; the sharp-
ened conflict and defined despair instead of vague anxiety and invisible enemies. "Life, my friends, is boring. We must not say so," wrote John Berryman,
and many years and "Dream Songs" later he leapt from a bridge in Minneapolis. But there is a devastating corollary to that statement: Life, friends, is ragged.
Loose ends are the rule.
    What happens when my daughter tells the television's story better than her 10
own is simply this: the TV offers an easier tale to tell. The salient points are there
for the plucking—indeed, they're the only points presented—and all she has to
do is to recall them. Instant Nostalgia! Life, on the other hand, slithers about
and runs down blind alleys and sometimes just fizzles at the climax. "The world

is ugly, / And the people are sad," sings the country bumpkin in Wallace Stevens's "Gubinnal." Who isn't tempted to ignore the inexorable fact of our insignificance on a dying planet? We all yearn for our private patch of blue.

COMPREHENSION

1. What is the thesis of the essay?
2. What is the meaning of the title as it pertains to the essay's argument?
3. Define the following terms: *subterfuge* (paragraph 6), *existential* (paragraph 9), *corollary* (paragraph 9), and *salient* (paragraph 10).

RHETORIC

1. What is the tone of the essay? How does Dove's use of language suggest this tone?
2. Where and how does Dove make the transition from the opening anecdote to her more general conclusions about television?
3. The author uses dashes, semicolons, and colons in her essay. Locate them. What is their stylistic effect and function?
4. Dove talks of television shows as exhibiting only the "salient points." How can this same observation be made of her writing?
5. Dove states of television that "we prefer it to reality" (paragraph 9). Who is the "we" in that statement?

WRITING

1. Observe a family watching a television drama or comedy. Observe their body language, demeanor, responses, and gestures. Write an essay describing your observations.
2. Interview five students on your campus and ask them to write one-paragraph responses to the question, "What is the function of television, and what is your purpose for watching television?" Compare and contrast them with Dove's argument regarding the purpose and function of television.
3. **WRITING AN ARGUMENT:** Argue for or against the proposition that watching television is a passive activity.

✳

The Plug-In Drug

Marie Winn

Marie Winn (b. 1936) was born in Prague, Czechoslovakia, and came to the United States in 1939. She has been a prolific author of books for children. She has contributed many articles to publications such as The New York Times Magazine, The New York Times Book Review, *and* Parade. *But she gained national fame with her book about the hazards of television,* The Plug-In Drug, *in 1977, followed by* Children without Childhood *in 1983. More recently, Winn has written articles and books on nature and birdwatching, including* Red-Tails in Love *(1998). The following is an excerpt from a chapter of* The Plug-In Drug, *one of the first books to alert parents to the effects of the mass media on their children.*

Cookies or Heroin?

The word "addiction" is often used loosely and wryly in conversation. People 1 will refer to themselves as "mystery-book addicts" or "cookie addicts." E. B. White writes of his annual surge of interest in gardening: "We are hooked and are making an attempt to kick the habit." Yet nobody really believes that reading mysteries or ordering seeds by catalogue is serious enough to be compared with addictions to heroin or alcohol. In these cases the word "addiction" is used jokingly to denote a tendency to overindulge in some pleasurable activity.

People often refer to being "hooked on TV." Does this, too, fall into the 2 lighthearted category of cookie eating and other pleasures that people pursue with unusual intensity? Or is there a kind of television viewing that falls into the more serious category of destructive addiction?

Not unlike drugs or alcohol, the television experience allows the participant 3 to blot out the real world and enter into a pleasurable and passive mental state. To be sure, other experiences, notably reading, also provide a temporary respite from reality. But it's much easier to stop reading and return to reality than to stop watching television. The entry into another world offered by reading includes an easily accessible return ticket. The entry via television does not. In this way television viewing, for those vulnerable to addiction, is more like drinking or taking drugs—once you start it's hard to stop.

Just as alcoholics are only vaguely aware of their addiction, feeling that they 4 control their drinking more than they really do ("I can cut it out any time I want—I just like to have three or four drinks before dinner"), many people overestimate their control over television watching. Even as they put off other activities to spend hour after hour watching television, they feel they could easily resume living in a different, less passive style. But somehow or other while the television set is present in their homes, it just stays on. With television's easy gratifications available, those other activities seem to take too much effort.

A heavy viewer (a college English instructor) observes: 5

> I find television almost irresistible. When the set is on, I cannot ignore it. I can't
> turn it off. I feel sapped, will-less, enervated. As I reach out to turn off the set,
> the strength goes out of my arms. So I sit there for hours and hours.

Self-confessed television addicts often feel they "ought" to do other 6
things—but the fact that they don't read and don't plant their garden or sew or
crochet or play games or have conversations means that those activities are no
longer as desirable as television viewing. In a way the lives of heavy viewers are
as unbalanced by their television "habit" as drug addicts' or alcoholics' lives.
They are living in a holding pattern, as it were, passing up the activities that
lead to growth or development or a sense of accomplishment. This is one reason
people talk about their television viewing so ruefully, so apologetically. They
are aware that it is an unproductive experience, that by any human measure almost any other endeavor is more worthwhile.

It is the adverse effect of television viewing on the lives of so many people 7
that makes it feel like a serious addiction. The television habit distorts the sense
of time. It renders other experiences vague and curiously unreal while taking on
a greater reality for itself. It weakens relationships by reducing and sometimes
eliminating normal opportunities for talking, for communicating.

And yet television does not satisfy, else why would the viewer continue to 8
watch hour after hour, day after day? "The measure of health," wrote the psychiatrist Lawrence Kubie, "is flexibility . . . and especially the freedom to cease
when sated." But heavy television viewers can never be sated with their television experiences. These do not provide the true nourishment that satiation requires, and thus they find that they cannot stop watching.

COMPREHENSION

1. Why does Winn consider television watching a true addiction?
2. Why does Winn consider television viewing hazardous to one's well-being?
3. What implicit assumption does Winn make concerning the purpose of
   human experience that leads her to conclude that television watching is
   harmful?

RHETORIC

1. Study the introductory paragraph. Is it truly needed? Does it add strength
   to the author's argument? What is its function, if any?
2. What function does the question Winn poses in paragraph 2 serve in setting
   up her argument?
3. In paragraph 3, what is the purpose of using comparison and contrast
   within the context of her argument?

4. The author states several effects of television viewing. Are these based on fact or opinion? Does it matter for the sake of her argument?
5. In her concluding paragraph, the author cites the work of a psychologist. Does this support her main argument? Explain.
6. The author presents television addiction as a serious issue. How does the tone of her essay communicate how seriously she regards the subject?

WRITING

1. In an essay, explain how watching television critically can be a positive educational experience.
2. Develop a manual for people who want to cut down on their television viewing. Model it after a weight-loss or smoking-cessation program.
3. **WRITING AN ARGUMENT:** It seems these days as though television is blamed for everything. Argue for the proposition that watching television is good for you.

www.mhhe.com/
mhreader

For more information on Marie Winn, go to:
More Resources > Ch. 9 Media & Pop Culture

2 Live Crew, Decoded

Henry Louis Gates Jr.

Henry Louis Gates Jr. (b. 1950) was born in Keyser, West Virginia, and was educated at Yale University and Clare College, Cambridge, where he received his PhD in 1979. He now teaches at Harvard University. Gates has edited numerous books addressing the issues of race, identity, and African American history and has contributed to over a dozen periodicals and journals, including Critical Inquiry, Black World, Yale Review, *and* Antioch Review, *among others. His work attempts to apply contemporary literary theories, such as structuralism and poststructuralism, to African and African American literature so that readers can develop a deep understanding of the structure, significance, methods, and meanings of this body of work. Much of his theoretical insights are summed up in his book* The Signifying Monkey: Towards a Theory of Afro-American Literary Criticism *(1988). Among his awards and honors have been a Carnegie Foundation fellowship, a MacArthur Prize fellowship, and a Mellon fellowship from Yale University. In the following essay, published in the* New York Times *in 1990, Gates offers a keen analysis of the rap music phenomenon.*

The rap group 2 Live Crew and their controversial hit recording, "As Nasty as   1
They Wanna Be," may well earn a signal place in the history of First Amendment
rights. But just as important is how these lyrics will be interpreted and by whom.

For centuries, African Americans have been forced to develop coded ways   2
of communicating to protect them from danger. Allegories and double meanings, words redefined to mean their opposites ("bad" meaning "good," for instance), even neologisms ("bodacious") have enabled blacks to share messages
only the initiated understand.

Many blacks were amused by the transcripts of Marion Barry's sting opera-   3
tion which reveals that he used the traditional black expression about one's
"nose being opened." This referred to a love affair and not, as Mr. Barry's prose-
cutors have suggested, to the inhalation of drugs. Understanding this phrase
could very well spell the difference (for the Mayor) between prison and freedom.

2 Live Crew is engaged in heavy-handed parody, turning the stereotypes of   4
black and white American culture on their heads. These young artists are acting
out, to lively dance music, a parodic exaggeration of the age-old stereotypes of
the oversexed black female and male. Their exuberant use of hyperbole (phan-
tasmagoric sexual organs, for example) undermines—for anyone fluent in black
cultural codes—a too literal-minded hearing of the lyrics.

This is the street tradition called "signifying" or "playing the dozens,"   5
which has generally been risqué, and where the best signifier or "rapper" is the
one who invents the most extravagant images, the biggest "lies," as the culture
says. (H. "Rap" Brown earned his nickname in just this way.) In the face of racist
stereotypes about black sexuality, you can do one of two things: you can disavow them or explode them with exaggeration.

2 Live Crew, like many "hip-hop" groups, is engaged in sexual carniva-   6
lesque. Parody reigns supreme, from a take-off of standard blues to a spoof of
the black power movement, their off-color nursery rhymes are part of a venera-
ble Western tradition. The group even satirizes the culture of commerce when it
appropriates popular advertising slogans ("Tastes great!" "Less filling!") and
puts them in a bawdy context.

2 Live Crew must be interpreted within the context of black culture gener-   7
ally and of signifying specifically. Their novelty, and that of other adventure-
some rap groups, is that their defiant rejection of euphemism now voices for the
mainstream what before existed largely in the "race record" market—where the
records of Redd Foxx and Rudy Ray Moore once were forced to reside.

Rock songs have always been about sex but have used elaborate sub-   8
terfuges to convey that fact. 2 Live Crew uses Anglo-Saxon words and is self-
conscious about it: a parody of a white voice in one song refers to "private
personal parts," as a coy counterpart to the group's bluntness.

Much more troubling than its so-called obscenity is the group's overt sex-   9
ism. Their sexism is so flagrant, however, that it almost cancels itself out in a
hyperbolic war between the sexes. In this, it recalls the inter-sexual jousting
in Zora Neale Hurston's novels. Still, many of us look toward the emergence
of more female rappers to redress sexual stereotypes. And we must not allow

ourselves to sentimentalize street culture: the appreciation of verbal virtuosity 10 does not lessen one's obligation to critique bigotry in all of its pernicious forms.

Is 2 Live Crew more "obscene" than, say, the comic Andrew Dice Clay? Clearly, this rap group is seen as more threatening than others that are just as sexually explicit. Can this be completely unrelated to the specter of the young black male as a figure of sexual and social disruption, the very stereotypes 2 Live Crew seem determined to undermine?

This question—and the very large question of obscenity and the First 11 Amendment—cannot even be addressed until those who would answer them become literate in the vernacular traditions of African Americans. To do less is to censor through the equivalent of intellectual prior restraint—and censorship is to art what lynching is to justice.

COMPREHENSION

1. What is the author's thesis?
2. According to Gates, what must one know before engaging in a critique of 2 Live Crew?
3. Does Gates consider 2 Live Crew's music obscene? Why or why not?

RHETORIC

1. The paragraphs in this essay are fairly short. How does this affect Gates's argument?
2. How does the author use definition to decode certain aspects of African American culture? Why is definition an important strategy in his argument?
3. Gates uses the word *hyperbole* in paragraph 4 and the word *hyperbolic* in paragraph 9. Why is it necessary for him to emphasize this concept to develop his argument?
4. Does the author appear to use a particular tone toward his subject matter? Does he appear to support the art of his subject, condemn it, explain it, or provide a mixture of all three approaches?
5. For whom is this essay written? What is its intended purpose? Explain your view.
6. Examine the final sentence of the essay. Does it provide an effective closure? Why is it particularly pertinent considering 2 Live Crew is an African American music group? Explain your view.

WRITING

1. In an essay, explain your position on rap music.
2. For a research project, write a paper on your favorite singer or music group.

3. **WRITING AN ARGUMENT:** Argue for or against the proposition that the music of 2 Live Crew or any more contemporary rap artist or group is obscene, basing your argument on the points raised in the article by Gates.

✳ www.mhhe.com/ **mhreader** | For more information on Henry Louis Gates Jr., go to: **More Resources > Ch. 9 Media & Pop Culture**

✳

My Creature from the Black Lagoon

Stephen King

Stephen King (b. 1947) was born in Portland, Maine. Raised by his mother, he spent parts of his childhood in Indiana, Connecticut, Massachusetts, and Maine. He graduated from the University of Maine at Orono in 1970 with a degree in English. During his early writing career, he sold several stories to mass market men's magazines and taught English in Hampden, Maine. In 1973, his novel Carrie *sold enough copies that he could devote his energies to writing full-time. He is the author of about 100 books, most focusing on horror and the occult. A number have been adapted for film and television, including* Carrie, The Dead Zone, The Shining, Christine, Pet Sematary, *and* Stand By Me, *among others. Besides writing, he belongs to an all-writers rock and roll band (with Dave Barry and Amy Tan) and is a major contributor to local and national charities. In the following selection, taken from* Danse Macabre *(1981), King compares and contrasts the responses of adults and children to horror movies.*

The first movie I can remember seeing as a kid was *Creature from the Black La-* 1 *goon*. It was at the drive-in, and unless it was a second-run job I must have been about seven, because the film, which starred Richard Carlson and Richard Denning, was released in 1954. It was also originally released in 3-D, but I cannot remember wearing the glasses, so perhaps I did see a rerelease.

I remember only one scene clearly from the movie, but it left a lasting im- 2 pression. The hero (Carlson) and the heroine (Julia Adams, who looked absolutely spectacular in a one-piece white bathing suit) are on an expedition somewhere in the Amazon basin. They make their way up a swampy, narrow waterway and into a wide pond that seems an idyllic South American version of the Garden of Eden.

But the creature is lurking—naturally. It's a scaly, batrachian monster that is 3 remarkably like Lovecraft's half-breed, degenerate aberrations—the crazed and blasphemous results of liaisons between gods and human women (It's difficult to get away from Lovecraft). This monster is slowly and patiently barricading the mouth of the stream with sticks and branches, irrevocably sealing the party of anthropologists in.

I was barely old enough to read at that time, the discovery of my father's   4
box of weird fiction still years away. I have a vague memory of boyfriends in my mom's life during that period—from 1952 until 1958 or so; enough of a memory to be sure she had a social life, not enough to even guess if she had a sex life. There was Norville, who smoked Luckies and kept three fans going in his two-room apartment during the summer; and there was Milt, who drove a Buick and wore gigantic blue shorts in the summertime; and another fellow, very small, who was, I believe, a cook in a French restaurant. So far as I know, my mother came close to marrying none of them. She'd gone that route once. Also, that was a time when a woman, once married, became a shadow figure in the process of decision-making and bread-winning. I think my mom, who could be stubborn, intractable, grimly persevering and nearly impossible to discourage, had gotten a taste for captaining her own life. And so she went out with guys, but none of them became permanent fixtures.

It was Milt we were out with that night, he of the Buick and the large blue   5
shorts. He seemed to genuinely like my brother and me, and to genuinely not mind having us along in the back seat from time to time (it may be that when you have reached the calmer waters of your early forties, the idea of necking at the drive-in no longer appeals so strongly . . . even if you have a Buick as large as a cabin cruiser to do it in). By the time the Creature made his appearance, my brother had slithered down onto the floor of the back and had fallen asleep. My mother and Milt were talking, perhaps passing a Kool back and forth. They don't matter, at least not in this context; nothing matters except the big black-and-white images up on the screen, where the unspeakable Thing is walling the handsome hero and the sexy heroine into . . . into . . . the Black Lagoon!

I knew, watching, that the Creature had become *my* Creature; I had bought   6
it. Even to a seven-year-old, it was not a terribly convincing Creature. I did not know then it was good old Ricou Browning, the famed underwater stuntman, in a molded latex suit, but I surely knew it was some guy in some kind of a monster suit . . . just as I knew that, later on that night, he would visit me in the black lagoon of my dreams, looking much more realistic. He might be waiting in the closet when we got back; he might be standing slumped in the blackness of the bathroom at the end of the hall, stinking of algae and swamp rot, all ready for a post-midnight snack of small boy. Seven isn't old, but it is old enough to know that you get what you pay for. You own it, you bought it, it's yours. It is old enough to feel the dowser suddenly come alive, grow heavy, and roll over in your hands, pointing at hidden water.

My reaction to the Creature on that night was perhaps the perfect reaction,   7
the one every writer of horror fiction or director who has worked in the field hopes for when he or she uncaps a pen or a lens: total emotional involvement, pretty much undiluted by any real thinking process—and you understand, don't you, that when it comes to horror movies, the only thought process really

necessary to break the mood is for a friend to lean over and whisper, "See the zipper running down his back?"

I think that only people who have worked in the field for some time truly   8
understand how fragile this stuff really is, and what an amazing commitment it imposes on the reader or viewer of intellect and maturity. When Coleridge spoke of "the suspension of disbelief" in his essay on imaginative poetry, I believe he knew that disbelief is not like a balloon, which may be suspended in air with a minimum of effort; it is like a lead weight, which has to be hoisted with a clean and a jerk and held up by main force. Disbelief isn't light; it's heavy. The difference in sales between Arthur Hailey and H. P. Lovecraft may exist because everyone believes in cars, and banks, but it takes a sophisticated and muscular intellectual act to believe, even for a little while, in Nyarlathotep, the Blind Faceless One, the Howler in the Night. And whenever I run into someone who expresses a feeling along the lines of, "I don't read fantasy or go to any of those movies; none of it's real," I feel a kind of sympathy. They simply can't lift the weight of fantasy. The muscles of the imagination have grown too weak.

In this sense, kids are the perfect audience for horror. The paradox is this: 9
children, who are physically quite weak, lift the weight of unbelief with ease. They are the jugglers of the invisible world—a perfectly understandable phenomenon when you consider the perspective they must view things from. Children deftly manipulate the logistics of Santa Claus's entry on Christmas Eve (he can get down small chimneys by making himself small, and if there's no chimney there's the letter slot, and if there's no letter slot there's always the crack under the door), the Easter Bunny, God (big guy, sorta old, white beard, throne), Jesus ("How do you think he turned the water into wine?" I asked my son Joe when he—Joe, not Jesus—was five; Joe's idea was that he had something "kinda like magic Kool-Aid, you get what I mean?"), the devil (big guy, red skin, horse feet, tail with an arrow on the end of it, Snidely Whiplash moustache), Ronald McDonald, the Burger King, the Keebler Elves, Dorothy and Toto, the Lone Ranger and Tonto, a thousand more.

Most parents think they understand this openness better than, in many   10
cases, they actually do, and try to keep their children away from anything that smacks too much of horror and terror—"Rated PG (or G in the case of *The Andromeda Strain*), but may be too intense for younger children," the ads for *Jaws* read—believing, I suppose, that to allow their kids to go to a real horror movie would be tantamount to rolling a live hand grenade into a nursery school.

But one of the odd Döppler effects that seems to occur during the selective   11
forgetting that is so much a part of "growing up" is the fact that almost *everything* has a scare potential for the child under eight. Children are literally afraid of their own shadows at the right time and place. There is the story of the four-year-old who refused to go to bed at night without a light on in his closet. His parents at last discovered he was frightened of a creature he had heard his father speak of often; this creature, which had grown large and dreadful in the child's imagination, was the "twi-night double-header."

Seen in this light, even Disney movies are minefields of terror, and the ani- 12
mated cartoons, which will apparently be released and rereleased even unto the
end of the world,* are usually the worst offenders. There are adults today, who,
when questioned, will tell you that the most frightening thing they saw at the
movies as children was Bambi's father shot by the hunter, or Bambi and his
mother running before the forest fire. Other Disney memories which are right
up there with the batrachian horror inhabiting the Black Lagoon include the
marching brooms that have gone totally out of control in *Fantasia* (and for the
small child, the real horror inherent in the situation is probably buried in the implied father-son relationship between Mickey Mouse and the old sorcerer; those
brooms are making a terrible mess, and when the sorcerer/father gets home,
there may be PUNISHMENT. . . . This sequence might well send the child of strict
parents into an ecstasy of terror); the night on Bald Mountain from the same
film; the witches in *Snow White* and *Sleeping Beauty,* one with her enticingly red
poisoned apple (and what small child is not taught early to fear the idea of POI-
SON?), the other with her deadly spinning wheel; this holds all the way up to the
relatively innocuous *One Hundred and One Dalmatians* which features the logical
granddaughter of those Disney witches from the thirties and forties—the evil
Cruella DeVille, with her scrawny, nasty face, her loud voice (grownups sometimes forget how terrified young children are of loud voices, which come from
the giants of their world, the adults), and her plan to kill all the dalmatian pup-
pies (read "children," if you're a little person) and turn them into dogskin coats.

Yet it is the parents, of course, who continue to underwrite the Disney pro- 13
cedure of release and rerelease, often discovering goosebumps on their own
arms as they rediscover what terrified them as children . . . because what the
good horror film (or horror sequence in what may be billed a "comedy" or an
"animated cartoon") does above all else is to knock the adult props out from under us and tumble us back down the slide into childhood. And there our own
shadow may once again become that of a mean dog, a gaping mouth, or a beck-
oning dark figure.

Perhaps the supreme realization of this return to childhood comes in David 14
Cronenberg's marvelous horror film *The Brood,* where a disturbed woman is literally producing "children of rage" who go out and murder the members of her
family, one by one. About halfway through the film, her father sits dispiritedly
on the bed in an upstairs room, drinking and mourning his wife, who has been
the first to feel the wrath of the brood. We cut to the bed itself . . . and clawed
hands suddenly reach out from beneath it and dig into the carpeting near the

---

*In one of my favorite Arthur C. Clarke stories, this actually happens. In this vignette, aliens from
space land on earth after the Big One has finally gone down. As the story closes, the best brains of
this alien culture are trying to figure out the meaning of a film they have found and learned how
to play back. The film ends with the words *A Walt Disney Production.* I have moments when I really believe that there would be no better epitaph for the human race, or for a world where the
only sentient being absolutely guaranteed of immortality is not Hitler, Charlemagne, Albert
Schweitzer, or even Jesus Christ—but is, instead, Richard M. Nixon, whose name is engraved on a
plaque placed on the airless surface of the moon. [This note is the author's.]

AN ALBUM OF ADVERTISEMENTS: IMAGES OF CULTURE

The advertisements that appear in this album reflect various ideas in the popular culture, what we value and what we desire. Advertising clearly is a powerful force in molding popular culture and taste. Consider what the advertisements in this album say about popular culture and how they embody forms of argument and persuasion that promote a product, person, or idea.

Advocacy Advertising: The "Get caught reading" campaign of the Association of American Publishers

CONSIDERING THE ADVERTISEMENT

1. What is the effect of using a celebrity—Alicia Keys—as an advocate in this advertisement? How does the designer present her within the frame of the ad?
2. Search in magazines or online for three other ads that use celebrities to promote a product or idea. Write an essay explaining why these ads are effective. If it is possible, download or scan these images into your essay.

Classic Advertisement: Campus Togs, Clothes for Men and Young Men—
a 1924 advertisement for young men's clothing

Contemporary Advertisement: Sean Jean—advertisement for contemporary men's clothing

CONSIDERING THE ADVERTISEMENTS

1. Describe the figures in each advertisement: How are the models similar? How do they differ? Who is the audience for each ad? What cultural assumptions does each advertisement build on? What forms of appeal does each ad make?
2. Using these classic and contemporary ads as a reference point, write an essay arguing for or against the proposition that the methods and goals of advertising have not changed or improved over time, even though the culture might have changed.

International Advertising: Budweiser in China

CONSIDERING THE ADVERTISEMENT

1. Explain the way in which this photograph captures the power of advertising to promote an American product overseas. What details hold your attention? Why is the poster advertising Budweiser so large?
2. In an essay, argue for or against the idea that advertising American products like Budweiser, Coke, or McDonald's overseas is a form of "cultural imperialism" that harms the societies of other countries.

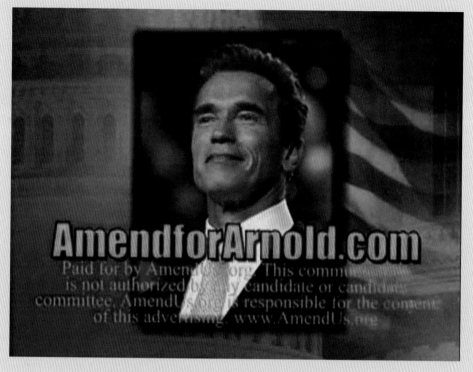

Political Advertising: AmendforArnold.com

CONSIDERING THE ADVERTISEMENT

1. Explain the various details and visual elements that create the "message" in this video still taken from an online advertisement advocating a movement to amend the U.S. Constitution so that the current governor of California can become president.
2. Search online for other Web sites that promote changes in local or national laws—for instance, banning same-sex marriage, supporting stem-cell research, or promoting "intelligent design" in school textbooks. In an essay, analyze the argumentative techniques underlying the verbal and nonverbal elements in the text you have selected. Attach or integrate the image into your essay.

Mobile Outdoor Advertising: NotinLA.com

CONSIDERING THE ADVERTISEMENT

1. What seems to be the purpose behind this mobile ad? What evidence can you cite to support your response? Why would it be helpful to go to the Web site to decipher the ad's message?
2. In an argumentative essay, explain why you think this NotinLA.com ad is or is not successful.

Thirty-five years ago,
Dave Thomas® opened the front door
and welcomed you inside.

Spokesperson Advertising: Wendy's ad featuring its founder

CONSIDERING THE ADVERTISEMENT

1. This advertisement features Dave Thomas, the deceased founder of Wendy's. How does the ad exploit the argumentative technique of *pathos* (appeal to emotion) and *ethos* (appeal to values and character)?
2. Write an essay in which you argue from one side or the other about the impact of fast-food advertising on American culture. In other words, do you think that the ads promoting the fast-food industry do more good or harm? Use the Wendy's ad and any others you want to locate to support your claim.

Anti-Advertising: God billboard

CONSIDERING THE ADVERTISEMENT

1. What is your response to this billboard ad? Are you surprised, shocked, or amused, and why? Who has paid for the billboard ad? What is the ad's purpose? Why might some viewers find the ad controversial? Why might the ad be considered a humorous form of anti-advertising?
2. Write an essay in which you contend that various expressions of faith should or should not be used in advertising, or that God should or should not be used to promote a product, person, or idea.

doomed father's shoes. And so Cronenberg pushes us down the slide; we are four again, and all of our worst surmises about what might be lurking under the bed have turned out to be true.

The irony of all this is that children are better able to deal with fantasy and 15 terror *on its own terms* than their elders are. You'll note I've italicized the phrase "on its own terms." An adult is able to deal with the cataclysmic terror of something like *The Texas Chain Saw Massacre* because he or she understands that it is all make-believe, and that when the take is done the dead people will simply get up and wash off the stage blood. The child is not so able to make this distinction, and *Chainsaw Massacre* is quite rightly rated R. Little kids do not need this scene, any more than they need the one at the end of *The Fury* where John Cassavetes quite literally blows apart. But the point is, if you put a little kid of six in the front row at a screening of *The Texas Chainsaw Massacre* along with an adult who was temporarily unable to distinguish between make-believe and "real things" (as Danny Torrance, the little boy in *The Shining* puts it)—if, for instance, you had given the adult a hit of Yellow Sunshine LSD about two hours before the movie started—my guess is that the kid would have maybe a week's worth of bad dreams. The adult might spend a year or so in a rubber room, writing home with Crayolas.

A certain amount of fantasy and horror in a child's life seems to me a per- 16 fectly okay, useful sort of thing. Because of the size of their imaginative capacity, children are able to handle it, and because of their unique position in life, they are able to put such feelings to work. They understand their position very well, too. Even in such a relatively ordered society as our own, they understand that their survival is a matter almost totally out of their hands. Children are "dependents" up until the age of eight or so in every sense of the word; dependent on mother and father (or some reasonable facsimile thereof) not only for food, clothing, and shelter, but dependent on them not to crash the car into a bridge abutment, to meet the school bus on time, to walk them home from Cub Scouts or Brownies, to buy medicines with childproof caps, dependent on them to make sure they don't electrocute themselves while screwing around with the toaster or while trying to play with Barbie's Beauty Salon in the bathtub.

Running directly counter to this necessary dependence is the survival direc- 17 tive built into all of us. The child realizes his or her essential lack of control, and I suspect it is this very realization which makes the child uneasy. It is the same sort of free-floating anxiety that many air travelers feel. They are not afraid because they believe air travel to be unsafe; they are afraid because they have surrendered control, and if something goes wrong all they can do is sit there clutching airsick bags or the in-flight magazine. To surrender control runs counter to the survival directive. Conversely, while a thinking, informed person may understand intellectually that travel by car is much more dangerous than flying, he or she is still apt to feel much more comfortable behind the wheel, because she/he has control . . . or at least an illusion of it.

This hidden hostility and anxiety toward the airline pilots of their lives may 18 be one explanation why, like the Disney pictures which are released during school vacations in perpetuity, the old fairy tales also seem to go on forever. A parent who would raise his or her hands in horror at the thought of taking his/her child to see *Dracula* or *The Changeling* (with its pervasive imagery of the drowning child) would be unlikely to object to the baby sitter reading "Hansel and Gretel" to the child before bedtime. But consider: the tale of Hansel and Gretel begins with deliberate abandonment (oh yes, the stepmother masterminds that one, but she is the symbolic mother all the same, and the father is a spaghetti-brained nurd who goes along with everything she suggests even though he know it's wrong—thus we can see her as amoral, him as actively evil in the Biblical and Miltonian sense), it progresses to kidnapping (the witch in the candy house), enslavement, illegal detention, and finally justifiable homicide and cremation. Most mothers and fathers would never take their children to see *Survive,* that quickie Mexican exploitation flick about the rugby players who survived the aftermath of a plane crash in the Andes by eating their dead teammates, but these same parents find little to object to in "Hansel and Gretel," where the witch is fattening the children up so she can eat them. We give this stuff to the kids almost instinctively, understanding on a deeper level, perhaps, that such fairy stories are the perfect points of crystallization for those fears and hostilities.

Even anxiety-ridden air travelers have their own fairy tales—all those *Airport* movies, which, like "Hansel and Gretel" and all those Disney cartoons, show every sign of going on forever . . . but which should only be viewed on Thanksgivings, since all of them feature a large cast of turkeys. [19]

My gut reaction to *Creature from the Black Lagoon* on that long-ago night was a kind of terrible, waking swoon. The nightmare was happening right in front of me; every hideous possibility that human flesh is heir to was being played out on that drive-in screen. [20]

Approximately twenty-two years later, I had a chance to see *Creature from the Black Lagoon* again—not on TV, with any kind of dramatic build and mood broken up by adverts for used cars, K-Tel disco anthologies, and Underalls pantyhose, thank God, but intact, uncut . . . and even in 3-D. Guys like me who wear glasses have a hell of a time with 3-D, you know; ask anyone who wears specs how they like those nifty little cardboard glasses they give you when you walk in the door. If 3-D ever comes back in a big way, I'm going to take myself down to the local Pearle Vision Center and invest seventy bucks in a special pair of prescription lenses: one red, one blue. Annoying glasses aside, I should add that I took my son Joe with me—he was then five, about the age I had been myself, that night at the drive-in (and imagine my surprise—my *rueful* surprise—to discover that the movie which had so terrified me on that long-ago night had been rated G by the MPAA . . . just like the Disney pictures). [21]

As a result, I had a chance to experience that weird doubling back in time that I believe most parents only experience at the Disney films with their children, or when reading them the Pooh books or perhaps taking them to the Shrine or the Barnum & Bailey circus. A popular record is apt to create a particular "set" in a listener's mind, precisely because of its brief life of six weeks to [22] three months, and "golden oldies" continue to be played because they are the emotional equivalent of freeze-dried coffee. When the Beach Boys come on the radio singing "Help Me, Rhonda," there is always that wonderful second or two when I can re-experience the wonderful, guilty joy of copping my first feel (and if you do the mental subtraction from my present age of thirty-three, you'll see that I was a little backward in that respect). Movies and books do the same thing, although I would argue that the mental set, its depth and texture, tends to be a little richer, a little more complex, when re-experiencing films and a lot more complex when dealing with books.

With Joe that day I experienced *Creature from the Black Lagoon* from the other 23 end of the telescope, but this particular theory of set identification still applied; in fact, it prevailed. Time and age and experience have all left their marks on me, just as they have on you; time is not a river, as Einstein theorized—it's a big . . . buffalo herd that runs us down and eventually mashes us into the ground, dead and bleeding, with a hearing-aid plugged into one ear and a colostomy bag instead of a .44 clapped on one leg. Twenty-two years later I knew that the Creature was really good old Ricou Browning, the famed underwater stuntman, in a molded latex suit, and the suspension of disbelief, that mental clean-and-jerk, had become a lot harder to accomplish. But I did it, which may mean nothing, or which may mean (I hope!) that the buffalo haven't got me yet. But when that weight of disbelief was finally up there, the old feelings came flooding in, as they flooded in some five years ago when I took Joe and my daughter Naomi to their first movie, a reissue of *Snow White and the Seven Dwarfs.* There is a scene in that film where, after Snow White has taken a bite from the poisoned apple, the dwarfs take her into the forest, weeping copiously. Half the audience of little kids was also in tears; the lower lips of the other half were trembling. The set identification in that case was strong enough so that I was also surprised into tears. I hated myself for being so blatantly manipulated, but manipulated I was, and there I sat, blubbering into my beard over a bunch of cartoon characters. But it wasn't Disney that manipulated me; I did it myself. It was the kid inside who wept, surprised out of dormancy and into schmaltzy tears . . . but at least awake for awhile.

During the final two reels of *Creature from the Black Lagoon,* the weight of 24 disbelief is nicely balanced somewhere above my head, and once again director Jack Arnold places the symbols in front of me and produces the old equation of the fairy tales, each symbol as big and as easy to handle as a child's alphabet block. Watching, the child awakes again and knows that this is what dying is like. Dying is when the Creature from the Black Lagoon dams up the exit. Dying is when the monster gets you.

In the end, of course, the hero and heroine, very much alive, not only sur- 25 vive but triumph—as Hansel and Gretel do. As the drive-in floodlights over the screen came on and the projector flashed its GOOD NIGHT, DRIVE SAFELY slide on that big white space (along with the virtuous suggestion that you ATTEND THE CHURCH OF YOUR CHOICE), there was a brief feeling of relief, almost of resurrection. But the feeling that stuck longest was the swooning sensation that good old Richard Carlson and Julia Adams were surely going down for the third time, and the image that remains forever after is of the creature slowly and patiently walling its victims into the Black Lagoon; even now I can see it peering over that growing wall of mud and sticks.

Its eyes. Its ancient eyes.   26

COMPREHENSION

1. Why does King claim that it is harder for an author to successfully bring a horror tale to life than a standard "realistic" one? What special skills does the horror writer need?
2. Why are children the "perfect audience for horror" (paragraph 9)? What exists in the structure of most horror films that make them suitable for children?
3. King titles his essay "My Creature from the Black Lagoon" rather than using the original title *The Creature from the Black Lagoon*. Why?
4. Why does King think it is ironic that many Disney movies are "G" rated while "horror" movies often contain warnings about content for children?
5. In paragraph 23, King remarks that he is pleased that he is still able to get a thrill from watching a horror movie even though he is an adult and understands the artifice behind the monster. Why does he feel this is a positive response? Why does he believe it would be beneficial for most adults to react this way?

RHETORIC

1. In paragraph 9, King attempts to reproduce the sense of what it is like to think like a child. How does he achieve this effect? What is his purpose?
2. How does King structure paragraph 18 to compare and contrast horror movies with "fairy tales"? What is his rhetorical intent?
3. What vocabulary choices does King use in his introduction to set up his conversational style of writing? What relationship does King intend to create between the writer and reader by employing this type of discourse?
4. In paragraphs 4 and 5, King recounts a childhood anecdote. What is the purpose of describing the outing to the drive-in theatre with his mother's boyfriend, Milt? How does King structure these two paragraphs so that they culminate rhetorically in a device similar to that employed in horror movies?
5. Although much of King's writing is informal, he does use references to popular culture, literature, and science in his writing. What is the significance and meaning of the following terms: *batrachian* (paragraph 3), *suspension of disbelief* (paragraph 8), *Döppler effects* (paragraph 11), *twi-night doubleheader* (paragraph 11), *possibility that human flesh is heir to* (paragraph 20), and *golden oldies* (paragraph 22).

6. The conclusion is only five words: two sentence fragments. Why has King chosen to end his essay this way?
7. King uses irony in his essay for comic effect; for example, in paragraph 25, what is the irony in the "sign-off" at the drive-in movie theater that reads: ATTEND THE CHURCH OF YOUR CHOICE?

WRITING

1. Think of the first horror movie you recall vividly from your childhood. Write an expository essay about how the film scared you. Include both the dramatic elements on the screen and your own state of mind while you watched.
2. Select a horror or science fiction book you've read that has been adapted for the screen. Compare and contrast the effects of each version. Which was more captivating? More engaging? More horrifying? More believable? Explain your view.
3. Compare and contrast the benefits or drawbacks, or both, of an adult reading a story to a child versus taking a child to the movies.
4. **WRITING AN ARGUMENT:** In an essay, argue for or against the proposition that horror movies are scarier when viewed at the movie theater than on home video.

www.mhhe.com/ **mhreader**

For more information on Stephen King, go to: **More Resources > Ch. 9 Media & Pop Culture**

✳

Red, White, and Beer

Dave Barry

Dave Barry (b. 1947) was born in Armonk, New York. He graduated from Haverford College in 1969 and was a reporter and editor at the Daily Local News *from 1971 to 1975. Since 1983, he has been a columnist for* The Miami Herald. *Besides writing his columns, Barry has written numerous books, all with his unique, amusing point of view. His books include* Stay Fit and Healthy Until You're Dead *(1985),* Dave Barry's Greatest Hits *(1988),* Dave Barry Turns 40 *(1990),* Dave Barry's Only Travel Guide You'll Ever Need *(1991), and* Boogers Are My Beat *(2002). Barry won the 1988 Pulitzer Prize for commentary. In the piece below, he comments on the relation between television commercials and patriotism.*

Lately I've been feeling very patriotic, especially during commercials. Like, 1
when I see those strongly pro-American Chrysler commercials, the ones where
the winner of the Bruce Springsteen Sound-Alike Contest sings about how The
Pride Is Back, the ones where Lee Iacocca himself comes striding out and practically challenges the president of Toyota to a knife fight, I get this warm, proud
feeling inside, the same kind of feeling I get whenever we hold routine naval
maneuvers off the coast of Libya.

But if you want to talk about *real* patriotism, of course, you have to talk 2
about beer commercials. I would have to say that Miller is the most patriotic
brand of beer. I grant you it tastes like rat saliva, but we are not talking about
taste here. What we are talking about, according to the commercials, is that
Miller is by God an *American* beer, "born and brewed in the U.S.A.," and the
men who drink it are American men, the kind of men who aren't afraid to per-
spire freely and shake a man's hand. That's mainly what happens in Miller
commercials: Burly American men go around, drenched in perspiration, shaking each other's hands in a violent and patriotic fashion.

You never find out exactly why these men spend so much time shaking 3
hands. Maybe shaking hands is just their simple straightforward burly masculine American patriotic way of saying to each other: "Floyd, I am truly sorry I
drank all that Miller beer last night and went to the bathroom in your glove
compartment." Another possible explanation is that, since there are never any
women in the part of America where beer commercials are made, the burly men
have become lonesome and desperate for any form of physical contact. I have
noticed that sometimes, in addition to shaking hands, they hug each other.
Maybe very late at night, after the David Letterman show, there are Miller commercials in which the burly men engage in slow dancing. I don't know.

I do know that in one beer commercial, I think this is for Miller—although it 4
could be for Budweiser, which is also a very patriotic beer—the burly men build
a house. You see them all getting together and pushing up a brand-new wall.
Me, I worry some about a house built by men drinking beer. In my experience,
you run into trouble when you ask a group of beer-drinking men to perform any
task more complex than remembering not to light the filter ends of cigarettes.

For example, in my younger days, whenever anybody in my circle of 5
friends wanted to move, he'd get the rest of us to help, and, as an inducement,
he'd buy a couple of cases of beer. This almost always produced unfortunate results, such as the time we were trying to move Dick "The Wretch" Curry from a
horrible fourth-floor walk-up apartment in Manhattan's Lower East Side to another horrible fourth-floor walk-up apartment in Manhattan's Lower East Side,
and we hit upon the labor-saving concept of, instead of carrying The Wretch's
possessions manually down the stairs, simply dropping them out the window,
down onto the street, where The Wretch was racing around, gathering up the
broken pieces of his life and shrieking at us to stop helping him move, his emotions reaching a fever pitch when his bed, which had been swinging wildly
from a rope, entered the apartment two floors below his through what had until seconds earlier been a window.

This is the kind of thinking you get, with beer. So I figure what happens, in 6 the beer commercial where the burly men are building the house, is they push the wall up so it's vertical, and then, after the camera stops filming them, they just keep pushing, and the wall crashes down on the other side, possibly onto somebody's pickup truck. And then they all shake hands.

But other than that, I'm in favor of the upsurge in retail patriotism, which is 7 lucky for me because the airwaves are saturated with pro-American commercials. Especially popular are commercials in which the newly restored Statue of Liberty—and by the way, I say Lee Iacocca should get some kind of medal for that, or at least be elected president—appears to be endorsing various products, as if she were Mary Lou Retton or somebody. I saw one commercial strongly suggesting that the Statue of Liberty uses Sure brand underarm deodorant.

I have yet to see a patriotic laxative commercial, but I imagine it's only a 8 matter of time. They'll show some actors dressed up as hard-working country folk, maybe at a church picnic, smiling at each other and eating pieces of pie. At least one of them will be a black person. The Statue of Liberty will appear in the background. Then you'll hear a country-style singer singing:

Folks 'round here they love this land;
They stand by their beliefs;
An' when they git themselves stopped up;
They want some quick relief.

Well, what do you think? Pretty good commercial concept, huh? 9

Nah, you're right. They'd never try to pull something like that. They'd put 10 the statue in the *foreground.*

COMPREHENSION

1. What does Barry mean by "retail patriotism" (paragraph 7)? How does the essay's title illustrate this concept?
2. According to Barry, what makes beer commercials, especially those for Miller, patriotic?
3. In Barry's opinion, what do sexism, patriotism, and beer have in common?

RHETORIC

1. Barry doesn't explicitly state his thesis anywhere in the essay. In your own words, what is his implied thesis? Use evidence from the essay to support your view.
2. Barry uses irony and humor very effectively in this piece. Cite some examples of his humor, and analyze how he achieves the desired effect.
3. The writer uses specific brand names in his essay. How does this device help strengthen his argument? Would eliminating them make the essay less persuasive? Why or why not?

4. Barry seems to digress from his point in paragraphs 4, 5, and 6. Why does he do this? How does this digression serve the purpose of the piece?
5. Does the anecdote Barry uses in paragraph 5 ring true? Why or why not? What purpose does it serve in the essay? Does its plausibility affect the strength of Barry's argument?
6. How does paragraph 10 function as a conclusion? Is it in keeping with the essay's tone and style? Is it an effective device? Justify your response.

WRITING

1. Barry's essay examines how television sells patriotism. Write an essay analyzing how television sells other abstract ideas, such as success, love, freedom, democracy. Pattern your essay after Barry's, using humor. Also, use specific television commercials you have seen as examples.
2. Write an essay entitled "Patriotism," using both denotative and connotative definitions of the word.
3. **WRITING AN ARGUMENT:** In an essay, argue for or against the claim that television advertising has had a harmful impact on American and global consumers.

www.mhhe.com/
mhreader

For more information on Dave Barry, go to:
More Resources > Ch. 9 Media & Pop Culture

✳
Wonder Woman

Gloria Steinem

Gloria Steinem (b. 1934) was born and raised in Toledo, Ohio; she attended Smith College, receiving a BA in government in 1956. A noted feminist and political activist, Steinem in 1968 helped to found New York magazine; in 1971 she cofounded Ms. magazine and served as its editor. Whether campaigning for Robert Kennedy, defending raising money for the United Farm Workers, or championing women's reproductive rights, Steinem has been on the cutting edge of American politics and social activism for almost four decades. Her books include The Thousand Indias (1957), Outrageous Acts and Everyday Rebellions (1983), Marilyn: Norma Jean (1986), Revolution from Within (1992), and Moving beyond Words (1994). In the following essay, Steinem explains why the comic book heroine, Wonder Woman (who was on the first cover of Ms.), was such a formative influence during her childhood.

Wonder Woman is the only female super-hero to be published continuously 1
since comic books began—indeed, she is one of the few to have existed at all or
to be anything other than part of a male super-hero group—but this may strike
many readers as a difference without much distinction. After all, haven't comic
books always been a little disreputable? Something that would never have been
assigned in school? The answer to those questions is yes, which is exactly why
they are important. Comic books have power—including over the child who
still lives within each of us—because they are *not* part of the "serious" grown-
up world.

I remember hundreds of nights reading comic books under the covers with 2
a flashlight; dozens of car trips while my parents told me I was ruining my eyes
and perhaps my mind ("brain-deadeners" was what my mother called them);
and countless hours spent hiding in a tree or some other inaccessible spot where
I could pore over their pages in sweet freedom. Because my family's traveling
meant I didn't go to school regularly until I was about twelve, comic books
joined cereal boxes and ketchup labels as the primers that taught me how to
read. They were even cheap enough to be the first things I bought on my own—
a customer who couldn't see over the countertop but whose dignity was greatly
enhanced by making a choice, counting out carefully hoarded coins, and completing a grown-up exchange.

I've always wondered if this seemingly innate drive toward independence in 3
children isn't more than just "a movement toward mastery," as psychologists
say. After all, each of us is the result of millennia of environment and heredity, a
unique combination that could never happen before—or again. Like a seed that
contains a plant, a child is already a unique person; an ancient spirit born into a
body too small to express itself, or even cope with the world. I remember feeling the greatest love for my parents whenever they allowed me to express my
own will, whether that meant wearing an inappropriate hat for days on end, or
eating dessert before I had finished dinner.

Perhaps it's our memories of past competence and dreams for the future 4
that create the need for super-heroes in the first place. Leaping skyscrapers in a
single bound, seeing through walls, and forcing people to tell the truth by encir-
cling them in a magic lasso—all would be satisfying fantasies at any age, but
they may be psychological necessities when we have trouble tying our shoes,
escaping a worldview composed mainly of belts and knees, and getting
grownups to *pay attention.*

The problem is that the super-heroes who perform magical feats—indeed, 5
even mortal heroes who are merely competent—are almost always men. A female child is left to believe that, even when her body is as big as her spirit, she
will still be helping with minor tasks, appreciating the accomplishments of others, and waiting to be rescued. Of course, pleasure is to be found in all these experiences of helping, appreciating, and being rescued; pleasure that should be
open to boys, too. Even in comic books, heroes sometimes work in groups or are
called upon to protect their own kind, not just helpless females. But the truth is

that a male super-hero is more likely to be vulnerable, if only to create suspense, than a female character is to be powerful or independent. For little girls, the only alternative is suppressing a crucial part of ourselves by transplanting our consciousness into a male character—which usually means a white one, thus penalizing girls of color doubly, and boys of color, too. Otherwise, choices remain limited: in the case of girls, to an "ideal" life of sitting around like a Technicolor clotheshorse, getting into jams with villains, and saying things like, "Oh, Superman! I'll always be grateful to you"; in the case of boys of color, to identifying with villains who may be the only ethnic characters with any power; and in the case of girls of color, to making an impossible choice between parts of their identity. It hardly seems worth learning to tie our shoes.

I'm happy to say that I was rescued from this dependent fate at the age of seven or so; rescued (Great Hera!) by a woman. Not only did she have the wisdom of Athena and Aphrodite's power to inspire love, she was also faster than Mercury and stronger than Hercules. In her all-woman home on Paradise Island, a refuge of ancient Amazon culture protected from nosy travelers by magnetic thought-fields that created an area known to the world as the Bermuda Triangle, she had come to her many and amazing powers naturally. Together with her Amazon sisters, she had been trained in them from infancy and perfected them in Greek-style contests of dexterity, strength, and speed. The lesson was that each of us might have unknown powers within us, if we only believed and practiced them. (To me, it always seemed boring that Superman had bullet-proof skin, X-ray vision, and the ability to fly. Where was the contest?) Though definitely white, as were all her Amazon sisters, she was tall and strong, with dark hair and eyes—a relief from the weak, bosomy, blonde heroines of the 1940s.

Of course, this Amazon did need a few fantastic gadgets to help her once she entered a modern world governed by Ares, God of War, not Aphrodite, Goddess of Love: a magic golden lasso that compelled all within its coils to obey her command, silver bracelets that repelled bullets, and an invisible plane that carried her through time as well as space. But she still had to learn how to throw the lasso with accuracy, be agile enough to deflect bullets from her silver-encased wrists, and navigate an invisible plane.

Charles Moulton, whose name appeared on each episode as Wonder Woman's writer and creator, had seen straight into my heart and understood the fears of violence and humiliation hidden there. No longer did I have to pretend to like the "POW!" and "SPLAT!" of boys' comic books, from Captain Marvel to the Green Hornet. No longer did I have nightmares after looking at ghoulish images of torture and murder, bloody scenes made all the more realistic by steel-booted Nazis and fang-toothed Japanese who were caricatures of World War II enemies then marching in every newsreel. (Eventually, the sadism of boys' comic books was so extreme that it inspired Congressional hearings, and publishers were asked to limit the number of severed heads and dripping entrails—a reminder that television wasn't the first popular medium selling sadism to boys.) Best of all, I could stop pretending to enjoy the ridicule, bossing-around, and constant endangering of female characters. In these Amazon adventures, only the villains bought the idea that "masculine" meant aggression and "feminine" meant submission. Only the occasional female accomplice said things like "Girls want superior men to boss them around," and even they were usually converted to the joys of self-respect by the story's end.

This was an Amazon super-hero who never killed her enemies. Instead, she 9 converted them to a belief in equality and peace, to self-reliance, and respect for the rights of others. If villains destroyed themselves, it was through their own actions or some unbloody accident. Otherwise, they might be conquered by force, but it was a force tempered by love and justice.

In short, she was wise, beautiful, brave, and explicitly out to change "a 10 world torn by the hatreds and wars of men."

She was Wonder Woman. 11

Only much later, when I was in my thirties and modern feminism had be- 12 gun to explain the political roots of women's status—instead of accepting some "natural" inferiority decreed by biology, God, or Freud—did I realize how hard Charles Moulton had tried to get an egalitarian worldview into comic book form. From Wonder Woman's birth myth as Princess Diana of Paradise Island, "that enlightened land," to her adventures in America disguised as Diana Prince, a be-spectacled army nurse and intelligence officer (a clear steal from Superman's Clark Kent), this female super-hero was devoted to democracy, peace, justice, and "liberty and freedom for all womankind."

One typical story centers on Prudence, a young pioneer in the days of the 13 American Frontier, where Wonder Woman has been transported by the invisible plane that doubles as a time machine. After being rescued from a Perils of Pauline life, Prudence finally realizes her own worth, and also the worth of all women. "From now on," she says proudly to Wonder Woman, "I'll rely on myself, not on a man." Another story ends with Wonder Woman explaining her own long-running romance with Captain Steve Trevor, the American pilot whose crash-landing on Paradise Island was Aphrodite's signal that the strongest and wisest of all the Amazons must answer the call of a war-torn world. As Wonder Woman says of this colleague whom she so often rescues: "I can never love a dominant man."

The most consistent villain is Ares, God of War, a kind of metavillain who 14 considers women "the natural spoils of war" and insists they stay home as the slaves of men. Otherwise, he fears women will spread their antiwar sentiments, create democracy in the world, and leave him dishonored and unemployed. That's why he keeps trying to trick Queen Hippolyte, Princess Diana's mother, into giving up her powers as Queen of the Amazons, thus allowing him to conquer Paradise Island and destroy the last refuge of ancient feminism. It is in memory of a past time when the Amazons did give in to the soldiers of Ares, and were enslaved by them, that Aphrodite requires each Amazon to wear a pair of cufflike bracelets. If captured and bound by them (as Wonder Woman sometimes is in particularly harrowing episodes), an Amazon loses all her power. Wearing them is a reminder of the fragility of female freedom.

In America, however, villains are marked not only by their violence, but by 15
their prejudice and lust for money. Thomas Tighe, woman-hating industrialist,
is typical. After being rescued by Wonder Woman from accidental imprisonment in his own bank vault, he refuses to give her the promised reward of a million dollars. Though the money is needed to support Holliday College, the
home of the band of college girls who aid Wonder Woman, Tighe insists that its
students must first complete impossible tests of strength and daring. Only after
Wonder Woman's powers allow them to meet every challenge does Tighe finally admit: "You win, Wonder Woman! . . . I am no longer a woman hater." She
replies: "Then you're the real winner, Mr. Tighe! Because when one ceases to
hate, he becomes stronger!"

Other villains are not so easily converted. Chief among them is Dr. Psycho, 16
perhaps a parody of Sigmund Freud. An "evil genius" who "abhors women,"
the mad doctor's intentions are summed up in this scene-setting preface to an
episode called "Battle for Womanhood": "With weird cunning and dark, forbidden knowledge of the occult, Dr. Psycho prepares to change the independent
status of modern American women back to the days of the sultans and slave
markets, clanking chains and abject captivity. But sly and subtle Psycho reckons
without Wonder Woman!"

When I looked into the origins of my proto-feminist super-hero, I discov- 17
ered that her pseudonymous creator had been a very non-Freudian psychologist named William Moulton Marston. Also a lawyer, businessman, prison
reformer, and inventor of the lie-detector test (no doubt the inspiration for Wonder Woman's magic lasso), he had invented Wonder Woman as a heroine for little girls, and also as a conscious alternative to the violence of comic books for
boys. In fact, Wonder Woman did attract some boys as readers, but the integrated world of comic book trading revealed her true status: at least three Wonder Woman comic books were necessary to trade for one of Superman. Among
the many male super-heroes, only Superman and Batman were to be as long-
lived as Wonder Woman, yet she was still a second-class citizen.

Of course, it's also true that Marston's message wasn't as feminist as it 18
might have been. Instead of portraying the goal of full humanity for women
and men, which is what feminism has in mind, he often got stuck in the sub-
ject/object, winner/loser paradigm of "masculine" versus "feminine," and
came up with female superiority instead. As he wrote: "Women represent love;
men represent force. Man's use of force without love brings evil and unhappiness. Wonder Woman proves that women are superior to men because they
have love in addition to force." No wonder I was inspired but confused by the
isolationism of Paradise Island: Did women have to live separately in order to
be happy and courageous? No wonder even boys who could accept equality
might have felt less than good about themselves in some of these stories: Were
there *any* men who could escape the cultural instruction to be violent?

Wonder Woman herself sometimes got trapped in this either/or choice. As 19
she muses to herself: "Some girls love to have a man stronger than they are to
make them do things. Do I like it? I don't know, it's sort of thrilling. But isn't it

more fun to make a man obey?" Even female villains weren't capable of being evil on their own. Instead, they were hyperfeminine followers of men's commands. Consider Priscilla Rich, the upper-class antagonist who metamorphoses into the Cheetah, a dangerous she-animal. "Women have been submissive to men," wrote Marston, "and taken men's psychology [force without love] as their own."

In those wartime years, stories could verge on a jingoistic, even racist patri- 20 otism. Wonder Woman sometimes forgot her initial shock at America's unjust patriarchal system and confined herself to defeating a sinister foreign threat by proving that women could be just as loyal and brave as men in service of their country. Her costume was a version of the Stars and Stripes. Some of her adversaries were suspiciously short, ugly, fat, or ethnic as a symbol of "un-American" status. In spite of her preaching against violence and for democracy, the good guys were often in uniform, and no country but the United States was seen as a bastion of freedom.

But Marston didn't succumb to stereotypes as often as most comic book writ- 21 ers of the 1940s. Though Prudence, his frontier heroine, is threatened by monosyllabic Indians, Prudence's father turns out to be the true villain, who has been cheating the Indians. And the irrepressible Etta Candy, one of Wonder Woman's band of college girls, is surely one of the few fat-girl heroines in comics.

There are other unusual rewards. Queen Hippolyte, for instance, is a rare 22 example of a mother who is good, powerful, and a mentor to her daughter. She founds nations, fights to protect Paradise Island, and is a source of strength to Wonder Woman as she battles the forces of evil and inequality. Mother and daughter stay in touch through a sort of telepathic TV set, and the result is a team of equals who are separated only by experience. In the flashback episode in which Queen Hippolyte succumbs to Hercules, she is even seen as a sexual being. How many girl children grew to adulthood with no such example of a strong, sensual mother—except for these slender stories? How many mothers preferred sons, or believed the patriarchal myth that competition is "natural" between mothers and daughters, or tamed their daughters instead of encouraging their wildness and strength? We are just beginning to realize the sense of anger and loss in girls whose mothers had no power to protect them, or forced them to conform out of fear for their safety, or left them to identify only with their fathers if they had any ambition at all.

Finally, there is Wonder Woman's ability to unleash the power of self- 23 respect within the women around her; to help them work together and support each other. This may not seem revolutionary to male readers accustomed to stories that depict men working together, but for females who are usually seen as competing for the favors of men—especially little girls who may just be getting to the age when girlfriends betray each other for the approval of boys—this discovery of sisterhood can be exhilarating indeed. Women get a rare message of independence, of depending on themselves, not even on Wonder Woman. "You saved yourselves," as she says in one of her inevitable morals at story's end. "I only showed you that you could."

Whatever the shortcomings of William Marston, his virtues became clear after 24
his death in 1947. Looking back at the post-Marston stories I had missed the
first time around—for at twelve or thirteen, I thought I had outgrown Wonder
Woman and had abandoned her—I could see how little her later writers understood her spirit. She became sexier-looking and more submissive, violent
episodes increased, more of her adversaries were female, and Wonder Woman
herself required more help from men in order to triumph. Like so many of her
real-life sisters in the postwar era of conservatism and "togetherness" of the
1950s, she had fallen on very hard times.

By the 1960s, Wonder Woman had given up her magic lasso, her bullet- 25
deflecting bracelets, her invisible plane, and all her Amazonian powers. Though
she still had adventures and even practiced karate, any attractive man could
disarm her. She had become a kind of female James Bond, though much more
boring because she was denied his sexual freedom. She was Diana Prince, a
mortal who walked about in boutique, car-hop clothes and took the advice of a
male mastermind named "I Ching."

It was in this sad state that I first rediscovered my Amazon super-hero in 26
1972. *Ms.* magazine had just begun, and we were looking for a cover story for
its first regular issue to appear in July. Since Joanne Edgar and other of its
founding editors had also been rescued by Wonder Woman in their childhoods,
we decided to rescue Wonder Woman in return. Though it wasn't easy to persuade her publishers to let us put her original image on the cover of a new and
unknown feminist magazine, or to reprint her 1940s Golden Age episodes inside, we finally succeeded. Wonder Woman appeared on newsstands again in
all her original glory, striding through city streets like a colossus, stopping
planes and bombs with one hand and rescuing buildings with the other.

Clearly, there were many nostalgic grown-ups and heroine-starved readers 27
of all ages. The consensus of response seemed to be that if we had all read more
about Wonder Woman and less about Dick and Jane, we might have been a lot
better off. As for her publishers, they, too, were impressed. Under the direction
of Dorothy Woolfolk, the first woman editor of Wonder Woman in all her long
history, she was returned to her original Amazon status—golden lasso,
bracelets, and all.

One day some months after her rebirth, I got a phone call from one of Won- 28
der Woman's tougher male writers. "Okay," he said, "she's got all her Amazon
powers back. She talks to the Amazons on Paradise Island. She even has a Black
Amazon sister named Nubia. Now will you leave me alone?"

I said we would.   29

In the 1970s, Wonder Woman became the star of a television series. As played 30
by Lynda Carter, she was a little blue of eye and large of breast, but she still retained her Amazon powers, her ability to convert instead of kill, and her appeal
for many young female viewers. There were some who refused to leave their
TV sets on Wonder Woman night. A few young boys even began to dress up as
Wonder Woman on Halloween—a true revolution.

In the 1980s, Wonder Woman's story line was revamped by DC Comics, 31 which reinvented its male super-heroes Superman and Batman at about the same time. Steve Trevor became a veteran of Vietnam; he remained a friend, but was romantically involved with Etta Candy. Wonder Woman acquired a Katharine Hepburn–Spencer Tracy relationship with a street-smart Boston detective named Ed Indelicato, whose tough-guy attitude played off Wonder Woman's idealism. She also gained a friend and surrogate mother in Julia Kapatelis, a leading archaeologist and professor of Greek culture at Harvard University who can understand the ancient Greek that is Wonder Woman's native tongue, and be a model of a smart, caring, single mother for girl readers. Julia's teenage daughter, Vanessa, is the age of many readers and goes through all of their uncertainties, trials, and tribulations, but has the joy of having a powerful older sister in Wonder Woman. There is even Myndi Mayer, a slick Hollywood public relations agent who turns Wonder Woman into America's hero, and is also in constant danger of betraying Diana's idealistic spirit. In other words, there are many of the currents of society today, from single mothers to the worries of teenage daughters and a commercial culture, instead of the simpler plots of America's dangers in World War II.

You will see whether Wonder Woman carries her true Amazon spirit into 32 the present. If not, let her publishers know. She belongs to you.

Since Wonder Woman's beginnings more than a half century ago, however, a 33 strange thing has happened: the Amazon myth has been rethought as archaeological relics have come to light. Though Amazons had been considered figments of the imagination, perhaps the mythological evidence of man's fear of woman, there is a tentative but growing body of evidence to support the theory that some Amazon-like societies did exist. In Europe, graves once thought to contain male skeletons—because they were buried with weapons or were killed by battle wounds—have turned out to hold skeletons of females after all. In the jungles of Brazil, scientists have found caves of what appears to have been an all-female society. The caves are strikingly devoid of the usual phallic design and theme; they feature, instead, the triangular female symbol, and the only cave that does bear male designs is believed to have been the copulatorium, where Amazons mated with males from surrounding tribes, kept only the female children, and returned male infants to the tribe. Such archaeological finds have turned up not only along the Amazon River in Brazil, but at the foot of the Atlas Mountains in northwestern Africa, and on the European and Asiatic sides of the Black Sea.

There is still far more controversy than agreement, but a shared supposition 34 of these myths is this: imposing patriarchy on the gynocracy of pre-history took many centuries and great cruelty. Rather than give up freedom and worship only male gods, some bands of women resisted. They formed all-woman cultures that survived by capturing men from local tribes, mating with them, and raising their girl children to have great skills of body and mind. These bands became warriors and healers who were sometimes employed for their skills by patriarchal cultures around them. As a backlash culture, they were doomed, but they may also have lasted for centuries.

Perhaps that's the appeal of Wonder Woman, Paradise Island, and this comic book message. It's not only a child's need for a lost independence, but an adult's need for a lost balance between women and men, between humans and nature. As the new Wonder Woman says to Vanessa, "Remember your *power*, little sister." 35

However simplified, that is Wonder Woman's message: Remember Our Power. 36

COMPREHENSION

1. According to the writer, why are children drawn to comic books and super-heroes?
2. Why did Wonder Woman appeal especially to Steinem? What distinctions does she draw between the way boys and girls view action heroes?
3. The writer traces the development of Wonder Woman from her inception during the 1940s to the 1980s. How did Wonder Woman change over the years? How did she remain true to her creator's (William Marston) conception of her? What does Steinem think about these changes?

RHETORIC

1. What is this essay's persuasive thesis?
2. At whom is this essay aimed—lovers of comic books, women, a general audience? On what do you base your conclusion?
3. In part, this is a personal essay. How does Steinem create her persona or self-image? Does the personal element enhance or detract from the analysis? Explain your response.
4. Sort out the complex cause-and-effect relationships in this essay. How does the comparative method reinforce the writer's analysis?
5. What types of evidence does the writer provide? Is it sufficient to convince readers? Where, if anywhere, would more detail be helpful?
6. Steinem divides the essay into five sections. What is her purpose? How successful is she in maintaining the essay's unity by employing this method?
7. What paragraphs form the writer's conclusion? How do they recapitulate and add to the substance of the overall essay?

WRITING

1. Write a personal essay about your favorite action hero or heroine—drawn from the comics, television cartoons, or computer games. Explain why this figure appeals to you, and what it reveals about the broader culture.

2. Compare and contrast the ways in which females and males approach action heroes. Refer to specific icons like the one in the *Terminator* film series or Buffy the Vampire Slayer on television to support your assessment.
3. **WRITING AN ARGUMENT:** Think about the numerous action or super-heroes that young children and adolescents encounter today in various media forms. Write an essay in which you contend that exposure to such super-heroes either does or does not encourage violent behavior in young people.

---

www.mhhe.com/
mhreader

For more information on Gloria Steinem, go to:
More Resources > Ch. 9 Media & Pop Culture

Supersaturation, or, The Media Torrent and Disposable Feeling

Todd Gitlin

Todd Gitlin (b. 1943) was born and grew up in New York City. He received a PhD in sociology from the University of California at Berkeley and was president of Students for a Democratic Society (SDS) in the 1960s. Gitlin is professor of culture, journalism, and sociology at New York University and has held the chair in American civilization at the Ecole des Hautes Etudes en Sciences Sociales in Paris. Gitlin also lectures at home and abroad on contemporary culture and history. He is the North American editor of the Web site openDemocracy.net. Among his notable books are Inside Prime Time *(1983),* The Twilight of Common Dreams: Why America Is Wracked by Culture Wars *(1995), and* Media Unlimited: How the Torrent of Images and Sounds Overwhelms Our Lives *(2001). In the selection from* Media Unlimited *that follows, Gitlin offers an overview of the ways in which the media influence our contemporary lives.*

On my bedroom wall hangs a print of Vermeer's *The Concert*, painted around   1
1660. A young woman is playing a spinet. A second woman, probably her maid, holds a letter. A cavalier stands between them, his back to us. A landscape is painted on the raised lid of the spinet, and on the wall hang two paintings, a landscape and *The Procuress*, a work by Baburen, another Dutch artist, depicting a man and two women in a brothel. As in many seventeenth-century Dutch paintings, the domestic space is decorated by paintings. In wealthy Holland, many homes, and not only bourgeois ones, featured such renderings of the outer world. These pictures were pleasing, but more: they were proofs of taste and prosperity, amusements and news at once.

Vermeer froze instants, but instants that spoke of the relative constancy of 2
the world in which his subjects lived. If he had painted the same room in the
same house an hour, a day, or a month later, the letter in the maid's hand would
have been different, and the woman might have been playing a different selection, but the paintings on the far wall would likely have been the same. There
might have been other paintings, etchings, and prints elsewhere in the house,
but they would not have changed much from month to month, year to year.

In what was then the richest country in the world, "everyone strives to em- 3
bellish his house with precious pieces, especially the room toward the street," as
one English visitor to Amsterdam wrote in 1640, noting that he had observed
paintings in bakeries, butcher's shops, and the workshops of blacksmiths and
cobblers.[1] Of course, the number of paintings, etchings, and prints in homes
varied considerably. One tailor owned five paintings, for example, while at the
high end, a 1665 inventory of a lavish patrician's house in Amsterdam held two
maps and thirteen paintings in one grand room, twelve paintings in his
widow's bedroom, and seven in the maid's room. Still, compared with today's
domestic imagery, the grandest Dutch inventories of that prosperous era were
tiny.[2] Even in the better-off households depicted by Vermeer, the visual field inhabited by his figures was relatively scanty and fixed.[3]

Today, Vermeer's equivalent, if he were painting domestic scenes, or shoot- 4
ing a spread for *Vanity Fair,* or directing commercials or movies, would also display his figures against a background of images; and if his work appeared
on-screen, there is a good chance that he would mix in a soundtrack as well.
Most of the images would be portraits of individuals who have never walked in
the door—not in the flesh—and yet are recognized and welcomed, though not
like actual persons. They would rapidly segue into others—either because they
had been edited into a video montage, or because they appear on pages meant
to be leafed through. Today's Vermeer would discover that the private space of
the home offers up vastly more impressions of the larger world than was possible in 1660. In seventeenth-century Delft, painters did not knock on the door
day and night offering fresh images for sale. Today, though living space has
been set apart from working space, as would have been the case only for the
wealthier burghers of Vermeer's time, the outside world has entered the home
with a vengeance—in the profusion of media.

The flow of images and sounds through the households of the rich world, 5
and the richer parts of the poor world, seems unremarkable today. Only a visi-

---

[1]Peter Mundy, quoted by Geert Mak, *Amsterdam,* trans. Philipp Blom (Cambridge, Mass: Harvard
University Press, 2000), p. 109. [This and subsequent notes in the selection are the author's.]
[2]Simon Schama, *The Embarrassment of Riches: An Interpretation of Dutch Culture in the Golden Age*
(New York: Knopf, 1987), pp. 313–19. Schama notes that research in the relevant archives is "still
in its early days" (p. 315).
[3]Many bourgeois Dutch houses also featured a camera lucida, a mounted magnifying lens trained
on objects in the vicinity. Because the lens was movable, motion could be simulated—distant objects being brought nearer and sent farther away. But because the apparatus was mounted in a
fixed location, the range of objects in motion was limited to those actually visible from the window. (Svetlana Alpers, personal communication, October 8, 1999.)

tor from an earlier century or an impoverished country could be startled by the fact that life is now played out against a shimmering multitude of images and sounds, emanating from television, videotapes, videodiscs, video games, VCRs, computer screens, digital displays of all sorts, always in flux, chosen partly at will, partly by whim, supplemented by words, numbers, symbols, phrases, fragments, all passing through screens that in a single minute can display more pictures than a prosperous seventeenth-century Dutch household contained over several lifetimes, portraying in one day more individuals than the Dutch burgher would have beheld in the course of years, and in one week more bits of what we have come to call "information" than all the books in all the households in Vermeer's Delft. And this is not yet to speak of our sonic surroundings: the music, voices, and sound effects from radios, CD players, and turntables. Nor is it to speak of newspapers, magazines, newsletters, and books. Most of the faces we shall ever behold, we shall behold in the form of images.

Because they arrive with sound, at home, in the car, the elevator, or the   6 waiting room, today's images are capable of attracting our attention during much of the day. We may ignore most of them most of the time, take issue with them or shrug them off (or think we are shrugging them off), but we must do the work of dispelling them—and even then, we know we can usher them into our presence whenever we like. Iconic plenitude is the contemporary condition, and it is taken for granted. To grow up in this culture is to grow into an expectation that images and sounds will be there for us on command, and that the stories they compose will be succeeded by still other stories, all bidding for our attention, all striving to make sense, all, in some sense, *ours*. Raymond Williams, the first analyst to pay attention to the fact that television is not just pictures but flow, and not just flow but drama upon drama, pointed out more than a quarter century ago, long before hundred-channel cable TV and VCRs, that

> we have never as a society acted so much or watched so many others acting. . . .
> [W]hat is really new . . . is that drama . . . is built into the rhythms of everyday
> life. In earlier periods drama was important at a festival, in a season, or as a
> conscious journey to a theater; from honouring Dionysus or Christ to taking in
> a show. What we have now is drama as habitual experience: more in a week, in
> many cases, than most human beings would previously have seen in a lifetime.[4]

Around the time Vermeer painted *The Concert,* Blaise Pascal, who worried   7 about the seductive power of distraction among the French royalty, wrote that "near the persons of kings there never fail to be a great number of people who see to it that amusement follows business, and who watch all the time of their leisure to supply them with delights and games, so that there is no blank in it."[5] In this one respect, today almost everyone—even the poor—in the rich countries resembles a king, attended by the courtiers of the media offering a divine right of choice.

---

[4]"Drama in a Dramatised Society," in Alan O'Connor, ed., *Raymond Williams on Television* (Toronto: Between the Lines, 1989 [1974]), pp. 3–5. *Flow* comes up in Williams's *Television: Technology and Cultural Form* (New York: Schocken, 1975), pp. 86 ff.

[5]*Pensées,* trans. W. F. Trotter (www.eserver.org/philosophy/pascal-pensees.txt), sec. 2, par. 142.

Measures of Magnitude

Statistics begin—but barely—to convey the sheer magnitude of this in-   8
touchness, access, exposure, plenitude, glut, however we want to think of it.

In 1999, a television set was on in the average American household more   9
than seven hours a day, a figure that has remained fairly steady since 1983. According to the measurements of the A. C. Nielsen Company, the standard used
by advertisers and the television business itself, the average individual watched
television about four hours a day, not counting the time when the set was on but
the individual in question was not watching. When Americans were asked to
keep diaries of how they spend their time, the time spent actually watching
dropped to a still striking three hours a day—probably an undercount. In 1995,
of those who watched, the percentage who watched "whatever's on," as opposed to any specific program, was 43 percent, up from 29 percent in 1979.[6]
Though cross-national comparisons are elusive because of differences in measurement systems, the numbers in other industrialized nations seem to be
comparable—France, for example, averaging three and a half hours per person.[7]
One survey of forty-three nations showed the United States ranking third in
viewing hours, after Japan and Mexico. None of this counts time spent discussing programs, reading about their stars, or thinking about either.[8]

Overall, wrote one major researcher in 1990, "watching TV is the dominant   10
leisure activity of Americans, consuming 40 percent of the average person's free
time as a primary activity [when people give television their undivided atten-
tion]. Television takes up more than half of our free time if you count . . . watching TV while doing something else like eating or reading . . . [or] when you have
the set on but you aren't paying attention to it."[9] Sex, race, income, age, and
marital status make surprisingly little difference in time spent.[10] Neither, at this
writing, has the Internet diminished total media use, even if you don't count the
Web as part of the media. While Internet users do watch 28 percent less television, they spend more time than nonusers playing video games and listening to
the radio and recorded music—obviously a younger crowd. Long-term users
(four or more years) say they go on-line for more than two hours a day, and
boys and girls alike spend the bulk of their Internet time entertaining themselves with games, hobbies, and the like.[11] In other words, the Internet redistrib-

---

[6]Robert D. Putnam, *Bowling Alone: The Collapse and Revival of American Community* (New York:
Simon and Schuster, 2000), p. 222, citing John P. Robinson and Geoffrey Godbey, *Time for Life:
The Surprising Ways Americans Use Their Time*, 2nd ed. (University Park: Pennsylvania State
University Press, 1999), pp. 136–53, 340–41, 222.
[7]This April 2001 figure for individuals fifteen and older comes from Mediamat (Mediame-
triewww.mediametria.fr/television/mediamat_mensuel/2001/avril.html).
[8]Putnam. *Bowling Alone*, p. 480, citing Eurodata TV (*One Television Year in the World: Audience
Report*, April 1999).
[9]John P. Robinson, "I Love My TV," *American Demographics*, September 1990, p. 24.
[10]Robert Kubey and Mihaly Csikszentmihalyi, *Television and the Quality of Life: How Viewing Shapes
Everyday Experience* (Hillsdale, N.J.: Lawrence Erlbaum Associates, 1990), pp. 71–73.
[11]UCLA Center for Communication Policy, *The UCLA Internet Report: Surveying the Digital Future*,
November 2000, pp. 10, 17, 18, 14 (www.ccp.ucla.edu).

utes the flow of unlimited media but does not dry it up. When one considers the overlapping and additional hours of exposure to radio, magazines, newspapers, compact discs, movies (available via a range of technologies as well as in theaters), and comic books, as well as the accompanying articles, books, and chats about what's on or was on or is coming up via all these means, it is clear that the media flow into the home—not to mention outside—has swelled into a torrent of immense force and constancy, an accompaniment *to* life that has become a central experience *of* life.

The place of media in the lives of children is worth special attention—not 11 simply because children are uniquely impressionable but because their experience shapes everyone's future; if we today take a media-soaked environment for granted, surely one reason is that we grew up in it and can no longer see how remarkable it is. Here are some findings from a national survey of media conditions among American children aged two through eighteen. The average American child lives in a household with 2.9 televisions, 1.8 VCRs, 3.1 radios, 2.6 tape players, 2.1 CD players, 1.4 video game players, and 1 computer. Ninety-nine percent of these children live in homes with one or more TVs, 97 percent with a VCR, 97 percent with a radio, 94 percent with a tape player, 90 percent with a CD player, 70 percent with a video game player, 69 percent with a computer. Eighty-eight percent live in homes with two or more TVs, 60 percent in homes with three or more. Of the 99 percent with a TV, 74 percent have cable or satellite service.[12] And so on, and on, and on.

The uniformity of this picture is no less astounding. A great deal about the 12 lives of children depends on their race, sex, and social class, but access to major media does not. For TV, VCR, and radio ownership, rates do not vary significantly among white, black, and Hispanic children, or between girls and boys. For television and radio, rates do not vary significantly according to the income of the community.[13]

How accessible, then, is the media cavalcade at home? Of children eight to 13 eighteen, 65 percent have a TV in their bedrooms, 86 percent a radio, 81 percent a tape player, 75 percent a CD player. Boys and girls are not significantly different in possessing this bounty, though the relative usages do vary by medium. Researchers also asked children whether the television was "on in their homes even if no one is watching 'most of the time,' 'some of the time,' 'a little of the time,' or 'never.'" Homes in which television is on "most of the time" are

---

[12]Donald F. Roberts, *Kids and Media @ the New Millennium* (Menlo Park, Calif.: Henry J. Kaiser Family Foundation, 1999), p. 9, table 1. There were 3,155 children in the sample, including oversamples of black and Hispanic children, to ensure that results in these minority populations would also be statistically significant. As best as a reader can discern, this was a reliable study, with a margin of error of no more than plus-or-minus five percentage points. Since the results for younger children, ages two to seven, come from parents' reports, they may well be conservative, since parents may be uninformed of the extent of their children's viewing or may be underplaying it in order not to feel ashamed before interviewers.
[13]Ibid., p. 11, tables 3-A, 3-B, 3-C.

termed *constant television households*. By this measure, 42 percent of all American households with children are constant television households. Blacks are more likely than whites or Hispanics to experience TV in their lives: 56 percent of black children live in constant television households (and 69 percent have a TV in their bedrooms, compared to 48 percent of whites). The lower the family education and the median income of the community, the greater the chance that a household is a constant television household.[14]

As for time, the average child spent six hours and thirty-two minutes per day exposed to media of all kinds, of which the time spent reading books and magazines—not counting schoolwork—averaged about forty-five minutes. For ages two to seven, the average for total media was four hours and seventeen minutes; for ages eight to thirteen, eight hours and eight minutes, falling to seven hours and thirty-five minutes for ages fourteen to eighteen.[15] Here, race and social class do count. Black children are most exposed, followed by Hispanics, than whites. At all age levels, the amount of exposure to all media varies inversely with class, from six hours and fifty-nine minutes a day for children in households where the median income for the zip code is under $25,000 to six hours and two minutes for children whose zip code median income is over $40,000. The discrepancy for TV exposure is especially pronounced, ranging from three hours and six minutes a day for children whose zip code incomes are under $25,000 to two hours and twenty-nine minutes for children whose zip code incomes are over $40,000.[16] Still, these differences are not vast. Given everything that divides the rich from the poor, the professional from the working class—differences in physical and mental health, infant mortality, longevity, safety, vulnerability to crime, prospects for stable employment, and so on—the class differences in media access and use are surprisingly slender. So are the differences between American and western European children, the latter averag-

---

[14]Ibid., pp. 13–15, tables 4, 5-A, 5-B, 6. In general, fewer western European or Israeli children than Americans have TVs in their bedrooms, but 70 percent in Great Britain do. Next highest in Europe is 64 percent in Denmark. The lows are 31 percent in Holland and 24 percent in Switzerland. Leen d'Haenens, "Old and New Media: Access and Ownership in the Home," in Sonia Livingstone and Moira Bovill, eds., *Children and Their Changing Media Environment: A European Comparative Study* (London: Lawrence Erlbaum Associates, 2001), p. 57.

[15]Roberts, *Kids and Media*, pp. 21–23, tables 8-C, 8-D.

[16]The same point applies to differences in media use throughout the prosperous world. As the economist Adair Turner writes: "European Internet penetration lags the US by 18 to 24 months. When cars or television sets were first introduced, the lag was more like 15 years. . . . The shortness of the lag also suggests that social concern about a 'digital divide,' whether within or between nations, is largely misplaced. . . . Time lags between different income groups in the penetration of personal computers, Internet connections or mobile phones are much shorter, once again because all these products are cheap. . . . At the global level the same scepticism about a digital divide should prevail. Africa may lag 15 years or so behind US levels of PC and Internet penetration, but it lags more like a century behind in basic literacy and health care." Adair Turner, "Not the e-conomy," *Prospect* (London), April 2001 (www.prospect-magazine.co.uk/highlights/essay_turner_april01).

ing six hours a day total, though in Europe only two and a quarter of those hours are spent with TV.[17]

All such statistics are crude, of course. Most of them register the time that people *say* they spend. They are—thankfully—not checked by total surveillance. Moreover, the meaning of *exposure* is hard to assess, since the concept encompasses rapt attention, vague awareness, oblivious coexistence, and all possible shadings in between. As the images glide by and the voices come and go, how can we assess what goes on in people's heads? Still, the figures do convey some sense of the media saturation with which we live—and so far we have counted only what can be counted at home. These numbers don't take into account the billboards, the TVs at bars and on planes, the Muzak in restaurants and shops, the magazines in the doctor's waiting room, the digital displays at the gas pump and over the urinal, the ads, insignias, and logos whizzing by on the sides of buses and taxis, climbing the walls of buildings, making announcements from caps, bags, T-shirts, and sneakers. To vary our experience, we can pay to watch stories about individuals unfold across larger-than-life-size movie screens, or visit theme parks and troop from image to image, display to display. Whenever we like, on foot or in vehicles, we can convert ourselves into movable nodes of communication, thanks to car radios, tape, CD, and game players, cell phones, beepers, Walkmen, and the latest in "personal communication systems"—and even if we ourselves refrain, we find ourselves drawn willy-nilly into the soundscape that others broadcast around us.

Crucially, who we are is how we live our time—or *spend* it, to use the term that registers its intrinsic scarcity. What we believe, or say we believe, is less important. We vote for a way of life with our time. And increasingly, when we are not at work or asleep, we are in the media torrent. (Sometimes at work, we are also there, listening to the radio or checking out sports scores, pin-ups, or headlines on the Internet.) Steadily more inhabitants of the wealthy part of the world have the means, incentives, and opportunities to seek private electronic companionship. The more money we have to spend, the more personal space each household member gets. With personal space comes solitude, but this solitude is instantly crowded with images and soundtracks. To a degree that was unthinkable in the seventeenth century, life experience has become an experience in the presence of media.

[17]Johannes W.J. Beentjes et al., "Children's Use of Different Media: For How Long and Why?" in Livingstone and Bovill, eds., *Children and Their Changing Media Environment,* p. 96.

COMPREHENSION

1. What does the author's title mean? How are the concepts of "supersaturation" and "disposable feeling" reflected in the essay?
2. Summarize Gitlin's treatment of Vermeer. Who was Vermeer? How, according to the writer, would his art be produced today?

3. List some of the facts and statistics that the writer presents to support his idea that we are caught in a "media torrent."

RHETORIC

1. What is Gitlin's argument? Where does he state his claim most clearly? What appeals to logic, ethics, and emotion does he make? Does he rely on his own opinions? Justify your answer.
2. How does the writer maintain unity between the two parts of this essay? Why, for example, does he open his essay with the story of Vermeer? How does Vermeer serve as a unifying element? What other unifying motifs can you find?
3. What varieties of evidence does the writer provide to bolster his argument? Does he rely on anecdotal or actual evidence? How do you know?
4. The writer employs a range of rhetorical strategies in this essay. Point to places where he uses description, comparison and contrast, classification, definition, and causal analysis.
5. Is the author's style personal, informal, or formal? How does this style explain Gitlin's relationship to his audience and the expectations he holds of his readers?
6. What strategy does the writer use in his conclusion? Is this strategy effective? Why or why not?
7. What do Gitlin's footnotes add to the essay? Why are they important?

WRITING

1. Write an essay in which you tell of how one medium—television or the Internet, for example—has affected or changed your life. Make sure that you provide adequate detail or evidence.
2. Write a comparative essay in which you analyze the similarities and differences between an "old" and a "new" component of the media—actual books and recordings of books, or telephones and cell phones, for example.
3. **WRITING AN ARGUMENT:** Write an essay in which you either agree or disagree with Gitlin's claim that increasingly "we are in a media torrent." Use appeals to logic, ethics, and emotion to advance your claim. Make certain that you have adequate evidence to support your major and minor propositions.

www.mhhe.com/
mhreader

For more information on Todd Gitlin, go to:
More Resources > Ch. 9 Media & Pop Culture

CONNECTIONS FOR CRITICAL THINKING

1. Examine the role of the media in society and the responsibilities or duties to humanity of individuals associated with the media. Use at least three essays from this chapter to illustrate or support your thesis.
2. Define *popular culture,* using the essays of Barry, Gates, and Steinem as reference points, along with any additional essays that you consider relevant.
3. Compare and contrast Dove's views on television with those of Winn.
4. Use the essays of Warshow, Willis, and Gitlin to explore the connections of media representations to American cultural experience. What strategies do these writers use? Are their goals similar?
5. Use the essays of Gates and Barry to explore the importance of both the causes and effects of the media promoting particular lifestyles to the public.
6. Gates refers to the African American "style" of communicating through music; and Barry presents beer commercials as communicating the traditional "patriotic symbols" of America. Do these authors have similar or differing points of view regarding the issues they address? Refer specifically to selections in each essay to support your view.
7. Search the Internet using the keywords *television* and *teenagers.* Select three or four sites, and write an expository paper describing the various ways the authors interpret any of the major themes.
8. After reading the essays of Warshow, Willis, Gates, and King, research the issue of the difference between popular entertainment and art. On the basis of your research, discuss whether there are legitimate criteria that distinguish the two forms. Apply these criteria to gangster films, rap music, and horror films.
9. Select several images of real "gangsters" from magazines or print out images of true gangsters from the Internet. Compare and contrast them with advertisements depicting gangsters from contemporary crime movies such as *Pulp Fiction.* What are the similarities and differences in the subjects' dress, demeanor, facial expression, and so on? What can you conclude from your comparisons?

chapter *10*

Literature and the Arts
Why Do They Matter?

Imagine a world without fiction, poetry, or drama; music, art, or other fine arts. We are so accustomed to taking the arts in their totality for granted that it is hard for us to conceive of contemporary culture without them. Our fondness for stories or paintings or any other creative form might help us understand our culture or might even move us to action. Yet the value of various artistic forms doesn't derive exclusively from their ability to tell us something about life. The arts can also take us into an imaginative realm offering perhaps more intense experiences than anything we encounter in the "real" world.

Think of literature and the arts as an exercise in imaginative freedom. You are free to select the books you read, the music that appeals to you, the exhibitions and concerts you attend, and the entertainment software with which you interact. Some of your decisions might be serious and consequential to your education. Other decisions, perhaps to watch a few soap operas on a rainy afternoon or to buy the latest potboiler, are less important. The way you view the arts—whether as a temporary escape from conventional reality or as a way to learn something about the temper of civilization—is entirely a matter of taste. Regardless of your purpose or intent, you approach literature and the arts initially for the sheer exhilaration and pleasure they provide. Art, as Plato observed, is a dream for awakened minds.

The arts awaken you to the power and intensity of the creative spirit. At the same time, you make judgments and evaluations of the nature of your creative encounter. When you assert that you like this painting or dislike that poem, you are assessing the work or value of the artistic experience. It is clear that you develop taste and become more equipped to discern the more subtle elements of art the more you are exposed to it. Perhaps you prefer to keep your experience of literature and the other arts a pleasurable pastime or an escape from reality. Or you may wish to participate in them as a creative writer, musician, painter, or photographer. Ultimately you may come to view literature and the arts as a transformational experience, a voyage of discovery in which you encounter diverse peoples and cultures, learn to see the world in creative terms, and begin to perceive your own creative potential in a new light.

Previewing the Chapter

As you read the essays in this chapter and respond to them in discussion and writing, consider the following questions:

- According to the author, what is the value of the art or literary form under discussion?
- What function does literature or art serve?
- Is the writer's perspective subjective or objective, and why?
- How does the author define his or her subject—whether it is poetry, fiction, art, or photography?
- Is the writer's experience of literature or art similar to or different from your own?
- In what ways do gender and race influence the writer's perspective on the subject?
- What is the main idea that the author wants to present about literature or the arts? Do you agree or disagree with this key concept?
- What have you learned about the importance of literature and the arts from reading these essays?

Classic and Contemporary Images
How Do We Evaluate a Work of Art?

Using a Critical Perspective   Although "greatness" in art and literature might be in the mind of the beholder, it could be argued that you need certain standards of excellence or judgment to determine the quality of any work. The artist's or writer's control of the medium, the projection of a unique vision, the evidence of a superlative style, and more: all enter into the evaluation process. As you consider the following sculptures by Auguste Rodin and Jeff Koons, try to evaluate their relative worth. Which work reflects greater artistic control? What makes the sculpture appealing, and why? Which work strikes you as "new" or original, or modern? Explain your response and criteria for evaluation.

Auguste Rodin (1840–1917), a French sculptor famous for his bronze and
marble figures, is thought by some critics to be one of the greatest
portraitists in the history of sculpture. Yet he was also criticized in
his time for the excessive realism and sensuousness of his figures.
Walking Man hints at some of the objections contemporary
critics lodged against Rodin's work.

568

Jeff Koons (b. 1955) is an American artist who, like Rodin, has had his admirers and detractors. Koons studied at the Art Institute of Chicago and elsewhere before becoming a commodity trader in New York City, which helped fund the materials for his art. *Rabbit* reflects Koons's fondness for popular culture and the way in which he takes consumer goods and repositions them as art objects.

Classic and Contemporary Essays
How Do We Know It's Good?

E. M. Forster and Fran Lebowitz, one English and the other American, represent two generations and two different literary and cultural styles. Forster, educated at Kings College, Cambridge, was associated with the famous Bloomsbury Group of writers, artists, and bohemians. Lebowitz, a high school dropout, removed by geography and time from her predecessor, came to maturity in the trendy artistic milieu of New York City. Yet Forster and Lebowitz share a hearty skepticism about the intellectual and artistic universe they inhabit. Each employs an ironic/satiric style—a common voice—to describe their respective artistic realms. Forster is more self-referential, adopting a self-effacing persona to confess his inability to look at pictures. Lebowitz, although she also is an actor in her essay, is more other-directed, creating an imaginary "Mr. Art" to accompany her on her rounds of Soho's avant-garde art scene. Forster roots his observations in classical art: Michelangelo, Velasquez, Van Gogh. On the other hand, Lebowitz focuses her eye on kitsch and pure junk, wondering why it is worth anyone's time to survey the contemporary art scene. However, both writers have the same purpose in mind—an objective shared by all the writers in this chapter: They want to establish those criteria by which we can determine if any work of art is good or bad, why literature and the arts matter, and what the arts tell us about ourselves, our culture, and our world.

✳

Not Looking at Pictures

E. M. Forster

Edward Morgan Forster (1879–1970), English essayist, novelist, biographer, and literary critic, wrote several notable works of fiction dealing with the constrictive effects of social and national conventions on human relationships. These novels include A Room with a View *(1908),* Howard's End *(1910), and* A Passage to India *(1924). In addition, his lectures on fiction, collected as* Aspects of the Novel *(1927), remain graceful elucidations of the genre. In "Not Looking at Pictures," an essay taken from* Two Cheers for Democracy *(1939), Forster offers a whimsical account of difficulties when trying to evaluate art.*

Pictures are not easy to look at. They generate private fantasies, they furnish 1
material for jokes, they recall scraps of historical knowledge, they show landscapes where one would like to wander and human beings whom one would
like to resemble or adore, but looking at them is another matter, yet they must
have been painted to be looked at. They were intended to appeal to the eye, but
almost as if it were gazing at the sun itself the eye often reacts by closing as soon
as it catches sight of them. The mind takes charge instead and goes off on some
alien vision. The mind has such a congenial time that it forgets what set it going.
Van Gogh and Corot and Michelangelo are three different painters, but if the
mind is undisciplined and uncontrolled by the eye, they may all three induce
the same mood; we may take just the same course through dreamland or fun-
land from them, each time, and never experience anything new.

I am bad at looking at pictures myself, and the late Roger Fry enjoyed going 2
to a gallery with me now and then, for this very reason. He found it an amusing
change to be with someone who scarcely ever saw what the painter had
painted. "Tell me, why do you like this, why do you prefer it to that?" he would
ask, and listen agape for the ridiculous answer. One day we looked at a fif-
teenth-century Italian predella, where a St. George was engaged in spearing a
dragon of the plesiosaurus type. I laughed. "Now, *what* is there funny in this?"
pounced Fry. I readily explained. The fun was to be found in the expression
upon the dragon's face. The spear had gone through its hooped-up neck once,
and now startled it by arriving at a second thickness. "Oh dear, here it comes
again, I hoped that was all" it was thinking. Fry laughed too, but not at the misfortunes of the dragon. He was amazed that anyone could go so completely off
the lines. There was no harm in it—but really, really! He was even more amazed
when our enthusiasms coincided: "I fancy we are talking about different
things," he would say, and we always were; I liked the mountain-back because
it reminded me of a peacock, he because it had some structural significance,
though not as much as the sack of potatoes in the foreground.

Long years of wandering down miles of galleries have convinced me that   3
there must be something rare in those colored slabs called "pictures," something which I am incapable of detecting for myself, though glimpses of it are to
be had through the eyes of others. How much am I missing? And what? And
are other modern sightseers in the same fix? Ours is an aural rather than a visual age, we do not get so lost in the concert hall, we seem able to hear music for
ourselves, and to hear it as music, but in galleries so many of us go off at once
into a laugh or a sigh or an amorous day-dream. In vain does the picture recall
us. "What have your obsessions got to do with me?" it complains. "I am neither
a theatre of varieties nor a spring-mattress, but paint. Look at my paint." Back
we go—the picture kindly standing still meanwhile, and being to that extent
more obliging than music—and resume the looking-business. But something is
sure to intervene—a tress of hair, the half-open door of a summer-house, a Crivelli dessert, a Bosch fish-and-fiend salad—and to draw us away.

One of the things that helps us keep looking is composition. For many years   4
now I have associated composition with a diagonal line, and when I find such a
line I imagine I have gutted the picture's secret. Giorgione's Castelfranco
Madonna has such a line in the lance of the warrior-saint, and Titian's Entombment at Venice has a very good one indeed. Five figures contribute to make up
the diagonal; beginning high on the left with the statue of Moses, it passes
through the heads of the Magdalene, Mary, and the dead Christ, and plunges
through the body of Joseph of Arimathea into the ground. Making a right angle
to it, flits the winged Genius of Burial. And to the right, apart from it, and per-
pendicular, balancing the Moses, towers the statue of Faith. Titian's Entombment is one of my easiest pictures. I look at photographs of it intelligently, and
encourage the diagonal and the pathos to reinforce one another. I see, with more
than usual vividness, the grim alcove at the back and the sinister tusked
pedestals upon which the two statues stand. Stone shuts in flesh; the whole picture is a tomb. I hear sounds of lamentation, though not to the extent of shatter-
ing the general scheme; that is held together by the emphatic diagonal, which
no emotion breaks. Titian was a very old man when he achieved this master-
piece; that too I realize, but not immoderately. Composition here really has been
a help, and it is a composition which no one can miss: the diagonal slopes as obviously as the band on a threshing-machine, and vibrates with power.

Unfortunately, having no natural esthetic aptitude, I look for diagonals   5
everywhere, and if I cannot find one thing the composition must be at fault. It is
a word which I have learnt—a solitary word in a foreign language. For instance,
I was completely baffled by Velasquez's Las Meninas. Wherever was the diagonal? Then the friend I was with—Charles Mauron, the friend who, after Roger
Fry, has helped me with pictures most—set to work on my behalf, and cautiously underlined the themes. There is a wave. There is a half-wave. The wave
starts up on the left, with the head of the painter, and curves down and up
through the heads of the three girls. The half-wave starts with the head of Isabel
de Velasco, and sinks out of the canvas through the dwarfs. Responding to these
great curves, or inverting them, are smaller ones on the women's dresses or

elsewhere. All these waves are not merely pattern; they are doing other work too—e.g., helping to bring out the effect of depth in the room, and the effect of air. Important too is the pushing forward of objects in the extreme left and right foregrounds, the easel of the painter in the one case, the paws of a placid dog in the other. From these, the composition curves back to the central figure, the lovely child-princess. I put it more crudely than did Charles Mauron, nor do I suppose that his account would have been Velasquez's, or that Velasquez would have given any account at all. But it is an example of the way in which pictures should be tackled for the benefit of us outsiders: coolly and patiently, as if they were designs, so that we are helped at last to the appreciation of something non-mathematical. Here again, as in the case of the Entombment, the composition and the action reinforced one another. I viewed with increasing joy that adorable party, which had been surprised not only by myself but by the King and Queen of Spain. There they were in the looking-glass! Las Meninas has a snapshot quality. The party might have been taken by Philip IV, if Philip IV had had a Kodak. It is all so casual—and yet it is all so elaborate and sophisticated, and I suppose those curves and the rest of it help to bring this out, and to evoke a vanished civilization.

Besides composition there is color. I look for that, too, but with even less 6 success. Color is visible when thrown in my face—like the two cherries in the great grey Michael Sweertz group in the National Gallery. But as a rule it is only material for dream.

On the whole, I am improving, and after all these years, I am learning to get 7 myself out of the way a little, and to be more receptive, and my appreciation of pictures does increase. If I can make any progress at all, the average outsider should do better still. A combination of courage and modesty is what he wants. It is so unenterprising to annihilate everything that's made to a green thought, even when the thought is an exquisite one. Not looking at art leads to one goal only. Looking at it leads to so many.

COMPREHENSION

1. Why does the author declare, "Pictures are not easy to look at" (paragraph 1)? Why is Forster himself "bad at looking at pictures" (paragraph 2)?
2. What does the author seem to like about art? Why does he persist in viewing artworks, despite his difficulties?
3. What does Forster say about "composition" and "color"?

RHETORIC

1. What is Forster's purpose in writing this essay? What response does he expect of his audience? How do you know?
2. Identify these allusions in the essay: Van Gogh, Corot, Michelangelo (paragraph 1); Roger Fry (paragraph 2); Crivelli, Bosch (paragraph 3); Giorgione,

Titian (paragraph 4); and Velasquez (paragraph 5). What do these allusions tell us about the degree of Forster's expertise?

3. Analyze the material presented in the introduction and the strategies employed.

4. How does Forster employ illustration to help structure each paragraph in this essay?

5. Explain the author's use of comparison and contrast in this essay.

6. What elements contribute to the gently humorous tone of this essay?

WRITING

1. Forster mentions "composition" and "color" as two aspects of art appreciation. Elaborate on these qualities and others in an essay explaining how you evaluate pictures.

2. Analyze a particular work of art based on a museum trip or an illustration.

3. **WRITING AN ARGUMENT:** Forster declares, "Ours is an aural rather than visual age" (paragraph 3). Do you agree or disagree with his assertion, and why?

www.mhhe.com/
mhreader

For more information on E. M. Forster, go to:
More Resources > Ch. 10 Literature & Arts

✳

Soho: Or Not at Home with Mr. Art

Fran Lebowitz

Fran Lebowitz (b. 1950), born in Morristown, New Jersey, is a humorist best known for her satirical essays on trendy urban lifestyles. A high school dropout (she ultimately obtained an equivalency degree), Lebowitz held self-described "colorful and picturesque" jobs in New York City, driving a taxi, cleaning apartments, and selling advertising, before breaking into journalism as a writer for Andy Warhol's Interview *magazine. Lebowitz has written* Metropolitan Life *(1978),* Social Studies *(1981),* Mr. Chas and Lisa Sue Met the Pandas *(1994), and* The Fran Lebowitz Reader *(1994). In the essay that follows, Lebowitz offers a deft, amusing portrait of the art scene in Soho, in New York City.*

Soho is a real place. By real I mean that it exists materially. Mr. Art is not a real   1
person. By not a real person I mean that he does not exist materially. Nevertheless, where Soho is concerned I wouldn't consider excluding Mr. Art from my

observations, for in matters such as this he is my most trusted adviser and confidant. He is a dapper little fellow, perhaps a bit wry—and while it has been said of him that his manner cannot really support his mannerisms, he is, I can assure you, a welcome relief from some of these other types we get in here.

I will introduce you to Soho slowly by noting that this area of downtown 2 Manhattan shares with Mr. Art nary a single characteristic. Up until a few years ago Soho was an obscure district of lofts used chiefly for storage and light manufacturing. It wasn't called Soho then—it wasn't called anything because no one ever went there except the people who make Christmas tree ornaments out of styrofoam and glitter or fabric trimmings out of highly colored stretch felt. And say what you will about members of these professions, they are generally, I am sure, very nice people who not only don't make those things out of choice but also don't go around calling obscure districts of Manhattan things like Soho. Ostensibly, Soho is called Soho because it begins *So*uth of *Ho*uston Street, but if you want my opinion I wouldn't be too terribly surprised to discover that the person who thought up this name is a person whose circle of friends in 1967 included at least one too many English photographers. It was, of course, a combination of many unattractive things that led to the Soho of today, but quite definitely the paramount factor was the advent of Big Art. Before Big Art came along, painters lived, as God undoubtedly intended them to, in garrets or remodeled carriage houses, and painted paintings of a reasonable size. A painting of a reasonable size is a painting that one can easily hang over a sofa. If a painting cannot be easily hung over a sofa it is obviously a painting painted by a painter who got too big for his brushes and is in fact the very sort of painting responsible for Mr. Art's chronically curled upper lip. Painters, however, are not the only ones involved here. Modern sculptors, or *those chiselers* as Mr. Art is wont to call them, must bear a good part of the blame, for when clay and marble went out and demolished tractor-trailer trucks came in, Big Art was here to stay.

One day a Big Artist realized that if he took all of the sewing machines and 3 bales of rags out of a three-thousand-square-foot loft and put in a bathroom and kitchen he would be able to live and make Big Art in the same place. He was quickly followed by other Big Artists and they by Big Lawyers, Big Boutique Owners, and Big Rich Kids. Soon there was a Soho and it was positively awash in hardwood floors, talked-to plants, indoor swings, enormous record collections, hiking boots, Conceptual Artists, video communes, Art book stores, Art grocery stores, Art restaurants, Art bars, Art galleries, and boutiques selling tie-dyed raincoats, macramé flower pots, and Art Deco salad plates.

Since the beginning of the Soho of today the only people in New York who 4 have been able to get through a Saturday afternoon without someone calling them on the telephone to suggest that they go down to Soho and look at the Art are those who belong to Black Nationalist organizations. As neither myself nor Mr. Art is a member of such a group, we consider it quite a feather in our mutual cap that we have succumbed to these ofttimes strongly worded suggestions so infrequently, and that on the rare occasions that we have we certainly have not been gracious about it.

A recent Saturday was just such an occasion and here is what we saw: 5

Art Gallery Number One

A girl who would probably have been a welcome addition to the teaching staff   6
of any progressive nursery school in the country had instead taken it upon herself to create out of ceramic clay exact replicas of such leather objects as shoes,
boots, suitcases, and belts. There was no question but that she had achieved her
goal—one had literally to snap one's fingernail against each object and hear it
ring before one was convinced that what one was snapping one's fingernail
against was indeed ceramic clay and not leather. And one could, of course,
choose to ignore Mr. Art as he hissed, "Why bother?" and struck a match across
a pair of gloves in order to light one of his aromatic foreign cigarettes.

Art Gallery Number Two

A young man who had apparently been refused admission to the Boy Scouts on   7
moral grounds had arranged on a shiny oak floor several groups of rocks. He
had then murdered a number of adolescent birch trees in order to bend them
into vaguely circular shapes and hang them on the wall. These things were all
for sale at prices that climbed well into the thousands. "First of all, imagine actually wanting to *own* any of this stuff," sneered Mr. Art, "and then imagine not
being able to figure out that with an ax and a wheelbarrow you could make it
all yourself in a single morning and still have time to talk to your plants."

Art Gallery Number Three

Two boys who were really good friends had taken a trip to North Africa. They   8
took a lot of color photographs of bowls, skies, pipes, animals, water, and each
other. They had arranged the photographs alphabetically—i.e., A—Ashes, B—
Bright sunny day—pasted them to pieces of varnished plywood, written intri-
cately simple little explanations beneath each photograph, and hung them
under their appropriate letters. I am compelled to admit that upon viewing this
work Mr. Art had to be forcibly restrained from doing bodily harm to himself
and those around him.

Art Gallery Number Four

Someone who had spent a deservedly lonely childhood in movie theaters had   9
gotten hold of a lot of stills from forties films, cut out the faces of the stars, hand-
colored them, and pasted them to blow-ups of picture postcards from Hollywood and Las Vegas. "Too camp," said Mr. Art testily upon being awakened;
"they oughta lock 'em all up."

Art Galleries Number Five through Sixteen

Scores of nine-by-twelve photo-realist renderings of gas stations, refrigerators,   10
pieces of cherry pie, art collectors, diners, '59 Chevys, and Mediterranean-style
dining room sets.

Mr. Art and I are presently seeking membership in a Black Nationalist or- 11
ganization. In the meantime we have taken our phone off the hook.

COMPREHENSION

1. How does the author describe Soho?
2. Who is the writer's friend, Mr. Art?
3. According to the author, what kinds of art are exhibited in Soho galleries? Who are the artists?

RHETORIC

1. How does Lebowitz establish a humorous tone in this piece? What is her purpose?
2. Does the author present a thesis or claim in this essay? Justify your answer.
3. What strategies does the author employ to introduce her subject in paragraphs 1–2?
4. Explain the writer's use of classification to organize part of the essay.
5. Does the writer have an adequate conclusion? Why or why not?

WRITING

1. Write a humorous essay on some aspect of literature or art—or on a specific work—that you consider bad, foolish, or pretentious.
2. Write an extended definition of *avant-garde*. Apply the term to art, music, literature, or any other medium you consider appropriate.
3. **WRITING AN ARGUMENT:** In an argumentative essay, take issue with Lebowitz, claiming that avant-garde art, music, or poetry enriches culture and has intrinsic value.

---

www.mhhe.com/
mhreader

For more information on Fran Lebowitz, go to:
More Resources > Ch. 10 Literature & Arts

Classic and Contemporary: Questions for Comparison

1. Explain how Forster and Lebowitz use humor—including irony and satire—in their essays to express their opinions. Use specific examples from their essays, and discuss why humor is particularly effective for the topic.
2. According to both writers, it is hard to appreciate art. What reasons do they give to support their shared proposition?

3. How are Forster and Mr. Art similar and dissimilar?
4. Compare the language used in both essays. How do they reflect the times in which the essays were written and the writers' respective backgrounds? Which essay do you find more effective stylistically, and why?

✳

Hearing Voices

Linda Hogan

Linda Hogan (b. 1947) is a Chickasaw poet, novelist, short-story writer, and essayist. She was born in Denver, Colorado, and earned her MA at the University of Colorado. She has been a professor of American and American Indian Studies at the University of Minnesota and the University of Colorado. She is one of three Indian writers who were commissioned to co-author a book celebrating the grand opening of the Smithsonian National Museum of the American Indian in 2004. Among her many books are the novel Mean Spirit *(1990), which was a finalist for the Pulitzer Prize, and the poetry collection* The Book of Medicines *(1993), a finalist for the National Book Critics Circle Award. A versatile writer, she has also published* The Woman Who Watches over the World: A Native Memoir *(2000) and* The Mysterious Journey of the Gray Whale *(2000). In the essay that follows, Hogan offers reflections on the complex interactions of tribal stories, poetry, landscape, and the human condition.*

When Barbara McClintock was awarded a Nobel Prize for her work on gene transposition in corn plants, the most striking thing about her was that she made her discoveries by listening to what the corn spoke to her, by respecting the life of the corn and "letting it come." 1

McClintock says she learned "the stories" of the plants. She "heard" them. She watched the daily green journeys of growth from earth toward sky and sun. She knew her plants in the way a healer or mystic would have known them, from the inside, the inner voices of corn and woman speaking to one another. 2

As an Indian woman, I come from a long history of people who have listened to the language of this continent, people who have known that corn grows with the songs and prayers of the people, that it has a story to tell, that the world is alive. Both in oral traditions and in mythology—the true language of inner life—account after account tells of the stones giving guidance, the trees singing, the corn telling of inner earth, the dragonfly offering up a tongue. This is true in the European traditions as well: Psyche received direction from the reeds and the ants, Orpheus knew the languages of earth, animals, and birds. 3

This intuitive and common language is what I seek for my writing, work in touch with the mystery and force of life, work that speaks a few of the many voices around us, and it is important to me that McClintock listened to the 4 voices of corn. It is important to the continuance of life that she told the truth of her method and that it reminded us all of where our strength, our knowing, and our sustenance come from.

It is also poetry, this science, and I note how often scientific theories lead to   5
the world of poetry and vision, theories telling us how atoms that were stars have been transformed into our living, breathing bodies. And in these theories, or maybe they should be called stories, we begin to understand how we are each many people, including the stars we once were, and how we are in essence the earth and the universe, how what we do travels clear around the earth and returns. In a single moment of our living, there is our ancestral and personal history, our future, even our deaths planted in us and already growing toward their fulfillment. The corn plants are there, and like all the rest we are forever merging our borders with theirs in the world collective.

Our very lives might depend on this listening. In the Chernobyl nuclear ac-   6
cident, the wind told the story that was being suppressed by the people. It gave away the truth. It carried the story of danger to other countries. It was a poet, a prophet, a scientist.

Sometimes, like the wind, poetry has its own laws speaking for the life of   7
the planet. It is a language that wants to bring back together what the other words have torn apart. It is the language of life speaking through us about the sacredness of life.

This life speaking life is what I find so compelling about the work of poets   8
such as Ernesto Cardenal, who is also a priest and was the Nicaraguan Minister of Culture. He writes: "The armadilloes are very happy with this government. . . . Not only humans desired liberation/the whole ecology wanted it." Cardenal has also written "The Parrots," a poem about caged birds who were being sent to the United States as pets for the wealthy, how the cages were opened, the parrots allowed back into the mountains and jungles, freed like the people, "and sent back to the land we were pulled from."

How we have been pulled from the land! And how poetry has worked hard   9
to set us free, uncage us, keep us from split tongues that mimic the voices of our captors. It returns us to our land. Poetry is a string of words that parades without a permit. It is a lockbox of words to put an ear to as we try to crack the safe of language, listening for the right combination, the treasure inside. It is life resonating. It is sometimes called Prayer, Soothsaying, Complaint, Invocation, Proclamation, Testimony, Witness. Writing is and does all these things. And like that parade, it is illegitimately insistent on going its own way, on being part of the miracle of life, telling the story about what happened when we were cosmic dust, what it means to be stars listening to our human atoms.

But don't misunderstand me. I am not just a dreamer. I am also the practi-   10
cal type. A friend's father, watching the United States stage another revolution in another Third World country, said, "Why doesn't the government just feed people and then let the political chips fall where they may?" He was right. It was easy, obvious, even financially more reasonable to do that, to let democracy be chosen because it feeds hunger. I want my writing to be that simple, that

clear and direct. Likewise, I feel it is not enough for me just to write, but I need to live it, to be informed by it. I have found over the years that my work has more courage than I do. It has more wisdom. It teaches me, leads me places I never knew I was heading. And it is about a new way of living, of being in the world.

I was on a panel recently where the question was raised whether we 11 thought literature could save lives. The audience, book people, smiled expectantly with the thought. I wanted to say, Yes, it saves lives. But I couldn't speak those words. It saves spirits maybe, hearts. It changes minds, but for me writing is an incredible privilege. When I sit down at the desk, there are other women who are hungry, homeless. I don't want to forget that, that the world of matter is still there to be reckoned with. This writing is a form of freedom most other people do not have. So, when I write, I feel a responsibility, a commitment to other humans and to the animal and plant communities as well.

Still, writing has changed me. And there is the powerful need we all have to 12 tell a story, each of us with a piece of the whole pattern to complete. As Alice Walker says, We are all telling part of the same story, and as Sharon Olds has said, Every writer is a cell on the body politic of America.

Another Nobel Prize laureate is Betty William, a Northern Ireland co- 13 winner of the 1977 Peace Prize. I heard her speak about how, after witnessing the death of children, she stepped outside in the middle of the night and began knocking on doors and yelling, behaviors that would have earned her a diagnosis of hysteria in our own medical circles. She knocked on doors that might have opened with weapons pointing in her face, and she cried out, "What kind of people have we become that we would allow children to be killed on our streets?" Within four hours the city was awake, and there were sixteen thousand names on petitions for peace. Now, that woman's work is a lesson to those of us who deal with language, and to those of us who are dealt into silence. She used language to begin the process of peace. This is the living, breathing power of the word. It is poetry. So are the names of those who signed the petitions. Maybe it is this kind of language that saves lives.

Writing begins for me with survival, with life and with freeing life, saving 14 life, speaking life. It is work that speaks what can't be easily said. It originates from a compelling desire to live and be alive. For me, it is sometimes the need to speak for other forms of life, to take the side of human life, even our sometimes frivolous living, and our grief-filled living, our joyous living, our violent living, busy living, our peaceful living. It is about possibility. It is based in the world of matter. I am interested in how something small turns into an image that is large and strong with resonance, where the ordinary becomes beautiful. I believe the divine, the magic, is here in the weeds at our feet, unacknowledged. What a world this is. Where else could water rise up to the sky, turn into snow crystals, magnificently brought together, fall from the sky all around us, pile up billions deep, and catch the small sparks of sunlight as they return again to water?

These acts of magic happen all the time; in Chaco Canyon, my sister has 15
seen a kiva, a ceremonial room in the earth, that is in the center of the canyon.
This place has been uninhabited for what seems like forever. It has been without water. In fact, there are theories that the ancient people disappeared when
they journeyed after water. In the center of it a corn plant was growing. It was
all alone and it had been there since the ancient ones, the old ones who came before us all, those people who wove dog hair into belts, who witnessed the painting of flute players on the seeping canyon walls, who knew the stories of corn.
And there was one corn plant growing out of the holy place. It planted itself
yearly. With no water, no person to care for it, no overturning of the soil, this
corn plant rises up to tell its story, and that's what this poetry is.

COMPREHENSION

1. Why does the writer specifically identify herself as an "Indian woman"
   (paragraph 3)? How does this identity influence Hogan's perception of the
   place of poetry in the life of a community or nation?
2. Where does the writer refer to "voices"? What types of voices does she hear,
   and why are they important?
3. What is the writer's definition of poetry? Why does she call poetry "life resonating" (paragraph 9)?

RHETORIC

1. What is the significance of the title? How does the title resonate throughout
   the essay?
2. What persona does the writer create for herself in this essay? What is her
   purpose in presenting herself in the way she does? What tone does she
   adopt?
3. As we might expect of a poet, Hogan employs highly figurative language in
   this essay. Locate examples of allusion, imagery, symbolism, and metaphor,
   and explain their meaning.
4. Where does the writer use illustration to help explain both the writing and
   nature of poetry? What do these concrete examples add to her extended
   definition of poetry?
5. How might this essay be considered an argument? What is the writer's
   claim? Where does she appeal to ethics and emotion? Why does she refer to
   writers such as Ernesto Cardenal and Alice Walker to support her claim?
   Why does the writer refer to the Chernobyl nuclear accident (paragraph 6)
   and the United States staging revolutions in Third World nations (paragraph 10)?
6. How does the concluding paragraph serve as a coda for the essay?

WRITING

1. Write your own definition of poetry, basing it on your personal experience, ethnic or racial background, and understanding of the world today.
2. Conduct additional research on Linda Hogan's poetry. Select one or two of her poems, and write an analysis of them.
3. **WRITING AN ARGUMENT:** Argue for or against the proposition, advanced by Hogan, that poetry—or, more generally, literature—has the power to change minds and even save lives. Provide evidence to support your claim.

www.mhhe.com/ **mhreader**   For more information on Linda Hogan, go to:
More Resources > Ch. 10 Literature & Arts

✳

What's American about American Poetry?

Billy Collins

Billy Collins (b. 1941) was appointed United States Poet Laureate for 2001–03. Born in New York City, he was educated at the College of the Holy Cross (BA, 1963) and the University of California, Riverside (PhD, 1971). Collins is known as a superb reader of poetry, connecting with his audience at a visceral level. "I have one reader in mind," he observes, "someone who is in the room with me, and who I'm talking to." He has taught at Lehman College of the City University of New York, Sarah Lawrence College, and Columbia University. Collins has received major awards for his verse, which has been collected in The Apple That Astonished Paris *(1988),* Questions about Angels *(1991),* The Art of Drowning *(1995),* Picnic, Lightning *(1998),* Sailing Alone around the Room *(2001), and* Nine Horses *(2002). In the following essay, Collins explains the "American-ness" of American poetry.*

I never really considered myself a particularly American poet until I went to   1
England some years ago to give a series of readings. I had put the tour together myself, and it looked it. The odd range of venues included a sixth form class, a jazz club in Brighton, a college of Sheffield University, and a community center in a small Yorkshire village. It was at this last site, by the way, that an elderly, agrarian-looking man rose from the audience during a question-and-answer

session to ask: "Mr. Collins, are *all* your poems written in prose?" But regardless of the audience or the venue, each reading left me with the same small but nagging realization: that my poems were written not in English but in American. At every reading I could sense dead spots occurring when I would utter a phrase such as "eggs over easy" or "sweat the final." I became convinced that the mention of "a state flower" in one of my poems must sound to the British ear like "estate flower." I was discovering that idiomatic American is difficult to translate not only into French or German, but into English. Just as one cannot understand what it is to be an American until one leaves the country, I was not aware of my own American voice—my written accent, so to speak—until I had faced several audiences of British listeners.

I was especially surprised to discover how steeped many of my poems were 2 in the American idiom, because for years I had consciously avoided using fad dialects or making references to contemporary culture. I knew that a phrase such as "frequent flyer," "hatch-back," or "Jello shot" would in time make a poem sound dated and thus could drastically shorten its shelf life. "Shelf life" is probably another example. I had tried to favor a more universal vocabulary, not a purely elemental diction of "rock," "cloud," "sky," and "tree," but a diction that leaned in that direction and was reluctant to allow in the linguistic news of the day. Ezra Pound put it most succinctly when he defined poetry as "the news that stays new." And I admired Mary Oliver's advice regarding a poet's notion of an audience: "write for a stranger born in a distant country hundreds of years from now." I wanted to include that stranger of the future in my audience, and I did not want him to have to consult a footnote for "Wonder Bread" or "Big Mac."

America, of course, is greater than the sum of its idioms, but if you selected 3 a few poets from an international pool and asked them about the relationship of their poetry to their nationality, most would place their mother tongue at the center of their responses. Czeslaw Milosz might cite the expressive possibilities of Polish; Yannis Ritsos might discuss the feel of writing in demotic Greek. But American poets can claim no exclusive, nationalistic rights to a mother tongue, for the language they write in is shared by the rest of the English-speaking world, which at this time is the most rapidly expanding language community in the world.

So where does American-ness lie for a writer if not in his native tongue? 4 D.H. Lawrence opens his seminal *Studies in Classic American Literature* by putting that question in the form of a challenge: "Where *is* this new bird called the true American? Show us the homunculus of the new era. Go on, show us him. Because all that is visible to the naked European eye, in America, is a sort of miscreant European." I find it odd that Lawrence calls the European eye "naked," for, if anything, compared to the bookish lenses covering the European eye, the American eye was the naked one; and the first poet to look at America with that naked eye—and, indeed, to appear naked before us—was Walt Whitman.

Lawrence recognized Whitman as the pioneer of a new American literature. 5
He called him "the greatest and the first and the only American teacher . . . the
first white aboriginal" though in the same breath he mocks Whitman's universal gesturing and accuses him of bogus sympathy. Surely, Whitman was the first
poet to try to get his arms around the continent so as to hold the lumberjack and
the secretary and the Eskimo in one loving cosmic embrace. A Long Islander
and a New Yorker, he refused to define himself as regional the way some American poets and ever more American novelists have done ever since. But the true
aboriginal stroke was Whitman's breaking loose from the iambic collar of traditional English poetry. *Leaves of Grass* moves to the cadence of the Bible, not the
British iambic two-step. The long poem was such a radical departure from customary meter and form that it triggered a critical debate as to whether it was really poetry, a debate which should have ended when one professor observed,
"If this is not poetry, it is something greater than poetry."

Strangely, it took a long time for anyone to follow Whitman's liberating lead. 6
As Lawrence put it, "Ahead of Whitman, nothing. Ahead of all poets, pioneering
into the wilderness of unopened life, Whitman." Eventually, American poetry
caught up with Whitman but not until his century had run out. By the early
1920s when Lawrence was making his assessments, many of the now canonical
modernist poems were appearing, and whatever else defined their veerings
away from convention, their freedom from the box of the stanza and the harness
of the iambic was the most common evidence of their experimentations.

These days, of course, "free verse" is not the exciting license it once was; 7
more often than not, it is simply an excuse to produce untidy, flat-footed poems,
an excuse in no way limited to poets in America. The more powerful, more difficult, yet abiding lesson of Whitman lies in his outrageousness. The audacity of
lines like "It is time to explain myself—let us stand" and "I sound my barbaric
yawp over the rooftops of the world" make possible Ginsberg's "American, I
am putting my queer shoulder to the wheel" and with some added coyness,
Frank O'Hara's "ah lunch! I think I am going crazy." Whitman's fearless, unheard-of voice shattered the glass of European gentility and eventually emboldened later generations of American poets to speak out in wilder tones.

If a writer is the sum of his or her influences, then my own poems are un- 8
avoidably the result of my exposure to the sounds and styles of both British and
American poetry. I even find myself playing one diction off against another,
usually for ironic effect. But more specifically, in thinking about myself as an
"American poet," and thus committing the dangerous act of auto-literary criticism, I find that a number of my poems seem determined to establish an American rootedness distinct from European influence. "American Sonnet," for
example, is a rejection of the Italian and English sonnet models in favor of the
American postcard which, like the sonnet, limits expression to a confined space
and, in addition, combines the verbal on one side with the pictorial on the other.
Like the traditional love sonnet, the traveler's postcard has acquired its own rit-
ualized conventions. The poem opens with an uncharacteristic "we," as if I
were speaking for all American poets.

American Sonnet

We do not speak like Petrarch or wear a hat like Spenser
and it is not fourteen lines
like furrows in a small, carefully plowed field

but the picture postcard, a poem on vacation,
that forces us to sing our songs in little rooms
or pour our sentiments into measuring cups.

We write on the back of a waterfall or lake,
adding to the view a caption as conventional
as an Elizabethan woman's heliocentric eyes.

We locate an adjective for weather.
We announce that we are having a wonderful time.
We express the wish that you were here

and hide the wish that we were where you are,
walking back from the mailbox, your head lowered
as you read and turn the thin message in your hands.

A slice of this faraway place, a width of white beach,
a piazza or carved spires of a cathedral
will pierce the familiar place where you remain,

and you will toss on the table this reversible display;
a few square inches of where we have strayed
and a compression of what we feel.

The ironic literary play of the first part of the poem gives way to a small   9
drama of separation, distance, and longing. The poem tries, but of course fails,
to mix irony and emotion with such equality as to achieve a perfectly ambiguous tone.

Another poem titled "Consolation" pretends to celebrate the pleasures of   10
spending the summer at home in the States rather than embarking on the traditional European holiday. "How agreeable it is not to be touring Italy," the poem
opens; then goes on to express the ease of staying put on native soil, cruising
"these local, familiar streets,/fully grasping the meaning of every road sign and
billboard/and all the sudden hand gestures of my compatriots." "Instead of
slouching in a cafe ignorant of the word for ice," the speaker prefers "the coffee
shop and the waitress known as Dot" where he will not have to have his photograph taken with the owner or figure out the exchange rate when the bill arrives. For him, "It is enough to climb back into the car/as if it were the great car
of English itself/and sounding my loud vernacular horn, speed off/down a
road that will never lead to Rome, not even Bologna." The poem is a mock-
rejection of literary Euro-centricism delivered by a speaker whose modest tastes
echo the sweet provincialism of the Wallace Shawn character in the film *My
Dinner with Andre*.

"Lines Written Over Three Thousand Miles from Tintern Abbey," as the ti- 11
tle implies, provides another example of this process of "Americanization," as
Wordsworth's famous autobiographical lyric is imported into the speaker's
American, and again, domestic, life.

> I was here before, a long time ago,
> and now I am here again
> is an observation that occurs in poetry
> as frequently as rain occurs in life.
>
> The fellow may be gazing
> over an English landscape,
> hillsides dotted with sheep,
> a row of tall trees topping the downs,
>
> or he could be moping through the shadows
> of a dark Bavarian forest,
> a wedge of cheese and a volume of fairy tales
> tucked into his rucksack.
>
> But the feeling is always the same:
> it was better the first time.
> This time is not nearly as good.
> I'm not feeling as chipper as I did back then.
>
> Something is always missing—
> Swans, a glint on the surface of a lake,
> some minor but essential touch.
> Or the quality of things has diminished.
>
> The sky was a deeper, more dimensional blue,
> clouds were more cathedral-like,
> and water rushed over rock
> with greater effervescence.
>
> From our chairs we have watched
> the poor author in his waistcoat
> as he recalls the dizzying icebergs of childhood
> and mills around in a field of weeds.
>
> We have heard the poets long dead
> declaim their dying
> from a promontory, a riverbank,
> next to a haycock, within a shadowy copse.
>
> We have listened to their dismay,
> the kind that issues from poems
> the way water issues forth from hoses,
> the way the match always gives its little speech on fire.

And when we put down the book at last,
lean back, close our eyes,
stinging with print,
and slip in the bookmark of sleep,

we will be schooled enough to know
that when we wake up
a little before dinner
things will not be nearly as good as they once were.

Something will be missing
from this long, coffin-shaped room,
the walls and windows now
only two different shades of gray,

the glossy gardenia drooping
in its chipped terra-cotta pot.
And on the floor, shoes, socks,
the browning core of an apple.

Nothing will be as it was
a few hours ago, back in the glorious past
before our naps, back in that Golden Age
that drew to a close sometime shortly after lunch.

The revisionist speaker's disenchantment with the Romantic theme of loss 12
is evident in his lumping together all the complaining poets of the 19th century,
both English and German. The domestication of this pattern of loss begins with
the homely images of the garden hose and the match. Time is compressed from
an autobiographical span to a few hours between lunch and dinner, and the
dated landscape of "promontory," "haycock," and "copse" is compressed into
an ordinary room-scape with its drooping flower and strewing of shoes and
socks. Romantic agony is reduced to reader fatigue. The Golden Age lies irre-
trievably behind us in an earlier part of the afternoon.

What makes poetry American can be measured in the kind of steps it makes 13
away from the poetry of the "Old World" as the schoolbooks used to say. Poetry
can also be American because of its idioms, its landscape, its irreverence toward
the European past, its audacious egotism, its ironic stances, its freedom of fixed
cadences, but most of all because of its immense variety. This last quality—its
democratic expansiveness and inclusiveness—was best expressed in a short
poem by Louis Simpson, who, for the moment, deserves to have the last word
on the subject.

American Poetry

Whatever it is, it must have
A stomach that can digest
Rubber, coal, uranium, moons, poems.

Like the shark, it contains a shoe.
It must swim for miles through the desert
Uttering cries that are almost human.

COMPREHENSION

1. How did the writer's visit to England help him understand what is American about American poetry?
2. According to the writer, what is distinctive about American poetry?
3. Paraphrase the two poems by Collins that he incorporates in this essay.

RHETORIC

1. What is the writer's thesis? Is it explicit or implied? Explain. Is his purpose to inform, argue, persuade, or analyze, or perhaps some combination of these motives?
2. At what sort of audience does Collins aim his essay? Do his allusions to other writers limit his audience? Why or why not?
3. Does Collins seem optimistic or pessimistic about the state of American poetry? How do you know?
4. Why does Collins insert his own poetry—as well as that of Louis Simpson—into the essay? Does it add to or detract from the unity and substance of the essay? Justify your response.
5. What strategies go into the writer's extended definition of American poetry?
6. Examine the writer's conclusion. How effective is his decision to end the essay with lines from a poem by Lewis Simpson? Explain your answer.

WRITING

1. Using the two poems by Collins that appear in this selection, write an essay explaining why they are "American."
2. Conduct research on Walt Whitman, one of the poets Collins alludes to. What makes Whitman a uniquely American poet? Write an essay in response to this question.
3. **WRITING AN ARGUMENT:** Write an essay arguing for or against the idea that the greatest literature is not American, English, Greek, or Polish (to cite four representative poets mentioned by Collins) but rather "universal" in its appeal. Select one famous writer to substantiate your claim. Conduct research if necessary.

www.mhhe.com/
mhreader

For more information on Billy Collins, go to:
More Resources > Ch. 10 Literature & Arts

✳

Imprisoning Time in a Rectangle

Lance Morrow

Lance Morrow (b. 1939), an American journalist and nonfiction writer, was born in Philadelphia, Pennsylvania. After getting his undergraduate degree from Harvard University in 1963, he worked as a reporter for the Washington Star *before joining the staff of* Time *magazine in 1965 where he currently is senior writer and essayist. His books include* The Chief: A Memoir of Fathers and Sons *(1984),* America: A Rediscovery *(1987),* Fishing the Tiber *(1988), and* Safari: Experiencing the World *(1992). After surviving a heart attack at the age of 53, he recorded his experience in* Heart: A Memoir *(1995). In the following essay, first printed in a special issue of* Time *(Fall 1989), Morrow offers insights into the art of photography and the importance of photojournalism.*

Balzac had a "vague dread" of being photographed. Like some primitive peoples, he thought the camera steals something of the soul—that, as he told a friend "every body in its natural state is made up of a series of ghostly images superimposed in layers to infinity, wrapped in infinitesimal films." Each time a photograph was made, he believed, another thin layer of the subject's being would be stripped off to become not life as before but a membrane of memory in a sort of translucent antiworld.

If that is what photography is up to, then the onion of the world is being peeled away, layer by layer—lenses like black holes gobbling up life's emanations. Mere images proliferate, while history pares down to a phosphorescence of itself.

The idea catches something of the superstition (sometimes justified, if you think about it) and the spooky metaphysics that go ghosting around photography. Taking pictures is a transaction that snatches instants away from time and imprisons them in rectangles. These rectangles become a collective public memory and an image-world that is located usually on the verge of tears, often on the edge of a moral mess.

It is possible to be entranced by photography and at the same time disquieted by its powerful capacity to bypass thought. Photography, as the critic Susan Sontag has pointed out, is an elegiac, nostalgic phenomenon. No one photographs the future. The instants that the photographer freezes are ever the past, ever receding. They have about them the brilliance or instancy of their moment but also the cello sound of loss that life makes when going irrecoverably away and lodging at last in the dreamworks.

The pictures made by photojournalists have the legitimacy of being news, fresh information. They slice along the hard edge of the present. Photojournalism is not self-conscious, since it first enters the room (the brain) as a battle report from the far-flung Now. It is only later that the artifacts of photojournal-ism sink into the textures of the civilization and tincture its memory: Jack Ruby shooting Lee Harvey Oswald, an image so raw and shocking, subsides at last into the ecology of memory where we also find thousands of other oddments from the time—John John saluting at the funeral, Jack and Jackie on Cape Cod, who knows?—bright shards that stimulate old feelings (ghost pangs, ghost ten-dernesses, wistfulness) but not thought really. The shocks turn into dreams. The memory of such pictures, flipped through like a disordered Rolodex, makes at last a cultural tapestry, an inventory of the kind that brothers and sisters and distant cousins may rummage through at family reunions, except that the greatest photojournalism has given certain memories the emotional prestige of icons.

If journalism—the kind done with words—is the first draft of history, what 6 is photojournalism? Is it the first impression of history, the first graphic flash? Yes, but it is also (and this is the disturbing thing) history's lasting visual im-pression. The service that the pictures perform is splendid, and so powerful as to seem preternatural. But sometimes the power they possess is more than they deserve.

Call up Eddie Adams' 1968 photo of General Nguyen Ngoc Loan, the police 7 chief of Saigon, firing his snub-nosed revolver into the temple of a Viet Cong of-ficer. Bright sunlight, Saigon: the scrawny police chief's arm, outstretched, goes by extension through the trigger finger into the V.C.'s brain. That photograph, and another in 1972 showing a naked young Vietnamese girl running in arms-outstretched terror up a road away from American napalm, outmanned the force of three U.S. Presidents and the most powerful Army in the world. The photo-graphs were considered, quite ridiculously, to be a portrait of America's moral disgrace. Freudians spend years trying to call up the primal image-memories, turned to trauma, that distort a neurotic patient's psyche. Photographs some-times have a way of installing the image and legitimizing the trauma: the very vividness of the image, the greatness of the photograph as journalism or even as art, forestalls examination.

Adams has always felt uncomfortable about his picture of Loan executing 8 the Viet Cong officer. What the picture does not show is that a few moments earlier the Viet Cong had slaughtered the family of Loan's best friend in a house just up the road. All this occurred during the Tet offensive, a state of general mayhem all over South Viet Nam. The Communists in similar circumstances would not have had qualms about summary execution.

But Loan shot the man; Adams took the picture. The image went firing 9 around the world and lodged in the conscience. Photography is the very dream of the Heisenberg uncertainty principle, which holds that the act of observing a physical event inevitably changes it. War is merciless, bloody, and by definition it occurs outside the orbit of due process. Loan's Viet Cong did not have a trial. He did have a photographer. The photographer's picture took on a life of its own and changed history.

All great photographs have lives of their own, but they can be as false as 10
dreams. Somehow the mind knows that and sorts out the matter, and permits itself to enjoy the pictures without getting sunk in the really mysterious business
that they involve.

Still, a puritan conscience recoils a little from the sheer power of photo- 11
graphs. They have lingering about them the ghost of the golden calf—the bright
object too much admired, without God's abstract difficulties. Great photographs bring the mind alive. Photographs are magic things that traffic in mystery. They float on the surface, and they have a strange life in the depths of the
mind. They bear watching.

COMPREHENSION

1. What does the writer mean by "the superstition . . . and the spooky metaphysics that go ghosting around photography" (paragraph 3)?
2. According to the author, how does photography deal with time? What is
   the relationship of photojournalism to truth?
3. Much of the essay deals with a famous photograph by Eddie Adams. (See
   page 133 for a reproduction of this photo.) Summarize the writer's analysis
   of this photograph.

RHETORIC

1. Morrow's introductory paragraph contains an allusion to the French novelist and short-story writer Honoré de Balzac (1799–1850). What is his purpose? How does this opening paragraph set the stage for the body of the
   writer's essay?
2. What claim does the writer make about the impact of photography and history? What evidence does he provide to support this claim?
3. Identify the many instances of figurative language—images, metaphors,
   and symbols—employed by the writer. What is he attempting to achieve
   with this highly figurative style? How effective is this language in advancing the writer's key propositions?
4. How does the writer apply a comparative method to advance his
   argument?
5. At what point does Morrow begin his conclusion? How does the conclusion
   reflect earlier motifs and ideas? Is the ending effective? Why or why not?

WRITING

1. Select any famous photograph, and write an essay explaining how it captures a moment in time and history.

2. Compare and contrast photography with pictorial art. Analyze examples of paintings and photographs; integrate these images into your essay.
3. **WRITING AN ARGUMENT:** Argue for or against the well-known adage, "The camera doesn't lie." Refer to Morrow's commentary to support your position on this proposition.

www.mhhe.com/
mhreader

For more information on Lance Morrow, go to:
More Resources > Ch. 10 Literature & Arts

＊

Saving the Life That Is Your Own: The Importance of Models in the Artist's Life

Alice Walker

Alice Walker (b. 1941) was born in Eatonton, Georgia, and now lives in San Francisco and Mendocino County, California. She attended Spelman College and graduated from Sarah Lawrence College. A celebrated and prolific novelist, short-story writer, poet, and essayist, she has also been active in the civil rights movement. Walker often draws on both her own personal experience and historical records to reflect on the African American experience. Her books include The Color Purple *(1976), which won the American Book Award and the Pulitzer Prize,* You Can't Keep a Good Woman Down *(1981),* Living in the World: Selected Essays, 1973–1987 *(1987),* The Temple of My Familiar *(1989),* By the Light of My Father's Smile *(1999), and* The Way Forward Is with a Broken Heart *(2001). The following essay, from* In Search of Our Mother's Gardens *(1983), offers a highly personalized and perceptive analysis of the importance of influence on both art and life.*

There is a letter Vincent Van Gogh wrote to Emile Bernard that is very meaningful to me. A year before he wrote the letter, Van Gogh had had a fight with his domineering friend Gauguin, left his company, and cut off, in desperation and anguish, his own ear. The letter was written in Saint-Remy, in the South of France, from a mental institution to which Van Gogh had voluntarily committed himself.

I imagine Van Gogh sitting at a rough desk too small for him, looking out at the lovely Southern light, and occasionally glancing critically next to him at his own paintings of the landscape he loved so much. The date of the letter is December 1889. Van Gogh wrote:

However hateful painting may be, and however cumbersome in the times we are living in, if anyone who has chosen this handicraft pursues it zealously, he is a man of duty, sound and faithful.

Society makes our existence wretchedly difficult at times, hence our impotence and the imperfection of our work.

. . . I myself am suffering under an absolute lack of models.

But on the other hand, there are beautiful spots here. I have just done five size 30 canvasses, olive trees. And the reason I am staying on here is that my health is improving a great deal.

What I am doing is hard, dry, but that is because I am trying to gather new strength by doing some rough work, and I'm afraid abstractions would make me soft.

Six months later, Van Gogh—whose health was "improving a great deal"— committed suicide. He had sold one painting during his lifetime. Three times was his work noticed in the press. But these are just details.   3

The real Vincent Van Gogh is the man who has "just done five size 30 canvasses, olive trees." To me, in context, one of the most moving and revealing descriptions of how a real artist thinks. And the knowledge that when he spoke of "suffering under an absolute lack of models" he spoke of that lack in terms of both the intensity of his commitment and the quality and singularity of his work, which was frequently ridiculed in his day.   4

The absence of models, in literature as in life, to say nothing of painting, is an occupational hazard for the artist, simply because models in art, in behavior, in growth of spirit and intellects—even if rejected—enrich and enlarge one's view of existence. Deadlier still, to the artist who lacks models, is the curse of ridicule, the bringing to bear on an artist's best work, especially his or her most original, most strikingly deviant, only a fund of ignorance and the presumption that, as an artist's critic, one's judgment is free of the restrictions imposed by prejudice, and is well informed, indeed, about all the art in the world that really matters.   5

What is always needed in the appreciation of art, or life, is the larger perspective. Connections made, or at least attempted, where none existed before, the straining to encompass in one's glance at the varied world the common thread, the unifying theme through immense diversity, a fearlessness of growth, of search, of looking, that enlarges the private and the public world. And yet, in our particular society, it is the narrowed and narrowing view of life that often wins.   6

Recently, I read at a college and was asked by one of the audience what I considered the major difference between the literature written by black and by white Americans. I had not spent a lot of time considering this question, since it is not the difference between them that interests me, but, rather, the way black writers and white writers seem to me to be writing one immense story—the same story, for the most part—with different parts of this immense story coming from a multitude of different perspectives. Until this is generally recognized, literature will always be broken into bits, black and white, and there will always be questions, wanting neat answers, such as this.   7

Still, I answered that I thought, for the most part, white American writers   8
tended to end their books and their characters' lives as if there were no better
existence for which to struggle. The gloom of defeat is thick.

By comparison, black writers seem always involved in a moral and/or   9
physical struggle, the result of which is expected to be some kind of larger freedom. Perhaps this is because our literary tradition is based on the slave narratives, where escape for the body and freedom for the soul went together, or
perhaps this is because black people have never felt themselves guilty of global,
cosmic sins.

This comparison does not hold up in every case, of course, and perhaps   10
does not really hold up at all. I am not a gatherer of statistics, only a curious
reader, and this has been my impression from reading many books by black and
white writers.

There are, however, two books by American women that illustrate what I   11
am talking about: *The Awakening,* by Kate Chopin, and *Their Eyes Were Watching
God,* by Zora Neale Hurston.

The plight of Mme Pontellier is quite similar to that of Janie Crawford. Each   12
woman is married to a dull, society-conscious husband and living in a dull,
propriety-conscious community. Each woman desires a life of her own and a
man who loves her and makes her feel alive. Each woman finds such a man.

Mme Pontellier, overcome by the strictures of society and the existence of   13
her children (along with the cowardice of her lover), kills herself rather than
defy the one and abandon the other. Janie Crawford, on the other hand, refuses
to allow society to dictate behavior to her, enjoys the love of a much younger,
freedom-loving man, and lives to tell others of her experience.

When I mentioned these two books to my audience, I was not surprised to   14
learn that only one person, a young black poet in the first row, had ever heard
of *Their Eyes Were Watching God* (*The Awakening* they had fortunately read in
their "Women in Literature" class), primarily because it was written by a black
woman, whose experience—in love and life—was apparently assumed to be
unimportant to the students (and the teachers) of a predominantly white school.

Certainly, as a student, I was not directed toward this book, which would   15
have urged me more toward freedom and experience than toward comfort and
security, but was directed instead toward a plethora of books by mainly white
male writers who thought most women worthless if they didn't enjoy bullfight-
ing or hadn't volunteered for the trenches in World War I.

Loving both these books, knowing each to be indispensable to my own   16
growth, my own life, I choose the model, the example, of Janie Crawford. And
yet this book, as necessary to me and to other women as air and water, is again
out of print.* But I have distilled as much as I could of its wisdom in this poem
about its heroine, Janie Crawford:

I love the way Janie Crawford
left her husbands
the one who wanted to change her

into a mule
and the other who tried to interest her
in being a queen.
A woman, unless she submits,
is neither a mule
nor a queen
though like a mule she may suffer
and like a queen pace the floor.

It has been said that someone asked Toni Morrison why she writes the kind 17
of books she writes, and that she replied: Because they are the kind of books I
want to read.

This remains my favorite reply to that kind of question. As if anyone read- 18
ing the magnificent, mysterious *Sula* or the grim, poetic *The Bluest Eye* would require more of a reason for their existence than for the brooding, haunting
Wuthering Heights, for example, or the melancholy, triumphant *Jane Eyre*. (I am
not speaking here of the most famous short line of that book, "Reader, I married
him," as the triumph, but, rather, of the triumph of Jane Eyre's control over her
own sense of morality and her own stout will, which are but reflections of her
creator's, Charlotte Brontë, who no doubt wished to write the sort of books *she*
wished to read.)

Flannery O'Connor has written that more and more the serious novelist 19
will write, not what other people want, and certainly not what other people expect, but whatever interests her or him. And that the direction taken, therefore,
will be away from sociology, away from the "writing of explanation," of statistics, and further into mystery, into poetry, and into prophecy. I believe this is
true, *fortunately true;* especially for "Third World Writers"; Morrison, Marquez,
Ahmadi, Camara Laye make good examples. And not only do I believe it is true
for serious writers in general, but I believe, as firmly as did O'Connor, that this
is our only hope—in a culture so in love with flash, with trendiness, with super-
ficiality, as ours—of acquiring a sense of essence, of timelessness, and of vision.
Therefore, to write the books one wants to read is both to point in the direction
of vision and, at the same time, to follow it.

When Toni Morrison said she writes the kind of books she wants to read, 20
she was acknowledging the fact that in a society in which "accepted literature"
is so often sexist and racist and otherwise irrelevant or offensive to so many
lives, she must do the work of two. She must be her own model as well as the
artist attending, creating, learning from, realizing the model, which is to say,
herself.

(It should be remembered that, as a black person, one cannot completely 21
identify with a Jane Eyre, or with her creator, no matter how much one admires
them. And certainly, if one allows history to impinge on one's reading pleasure,
one must cringe at the thought of how Heathcliff, in the New World far from
Wuthering Heights, amassed his Cathy-dazzling fortune.) I have often been
asked why, in my own life and work, I have felt such a desperate need to know

and assimilate the experiences of earlier black women writers, most of them un-
heard of by you and by me, until quite recently; why I felt a need to study them
and to teach them.

I don't recall the exact moment I set out to explore the works of black   22
women, mainly those in the past, and certainly, in the beginning, I had no desire
to teach them. Teaching being for me, at that time, less rewarding than star-
gazing on a frigid night. My discovery of them—most of them out of print, abandoned, discredited, maligned, nearly lost—came about, as many things of value
do, almost by accident. As it turned out—and this should not have surprised
me—I found I was in need of something that only one of them could provide.

Mindful that throughout my four years at a prestigious black and then a   23
prestigious white college I had heard not one word about early black women
writers, one of my first tasks was simply to determine whether they had existed.
After this, I could breathe easier, with more assurance about the profession I
myself had chosen.

But the incident that started my search began several years ago: I sat down   24
at my desk one day, in a room of my own, with key and lock, and began prepa-
rations for a story about voodoo, a subject that had always fascinated me. Many
of the elements of this story I had gathered from a story my mother several
times told me. She had gone, during the Depression, into town to apply for
some government surplus food at the local commissary, and had been turned
down, in a particularly humiliating way, by the white woman in charge.

My mother always told this story with a most curious expression on her   25
face. She automatically raised her head higher than ever—it was always high—
and there was a look of righteousness, a kind of holy *heat* coming from her eyes.
She said she had lived to see this same white woman grow old and senile and
so badly crippled she had to get about on *two* sticks.

To her, this was clearly the working of God, who, as in the old spiritual,   26
". . . may not come when you want him, but he's right on time!" To me, hearing
the story for about the fiftieth time, something else was discernible: the possibilities of the story, for fiction.

What, I asked myself, would have happened if, after the crippled old lady   27
died, it was discovered that someone, my mother perhaps (who would have
been mortified at the thought, Christian that she is), had voodooed her?

Then, my thoughts sweeping me away into the world of hexes and conjur-   28
ings of centuries past, I wondered how a larger story could be created out of my
mother's story; one that would be true to the magnitude of her humiliation and
grief, and to the white woman's lack of sensitivity and compassion.

My third quandary was: How could I find out all I needed to know in order   29
to write a story that used *authentic* black witchcraft?

Which brings me back, almost, to the day I became really interested in black   30
women writers. I say "almost" because one other thing, from my childhood,
made the choice of black magic a logical and irresistible one for my story. Aside
from my mother's several stories about root doctors she had heard of or known,
there was the story I had often heard about my "crazy" Walker aunt.

Many years ago, when my aunt was a meek and obedient girl growing up 31
in a strict, conventionally religious house in the rural South, she had suddenly
thrown off her meekness and had run away from home, escorted by a rogue of
a man permanently attached elsewhere.

When she was returned home by her father, she was declared quite mad. In 32
the backwoods South at the turn of the century, "madness" of this sort was
cured not by psychiatry but by powders and by spells. (One can see Scott
Joplin's *Treemonisha* to understand the role voodoo played among black people
of that period.) My aunt's madness was treated by the community conjurer,
who promised, and delivered, the desired results. His treatment was a bag of
white powder, bought for fifty cents, and sprinkled on the ground around her
house, with some of it sewed, I believe, into the bodice of her nightgown.

So when I sat down to write my story about voodoo, my crazy Walker aunt 33
was definitely on my mind.

But she had experienced her temporary craziness so long ago that her story 34
had all the excitement of a might-have-been. I needed, instead of family memories, some hard facts about the *craft* of voodoo, as practiced by Southern blacks
in the nineteenth century. (It never once, fortunately, occurred to me that
voodoo was not worthy of the interest I had in it, or was too ridiculous to study
seriously.)

I began reading all I could find on the subject of "The Negro and His Folk- 35
ways and Superstitions." There were Botkin and Puckett and others, all white,
most racist. How was I to believe anything they wrote, since at least one of
them, Puckett, was capable of wondering, in his book, if "The Negro" had a
large enough brain?

Well, I thought, where are the *black* collectors of folklore? Where is the *black* 36
anthropologist? Where is the *black* person who took the time to travel the back
roads of the South and collect the information I need: how to cure heat trouble,
treat dropsy, hex somebody to death, lock bowels, cause joints to swell, eyes to
fall out, and so on. Where was this black person?

And that is when I first saw, in a *footnote* to the white voices of authority, the 37
name Zora Neale Hurston.

Folklorist, novelist, anthropologist, serious student of voodoo, also all- 38
around black woman, with guts enough to take a slide rule and measure random black heads in Harlem; not to prove their inferiority, but to prove that
whatever their size, shape, or present condition of servitude, those heads contained all the intelligence anyone could use to get through this world.

Zora Hurston, who went to Barnard to learn how to study what she really 39
wanted to learn: the ways of her own people, and what ancient rituals, customs,
and beliefs had made them unique.

Zora, of the sandy-colored hair and the daredevil eyes, a girl who escaped 40
poverty and parental neglect by hard work and a sharp eye for the main chance.

Zora, who left the South only to return to look at it again. Who went to root 41
doctors from Florida to Louisiana and said, "Here I am. I want to learn your
trade."

Zora, who had collected all the black folklore I could ever use.   42
That Zora.   43

And having found *that Zora* (like a golden key to a storehouse of varied   44
treasure), I was hooked.

What I had discovered, of course, was a model. A model, who, as it hap-   45
pened, provided more than voodoo for my story, more than one of the greatest
novels America had produced—though, being America, it did not realize this.
She had provided, as if she knew someday I would come along wandering in
the wilderness, a nearly complete record of her life. And though her life
sprouted an occasional wart, I am eternally grateful for that life, warts and all.

It is not irrelevant, nor is it bragging (except perhaps to gloat a little on the   46
happy relatedness of Zora, my mother and me), to mention here that the story I
wrote, called "the Revenge of Hannah Kemhuff," based on my mother's experiences during the Depression, and on Zora Hurston's folklore collection of the
1920s, and on my own response to both out of a contemporary existence, was
immediately published and was later selected, by a reputable collector of short
stories, as one of the *Best Short Stories of 1974.*

I mention it because this story might never have been written, because the   47
very bases of its structure, authentic black folklore, viewed from a black perspective, might have been lost.

Had it been lost, my mother's story would have had no historical underpin-   48
ning, none I could trust, anyway. I would not have written the story, which I enjoyed writing as much as I've enjoyed writing anything in my life, had I not
known that Zora had already done a thorough job of preparing the ground over
which I was then moving.

In that story I gathered up the historical and psychological threads of the   49
life my ancestors lived, and in the writing of it I felt joy and strength and my
own continuity. I had that wonderful feeling writers get sometimes, not very often, of being *with* a great many people, ancient spirits, all very happy to see me
consulting and acknowledging them, and eager to let me know, through the joy
of their presence, that, indeed, I am not alone.

To take Toni Morrison's statement further, if that is possible, in my own   50
work I write not only what I want to read—understanding fully and indelibly
that if I don't do it no one else is so vitally interested, or capable of doing it to
my satisfaction—I write all the things *I should have been able to read.* Consulting,
as belatedly discovered models, those writers—most of whom, not surprisingly,
are women—who understood that their experience as ordinary human beings
was also valuable, and in danger of being misrepresented, distorted, or lost:

Zora Hurston—novelist, essayist, anthropologist, autobiographer;

Jean Toomer—novelist, poet, philosopher, visionary, a man who cared
what women felt;

Colette—whose crinkly hair enhances her French, part-black face; novelist,
playwright, dancer, essayist, newspaperwoman, lover of women, men,
small dogs; fortunate not to have been born in America;

Anaïs Nin—recorder of everything, no matter how minute;

Tillie Olson—a writer of such generosity and honesty, she literally saves lives;

Virginia Woolf—who has saved so many of us.

It is, in the end, the saving of lives that we writers are about. Whether we 51 are "minority" writers or "majority." It is simply in our power to do this.

We do it because we care. We care that Vincent Van Gogh mutilated his ear. 52 We care that behind a pile of manure in the yard he destroyed his life. We care that Scott Joplin's music *lives!* We care because we know this: *the life we save is our own.*

COMPREHENSION

1. Explain the significance of Walker's title. How does it serve her purpose and guide readers to her thesis? What is her thesis?
2. According to the author, what is the importance of models in art? What is the relationship of models to life? List the models in Walker's life. Which of them stand out?
3. Paraphrase Walker's remarks on the relationship between black American and white American writing.

RHETORIC

1. Walker uses many allusions in this essay. Identify as many as you can. What is the allusion in the title? Comment on the general effectiveness of her allusions.
2. Is the author's style and choice of diction suitable to her subject matter and to her audience? Why or why not?
3. Why does the author personalize her treatment of the topic? What does she gain? Is there anything lost?
4. Walker employs several unique structuring devices in this essay. Cite at least three, and analyze their utility.
5. Explain Walker's use of examples to reinforce her generalizations and to organize the essay.
6. Which paragraphs constitute Walker's conclusion? What is their effect?

WRITING

1. Write an essay expanding the meaning of Walker's remark, "What is always needed in the appreciation of art, or life, is the larger perspective" (paragraph 6).
2. If you were planning on a career as a writer, artist, actor, or musician, who would your models be, and why?

3. **WRITING AN ARGUMENT:** Argue for or against Walker's proposition that the absence of models in art and life is an "occupational hazard" (paragraph 5).

www.mhhe.com/
mhreader

For more information on Alice Walker, go to:
More Resources > Ch. 10 Literature & Arts

✳

The Beatles

Ned Rorem

Ned Rorem (b. 1923), an acclaimed composer and writer, was born in Richmond, Indiana. He studied at several conservatories before graduating from Juilliard School of Music in New York City with BA (1946) and MA (1948) degrees. Rorem also studied privately with the famous composers Aaron Copland and Virgil Thomson. During a long, distinguished career, Rorem has composed three symphonies, six operas, numerous concertos and chamber pieces, and more than 500 songs. His music has received many awards, including the Pulitzer Prize for Air Music for Orchestra *(1976) and the Gold Medal from the American Academy of Arts and Letters. Rorem is the author of 17 books, among them five diaries, starting with* The Paris Diary *(1966). He has also written* Music from Inside Out *(1967) and* Other Entertainment *(1996). Rorem's knowledge of and fondness for music in all its forms—from classical to pop—is revealed in the following essay, written when the Beatles were at the height of their popularity in 1968.*

I never go to classical concerts any more, and I don't know anyone who does.   1
It's hard still to care whether some virtuoso tonight will perform the *Moonlight Sonata* a bit better or a bit worse than another virtuoso performed it last night.

I do often attend what used to be called avant-garde recitals, though seldom   2
with delight, and inevitably I look around and wonder: What am I doing here? What am I learning? Where are the poets and painters and even composers who used to flock to these things? Well, perhaps what I'm doing here is a duty, keeping an ear on my profession so as to justify the joys of resentment, to steal an idea or two, or just to show charity toward some friend on the program. But I learn less and less. Meanwhile the absent artists are home playing records; they are *reacting* again, finally, to something they no longer find at concerts.

Reacting to what? To the Beatles, of course—the Beatles, whose arrival has   3
proved one of the most healthy events in music since 1950. They and their off-shoots represent—as any nonspecialized intellectual will tell you—the finest communicable music of our time.

This music was already sprouting a decade ago through such innocent male ₄ sex symbols as Presley in America and Johnny Halliday in France, both of whom were then caricatured by the English in a movie called *Expresso Bongo*, a precursor of *Privilege*, about a none-too-bright rock singer. These young soloists (still functioning and making lots of money) were the parents of more sophisticated, more *committed*, soloists like Dylan and Donovan, who in turn spawned a horde of masculine offspring including twins (Simon and Garfunkel, the most cultured), quintuplets (Country Joe & The Fish, the most exotic), sextuplets (The Association, the most nostalgic), even septuplets (Mothers of Invention, the most madly satirical). With much less frequency were born female descendants such as Janis Ian or Bobbie Gentry (each of whom has produced one, and only one, good song—and who may be forgotten or immortal by the time this is read) and the trio of Supremes. Unlike their "grandparents," all of these groups, plus some twenty other fairly good ones, write most of their own material, thus combining the traditions of 12th-century troubadours, 16th-century madrigalists, and 18th-century musical artisans who were always composer-performers—in short, combining all sung expression (except opera) as it was before the twentieth century.

Why are the Beatles superior? It is easy to say that most of their competition ₅ (like most everything everywhere) is junk; more important, their betterness is consistent: each of the songs from their last three albums is memorable. The best of these memorable tunes—and the best is a large percentage ("Here, There and Everywhere," "Good Day Sunshine," "Michelle," "Norwegian Wood" are already classics)—compare with those by composers from great eras of song: Monteverdi, Schumann, Poulenc.

Good melody—even perfect melody—can be both defined and taught, as ₆ indeed can the other three "dimensions" of music: rhythm, harmony, counterpoint (although rhythm is the only one that can exist alone). Melody may be described thus: a series of notes of varying pitch and length, which evolve into a recognizable musical shape. In the case of a melody (*tune* means the same thing) which is set to words, the musical line will flow in curves relating to the verse that propels it inevitably toward a "high" point, usually called climax, and thence to the moment of culmination. The *inevitable* element is what makes the melody good—or perfect. But perfection can be sterile, as witness the thousands of 32-bar models turned out yesterday in Tin Pan Alley, or today by, say, Jefferson Airplane. Can we really recall such tunes when divorced from their words?

Superior melody results from the same recipe, with the difference that certain of the ingredients are blessed with the Distortion of Genius. The Beatles' ₇ words often go against the music (the crushing poetry that opens "A Day in the Life" intoned to the blandest of tunes), even as Martha Graham's music often contradicts her dance (she gyrates hysterically to utter silence, or stands motionless while all hell breaks loose in the pit). Because the Beatles pervert with naturalness they usually build solid structures, whereas their rivals pervert with affectation, aping the gargoyles but not the cathedral.

The unexpected in itself, of course, is no virtue, though all great works seem ₈ to contain it. For instance, to cite as examples only the above four songs: "Here,

There, and Everywhere" would seem at mid-hearing to be no more than a charming college show ballad, but once concluded it has grown immediately memorable. Why? Because of the minute harmonic shift on the words "wave of her hand," as surprising, yet as satisfyingly *right* as that in a Monteverdi madrigal like "A un giro sol." The notation of the hyper-exuberant rhythms in "Good Day Sunshine" was as aggravatingly elusive to me as some by Charles Ives, until I realized it was made by *triplets over the bar;* the "surprise" here was that the Beatles had made so simple a process *sound* so complex to a professional ear, and yet (by a third convolution) be instantly imitable by any amateur "with a beat." "Michelle" changes key on the very second measure (which is also the second word): in itself this is "allowed"—Poulenc often did it, and certainly he was the most derivative and correct composer who ever lived; the point is that he *chose* to do it on just the second measure, and that the choice worked. Genius doesn't lie in not being derivative, but in making right choices instead of wrong ones. As for "Norwegian Wood," again it is the arch of the tune—a movement growing increasingly disjunct, an inverted pyramid formed by a zigzag—which proves the song unique and memorable, rather than merely original.

Newness per se has never been the basis—or even especially an ingredient—of the Beatles' work. On the contrary, they have revitalized music's basics (harmony, counterpoint, rhythm, melody) by using them again in the simplest manner, a manner directed away from intellectualism and toward the heart. The Beatles' instrumentation may superficially sound far-out, but it apes the flashier element of electronic background no more advanced than the echo-chamber sound tracks of 1930s horror movies. Their "newest" thing is probably a kind of prosodic liberty; their rendition—their *realization*—often sounds contrary to the verses' predictable look on paper. Yet even at that, are they much different from our definitive songwriters of the past? From Purcell, say, or Debussy? It is not in their difference but in their betterness that their superiority lies.   9

But their betterness is not always apparent. Again, like Stravinsky, they are   10
already classifiable with retrospective periods. Inasmuch as they try to surpass or even consciously to redefine themselves with each period, they fail, as they mostly have with their *Magical Mystery Tour.* This isn't surprising with persons so public and hence so vulnerable. But where, from this almost complacent "civilization," can they go from here?

Well, where does any artist go? Merely on. Still, it should now be clear that   11
they are not the sum of their parts, but four distinct entities. Paul, I guess, is a genius with tunes; though what, finally, is genius without training? John, it seems, is no less clever than James Joyce; though where, ultimately, can that lead, when he is no *more* clever? George, they say, has brought East to West; but what, really, can that prove, when even Kipling realized it's not the twain of deeds but of concepts which never seem to meet? And Ringo, to at least one taste, is cute as a bug; though anyone, actually, can learn quick to play percussion, as our own George Plimpton now is demonstrating.

We've become so hung up on what they *mean,* we can no longer hear what 12
they're performing. Nor was Beethoven ever so Freudianized.

Just as twenty years ago one found oneself reading more books about Kafka 13
than reading Kafka himself, so today one gets embarrassed at being overheard
in deep discussion of the Beatles. I love them. But I love them not as symbolic
layers of "the scene" (or whatever it's called), and even less as caricatures of
themselves (which, like Mae West, they're inclined to become). I love them as
the hearty barbaric troubadours they essentially are. As such I hope they will
continue to develop, together or apart, for they represent the most invigorating
music of an era so civilized that it risks extinction less from fallout than from
boredom.

COMPREHENSION

1. Why does Rorem, a classically trained musician, like the Beatles? Why, according to the writer, are they superior to other pop groups?
2. Explain what the author means by *harmony, counterpoint, rhythm,* and *melody*
   (paragraph 6). Which songs does he cite as examples of these elements?
3. Is this an essay exclusively about the Beatles? Why or why not?

RHETORIC

1. Does "The Beatles" have an explicitly stated thesis or claim? Why or why
   not? Is his purpose to inform, persuade, or both? Justify your answer.
2. How much do you think Rorem expects readers to know about music composition, as well as both popular and classical music? How can you tell? In
   what ways does the writer establish himself as an authority on these matters? What is his tone?
3. Which paragraphs constitute the writer's introduction? What is the organizing principle in these paragraphs? How do subsequent paragraphs come
   back to ideas first expressed in the introduction?
4. Why does Rorem divide the essay into two sections? What is the relationship between these sections?
5. How does the writer develop an extended definition of what constitutes
   good or memorable music? What rhetorical strategies does he employ?
6. Evaluate the writer's conclusion. Why does he refer to Kafka and Mae West
   in an essay ostensibly about music?

WRITING

1. Select a musician or musical group. Explain what you like about the music
   and, if you can, what compositional elements the music reveals. Refer to
   other musicians to establish a comparative framework.

2. What types of music do you like? Write a classification essay answering this question.

3. **WRITING AN ARGUMENT:** Does Rorem convince you that the Beatles composed some of "the finest communicable music of our time" (paragraph 3)? Write an argumentative essay in which you either support or reject his claim.

| www.mhhe.com/ mhreader | For more information on Ned Rorem, go to: **More Resources > Ch. 10 Literature & Arts** |

✳

The Ocean, the Bird, and the Scholar

Helen Vendler

Helen Vendler (b. 1933), born in Boston, is regarded by many as one of America's finest critics of poetry. She has degrees from Emmanuel College (BA, 1954) and Harvard University (PhD, 1960). Vendler has taught widely in the United States and Europe, including Smith College, Boston University, and the University of Bordeaux; she has been a professor of English at Harvard since 1985. She has written studies of Wallace Stevens, William Butler Yeats, George Herbert, John Keats, and Seamus Heaney. Her books include Part of Nature, Part of Us: Modern American Poets *(1980),* Wallace Stevens: Words Chosen Out of Desire *(1986), and* The Art of Shakespeare's Sonnets *(1997). Vendler is also poetry reviewer for the* New Yorker *and other magazines. In the selection that follows, published in the* New Republic *in 2004, Vendler evaluates the ways in which the arts help us to live.*

When it became useful in educational circles in the United States to group various university disciplines under the name "The Humanities," it seems to have been tacitly decided that philosophy and history would be cast as the core of this grouping, and that other forms of learning—the study of languages, literatures, religion, and the arts—would be relegated to subordinate positions. Philosophy, conceived of as embodying truth, and history, conceived of as a factual record of the past, were proposed as the principal embodiments of Western culture, and given pride of place in general education programs.   1

But this confidence in a reliable factual record, not to speak of faith in a reliable philosophical synthesis, has undergone considerable erosion. Historical and philosophical assertions issue, it seems, from particular vantage points, and are no less contestable than the assertions of other disciplines. The day of limiting cultural education to Western culture alone is over. There are losses here, of   2 course—losses in depth of learning, losses in coherence—but these very changes have thrown open the question of how the humanities should now be conceived, and how the study of the humanities should, in this moment, be encouraged.

I want to propose that the humanities should take, as their central objects of 3 study, not the texts of historians or philosophers, but the products of aesthetic endeavor: art, dance, music, literature, theater, architecture, and so on. After all, it is by their arts that cultures are principally remembered. For every person who has read a Platonic dialogue, there are probably ten who have seen a Greek marble in a museum; or if not a Greek marble, at least a Roman copy; or if not a Roman copy, at least a photograph. Around the arts there exist, in orbit, the commentaries on art produced by scholars: musicology and music criticism, art history and art criticism, literary and linguistic studies. At the periphery we might set the other humanistic disciplines—philosophy, history, the study of re-ligion. The arts would justify a broad philosophical interest in ontology, phe-nomenology, and ethics; they would bring in their train a richer history than one which, in its treatment of mass phenomena, can lose sight of individual hu-man uniqueness—the quality most prized in artists, and most salient, and most valued, in the arts.

What would be the advantage of centering humanistic study on the arts? 4 The arts present the whole uncensored human person—in emotional, physical, and intellectual being, and in single and collective form—as no other branch of human accomplishment does. In the arts we see both the nature of human predicaments—in Job, in Lear, in Isabel Archer—and the evolution of represen-tation over long spans of time (as the taste for the Gothic replaces the taste for the Romanesque, as the composition of opera replaces the composition of plain-chant). The arts bring into play historical and philosophical questions without implying the prevalence of a single system or of universal solutions. Artworks embody the individuality that fades into insignificance in the massive canvas of history and is suppressed in philosophy by the desire for impersonal assertion. The arts are true to the way we are and were, to the way we actually live and have lived—as singular persons swept by drives and affections, not as collective entities or sociological paradigms. The case histories developed within the arts are in part idiosyncratic, but in part they are applicable by analogy to a class larger than the individual entities they depict. Hamlet is a very specific figure—a Danish prince who has been to school in Germany—but when Prufrock says, "I am not Prince Hamlet," he is in a way testifying to the fact that Hamlet means something to everyone who knows about the play.

If the arts are so satisfactory an embodiment of human experience, why do we 5 need studies commenting on them? Why not merely take our young people to museums, to concerts, to libraries? There is certainly no substitute for hearing Mozart, reading Dickinson, or looking at the boxes of Joseph Cornell. Why should we support a brokering of the arts? Why not rely on their direct impact? The simplest answer is that reminders of art's presence are constantly necessary.

As art goes in and out of fashion, some scholar is always necessarily reviving Melville, or editing Monteverdi, or recommending Jane Austen. Critics and scholars are evangelists, plucking the public by the sleeve, saying, "Look at this," or "Listen to this," or "See how this works." It may seem hard to believe, but there was a time when almost no one valued Gothic art, or, to come closer to our own time, *Moby-Dick* and *Billy Budd.*

A second reason to encourage scholarly studies of the arts is that such stud- 6
ies establish in human beings a sense of cultural patrimony. We in the United States are the heirs of several cultural patrimonies: a world patrimony (of which we are becoming increasingly conscious); a Western patrimony (from which we derive our institutions, civic and aesthetic); and a specifically American patrimony (which, though great and influential, has, bafflingly, yet to be established securely in our schools). In Europe, although the specifically national patrimony was likely to be urged as pre-eminent—Italian pupils studied Dante, French pupils studied Racine—most nations felt obliged to give their students an idea of the Western inheritance extending beyond native production. As time passed, colonized nations, although instructed in the culture of the colonizer, found great energy in creating a national literature and culture of their own with and against the colonial model. (We can see this, for instance, in the example of nineteenth- and twentieth-century Ireland.) For a long time, American schooling paid homage, culturally speaking, to Europe and to England; but increasingly we began to cast off European and English influence in arts and letters without, unfortunately, filling the consequent cultural gap in the schools with our own worthy creations in art and literature. Our students leave high school knowing almost nothing about American art, music, architecture, and sculpture, and having only a superficial acquaintance with a few American writers.

We will ultimately want to teach, with justifiable pride, our national patrimony 7
in arts and letters—by which, if by anything, we will be remembered—and we hope, of course, to foster young readers and writers, artists and museumgoers, composers and music enthusiasts. But these patriotic and cultural aims alone are not enough to justify putting the arts and the studies of the arts at the center of our humanistic and educational enterprise. What, then, might lead us to recommend the arts and their commentaries as the center of the humanities? Art, said Wallace Stevens, helps us to live our lives. I'm not sure we are greatly helped to live our lives by history (since, whether or not we remember it, we seem doomed to repeat it), or by philosophy (the consolations of philosophy have never been very widely received). Stevens's assertion is a large one, and we have a right to ask how he would defend it. How do the arts, and the scholarly studies attendant on them, help us to live our lives?

Stevens was a democratic author, and he expected his experience, and his 8
reflections on it, to apply widely. For him, as for any other artist, "to live our lives" means to live in the body as well as in the mind, on the sensual earth as well as in the celestial clouds. The arts exist to relocate us in the body by means

of the work of the mind in aesthetic creation; they situate us on the earth, para-doxically, by means of a mental paradigm of experience embodied, with symbolic concision, in a physical medium. It distressed Stevens that most of the human beings he saw walked about blankly, scarcely seeing the earth on which they lived, filtering it out from their pragmatic urban consciousness. Even when he was only in his twenties, Stevens was perplexed by the narrowness of the way in which people inhabit the Earth:

> I thought, on the train, how utterly we have forsaken the Earth, in the sense of excluding it from our thoughts. There are but few who consider its physical hugeness, its rough enormity. It is still a disparate monstrosity, full of solitudes & barrens & wilds. It still dwarfs & terrifies & crushes. The rivers still roar, the mountains still crash, the winds still shatter. Man is an affair of cities. His gardens & orchards & fields are mere scrapings. Somehow, however, he has managed to shut out the face of the giant from his windows. But the giant is there, nevertheless.

The arts and their attendant disciplines restore human awareness by releasing it into the ambience of the felt world, giving a habitation to the tongue in newly coined language, to the eyes and ears in remarkable re-creations of the physical world, to the animal body in the kinesthetic flex and resistance of the artistic medium. Without an alert sense of such things, one is only half alive. Stevens reflected on this function of the arts—and on the results of its absence—in three poems that I will take up as proof-texts for what follows. Although Stevens speaks in particular about poetry, he extends the concept to *poesis*—the Greek term for making, widely applicable to all creative effort.

Like geography and history, the arts confer a patina on the natural world. A vacant stretch of grass becomes humanly important when one reads the sign "Gettysburg." Over the grass hangs an extended canopy of meaning—struggle, corpses, tears, glory—shadowed by a canopy of American words and works, from the Gettysburg Address to the Shaw Memorial. The vacant plain of the sea becomes human when it is populated by the ghosts of Ahab and Moby-Dick. An unremarkable town becomes "Winesburg, Ohio"; a rustic bridge becomes "the rude bridge that arched the flood" where Minutemen fired "the shot heard round the world." One after the other, cultural images suspend themselves, invisibly, in the American air, as—when we extend our glance—the Elgin marbles, wherever they may be housed, hover over the Parthenon, once their home; as Michelangelo's Adam has become, to the Western eye, the Adam of Genesis. The patina of culture has been laid down over centuries, so that in an English field one can find a Roman coin, in an Asian excavation an emperor's stone army, in our Western desert the signs of the mound builders. Over Stevens's giant earth, with its tumultuous motions, there floats every myth, every text, every picture, every system, that creators—artistic, religious, philosophical— have conferred upon it. The Delphic oracle hovers there next to Sappho, Luther's theses hang next to the Grünewald altar, China's Cold Mountain neighbors Sinai, the B minor Mass shares space with Rabelais.

If there did not exist, floating over us, all the symbolic representations that 11
art and music, religion, philosophy, and history, have invented, and all the interpretations and explanations of them that scholarly effort has produced, what sort
of people would we be? We would, says Stevens, be sleepwalkers, going about
like automata, unconscious of the very life we were living: this is the import of
Stevens's poem "Somnambulisma," written in 1943. The poem rests on three images, of which the first is the incessantly variable sea, the vulgar reservoir from
which the vulgate—the common discourse of language and art alike—is drawn.
The second image is that of a mortal bird, whose motions resemble those of the
water but who is ultimately washed away by the ocean. The subsequent generations of the bird, too, are always washed away. The third image is that of a
scholar, without whom ocean and bird alike would be incomplete.

> On an old shore, the vulgar ocean rolls
> Noiselessly, noiselessly, resembling a thin bird,
> That thinks of settling, yet never settles, on a nest.
>
> The wings keep spreading and yet are never wings.
> The claws keep scratching on the shale, the shallow shale,
> The sounding shallow, until by water washed away.
>
> The generations of the bird are all
> By water washed away. They follow after.
> They follow, follow, follow, in water washed away.
>
> Without this bird that never settles, without
> Its generations that follow in their universe,
> The ocean, falling and falling on the hollow shore,
> Would be a geography of the dead: not of that land
> To which they may have gone, but of the place in which
> They lived, in which they lacked a pervasive being,
>
> In which no scholar, separately dwelling,
> Poured forth the fine fins, the gawky beaks, the personalia,
> Which, as a man feeling everything, were his.

Without the bird and its generations, the ocean, says the poet, would be "a 12
geography of the dead"—not in the sense of their having gone to some other
world, but in the sense of their being persons who were emotionally and intellectually dead while alive, who lacked "a pervasive being." To lack a pervasive
being is to fail to live fully. A pervasive being is one that extends through the
brain, the body, the senses, and the will, a being that spreads to every moment,
so that one not only feels what Keats called "the poetry of earth" but responds
to it with creative motions of one's own.

Unlike Keats's nightingale, Stevens's bird does not sing; its chief functions 13
are to generate generations of birds, to attempt to sprout wings, and to try to
leave behind some painstakingly scratched record of its presence. The water
restlessly moves, sometimes noiselessly, sometimes in "sounding shallow[s]";
the bird "never settles." The bird tries to generate wings, but never quite suc-

ceeds; it tries to inscribe itself on the shale, but its scratchings are washed away. The ocean is "falling and falling," the mortal generations are following and following. Time obliterates birds and inscriptions alike.

Imagine being psychically dead during the very life you have lived. That, [14] says Stevens, would be the fate of the generations were it not for the scholar. Stevens does not locate his scholar in the ocean or on the shale, the haunts of the bird; the scholar, says the poet, dwells separately. But he dwells in immense fertility: things pour forth from him. He makes up for the wings that are never wings, for the impotent claws; he generates "fine fins," the essence of the ocean's fish; he creates "gawky beaks," opening in fledglings waiting to be fed so that they may rise into their element, the air; and he produces new garments for the earth, called not regalia (suitable for a monarchy) but "personalia," suitable for the members of a democracy. How is the scholar capable of such profusion? He is fertile both because he is a man who "feel[s] everything," and because every thing that he feels reifies itself in a creation. He gives form and definition both to the physical world (as its scientific observer) and to the inchoate aesthetic world (as the quickened responder to the bird's incomplete natural song). He is analogous to the God of Genesis; as he observes and feels finniness, he says, "Let there be fine fins," and fine fins appear.

Why does Stevens name this indispensable figure a "scholar"? (Elsewhere [15] he calls him a "rabbi"—each is a word connoting learning.) What does learning have to do with creation? Why are study and learning indispensable in reifying and systematizing the world of phenomena and their aesthetic representations? Just as the soldier is poor without the poet's lines (as Stevens says elsewhere), so the poet is poor without the scholar's cultural memory, his taxonomies and his histories. Our systems of thought—legal, philosophical, scientific, religious—have all been devised by "scholars" without whose aid widespread complex thinking could not take place and be debated, intricate texts and scores could not be accurately established and interpreted. The restless emotions of aesthetic desire, the wing-wish and inscription-yearning of the bird, perish without the arranging and creative powers of intellectual endeavor. The arts and the studies of the arts are for Stevens a symbiotic pair, each dependent on the other. Nobody is born understanding string quartets or reading Latin or creating poems; without the scholar and his libraries, there would be no perpetuation and transmission of culture. The mutual support of art and learning, the mutual delight each ideally takes in each, can be taken as a paradigm of how the humanities might be integrally conceived and educationally conveyed as inextricably linked to the arts.

"Somnambulisma" is the illustration of Stevens's adage "Poetry is the scholar's [16] art." What is necessary, asks "Somnambulisma," for creative effort? Emotion, desire, generative energy, and learned invention—these, replies the poem, are indispensable in the artist. But there is another way of thinking about art, focusing less on the creator of art than on those of us who make up art's audience. What do we gain in being the audience for the arts and their attendant disciplines? Let us, says Stevens, imagine ourselves deprived of all the products of aesthetic and humanistic effort, living in a world with no music, no art, no architecture, no books, no films, no choreography, no theater, no histories, no songs, no prayers, no images floating above the earth to keep it from being a geography of the dead. Stevens creates the desolation of that deprivation in a poem—the second of my three texts—called "Large Red Man Reading." The poem is like a painting by Matisse, showing us an earthly giant the color of the sun, reading aloud from great sky-sized tabulae which, as the day declines, darken from blue to purple. The poem also summons up the people of the giant's audience: they are ghosts, no longer alive, who now inhabit unhappily (having expected more from the afterlife) the remote "wilderness of stars." What does the giant describe to the ghosts as he reads from his blue tabulae? Nothing extraordinary—merely the normal furniture of life, the common and the beautiful, the banal, the ugly, and even the painful. But to the ghosts these are things achingly familiar from life and yet disregarded within it. Now they are achingly lost, things they never sufficiently prized when alive, but which they miss devastatingly in the vacancy of space, among the foreign stars.

> There were ghosts that returned to earth to hear his phrases,
> As he sat there reading, aloud, the great blue tabulae.
> They were those from the wilderness of stars that had expected more.
>
> There were those that returned to hear him read from the poem of life,
> Of the pans above the stove, the pots on the table, the tulips among them.
> They were those that would have wept to step barefoot into reality,
>
> They would have wept and been happy, have shivered in the frost
> And cried out to feel it again, have run fingers over leaves
> And against the most coiled thorn, have seized on what was ugly
>
> And laughed, as he sat there reading, from out of the purple tabulae.
> The outlines of being and its expressings, the syllables of its law:
> *Poesis, poesis,* the literal characters, the vatic lines,
>
> Which in those ears and in those thin, those spended hearts,
> Took on color, took on shape and the size of things as they are
> And spoke the feeling for them, which was what they had lacked.

The ghosts, while they were alive, had lacked feeling, because they had not [17] registered in their memory "the outlines of being and its expressings, the syllables of its law." It is a triple assertion that Stevens makes here: that being possesses not only outlines (as all bodies do) and expressings (in all languages) but also a law, which is stricter than mere "expressings." Expressings by themselves cannot exemplify the law of being: only *poesis*—the creator's act of replicating in symbolic form the structures of life—pervades being sufficiently to intuit and to embody its law. *Poesis* not only reproduces the content of life (its daily phenomena) but finds a manner (inspired, "vatic") for that content, and in the means of its medium—here, the literal characters of its language—embodies the structural laws that shape being to our understanding.

Stevens's anecdote-of-audience in "Large Red Man Reading" suggests how   18
ardently we would want to come back, as ghosts, in order to recognize and rel-
ish the parts of life we had insufficiently noticed and hardly valued when alive.
But we cannot—according to the poem—accomplish this by ourselves: it is only
when the earthly giant of vital being begins to read, using poetic and prophetic
syllables to express the reality, and the law, of being, that the experiences of life
can be reconstituted and made available as beauty and solace, to help us live
our lives.

How could our lives be different if we reconstituted the humanities around the   19
arts and the studies of the arts? Past civilizations are recalled in part, of course,
for their philosophy and their history, but for most of us it is the arts of the past
that preserve Egypt and Greece and Rome, India and Africa and Japan. The
names of the artists may be lost, the arts themselves in fragments, the scrolls incomplete, the manuscripts partial—but Anubis and the Buddha and *The Canterbury Tales* still populate our imaginative world. They come trailing their
interpretations, which follow them and are like water washed away. Scholarly
and critical interpretations may not outlast the generation to which they are relevant; as intellectual concepts flourish and wither, so interpretations are proposed and discarded. But we would not achieve our own grasp on Vermeer or
Horace, generation after generation, without the scholars' outpourings.

If we are prepared to recognize the centrality of artists and their interpreters   20
to every past culture, we might begin to reflect on what our own American culture has produced that will be held dear centuries from now. Which are the
paintings, the buildings, the novels, the musical compositions, the poems,
through which we will be remembered? What set of representations of life will
float above the American soil, rendering each part of it as memorable as Marin's
Maine or Langston Hughes's Harlem, as Cather's Nebraska or Lincoln's Gettysburg? How will the outlines and the expressings and the syllables of American
being glow above our vast geography? How will our citizens be made aware of
their cultural inheritance, and become proud of their patrimony? How will they
pass it on to their children as their own generation is by water washed away?
How will their children become capable of "feeling everything," of gaining "a
pervasive being," capable of helping the bird to spread its wings and the fish to
grow their "fine fins" and the scholar to pour forth his "personalia"?

To link, by language, feeling to phenomena has always been the poet's aim.   21
"Poetry," said Wordsworth in 1798, "is the breath and finer spirit of all knowledge; it is the impassioned expression which is in the countenance of all science."
Our culture cannot afford to neglect the thirst of human beings for the representations of life offered by the arts, the hunger of human beings for commentary on
those arts as they appear on the cultural stage. The training in subtlety of response (which used to be accomplished in large part by religion and the arts) cannot be responsibly left to commercial movies and television. Within education,
scientific training, which necessarily brackets emotion, needs to be complemented by the direct mediation—through the arts and their interpretations—

of feeling, vicarious experience, and interpersonal imagination. Art can often be trusted—once it is unobtrusively but ubiquitously present—to make its own impact felt. A set of Rembrandt self-portraits in a shopping mall, a group of still lifes in a subway, sonatas played in the lunchroom, spirituals sung chorally from kindergarten on—all such things, appearing entirely without commentary, can be offered in the community and the schools as a natural part of living. Students can be gently led, by teachers and books, from passive reception to active reflection. The arts are too profound and far-reaching to be left out of our children's patrimony: the arts have a right, within our schools, to be as serious an object of study as molecular biology or mathematics. Like other complex products of the mind, they ask for reiterated exposure, sympathetic exposition, and sustained attention.

The arts have the advantage, once presented, of making people curious not 22 only about aesthetic matters, but also about history, philosophy, and other cultures. How is it that pre-Columbian statues look so different from Roman ones? Why do some painters concentrate on portraits and others on landscapes? Why did great ages of drama arise in England and Spain and then collapse? Who first found a place for jazz in classical music, and why? Why do some writers become national heroes, and others do not? Who evaluates art, and how? Are we to believe what a piece of art says? Why does Picasso represent a full face and a profile at the same time? How small can art be and still be art? Why have we needed to invent so many subsets within each art—within literature the epic, drama, lyric, novel, dialogue, essay; within music everything from the solo partita to the chorales of Bach? Why do cultures use different musical instruments and scales? Who has the right to be an artist? How does one claim that right? The questions are endless, and the answers provocative; and both questions and answers require, and indeed generate, sensuous responsiveness, a trained eye, fine discrimination, and a hunger for learning, all qualities we would like to see in ourselves and in our children.

Best of all, the arts are enjoyable. The "grand elementary principle of pleas- 23 ure" (as Wordsworth called it) might be invoked more urgently than it now is to make the humanities, both past and present, mean something relevant to Americans. Once the appetite for an art has been awakened by pleasure, the nursery rhyme and the cartoon lead by degrees to Stevens and Eakins. A curriculum relying on the ocean, the bird, and the scholar, on the red man and his blue tabulae, would produce a love of the arts and humanities that we have not yet succeeded in generating in the population at large. When reality is freshly seen, through the artists and their commentators, something happens to the felt essence of life. As Stevens wrote in the third of my texts, "Angel Surrounded by Paysans," the angel of reality then briefly appears at our door, saying:

> . . . I am the necessary angel of earth,
> Since, in my sight, you see the earth again.
> Cleared of its stiff and stubborn, man-locked set,
> And, in my hearing, you hear its tragic drone

Rise liquidly in liquid lingerings,
Like watery words awash; like meanings said

By repetitions of half meanings. Am I not,
Myself, only half of a figure of a sort,

A figure half seen, or seen for a moment, a man
Of the mind, an apparition apparelled in

Apparels of such lightest look that a turn
Of my shoulder and quickly, too quickly, I am gone?

That art-angel of the earth, renewing our sense of life and of ourselves, is   24
only half meaning, because we provide the other half. Among us are the scholars who interpret those half-meanings into full ones, apparelling us anew in
their personalia. In the apparels of his messenger, Stevens is recalling these lines
from Wordsworth's great ode:

There was a time when meadow, grove, and stream,
The earth, and every common sight
To me did seem
Apparelled in celestial light,
The glory and the freshness of a dream.

The secular angel refreshing our sense of the world, apparelled in   25
Wordsworthian light, stays only for a moment, our moment of attention. But
that moment of mental acuity recalls us to being, the body, and the emotions,
which are, peculiarly, so easy for us to put to one side as we engage in purely intellectual or physical work. Just as art is only half itself without us—its audience, its analysts, its scholars—so we are only half ourselves without it. When,
in this country, we become fully ourselves, we will have balanced our great
accomplishments in progressive abstraction—in mathematics and the natural
sciences—with an equally great absorption in art, and in the disciplines ancil-
lary to art. The arts, though not progressive, aim to be eternal, and sometimes
are. And why should the United States not have as much eternity as any other
nation? As Marianne Moore said of excellence, "It has never been confined to
one locality."

COMPREHENSION

1. Consider the title. What is the significance of "the ocean, the bird, and the
   scholar"? Where in the essay does Vendler explain these connections?
2. How does the writer define "the humanities"? Why does she think the arts
   should be central to humanistic study?

3. Who was Wallace Stevens? Where do references to this individual appear in the essay, and what observations does Vendler make about him? Why, according to the author, is he important to our understanding of the arts?

RHETORIC

1. What is the author's purpose in this essay, and what is her claim? What appeals to logic and ethics does she make? Do you find her persuasive? Why or why not?
2. This essay is a version of the 33rd Jefferson Lecture in the Humanities delivered by Vendler. What can you infer about the original audience for the speech? What type of readership would find the printed version appealing or provocative? What assumptions does she make about her audience? How can you tell?
3. Why does the author divide the essay into several sections? What progression in her argument do you detect from section to section?
4. How do references to Wallace Stevens serve as an organizing principle for large segments of the essay? Why does the author quote sections from his poetry and analyze them?
5. What types of evidence does Vendler provide to support her points? How effective is this evidence in convincing you that her claim is justifiable? Explain.
6. How compelling is Vendler's conclusion? Is it appropriate for her to end with a quote from the poetry of Marianne Moore? Why or why not?

WRITING

1. Select a poem and write an analysis and evaluation of it. Explain why it reveals something important about life, culture, and society.
2. Select one writer, artist, or musician mentioned by Vendler in her essay. Conduct research on this figure, and then write a brief research paper on the way the artist reflects certain cultural or national values.
3. **WRITING AN ARGUMENT:** Take a position on the importance of the arts, as Vendler defines them, to a college education. Use material from Vendler's essay as well as materials from other essays in this chapter to support your points.

★ www.mhhe.com/ **mhreader**

For more information on Helen Vendler, go to:
More Resources > Ch. 10 Literature & Arts

CONNECTIONS FOR CRITICAL THINKING

1. Write an essay comparing and contrasting poetry and any other literary or art form. What merits does each form have? Are there any limitations in either form? Which do you find more satisfying? Which form is more accessible? Use at least three essays in this chapter to illustrate or support your thesis.

2. Write an essay exploring the importance of role models in art and literature. Refer to the essays by Walker, Rorem, and Vendler to address the issue.

3. Develop an extended definition of poetry, using the essays by Hogan, Collins, and Vendler to help develop your ideas. Explain in your own words why poetry is important. Refer to examples of poetry in popular music and culture if you wish.

4. Forster, Collins, Rorem, Morrow, and other writers in this chapter provide extended examples of art and artists. How do they develop these illustrations? What strategies do they use? Are their goals similar or not? Explain your response.

5. Use the essays by Forster, Lebowitz, and Rorem to explore the question of excellence in the arts. Answer the question: How do you know the work of art is good?

6. Examine the role of the artist in society and the artist's purpose or duty to society. How would the writers in this chapter address this issue?

7. Working as a group, discuss the two poems by Billy Collins that appear in his essay. Then explore the ways in which other writers in this chapter have helped the group to analyze and evaluate these poems.

8. After visiting several Web sites, write an essay in which you explain the importance of photography and photojournalism. Connect your findings to the ideas presented by Morrow in his essay.

chapter *11*

Philosophy, Ethics, and Religion
What Do We Believe?

You do not have to be an academician in an ivory tower to think about religion and the destiny of humankind or about questions of right and wrong. All of us possess beliefs about human nature and conduct, about "rival conceptions of God" (to use C. S. Lewis's phrase), about standards of behavior and moral duty. In fact, as Robert Coles argues in an essay appearing in this chapter, even children make ethical choices every day and are attuned to the "moral currents and issues in the large society."

Most of us have a system of ethical and religious beliefs, a philosophy of sorts, although it may not be a fully logical and systematic philosophy and we may not be conscious that it determines what we do in everyday life. This system of beliefs and values is transmitted to us by family members, friends, educators, religious figures, and representatives of social groups. Such a philosophical system is not unyielding or unchanging, because our typical conflicts and the choices that we make often force us to test our ethical assumptions and our values. For example, you may believe in nonviolence, but what would you do if someone threatened physical harm to you or a loved one? Or you may oppose the death penalty but encounter an essay that causes you to reassess your position. Our beliefs about nonviolence, capital punishment, abortion, cheating, equality, and so on are often paradoxical and place us in a universe of ethical dilemmas.

Your ability to resolve such dilemmas and make complex ethical decisions depends on your storehouse of knowledge and experience and on how well formulated your philosophy or system of beliefs is. When you know what is truly important in your life, you can make choices and decisions carefully and responsibly. Growing up in a world with competing views on morality often makes these choices that much harder, for constellations of cultures, beliefs, and influences contribute to our own personal development. As Plato observes in his classic "The Allegory of the Cave," the idea of what is truly good and correct never appears without wisdom and effort.

In this context, religion is also intrinsically connected to our sense of morality and ethics. Our personal code of ethics often has a religious grounding. Our religion often determines the way in which we apply our ethics—for instance, it may determine our attitudes toward contraception, equality of the races or the sexes, and evolution. In all instances, competing religious and secular values may force us to make hard decisions about our positions on significant cultural issues.

In one essay in this section, Virginia Woolf contemplates a seemingly insignificant creature—a moth—that tells her (and us) a great deal about life and death. All authors in this chapter seek the essence of the values and ideas that we develop during our brief moment on this planet and that lend meaning and vitality to our lives.

Previewing the Chapter

As you read the essays in this chapter and respond to them in discussion and writing, consider the following questions:

- On what ethical or religious problem or conflict does the author focus?
- Is the author's view of life optimistic or pessimistic? Why?
- Do you agree or disagree with the philosophical or religious perspective that the author adopts?
- Is there a clear solution to the issue the author investigates?
- Does the author present rational arguments or engage in emotional appeals and weak reasoning?
- Does the author approach ethical, theological, and philosophical issues in an objective or in a subjective way?
- How significant is the ethical or philosophical subject addressed by the author?
- What social, political, or racial issues are raised by the author?
- Are there religious dimensions to the essay? If so, how does religion reinforce the author's philosophical inquiry?
- How do these essays encourage you to examine your own attitudes and values? In reading them, what do you discover about your own system of beliefs and the beliefs of society at large?

Classic and Contemporary Images
HOW DO WE VIEW ANGELS AND DEVILS?

Using a Critical Perspective   Comment on the composition of each of these works of art. How does each artist present the supernatural beings depicted? What do you notice about the organization of the images? From what angle does the artist approach the depiction? What do the artists have in common? Is the overall impression or effect of each illustration the same or different? Explain.

Angels, supernatural beings who serve as messengers from God, are found in the literature and imagery of Judaism, Christianity, and Islam from ancient times to the present, as in the Islamic painting from India shown here.

618

In more recent times, the sculptor Jacob Epstein (1880–1959) created a bronze
statue of St. Michael for Coventry Cathedral in England. The ancient
cathedral at Coventry was destroyed by German bombs in 1940.
During the 1950s, a new cathedral was built near the ruins of the old one.
With his spear in hand and his wings outstretched, St. Michael
stands in triumph over the prone, chained figure of the devil.

Classic and Contemporary Essays
IS SUPERSTITION A FORM OF BELIEF?

Although most contemporary individuals who consider themselves "educated" deny any strong influence of superstition in their lives, both Margaret Mead and Letty Cottin Pogrebin suggest in their respective essays that neither contemporary ideas, with their reliance on science, nor higher education, with its focus on rational thinking, insulate us from allowing at least a small amount of superstition into our lives. Mead discounts the notion that superstition is relegated to "primitive" societies or to the uneducated who have not been enlightened by a firm grounding in empiricism. The famed anthropologist suggests that we need superstition to provide coherence to our lives when other forms of belief and thought are not competent to satisfy us. For example, in facing the unknown, we may turn to superstition as a welcome friend. How many of your fellow schoolmates, for example, will cross their fingers before an exam or keep a good-luck charm attached to their computer? Simply put, very few of us are so secure that we can rely on our own inner fortitude to ward off occasional fears or feelings of helplessness. Superstition, therefore, provides a framework to maintain a private sanctuary against the unknown. Pogrebin admits to being a "very rational person" who also "happen[s] to be superstitious." Although the particular rituals that inform her superstition were learned from her mother, who used them as a means of "imposing order," the function of superstition in her own life, Pogrebin claims, is to maintain coherence with the past: to feel the connection between herself and her mother. In other words, like Mead, Pogrebin contends superstition helps us maintain a sense of protection in an environment where we do not have complete control. By keeping the same rituals as her mother, Pogrebin senses her mother's protection. Does this mean that humans are flawed, weak creatures? If we consider the vicissitudes and uncertainties of modern life, perhaps the tendency of humans to have a bit of superstition in their worldview is a sign of intelligence.

✳

New Superstitions for Old

Margaret Mead

Margaret Mead (1901–1979), famed American anthropologist, was curator of ethnology at the American Museum of Natural History and a professor at Columbia University. Her field expeditions to Samoa, New Guinea, and Bali in the 1920s and 1930s produced several major studies, notably Coming of Age in Samoa *(1928),* Growing Up in New Guinea *(1930), and* Sex and Temperament in Three Primitive Societies *(1935). In this essay, first published in* A Way of Seeing *(1970), Mead discusses the role that superstition plays in our daily life.*

Once in a while there is a day when everything seems to run smoothly and even   1
the riskiest venture comes out exactly right. You exclaim, "This is my lucky day!" Then as an afterthought you say, "Knock on wood!" Of course, you do not really believe that knocking on wood will ward off danger. Still, boasting about your own good luck gives you a slightly uneasy feeling—and you carry out the little protective ritual. If someone challenged you at that moment, you would probably say, "Oh, that's nothing. Just an old superstition."

But when you come to think about it, what is superstition?   2

In the contemporary world most people treat old folk beliefs as superstitions—the belief, for instance, that there are lucky and unlucky days or numbers, that future events can be read from omens, that there are protective charms or that what happens can be influenced by casting spells. We have excluded magic from our current world view, for we know that natural events have natural causes.   3

In a religious context, where truths cannot be demonstrated, we accept   4
them as a matter of faith. Superstitions, however, belong to the category of beliefs, practices and ways of thinking that have been discarded because they are inconsistent with scientific knowledge. It is easy to say that other people are superstitious because they believe what we regard to be untrue. "Superstition" used in that sense is a derogatory term for the beliefs of other people that we do not share. But there is more to it than that. For superstitions lead a kind of half life in a twilight world where, sometimes, we partly suspend our disbelief and act as if magic worked.

Actually, almost every day, even in the most sophisticated home, something   5
is likely to happen that evokes the memory of some old folk belief. The salt spills. A knife falls to the floor. Your nose tickles. Then perhaps, with a slightly embarrassed smile, the person who spilled the salt tosses a pinch over his left shoulder. Or someone recites the old rhyme, "Knife falls, gentleman calls." Or as you rub your nose you think, That means a letter. I wonder who's writing?

No one takes these small responses very seriously or gives them more than a passing thought. Sometimes people will preface one of these ritual acts—walking around instead of under a ladder or hastily closing an umbrella that has been opened inside a house—with such remarks as "I remember my great-aunt used to . . . " or "Germans used to say you ought not . . . " And then, having placed the belief at some distance away in time or space, they carry out the ritual.

Everyone also remembers a few of the observances of childhood—wishing   6
on the first star; looking at the new moon over the right shoulder; avoiding the cracks in the sidewalk on the way to school while chanting, "Step on a crack, break your mother's back"; wishing on white horses, on loads of hay, on covered bridges, on red cars; saying quickly, "Bread-and-butter" when a post or a tree separated you from the friend you were walking with. The adult may not actually recite the formula "Star light, star bright . . . " and may not quite turn to look at the new moon, but his mood is tempered by a little of the old thrill that came when the observance was still freighted with magic.

Superstition can also be used with another meaning. When I discuss the re-   7
ligious beliefs of other peoples, especially primitive peoples, I am often asked, "Do they really have a religion, or is it all just superstition?" The point of contrast here is not between a scientific and a magical view of the world but between the clear, theologically defensible religious beliefs of members of civilized societies and what we regard as the false and childish views of the heathen who "bow down to wood and stone." Within the civilized religions, however, where membership includes believers who are educated and urbane and others who are ignorant and simple, one always finds traditions and practices that the more sophisticated will dismiss offhand as "just superstition" but that guide the steps of those who live by older ways. Mostly these are very ancient beliefs, some handed on from one religion to another and carried from country to country around the world.

Very commonly, people associate superstition with the past, with very old   8
ways of thinking that have been supplanted by modern knowledge. But new superstitions are continually coming into being and flourishing in our society. Listening to mothers in the park in the 1930s, one heard them say, "Now, don't you run out into the sun, or Polio will get you." In the 1940s elderly people explained to one another in tones of resignation, "It was the Virus that got him down." And every year the cosmetics industry offers us new magic—cures for baldness, lotions that will give every woman radiant skin, hair coloring that will restore to the middle-aged the charm and romance of youth—results that are promised if we will just follow the simple directions. Families and individuals also have their cherished, private superstitions. You must leave by the back door when you are going on a journey, or you must wear a green dress when you are taking an examination. It is a kind of joke, of course, but it makes you feel safe.

These old half-beliefs and new half-beliefs reflect the keenness of our wish   9
to have something come true or to prevent something bad from happening. We

do not always recognize new superstitions for what they are, and we still follow the old ones because someone's faith long ago matches our contemporary hopes and fears. In the past people "knew" that a black cat crossing one's path was a bad omen, and they turned back home. Today we are fearful of taking a journey and would give anything to turn back—and then we notice a black cat running across the road in front of us.

Child psychologists recognize the value of the toy a child holds in his hand 10 at bedtime. It is different from his thumb, with which he can close himself in from the rest of the world, and it is different from the real world to which he is learning to relate himself. Psychologists call these toys—these furry animals and old, cozy baby blankets—"transitional objects"; that is, objects that help the child move back and forth between the exactions of everyday life and the world of wish and dream.

Superstitions have some of the qualities of these transitional objects. They 11 help people pass between the areas of life where what happens has to be accepted without proof and the areas where sequences of events are explicable in terms of cause and effect, based on knowledge. Bacteria and viruses that cause sickness have been identified; the cause of symptoms can be diagnosed and a rational course of treatment prescribed. Magical charms no longer are needed to treat the sick; modern medicine has brought the whole sequence of events into the secular world. But people often act as if this change had not taken place. Laymen still treat germs as if they were invisible, malign spirits, and physicians sometimes prescribe antibiotics as if they were magic substances.

Over time, more and more of life has become subject to the controls of 12 knowledge. However, this is never a one-way process. Scientific investigation is continually increasing our knowledge. But if we are to make good use of this knowledge, we must not only rid our minds of old, superseded beliefs and fragments of magical practice, but also recognize new superstitions for what they are. Both are generated by our wishes, our fears and our feeling of helplessness in difficult situations.

Civilized peoples are not alone in having grasped the idea of superstitions—beliefs and practices that are superseded but that still may evoke the different worlds in which we live—the sacred, the secular and the scientific. They allow us to keep a private world also, where, smiling a little, we can banish danger with a gesture and summon luck with a rhyme, make the sun shine in spite of storm clouds, force the stranger to do our bidding, keep an enemy at bay and straighten the paths of those we love.

COMPREHENSION

1. Explain in your own words the religious context for this essay.
2. What point is Mead making about superstition in modern life? Where does she state her main idea?
3. Where does Mead define *superstition?* How does it differ from folk beliefs?

RHETORIC

1. Explain what Mead means by "transitional objects" (paragraph 10). Why does she mention them?
2. Discuss the author's use of the pronouns *we* and *us* in the conclusion. Why does she state the conclusion in personal terms?
3. How does Mead use definition to differentiate *superstition* from *faith?* Explain the logic behind her distinction.
4. How does Mead use classification to describe the "worlds in which we live"? What are these worlds? What examples does she give of superstition in each of these worlds?
5. Look at paragraph 10. What is the purpose of this example? How does it figure in the context of Mead's essay?
6. Discuss the term "theologically defensible" as used in paragraph 7. Does Mead support this concept by example or evidence? Why?

WRITING

1. Write an essay about beliefs you once held that you have since abandoned. Why did you abandon them? What was the practical result?
2. Select a saying or phrase based in superstition or folk belief that you or a friend is fond of. Analyze its appeal.
3. **WRITING AN ARGUMENT:** This article was published in 1966. Have we made any progress toward banishing superstition since then? Will we ever live in a culture free of superstition? Do we want to? Answer these questions in an argumentative essay.

---

www.mhhe.com/ **mhreader**

For more information on Margaret Mead, go to:
More Resources > Ch. 11 Ethics & Religion

✳

Superstitious Minds

Letty Cottin Pogrebin

Letty Cottin Pogrebin (b. 1939) is deeply committed to women's issues, family politics, and the nonsexist rearing and education of children. A native of New York, she graduated from Brandeis University and from 1971 to 1987 was the editor of Ms. *magazine, for which she remains a contributing editor. She has also contributed to such publications as* The New York Times *and* The Nation *and has written a number of books,*

including Among Friends *(1986)*; Debra Golda and Me: Being Female and Jewish in America *(1991); and* Getting Over Getting Older *(1996). Pogrebin lectures
frequently and is a founder of the Women's Political Caucus as well as president of the
Authors' Guild. In the following essay, she reminisces about her fearful, superstitious
mother, whom she understands much better since becoming a mother herself.*

I am a very rational person. I tend to trust reason more than feeling. But I also   1
happen to be superstitious—in my fashion. Black cats and rabbits' feet hold no
power for me. My superstitions are my mother's superstitions, the amulets and
incantations she learned from her mother and taught me.

I don't mean to suggest that I grew up in an occult atmosphere. On the con-   2
trary, my mother desperately wanted me to rise above her immigrant ways and
become an educated American. She tried to hide her superstitions, but I came to
know them all: Slap a girl's cheeks when she first gets her period. Never take a
picture of a pregnant woman. Knock wood when speaking about your good
fortune. Eat the ends of bread if you want to have a boy. Don't leave a bride
alone on her wedding day.

When I was growing up, my mother often would tiptoe in after I seemed to   3
be asleep and kiss my forehead three times, making odd noises that sounded
like a cross between sucking and spitting. One night I opened my eyes and demanded an explanation. Embarrassed, she told me she was excising the "Evil
Eye"—in case I had attracted its attention that day by being especially wonderful. She believed her kisses could suck out any envy or ill will that those less fortunate may have directed at her child.

By the time I was in my teens, I was almost on speaking terms with the Evil   4
Eye, a jealous spirit that kept track of those who had "too much" happiness and
zapped them with sickness and misery to even the score. To guard against his
mischief, my mother practiced rituals of interference. evasion, deference, and
above all, avoidance of situations where the Evil Eye might feel at home.

This is why I wasn't allowed to attend funerals. This is also why my mother   5
hated to mend my clothes while I was wearing them. The only garment one
should properly get sewn *into* is a shroud. To ensure that the Evil Eye did not
confuse my pinafore with a burial outfit, my mother insisted that I chew a
thread while she sewed, thus proving myself very much alive. Outwitting the
Evil Eye also accounted for her closing the window shades above my bed
whenever there was a full moon. The moon should only shine on cemeteries,
you see; the living need protection from the spirits.

Because we were dealing with a deadly force, I also wasn't supposed to say   6
any words associated with mortality. This was hard for a 12-year-old who punctuated every anecdote with the verb "to die," as in "You'll die when you hear
this!" or "If I don't get home by ten, I'm dead." I managed to avoid using such
expressions in the presence of my mother until the day my parents brought
home a painting I hated and we were arguing about whether it should be displayed on our walls. Unthinking, I pressed my point with a melodramatic idiom: "That picture will hang over my dead body!" Without a word, my mother
grabbed a knife and slashed the canvas to shreds.

I understand all this now. My mother emigrated in 1907 from a small Hun- 7
garian village. The oldest of seven children, she had to go out to work before
she finished the eighth grade. Experience taught her that life was unpredictable
and often incomprehensible. Just as an athlete keeps wearing the same T-shirt
in every game to prolong a winning streak, my mother's superstitions gave her
a means of imposing order on a chaotic system. Her desire to control the fates
sprung from the same helplessness that makes the San Francisco 49ers' defensive more superstitious than its offensive team. Psychologists speculate this is
because the defense has less control; they don't have the ball.

Women like my mother never had the ball. She died when I was 15, leaving 8
me with deep regrets for what she might have been—and a growing understanding of who she was. *Superstitious* is one of the things she was. I wish I had
a million sharp recollections of her, but when you don't expect someone to die,
you don't store up enough memories. Ironically, her mystical practices are
among the clearest impressions she left behind. In honor of this matrilineal
heritage—and to symbolize my mother's effort to control her life as I in my way
try to find order in mine—knock on wood and I do not let the moon shine on
those I love. My children laugh at me, but they understand that these tiny rituals have helped keep my mother alive in my mind.

A year ago, I awoke in the night and realized that my son's window blinds 9
had been removed for repair. Smiling at my own compulsion, I got a bed sheet
to tack up against the moonlight and I opened his bedroom door. What I saw
brought tears to my eyes. There, hopelessly askew, was a blanket my son, then
18, had taped to his window like a curtain.

My mother never lived to know David, but he knew she would not want 10
the moon to shine upon him as he slept.

COMPREHENSION

1. What is the function of superstition in the writer's life? What purpose did it
   serve in her mother's life?
2. What was Pogrebin's reaction to her mother's behavior while she was
   growing up? How does the adult feel?
3. How does the writer use superstitions now as an adult? Has she passed on
   these beliefs to her children? Explain.

RHETORIC

1. Examine Pogrebin's first sentence. How does it prepare the reader for the
   content of the essay? How does its simplicity add to its force?
2. How do the accumulated examples in paragraph 2 illustrate the point of the
   paragraph?
3. What is the writer's tone? Justify your answer.
4. What is the point of paragraph 7? How does the metaphor work to support
   Pogrebin's point?

5. What is the purpose of the essay? Where does it become apparent? How do the other paragraphs reinforce it?
6. Comment on the author's final sentence. What effect does it have on the reader? How does it help to hold the essay together?

WRITING

1. Children are often annoyed or embarrassed by their parents' behavior or beliefs. Write an essay describing something your parents repeatedly said or did that caused you discomfort or confusion. Include how you now feel about their actions and any insight you may have gained about their motives or feelings.
2. Write an essay about superstition. Consider the meaning of the word. What connection, if any, does it have with religion? How do superstitions affect the people who believe in them? Why do they believe? What is the role of superstition in your family? Provide examples of superstitions.
3. Write an essay in which you consider how your parents raised you—the values, opinions, beliefs they instilled in you. Would you want to pass these on to your children? Why, or why not?
4. **WRITING AN ARGUMENT:** Argue for or against the proposition that children need superstition—Santa Claus, the Tooth Fairy, and so forth—in their lives.

 www.mhhe.com/ **mhreader**   For more information on Letty Cottin Pogrebin, go to: **More Resources > Ch. 11 Ethics & Religion**

Classic and Contemporary: Questions for Comparison

1. Mead presents her argument regarding the benefits of superstition through an anthropological analysis of the subject, whereas Pogrebin employs a more personal approach, focusing on the role of superstition in her own life. What are the merits of each approach? Do the two essays together contribute to a greater understanding of the nature of superstition than either one alone?
2. Both Mead and Pogrebin discuss the value of superstition in childhood. Why is childhood in particular a time in life when superstition can prove valuable? What is Mead's answer to this issue? How does it differ from Pogrebin's?
3. The tone of Mead's essay is objective, scientific, and critical. In addition, she never refers to personal experience, and we can assume she is writing from the perspective of someone who needs to maintain her professional status.

Why would these factors influence her decision to contour her style so that she seems an observer of superstition rather than someone with superstitious leanings? On the other hand, why would Pogrebin choose to "personalize" her essay by referring to personal experience? Are these choices matters of style, audience, purpose, or a combination of these?

4. Mead seems more concerned with explicating her subject matter in academic terms. Note how she articulates the meaning of *folk beliefs*, *religion*, and *transitional objects*. How would you characterize the differences between an academically oriented argument such as Mead's and one intended for a more general audience such as Pogrebin's?

✳

I Listen to My Parents and I Wonder What They Believe

Robert Coles

Robert Coles (b. 1929), author and psychologist, won the Pulitzer Prize in general nonfiction for volumes 1 and 2 of Children of Crisis, *in which he examines with compassion and intelligence the effects of the controversy over integration on children in the South. Walker Percy praised Coles because he "spends his time listening to people and trying to understand them." In its final form,* Children of Crisis *has five volumes, and Coles has widened its focus to include the children of the wealthy and the poor, the exploited and the exploiters. In collaboration with Jane Coles, he completed* Women in Crisis II *(1980). He has also written* The Secular Mind *(1999) and* Lives of Moral Leadership *(2000). Below, Coles demonstrates his capacity to listen to and to understand children.*

Not so long ago children were looked upon in a sentimental fashion as "angels," or as "innocents." Today, thanks to Freud and his followers, boys and girls are understood to have complicated inner lives; to feel love, hate, envy and rivalry in various and subtle mixtures; to be eager participants in the sexual and emotional politics of the home, neighborhood and school. Yet some of us parents still cling to the notion of childhood innocence in another way. We do not see that our children also make ethical decisions every day in their own lives, or realize how attuned they may be to moral currents and issues in the larger society.

In Appalachia I heard a girl of eight whose father owns coal fields (and gas stations, a department store and much timberland) wonder about "life" one day: "I'll be walking to the school bus, and I'll ask myself why there's some who are poor and their daddies can't find a job, and there's some who are lucky like me. Last month there was an explosion in a mine my daddy owns, and everyone became upset. Two miners got killed. My daddy said it was their own fault, because they'll be working and they get careless. When my mother asked if there was anything wrong with the safety down in the mine, he told her no and she shouldn't ask questions like that. Then the Government people came and they said it was the owner's fault—Daddy's. But he has a lawyer and the lawyer is fighting the Government and the union. In school, kids ask me what I think, and I sure do feel sorry for the two miners and so does my mother—I know that. She told me it's just not a fair world and you have to remember that. Of course, there's no one who can be sure there won't be trouble; like my daddy says, the rain falls on the just and the unjust. My brother is only six and he asked Daddy awhile back who are the 'just' and the 'unjust,' and Daddy said there are people who work hard and they live good lives, and there are lazy people and they're always trying to sponge off others. But I guess you have to feel sorry for anyone who has a lot of trouble, because it's poured-down, heavy rain."

Listening, one begins to realize that an elementary-school child is no 3 stranger to moral reflection—and to ethical conflict. This girl was torn between her loyalty to her particular background, its values and assumptions, and to a larger affiliation—her membership in the nation, the world. As a human being whose parents were kind and decent to her, she was inclined to be thoughtful and sensitive with respect to others, no matter what their work or position in society. But her father was among other things a mineowner, and she had already learned to shape her concerns to suit that fact of life. The result: a moral oscillation of sorts, first toward nameless others all over the world and then toward her own family. As the girl put it later, when she was a year older: "You should try to have 'good thoughts' about everyone, the minister says, and our teacher says that too. But you should honor your father and mother most of all; that's why you should find out what they think and then sort of copy them. But sometimes you're not sure if you're on the right track."

Sort of copy them. There could be worse descriptions of how children acquire 4 moral values. In fact, the girl understood how girls and boys all over the world "sort of" develop attitudes of what is right and wrong, ideas of who the just and the unjust are. And they also struggle hard and long, and not always with success, to find out where the "right track" starts and ends. Children need encouragement or assistance as they wage that struggle.

In home after home that I have visited, and in many classrooms, I have met 5 children who not only are growing emotionally and intellectually but also are trying to make sense of the world morally. That is to say, they are asking themselves and others about issues of fair play, justice, liberty, equality. Those last words are abstractions, of course—the stuff of college term papers. And there are, one has to repeat, those in psychology and psychiatry who would deny elementary-school children access to that "higher level" of moral reflection. But any parent who has listened closely to his or her child knows that girls and boys are capable of wondering about matters of morality, and knows too that often it is their grown-up protectors (parents, relatives, teachers, neighbors) who are made uncomfortable by the so-called "innocent" nature of the questions children may ask or the statements they may make. Often enough the issue is not the moral capacity of children but the default of us parents who fail to respond to inquiries put to us by our daughters and sons—and fail to set moral standards for both ourselves and our children.

Do's and don't's are, of course, pressed upon many of our girls and boys. ₆ But a moral education is something more than a series of rules handed down, and in our time one cannot assume that every parent feels able—sure enough of her own or his own actual beliefs and values—to make even an initial explanatory and disciplinary effect toward a moral education. Furthermore, for many of us parents these days it is a child's emotional life that preoccupies us.

In 1963, when I was studying school desegregation in the South, I had ex- ₇ tended conversations with Black and white elementary-school children caught up in a dramatic moment of historical change. For longer than I care to remember, I concentrated on possible psychiatric troubles, on how a given child was managing under circumstances of extreme stress, on how I could be of help— with "support," with reassurance, with a helpful psychological observation or interpretation. In many instances I was off the mark. These children weren't "patients"; they weren't even complaining. They were worried, all right, and often enough they had things to say that were substantive—that had to do not so much with troubled emotions as with questions of right and wrong in the real-life dramas taking place in their worlds.

Here is a nine-year-old white boy, the son of ardent segregationists, telling ₈ me about his sense of what desegregation meant to Louisiana in the 1960s: "They told us it wouldn't happen—never. My daddy said none of us white people would go into schools with the colored. But then it did happen, and when I went to school the first day I didn't know what would go on. Would the school stay open or would it close up? We didn't know what to do; the teacher kept telling us that we should be good and obey the law, but my daddy said the law was wrong. Then my mother said she wanted me in school even if there were some colored kids there. She said if we all stayed home she'd be a 'nervous wreck.' So I went.

"After a while I saw that the colored weren't so bad. I saw that there are dif- ₉ ferent kinds of colored people, just like with us whites. There was one of the colored who was nice, a boy who smiled, and he played real good. There was another one, a boy, who wouldn't talk with anyone. I don't know if it's right that we all be in the same school. Maybe it isn't right. My sister is starting school next year, and she says she doesn't care if there's 'mixing of the races.' She says they told her in Sunday school that everyone is a child of God, and then a kid asked if that goes for the colored too and the teacher said yes, she thought so. My daddy said that it's true, God made everyone—but that doesn't mean we all have to be living together under the same roof in the home or the school. But my mother said we'll never know what God wants of us but we have to try to read His mind, and that's why we pray. So when I say my prayers I ask God to tell me what's the right thing to do. In school I try to say hello to the colored, because they're kids, and you can't be mean or you'll be 'doing wrong,' like my grandmother says."

Children aren't usually long-winded in the moral discussions they have 10 with one another or with adults, and in quoting this boy I have pulled together comments he made to me in the course of several days. But everything he said was of interest to me. I was interested in the boy's changing racial attitudes. It was clear he was trying to find a coherent, sensible moral position too. It was also borne in on me that if one spends days, weeks in a given home, it is hard to escape a particular moral climate just as significant as the psychological one.

In many homes parents establish moral assumptions, mandates, priorities. 11 They teach children what to believe in, what not to believe in. They teach children what is permissible or not permissible—and why. They may summon up the Bible, the flag, history, novels, aphorisms, philosophical or political sayings, personal memories—all in an effort to teach children how to behave, what and whom to respect and for which reasons. Or they may neglect to do so, and in so doing teach their children *that*—a moral abdication, of sorts—and in this way fail their children. Children need and long for words of moral advice, instruction, warning, as much as they need words of affirmation or criticism from their parents about other matters. They must learn how to dress and what to wear, how to eat and what to eat; and they must also learn how to behave under X or Y or Z conditions, and why.

All the time, in 20 years of working with poor children and rich children, 12 Black children and white children, children from rural areas and urban areas and in every region of this county, I have heard questions—thoroughly intelligent and discerning questions—about social and historical matters, about personal behavior, and so on. But most striking is the fact that almost all those questions, in one way or another, are moral in nature: Why did the Pilgrims leave England? Why didn't they just stay and agree to do what the king wanted them to do? . . . Should you try to share all you've got or should you save a lot for yourself? . . . What do you do when you see others fighting—do you try to break up the fight, do you stand by and watch or do you leave as fast as you can? . . . Is it right that some people haven't got enough to eat? . . . I see other kids cheating and I wish I could copy the answers too; but I won't cheat, though sometimes I feel I'd like to and I get all mixed up. I go home and talk with my parents, and I ask them what should you do if you see kids cheating—pay no attention, or report the kids or do the same thing they are doing?

Those are examples of children's concerns—and surely millions of Ameri- 13 can parents have heard versions of them. Have the various "experts" on childhood stressed strongly enough the importance of such questions—and the importance of the hunger we all have, no matter what our age or background, to examine what we believe in, are willing to stand up for, and what we are determined to ask, likewise, of our children?

Children not only need our understanding of their complicated emotional 14 lives; they also need a constant regard for the moral issues that come their way as soon as they are old enough to play with others and take part in the politics of the nursery, the back yard and the schoolroom. They need to be told what they must do and what they must not do. They need control over themselves and a sense of what others are entitled to from them—cooperation, thoughtfulness, an attentive ear and eye. They need discipline not only to tame their excesses of emotion but discipline also connected to stated and clarified moral values. They need, in other words, something to believe in that is larger than their own appetites and urges and, yes, bigger than their "psychological drives." They need a larger view of the world, a moral context, as it were—a faith that addresses itself to the meaning of this life we all live and, soon enough, let go of.

Yes, it is time for us parents to begin to look more closely at what ideas our   15 children have about the world; and it would be well to do so before they become teenagers and young adults and begin to remind us, as often happens, of how little attention we did pay to their moral development. Perhaps a nine-year-old girl from a well-off suburban home in Texas put it better than anyone else I've met:

> I listen to my parents, and I wonder what they believe in more than anything else. I asked my mom and my daddy once: What's the thing that means most to you? They said they didn't know but I shouldn't worry my head too hard with questions like that. So I asked my best friend, and she said she wonders if there's a God and how do you know Him and what does He want you to do— I mean, when you're in school or out playing with your friends. They talk about God in church, but is it only in church that He's there and keeping an eye on you? I saw a kid steal in a store, and I know her father has a lot of money—because I hear my daddy talk. But stealing's wrong. My mother said she's a "sick girl," but it's still wrong what she did. Don't you think?

There was more—much more—in the course of the months I came to know   16 that child and her parents and their neighbors. But those observations and questions—a "mere child's"—reminded me unforgettably of the aching hunger for firm ethical principles that so many of us feel. Ought we not begin thinking about this need? Ought we not all be asking ourselves more intently what standards we live by—and how we can satisfy our children's hunger for moral values?

COMPREHENSION

1. How does the author's title capture the substance of his essay? What is his thesis?
2. According to Coles, why do parents have difficulty explaining ethics to their children? On what aspects of their children's development do they tend to concentrate? Why?
3. There is an implied contrast between mothers' and fathers' attitudes toward morality in Coles's essay. Explain this contrast, and cite examples for your explanation.

RHETORIC

1. What point of view does Coles use here? How does that viewpoint affect the tone of the essay?
2. Compare Coles's sentence structure with the sentence structure of the children he quotes. How do they differ?
3. Does this essay present an inductive or a deductive argument? Give evidence for your answer.
4. How does paragraph 13 differ from paragraphs 3, 10, and 16? How do all four paragraphs contribute to the development of the essay?
5. Explain the line of reasoning in the first paragraph. Why does Coles allude to Freud? How is that allusion related to the final sentence of the paragraph?
6. What paragraphs constitute the conclusion of the essay? Why? How do they summarize Coles's argument?

WRITING

1. Write an essay describing conflict between your parents' ethical views and your own.
2. Gather evidence, from conversations with your friends and relatives, about an ethical issue such as poverty, world starvation, abortion, or capital punishment. Incorporate their opinions in your essay through direct and indirect quotation.
3. **WRITING AN ARGUMENT:** Coles asserts the need for clear ethical values. How have your parents provided such values? What kind of values will you give your children? Answer these questions in a brief argumentative essay.

---

www.mhhe.com/ **mhreader** | For more information on Robert Coles, go to: **More Resources > Ch. 11 Ethics & Religion**

Salvation

Langston Hughes

James Langston Hughes (1902–1967), poet, playwright, fiction writer, biographer, and essayist, was for more than 50 years one of the most productive and significant modern American authors. In The Weary Blues *(1926),* Simple Speaks His Mind *(1950),* The Ways of White Folks *(1940),* Selected Poems *(1959), and dozens of other books, he strove, in his own words, "to explain the Negro condition in America."*

This essay, from his 1940 autobiography, The Big Sea, *reflects the sharp, humorous, often bittersweet insights contained in Hughes's examination of human behavior.*

I was saved from sin when I was going on thirteen. But not really saved. It happened like this. There was a big revival at my Auntie Reed's church. Every night for weeks there had been much preaching, singing, praying, and shouting, and some very hardened sinners had been brought to Christ, and the membership of the church had grown by leaps and bounds. Then just before the revival ended, they held a special meeting for children, "to bring the young lambs to the fold." My aunt spoke of it for days ahead. That night I was escorted to the front row and placed on the mourners' bench with all the other young sinners, who had not yet been brought to Jesus.

My aunt told me that when you were saved you saw a light, and something happened to you inside! And Jesus came into your life! And God was with you from then on! She said you could see and hear and feel Jesus in your soul. I believed her. I had heard a great many old people say the same thing and it seemed to me they ought to know. So I sat there calmly in the hot, crowded church, waiting for Jesus to come to me.

The preacher preached a wonderful rhythmical sermon, all moans and shouts and lonely cries and dire pictures of hell, and then he sang a song about the ninety and nine safe in the fold, but one little lamb was left out in the cold. Then he said: "Won't you come? Won't you come to Jesus? Young lambs, won't you come?" And he held out his arms to all us young sinners there on the mourners' bench. And the little girls cried. And some of them jumped up and went to Jesus right away. But most of us just sat there.

A great many old people came and knelt around us and prayed, old women with jet-black faces and braided hair, old men with work-gnarled hands. And the church sang a song about the lower lights are burning, some poor sinners to be saved. And the whole building rocked with prayer and song.

Still I kept waiting to *see* Jesus.

Finally all the young people had gone to the altar and were saved, but one boy and me. He was a rounder's son named Westley. Westley and I were surrounded by sisters and deacons praying. It was very hot in the church, and getting late now. Finally Westley said to me in a whisper: "God damn! I'm tired o' sitting here. Let's get up and be saved." So he got up and was saved.

Then I was left all alone on the mourners' bench. My aunt came and knelt at my knees and cried, while prayers and song swirled all around me in the little church. The whole congregation prayed for me alone, in a mighty wail of moans and voices. And I kept waiting serenely for Jesus, waiting, waiting—but he didn't come. I wanted to see him, but nothing happened to me. Nothing! I wanted something to happen to me, but nothing happened.

I heard the songs and the minister saying: "Why don't you come? My dear child, why don't you come to Jesus? Jesus is waiting for you. He wants you. Why don't you come? Sister Reed, what is this child's name?"

"Langston," my aunt sobbed. 9

"Langston, why don't you come? Why don't you come and be saved? Oh, 10 Lamb of God! Why don't you come?"

Now it was really getting late. I began to be ashamed of myself, holding 11 everything up so long. I began to wonder what God thought about Westley, who certainly hadn't seen Jesus either, but who was now sitting proudly on the platform, swinging his knickerbockered legs and grinning down at me, surrounded by deacons and old women on their knees praying. God had not struck Westley dead for taking his name in vain or for lying in the temple. So I decided that maybe to save further trouble, I'd better lie, too, and say that Jesus had come, and get up and be saved.

So I got up. 12

Suddenly the whole room broke into a sea of shouting, as they saw me rise. 13 Waves of rejoicing swept the place. Women leaped in the air. My aunt threw her arms around me. The minister took me by the hand and led me to the platform.

When things quieted down, in a hushed silence, punctuated by a few ec- 14 static "Amens," all the new young lambs were blessed in the name of God. Then joyous singing filled the room.

That night, for the last time in my life but one—for I was a big boy twelve 15 years old—I cried. I cried, in bed alone, and couldn't stop. I buried my head under the quilts, but my aunt heard me. She woke up and told my uncle I was crying because the Holy Ghost had come into my life, and because I had seen Jesus. But I was really crying because I couldn't bear to tell her that I had lied, that I had deceived everybody in the church, that I hadn't seen Jesus, and that now I didn't believe there was a Jesus any more, since he didn't come to help me.

COMPREHENSION

1. What does the title tell you about the subject of this essay? How would you state, in your own words, the thesis that emerges from the title and the essay?
2. How does Hughes recount the revival meeting he attended? What is the dominant impression?
3. Explain Hughes's shifting attitude toward salvation in this essay. Why is he disappointed in the religious answers provided by his church? What does he say about salvation in the last paragraph?

RHETORIC

1. Key words and phrases in this essay relate to the religious experience. Locate five of these words and expressions, and explain their connotations.
2. Identify the level of language in the essay. How does Hughes employ language effectively?

3. Where is the thesis statement in the essay? Consider the following: the use of dialogue, the use of phrases familiar to you (idioms), and the sentence structure. Cite examples of these elements.
4. How much time elapses, and why is this important to the effect? How does the author achieve narrative coherence?
5. Locate details and examples in the essay that are especially vivid and interesting. Compare your list with what others have listed. What are the similarities? The differences?
6. What is the tone of the essay? What is the relationship between tone and point of view?

WRITING

1. Describe a time in your life when you suppressed your feelings about religion before friends or adults because you thought they would misunderstand.
2. Write a narrative account of the most intense religious experience in your life.
3. **WRITING AN ARGUMENT:** In an argumentative essay, explain why you think or do not think that politicians today often profess their religious beliefs simply in order to satisfy voters and not because of firmly held religious sentiments.

| www.mhhe.com/ **mhreader** | For more information on Langston Hughes, go to: **More Resources > Ch. 11 Ethics & Religion** |

✳

The Divine Revolution

Václav Havel

Václav Havel (b. 1936) was born into a well-to-do family in Prague. Because of his "bourgeois" background he was denied entrance to a university and instead studied at a technical college. In the 1960s Havel became interested in the theatre. He subsequently enrolled in the Academy of Dramatic Arts and graduated in 1967. During the 1970s and 1980s, Havel was repeatedly arrested for his dissident activities. In November 1989, he formed a political opposition group, the Civic Forum, and, with the fall of communism, was elected by popular vote as the president of the Czech and Slovak Federal Republic. He resigned in 1992 but was elected president of the New Czech Republic in 1993. Among his plays are The Beggar's Opera *(1976),* Largo Desolato *(1985), and* Temptation *(1986). He has also published numerous essays; a collection of letters to his wife while he was in prison,* Letters to Olga *(1988); and a collection of thoughts on life, literature, and politics titled* Disturbing the Peace *(1990). He has won numerous international prizes for his writing and humanitarian efforts. In the following essay, he disputes the importance of the "God of Technology" as an answer to our current problems, and instead calls for a new spiritual vision for the future.*

Humankind today is well aware of the spectrum of threats looming over its head. We know that the number of people living on our planet is growing at a soaring rate and that within a relatively short time we can expect it to total in the tens of billions. We know that the already-deep abyss separating the planet's poor and rich could deepen further, and more and more dangerously, because of this rapid population growth. We also know that we've been destroying the environment on which our existence depends and that we are headed for disaster by producing weapons of mass destruction and allowing them to proliferate. 1

And yet, even though we are aware of these dangers, *we do almost nothing to avert* them. It's fascinating to me how preoccupied people are today with catastrophic prognoses, how books containing evidence of impending crises become bestsellers, but how very little account we take of these threats in our everyday activities. Doesn't every schoolchild know that the resources of this planet are limited and that if they are expended faster than they are recovered, we are doomed? And still we continue in our wasteful ways and don't even seem perturbed. Quite the contrary: Rising *production is considered to be the main sign of national success,* not only in poor states where such a position could be justified, but also in wealthy ones, which are cutting the branch on which they sit with their ideology of indefinitely prolonged and senseless growth. 2

The most important thing we can do today is to study the reasons why humankind does little to address these threats and why it allows itself to be carried onward by some kind of perpetual motion, unaffected by self-awareness or a sense of future options. It would be unfair to ignore the existence of numerous projects for averting these dangers, or to deny that a lot already has been done. However, all attempts of this kind have one thing in common: *They do not touch the seed from which the threats I'm speaking of sprout,* but merely try to diminish their impact. (A typical example is the list of legal acts, ordinances, and international treaties stipulating how much toxic matter this or that plant may discharge into the environment.) I'm not criticizing these safeguards; I'm only saying that they are technical tricks that have no real effect on the substance of the matter. 3

What, then, is the substance of the matter? What could change the direction of today's civilization? 4

It is my deep conviction that the only option is a change in the sphere of the spirit, in the sphere of human conscience. It's not enough to invent new machines, new regulations, new institutions. We must develop a new understanding of the true purpose of our existence on this earth. Only by making such a fundamental shift will we be able to create new models of behavior and a new set of values for the planet. In short, it appears to me that it would be better to start from the head rather than the tail. 5

Whenever I've gotten involved in a major global problem—the logging of rainforests, ethnic or religious intolerance, the brutal destruction of indigenous cultures—I've always discovered somewhere in the long chain of events that gave rise to it a basic lack of responsibility for the planet.

There are countless types of responsibility—more or less pressing, depending on who's involved. We feel responsible for our personal welfare, our families, our companies, our communities, our nations. And somewhere in the background there is, in every one of us, a small feeling of responsibility for the planet and its future. It seems to me that this last and deepest responsibility has become a very low priority—dangerously low, considering that the world today is more interlinked than ever before and that we are, for all intents and purposes, living one global destiny.

At the same time, our world is dominated by several great religious systems, whose differences seem to be coming to the fore with increasing sharpness and setting the stage for innumerable political and armed conflicts. In my opinion, this fact—which is attracting, understandably, a great deal of media attention—partly conceals a more important fact: that the civilization within which this religious tension is taking place is, in essence, a deeply atheistic one. Indeed, it is the first atheistic civilization in the history of humankind.

Perhaps the real issue is a crisis of respect for the moral order extended to us from above, or simply a crisis of respect for any kind of authority higher than our own earthly being, with its material and thoroughly ephemeral interests. Perhaps our lack of responsibility for the planet is only the logical consequence of the modern conception of the universe as a complex of phenomena controlled by certain scientifically identifiable laws, formulated for God-knows-what purpose. This is a conception that does not inquire into the meaning of existence and renounces any kind of metaphysics, including its own metaphysical roots.

In the process, we've lost our certainty that the universe, nature, existence, our own lives are works of creation that have a definite meaning and purpose. This loss is accompanied by loss of the feeling that whatever we do must be seen in the light of a higher order of which we are part and whose authority we must respect.

In recent years the great religions have been playing an increasingly important role in global politics. Since the fall of communism, the world has become multipolar instead of bipolar, and many countries outside the hitherto dominant Euro-American cultural sphere have grown in self-confidence and influence. But the more closely tied we are by the bonds of a single global civilization, the more the various religious groups emphasize all the ways in which they differ from each other. This is an epoch of accentuated spiritual, religious, and cultural "otherness."

How can we restore in the human mind a shared attitude to what is above if people everywhere feel the need to stress their otherness? Is there any sense in trying to turn the human mind to the heavens when such a turn would only aggravate the conflict among our various deities?

I'm not, of course, an expert on religion, but it seems to me that the major 13
faiths have much more in common than they are willing to admit. They share a
basic point of departure—that this world and our existence are not freaks of
chance but rather part of a mysterious, yet integral, act whose sources, direction, and purpose are difficult for us to perceive in their entirety. And they share
a large complex of moral imperatives that this mysterious act implies. In my
view, whatever differences these religions might have are not as important as
these fundamental similarities.

Perhaps the way out of our current bleak situation could be found by 14
searching for what unites the various religions—a purposeful search for common principles. Then we could cultivate human coexistence while, at the same
time, cultivating the planet on which we live, suffusing it with the spirit of this
religious and ethical common ground—what I would call the common spiritual
and moral minimum.

Could this be a way to stop the blind perpetual motion dragging us toward 15
hell? Can the persuasive words of the wise be enough to achieve what must be
done? Or will it take an unprecedented disaster to provoke this kind of existential revolution—a universal recovery of the human spirit and renewed responsibility for the world?

COMPREHENSION

1. What is the thesis of the essay?
2. What does Havel mean by the "divine revolution"? How does it differ from
   what we usually have in mind when we think of a revolution?
3. According to Havel, what is the major shortcoming of humankind today?

RHETORIC

1. Havel's first paragraph contains the phrase "we know" three times. What
   function does the use of this repetition serve in introducing his subject?
2. Paragraph 4 is composed of two short questions. How does Havel employ
   these questions to make the paragraph a major transitional point in his essay? What argument grows from these questions?
3. Study paragraph 8. Where is the topic sentence located? How does Havel
   develop the ideas in this paragraph so that its placement is so effective?
4. Paragraph 12 is similar in form and rhetorical purpose to paragraph 4. It too
   is composed of two questions. How does paragraph 12 create another transition in logical development in the essay, and how do paragraphs 4 and 12
   help create an overall rhetorical structure for the entire essay?
5. Havel makes liberal use of the words *could* and *perhaps* as he makes suggestions regarding how to improve the current state of civilization. How do
   these words affect the tone of his argument? What do they imply about the
   author's voice?

6. Havel refers to the contemporary world as being "senseless" (paragraph 2); "atheistic" (paragraph 8); "bleak" (paragraph 14); and "blind" (paragraph 15). Are these prognoses fact or opinion? Explain your view.
7. Havel concludes his essay with three questions. How do they reinforce the tone of the essay?

WRITING

1. Write an expository essay exploring how the media exploits our spiritual crisis by obsessing over issues such as political scandal, marginal social behavior (as depicted on such shows as *The Jerry Springer Show*) and celebrity tragedies.
2. Write a 300-word summary of Havel's essay.
3. **WRITING AN ARGUMENT:** Argue for or against the view that Havel is much too pessimistic about the world today given the rise in awareness and assistance toward homelessness, foster care, adoption, and other social and cultural issues.

www.mhhe.com/
mhreader

For more information on Václav Havel, go to:
More Resources > Ch. 11 Ethics & Religion

The Death of the Moth

Virginia Woolf

Virginia Woolf (1882–1941), English novelist and essayist, was the daughter of Sir Leslie Stephen, a famous critic and writer on economics. An experimental novelist, Woolf attempted to portray consciousness through a poetic, symbolic, and concrete style. Her novels include Jacob's Room *(1922),* Mrs. Dalloway *(1925),* To the Lighthouse *(1927), and* The Waves *(1931). She was also a perceptive reader and critic, and her criticism appears in* The Common Reader *(1925) and* The Second Common Reader *(1933). The following essay, which demonstrates Woolf's capacity to find profound meaning even in commonplace events, appeared in* The Death of the Moth and Other Essays *(1942).*

Moths that fly by day are not properly to be called moths; they do not excite that   1
pleasant sense of dark autumn nights and ivy-blossom which the commonest yellow-underwing asleep in the shadow of the curtain never fails to rouse in us. They are hybrid creatures, neither gay like butterflies nor somber like their own

species. Nevertheless the present specimen, with his narrow hay-colored wings, fringed with a tassel of the same color, seemed to be content with life. It was a pleasant morning, mid-September, mild, benignant, yet with a keener breath than that of the summer months. The plough was already scoring the field opposite the window, and where the share had been, the earth was pressed flat and gleamed with moisture. Such vigor came rolling in from the fields and the down beyond that it was difficult to keep the eyes strictly turned upon the book. The rooks too were keeping one of their annual festivities; soaring round the tree tops until it looked as if a vast net with thousands of black knots in it had been cast up into the air; which, after a few moments sank slowly down upon the trees until every twig seemed to have a knot at the end of it. Then, suddenly, the net would be thrown into the air again in a wider circle this time, with the utmost clamor and vociferation, as though to be thrown into the air and settle down upon the tree tops were a tremendously exciting experience.

The same energy which inspired the rooks, the ploughmen, the horses, and   2
even, it seemed, the lean bare-backed downs, sent the moth fluttering from side to side of his square of the windowpane. One could not help watching him. One was, indeed, conscious of a queer feeling of pity for him. The possibilities of pleasure seemed that morning so enormous and so various that to have only a moth's part in life, and a day moth's at that, appeared a hard fate, and his zest in enjoying his meager opportunities to the full, pathetic. He flew vigorously to one corner of his compartment, and, after waiting there a second, flew across to the other. What remained for him but to fly to a third corner and then to a fourth? That was all he could do, in spite of the size of the downs, the width of the sky, the far-off smoke of houses, and the romantic voice, now and then, of a steamer out at sea. What he could do he did. Watching him, it seemed as if a fibre, very thin but pure, of the enormous energy of the world had been thrust into his frail and diminutive body. As often as he crossed the pane, I could fancy that a thread of vital light became visible. He was little or nothing but life.

Yet, because he was so small, and so simple a form of the energy that was   3
rolling in at the open window and driving its way through so many narrow and intricate corridors in my own brain and in those of other human beings, there was something marvelous as well as pathetic about him. It was as if someone had taken a tiny bead of pure life and decking it as lightly as possible with down and feathers, had set it dancing and zigzagging to show us the true nature of life. Thus displayed one could not get over the strangeness of it. One is apt to forget all about life, seeing it humped and bossed and garnished and cumbered so that it has to move with the greatest circumspection and dignity. Again, the thought of all that life might have been had he been born in any other shape caused one to view his simple activities with a kind of pity.

After a time, tired by his dancing apparently, he settled on the window   4
ledge in the sun, and, the queer spectacle being at an end, I forgot about him. Then, looking up, my eye was caught by him. He was trying to resume his dancing, but seemed either so stiff or so awkward that he could only flutter to the bottom of the windowpane; and when he tried to fly across it he failed. Being intent on other matters I watched these futile attempts for a time without thinking,

unconsciously waiting for him to resume his flight, as one waits for a machine, that has stopped momentarily, to start again without considering the reason of its failure. After perhaps a seventh attempt he slipped from the wooden ledge and fell, fluttering his wings, on to his back on the window sill. The helplessness of his attitude roused me. It flashed upon me that he was in difficulties; he could no longer raise himself; his legs struggled vainly. But, as I stretched out a pencil, meaning to help him to right himself, it came over me that the failure and awkwardness were the approach of death. I laid the pencil down again.

The legs agitated themselves once more. I looked as if for the enemy against   5
which he struggled. I looked out of doors. What had happened there? Presumably it was midday, and work in the fields had stopped. Stillness and quiet had replaced the previous animation. The birds had taken themselves off to feed in the brooks. The horses stood still. Yet the power was there all the same, massed outside, indifferent, impersonal, not attending to anything in particular. Somehow it was opposed to the little hay-colored moth. It was useless to try to do anything. One could only watch the extraordinary efforts made by those tiny legs against an oncoming doom which could, had it chosen, have submerged an entire city, not merely a city, but masses of human beings; nothing, I knew, had any chance against death. Nevertheless after a pause of exhaustion the legs fluttered again. It was superb this last protest, and so frantic that he succeeded at last in righting himself. One's sympathies, of course, were all on the side of life. Also, when there was nobody to care or to know, this gigantic effort on the part of an insignificant little moth, against a power of such magnitude, to retain what no one else valued or desired to keep, moved one strangely. Again, somehow, one saw life, a pure bead. I lifted the pencil again, useless though I knew it to be. But even as I did so, the unmistakable tokens of death showed themselves. The body relaxed, and instantly grew stiff. The struggle was over. The insignificant little creature now knew death. As I looked at the dead moth, this minute wayside triumph of so great a force over so mean an antagonist filled me with wonder. Just as life had been strange a few minutes before, so death was now as strange. The moth having righted himself now lay most decently and uncomplainingly composed. O yes, he seemed to say, death is stronger than I am.

COMPREHENSION

1. Why is Woolf so moved by the moth's death? Why does she call the moth's protest "superb" (paragraph 5)?
2. What, according to Woolf, is the "true nature of life" (paragraph 3)?
3. What paradox is inherent in the death of the moth?

RHETORIC

1. Examine Woolf's use of simile in paragraph 1. Where else in the essay does she use similes? Are any of them similar to the similes used in paragraph 1?

2. Why does the author personify the moth?
3. What sentences constitute the introduction of this essay? What rhetorical device do they use?
4. Divide the essay into two parts. Now explain why you divided the essay where you did. How are the two parts different? How are they similar?
5. Explain the importance of description in this essay. Where, particularly, does Woolf describe the setting of her scene? How does that description contribute to the development of her essay? How does she describe the moth, and how does this description affect tone?
6. How is narration used to structure the essay?

WRITING

1. Woolf implicitly connects insect and human life. What else can we learn about human development by looking at other forms of life? Analyze this connection in an essay.
2. Write a detailed description of a small animal. Try to invest it with the importance that Woolf gives her moth.
3. **WRITING AN ARGUMENT:** Write a rebuttal of this essay, explaining why a moth, or any other insect or animal, can tell us nothing about the human condition.

---

www.mhhe.com/
mhreader

For more information on Virginia Woolf, go to:
More Resources > Ch. 11 Ethics & Religion

The Allegory of the Cave

Plato

Plato (427–347 B.C.), *pupil and friend of Socrates, was one of the greatest philosophers of the ancient world. Plato's surviving works are all dialogues and epistles, many of the dialogues purporting to be conversations of Socrates and his disciples. Two key aspects of his philosophy are the dialectical method—represented by the questioning and probing of the particular event to reveal the general truth—and the existence of Forms. Plato's best-known works include the* Phaedo, Symposium, Phaedrus, *and* Timaeus. *The following selection, from the* Republic, *is an early description of the nature of Forms.*

And now, I said, let me show in a figure how far our nature is enlightened or   1
unenlightened: Behold! human beings living in an underground den, which has
a mouth open towards the light and reaching all along the den; here they have
been from their childhood, and have their legs and necks chained so that they
cannot move, and can only see before them, being prevented by the chains from
turning round their heads. Above and behind them a fire is blazing at a distance, and between the fire and the prisoners there is a raised way; and you will
see, if you look, a low wall built along the way, like the screen which marionette
players have in front of them, over which they show the puppets.

I see.   2

And do you see, I said, men passing along the wall carrying all sorts of   3
vessels, and statues and figures of animals made of wood and stone and various materials, which appear over the wall? Some of them are talking, others
silent.

You have shown me a strange image, and they are strange prisoners.   4

Like ourselves, I replied; and they see only their own shadows, or the shad-   5
ows of one another, which the fire throws on the opposite wall of the cave?

True, he said; how could they see anything but the shadows if they were   6
never allowed to move their heads?

And of the objects which are being carried in like manner they would only   7
see the shadows?

Yes, he said.   8

And if they were able to converse with one another, would they not sup-   9
pose that they were naming what was actually before them?

Very true.   10

And suppose further that the prison had an echo which came from the   11
other side, would they not be sure to fancy when one of the passersby spoke
that the voice which they heard came from the passing shadow?

No question, he replied.   12

To them, I said, the truth would be literally nothing but the shadows of the   13
images.

That is certain.   14

And now look again, and see what will naturally follow if the prisoners are   15
released and disabused of their error. At first, when any of them is liberated and
compelled suddenly to stand up and turn his neck round and walk and look towards the light, he will suffer sharp pains; the glare will distress him and he will
be unable to see the realities of which in his former state he had seen the shadows; and then conceive some one saying to him, that what he saw before was an
illusion, but that now, when he is approaching nearer to being and his eye is
turned towards more real existence, he has a clearer vision—what will be his reply? And you may further imagine that his instructor is pointing to the objects
as they pass and requiring him to name them—will he not be perplexed? Will
he not fancy that the shadows which he formerly saw are truer than the objects
which are now shown to him?

Far truer.   16

And if he is compelled to look straight at the light, will he not have a pain 17
in his eyes which will make him turn away to take refuge in the objects of vision
which he can see, and which he will conceive to be in reality clearer than the
things which are now being shown to him?

True, he said.   18

And suppose once more, that he is reluctantly dragged up a steep and 19
rugged ascent, and held fast until he is forced into the presence of the sun himself, is he not likely to be pained and irritated? When he approaches the light his
eyes will be dazzled and he will not be able to see anything at all of what are
now called realities.

Not all in a moment, he said.   20

He will require to grow accustomed to the sight of the upper world. And 21
first he will see the shadows best, next the reflections of men and other objects
in the water, and then the objects themselves; then he will gaze upon the light of
the moon and the stars and the spangled heaven; and he will see the sky and the
stars by night better than the sun or the light of the sun by day?

Certainly.   22

Last of all he will be able to see the sun, and not mere reflections of him in 23
the water, but he will see him in his own proper place, and not in another; and
he will contemplate him as he is.

Certainly.   24

He will then proceed to argue that this is he who gives the season and the 25
years, and is the guardian of all that is in the visible world, and in a certain way
the cause of all things which he and his fellows have been accustomed to
behold?

Clearly, he said, he would first see the sun and then reason about him.   26

And when he remembered his old habitation, and the wisdom of the den 27
and his fellow-prisoners, do you not suppose that he would felicitate himself on
the change, and pity them?

Certainly, he would.   28

And if they were in the habit of conferring honors among themselves on 29
those who were quickest to observe the passing shadows and to remark which of
them went before, and which followed after, and which were together; and who
were therefore best able to draw conclusions as to the future, do you think that he
would care for such honors and glories, or envy the possessors of them? Would
he not say with Homer, Better to be the poor servant of a poor master, and to endure anything, rather than think as they do and live after their manner?

Yes, he said, I think that he would rather suffer anything than entertain 30
these false notions and live in this miserable manner.

Imagine once more, I said, such an one coming suddenly out of the sun to 31
be replaced in his old situation; would he not be certain to have his eyes full of
darkness?

To be sure, he said.   32

And if there were a contest, and he had to compete in measuring the shad- 33
ows with the prisoners who had never moved out of the den, while his sight

was still weak, and before his eyes had become steady (and the time which would be needed to acquire this new habit of sight might be very considerable) would he not be ridiculous? Men would say of him that up he went and down he came without his eyes; and that it was better not even to think of ascending; and if any one tried to loose another and lead him up to the light, let them only catch the offender, and they would put him to death.

No question, he said. 34

This entire allegory, I said, you may now append, dear Glaucon, to the pre- 35 vious argument; the prison-house is the world of sight, the light of fire is the sun, and you will not misapprehend me if you interpret the journey upwards to be the ascent of the soul into the intellectual world according to my poor belief, which, at your desire, I have expressed—whether rightly or wrongly God knows. But, whether true or false, my opinion is that in the world of knowledge the idea of good appears last of all, and is seen only with an effort; and, when seen, is also inferred to be the universal author of all things beautiful and right, parent of light and of the lord of light in this visible world, and the immediate source of reason and truth in the intellectual; and that this is the power upon which he who would act rationally either in public or private life must have his eye fixed.

I agree, he said, as far as I am able to understand you. 36

Moreover, I said, you must not wonder that those who attain to this beauti- 37 ful vision are unwilling to descend to human affairs; for their souls are ever hastening into the upper world where they desire to dwell; which desire of theirs is very natural, if our allegory may be trusted.

Yes, very natural. 38

And is there anything surprising in one who passes from divine contempla- 39 tions to the evil state of man, misbehaving himself in a ridiculous manner; if, while his eyes are blinking and before he has become accustomed to the surrounding darkness, he is compelled to fight in courts of law, or in other places, about the images or the shadows of images of justice, and is endeavoring to meet the conceptions of those who have never yet seen absolute justice?

Anything but surprising, he replied. 40

Any one who has common sense will remember that the bewilderments of 41 the eyes are of two kinds, and arise from two causes, either from coming out of the light or from going into the light, which is true of the mind's eye, quite as much as of the bodily eye; and he who remembers this when he sees any one whose vision is perplexed and weak, will not be too ready to laugh; he will first ask whether that soul of man has come out of the brighter light, and is unable to see because unaccustomed to the dark, or having turned from darkness to the day is dazzled by excess of light. And he will count the one happy in his condition and state of being, and he will pity the other; or, if he have a mind to laugh at the soul which comes from below into the light, there will be more reason in this than in the laugh which greets him who returns from above out of the light into the den.

That, he said, is a very just distinction. 42

COMPREHENSION

1. What does Plato hope to convey to readers of his allegory?
2. According to Plato, do human beings typically perceive reality? To what does he compare the world?
3. According to Plato, what often happens to people who develop a true idea of reality? How well do they compete with others? Who is usually considered superior? Why?

RHETORIC

1. Is the conversation portrayed here realistic? How effective is this conversational style at conveying information?
2. How do you interpret such details of this allegory as the chains, the cave, and the fire? What connotations do such symbols have?
3. How does Plato use conversation to develop his argument? What is Glaucon's role in the conversation?
4. Note examples of transition words that mark contrasts between the real and the shadow world. How does Plato use contrast to develop his idea of the true real world?
5. Plato uses syllogistic reasoning to derive human behavior from his allegory. Trace his line of reasoning, noting transitional devices and the development of ideas in paragraphs 5 to 14. Find and describe a similar line of reasoning.
6. In what paragraph does Plato explain his allegory? Why do you think he locates his explanation where he does?

WRITING

1. Are Plato's ideas still influencing contemporary society? How do his ideas affect our evaluation of materialism, sensuality, sex, and love?
2. Write an allegory based on a sport, business, or space flight to explain how we act in the world.
3. **WRITING AN ARGUMENT:** In an extended essay try to convince your audience that *The Matrix* films are based on Plato's essay.

✳

November 2001: Not about Islam?

Salman Rushdie

Salman Rushdie (b. 1947), a well-known novelist, essayist, and critic, was born in Bombay, India, into a middle-class Muslim family that relocated to Pakistan following the bloody Partition. He attended public school in Pakistan and England and graduated from Kings College at Cambridge University. Rushdie first received critical acclaim for Midnight's Children *(1981) and* Shame *(1983). With the publication of his controversial novel* Satanic Verses *(1989) and the subsequent* fatwa, *or religious edict, issued by Ayatollah Khomeini ordering his death for blasphemy against Islam and his depiction of Muhammad, Rushdie went into hiding for several years. In 1998, the Islamic Republic of Iran announced that it would not carry out Rushdie's death sentence, but the* fatwa *remains in force. Rushdie's more recent work includes* Imaginary Homelands: Essays and Criticism *(1991),* The Moor's Last Sigh *(1995), and* Fury *(2001). In the selection that follows, published in* The New York Times *in 2001, shortly after the 9/11 attacks, Rushdie confronts the issue of Islamic terrorism.*

"This isn't about Islam." The world's leaders have been repeating this mantra   1
for weeks, partly in the virtuous hope of deterring reprisal attacks on innocent Muslims living in the West, partly because if the United States is to maintain its coalition against terror it can't afford to allege that Islam and terrorism are in any way related.

The trouble with this necessary disclaimer is that it isn't true. If this isn't   2
about Islam, why the worldwide Muslim demonstrations in support of Osama bin Laden and Al-Qaida? Why did those ten thousand men armed with swords and axes mass on the Pakistan-Afghanistan frontier, answering some mullah's call to jihad? Why are the war's first British casualties three Muslim men who died fighting on the Taliban side?

Why the routine anti-Semitism of the much-repeated Islamic slander that   3
"the Jews" arranged the hits on the World Trade Center and Pentagon, with the oddly self-deprecating explanation offered by the Taliban leadership among others; that Muslims could not have the technological know-how or organizational sophistication to pull off such a feat? Why does Imran Khan, the Pakistani ex–sports star turned politician, demand to be shown the evidence of Al-Qaida's guilt while apparently turning a deaf ear to the self-incriminating statements of Al-Qaida's own spokesmen (there will be a rain of aircraft from the skies, Muslims in the West are warned not to live or work in tall buildings, et cetera)? Why all the talk about U.S. military infidels desecrating the sacred soil of Saudi Arabia, if some sort of definition of what is sacred is not at the heart of the present discontents?

Let's start calling a spade a spade. Of course this is "about Islam." The ques- 4
tion is, what exactly does that mean? After all, most religious belief isn't very
theological. Most Muslims are not profound Quranic analysts. For a vast number of "believing" Muslim men, "Islam" stands, in a jumbled, half-examined
way, not only for the fear of God—the fear more than the love, one suspects—
but also for a cluster of customs, opinions, and prejudices that include their dietary practices; the sequestration or near-sequestration of "their" women; the
sermons delivered by their mullah of choice; a loathing of modern society in
general, riddled as it is with music, godlessness, and sex; and a more particularized loathing (and fear) of the prospect that their own immediate surroundings
could be taken over—"Westoxicated"—by the liberal Western-style way of life.

Highly motivated organizations of Muslim men (oh, for the voices of Mus- 5
lim women to be heard) have been engaged, over the last thirty years or so,
on growing radical political movements out of this mulch of "belief." These
Islamists—we must get used to this word, "Islamists," meaning those who are
engaged upon such political projects, and learn to distinguish it from the more
general and politically neutral "Muslim"—include the Muslim Brotherhood in
Egypt, the blood-soaked combatants of the FIS and GIA in Algeria, the Shia rev-
olutionaries of Iran, and the Taliban. Poverty is their great helper, and the fruit
of their efforts is paranoia. This paranoid Islam, which blames outsiders, "infidels," for all the ills of Muslim societies, and whose proposed remedy is the
closing of those societies to the rival project of modernity, is presently the
fastest-growing version of Islam in the world.

This is not really to go along with Samuel Huntington's thesis about the 6
"clash of civilizations," for the simple reason that the Islamists' project is turned
not only against the West and "the Jews" but also against their fellow Islamists.
Whatever the public rhetoric, there's little love lost between the Taliban and
Iranian regimes. Dissensions between Muslim nations run at least as deep as, if
not deeper than, those nations' resentment of the West. Nevertheless, it would
be absurd to deny that this self-exculpatory, paranoiac Islam is an ideology with
widespread appeal.

Twenty years ago, when I was writing a novel about power struggles in a 7
fictionalized Pakistan, it was already de rigueur in the Muslim world to blame
all its troubles on the West and, in particular, the United States. Then as now,
some of these criticisms were well-founded; no room here to rehearse the
geopolitics of the Cold War, and America's frequently damaging foreign policy
"tilts," to use the Kissinger term, toward (or away from) this or that temporarily useful (or disapproved-of) nation-state, or America's role in the installation
and deposition of sundry unsavory leaders and regimes. But I wanted then to
ask a question which is no less important now: suppose we say that the ills of
our societies are not primarily America's fault—that we are to blame for our
own failings? How would we understand them then? Might we not, by accepting our own responsibility for our problems, begin to learn to solve them for
ourselves?

It is interesting that many Muslims, as well as secularist analysts with roots 8
in the Muslim world, are beginning to ask such questions now. In recent weeks
Muslim voices have everywhere been raised against the obscurantist "hijack" of
their religion. Yesterday's hotheads (among them Yusuf Islam, a.k.a. Cat
Stevens) are improbably repackaging themselves as today's pussycats. An Iraqi
writer quotes an earlier Iraqi satirist: "The disease that is in us, is from us." A
British Muslim writes that "Islam has become its own enemy." A Lebanese
writer friend, returning from Beirut, tells me that, in the aftermath of September
11, public criticism of Islamism has become much more outspoken. Many commentators have spoken of the need for a Reformation in the Muslim world. I'm
reminded of the way non-communist socialists used to distance themselves
from the tyrannous "actually existing" socialism of the Soviets; nevertheless,
the first stirrings of this counterproject are of great significance. If Islam is to be
reconciled with modernity, these voices must be encouraged until they swell
into a roar.

Many of them speak of another Islam, their personal, private faith, and the 9
restoration of religion to the sphere of the personal, its de-politicization, is the
nettle that all Muslim societies must grasp in order to become modern. The only
aspect of modernity in which the terrorists are interested is technology, which
they see as a weapon that can be turned against its makers. If terrorism is to be
defeated, the world of Islam must take on board the secularist-humanist principles on which the modern is based, and without which their countries' freedom
will remain a distant dream.

COMPREHENSION

1. What is Rushdie's response to statements that 9/11 was not "about Islam"?
2. What distinction does Rushdie draw between "Muslims" and "Islamists"?
   What, according to the writer, is the proper role of religion in the contemporary world?
3. Explain Rushdie's solution to some of the problems confronting Islamic nations today.

RHETORIC

1. How would you describe the writer's tone? Identify words, phrases, and
   sentences that capture the writer's attitude toward his subject. How does
   the fact that Rushdie writes in the immediate aftermath of the events of September 11, 2001, affect the tone? How do his personal difficulties bear on his
   approach to the subject?
2. What is the key claim that the writer makes in this essay? Is it stated or
   implied?
3. Where does Rushdie engage in rebuttal of his opponents' points? Does he
   refute these points clearly and adequately? Why or why not?

4. What reasons and evidence does the writer offer to support his contention that if Muslim nations accepted responsibility for their internal conditions rather than blaming the West, they could solve their own problems?
5. Where does the writer apply the comparative method to advance his argument?
6. How does Rushdie support his premise that Islamic societies want to become modernized and can do so if their religion "returns to the sphere of the personal" (paragraph 11)?

WRITING

1. Write a personal essay in which you explain what you think the proper role of religion should be in the post–September 11 world.
2. Write a comparative essay in which you distinguish between "Islam" and "Islamists." Conduct research if necessary.
3. **WRITING AN ARGUMENT:** Take issue with Rushdie's claim that September 11, 2001, is "about Islam." Rebut his reasons, offering ideas and support for your alternative explanation.

---

www.mhhe.com/
mhreader

For more information on Salman Rushdie, go to:
More Resources > Ch. 11 Ethics & Religion

＊

The Rival Conceptions of God

C. S. Lewis

Clive Staples Lewis (1898–1963) was born in Belfast, Ireland, but spent the most important years of his life as a lecturer in English at Oxford. His first book, Dymer, *was published in 1926, but it was not until the publication of* The Pilgrim's Regress *in 1933 that he addressed the central work of his life: a passionate defense of the Christian faith. Lewis's immense output embraces science fiction, fantasy, children's books, theology, and literary criticism. Among his best-known works are* The Screwtape Letters *(1942),* The Lion, the Witch, and the Wardrobe *(1950), and* The Chronicles of Narnia *(1956). In this essay, Lewis describes the reasoning that led to his conversion.*

I have been asked to tell you what Christians believe, and I am going to begin   1
by telling you one thing that Christians do not need to believe. If you are a Christian you do not have to believe that all the other religions are simply wrong all through. If you are an atheist you do have to believe that the main

point in all the religions of the whole world is simply one huge mistake. If you are a Christian, you are free to think that all these religions, even the queerest ones, contain at least some hint of the truth. When I was an atheist I had to try to persuade myself that most of the human race have always been wrong about the question that mattered to them most; when I became a Christian I was able to take a more liberal view. But, of course, being a Christian does mean thinking that where Christianity differs from other religions, Christianity is right and they are wrong. As in arithmetic—there is only one right answer to a sum, and all other answers are wrong: but some of the wrong answers are much nearer being right than others.

The first big division of humanity is into the majority, who believe in some   2
kind of God or gods, and the minority who do not. On this point, Christianity lines up with the majority—lines up with ancient Greeks and Romans, modern savages, Stoics, Platonists, Hindus, Mohammedans, etc., against the modern Western European materialist.

Now I go on to the next big division. People who all believe in God can be   3
divided according to the sort of God they believe in. There are two very different ideas on this subject. One of them is the idea that He is beyond good and evil. We humans call one thing good and another thing bad. But according to some people that is merely our human point of view. These people would say that the wiser you become the less you would want to call anything good or bad, and the more clearly you would see that everything is good in one way and bad in another, and that nothing could have been different. Consequently, these people think that long before you got anywhere near the divine point of view the distinction would have disappeared altogether. We call a cancer bad, they would say, because it kills a man; but you might just as well call a successful surgeon bad because he kills a cancer. It all depends on the point of view. The other and opposite idea is that God is quite definitely "good" or "righteous," a God who takes sides, who loves love and hates hatred, who wants us to behave in one way and not in another. The first of these views—the one that thinks God beyond good and evil—is called Pantheism. It was held by the great Prussian philosopher Hegel and, as far as I can understand them, by the Hindus. The other view is held by Jews, Mohammedans and Christians.

And with this big difference between Pantheism and the Christian idea of   4
God, there usually goes another. Pantheists usually believe that God, so to speak, animates the universe as you animate your body: that the universe almost *is* God, so that if it did not exist He would not exist either, and anything you find in the universe is a part of God. The Christian idea is quite different. They think God invented and made the universe—like a man making a picture or composing a tune. A painter is not a picture, and he does not die if his picture is destroyed. You may say, "He's put a lot of himself into it," but you only mean that all its beauty and interest has come out of his head. His skill is not in the picture in the same way that it is in his head, or even in his hands. I expect you see how this difference between Pantheists and Christians hangs together with

the other one. If you do not take the distinction between good and bad very seriously, then it is easy to say that anything you find in this world is a part of God. But, of course, if you think some things really bad, and God really good, then you cannot talk like that. You must believe that God is separate from the world and that some of the things we see in it are contrary to His will. Confronted with a cancer or a slum the Pantheist can say, "If you could only see it from the divine point of view, you would realize that this also is God." The Christian replies, "Don't talk damned nonsense."[1] For Christianity is a fighting religion. It thinks God made the world—that space and time, heat and cold, and all the colors and tastes, and all the animals and vegetables, are things that God "made up out of His head" as a man makes up a story. But it also thinks that a great many things have gone wrong with the world that God made and that God insists, and insists very loudly, on our putting them right again.

And, of course, that raises a very big question. If a good God made the world why has it gone wrong? And for many years I simply refused to listen to the Christian answers to this question, because I kept on feeling "whatever you say, and however clever your arguments are, isn't it much simpler and easier to say that the world was not made by any intelligent power? Aren't all your arguments simply a complicated attempt to avoid the obvious?" But then that threw me back into another difficulty.

My argument against God was that the universe seemed so cruel and unjust. But how had I got this idea of *just* and *unjust*? A man does not call a line crooked unless he has some idea of a straight line. What was I comparing this universe with when I called it unjust? If the whole show was bad and senseless from A to Z, so to speak, why did I, who was supposed to be part of the show, find myself in such violent reaction against it? A man feels wet when he falls into water, because man is not a water animal: a fish would not feel wet. Of course I could have given up my idea of justice by saying it was nothing but a private idea of my own. But if I did that, then my argument against God collapsed too—for the argument depended on saying that the world was really unjust, not simply that it did not happen to please my private fancies. Thus in the very act of trying to prove that God did not exist—in other words, that the whole of reality was senseless—I found I was forced to assume that one part of reality—namely my idea of justice—was full of sense. Consequently atheism turns out to be too simple. If the whole universe has no meaning, we should never have found out that it has no meaning: just as, if there were no light in the universe and therefore no creature with eyes, we should never know it was dark. *Dark* would be without meaning.

[1]One listener complained of the word *damned* as frivolous swearing. But I mean exactly what I say—nonsense that is *damned* is under God's curse, and will (apart from God's grace) lead those who believe it to eternal death. [This note is the author's.]

COMPREHENSION

1. Who is Lewis's audience? What is his purpose? How do you know?
2. Lewis divides humanity into a number of distinct categories. Name them, and discuss his purpose in establishing these categories.
3. What is Lewis's purpose in likening Christianity to arithmetic? In what sense is this apt? Where does he use a similar image?

RHETORIC

1. Look up the following words from paragraph 2 in a dictionary or an encyclopedia: *Stoics, Platonists, Hindus,* and *Mohammedans.* What are the major tenets of their beliefs?
2. Explain Lewis's use of the word *damned* (paragraph 4). What is specific about his use of this word? Is it appropriate?
3. How does Lewis develop his argument? What line of reasoning does he follow? What transition markers does Lewis use?
4. How does Lewis use definition to structure certain parts of his argument?
5. In which paragraph is Lewis making what he considers the one irrefutable argument in favor of the existence of God? Is this paragraph coherently reasoned in terms of the whole essay? Explain.
6. Why is Lewis's idea of justice critical to the evaluation of his thought? Is his use of the word *justice* idiosyncratic or objective? How does accepting his definition make an important difference to the response that a reader would give to this piece?

WRITING

1. The Western tradition is based, in large part, on the belief that Christianity is right and other religions are wrong. Is this belief as strong today as it was in the past? Does it still cohere as an argument?
2. Write an essay describing your religious beliefs and how they originated.
3. **WRITING AN ARGUMENT:** Argue for or against atheism.

www.mhhe.com/
mhreader

For more information on C. S. Lewis, go to:
More Resources > Ch. 11 Ethics & Religion

✳

The Culture of Disbelief

Stephen L. Carter

Stephen L. Carter (b. 1954) received a BA from Stanford University in 1976 and graduated from Yale University Law School in 1979. He served as a law clerk for the U.S. Supreme Court and as a lawyer in private practice before becoming a professor of law at the Yale University Law School in 1982. An African American who is opposed to affirmative action, he has become a controversial figure among proponents of the policy. His first book, Reflections of an Affirmative Action Baby *(1991), outlines his views on the subject and draws on personal experience to show that, even though he was a beneficiary of affirmative action, such preference ultimately makes successful African Americans seem to have received preferential treatment. His second book,* The Culture of Disbelief: How American Law and Politics Trivialize Religious Devotion *(1993), addresses the ways he believes the law has recently operated against the spirit of American values in its effort to ban religion from political discourse and expression. The following essay succinctly sums up this argument.*

Contemporary American politics faces few greater dilemmas than deciding how to deal with the resurgence of religious belief. On the one hand, American ideology cherishes religion, as it does all matters of private conscience, which is why we justly celebrate a strong tradition against state interference with private religious choice. At the same time, many political leaders, commentators, scholars, and voters are coming to view any religious element in public moral discourse as a tool of the radical right for reshaping American society. But the effort to banish religion for politics' sake has led us astray: In our sensible zeal to keep religion from dominating our politics, we have created a political and legal culture that presses the religiously faithful to be other than themselves, to act publicly, and sometimes privately as well, as though their faith does not matter to them.

Recently, a national magazine devoted its cover story to an investigation of prayer: how many people pray, how often, why, how, and for what. A few weeks later came the inevitable letter from a disgruntled reader, wanting to know why so much space had been dedicated to such nonsense.[1]

Statistically, the letter writer was in the minority: by the magazine's figures, better than nine out of ten Americans believe in God and some four out of five

---

[1]"Talking to God," *Newsweek*, Jan. 6, 1992, p. 38; Letter to the Editor, *Newsweek,* Jan. 1992, p. 10. The letter called the article a "theocratic text masquerading as a news article." [This and subsequent notes in the selection are the author's.]

pray regularly.[2] Politically and culturally, however, the writer was in the American mainstream, for those who do pray regularly—indeed, those who believe in God—are encouraged to keep it a secret, and often a shameful one at that. Aside from the ritual appeals to God that are expected of our politicians, for Americans to take their religions seriously, to treat them as ordained rather than chosen, is to risk assignment to the lunatic fringe.

Yet religion matters to people, and matters a lot. Surveys indicate that 4 Americans are far more likely to believe in God and to attend worship services regularly than any other people in the Western world. True, nobody prays on prime-time television unless religion is a part of the plot, but strong majorities of citizens tell pollsters that their religious beliefs are of great importance to them in their daily lives. Even though some popular histories wrongly assert the contrary, the best evidence is that this deep religiosity has always been a facet of the American character and that it has grown consistently through the nation's history.[3] And today, to the frustration of many opinion leaders in both the legal and political cultures, religion, as a moral force and perhaps a political one too, is surging. Unfortunately, in our public life, we prefer to pretend that it is not.

Consider the following events: 5

- When Hillary Rodham Clinton was seen wearing a cross around her neck at some of the public events surrounding her husband's inauguration as President of the United States, many observers were aghast, and one television commentator asked whether it was appropriate for the First Lady to display so openly a religious symbol. But if the First Lady can't do it, then certainly the President can't do it, which would bar from ever holding the office an Orthodox Jew under a religious compulsion to wear a yarmulke.
- Back in the mid-1980s, the magazine *Sojourners*—published by politically liberal Christian evangelicals—found itself in the unaccustomed position of defending the conservative evangelist Pat Robertson against secular liberals who, a writer in the magazine sighed, "see[m] to consider Robertson a dangerous neanderthal because he happens to believe that God can heal diseases."[4] The point is that the editors of *Sojourners*, who are no great admirers of Robertson, also believe that God can heal diseases. So do tens of millions of Americans. But they are not supposed to say so.

---

[2]"Talking to God," p. 39. The most recent Gallup data indicate that 96 percent of Americans say they believe in God, including 82 percent who describe themselves as Christians (56 percent Protestant, 25 percent Roman Catholic) and 2 percent who describe themselves as Jewish. (No other faith accounted for as much as 1 percent.) See Ari L. Goldman, "Religion Notes," *New York Times*, Feb. 27, 1993, p. 9.

[3]See, for example, Jon Butler, *Awash in a Sea of Faith* (Cambridge: Harvard University Press, 1990).

[4]Collum, "The Kingdom and the Power," *Sojourners*, Nov. 1986, p. 4. Some 82 percent of Americans believe that God performs miracles today. George Gallup, Jr., and Jim Castelli, *The People's Religion: American Faith in the '90s* (New York: Macmillan, 1989), p. 58.

- In the early 1980s, the state of New York adopted legislation that, in effect, requires an Orthodox Jewish husband seeking a divorce to give his wife a *get*—a religious divorce—without which she cannot remarry under Jewish law. Civil libertarians attacked the statute as unconstitutional. Said one critic, the "barriers to remarriage erected by religious law . . . only exist in the minds of those who believe in the religion."[5] If the barriers are religious, it seems, then they are not real barriers, they are "only" in the woman's mind—perhaps even a figment of the imagination.
- When the Supreme Court of the United States, ostensibly the final refuge of religious freedom, struck down a Connecticut statute requiring employers to make efforts to allow their employees to observe the sabbath, one Justice observed that the sabbath should not be singled out because all employees would like to have "the right to select the day of the week in which to refrain from labor."[6] Sounds good, except that, as one scholar has noted, "It would come as some surprise to a devout Jew to find that he has 'selected the day of the week in which to refrain from labor,' since the Jewish people have been under the impression for some 3,000 years that this choice was made by God."[7] If the sabbath is just another day off, then religious choice is essentially arbitrary and unimportant; so if one sabbath day is inconvenient, the religiously devout employee can just choose another.
- When President Ronald Reagan told religious broadcasters in 1983 that all laws passed since biblical times "have not improved on the Ten Commandments one bit," which might once have been considered a pardonable piece of rhetorical license, he was excoriated by political pundits, including one who charged angrily that Reagan was giving "short shrift to the secular laws and institutions that a president is charged with protecting."[8] And as for the millions of Americans who consider the Ten Commandments the fundaments on which they build their lives, well, they are no doubt subversive of these same institutions.

These examples share a common rhetoric that refuses to accept the notion that rational, public-spirited people can take religion seriously. It might be argued that such cases as these involve threats to the separation of church and state, the durable and vital doctrine that shields our public institutions from religious domination and our religious institutions from government domination.

---

[5]Madeline Kochen, "Constitutional Implications of New York's 'Get' Statute," *New York Law Journal,* Oct. 27, 1983, p. 32.
[6]*Estate of Thornton v. Caldor, Inc.,* 472 U.S. 703, 711 (1985) (Justice Sandra Day O'Connor, concurring).
[7]Michael W. McConnell, "Religious Freedom at a Crossroads," *University of Chicago Law Review* 59 (1992):115.
[8]Robert G. Kaiser, "Hypocrisy: This Puffed-Up Piety Is Perfectly Preposterous," *Washington Post,* March 18, 1984, p. C1.

I am a great supporter of the separation of church and state . . . but that is not what these examples are about.

What matters about these examples is the *language* chosen to make the points. In each example, as in many more that I shall discuss, one sees a trend in our political and legal cultures toward treating religious beliefs as arbitrary and unimportant, a trend supported by a rhetoric that implies that there is something wrong with religious devotion. More and more, our culture seems to take the position that believing deeply in the tenets of one's faith represents a kind of mystical irrationality, something that thoughtful, public-spirited American citizens would do better to avoid. If you must worship your God, the lesson runs, at least have the courtesy to disbelieve in the power of prayer; if you must observe your sabbath, have the good sense to understand that it is just like any other day off from work. 7

The rhetoric matters. A few years ago, my wife and I were startled by a teaser for a story on a network news program, which asked what was meant to be a provocative question: "When is a church more than just a place of worship?" For those to whom worship is significant, the subtle arrangement of words is arresting: *more than* suggests that what follows ("just a place of worship") is somewhere well down the scale of interesting or useful human activities, and certainly that whatever the story is about is *more than* worship; and *just*—suggests that what follows ("place of worship") is rather small potatoes. 8

A friend tells the story of how he showed his résumé to an executive search consultant—in the jargon, a corporate headhunter—who told him crisply that if he was serious about moving ahead in the business world, he should remove from the résumé any mention of his involvement with a social welfare organization that was connected with a church, but not one of the genteel mainstream denominations. Otherwise, she explained, a potential employer might think him a religious fanatic. 9

How did we reach this disturbing pass, when our culture teaches that religion is not to be taken seriously, even by those who profess to believe in it? Some observers suggest that the key moment was the Enlightenment, when the Western tradition sought to sever the link between religion and authority. One of the playwright Tom Stoppard's characters observes that there came "a calendar date—*a moment*—when the onus of proof passed from the atheist to the believer, when, quite suddenly, the noes had it."[9] To which the philosopher Jeffrey Stout appends the following comment: "If so, it was not a matter of majority rule."[10] Maybe not—but a strong undercurrent of contemporary American politics holds that religion must be kept in its proper place and, still more, in proper perspective. There are, we are taught by our opinion leaders, religious matters and important matters, and disaster arises when we confuse the two. Rationality, it seems, consists in getting one's priorities straight. (Ignore your 10

[9]Tom Stoppard, *Jumpers*, quoted in Jeffrey Stout, *The Flight from Authority: Religion, Morality and the Quest for Autonomy* (South Bend, Indiana: University of Notre Dame Press, 1981), p.150.
[10]Ibid.

religious law and marry at leisure.) Small wonder, then, that we have recently been treated to a book, coauthored by two therapists, one of them an ordained minister, arguing that those who would put aside, say, the needs of their families in order to serve their religions are suffering from a malady the authors called "toxic faith"—for no normal person, evidently, would sacrifice the things that most of us hold dear just because of a belief that God so intended it.[11] (One wonders how the authors would have judged the toxicity of the faith of Jesus, Moses, or Mohammed.)

We are trying, here in America, to strike an awkward but necessary balance, 11 one that seems more and more difficult with each passing year. On the one hand, a magnificent respect for freedom of conscience, including the freedom of religious belief, runs deep in our political ideology. On the other hand, our understandable fear of religious domination of politics presses us, in our public personas, to be wary of those who take their religion too seriously. This public balance reflects our private selves. We are one of the most religious nations on earth, in the sense that we have a deeply religious citizenry; but we are also perhaps the most zealous in guarding our public institutions against explicit religious influences. One result is that we often ask our citizens to split their public and private selves, telling them in effect that it is fine to be religious in private, but there is something askew when those private beliefs become the basis for public action.

We teach college freshmen that the Protestant Reformation began the 12 process of freeing the church from the state, thus creating the possibility of a powerful independent moral force in society. As defenders of the separation of church and state have argued for centuries, autonomous religions play a vital role as free critics of the institutions of secular society. But our public culture more and more prefers religion as something without political significance, less an independent moral force than a quietly irrelevant moralizer, never heard, rarely seen. "[T]he public sphere," writes the theologian Martin Marty, "does not welcome explicit Reformed witness—or any other particularized Christian witness."[12] Or, for that matter, any religious witness at all.

Religions that most need protection seem to receive it least. Contemporary 13 America is not likely to enact legislation aimed at curbing the mainstream Protestant, Roman Catholic, or Jewish faiths. But Native Americans, having once been hounded from their lands, are now hounded from their religions, with the complicity of a Supreme Court untroubled when sacred lands are taken for road building or when Native Americans under a bona fide religious compulsion to use *peyote* in their rituals are punished under state antidrug regulations.[13] (Imagine the brouhaha if New York City were to try to take St.

---

[11]Stephen Arterburn and Jack Felton, *Toxic Faith: Understanding and Overcoming Religious Addiction* (Nashville, Tenn.: Oliver-Nelson Books, 1991).
[12]Martin E. Marty, "Reformed America and America Reformed," *Reformed Journal* (March 1989): 8, 10.
[13]*Employment Division, Department of Human Resources v. Smith*, 494 U.S. 872 (1990).

Patrick's Cathedral by eminent domain to build a new convention center, or if Kansas, a dry state, were to outlaw the religious use of wine.) And airports, backed by the Supreme Court, are happy to restrict solicitation by devotees of Krishna Consciousness, which travelers, including this one, find irritating.[14] (Picture the response should the airports try to regulate the wearing of crucifixes or yarmulkes on similar grounds of irritation.)

The problem goes well beyond our society's treatment of those who simply 14 want freedom to worship in ways that most Americans find troubling. An analogous difficulty is posed by those whose religious convictions move them to action in the public arena. Too often, our rhetoric treats the religious impulse to public action as presumptively wicked—indeed, as necessarily oppressive. But this is historically bizarre. Every time people whose vision of God's will moves them to oppose abortion rights are excoriated for purportedly trying to impose their religious views on others, equal calumny is implicitly heaped upon the mass protest wing of the civil rights movement, which was openly and unashamedly religious in its appeals as it worked to impose its moral vision on, for example, those who would rather segregate their restaurants.

One result of this rhetoric is that we often end up fighting the wrong battles. 15 Consider what must in our present day serve as the ultimate example of religion in the service of politics: the 1989 death sentence pronounced by the late Ayatollah Ruhollah Khomeini upon the writer Salman Rushdie for his authorship of *The Satanic Verses*, which was said to blaspheme against Islam. The death sentence is both terrifying and outrageous, and the Ayatollah deserved all the fury lavished upon him for imposing it. Unfortunately, for some critics the facts that the Ayatollah was a religious leader and that the "crime" was a religious one lends the sentence a particular monstrousness; evidently they are under the impression that writers who are murdered for their ideas are choosy about the motivations of their murderers, and that those whose writings led to their executions under, say, Stalin, thanked their lucky stars at the last instant of their lives that Communism was at least godless.

To do battle against the death sentence for Salman Rushdie—to battle 16 against the Ayatollah—one should properly fight against official censorship and intimidation, not against religion. We err when we presume that religious motives are likely to be illiberal, and we compound the error when we insist that the devout should keep their religious ideas—whether good or bad—to themselves. We do no credit to the ideal of religious freedom when we talk as though religious belief is something of which public-spirited adults should be ashamed.

The First Amendment to the Constitution, often cited as the place where 17 this difficulty is resolved, merely restates it. The First Amendment guarantees the "free exercise" of religion but also prohibits its "establishment" by the government. There may have been times in our history when we as a nation have tilted too far in one direction, allowing too much religious sway over politics. But in late-twentieth-century America, despite some loud fears about the influ-

---

[14]*International Society for Krishna Consciousness v. Lee*, 112 S. Ct. 2701 (1992).

ence of the weak and divided Christian right, we are upsetting the balance afresh by tilting too far in the other direction—and the courts are assisting in the effort. For example, when a group of Native Americans objected to the Forest Service's plans to allow logging and road building in a national forest area traditionally used by the tribes for sacred rituals, the Supreme Court offered the back of its hand. True, said the Justices, the logging "could have devastating effects on traditional Indian religious practices." But that was just too bad: "government simply could not operate if it were required to satisfy every citizen's religious needs and desires."[15]

A good point: but what, exactly, are the protesting Indians left to do? Pre- 18 sumably, now that their government has decided to destroy the land they use for their sacred rituals, they are free to choose new rituals. Evidently, a small matter like the potential destruction of a religion is no reason to halt a logging project. Moreover, had the government decided instead to prohibit logging in order to preserve the threatened rituals, it is entirely possible that the decision would be challenged as a forbidden entanglement of church and state. Far better for everyone, it seems, for the Native Americans to simply allow their rituals to go quietly into oblivion. Otherwise, they run the risk that somebody will think they actually take their rituals seriously.

The Price of Faith

When citizens do act in their public selves as though their faith matters, they 19 risk not only ridicule, but actual punishment. In Colorado, a public school teacher was ordered by his superiors, on pain of disciplinary action, to remove his personal Bible from his desk where students might see it. He was forbidden to read it silently when his students were involved in other activities. He was also told to take away books on Christianity he had added to the classroom library, although books on Native American religious traditions, as well as on the occult, were allowed to remain. A federal appeals court upheld the instruction, explaining that the teacher could not be allowed to create a religious atmosphere in the classroom, which, it seems, might happen if the students knew he was a Christian.[16] One wonders what the school, and the courts, might do if, as many Christians do, the teacher came to school on Ash Wednesday with ashes in the shape of a cross imposed on his forehead—would he be required to wash them off? He just might. Early in 1993, a judge required a prosecutor arguing a case on Ash Wednesday to clean the ashes from his forehead, lest the jury be influenced by its knowledge of the prosecutor's religiosity.

Or suppose a Jewish teacher were to wear a yarmulke in the classroom. If the 20 school district tried to stop him, it would apparently be acting within its authority. In 1986, after a Jewish Air Force officer was disciplined for wearing a yarmulke while on duty, in violation of a military rule against wearing head-gear

[15]*Lyng v. Northwest Indian Cemetery Protective Association*, 485 U.S. 439 (1988).
[16]*Roberts v. Madigan*, 921 F. 2d 1047 (10th Cir. 1990).

indoors, the Supreme Court shrugged: "The desirability of dress regulations in the military is decided by the appropriate military officials," the justices explained, "and they are under no constitutional mandate to abandon their considered professional judgment."[17] The Congress quickly enacted legislation permitting the wearing of religious apparel while in uniform as long as "the wearing of the item would [not] interfere with the performance of the member's military duties," and—interesting caveat—as long as the item is "neat and conservative."[18] Those whose faiths require them to wear dreadlocks and turbans, one supposes, need not apply to serve their country, unless they are prepared to change religions.

Consider the matter of religious holidays. One Connecticut town recently   21 warned Jewish students in its public schools that they would be charged with *six* absences if they missed two days instead of the officially allocated one for Yom Kippur, the holiest observance in the Jewish calendar. And Alan Dershowitz of Harvard Law School, in his controversial book *Chutzpah*, castigates Harry Edwards, a Berkeley sociologist, for scheduling an examination on Yom Kippur, when most Jewish students would be absent. According to Dershowitz's account, Edwards answered criticism by saying: "That's how I'm going to operate. If the students don't like it, they can drop the class." For Dershowitz, this was evidence that "Jewish students [are] second-class citizens in Professor Edwards's classes."[19] Edwards has heatedly denied Dershowitz's description of events, but even if it is accurate, it is possible that Dershowitz has identified the right crime and the wrong villain. The attitude that Dershowitz describes, if it exists, might reflect less a personal prejudice against Jewish students than the society's broader prejudice against religious devotion, a prejudice that masquerades as "neutrality." If Edwards really dared his students to choose between their religion and their grade, and if that meant that he was treating them as second-class citizens, he was still doing no more than the courts have allowed all levels of government to do to one religious group after another—Jews, Christians, Muslims, Sikhs, it matters not at all. The consistent message of modern American society is that whenever the demands of one's religion conflict with what one has to do to get ahead, one is expected to ignore the religious demands and act . . . well . . . *rationally.*

Consider Jehovah's Witnesses, who believe that a blood transfusion from   22 one human being to another violates the biblical prohibition on ingesting blood. To accept the transfusion, many Witnesses believe, is to lose, perhaps forever, the possibility of salvation. As the Witnesses understand God's law, moreover, the issue is not whether the blood transfusion is given against the recipient's will, but whether the recipient is, at the time of the transfusion, actively protesting. This is the reason that Jehovah's Witnesses sometimes try to impede the physical access of medical personnel to an unconscious Witness: lack of consciousness is no de-

[17]*Goldman v Weinberger,* 475 U.S. 503 (1986).
[18]45 U.S.C. 774, as amended by Pub. L. No. 100-80, Dec. 4, 1987.
[19]Alan M. Dershowitz, *Chutzpah* (Boston: Little, Brown, 1991), pp. 329–30.

fense. This is also the reason that Witnesses try to make the decisions on behalf of their children: a child cannot be trusted to protest adequately.

The machinery of law has not been particularly impressed with these argu- 23 ments. There are many cases in which the courts have allowed or ordered transfusions to save the lives of unconscious Witnesses, even though the patient might have indicated a desire while conscious not to be transfused.[20] The machinery of modern medicine has not been impressed, either, except with the possibility that the Witnesses have gone off the deep end; at least one hospital's protocol apparently requires doctors to refer protesting Witnesses to psychiatrists.[21] Although the formal text of this requirement states as the reason the need to be sure that the Witness knows what he or she is doing, the subtext is a suspicion that the patient was not acting rationally in rejecting medical advice for religious reasons. After all, there is no protocol for packing *consenting* patients off to see the psychiatrist. But then, patients who consent to blood transfusions are presumably acting rationally. Perhaps, with a bit of gentle persuasion, the dissenting Witness can be made to act rationally too—even if it means giving up an important tenet of the religion.

And therein lies the trouble. In contemporary American culture, the reli- 24 gions are more and more treated as just passing beliefs—almost as fads, older, stuffier, less liberal versions of so-called New Age—rather than as the fundaments upon which the devout build their lives. (The noes have it!) And if religions *are* fundamental, well, too bad—at least if they're the *wrong* fundaments—if they're inconvenient, give them up! If you can't remarry because you have the wrong religious belief, well, hey, believe something else! If you can't take your exam because of a Holy Day, get a new Holy Day! If the government decides to destroy your sacred lands, just make some other lands sacred! If you must go to work on your sabbath, it's no big deal! It's just a day off! Pick a different one! If you can't have a blood transfusion because you think God forbids it, no problem! Get a new God! And through all of this trivializing rhetoric runs the subtle but unmistakable message: pray if you like, worship if you must, but whatever you do, do not on any account take your religion seriously.

[20]In every decided case that I have discovered involving efforts by Jehovah's Witness parents to prevent their children from receiving blood transfusions, the court has allowed the transfusion to proceed in the face of parental objection. I say more about transfusions of children of Witnesses, and about the rights of parents over their children's religious lives, in chapter 11 [of my book].
[21]See Ruth Macklin, "The Inner Workings of an Ethics Committee: Latest Battle over Jehovah's Witnesses," *Hastings Center Report* 18 (February/March 1988): 15.

COMPREHENSION

1. Where does the author articulate the thesis of his essay?
2. The author cites the First Amendment as being a significant historical reference in raising the debate regarding the relationship between government

and religion in the United States. What is the First Amendment to the Constitution? What does it mean that the Constitution was amended?

3. In your own words, what is the meaning of the essay's title?

RHETORIC

1. How does the opening line of the essay draw the reader into the concerns of the author?
2. What is the rhetorical function of the bulleted examples the author uses in paragraph 5?
3. In paragraphs 8 and 9, the author introduces a personal tone to his essay. Does this add to or diminish his argument?
4. In paragraph 7, Carter places the word *language* in italics; while in other places, he refers to the use of rhetoric as a way of demeaning the religious impulse. For example, in paragraph 14, where he states, "Too often, our rhetoric treats the religious impulse to public action as presumptively wicked." Why does Carter focus so much on the use of language as a tool in the attack on religion?
5. The author uses mainly anecdotal evidence to support his views, yet most social sciences claim that anecdotes are a poor form of evidence because they refer only to individual cases, and not to general trends. To what degree does Carter's strategy in using anecdotes strengthen or weaken his argument?
6. Carter devotes one section of his essay to "The Price of Faith." Why has he emphasized this religious issue by placing it in a separate category?
7. How does Carter use irony in his final paragraph? Why is this an effective way of both summing up his main points and drawing attention to them?

WRITING

1. In a research paper, compare and contrast court rulings regarding perceived governmental infringements on Christian rights of worship versus Native American rights of worship.
2. Assume the role of the CEO of a corporation. Write a policy statement in which you provide guidelines for acceptable and unacceptable displays of religious behavior and symbols.
3. **WRITING AN ARGUMENT:** Argue for or against the view that the strength of religious toleration among the American people renders any specific legislation regarding religion merely an academic exercise, with no true social effect.

★ www.mhhe.com/ **mhreader**   For more information on Stephen L. Carter, go to:
More Resources > Ch. 11 Ethics & Religion

CONNECTIONS FOR CRITICAL THINKING

1. How do writers like Plato, Havel, and Woolf use figurative language to make philosophical points? Use specific examples from these authors' works to formulate your answer.

2. Compare and contrast the way Plato presents his allegory to help the reader arrive at wisdom with the way Woolf uses the "mystery" of the moth to arrive at a unique form of understanding. What is the main difference between the two? Is one preferable over another?

3. Explore the connection between Plato, the philosopher, and Coles, the psychiatrist. How do their essays complement each other? How does Coles's attitude toward existence reflect Plato's philosophy of the cave?

4. What distinguishes a "true" religious belief from a superstition? What are their various functions? Is one more valid than the other? Explain your answer with reference to Mead, Pogrebin, and Hughes.

5. Coles argues that the moral education of children is essential to a well-functioning society. What function does superstition serve in the lives of children that a pure moral education may fail to provide?

6. Based on your reading of Lewis, explain whether you think he would agree or disagree with Rushdie's observations about Islam?

7. What is the difference between philosophy and religion? Is it merely a matter of belief? Address this question in an essay, using support from writers in this chapter.

8. Write an essay entitled "The Purpose of Life." Using examples and evidence from their works, choose three writers in this chapter to develop this theme.

9. Join two religious newsgroups. Spend two weeks monitoring their messages. Compare and contrast their concerns, questions, perspectives, and beliefs.

10. Work with other class members in creating your own interactive Web site displaying an excerpt from Rushdie's essay. Ask for personal responses from all its visitors, and report your findings.

11. Visit a Web site with summaries of judicial rulings by the Supreme Court. Using the keyword *religion,* study three case histories and the Court's ruling on each.

12. Research the role of cults in American society, particularly among young people. Focus on finding specific superstitions they have that can inflict self-harm or harm on others. Using the essays by Mead and Pogrebin as sources, explore the differences between "good" superstitions and "bad" superstitions.

 chapter *12*

Health and Medicine
What Are the Challenges?

Today, medicine and the health sciences are recasting our lives and the world we know. From stem-cell research, to the abortion debate, to the AIDS epidemic, we are dealing with enormous medical challenges and controversies. At the same time, commonplace conditions ranging from starvation to the common cold continue to defy solutions. There are surely medical breakthroughs—new drugs, therapies, technologies, and delivery systems—that offer some cause for optimism. However, we must acknowledge the ongoing reality of illness, both physical and psychological, and the serious imbalance in a person's access to health care.

It could be argued that medical science presents an unequal playing field to Americans and people worldwide, for health care clearly is a privilege rather than a right. For millions of people in the United States and billions around the world, health care is rudimentary or nonexistent. College students, of course, are among the privileged. They typically enjoy health insurance, access to campus clinics or affiliated hospitals, counseling and psychiatric intervention, and an entire network of other health care support systems. Health care at an American college or university is a model that we would wish for any culture, society, or nation.

Medicine and health care are also subjects for civic discourse, cultural argument, and political debate. Disputes over medicine—abortion, cloning, drug addiction, and more—are also part of everyday life. The subject was an integral part of the most recent U.S. presidential election and assuredly will resurface in future election cycles. And of course the media (as we discussed in an earlier chapter) exploit our fascination with health and medicine. Television shows feature extreme surgical makeovers and contests among oversized people competing to lose the most weight. Magazines and television promote potentially dangerous body images. On the Internet, people suffering from anorexia and other dietary disorders can find solace and support.

The writers in this chapter contend with some of the most pressing issues confronting the medical sciences today. Some of the writers are physicians. Some authors personalize their subject; others offer objective analysis or compelling arguments. All raise moral and ethical issues as they deal with ways in which medicine is shaping our personalities and our lives.

Previewing the Chapter

As you read the essays in this chapter and respond to them in discussion and writing, consider the following questions:

- What is the writer's subject? What perspective on medicine or health does the writer take?
- What is the writer's purpose: to explain, narrate an event, argue, or persuade?
- Do you find the writer's tone to be subjective or objective? Does the author have a personal motive in addressing the topic in the way he or she does?
- What moral, ethical, or religious issues does the writer raise in connection with medicine or health science?
- What logical, emotional, and ethical appeals does the author make to his or her audience?
- Which level of specialized knowledge—history, science, medicine, or other area—does the author bring to bear on the subject? What level of authority does the writer bring to the topic?
- What cultural, economic, or political problems does the author connect to the medical topic under consideration?
- Do you agree or disagree with the author's thesis or claim, and why?
- Which essays appear similar in subject, thesis, or perspective?
- Which essays did you find most compelling or convincing, and why? Which ones changed your opinion or altered your thinking on the subject?

Classic and Contemporary Images
WHAT DOES MEDICAL RESEARCH TELL US?

Using a Critical Perspective   The medical universe has changed radically since the time in 1632 when Rembrandt painted *The Anatomy Lesson of Professor Nicolaes Tulp*. Switch forward to 2004 and move from Holland to New York City, where photographer Gary Bramnick captures the release of conjoined twins after successful surgery. As you consider these visual texts, answer these questions: What is the main purpose of the artist or the photographer? What elements in each image contribute to the overall effect? How is the human subject portrayed, and which scene evokes the strongest emotional reaction? How does the much earlier scene relate to the contemporary one?

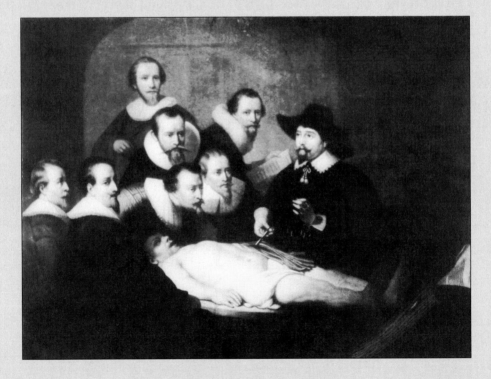

Rembrandt van Rijn (1606–1669) was the most gifted painter, draftsman, and etcher of Holland's Golden Age. *The Anatomy Lesson* (1632) is a group portrait of the Amsterdam surgeon's guild whose members, led by Dr. Nicolaes Tulp, were unsurpassed in the surgical techniques of the period.

Clarence and Carl Aguirre, formerly conjoined twins who were separated by surgery, are followed by their mother Arlene Aguirre, center left, and grandmother Evelyn, center right, as they leave Children's Hospital at Montefiore Tuesday, August 30, 2004, in New York, flanked by the doctors who performed the surgery, Dr. James T. Goodrich, left, and Dr. David A. Staffenberg. Nurses loaded the boys into separate ambulances, which took off with a police escort for Blythedale Children's Hospital in Valhalla, where they and their mother have lived between operations at Montefiore.

Classic and Contemporary Essays

We would like to think that the 21st century will avoid a plague like the one that swept through Asia and Europe during the 14th century. Yet the AIDS epidemic proves to be intractable and growing in sub-Saharan Africa, Asia, Russia, and the Indian subcontinent, while outbreaks of new potential epidemics like Ebola and SARS pose serious challenges for medical researchers and dangers for humankind. Plagues seem to be as old and persistent as civilization itself. Looking back to the 14th century, the noted historian Barbara Tuchman tells the story of the Black Death—the bubonic plague—that devastated Europe, resulting in the extinction of one-third of the population. With a historian's eye for narrative and detail, she describes the symptoms associated with plague, the process by which it spread inexorably from one nation to the next, and the religious, political, and cultural impact of the disease on the continent. Unlike Tuchman, the physician and writer Ronald J. Glasser deals with the present and with the possibility that we are facing new plague years. Like Tuchman, he traces the origins of "the emerging plagues" and their consequences for peoples around the world. But he also provides a pointed argument about the failure of the American health care system, as well as politicians, to recognize and deal rationally with diseases of potentially epidemic proportions. As you read these two essays, one looking backward in history and the other forward in time, consider the rhetorical strategies that Tuchman and Glasser employ to render their vision of the plague years in vivid and compelling ways.

✳

"This Is the End of the World": The Black Death

Barbara Tuchman

Barbara Tuchman (1912–1989) was born in New York City and graduated from Rad-cliffe College. A self-taught historian, she worked as a writer for The Nation *magazine and during World War II served as an editor at the U.S. Office of War Information. Her book* The Guns of August *(1960), a narrative history of the outbreak of World War I, won the Pulitzer Prize. She won it again for her book* Stilwell and the American Ex-perience in China: 1911–45 *(1971). Her other books included such best sellers as* A Distant Mirror: The Calamitous 14th Century *(1978) and* The First Salute *(1989). In her later years, she was a lecturer at Harvard University and at the U.S. Naval War College. In this selection, excerpted from* A Distant Mirror, *Tuchman explains in her vivid narrative style the effects of the bubonic plague on Western Europe.*

In October 1347, two months after the fall of Calais, Genoese trading ships put   1
into the harbor of Messina in Sicily with dead and dying men at the oars. The ships had come from the Black Sea port of Caffa (now Feodosiya) in the Crimea, where the Genoese maintained a trading post. The diseased sailors showed strange black swellings about the size of an egg or an apple in the armpits and groin. The swellings oozed blood and pus and were followed by spreading boils and black blotches on the skin from internal bleeding. The sick suffered severe pain and died quickly within five days of the first symptoms. As the disease spread, other symptoms of continuous fever and spitting of blood appeared in-stead of the swellings or buboes. These victims coughed and sweated heavily and died even more quickly, within three days or less, sometimes in 24 hours. In both types everything that issued from the body—breath, sweat, blood from the buboes and lungs, bloody urine, and blood-blackened excrement—smelled foul. Depression and despair accompanied the physical symptoms, and before the end "death is seen seated on the face."

The disease was bubonic plague, present in two forms: one that infected the   2
bloodstream, causing the buboes and internal bleeding, and was spread by con-tact; and a second, more virulent pneumonic type that infected the lungs and was spread by respiratory infection. The presence of both at once caused the high mortality and speed of contagion. So lethal was the disease that cases were known of persons going to bed well and dying before they woke, of doctors catching the illness at a bedside and dying before the patient. So rapidly did it spread from one to another that to a French physician, Simon de Covino, it seemed as if one sick person "could infect the whole world." The malignity of the pestilence appeared more terrible because its victims knew no prevention and no remedy.

The physical suffering of the disease and its aspect of evil mystery were ex-   ₃
pressed in a strange Welsh lament which saw "death coming into our midst like
black smoke, a plague which cuts off the young, a rootless phantom which has no
mercy for fair countenance. Woe is me of the shilling in the armpit! It is seething,
terrible . . . a head that gives pain and causes a loud cry . . . a painful angry knob
. . . Great is its seething like a burning cinder . . . a grievous thing of ashy color."
Its eruption is ugly like the "seeds of black peas, broken fragments of brittle sea-
coal . . . the early ornaments of black death, cinders of the peelings of the cockle
weed, a mixed multitude, a black plague like halfpence, like berries. . . ."

Rumors of a terrible plague supposedly arising in China and spreading   ₄
through Tartary (Central Asia) to India and Persia, Mesopotamia, Syria, Egypt,
and all of Asia Minor had reached Europe in 1346. They told of a death toll so
devastating that all of India was said to be depopulated, whole territories covered by dead bodies, other areas with no one left alive. As added up by Pope
Clement VI at Avignon, the total of reported dead reached 23,840,000. In the absence of a concept of contagion, no serious alarm was felt in Europe until the
trading ships brought their black burden of pestilence into Messina while other
infected ships from the Levant carried it to Genoa and Venice.

By January 1348 it penetrated France via Marseille, and North Africa via Tu-   ₅
nis. Shipborne along coasts and navigable rivers, it spread westward from Marseille through the ports of Languedoc to Spain and northward up the Rhône to
Avignon, where it arrived in March. It reached Narbonne, Montpellier, Carcassonne, and Toulouse between February and May, and at the same time in Italy
spread to Rome and Florence and their hinterlands. Between June and August
it reached Bordeaux, Lyon, and Paris, spread to Burgundy and Normandy, and
crossed the Channel from Normandy into southern England. From Italy during
the same summer it crossed the Alps into Switzerland and reached eastward to
Hungary.

In a given area the plague accomplished its kill within four to six months   ₆
and then faded, except in the larger cities, where, rooting into the close-
quartered population, it abated during the winter, only to reappear in spring
and rage for another six months.

In 1349 it resumed in Paris, spread to Picardy, Flanders, and the Low Coun-   ₇
tries, and from England to Scotland and Ireland as well as to Norway, where a
ghost ship with a cargo of wool and a dead crew drifted offshore until it ran
aground near Bergen. From there the plague passed into Sweden, Denmark,
Prussia, Iceland, and as far as Greenland. Leaving a strange pocket of immunity
in Bohemia, and Russia unattacked until 1351, it had passed from most of Europe by mid-1350. Although the mortality rate was erratic, ranging from one
fifth in some places to nine tenths or almost total elimination in others, the overall estimate of modern demographers has settled—for the area extending from
India to Iceland—around the same figure expressed in Froissart's casual words:
"a third of the world died." His estimate, the common one at the time, was not
an inspired guess but a borrowing of St. John's figure for mortality from plague
in Revelation, the favorite guide to human affairs of the Middle Ages.

A third of Europe would have meant about 20 million deaths. No one 8
knows in truth how many died. Contemporary reports were an awed impression, not an accurate count. In crowded Avignon, it was said, 400 died daily; 7,000 houses emptied by death were shut up; a single graveyard received 11,000 corpses in six weeks; half the city's inhabitants reportedly died, including 9 cardinals or one third of the total, and 70 lesser prelates. Watching the endlessly passing death carts, chroniclers let normal exaggeration take wings and put the Avignon death toll at 62,000 and even at 120,000, although the city's total population was probably less than 50,000.

When graveyards filled up, bodies at Avignon were thrown into the Rhône 9
until mass burial pits were dug for dumping the corpses. In London in such pits corpses piled up in layers until they overflowed. Everywhere reports speak of the sick dying too fast for the living to bury. Corpses were dragged out of homes and left in front of doorways. Morning light revealed new piles of bodies. In Florence the dead were gathered up by the Compagnia della Misericordia—founded in 1244 to care for the sick—whose members wore red robes and hoods masking the face except for the eyes. When their efforts failed, the dead lay putrid in the streets for days at a time. When no coffins were to be had, the bodies were laid on boards, two or three at once, to be carried to graveyards or common pits. Families dumped their own relatives into the pits, or buried them so hastily and thinly "that dogs dragged them forth and devoured their bodies."

Amid accumulating death and fear of contagion, people died without last 10
rites and were buried without prayers, a prospect that terrified the last hours of

Burial of the plague victims. From Annales de Gilles de Muisit.

the stricken. A bishop in England gave permission to laymen to make confession to each other as was done by the Apostles, "or if no man is present then even to a woman," and if no priest could be found to administer extreme unction, "then faith must suffice." Clement VI found it necessary to grant remissions of sin to all who died of the plague because so many were unattended by priests. "And no bells tolled," wrote a chronicler of Siena, "and nobody wept no matter what his loss because almost everyone expected death. . . . And people said and believed, 'This is the end of the world.'"

In Paris, where the plague lasted through 1349, the reported death rate was 800 a day, in Pisa 500, in Vienna 500 to 600. The total dead in Paris numbered 50,000 or half the population. Florence, weakened by the famine of 1347, lost three to four fifths of its citizens, Venice two thirds, Hamburg and Bremen, though smaller in size, about the same proportion. Cities, as centers of transportation, were more likely to be affected than villages, although once a village was infected, its death rate was equally high. At Givry, a prosperous village in Burgundy of 1,200 to 1,500 people, the parish register records 615 deaths in the space of fourteen weeks, compared to an average of thirty deaths a year in the previous decade. In three villages of Cambridgeshire, manorial records show a death rate of 47 percent, 57 percent, and in one case 70 percent. When the last survivors, too few to carry on, moved away, a deserted village sank back into the wilderness and disappeared from the map altogether, leaving only a grass-covered ghostly outline to show where mortals once had lived. 11

In enclosed places such as monasteries and prisons, the infection of one person usually meant that of all, as happened in the Franciscan convents of Carcassonne and Marseille, where every inmate without exception died. Of the 140 Dominicans at Montpellier only seven survived. Petrarch's brother Gherardo, member of a Carthusian monastery, buried the prior and 34 fellow monks one by one, sometimes three a day, until he was left alone with his dog and fled to look for a place that would take him in. Watching every comrade die, men in such places could not but wonder whether the strange peril that filled the air had not been sent to exterminate the human race. In Kilkenny, Ireland, Brother John Clyn of the Friars Minor, another monk left alone among dead men, kept a record of what had happened lest "things which should be remembered perish with time and vanish from the memory of those who come after us." Sensing "the whole world, as it were, placed within the grasp of the Evil One," and waiting for death to visit him too, he wrote, "I leave parchment to continue this work, if perchance any man survive and any of the race of Adam escape this pestilence and carry on the work which I have begun." Brother John, as noted by another hand, died of the pestilence, but he foiled oblivion. 12

The largest cities of Europe, with populations of about 100,000, were Paris and Florence, Venice and Genoa. At the next level, more than 50,000 were Ghent and Bruges in Flanders, Milan, Bologna, Rome, Naples, and Palermo, and Cologne. London hovered below 50,000, the only city in England except York with more than 10,000. At the level of 20,000 to 50,000 were Bordeaux, Toulouse, Montpellier, Marseille, and Lyon in France, Barcelona, Seville, and Toledo in 13

Spain, Siena, Pisa, and other secondary cities in Italy, and the Hanseatic trading cities of the Empire. The plague raged through them all, killing anywhere from one third to two thirds of their inhabitants. Italy, with a total population of 10 to 11 million, probably suffered the heaviest toll. Following the Florentine bankruptcies, the crop failures and workers' riots of 1346–47, the revolt of Cola di Rienzi that plunged Rome into anarchy, the plague came as the peak of successive calamities. As if the world were indeed in the grasp of the Evil One, its first appearance on the European mainland in January 1348 coincided with a fearsome earthquake that carved a path of wreckage from Naples up to Venice. Houses collapsed, church towers toppled, villages were crushed, and the destruction reached as far as Germany and Greece. Emotional response, dulled by horrors, underwent a kind of atrophy epitomized by the chronicler who wrote, "And in these days was burying without sorrowe and wedding without friendschippe."

In Siena, where more than half the inhabitants died of the plague, work was 14 abandoned on the great cathedral, planned to be the largest in the world, and never resumed, owing to loss of workers and master masons and "the melancholy and grief" of the survivors. The cathedral's truncated transept still stands in permanent witness to the sweep of death's scythe. Angolo di Tura, a chronicler of Siena, recorded the fear of contagion that froze every other instinct. "Father abandoned child, wife husband, one brother another," he wrote, "for this plague seemed to strike through the breath and sight. And so they died. And no one could be found to bury the dead for money or friendship. . . . And I, Angolo di Tura, called the Fat, buried my five children with my own hands, and so did many others likewise."

There were many to echo his account of inhumanity and few to balance it, 15 for the plague was not the kind of calamity that inspired mutual help. Its loathsomeness and deadliness did not herd people together in mutual distress, but only prompted their desire to escape each other. "Magistrates and notaries refused to come and make the wills of the dying," reported a Franciscan friar of Piazza in Sicily; what was worse, "even the priests did not come to hear their confessions." A clerk of the Archbishop of Canterbury reported the same of English priests who "turned away from the care of their benefices from fear of death." Cases of parents deserting children and children their parents were reported across Europe from Scotland to Russia. The calamity chilled the hearts of men, wrote Boccaccio in his famous account of the plague in Florence that serves as introduction to the *Decameron*. "One man shunned another . . . kinsfolk held aloof, brother was forsaken by brother, oftentimes husband by wife; nay, what is more, and scarcely to be believed, fathers and mothers were found to abandon their own children to their fate, untended, unvisited as if they had been strangers." Exaggeration and literary pessimism were common in the 14th century, but the Pope's physician, Guy de Chauliac, was a sober, careful observer who reported the same phenomenon: "A father did not visit his son, nor the son his father. Charity was dead."

Yet not entirely. In Paris, according to the chronicler Jean de Venette, the 16 nuns of the Hôtel Dieu or municipal hospital, "having no fear of death, tended the sick with all sweetness and humility." New nuns repeatedly took the places of those who died, until the majority "many times renewed by death now rest in peace with Christ as we may piously believe."

When the plague entered northern France in July 1348, it settled first in Normandy and, checked by winter, gave Picardy a deceptive interim until the next summer. Either in mourning or warning, black flags were flown from church towers of the worst-stricken villages of Normandy. "And in that time," wrote a monk of the abbey of Fourcarment, "the mortality was so great among the people of Normandy that those of Picardy mocked them." The same unneighborly reaction was reported of the Scots, separated by a winter's immunity from the English. Delighted to hear of the disease that was scourging the "southrons," they gathered forces for an invasion, "laughing at their enemies." Before they could move, the savage mortality fell upon them too, scattering some in death and the rest in panic to spread the infection as they fled.

In Picardy in the summer of 1349 the pestilence penetrated the castle of Coucy to kill Enguerrand's mother, Catherine, and her new husband. Whether her nine-year-old son escaped by chance or was perhaps living elsewhere with one of his guardians is unrecorded. In nearby Amiens, tannery workers, responding quickly to losses in the labor force, combined to bargain for higher wages. In another place villagers were seen dancing to drums and trumpets, and on being asked the reason, answered that, seeing their neighbors die day by day while their village remained immune, they believed they could keep the plague from entering "by the jollity that is in us. That is why we dance." Further north in Tournai on the border of Flanders, Gilles li Muisis, Abbot of St. Martin's, kept one of the epidemic's most vivid accounts. The passing bells rang all day and all night, he recorded, because sextons were anxious to obtain their fees while they could. Filled with the sound of mourning, the city became oppressed by fear, so that the authorities forbade the tolling of bells and the wearing of black and restricted funeral services to two mourners. The silencing of funeral bells and of criers' announcements of deaths was ordained by most cities. Siena imposed a fine on the wearing of mourning clothes by all except widows.

Flight was the chief recourse of those who could afford it or arrange it. The rich fled to their country places like Boccaccio's young patricians of Florence, who settled in a pastoral palace "removed on every side from the roads" with "wells of cool water and vaults of rare wines." The urban poor died in their burrows, "and only the stench of their bodies informed neighbors of their death." That the poor were more heavily afflicted than the rich was clearly remarked at the time, in the north as in the south. A Scottish chronicler, John of Fordun, stated flatly that the pest "attacked especially the meaner sort and common people—seldom the magnates." Simon de Covino of Montpellier made the same observation. He ascribed it to the misery and want and hard lives that made the poor more susceptible, which was half the truth. Close contact and lack of sanitation was the unrecognized other half. It was noticed too that the young died in greater proportion than the old. Simon de Covino compared the disappearance of youth to the withering of flowers in the fields.

In the countryside peasants dropped dead on the roads, in the fields, in 20 their houses. Survivors in growing helplessness fell into apathy, leaving ripe wheat uncut and livestock untended. Oxen and asses, sheep and goats, pigs and chickens ran wild and they too, according to local reports, succumbed to the pest. English sheep, bearers of the precious wool, died throughout the country. The chronicler Henry Knighton, canon of Leicester Abbey, reported 5,000 dead in one field alone, "their bodies so corrupted by the plague that neither beast nor bird would touch them," and spreading an appalling stench. In the Austrian Alps wolves came down to prey upon sheep and then, "as if alarmed by some invisible warning, turned and fled back into the wilderness." In remote Dalmatia bolder wolves descended upon a plague-stricken city and attacked human survivors. For want of herdsmen, cattle strayed from place to place and died in hedgerows and ditches. Dogs and cats fell like the rest.

The dearth of labor held a fearful prospect because the 14th century lived 21 close to the annual harvest both for food and for next year's seed. "So few servants and laborers were left," wrote Knighton, "that no one knew where to turn for help." The sense of a vanishing future created a kind of dementia of despair. A Bavarian chronicler of Neuberg on the Danube recorded that "Men and women . . . wandered around as if mad" and let their cattle stray "because no one had any inclination to concern themselves about the future." Fields went uncultivated, spring seed unsown. Second growth with nature's awful energy crept back over cleared land, dikes crumbled, salt water reinvaded and soured the lowlands. With so few hands remaining to restore the work of centuries, people felt, in Walsingham's words, that "the world could never again regain its former prosperity."

Though the death rate was higher among the anonymous poor, the known 22 and the great died too. King Alfonso XI of Castile was the only reigning monarch killed by the pest, but his neighbor King Pedro of Aragon lost his wife, Queen Leonora, his daughter Marie, and a niece in the space of six months. John Cantacuzene, Emperor of Byzantium, lost his son. In France the lame Queen Jeanne and her daughter-in-law Bonne de Luxemburg, wife of the Dauphin, both died in 1349 in the same phase that took the life of Enguerrand's mother. Jeanne, Queen of Navarre, daughter of Louis X, was another victim. Edward III's second daughter, Joanna, who was on her way to marry Pedro, the heir of Castile, died in Bordeaux. Women appear to have been more vulnerable than men, perhaps because, being more housebound, they were more exposed to fleas. Boccaccio's mistress Fiammetta, illegitimate daughter of the King of Naples, died, as did Laura, the beloved—whether real or fictional—of Petrarch. Reaching out to us in the future, Petrarch cried, "Oh happy posterity who will not experience such abysmal woe and will look upon our testimony as a fable."

In Florence Giovanni Villani, the great historian of his time, died at 68 in the 23 midst of an unfinished sentence: ". . . *e dure questo pistolenza fino a* . . . (in the midst of this pestilence there came to an end . . .)." Siena's master painters, the brothers Ambrogio and Pietro Lorenzetti, whose names never appear after 1348, presumably perished in the plague, as did Andrea Pisano, architect and sculptor of Florence. William of Ockham and the English mystic Richard Rolle of Hampole both disappear from mention after 1349. Francisco Datini, merchant of Prato, lost both his parents and two siblings. Curious sweeps of mortality afflicted certain bodies of merchants in London. All eight wardens of the Company of Cutters, all six wardens of the Hatters, and four wardens of the Goldsmiths died before July 1350. Sir John Pulteney, master draper and four times Mayor of London, was a victim, likewise Sir John Montgomery, Governor of Calais.

Among the clergy and doctors the mortality was naturally high because of the nature of their professions. Out of 24 physicians in Venice, 20 were said to have lost their lives in the plague, although according to another account, some were believed to have fled or to have shut themselves up in their houses. At Montpellier, site of the leading medieval medical school, the physician Simon de Covino reported that, despite the great number of doctors, "hardly one of them escaped." In Avignon, Guy de Chauliac confessed that he performed his medical visits only because he dared not stay away for fear of infamy, but "I was in continual fear." He claimed to have contracted the disease but to have cured himself by his own treatment; if so, he was one of the few who recovered. [24]

Clerical mortality varied with rank. Although the one-third toll of cardinals reflects the same proportion as the whole, this was probably due to their concentration in Avignon. In England, in strange and almost sinister procession, the Archbishop of Canterbury, John Stratford, died in August 1348, his appointed successor died in May 1349, and the next appointee three months later, all three within a year. Despite such weird vagaries, prelates in general managed to sustain a higher survival rate than the lesser clergy. Among bishops the deaths have been estimated at about one in twenty. The loss of priests, even if many avoided their fearful duty of attending the dying, was about the same as among the population as a whole. [25]

Government officials, whose loss contributed to the general chaos, found, on the whole, no special shelter. In Siena four of the nine members of the governing oligarchy died, in France one third of the royal notaries, in Bristol 15 out of the 52 members of the Town Council or almost one third. Tax-collecting obviously suffered, with the result that Philip VI was unable to collect more than a fraction of the subsidy granted him by the Estates in the winter of 1347–48. [26]

Lawlessness and debauchery accompanied the plague as they had during the great plague of Athens of 430 B.C., when according to Thucydides, men grew bold in the indulgence of pleasure: "For seeing how the rich died in a moment and those who had nothing immediately inherited their property, they reflected that life and riches were alike transitory and they resolved to enjoy themselves while they could." Human behavior is timeless. When St. John had his vision of plague in Revelation, he knew from some experience or race memory that those who survived "repented not of the work of their hands. . . . Neither repented they of their murders, nor of their sorceries, nor of their fornication, nor of their thefts." [27]

COMPREHENSION

1. The title of this essay suggests a religious theme. Why did intellectuals and religious leaders associate the bubonic plague with biblical prophecy?
2. Does this essay have a thesis or does it merely record in detail a period in European history? If it does have a thesis, is it implied or expressed directly? Explain your answer.
3. Does Tuchman suggest that Europe was "fated" to endure the tragic consequences of the plague owing to a higher power, or does she attribute the disaster to a confluence of history and chance? Explain your answer.

RHETORIC

1. Tuchman begins her essay by describing in detail the physical symptoms of the plague. What strategy lies behind this rhetorical decision?
2. Tuchman has a reputation as a historian whose goal was to bring "history to life." What methods does she use to realize this goal? Is she successful? Why or why not? What does the illustration on page 673 contribute?
3. Contemporary authors and filmmakers often select morbid themes for their sensational value or for financial gain, or both. For example, there is a plethora of "true-crime" stories, "re-creations" of natural disasters, and profiles of aberrant and murderous personalities such as Jeffrey Dahmer, Ted Bundy, and the "Hillside Strangler." Is this Tuchman's purpose? Explain why or why not.
4. Note the particular parts of speech Tuchman uses to begin paragraphs 5 through 7, 9, 11, 12, 14, 16 through 18, 20, 22, and 23. All begin with either conjunctions or prepositions. How do these grammatical devices help maintain the flow of Tuchman's narrative?
5. Tuchman makes references to a vast number of historical figures and specific locations in 14th century Europe. What is her assumption about the educational level of her intended audience? About the specialization of her readership? Is it necessary to know something about the people and places she cites to appreciate the essay? Or is Tuchman writing a book of general interest, with the implicit supposition that different readers will extract their own level of appreciation from her narrative? Explain.
6. Tuchman uses direct quotations from the observers and chroniclers of the times. Examine the use of such sources in paragraphs 10, 13, 15, 16, 17, 18, 20, and 23, among others. How does Tuchman weave their observations into her own narrative so that the essay maintains unity and coherence? How does her use of these citations affect the strength of her writing?

WRITING

1. Write a 300-word summary of Tuchman's essay.

2. For a research project, study Tuchman's philosophy regarding how history should be reported. Apply your research to her treatment of the Black Death.
3. **WRITING AN ARGUMENT:** Argue for or against the proposition that an epidemic as severe as the one that Tuchman describes could not possibly occur in the 21st century.

www.mhhe.com/ **mhreader** | For more information on Barbara Tuchman, go to:
More Resources > Ch. 12 Health & Medicine

✳

We Are Not Immune

Ronald J. Glasser

Ronald J. Glasser (b. 1940) is a Minneapolis specialist in pediatric nephrology and rheumatology. Additionally, he is a nonfiction writer and novelist. His books include 365 Days *(1971), a collection of sketches about wounded and dying American soldiers fighting in the Vietnam War, and* Ward 402 *(1973), the story of a young girl dying of cancer. His nonfiction books include a study of cancer,* The Greatest Battle *(1976), and* The Light in the Skull: An Odyssey of Medical Discovery *(1997). In the following essay, which appeared in* Harper's Magazine *in 2004, Glasser discusses current epidemics and the collapse of public health.*

Death is inevitable, but not disease. The difference may be as simple as washing   1
our hands or keeping the wastes of industrialized farming out of the water supply, but it is often much more complicated. Bacteria and viruses are no mean adversaries, nor are they easily defeated. If we fail to be watchful or to protect those most at risk, a public-health catastrophe is inevitable, and yet somewhere within the span of the last thirty years the idea of the common good has disappeared from our national consciousness, giving way to the misconception that we no longer need concern ourselves with the welfare of our fellow citizens. It is a dangerous conceit, and it leads us toward a future infected with unprecedented and unnecessary disease.

We have grown not so much complacent as narcotized, lulled into a sense   2
of security by the almost daily pronouncements from corporate medicine and the pharmaceutical industry of ever better drugs and more "breakthrough" treatments. The spectacular progress of twentieth-century medicine, most recently the sequencing of the human genome, sponsors the widespread fancy that disease might someday be conquered, that genetic manipulation or nanotechnology or some other science-fiction marvel might bring with it a cure for

death. Long forgotten are the days when the loss of a child to diphtheria or whooping cough or yellow fever was a commonplace event, the days before widespread vaccination and government safety and health regulations; we no longer remember life before publicly funded sewage-treatment plants and the passage of the clean-air and -water acts. Public health is often invisible and unremarked when it works well; when it fails, our neighbors sicken and die.

A public-health system is only as strong as its weakest link; an epidemic en- 3 forces, in the most rigorous fashion, the American credo that all men are created equal. If we allow one segment of our society to suffer and perish from preventable disease, little stands in the way of collective doom. Yet today, 44 million people in the United States are without health insurance; those who can afford to pay for it generally receive inferior treatment, despite the fact that Americans spend $1.4 trillion annually for their health care. Public-health departments across the country have never recovered from decades of cutbacks, despite injections of funding in response to specific emergencies such as AIDS or the threat of bioterrorism. Purchases of newer and more reliable diagnostic-testing equipment have been deferred; technical staff and other employees needed to support epidemiologic and testing programs have been downsized; vital on-site bacteriological and viral laboratories have been closed and the testing outsourced to the lowest bidder or simply abandoned.* State and local early-childhood services, prenatal care, immunization campaigns for the poor, alcohol-abuse and smoking-awareness campaigns, monitoring programs for lead and arsenic levels, as well as HIV/AIDS treatment programs, have been curtailed as health departments shift around available monies and reassign what few permanent staff members they have left in an attempt to keep the most critical programs in operation. Prevention becomes secondary to simply keeping people alive. Nor must we concern ourselves simply with the state of American public health; as distances collapse and human populations grow ever more mobile, so also new and deadly diseases (among them Ebola and the Marburg virus) find their way across deserts and oceans. AIDS took decades to escape its origins in central Africa; we should not expect the next simian retrovirus to take so long. SARS made its way from Asia to Toronto in a matter of weeks.

Medical historians describe the last few decades as the age of "the emerging 4 plagues." Overpopulation, poverty, ecological devastation and global climate change, chemical pollution and industrial agriculture—all of these factors conspire to create the conditions for unprecedented death by infectious disease. Between 1977 and 1994, twenty-nine previously unknown human pathogens emerged, and it is estimated, moreover, that we have identified only 1 percent of the bacteria and 4 percent of the viruses on the planet. Tuberculosis, a disease that should have disappeared decades ago, has reemerged as an epidemic, and

---

*More than 3,000 hospital beds and 35 hospitals have been eliminated in Minnesota over the last twenty years; the state has 16,511 licensed beds, of which only about 7,000 are staffed, and most of these are occupied every day. Nationally, between 1980 and 2000, 1,000 hospitals shut their doors. [This and subsequent notes in the selection are the author's.]

drug-resistant strains continue to spread throughout our cities. In 1995, 1.7 million American patients contracted hospital-spread infections; 88,000 of these patients died; 70 percent of the infections were drug-resistant.** Each year an estimated 76 million Americans fall ill to food-borne illnesses resulting in approximately 325,000 hospitalizations and 5,000 deaths. Influenza infects 10 to 20 percent of the U.S. population every year and kills 36,000; a virulent avian flu could kill millions. Such numbers, a mere sampling of those available, paint a grim portrait, and the view does not improve if we narrow the perspective.

During a two-week period in 1993 one of Milwaukee's two water-treatment   5 plants malfunctioned. This waterworks supplied treated drinking water directly from Lake Michigan to at least half the population of Milwaukee and nine of its suburbs. The investigation that eventually followed revealed unprecedented increases in the density levels of the supposedly treated water during that two-week period. The gauges designed for continuous measurement of water purity had clearly not been functioning properly for periods as long as eight to twelve hours at a time. The precise reason for the failure remains obscure, but what is clear is that no alarms went off, no backup systems were brought online, and no one noticed the increases in turbidity that led to the largest waterborne epidemic ever to occur in the United States.

Within days of the plant's malfunction and continuing for an additional   6 month, more than 403,000 people in the Milwaukee area developed fever, vomiting, and diarrhea. One hundred people died. The cause of the illness was *Cryptosporidium*, a single-celled microorganism that survives in bodies of standing water and has been known as a cause of diarrhea, abdominal cramping, nausea, vomiting, and fever since the 1970s. There is no medical treatment for the infection, and in otherwise healthy individuals the disease is usually self-limiting, though in a minority of cases the disease can lead to weeks of disability. In patients on immunosuppressive medications, those undergoing chemotherapy, or those with AIDS, the infection can be ruthless, unrelenting, and fatal.

The seriousness of *Cryptosporidium* in an immune-suppressed patient be-   7 came clear at the beginning of the AIDS epidemic when physicians first found this strange and unexpected parasite in the blood and bone marrows of infected patients. Knowing that the organism was basically a disease of herd animals, the doctors contacted the preeminent expert on *Cryptosporidium* in the department of agriculture at the University of Iowa. When the professor was asked how infected sheep were treated, he hesitated. "There is no treatment," he answered. "We shoot them."

**The failure of antibiotics to control common infections such as staph and strep is one of the most chilling of recent developments, one hastened by the reckless overprescription of physicians and by massive application of antibiotics to livestock that would otherwise perish in the lethal miasma created by industrial agriculture.

There are well over 1,400 documented microorganisms that can infect hu- 8
mans, of which fully one half first caused disease in animals. *Cryptosporidium*
made the transfer to humans through the contamination of surface waters by
runoff from farmlands and drainage ditches. Unfortunately, the cysts that
spread the infection are highly resistant to chlorine and even remain viable in
the laboratory after exposures to full-strength household bleaches. You can't kill
the *Cryptosporidium* cysts, and while alive they remain astonishingly infectious.
Disease has been known to occur through ingestion of as few as thirty cysts,
and experimental data have shown that even a single cyst can pass on the disease to uninfected sheep as well as to humans. The only means of prevention
for a contaminated water supply are filters of less than one micron placed
within water-purification systems that physically remove the millions of heavy,
dense cysts before they reach the household taps of public water supplies.

The failure of the main safety and backup gauges in Milwaukee was clearly 9
a disaster, but the most unnerving aspect of the Wisconsin epidemic was not the
astronomical numbers of affected people or even the deaths; it was the fact that,
in the midst of the worsening epidemic, it was not the federal government or
any state or local health department or surveillance program or emergency-
room database or managed-healthcare reporting system that alerted the public
that an epidemic was in progress. It was a pharmacist, who happened to notice
unusual sales of over-the-counter diarrhea medication. The local media reported the outbreak days before the health department took action, almost a
week after the alarm was first raised.

The outbreak of *Cryptosporidium* in Milwaukee was more than the simple 10
malfunction of a few gauges in a midsized American city; to those concerned
about the nation's ability to treat its people and control disease, it was a clear
sign that our infectious-disease and medical surveillance and prevention programs were no longer working. Although the Milwaukee disaster was unusual
for its size, waterborne outbreaks of disease are not uncommon. The Centers for
Disease Control maintain a database on the subject, but the statistics are not
very reliable because they depend on the voluntary reporting of state and local
health officials. Some states choose not to make these reports; some states do
not even have active disease-surveillance systems. But local failures can often
have far broader consequences, as we learned from the 1999 outbreak of the
West Nile virus in New York City. Mosquito surveillance and control was a local budget casualty that led to a national epidemic, and by last year West Nile
had appeared in every state but Washington, Oregon, Alaska, and Hawaii;
14,163 people are known to have been infected and 564 have died. The discovery in 2002 that the virus was transmitted via organ transplantation, and possibly by blood transfusion, has led to fears that the national blood supply could
be contaminated. Testing for West Nile raises the cost of blood by $4 to $7 a unit.
And even as the virus spreads across the continent, federal funding is being cut;
in 2004 the Mosquito Abatement for Safety and Health Act received zero funding, and none has been requested for 2005.

The United States has no single agency responsible for public health and   11
thus no coherent policy. As Laurie Garrett suggests in her monumental study,
Betrayal of Trust: The Collapse of Global Public Health, it is no exaggeration to say
that we simply lack a public-health system per se; what we do have is best described as "a hodgepodge of programs, bureaucracies, and failings."

The great public-health victories of the nineteenth and early twentieth cen-   12
turies over yellow fever, cholera, encephalitis, smallpox, puerperal fever, and a
host of other infectious diseases were largely the result of preventive measures
enacted by visionary public officials: improved sanitation and nutrition (safe
water and food, decent housing, paved streets, sewers), vigorous powers of
quarantine to prevent contagion, mosquito control and the installation of window glass, and the creation of vaccination programs. Few advances were as important as the realization that merely washing one's hands could prevent the
spread of disease. Life expectancy in the eighteenth century for an average male
was about thirty years; by the early 1970s, it was seventy-five years. And as
Garrett points out, most of that progress occurred prior to the invention of antibiotics, and "less than 4 percent of the total improvement in life expectancy
since the 1700s can be credited to twentieth century advances in medical care."
Ironically, the medical revolutions of the twentieth century have contributed to
our overconfident and complacent neglect of the public-health infrastructure.
We spend vast sums to lengthen the lives of terminally ill patients by a few days
and refuse to make modest investments that would prevent millions of needless
illnesses and deaths.

The peculiar dynamics of American politics, with its periodic spasms of irra-   13
tional antigovernment hysteria, have ensured that few effective public-health
policies fail to attract powerful political enemies, enemies that more often than
not have succeeded in weakening the agencies charged by Congress with the responsibility for the health and well-being of the American people. Not even the
CDC is immune from the virus of partisan politics; despite an overwhelming
medical consensus, the agency has refused to take a position on the use of condoms to prevent AIDS and has curtailed the printing or distribution of any data
on the control or treatment of sexually transmitted diseases that might offend
the most conservative Christians. In response to political pressure from the
NRA and threats from Congress to withhold funding, the CDC has also discon-
tinued its definitive research documenting the public-health costs of handguns.

The Food and Drug Administration presents the same self-defeating pat-   14
tern of regulatory behavior. In May of this year, the agency refused to approve
a morning-after contraceptive pill for over-the-counter use, even after its own
expert advisory panel recommended it. Far worse is the degree to which the
FDA panders to its industrial constituency. Drugs receive approval without adequate testing; the agency dithers when patients begin to die; eventually it turns
out that adverse findings were ignored or suppressed. Often more concerned
for the well-being of the pharmaceutical industry than for the health of American citizens, the FDA challenges states that seek to purchase cheaper Canadian

drugs for their citizens and ignores the ongoing concentration of drug and vaccine production into the hands of fewer and larger companies, which has led to greater consumer costs and vaccine shortages. The agency has shown no inclination to pressure manufacturers into adopting new technologies that would allow the timely and safe development of new vaccines in response to emerging diseases. Not too long ago the FDA supported the pharmaceutical industry's wish to give antidepressant drugs to children despite the agency's own finding that such drugs might cause them to commit suicide.

Faced with alarming outbreaks of food-borne illness, the Department of   15 Agriculture has refused to enforce the use of any of the more definitive and reliable, though admittedly more costly, bacterial tests of meat and meat products to replace the pathetically ineffective "poke and sniff" test used in all government-monitored and -approved slaughterhouses and meat-processing plants. *E. coli,* salmonella, listeria, shigella, have all caused outbreaks of disease. What is astonishing is not that a million pounds of hamburger can be contaminated from one infected cow but that the federal government demands only "voluntary" recalls. Confronted with proof that mad cow disease has infected the American food supply, the agency has prohibited the routine testing of American cattle for the disease, using the newly available tests only in obviously diseased animals and then allowing the animals to be slaughtered and put into the food supply before the results of those tests are available. The USDA has dismissed the recommendations of some of the nation's most prominent professors of agriculture and veterinary medicine to institute a more rigorous and scientific method of testing for this disease, usually citing as an excuse the meat industry's concern that any testing will add an additional three to five cents a pound to consumer prices. The USDA is not so much a regulatory agency as it is an arm of the meat-industry lobby.

Americans, we know, pay too much for their health care, and compared with   16 other countries we receive a very poor return on our investment. The reasons are many, but they are not hard to understand: in essence, we have tended historically to view health care as a commodity like any other. But health is not a product; it is a public good. The evidence is clear that even when viewed through the reductive lens of purely economic self-interest, market-based, entrepreneurial medicine is a failure. Healing people after they fall ill is vastly more expensive than preventing the illness in the first place: every dollar spent preventing diphtheria, for instance, saves $27; every dollar spent on measles, mumps, and rubella saves $23. Yet policymakers have consistently preferred the most expensive and least efficient models of health care, proving once again that the apostles of privatization are motivated not by hard-nosed economics but by an incoherent ideology that is little more than a brittle mask concealing the most irrational species of self-interest.

For the last quarter century, especially after the election of Ronald Reagan   17 and his declaration that government itself is the problem that afflicts us, the public-health infrastructure of this country has been eviscerated. Between 1981 and 1993, public-health expenditures declined by 25 percent as a proportion of overall health spending; in 1992, less than 1 percent of all American health-care spending was devoted to public health. That trend has continued, even after the anthrax attacks of 2001, when politicians suddenly realized how vulnerable the nation was to biological attack.

Since then, it is true, the federal government has appropriated about $2 bil- 18 lion for bioterrorism response, an undertaking that if it were actually carried out would necessarily involve improving the public-health infrastructure. In theory, the bioterrorism money is channeled through the CDC, which distributes it to the states, which in turn disperse money to local health departments. Superficially, the gains are impressive: the CDC's budget for "public health preparedness and response for bioterrorism" increased from $49.9 million in 2001 to $918 million in 2002 and $870 million in 2003. Yet strangely enough, state and local public-health budgets have continued to decline. Public-health laboratories in California could lose 20 percent of their funding this year; the Alabama Department of Public Health expects to fire 250 people and to close regional labs and cut back on its flu-vaccination programs. State funding for AIDS prevention in Massachusetts has been cut by 40 percent over the last two years. Larimer County, Colorado, where last summer 500 people contracted the West Nile virus, received $100,000 in federal funds but lost $700,000 in state money. Overall, thirty-two states cut their public-health budgets between fiscal years 2002 and 2003. Michigan cut its spending by 24 percent, Massachusetts by 23 percent, and Montana, which received more federal bioterror money per capita than New York, cut its public-health budget by 19 percent. Many states, facing huge budget deficits, apparently took the federal money and simply cut their own appropriations. This should come as no surprise: in 2003 the states collectively faced a $66 billion shortfall, and in 2004 state deficits are estimated to be $78 billion. Federal investment will do no good if state politicians, struggling to cope with the economic effects of other federal policies, use those funds to reduce their own deficits.

The Trust for America's Health (TFAH), a nonprofit group that monitors 19 public-health policy, in December released a comprehensive study of what the state health departments have accomplished with their "increased" funding. TFAH found that only twenty-four states had spent at least 90 percent of their 2002 bioterror funds, and only seventeen states had passed at least 50 percent of the money along to local health departments. Much of the money is mired in bureaucracy. A February GAO report revealed that the states were not much better prepared for bioterrorism (and by extension, a natural epidemic) than they were in 2001.

Of course, state health departments can hardly be blamed for their inability 20 to correct a quarter century of neglect with what amounts to a mere $2,000 for every staffed hospital bed in America. Bioterrorism funds are being used simply to keep the lights on, and no one who has carefully observed the Bush Administration would expect it to follow through with its promises to rebuild the public-health system. In fact, the President's 2005 budget proposal calls for a $105 million decrease in state and local bioterrorism funding. The new budget also cuts $1.1 billion from the "Function 550" account, which finances disease-prevention programs and other public-health initiatives, and the federal Public Health Improvements Programs were cut by 64 percent.

Secretary of Health and Human Services Tommy Thompson has claimed 21 that preparing for bioterrorism will enable the government to respond to influenza and other infectious diseases; in fact, the reverse is true. Bioterrorism is a remote threat and a massive attack is very unlikely, but it captures the imagination of weak-minded politicians and a populace raised on movies starring Bruce Willis. The truly imminent biological threat, which all public-health experts agree will inevitably strike, is an influenza pandemic. The 1918 pandemic killed 550,000 Americans and 30 million worldwide. A virulent flu would thus be much worse than a bioterrorism attack, and it would strike every part of the country more or less simultaneously. These facts are well known and understood, yet TFAH found that only thirteen states have a plan or at least a draft of a plan to confront an influenza pandemic. Amazingly, the CDC itself has yet to release a federal plan for such a pandemic; nor does the CDC require states to report flu cases or even flu deaths.

Every year influenza epidemics emerge from areas such as the Guangdong re- 22 gion of China, where large populations of farmers, pigs, and poultry share their species' various strains of the influenza virus. When multiple strains of the virus infect the same host, they begin to share genes, creating new mutations; when a new strain emerges for which humans have no immunity, a pandemic can occur.

In response to a 1997 avian influenza outbreak that began to infect humans 23 but stopped short, for some reason, of becoming an epidemic, the World Health Organization significantly expanded its flu-prevention activities and set up its Global Agenda for Influenza Surveillance and Control, a program whose four main objectives are to monitor the spread of influenza in animals and humans, to identify each year's newest infective strain, to accelerate global pandemic awareness, and to increase usage and speed development of an effective vaccine. Each year the WHO surveillance program puts its infectious-disease teams along with its worldwide network of more than one hundred laboratories on alert, hoping to detect outbreaks before they spread around the globe. Such generalized surveillance is difficult and expensive, but the danger of emerging infections and the continuing influenza threat have left the world health community with little choice.

In February 2003 the WHO issued a report about a group of patients with 24 severe influenza in Hong Kong. The index case was a physician from Guangdong province in China. A global alert was soon issued concerning similar illnesses in Singapore and Hanoi. The WHO sent Dr. Carlo Urbani, an Italian infectious-disease specialist, to Hanoi to investigate. Urbani swiftly determined that the disease was something unusual and that it was highly contagious and virulent. Unlike influenza, which always begins with a runny nose, waves of generalized aches and pains, and weakness, followed by days of fever and an increasing cough before the onset of pneumonia, this disease progressed almost immediately to severe pneumonia, respiratory collapse, and, for many, death. We now know that these alerts were describing the SARS outbreak, which nearly became a global pandemic. Working closely with the Vietnamese authorities, Urbani and other specialists from the WHO, the CDC, and Doctors Without Borders were able to contain the disease in Hanoi, though tragically Urbani himself contracted SARS and died in a makeshift isolation ward in Bangkok. It was not long before the disease spread to Toronto. By late March, 6,800 people there had already been quarantined, with another 5,200 health-care staff working "in quarantine" at facilities that public-health officials had quickly set aside for treating suspected SARS cases. In the United States public-health officials were simply holding their breath and hoping for the best. Not only have cutbacks stripped rural areas of their hospitals and clinics but even the major cities now lack the number of acute-care and infectious-disease beds—not to mention the nursing staff, technicians, and isolation units—to deal with a bad year of influenza much less a full-fledged disease with what appeared to be the staggering demands of SARS.

What happened next was unprecedented: researchers quickly determined   25 that the disease was caused by a new type of virus and very rapidly isolated the cause as a previously unknown coronavirus that had apparently jumped from an animal species to humans. It was not lost on the world's infectious-disease experts that what had taken physicians and scientists almost four years in the case of AIDS was accomplished for SARS in less than four months. It is no exaggeration to say that the billions of dollars so reluctantly pushed into viral research as a result of the efforts of AIDS activists in the 1980s and 1990s enabled the WHO to quickly find the cause of another viral plague. And it was the ability to share accurate information in real time via email and the Internet that allowed the WHO to hold the disease in check.

In the midst of all the tracking of potential contacts, the increased hospital   26 izations, the thousands of people in quarantine, the disease simply vanished at virtually the same time all over the world. Coronaviruses thrive in cold weather, and, like influenza, they spread during the winter months, which accounts for the yearly outbreaks of colds and upper-respiratory infections. The realization that SARS is a cold-weather virus is troubling, because it means that there has been no real victory, only a reprieve. It has to be assumed that SARS is still out there waiting for another winter.

The lesson of the SARS outbreak was that preparation, surveillance, and de   27 cisive action from public officials can prevent epidemics. The WHO response was exemplary—training, staffing, equipment, and funding were all in place, ready for an emergency—but we still lack a truly global early-warning system. In the United States we continue to be without an effective national warning system. As *Lancet* editor Richard Horten writes in *Health Wars,* his scathing critique of contemporary medicine, "No single agency—CDC, WHO, the military, or a nongovernmental organization (such as Médecins Sans Frontières)—

currently has the resources, staff, or equipment to act as a rapid-response strike force during a civilian health emergency." If SARS had come to the United States, there is little hope that it could have been contained.

Today, we are no better prepared for a SARS epidemic than we were last 28 year. "Homeland security," curiously interpreted to exclude the most plausible and deadly threats facing our population, has remained the priority. The massive smallpox immunization program in 2002 was little more than a distraction and waste of precious funds. Meanwhile, we are afflicted with a government that has waged war all across the world to avenge the deaths of 3,000 terror victims, far fewer than die of influenza in a mild year; a government that insists on spending $50 billion to build a missile-defense system that does not work, a military-industrial make-work project designed to meet a threat that does not exist. The war in Iraq consumes almost $4 billion a month, twice the amount we have largely squandered on bioterrorism since 2001. We have grown so foolish and so incompetent that perhaps we do not deserve to survive. Perhaps it is simply time to die.

COMPREHENSION

1. What does the author mean when he writes, "We have grown not so much complacent as narcotized" about medicine and illness (paragraph 2)? What examples does he give to support this assertion?
2. According to the author, what are some of the main examples of the age of the "emerging plagues" (paragraph 4)?
3. Explain Glasser's opinion of the state of American health care. How is health care policy affected by politics? What does the author conclude?

RHETORIC

1. What is the writer's claim? Where does it appear most clearly? How does the tone of the essay reinforce the claim?
2. The writer divides his essay into five sections. What is the topic of each section? How does each section relate to the unit that follows? How does the writer maintain coherence?
3. What appeals to authority does the author make? How does he establish his own credibility?
4. What are the different types of evidence that the writer offers in this essay? What are the premises behind the evidence? How convincing is this evidence? Explain your answer.
5. What patterns of cause and effect does the author analyze? What is his purpose in using causal analysis?
6. Examine the first and last paragraphs. What similarities do you detect in content and tone? How do these beginning and ending paragraphs serve to frame the entire essay?

WRITING

1. Write a personal essay in which you describe an instance when you did not receive proper or adequate health care.
2. Research one disease that Glasser mentions—for example influenza, AIDS, or SARS—and write an essay defining and describing the illness and its epidemic nature either in the United States or worldwide.
3. **WRITING AN ARGUMENT:** Write a response to the author, either agreeing with his views or refuting them. Be certain to deal with his concluding remark, "Perhaps it is simply time to die."

www.mhhe.com/
mhreader

For more information on Ronald J. Glasser, go to:
More Resources > Ch. 12 Health & Medicine

Classic and Contemporary: Questions for Comparison

1. Discuss Tuchman's and Glasser's essays in terms of style, method, and tone. What level of language do they employ? What forms of authority do they bring to bear on their subject? What rhetorical strategies do they use? How does the fact that one writer is a historian and the other a doctor affect their approach to the subject? Does one argue and the other explain, or do they both make similar assertions?
2. Compare Tuchman's view of the bubonic plague of the 14th century with Glasser's presentation of the new plague years.
3. In groups of three or four, conduct research on epidemics throughout history not mentioned by either Tuchman or Glasser. How do these epidemics serve to reinforce the assertions that the two writers make?

✳

I Worked Hard for That Furrowed Brow

Ellen Goodman

Ellen Goodman (b. 1941) is an award-winning journalist and associate editor of the Boston Globe. *She was born in Newton, Massachusetts, and educated at Radcliffe College. She began working for the* Boston Globe *in 1967 and started writing a weekly column in 1971, which today is syndicated in more than 400 newspapers across the*

United States. Starting with Close to Home *(1979), Goodman has published five col-*
lections of her columns, dealing with a broad range of topics, including the status of
women, health and reproductive issues, politics, and the family. In 1980 her columns
were awarded the Pulitzer Prize for Distinguished Commentary. Here, in a column that
appeared in 2002, she dissents from the current obsession with Botox and other forms
of cosmetic surgery.

Just because the FDA has approved of Botox doesn't mean that I have to. In fact,   1
since 835,000 people have already had their foreheads injected with the paralyzing fluid that keeps them from being able to frown, I figure that somebody has
to frown for them.

When I first read about Botox as a cosmetic, I thought there was something   2
vaguely charming about the idea. After all, the microbe created by the U.S.
Army to inflict botulism poisoning on our enemies was now being used for domestic and aesthetic purposes. Talk about beating your swords into tweezers.

But even before the FDA gave the green light, we heard that Botox gather-   3
ings of women had become the Tupperware parties of the 21st century. Only
what's being preserved are the women, not the leftovers.

This is not, I promise you, a screed about the political incorrectness of plas-   4
tic surgery or vanity. Nor is it about how beauty is only skin deep.

Over the years, my attitudes—like my jaw line—have softened toward   5
women who choose to change their faces rather than live with them. I know
there's a line between those who "need" to be "fixed" and those who don't,
between those who need surgery—think burn victim—and those who need
therapy—think Michael Jackson. But I'm less inclined to draw it for anyone else.

When 47-year-old Greta Van Susteren became the poster anchor for plastic   6
surgery, I thought the criticism was way over the top. As she said, "Having
plastic surgery isn't shoplifting." If it were, nearly every female-and-fifty face
on TV would be behind bars. After all, most of us choose, um, some self-
improvement. Where is the unacceptable point on the aesthetic slope between
braces and face lifts? Aging gracefully does not mean that you have to age
grayly. So, you tell me the cut-off between hair color and collagen.

Nevertheless.   7

As a woman of a certain age—the age targeted by the hefty $53 million ad   8
campaign being launched by Allergan, the maker of Botox—every time someone I know, or watch, has some "work" done, I have a vague feeling of being
deserted. It's as if they'd left a threatened neighborhood, the endangered, natural species free range, and sided with the image-makers.

Remember back when Gloria Steinem turned 40? (If you do, it's probably   9
too late for Botox, anyway.) She said: "This is what 40 looks like." At that time it
was a statement that said proudly: We are not your grandmother's 40-year-old.

Of course, 40 never did necessarily look like Gloria. But what happens   10
when 50 is supposed to look like 40? Does that mean the whole standard of aging has changed? Do we think 60 should look like 50? Does, say, a 70-year-old
Barbara Walters actually change the future for older women on TV? Or is an
older woman only accepted if she doesn't look her age?

Chemical peels. Endoscopic lifts. Microfat injections. Eyelid lifts. Face lifts.   11
Botox marketed to women the way Viagra is to men (never mind). How long is
it before looking "your age" is regarded as a slatternly failure of effort? How
long before any woman who doesn't try one of the above is dismissed as someone who is "letting herself go"?

I have always loved the expression, "letting yourself go." Where do you go,   12
when you let yourself? To the recycle bin or to freedom? On Oscar night, in a
sea of nipped and tucked, siliconed and surgeried women, the only seamed
faces over 50 belonged to the likes of Judi Dench, Maggie Smith, and Helen Mir-
ren. They are all character actors. Is that where they let themselves go? Into
character?

In the past few years, I have found myself looking at older women as har-   13
bingers of the future. I'm looking for energy and confidence, and yes, attractive-
ness. Who do I want to be when I grow up? I am sure there are young women
searching for the same clues. But there's no way to find them on the Botox party
masks. This is the real symbolism of Botox. It eliminates lines temporarily by
paralyzing muscles. It offers an actual trade-off. You trade the ability, literally, to
express your emotions—furrow that brow, crinkle that eye—for a flawless appearance. In the search for approval from others, you hide what you are feeling.
Especially anger.

This seems to my cranky eye and creased eyebrow to be exactly the oppo-   14
site of my goal to become an outspoken, maybe even outrageous, laugh-out-
loud, nothing-left-to-lose old lady. Spare me the Botox. I plan to remain the kind
of character actor who wears her emotions, not on her sleeve or on her surgeon's bill, but on her face.

COMPREHENSION

1. What does Goodman think about Botox? What other forms of cosmetic surgery does she mention? What is her attitude toward them?
2. Why does the author mention Michael Jackson? Who are Greta Van Susteren and Gloria Steinem (see page 548), and how do they differ in their approach to beauty?
3. What, finally, is the author's attitude toward aging?

RHETORIC

1. A newspaper column places strict stylistic and formal demands on a writer.
   What journalistic elements do you find in this essay? What elements of style
   stand out?
2. Explain the tone of this essay. Where is the author serious? Where does she
   employ irony and satire? What is the overall effect?
3. What is the main idea of this essay? Which sentence serves as the thesis
   statement?

4. Analyze the essay as an argument and an attempt to persuade. What is the author's warrant, claim, and support? Does she engage in refutation? Why or why not?
5. What is the purpose and effect of the many questions that the author poses in this essay?
6. Where does Goodman use metaphors and other types of figurative language? What ideas do they convey?
7. How effective is the conclusion? Why?

WRITING

1. Have you had cosmetic surgery, or do you know someone who has? Write an essay telling about the procedure and the result.
2. Write an essay that compares and contrasts the ways that men and women view cosmetic surgery.
3. **WRITING AN ARGUMENT:** Compose an answer to Goodman in which you offer a humorous defense of cosmetic surgery.

✴ www.mhhe.com/ **mhreader**	For more information on Ellen Goodman, go to: **More Resources > Ch. 12 Health & Medicine**

✴

A New Look, an Old Battle

Anna Quindlen

Anna Quindlen (b. 1953) was born in Philadelphia and educated at Barnard College (BA, 1974). A journalist and novelist, she began her writing career as a reporter for the New York Post *and later moved on to* The New York Times, *where she was a syndicated columnist. Quindlen has written a number of books, including* Living Out Loud *(1986),* Object Lessons *(1991),* One True Thing *(1994),* Black and Blue *(1998),* Blessings *(2002), and* Loud and Clear *(2004). Quindlen received the Pulitzer Prize for Commentary in 1992. She is currently a columnist for* Newsweek *magazine. In this 2002 essay, Quindlen offers an argument about stem-cell research and the anti-abortion movement.*

Public personification has always been the struggle on both sides of the abor-   1
tion battle lines. That is why the people outside clinics on Saturday mornings carry signs with photographs of infants rather than of zygotes, why they wear lapel pins fashioned in the image of tiny feet and shout, "Don't kill your baby,"

rather than, more accurately, "Don't destroy your embryo." Those who support the legal right to an abortion have always been somewhat at a loss in the face of all this. From time to time women have come forward to speak about their decision to have an abortion, but when they are prominent, it seems a bit like grandstanding, and when they are not, it seems a terrible invasion of privacy when privacy is the point in the first place. Easier to marshal the act of presumptive ventriloquism practiced by the opponents, pretending to speak for those unborn unknown to them by circumstance or story.

But the battle of personification will assume a different and more sympathetic visage in the years to come. Perhaps the change in the weather was best illustrated when conservative Sen. Strom Thurmond invoked his own daughter to explain a position opposed by the anti-abortion forces. The senator's daughter has diabetes. The actor Michael J. Fox has Parkinson's disease. Christopher Reeve is in a wheelchair because of a spinal-cord injury, Ronald Reagan locked in his own devolving mind by Alzheimer's.* In the faces of the publicly and personally beloved lies enormous danger for the life-begins-at-conception lobby. 2

The catalytic issue is research on stem cells. These are versatile building blocks that may be coaxed into becoming any other cell type; they could therefore hold the key to endless mysteries of human biology, as well as someday help provide a cure for ailments as diverse as diabetes, Parkinson's, spinal-cord degeneration, and Alzheimer's. By some estimates, more than 100 million Americans have diseases that scientists suspect could be affected by research on stem cells. Scientists hope that the astonishing potential of this research will persuade the federal government to help fund it and allow the National Institutes of Health to help oversee it. This is not political, researchers insist. It is about science, not abortion. 3

And they are correct. Stem-cell research is typically done by using frozen embryos left over from in vitro fertilization. If these embryos were placed in the womb, they might eventually implant, become a fetus, then a child. Unused, they are the earliest undifferentiated collection of cells made by the joining of the egg and sperm, no larger than the period at the end of this sentence. One of the oft-used slogans of the anti-abortion movement is "abortion stops a beating heart." There is no heart in this preimplantation embryo, but there are stem cells that, in the hands of scientists, might lead to extraordinary work affecting everything from cancer to heart disease. 4

All of which leaves the anti-abortion movement trying desperately to hold its hard line, and failing. Judie Brown of the American Life League can refer to these embryos as "the tiniest person," and the National Right to Life organization can publish papers that refer to stem-cell research as the "destruction of life." But ordinary people with family members losing their mobility or their grasp on reality will be able to be more thoughtful and reasonable about the issues involved. 5

The anti-abortion activists know this, because they have already seen the defections. Some senators have abandoned them to support fetal-tissue re- 6 search, less promising than stem-cell work but still with significant potential for treating various ailments. Elected officials who had voted against abortion rights found themselves able to support procedures that used tissue from aborted fetuses; perhaps they were men who had fathers with heart disease, who had mothers with arthritis and whose hearts resonated with the possibilities for alleviating pain and prolonging life. Senator Thurmond was one, Senator McCain another. Former senator Connie Mack of Florida recently sent a letter to the president, who must decide the future role of the federal government in this area, describing himself "as a conservative pro-life now former member" of Congress, and adding that there "were those of us identified as such who supported embryonic stem-cell research."

When a recent test of fetal tissue in patients with Parkinson's had disastrous side effects, the National Right to Life Web site ran an almost gloating report: "horrific," "rips to shreds," "media cheerleaders," "defy description." The tone is a reflection of fear. It's the fear that the use of fetal tissue to produce cures for debilitating ailments might somehow launder the process of terminating a pregnancy, a positive result from what many people still see as a negative act. And it's the fear that thinking—really thinking—about the use of the earliest embryo for lifesaving research might bring a certain long-overdue relativism to discussions of abortion across the board.

The majority of Americans have always been able to apply that relativism to these issues. They are more likely to accept early abortions than later ones. They are more tolerant of a single abortion under exigent circumstances than multiple abortions. Some who disapprove of abortion in theory have discovered that they can accept it in fact if a daughter or a girlfriend is pregnant.

And some who believe that life begins at conception may look into the vacant eyes of an adored parent with Alzheimer's or picture a paralyzed child walking again, and take a closer look at what an embryo really is, at what stem-cell research really does, and then consider the true cost of a cure. That is what Senator Thurmond obviously did when he looked at his daughter and broke ranks with the true believers. It may be an oversimplification to say that real live loved ones trump the imagined unborn, that a cluster of undifferentiated cells due to be discarded anyway is a small price to pay for the health and welfare of millions. Or perhaps it is only a simple commonsensical truth.

COMPREHENSION

1. What is the "new look" and "old battle" mentioned by Quindlen in the title of her essay?
2. Why are anti-abortion activists opposed to stem-cell research? What is the author's response to them?
3. Explain the meaning of "public personification" (paragraph 1) and how it relates to the author's argument.

RHETORIC

1. Explain the argumentative strategies used by the author in the introductory paragraph. How do the phrases "public personification" and "presumptive ventriloquism" color Quindlen's argument?
2. What is the author's claim, and what tone does she adopt to advance it?
3. How does the author structure her argument?
4. Do the author's examples adequately support her view? Explain.
5. According to the author, what are the weaknesses in the opposition's arguments? Do you think that she fairly presents these arguments? How does she refute them?
6. What ethical and emotional appeals does the author make? How does the conclusion reinforce these appeals?

WRITING

1. Write an essay about the ways in which "public personification" obscures the debate over a specific medical or health issue.
2. Write an essay that takes an objective approach to the pros and cons of the stem-cell debate.
3. **WRITING AN ARGUMENT:** Write an essay supporting or opposing a woman's right to an abortion.

www.mhhe.com/ **mhreader**   For more information on Anna Quindlen, go to: **More Resources > Ch. 12 Health & Medicine**

Macho

Perri Klass

Perri Klass (b. 1958) was born in Trinidad of American parents. She attended Harvard University, receiving a BA (1979) and MD (1986). A pediatrician and mother of two, she is the author of many books, including two novels, Recombinations *(1985) and* Other Women's Children *(1990); two collections of short stories,* I Am Having an Adventure *(1986) and* Love and Modern Medicine *(2001); and a collection of essays,* A Not Entirely Benign Procedure: Four Years as a Medical Student *(1987). In the following essay from the latter book, she demonstrates that the medical profession and personality types have very strong ties.*

Purely by coincidence, our team has four women and one man. The two interns 1
and the two medical students are female, and the resident, who leads the team,
is male. We are clattering up the stairs one morning in approved hospital fashion, conveying by our purposeful demeanor: out of the way, doctors coming,
decisions to make, lives to save. (In fact, what we are actually trying to do is to
rush through morning rounds in time to get to breakfast before the cafeteria
stops serving hot food, but never mind that now.) We barrel through the door
into the intensive care unit, and some other resident, standing by, announces,
"Here comes The A-Team." Immediately our resident swings around to respond to some undertone he has detected:

"Are you saying my team is weak? Huh? You saying my team is weak?" 2

We continue on our rounds, the resident occasionally prompting one of the 3
interns, "You should be pushing me out of the way, you know. Go on, push me
out of the way." He means that since she is on call that day, she should be first
through every door, first to lay hands on every patient.

My fellow medical student and I trail along in the rear, a position that accu- 4
rately reflects our place in the hierarchy and also my energy level (this is, after
all, 7:00 A.M.) She whispers to me, "Straighten your back! Suck in your stomach!
This is war!" And as we start to clatter down the stairs to breakfast, our morning mission successfully accomplished, she and I are both singing under our
breaths, "Macho macho doc, I wanna be a macho doc. . . ."

Macho in medicine can mean a number of things. Everyone knows it's out 5
there as a style, either an ideal or an object of ridicule. You hear echoes of it in
the highest praise one can receive in the hospital, "Strong work," which may be
said to an intern who got a very sick patient through the night or to a medical
student who successfully fielded some obscure questions on rounds. And the
all-purpose term of disparagement is "Weak." They're being really weak down
in the emergency room tonight, admitting people who could just as well be sent
home. Dr. So-and-so is being weak with that patient—why doesn't he just tell
him he *has* to have the surgery? You were pretty weak this morning when they
were asking you about rheumatic heart disease—better read up on it.

Macho can refer to your willingness to get tough with your patients, to keep 6
them from pushing you around. It can refer to your eagerness to do invasive
procedures—"The hell with radiology, I wanna go for the biopsy." Talk like that
and they'll call you a cowboy, and generally mean it as a compliment. Macho
can mean territoriality: certain doctors resent calling in expert consultations
and, when they finally have to, await the recommendations with truculent
eagerness to disregard them. "These are *our* patients and *we* make all the
decisions," I heard over and over from one resident I worked under. The
essence of macho, any kind of macho, after all, is that life is a perpetual contest.
You must not let others intrude on your stamping grounds. You must not let
anyone tell you what to do. And of course, the most basic macho fear is the fear
of being laughed at; whatever you do, you must not let anyone mock you—or
our team.

A medical student once said to me when I teased him about not being able 7
to work the Addressograph machine after six weeks in the hospital, "That's secretarial work. I can draw a blood gas blindfolded, from thirty feet away!"

Life in the hospital is full of opportunities to prove yourself, if you want to 8
look at it that way. "I want you guys to be able to get blood from a stone," announced our new resident on his first day as our leader. The "guys," the other
female medical student and I, must have looked a little dubious, because he
continued, "Okay, it may mean the patient gets stuck a few extra times, but I
don't want you giving up just because of that." And sure enough, when I came
to tell him that I had stuck one particular woman six times without success, and
could he please come show me where he thought a decent vein might be, he
sent me back in to try her ankles. "Blood from a stone!" he called after me, and
when I finally got a tube of this unfortunate woman's blood, he patted me on
the back and said, "Strong work."

If we are at war, then who is the enemy? Rightly the enemy is disease, and 9
even if that is not your favorite metaphor, it is a rather common way to think of
medicine: we are combating these deadly processes for the bodies of our patients. They become battlefields, lying there passively in bed while the evil
armies of pathology and the resplendent forces of modern medicine fight it out.
Still, there are very good doctors who seem to think that way, who take disease
as a personal enemy and battle it with fury and dedication. The real problem
arises because all too often the patient comes to personify the disease, and
somehow the patient becomes the enemy.

We don't say, or think, "Mrs. Hawthorne's cancer is making her sicker." We 10
say, "Mrs. Hawthorne's crumping on me," and Mrs. Hawthorne represents the
challenge we cannot meet, the disease we cannot cure. And instead of hating
her cancer, it's not hard to start hating Mrs. Hawthorne—especially if she has an
irritating personality, and most especially of all if she somehow seems to be
blaming us. That is, if every day the doctor sees the challenge again in the patient's eye, hears it in the patient's voice: "You can't do anything for me, can
you, despite all the tests and all the medicines?"

The patient may want the doctor to continue fighting, may even take re- 11
newed hope as new therapies are instituted, but the doctor, knowing them to be
essentially futile, may become angrier and angrier. When the disease has essentially won and the patient continues to present the challenge, the macho doctor
is left with no appropriate response. He cannot sidestep the challenge by offering comfort rather than combat, because comfort is not in his repertoire. And
unable to do battle against the disease to any real effect, he may feel almost
ready to battle the patient.

I have been talking as if macho medicine is a male preserve, and to a large ex- 12
tent that's true. Certainly there are some female doctors who end up being fairly
macho and, much more important, many men who are not macho at all. Some
of the gentlest, most reasonable doctors I worked with were male, good teachers and superb healers. But there are also many macho docs, and certainly it is

pervasive as a style in the hospital. I don't believe that would be the case if the majority of doctors up to now had been female, and perhaps it will change over time as more women become doctors. The tradition of medical training is partly a tradition of hazing, boot camp, basic training. New buzzwords are now being muttered, like "nurturing" or "supportive," but there are many doctors riding the range out there to whom you wouldn't dare mutter any such words.

"Sup-por-tive," you can almost hear The Duke drawl as the doctor looks 13 down at the newest sissy in town. At which point you tuck your hypodermic needle back into its holster and march, on the double, back into that pesky varmint's room to let him know who's boss in this here hospital.

COMPREHENSION

1. What does Klass mean by the term *macho?* What actions constitute macho behavior?
2. What specific negative effects does Klass see as a result of doctors' macho behavior? Does she see any advantages?
3. Klass states that members of the medical profession have their own vernacular, such as "strong work," "cowboy," and "weak." What do they mean within the context of Klass's work environment? What do they say about the attitude of the doctors who use them?

RHETORIC

1. What is Klass's purpose in the essay? Where is the purpose best articulated? Explain.
2. Klass has chosen to quote some of her colleagues directly. What is the rhetorical purpose of using these quotations?
3. Why is the essay divided into two sections? What change in tone or mood occurs in the second section?
4. Klass uses irony in describing the world of her internship. For example, in paragraph 1, she states that the reason they are rushing through morning rounds is to get a hot breakfast before the cafeteria stops serving. What other examples of irony can you find in the essay? What is the overall purpose of Klass's irony?
5. Klass uses two references to popular culture: one in paragraph 4 where she parodies a popular song by using the lyrics "Macho macho doc" and the other, in paragraph 13, where she refers to "The Duke." What are the origins of these references, and what is their rhetorical purpose in the essay?
6. What seems to be Klass's attitude toward macho medical behavior? What is the relationship between her attitude and the tone of the essay?
7. Klass uses an extended metaphor in paragraph 9 by comparing the treatment of disease to waging war. Where else does she use this metaphor?

WRITING

1. Select a profession within medicine, psychology, or the health sciences, and find an attitude among members of that profession that can be used conceptually to explain how members of the profession perceive their roles.
2. Drawing upon personal experience, explain why women in the medical field might be gentler and more sympathetic than their male counterparts.
3. **WRITING AN ARGUMENT:** Write an essay in which you argue that medical practice has either changed for the better, changed for the worse, or remained the same since 1987, when Klass's essay was written. You may consult your parents and other older adults to assist you.

www.mhhe.com/
mhreader

For more information on Perri Klass, go to:
More Resources > Ch. 12 Health & Medicine

✳

The Embalming of Mr. Jones

Jessica Mitford

Jessica Mitford (1917–1996) was born in England and grew up with her brother and five sisters. She moved to the United States in 1939 and became a naturalized citizen in 1944. Her book The American Way of Death *(1963), a critical examination of the American funeral industry, became a best seller. Other books included* The Trial of Doctor Spock *(1969),* Kind and Usual Punishment: The Prison Business *(1973),* A Fine Old Conflict *(1977), an account of her experiences as a member of the U.S. Communist Party, and* The American Way of Birth *(1992). In the following essay, Mitford explains in ironic and painstaking terms the processes of embalming and restoring a cadaver, using meticulous detail to support her thesis.*

Embalming is indeed a most extraordinary procedure, and one must wonder at   1
the docility of Americans who each year pay hundreds of millions of dollars for
its perpetuation, blissfully ignorant of what it is all about, what is done, how it
is done. Not one in ten thousand has any idea of what actually takes place.
Books on the subject are extremely hard to come by. They are not to be found in
most libraries or bookshops.

In an era when huge television audiences watch surgical operations in the   2
comfort of their living rooms, when, thanks to the animated cartoon, the geography of the digestive system has become familiar territory even to the nursery
school set, in a land where the satisfaction of curiosity about almost all matters

is a national pastime, the secrecy surrounding embalming can, surely, hardly be attributed to the inherent gruesomeness of the subject. Custom in this regard has within this century suffered a complete reversal. In the early days of American embalming, when it was performed in the home of the deceased, it was almost mandatory for some relative to stay by the embalmer's side and witness the procedure. Today, family members who might wish to be in attendance would certainly be dissuaded by the funeral director. All others, except apprentices, are excluded by law from the preparation room.

A close look at what does actually take place may explain in large measure 3 the undertaker's intractable reticence concerning a procedure that has become his major *raison d'être*. Is it possible he fears that public information about embalming might lead patrons to wonder if they really want this service? If the funeral men are loath to discuss the subject outside the trade, the reader may, understandably, be equally loath to go on reading at this point. For those who have the stomach for it, let us part the formaldehyde curtain. . . .

The body is first laid out in the undertaker's morgue—or rather, Mr. Jones 4 is reposing in the preparation room—to be readied to bid the world farewell.

The preparation room in any of the better funeral establishments has the 5 tiled and sterile look of a surgery, and indeed the embalmer-restorative artist who does his chores there is beginning to adopt the term "dermasurgeon" (appropriately corrupted by some mortician-writers as "demisurgeon") to describe his calling. His equipment, consisting of scalpels, scissors, augers, forceps, clamps, needles, pumps, tubes, bowls and basins, is crudely imitative of the surgeon's as is his technique, acquired in a nine- or twelve-month post-high-school course in an embalming school. He is supplied by an advanced chemical industry with a bewildering array of fluids, sprays, pastes, oils, powders, creams, to fix or soften tissue, shrink or distend it as needed, dry it here, restore the moisture there. There are cosmetics, waxes and paints to fill and cover features, even plaster of Paris to replace entire limbs. There are ingenious aids to prop and stabilize the cadaver: a Vari-Pose Head Rest, the Edwards Arm and Hand Positioner, the Repose Block (to support the shoulders during the embalming), and the Throop Foot Positioner, which resembles an old-fashioned stocks.

Mr. John H. Eckels, president of the Eckels College of Mortuary Science, 6 thus describes the first part of the embalming procedure: "In the hands of a skilled practitioner, this work may be done in a comparatively short time and without mutilating the body other than by slight incision—so slight that it scarcely would cause serious inconvenience if made upon a living person. It is necessary to remove all the blood, and doing this not only helps in the disinfecting, but removes the principal cause of disfigurements due to discoloration."

Another textbook discusses the all-important time element: "The earlier this 7 is done, the better, for every hour that elapses between death and embalming will add to the problems and complications encountered. . . ." Just how soon should one get going on the embalming? The author tells us, "On the basis of such scanty information made available to this profession through its rudimentary and haphazard system of technical research, we must conclude that the best results are to be obtained if the subject is embalmed before life is completely extinct—that is, before cellular death has occurred. In the average case, this would mean within an hour after somatic death." For those who feel that there is something a little rudimentary, not to say haphazard, about this advice, a comforting thought is offered by another writer. Speaking of fears entertained in early days of premature burial, he points out, "One of the effects of embalming by chemical injection, however, has been to dispel fears of live burial." How true; once the blood is removed, chances of live burial are indeed remote.

To return to Mr. Jones, the blood is drained out through the veins and re- 8 placed by embalming fluid pumped in through the arteries. As noted in *The Principles and Practices of Embalming*, "every operator has a favorite injection and drainage point—a fact which becomes a handicap only if he fails or refuses to forsake his favorites when conditions demand it." Typical favorites are the carotid artery, femoral artery, jugular vein, subclavian vein. There are various choices of embalming fluid. If Flextone is used, it will produce a "mild, flexible rigidity. The skin retains a velvety softness, the tissues are rubbery and pliable. Ideal for women and children." It may be blended with B. and G. Products Company's Lyf-Lyk tint, which is guaranteed to reproduce "nature's own skin texture . . . the velvety appearance of living tissue." Suntone comes in three separate tints: Suntan; Special Cosmetic Tint, a pink shade "especially indicated for young female subjects"; and Regular Cosmetic Tint, moderately pink.

About three to six gallons of a dyed and perfumed solution of formalde- 9 hyde, glycerin, borax, phenol, alcohol and water is soon circulating through Mr. Jones, whose mouth has been sewn together with a "needle directed upward between the upper lip and gum and brought out through the left nostril," with the corners raised slightly "for a more pleasant expression." If he should be bucktoothed, his teeth are cleaned with Bon Ami and coated with colorless nail polish. His eyes, meanwhile, are closed with flesh-tinted eye caps and eye cement.

The next step is to have at Mr. Jones with a thing called a trocar. This is a 10 long, hollow needle attached to a tube. It is jabbed into the abdomen, poked around the entrails and chest cavity, the contents of which are pumped out and replaced with "cavity fluid." This is done, and the hole in the abdomen sewed up, Mr. Jones's face is heavily creamed (to protect the skin from burns which may be caused by leakage of the chemicals), and he is covered with a sheet and left unmolested for a while. But not for long—there is more, much more, in store for him. He has been embalmed, but not yet restored, and the best time to start restorative work is eight to ten hours after embalming, when the tissues have become firm and dry.

The object of all this attention to the corpse, it must be remembered, is to 11 make it presentable for viewing in an attitude of healthy repose. "Our customs require the presentation of our dead in the semblance of normality . . . unmarred by the ravages of illness, disease or mutilation," says Mr. J. Sheridan Mayer in his *Restorative Art*. This is rather a large order since few people die in the full bloom of health, unravaged by illness and unmarked by some disfigurement. The funeral industry is equal to the challenge: "In some cases the gruesome appearance of a mutilated or disease-ridden subject may be quite discouraging. The task of restoration may seem impossible and shake the confidence of the embalmer. This is the time for intestinal fortitude and determination. Once the formative work is begun and affected tissues are cleaned or removed, all doubts of success vanish. It is surprising and gratifying to discover the results which may be obtained."

The embalmer, having allowed an appropriate interval to elapse, returns to 12 the attack, but now he brings into play' the skill and equipment of sculptor and cosmetician. Is a hand missing? Casting one in plaster of Paris is a simple matter. "For replacement purposes, only a cast of the back of the hand is necessary; this is within the ability of the average operator and is quite adequate." If a lip or two, a nose or an ear should be missing, the embalmer has at hand a variety of restorative waxes with which to model replacements. Pores and skin texture are simulated by stippling with a little brush, and over this cosmetics are laid on. Head off? Decapitation cases are rather routinely handled. Ragged edges are trimmed, and head joined to torso with a series of splints, wires and sutures. It is a good idea to have a little something at the neck—a scarf or high collar—when time for viewing comes. Swollen mouth? Cut out tissue as needed from inside the lips. If too much is removed, the surface contour can easily be restored by padding with cotton. Swollen necks and cheeks are reduced by removing tissue through vertical incisions made down each side of the neck. "When the deceased is casketed, the pillow will hide the suture incisions . . . as an extra precaution against leakage, the suture may be painted with liquid sealer."

The opposite condition is more likely to be present itself—that of emacia- 13 tion. His hypodermic syringe now loaded with massage cream, the embalmer seeks out and fills the hollowed and sunken areas by injection. In this procedure the backs of the hands and fingers and the under-chin area should not be neglected.

Positioning the lips is a problem that recurrently challenges the ingenuity of 14 the embalmer. Closed too tightly, they tend to give a stern, even disapproving expression. Ideally, embalmers feel, the lips should give the impression of being ever so slightly parted, the upper lip protruding slightly for a more youthful appearance. This takes some engineering, however, as the lips tend to drift apart. Lip drift can sometimes be remedied by pushing one or two straight pins through the inner margin of the lower lip and then inserting them between the two front upper teeth. If Mr. Jones happens to have no teeth, the pins can just as easily be anchored in his Armstrong Face Former and Denture Replacer. Another method to maintain lip closure is to dislocate the lower jaw, which is then held in its new position by a wire run through holes which have been drilled through the upper jaws at the midline. As the French are fond of saying, il faut souffrir pour être belle.*

---

*ED'S. NOTE—It is necessary to suffer in order to be beautiful.

If Mr. Jones has died of jaundice, the embalming fluid will very likely turn   15
him green. Does this deter the embalmer? Not if he has intestinal fortitude.
Masking pastes and cosmetics are heavily laid on, burial garments and casket
interiors are color-correlated with particular care, and Jones is displayed beneath rose-colored lights. Friends will say, "How *well* he looks." Death by carbon monoxide, on the other hand, can be rather a good thing from the
embalmer's viewpoint: "One advantage is the fact that this type of discoloration
is an exaggerated form of a natural pink coloration." This is nice because the
healthy glow is already present and needs but little attention.

The patching and filling completed, Mr. Jones is now shaved, washed and   16
dressed. Cream-based cosmetic, available in pink, flesh, suntan, brunette and
blonde, is applied to his hands and face, his hair is shampooed and combed
(and, in the case of Mrs. Jones, set), his hands manicured. For the horny-handed
son of toil special care must be taken; cream should be applied to remove ingrained grime, and the nails cleaned. "If he were not in the habit of having them
manicured in life, trimming and shaping is advised for better appearance—
never questioned by kin."

Jones is now ready for casketing (this is the present participle of the verb "to   17
casket"). In this operation his right shoulder should be depressed slightly "to
turn the body a bit to the right and soften the appearance of lying flat on the
back." Positioning the hands is a matter of importance, and special rubber positioning blocks may be used. The hands should be cupped slightly for a more lifelike, relaxed appearance. Proper placement of the body requires a delicate sense
of balance. It should lie as high as possible in the casket, yet not so high that the
lid, when lowered, will hit the nose. On the other hand, we are cautioned, placing the body too low "creates the impression that the body is in a box."

Jones is next wheeled into the appointed slumber room where a few last   18
touches may be added—his favorite pipe placed in his hand or, if he was a great
reader, a book propped into position. (In the case of little Master Jones a Teddy
bear may be clutched.) Here he will hold open house for a few days, visiting
hours 10 A.M. to 9 P.M.

COMPREHENSION

1. What is Mitford's value judgment regarding the work of the contemporary
    American embalmer? By extension, what is her view of the American pub-
    lic that seeks out his or her services? Explain.
2. In paragraph 2, Mitford tells us that at one time a family member was pres-
    ent during the embalming procedure. Nowadays, however, embalmers
    would probably oppose nearly anyone who might wish to witness their
    craft. Why do you suppose this practice has changed? What does it suggest
    about changes in American culture?
3. Why does Mitford refer to the "generic" corpse as "Mr. Jones"?
4. Does this essay contain a thesis, or is it simply an example of "process
    analysis"? Explain your answer.

RHETORIC

1. At what point does the introduction to the essay end and its body paragraphs begin? How does Mitford signal this transition? How can this transition be considered as a metaphor for the embalming process?
2. Embalmers, embalming, and embalming products are referred to in a number of ways. For example, a practitioner is referred to as "dermasurgeon" or "demisurgeon" (paragraph 5). Embalming is referred to as "Mortuary Science" (paragraph 6). Products include such items as "Flextone," "Lyf-Lyk tint," "Suntan," and "Regular Cosmetic Tint" (paragraph 8). What is the irony behind the names and functions of these nomenclatures? What does the inclusion of these terms suggest about Mitford's attitude toward the embalming process?
3. Why does Mitford quote various "experts" in the field of embalming? What attitude does she seem to have toward these experts?
4. Mitford employs considerable irony in her essay. What is ironic about her explanation of the effects of two types of death (through "jaundice" and "carbon monoxide") in paragraph 15?
5. Mitford uses extensive detail to convey her attitude about the embalming of Mr. Jones. How does she convey her attitude through her tone, without directly stating it?
6. What is the ironic purpose for using a French expression at the conclusion of paragraph 14?
7. Define the following terms employed by the author: *perpetuation* (paragraph 1), *dissuaded* (paragraph 2), *pliable* (paragraph 8), *unravaged* (paragraph 11), and *stippling* (paragraph 12).
8. In paragraph 17, Mitford describes several methods embalmers use to position the body in the casket. What are these methods, and why does Mitford present them as ironic?

WRITING

1. Write an explanation of a medical process you personally find disgusting or delightful that seems to reflect aspects of American culture. Make your attitude clear by your choice of words.
2. Interview a fellow student who is from another culture about funeral practices and attitudes toward dying in his or her country. Compare them to those you have observed in America.
3. **WRITING AN ARGUMENT:** In the role of a loved one of someone recently deceased, write a letter to Mitford defending your decision to have the deceased embalmed for public viewing.

✳

The Terrifying Normalcy of AIDS

Stephen Jay Gould

Stephen Jay Gould (1941–2002), an acclaimed contemporary science writer, taught biology, geology, and the history of science at Harvard University, where he was Alexander Agassiz Professor of Zoology. Born in New York City, he was educated at Antioch College (BA, 1963) and Columbia University (PhD, 1967). He wrote a monthly column, "This View of Life," for Natural History *and was the author of* Ever Since Darwin *(1977),* Ontogeny and Phylogeny *(1977),* The Panda's Thumb *(1980),* Wonderful Life *(1989),* Bully for Brontosaurus *(1991),* The Structure of Evolutionary Theory *(2002), and other books. In this 1987 essay, Gould explains in clear, precise language why AIDS is a "natural phenomenon" and warns against viewing it in moral terms.*

Disney's Epcot Center in Orlando, Fla., is a technological tour de force and a con-   1
ceptual desert. In this permanent World's Fair, American industrial giants have built their versions of an unblemished future. These masterful entertainments convey but one message, brilliantly packaged and relentlessly expressed: progress through technology is the solution to all human problems. G.E. proclaims from Horizons: "If we can dream it, we can do it." A.T.&T. speaks from on high within its giant golf ball: We are now "unbounded by space and time." United Technologies bubbles from the depths of Living Seas: "With the help of modern technology, we feel there's really no limit to what can be accomplished."

Yet several of these exhibits at the Experimental Prototype Community of   2
Tomorrow, all predating last year's space disaster, belie their stated message from within by using the launch of the shuttle as a visual metaphor for technological triumph. The Challenger disaster may represent a general malaise, but it remains an incident. The AIDS pandemic, an issue that may rank with nuclear weaponry as the greatest danger of our era, provides a more striking proof that mind and technology are not omnipotent and that we have not canceled our bond to nature.

In 1984, John Platt, a biophysicist who taught at the University of Chicago   3
for many years, wrote a short paper for private circulation. At a time when most of us were either ignoring AIDS, or viewing it as a contained and peculiar affliction of homosexual men, Platt recognized that the limited data on the origin of AIDS and its spread in America suggested a more frightening prospect: we are all susceptible to AIDS, and the disease has been spreading in a simple exponential manner.

Exponential growth is a geometric increase. Remember the old kiddy prob-   4
lem: if you place a penny on square one of a checkerboard and double the num-

ber of coins on each subsequent square—2, 4, 8, 16, 32 . . .—how big is the stack by the sixty-fourth square? The answer: about as high as the universe is wide. Nothing in the external environment inhibits this increase, thus giving to exponential processes their relentless character. In the real, noninfinite world, of course, some limit will eventually arise, and the process slows down, reaches a steady state, or destroys the entire system: the stack of pennies falls over, the bacterial cells exhaust their supply of nutrients.

Platt noticed that data for the initial spread of AIDS fell right on an exponential curve. He then followed the simplest possible procedure of extrapolating the curve unabated into the 1990's. Most of us were incredulous, accusing Platt of the mathematical gamesmanship that scientists call "curve fitting." After all, aren't exponential models unrealistic? Surely we are not all susceptible to AIDS. Is it not spread only by odd practices to odd people? Will it not, therefore, quickly run its short course within a confined group?

Well, hello 1987—worldwide data still match Platt's extrapolated curve. This will not, of course, go on forever. AIDS has probably already saturated the African areas where it probably originated, and where the sex ratio of afflicted people is 1-to-1, male-female. But AIDS still has far to spread, and may be moving exponentially, through the rest of the world. We have learned enough about the cause of AIDS to slow its spread, if we can make rapid and fundamental changes in our handling of that most powerful part of human biology—our own sexuality. But medicine, as yet, has nothing to offer as a cure and precious little even for palliation.

This exponential spread of AIDS not only illuminates its, and our, biology, but also underscores the tragedy of our moralistic misperception. Exponential processes have a definite time and place of origin, an initial point of "inoculation"—in this case, Africa. We didn't notice the spread at first. In a population of billions, we pay little attention when one increases to two, or eight to sixteen, but when one million becomes two million, we panic, even though the *rate* of doubling has not increased.

The infection has to start somewhere, and its initial locus may be little more than an accident of circumstance. For a while, it remains confined to those in close contact with the primary source, but only by accident of proximity, not by intrinsic susceptibility. Eventually, given the power and lability of human sexuality, it spreads outside the initial group and into the general population. And now AIDS has begun its march through our own heterosexual community.

What a tragedy that our moral stupidity caused us to lose precious time, the greatest enemy in fighting an exponential spread, by down-playing the danger because we thought that AIDS was a disease of three irregular groups of minorities: minorities of life style (needle users), of sexual preference (homosexuals) and of color (Haitians). If AIDS had first been imported from Africa into a Park Avenue apartment, we would not have dithered as the exponential march began.

The message of Orlando—the inevitability of technological solutions—is 10
wrong, and we need to understand why.

Our species has not won its independence from nature, and we cannot do 11
all that we can dream. Or at least we cannot do it at the rate required to avoid
tragedy, for we are not unbounded from time. Viral diseases are preventable in
principle, and I suspect that an AIDS vaccine will one day be produced. But
how will this discovery avail us if it takes until the millennium, and by then
AIDS has fully run its exponential course and saturated our population, killing
a substantial percentage of the human race? A fight against an exponential enemy is primarily a race against time.

We must also grasp the perspective of ecology and evolutionary biology 12
and recognize, once we reinsert ourselves properly into nature, that AIDS represents the ordinary workings of biology, not an irrational or diabolical plague
with a moral meaning. Disease, including epidemic spread, is a natural phenomenon, part of human history from the beginning. An entire subdiscipline of
my profession, paleopathology, studies the evidence of ancient diseases preserved in the fossil remains of organisms. Human history has been marked by
episodic plagues. More native peoples died of imported disease than ever fell
before the gun during the era of colonial expansion. Our memories are short,
and we have had a respite, really, only since the influenza pandemic at the end
of World War I, but AIDS must be viewed as a virulent expression of an ordinary natural phenomenon.

I do not say this to foster either comfort or complacency. The evolutionary 13
perspective is correct, but utterly inappropriate for our human scale. Yes, AIDS
is a natural phenomenon, one of a recurring class of pandemic diseases. Yes,
AIDS may run through the entire population, and may carry off a quarter or
more of us. Yes, it may make no *biological* difference to Homo sapiens in the long
run: there will still be plenty of us left and we can start again. Evolution cares as
little for its agents—organisms struggling for reproductive success—as physics
cares for individual atoms of hydrogen in the sun. But we care. These atoms are
our neighbors, our lovers, our children and ourselves. AIDS is both a natural
phenomenon and, potentially, the greatest natural tragedy in human history.

The cardboard message of Epcot fosters the wrong attitudes: we must both rein- 14
sert ourselves into nature and view AIDS as a natural phenomenon in order to
fight properly. If we stand above nature and if technology is all-powerful, then
AIDS is a horrifying anomaly that must be trying to tell us something. If so, we
can adopt one of two attitudes, each potentially fatal. We can either become
complacent, because we believe the message of Epcot and assume that medicine
will soon generate a cure, or we can panic in confusion and seek a scapegoat for
something so irregular that it must have been visited upon us to teach us a
moral lesson.

But AIDS is not irregular. It is part of nature. So are we. This should galva- 15
nize us and give us hope, not prompt the worst of all responses: a kind of "new-

age" negativism that equates natural with what we must accept and cannot, or even should not, change. When we view AIDS as natural, and when we recognize both the exponential property of its spread and the accidental character of its point of entry into America, we can break through our destructive tendencies to blame others and to free ourselves of concern.

If AIDS is natural, then there is no message in its spread. But by all that science has learned and all that rationality proclaims, AIDS works by a *mechanism*—and we can discover it. Victory is not ordained by any principle of progress, or any slogan of technology, so we shall have to fight like hell, and be watchful. There is no message, but there is a mechanism. 16

COMPREHENSION

1. What does Gould mean when he defines AIDS as a "natural phenomenon" (paragraph 12)? How does the title support this definition?
2. What does Gould mean by "our moral stupidity" in paragraph 9?
3. What connection does Gould make between our reaction to the AIDS crisis and our alienation from nature?

RHETORIC

1. What is Gould's main idea? Where in the essay is it stated?
2. What is the purpose of paragraphs 1 and 2? How do they contribute to Gould's argument? How do they help establish the tone of the essay? What is the tone? What is the importance of Epcot Center to Gould's thesis?
3. Gould uses scientific terminology in his essay. Define the words *exponential* in paragraph 3 and *pandemic* and *phenomenon* in paragraph 13. Is this essay intended for a specialized audience? Justify your response.
4. Trace the progression of ideas in paragraphs 2, 3, 4, and 5. What transitions does Gould employ?
5. Does Gould use rhetorical strategies besides argument in his essay? Cite evidence of this varied rhetorical approach.
6. Explain the final sentence in Gould's conclusion. What is its relation to the paragraph as a whole?

WRITING

1. Gould states that we must "reinsert ourselves into nature" (paragraph 14). What does he mean by this? How would this affect the way in which we deal with disease and death in our society? Explore this issue in a brief essay.
2. Write an extended definition of HIV/AIDS, attempting to avoid moralizing about the subject. Conduct research if necessary.

3. **WRITING AN ARGUMENT:** Write an essay in which you expand on Gould's belief that our moral stupidity has not only hindered society's recognition of the AIDS threat but continues to impede AIDS research and treatment.

| ✸ www.mhhe.com/ **mhreader** | For more information on Stephen Jay Gould, go to: **More Resources > Ch. 12 Health & Medicine** |

✸

Sarcophagus

Richard Selzer

Richard Selzer (b. 1928) is a professional surgeon who started writing for several hours each night after he had already established a successful medical career. He was born in Troy, New York, and received degrees from Union College (BS, 1948) and Albany Medical College (MD, 1953). His first book of essays, Mortal Essays: Notes on the Art of Surgery *(1974), established him as a prominent essayist specializing in the world of medicine and surgery. Selzer employs his elegant prose style in describing the often tragic, unpleasant, and painful world of medical patients. He is a contributor to popular magazines, and his essays have been collected in several books, among them* Confessions of a Knife *(1979),* Letters to a Young Doctor *(1982),* Taking the World in for Repairs *(1997), and* The Exact Location of the Soul *(2001). He has also contributed to a number of anthologies and written a book of stories,* Imagine a Woman *(1997). The following essay demonstrates Selzer's experience and expertise as a surgeon as well as his unique ability to describe the world of medicine in poetic and graceful terms.*

We are six who labor here in the night. No . . . seven! For the man horizontal upon the table strives as well. But we do not acknowledge his struggle. It is our own that preoccupies us.

I am the surgeon.

David is the anesthesiologist. You will see how kind, how soft he is. Each patient is, for him, a preparation respectfully controlled. Blood pressure, pulse, heartbeat, flow of urine, loss of blood, temperature, whatever is measurable, David measures. And he is a titrator, adding a little gas, drug, oxygen, fluid, blood in order to maintain the dynamic equilibrium that is the only state compatible with life. He is in the very center of the battle, yet he is one step removed; he has not known the patient before this time, nor will he deal with the next of kin. But for him, the occasion is no less momentous.

Heriberto Paz is an assistant resident in surgery. He is deft, tiny, mercurial. I   4
have known him for three years. One day he will be the best surgeon in Mexico.

Evelyn, the scrub nurse, is a young Irish woman. For seven years we have   5
worked together. Shortly after her immigration, she led her young husband into
my office to show me a lump on his neck. One year ago he died of Hodgkin's
disease. For the last two years of his life, he was paralyzed from the waist down.
Evelyn has one child, a boy named Liam.

Brenda is a black woman of forty-five. She is the circulating nurse, who will   6
conduct the affairs of this room, serving our table, adjusting the lights, counting
the sponges, ministering to us from the unsterile world.

Roy is a medical student who is beginning his surgical clerkship. He has been   7
assigned to me for the next six weeks. This is his first day, his first operation.

David is inducing anesthesia. In cases where the stomach is not empty   8
through fasting, the tube is passed into the windpipe while the patient is awake.
Such an "awake" intubation is called crashing. It is done to avoid vomiting and
the aspiration of stomach contents into the lungs while the muscles that control
coughing are paralyzed.

We stand around the table. To receive a tube in the windpipe while fully   9
awake is a terrifying thing.

"Open your mouth wide," David says to the man. The man's mouth opens   10
slowly to its fullest, as though to shriek. But instead, he yawns. We smile down
at him behind our masks.

"OK. Open again. Real wide."   11

David sprays the throat of the man with a local anesthetic. He does this   12
three times. Then, into the man's mouth. David inserts a metal tongue depres-
sor which bears a light at the tip. It is called a laryngoscope. It is to light up the
throat, reveal the glottic chink through which the tube must be shoved. All this
while, the man holds his mouth agape, submitting to the hard pressure of the
laryngoscope. But suddenly, he cannot submit. The man on the table gags,
struggles to free himself, to spit out the instrument. In his frenzy his lip is
pinched by the metal blade.

There is little blood.   13

"Suction," says David.   14

Secretions at the back of the throat obscure the view. David suctions them   15
away with a plastic catheter.

"Open," commands David. More gagging. Another pass with the scope.   16
Another thrust with the tube. Violent coughing informs us that the tube is in the
right place. It has entered the windpipe. Quickly the balloon is inflated to snug
it against the wall of the trachea. A bolus of Pentothal is injected into a vein in
the man's arm. It takes fifteen seconds for the drug to travel from his arm to his
heart, then on to his brain. I count them. In fifteen seconds, the coughing stops,
the man's body relaxes. He is asleep.

"All set?" I ask David.   17
"Go ahead," he nods.   18

A long incision. You do not know how much room you will need. This part 19
of the operation is swift, tidy. Fat . . . muscle . . . fascia . . . the peritoneum is
snapped open and a giant shining eggplant presents itself. It is the stomach,
black from the blood it contains and that threatens to burst it. We must open
that stomach, evacuate its contents, explore.

Silk sutures are placed in the wall of the stomach as guidelines between 20
which the incision will be made. They are like the pitons of a mountaineer. I cut
again. No sooner is the cavity of the stomach achieved, than a columnar geyser
of blood stands from the small opening I have made. Quickly, I slice open the
whole front of the stomach. We scoop out handfuls of clot, great black gelati-
nous masses that shimmy from the drapes to rest against our own bellies as
though, having been evicted from one body, they must find another in which to
dwell. Now and then we step back to let them slither to the floor. They are under our feet. We slip in them. "Jesus," I say. "He is bleeding all over North
America." Now my hand is inside the stomach, feeling, pressing. There! A tumor spreads across the back wall of this stomach. A great hard craterous plain,
the dreaded linitis plastica (leather bottle) that is not content with seizing one
area, but infiltrates between the layers until the entire organ is stiff with cancer.
It is that, of course, which is bleeding. I stuff wads of gauze against the tumor. I
press my fist against the mass of cloth. The blood slows. I press harder. The
bleeding stops.

A quick glance at Roy. His gown and gloves, even his mask, are sprinkled 21
with blood. Now is he dipped; and I, his baptist.

David has opened a second line into the man's veins. He is pumping blood 22
into both tubings.

"Where do we stand?" I ask him. 23
"Still behind. Three units." He checks the blood pressure. 24
"Low, but coming up," he says. 25
"Shall I wait 'til you catch up?" 26
"No. Go ahead. I'll keep pumping." 27
I try to remove my fist from the stomach, but as soon as I do, there is a fresh 28
river of blood.
"More light," I say. "I need more light." 29
Brenda stands on a platform behind me. She adjusts the lamps. 30
"More light," I say, like a man going blind. 31
"That's it," she says. "There is no more light." 32
"We'll go around from the outside," I say. Heriberto nods agreement. "Free 33
up the greater curvature first, then the lesser, lift the stomach up and get some
control from behind."

I must work with one hand. The other continues as the compressor. It is the 34
tiredest hand of my life. One hand, then, inside the stomach, while the other
creeps behind. Between them . . . a ridge of tumor. The left hand fumbles,
gropes toward its mate. They swim together. I lift the stomach forward to find
that *nothing* separates my hands from each other. The wall of the stomach has
been eaten through by the tumor. One finger enters a large tubular structure. It

is the aorta. The incision in the stomach has released the tamponade of blood and brought us to this rocky place.

"Curved aortic clamp." 35

A blind grab with the clamp, high up at the diaphragm. The bleeding slack- 36 ens, dwindles. I release the pressure warily. A moment later there is a great bang of blood. The clamp has bitten through the cancerous aorta.

"Zero silk on a big Mayo needle." 37

I throw the heavy sutures, one after the other, into the pool of blood, hoping 38 to snag with my needle some bit of tissue to close over the rent in the aorta, to hold back the blood. There is no tissue. Each time, the needle pulls through the crumble of tumor. I stop. I repack the stomach. Now there is a buttress of packing both outside and inside the stomach. The bleeding is controlled. We wait. Slowly, something is gathering here, organizing. What had been vague and shapeless before is now declaring itself. All at once, I know what it is. There is nothing to do.

For what tool shall I ask? With what device fight off this bleeding? A knife? 39 There is nothing here to cut. Clamps? Where place the jaws of a hemostat? A scissors? Forceps? Nothing. The instrument does not exist that knows such deep red jugglery. Not all my clever picks, my rasp . . . A miner's lamp, I think, to cast a brave glow.

David has been pumping blood steadily. 40

"He is stable at the moment," he says. "'Where do we go from here?" 41

"No place. He's going to die. The minute I take away my pressure, he'll 42 bleed to death."

I try to think of possibilities, alternatives. I cannot; there are none. Minutes 43 pass. We listen to the cardiac monitor, the gassy piston of the anesthesia machine.

"More light!" I say. "Fix the light." 44

The light seems dim, aquarial, a dilute beam slanting through a green sea. 45 At such a fathom the fingers are clumsy. There is pressure. It is cold.

"Dave," I say, "stop the transfusion." I hear my voice coming as from a 46 great distance. "Stop it," I say again.

David and I look at each other, standing among the drenched rags, the 47 smeared equipment.

"I can't," he says. 48

"Then I will," I say, and with my free hand I reach across the boundary that 49 separates the sterile field from the outside world, and I close the clamp on the intravenous tubing. It is the act of an outlaw, someone who does not know right from wrong. But I know. I know that this is right to do.

"The oxygen." I say. "Turn it off." 50

"You want it turned off, you do it," he says. 51

"Hold this," I say to Heriberto, and I give over the packing to him. I step 52 back from the table, and go to the gas tanks.

"This one?" I have to ask him. 53

"Yes," David nods. 54

I turn it off. We stand there, waiting, listening to the beeping of the electro- 55
cardiograph. It remains even, regular, relentless. Minutes go by, and the sound
continues. The man will not die. At last, the intervals on the screen grow longer,
the shape of the curve changes, the rhythm grows wild, furious. The line
droops, flattens. The man is dead.

It is silent in the room. Now we are no longer a team, each with his circum- 56
scribed duties to perform. It is Evelyn who speaks first.

"It is a blessing," she says. I think of her husband's endless dying. 57

"No," says Brenda. "Better for the family if they have a few days . . . to get 58
used to the idea of it."

"But, look at all the pain he's been spared." 59

"Still, for the ones that are left, it's better to have a little time." 60

I listen to the two women murmuring, debating without rancor, speaking in 61
hushed tones of the newly dead as women have done for thousands of years.

"May I have the name of the operation?" It is Brenda, picking up her duties. 62
She is ready with pen and paper.

"Exploratory laparotomy. Attempt to suture malignant aorto-gastric 63
fistula."

"Is he pronounced?" 64

"What time is it?" 65

"Eleven-twenty." 66

"Shall I put that down?" 67

"Yes." 68

"Sew him up," I say to Heriberto. "I'll talk to the family." 69

To Roy I say, "You come with me." 70

Roy's face is speckled with blood. He seems to me a child with the measles. 71
What, in God's name, is he doing here?

From the doorway, I hear the voices of the others, resuming. 72

"Stitch," says Heriberto. 73

Roy and I go to change our bloody scrub suits. We put on long white coats. In 74
the elevator, we do not speak. For the duration of the ride to the floor where the
family is waiting, I am reasonable. I understand that in its cellular wisdom, the
body of this man had sought out the murderous function of my scalpel, and
stretched itself upon the table to receive the final stabbing. For this little time, I
know that it is not a murder committed but a mercy bestowed. Tonight's knife
is no assassin, but the kind scythe of time.

We enter the solarium. The family rises in unison. There are so many! How 75
ruthless the eyes of the next of kin.

"I am terribly sorry . . .," I begin. Their faces tighten, take guard. "There was 76
nothing we could do."

I tell them of the lesion, tell of how it began somewhere at the back of the 77
stomach; how, long ago, no one knows why, a cell lost the rhythm of the body,
fell out of step, sprang, furious, into rebellion. I tell of how the cell divided and

begat two of its kind, which begat four more and so on, until there was a whole race of lunatic cells, which is called cancer.

I tell of how the cancer spread until it had replaced the whole back of the 78 stomach, invading, chewing until it had broken into the main artery of the body. Then it was, I tell them, that the great artery poured its blood into the stomach. I tell of how I could not stop the bleeding, how my clamps bit through the crumbling tissue, how my stitches would not hold, how there was nothing to be done. All of this I tell.

A woman speaks. She has not heard my words, only caught the tone of my 79 voice.

"Do you mean he is dead?" 80

Should I say "passed away" instead of "died"? No. I cannot. 81

"Yes," I tell her, "he is dead." 82

Her question and my answer unleash their anguish. Roy and I stand among 83 the welter of bodies that tangle, grapple, rock, split apart to form new couplings. Their keening is exuberant, wild. It is more than I can stand. All at once, a young man slams his fist into the wall with great force.

"Son of a bitch!" he cries. 84

"Stop that!" I tell him sharply. Then, more softly, "Please try to control 85 yourself."

The other men crowd about him, patting, puffing, grunting. They are all fat, 86 with huge underslung bellies. Like their father's. A young woman in a nun's habit hugs each of the women in turn.

"Shit!" says one of the men. 87

The nun hears, turns away her face. Later, I see the man apologizing to her. 88

The women, too, are fat. One of them has a great pile of yellowish hair that 89 has been sprayed and rendered motionless. All at once, she begins to whine. A single note, coming louder and louder. I ask a nurse to bring tranquilizer pills. She does, and I hand them out, one to each, as though they were the wafers of communion. They urge the pills upon each other.

"Go on, Theresa, take it. Make her take one." 90

Roy and I are busy with cups of water. Gradually it grows quiet. One of the 91 men speaks.

"What's the next step?" 92

"Do you have an undertaker in mind?" 93

They look at each other, shrug. Someone mentions a name. The rest nod. 94

"Give the undertaker a call. Let him know. He'll take care of everything." 95

I turn to leave. 96

"Just a minute," one of the men calls. "Thanks, Doc. You did what you 97 could."

"Yes," I say. 98

Once again in the operating room. Blood is everywhere. There is a wild smell, 99 as though a fox had come and gone. The others, clotted about the table, work on. They are silent, ravaged.

"How did the family take it?"   100

"They were good, good."   101

Heriberto has finished reefing up the abdomen. The drapes are peeled back. 102
The man on the table seems more than just dead. He seems to have gone beyond that, into a state where expression is possible—reproach and scorn. I
study him. His baldness had advanced beyond the halfway mark. The remaining strands of hair had been gallantly dyed. They are, even now, neatly combed
and crenellated. A stripe of black moustache rides his upper lip. Once, he had
been spruce!

We all help lift the man from the table to the stretcher.   103

"On three," says David. "One . . . two . . . three."   104

And we heft him over, using the sheet as a sling. My hand brushes his 105
shoulder. It is cool. I shudder as though he were infested with lice. He has become something that I do not want to touch.

More questions from the women.   106

"Is a priest coming?"   107

"Does the family want to view him?"   108

"Yes. No. Don't bother me with these things."   109

"Come on," I say to Roy. We go to the locker room and sit together on a 110
bench. We light cigarettes.

"Well?" I ask him.   111

"When you were scooping out the clots, I thought I was going to swoon."   112

I pause over the word. It is too quaint, too genteel for this time. I feel, at that 113
moment, a great affection for him.

"But you fought it."   114

"Yes. I forced it back down. But, almost . . ."   115

"Good," I say. Who knows what I mean by it? I want him to know that I 116
count it for something.

"And you?" he asks me. The students are not shy these days.   117

"It was terrible, his refusal to die."   118

I want him to say that it was right to call it quits, that I did the best I could. 119
But he says nothing. We take off our scrub suits and go to the shower. There are
two stalls opposite each other. They are curtained. But we do not draw the curtains. We need to see each other's healthy bodies. I watch Roy turn his face directly upward into the blinding fall of water. His mouth is open to receive it. As
though it were milk flowing from the breasts of God. For me, too, this water is
like a well in a wilderness.

In the locker room, we dress in silence.   120

"Well, goodnight."   121

Awkwardly our words come out in unison.   122

"In the morning . . ."   123

"Yes, yes, later."   124

"Goodnight."   125

I watch him leave through the elevator door.   126

For the third time I go to that operating room. The others have long since fin- 127
ished and left. It is empty, dark. I turn on the great lamps above the table that
stands in the center of the room. The pediments of the table and the floor have
been scrubbed clean. There is no sign of the struggle. I close my eyes and see
again the great pale body of the man, like a white bullock, bled. The line of
stitches on his abdomen is a hieroglyph. Already, the events of this night are
hidden from me by these strange untranslatable markings.

COMPREHENSION

1. What has the author implied by choosing his title for the essay? How is the
   title reinforced by the final paragraph?
2. Based upon Selzer's description of the surgeon's work, to what other profession does the author draw analogies? Explain.
3. Why do the other members of the operating team refuse to tamper with the
   medical apparatus, even after being ordered to do so by the surgeon?

RHETORIC

1. What is the dramatic effect of telling the story in the present tense?
2. The author often eschews conventional sentence structure. For example in
   paragraph 16, he employs fragments: "Another pass with the scope. Another thrust with the tube." In paragraph 21, he uses odd syntax, "Now is
   he dipped; and I, his baptist." And some sentences are extremely short,
   such as paragraph 96: "I turn to leave." What is the cumulative effect of using such innovative sentence structure?
3. Why has the author divided his essay into six parts? What is the function of
   each part? How does the author create drama via the juxtaposition of one
   section to the next?
4. Dialogue is used frequently in the essay. What is the function and effect of
   the dialogue?
5. Selzer's imagery is often vivid and original. How do the following excerpts
   contribute to the tone of the essay: "I understand that in its cellular wisdom,
   the body of this man had sought out the murderous function of my
   scalpel"; "Tonight's knife is no assassin, but the kind scythe of time" (paragraph 74). "There is a wild smell, as though a fox had come and gone"
   (paragraph 99).
6. There are several references to religion in the essay. Locate them, and explain what their cumulative effect is on the tone of the essay.
7. Mystery plays a significant part in the author's mood; for example, the final
   sentence reads, "Already, the events of this night are hidden from me by
   these strange untranslatable markings." What other passages reflect this
   mood of mystery in the essay? How does this mood affect the description of
   the essay's events, which are supposedly based on science?

WRITING

1. We take for granted many things that are mysterious to us; for example, the act of reading, writing, and breathing. Write a descriptive essay in which you reflect upon some basic activity that you have never analyzed before.
2. Write a critique of Selzer's essay, entitling it "Religious Imagery in Selzer's 'Sarcophagus.'"
3. **WRITING AN ARGUMENT:** Explain why it is dangerous to treat physicians as supermen or superwomen, and why it is especially damaging to think that surgeons are gods or goddesses.

| www.mhhe.com/ **mhreader** | For more information on Richard Selzer, go to: **More Resources > Ch. 12 Health & Medicine** |

The Globalization of Eating Disorders

Susan Bordo

Susan Bordo *(b. 1947) was born in Newark, New Jersey, and was educated at Carleton University (BA, 1972) and the State University of New York at Stony Brook (PhD, 1982). She is the Singletary Chair in the Humanities and a professor of English and Women's Studies at the University of Kentucky. A feminist philosopher and interdisciplinary scholar who focuses on Western culture's attitudes toward gender and the body, Bordo has written* The Flight to Objectivity: Essays on Cartesianism and Culture *(1987),* Unbearable Weight: Feminism, Western Culture, and the Body *(1993),* Twilight Zones: The Hidden Life of Cultural Images from Plato to O.J. *(1997), and* The Male Body: A New Look at Men in Public and in Private *(1999). In this selection, Bordo offers an overview of a new kind of epidemic, fueled by Western media images, that is affecting cultures around the world.*

The young girl stands in front of the mirror. Never fat to begin with, she's been   1
on a no-fat diet for a couple of weeks and has reached her goal weight: 115 lb.,
at 5'4—exactly what she should weigh, according to her doctor's chart. But in
her eyes she still looks dumpy. She can't shake her mind free of the "Lady
Marmelade" video from Moulin Rouge. Christina Aguilera, Pink, L'il Kim, and
Mya, each one perfect in her own way: every curve smooth and sleek, lean-sexy,
nothing to spare. Self-hatred and shame start to burn in the girl, and envy tears

at her stomach, enough to make her sick. She'll never look like them, no matter how much weight she loses. Look at that stomach of hers, see how it sticks out? Those thighs—they actually jiggle. Her butt is monstrous. She's fat, gross, a dough girl.

As you read the imaginary scenario above, whom did you picture standing in front of the mirror? If your images of girls with eating and body image problems have been shaped by *People* magazine and Lifetime movies, she's probably white, North American, and economically secure. A child whose parents have never had to worry about putting food on the family table. A girl with money to spare for fashion magazines and trendy clothing, probably college-bound. If you're familiar with the classic psychological literature on eating disorders, you may also have read that she's an extreme "perfectionist" with a hyper-demanding mother, and that she suffers from "body-image distortion syndrome" and other severe perceptual and cognitive problems that "normal" girls don't share. You probably don't picture her as Black, Asian, or Latina.

Read the description again, but this time imagine twenty-something Tenisha Williamson standing in front of the mirror. Tenisha is black, suffers from anorexia, and feels like a traitor to her race. "From an African-American standpoint," she writes, "we as a people are encouraged to embrace our big, voluptuous bodies. This makes me feel terrible because I don't want a big, voluptuous body! I don't ever want to be fat—ever, and I don't ever want to gain weight. I would rather die from starvation than gain a single pound."[1] Tenisha is no longer an anomaly. Eating and body image problems are now not only crossing racial and class lines, but gender lines. They have also become a global phenomenon.

Fiji is a striking example. Because of their remote location, the Fiji islands did not have access to television until 1995, when a single station was introduced. It broadcasts programs from the United States, Great Britain, and Australia. Until that time, Fiji had no reported cases of eating disorders, and a study conducted by anthropologist Anne Becker showed that most Fijian girls and women, no matter how large, were comfortable with their bodies. In 1998, just three years after the station began broadcasting, 11 percent of girls reported vomiting to control weight, and 62 percent of the girls surveyed reported dieting during the previous months.[2]

Becker was surprised by the change; she had thought that Fijian cultural traditions, which celebrate eating and favor voluptuous bodies, would "withstand" the influence of media images. Becker hadn't yet understood that we live in an empire of images, and that there are no protective borders.

In Central Africa, for example, traditional cultures still celebrate voluptuous women. In some regions, brides are sent to fattening farms, to be plumped and massaged into shape for their wedding night. In a country plagued by AIDS,

---

[1]From the Colours of Ana website (http://coloursofana.com//ss8.asp). [This and subsequent notes in the selection are the author's.]
[2]Reported in Nancy Snyderman, *The Girl in the Mirror* (New York: Hyperion, 2002), p. 84.

the skinny body has meant—as it used to among Italian Jewish, and Black Americans—poverty, sickness, death. "An African girl must have hips," says dress designer Frank Osodi, "We have hips. We have bums. We like flesh in Africa." For years, Nigeria sent its local version of beautiful to the Miss World Competition. The contestants did very poorly. Then a savvy entrepreneur went against local ideals and entered Agbani Darego, a light-skinned, hyper-skinny beauty. (He got his inspiration from M-Net, the South African network seen across Africa on satellite television, which broadcasts mostly American movies and television shows.) Agbani Darego won the Miss World Pageant, the first Black African to do so. Now, Nigerian teenagers fast and exercise, trying to become "lepa"—a popular slang phrase for the thin "it" girls that are all the rage. Said one: "People have realized that slim is beautiful."[3]

How can mere images be so powerful? For one thing, they are never "just pictures," as the fashion magazines continually maintain (disingenuously) in their own defense. They speak to young people not just about how to be beautiful but also about how to become what the dominant culture admires, values, rewards. They tell them how to be cool, "get it together," overcome their shame. To girls who have been abused they may offer a fantasy of control and invulnerability, immunity from pain and hurt. For racial and ethnic groups whose bodies have been deemed "foreign," earthy, and primitive, and considered unattractive by Anglo-Saxon norms, they may cast the lure of being accepted as "normal" by the dominant culture.   7

In today's world, it is through images—much more than parents, teachers, or clergy—that we are taught how to be. And it is images, too, that teach us how to see, that educate our vision in what's a defect and what is normal, that give us the models against which our own bodies and the bodies of others are measured. Perceptual pedagogy: "How To Interpret Your Body 101." It's become a global requirement.   8

I was intrigued, for example, when my articles on eating disorders began to be translated, over the past few years, into Japanese and Chinese. Among the members of audiences at my talks, Asian women had been among the most insistent that eating and body image weren't problems for their people, and indeed, my initial research showed that eating disorders were virtually unknown in Asia. But when, this year, a Korean translation of *Unbearable Weight* was published, I felt I needed to revisit the situation. I discovered multiple reports on dramatic increases in eating disorders in China, South Korea, and Japan. "As many Asian countries become Westernized and infused with the Western aesthetic of a tall, thin, lean body, a virtual tsunami of eating disorders has swamped Asian countries," writes Eunice Park in *Asian Week* magazine. Older people can still remember when it was very different. In China, for example, where revolutionary ideals once condemned any focus on appearance and there have been several disastrous famines, "little fatty" was a term of endearment   9

[3]Norimitsu Onishi, "Globalization of Beauty Makes Slimness Trendy," *The New York Times,* Oct. 3, 2002.

for children. Now, with fast food on every corner, childhood obesity is on the rise, and the cultural meaning of fat and thin has changed. "When I was young," says Li Xiaojing, who manages a fitness center in Beijing, "people admired and were even jealous of fat people since they thought they had a better life. . . . But now, most of us see a fat person and think 'He looks awful.'"[4]

Clearly, body insecurity can be exported, imported, and marketed—just like 10 any other profitable commodity. In this respect, what's happened with men and boys is illustrative. Ten years ago men tended, if anything, to see themselves as better looking than they (perhaps) actually were. And then (as I chronicle in detail in my book *The Male Body*) the menswear manufacturers, the diet industries, and the plastic surgeons "discovered" the male body. And now, young guys are looking in their mirrors, finding themselves soft and ill defined, no matter how muscular they are. Now they are developing the eating and body image disorders that we once thought only girls had. Now they are abusing steroids, measuring their own muscularity against the oiled and perfected images of professional athletes, body-builders, and *Men's Health* models. Now the industries in body-enhancement—cosmetic surgeons, manufacturers of anti-aging creams, spas and salons—are making huge bucks off men, too.

What is to be done? I have no easy answers. But I do know that we need to 11 acknowledge, finally and decisively, that we are dealing here with a cultural problem. If eating disorders were biochemical, as some claim, how can we account for their gradual "spread" across race, gender, and nationality? And with mass media culture increasingly providing the dominant "public education" in our children's lives—and those of children around the globe—how can we blame families? Families matter, of course, and so do racial and ethnic traditions. But families exist in cultural time and space—and so do racial groups. In the empire of images, no one lives in a bubble of self-generated "dysfunction" or permanent immunity. The sooner we recognize that—and start paying attention to the culture around us and what it is teaching our children—the sooner we can begin developing some strategies for change.

COMPREHENSION

1. How does the author define the "body-image distortion syndrome" (paragraph 2)?
2. Why have body image and weight problems become a global phenomenon? What is the main cause of this phenomenon?
3. How, according to the author, should we deal with the globalization of eating disorders?

---

[4]Reported in Elizabeth Rosenthal, "Beijing Journal: China's Chic Waistline: Convex to Concave," *The New York Times*, Dec. 9, 1999.

RHETORIC

1. How does the author establish herself as an authority on her subject? Do you think that she succeeds? Why or why not?
2. What is the writer's claim? Where does she place it, and why?
3. Bordo begins with an imaginary situation. Does this strategy enhance or detract from the validity of her argument? Justify your response.
4. The writer uses several rhetorical strategies to advance her argument. Identify places where she employs description, illustration, comparison and contrast, and causal analysis.
5. The writer has been praised for her readable or accessible style. Do you think that this essay is well written and thought provoking? Explain.
6. How does Bordo develop this selection as a problem-solution essay? Where does the solution appear, and how effective is its placement within the essay?

WRITING

1. Write a causal essay analyzing young Americans' fascination with body image and the consequences of this preoccupation.
2. Why are women in the United States and around the world more susceptible to eating disorders than men? Answer this question in an analytical essay.
3. **WRITING AN ARGUMENT:** Write an essay entitled "Body Images, Eating Disorders, and Cultural Imperialism." In this essay, argue for or against the proposition that American media are exporting potentially unhealthy images of the human body.

| ✴ www.mhhe.com/ **mhreader** | For more information on Susan Bordo, go to: **More Resources > Ch. 12 Health & Medicine** |

CONNECTIONS FOR CRITICAL THINKING

1. Research the current status of public health in the United States and what is being done to prevent major outbreaks of disease. Refer to at least three essays in this chapter to support your findings and thesis.
2. Compare approaches to women's health issues as discussed by several writers in this chapter.
3. Explore the World Health Organization Web site (www.who.int/en), and summarize what you find about global pandemics.
4. Compare and contrast the strategies that Glasser, Goodman, Quindlen, and Bordo use to develop their arguments.
5. Referring to any three essays in this chapter, analyze some of the major moral, ethical, and religious issues raised by current medical research.
6. Compare and contrast the treatment of death in the essays by Tuchman, Glasser, Selzer, and Mitford.
7. Examine the use of irony and satire in the essays by Goodman and Mitford.
8. Research the subject of AIDS, and connect your findings to the essays by Tuchman, Glasser, and Gould.
9. Working with three other class members, develop a PowerPoint presentation informing your college about a health issue of campus concern.
10. Compare and contrast two Web sites devoted to some aspect of medicine and health—for example, stem-cell research, abortion, funeral practices, cosmetic surgery, or dieting.

chapter *13*

Nature and the Environment
How Do We Relate to the Natural World?

We are at a point in the history of civilization where consciousness of our fragile relationship to nature and the environment is high. Even as you spend an hour reading a few of the essays in this chapter, it is estimated that we are losing 3,000 acres of rain forest around the world and four species of plants or animals. From pollution to the population explosion to the depletion of the ozone layer, we seem to be confronted with ecological catastrophe. Nevertheless, as Rachel Carson reminds us, we have "an obligation to endure," to survive potential natural catastrophe by understanding and managing our relationship with the natural world.

Ecology, or the study of nature and the environment, as many of the essayists in this chapter attest, involves us in the conservation of the earth. It moves us to suppress our rapacious destruction of the planet. Clearly, the biological stability of the planet is increasingly precarious. More plants, insects, birds, and animals became extinct in the 20th century than in any era since the Cretaceous catastrophe more than 65 million years ago that led to the extinction of the dinosaurs. Within this ecological context, writers like Carson become our literary conscience, reminding us of how easily natural processes can break down unless we insist on a degree of ecological economy.

Of course, any modification of human behavior in an effort to conserve nature is a complex matter. To save the spotted owl in the Pacific Northwest, we must sacrifice the jobs of people in the timber industry. To reduce pollution, we must forsake gas and oil for alternate energy sources that are costly to develop. To reduce the waste stream, we must shift from a consumption to a conservation society. The ecological debate is complicated, but it is clear that the preservation of the myriad life cycles on earth is crucial, for we, too, could become an endangered species.

The language of nature is as enigmatic as the sounds of dolphins and whales communicating with their respective species. Writers like Barry Lopez and Rachel Carson and the vision found in the letter of Chief Seattle help us decipher the language of our environment. These writers encourage us to converse with nature, learn from it, and even revere it. All of us are guests on this planet; the natural world is our host. If we do not protect the earth, how can we guarantee the survival of global civilization?

Previewing the Chapter

As you read the essays in this chapter and respond to them in discussion and writing, consider the following questions:

- According to the author, what should our relationship to the natural world be?
- What claims or arguments does the author make about the importance of nature? Do you agree or disagree with these claims and arguments?
- What specific ecological problem does the author investigate?
- How does the author think that nature influences human behavior?
- What cultural factors are involved in our approach to the environment?
- Is the writer optimistic, pessimistic, or neutral in the assessment of our ability to conserve nature?
- Do you find that the author is too idealistic or sentimental in the depiction of nature? Why?
- Based on the author's essay, how does he or she qualify as a nature writer?
- How have you been challenged or changed by the essays in this chapter?

Classic and Contemporary Images:

ARE WE DESTROYING OUR NATURAL WORLD?

Using a Critical Perspective   Imagine yourself to be part of each of the scenes depicted in these two illustrations. How do you feel, and why? Now examine the purpose of each image. What details do the artists emphasize to convey their statement about our relationship to the natural world? What images does each artist create to capture your attention and direct your viewing and thinking toward a specific, dominant impression?

The painters of the Hudson River School such as John Frederick Kensett (1816–1872) celebrated American landscapes in their art, painting breathtaking scenes in meticulous detail. In *Along the Hudson* (1852), the beauty of the river is unspoiled.

Vehicles travel on the 405 Freeway where it intersects with the 10 West Freeway, Thursday, September 23, 2004, in Los Angeles. California could take the lead in the international effort to reduce global warming if the state's Air Resources Board gives approval to a package of regulations that would cut vehicle emissions by as much as 25 percent.

Classic and Contemporary Essays
DO WE OWN NATURE?

The simple yet passionate reflections of Chief Seattle regarding the destruction of a worldview are complemented by the more scholarly and learned meditation written nearly a century and a half later by the esteemed naturalist and writer Barry Lopez. Chief Seattle mourns the death of a way of life, a way of thinking, and a way of being as he tragically accepts that the cultural world of his people is doomed to disappear with the encroachment of "civilization." The white man exploits nature, uses nature, and perhaps most radically of all, perceives himself as apart from nature. This is in profound contrast to the ways of Chief Seattle's people, who saw themselves in harmony with nature, or more specifically inseparable from it, as inseparable perhaps as from a part of their own bodies. Chief Seattle's address is simple. And so perhaps is his message, although one should not confuse simplicity with lack of profundity. Lopez, writing from the perspective of a 21st century academic and naturalist, informs the reader of humans' ongoing "feud" with nature—not only by reflecting on his own experience and perceptions—but by drawing on his "book" knowledge as well. He relates many relevant tales from other cultures and mythologies, and excavates the historical record when reflecting on the "stone horse" he finds hidden in the wilderness, a symbol of that union with nature that Chief Seattle spoke of so eloquently. Although they speak in very different levels of discourse and from different perspectives, it is evident that the awe of nature has not been demolished despite many deliberate and random attempts to extinguish it.

✳

Letter to President Pierce, 1855

Chief Seattle

Chief Seattle (1786–1866) was the leader of the Dewamish and other Pacific North-west tribes. The city of Seattle, Washington, bears his name. In 1854, Chief Seattle re-luctantly agreed to sell tribal lands to the United States government and move to the government-established reservations. The authenticity of the following speech has been challenged by many scholars. However, most specialists agree it contains the substance and perspective of Chief Seattle's attitude toward nature and the white race.

We know that the white man does not understand our ways. One portion of the 1
land is the same to him as the next, for he is a stranger who comes in the night
and takes from the land whatever he needs. The earth is not his brother, but his
enemy, and when he has conquered it, he moves on. He leaves his fathers'
graves, and his children's birthright is forgotten. The sight of your cities pains
the eyes of the red man. But perhaps it is because the red man is a savage and
does not understand.

There is no quiet place in the white man's cities. No place to hear the leaves 2
of spring or the rustle of insect's wings. But perhaps because I am a savage and
do not understand, the clatter only seems to insult the ears. The Indian prefers
the soft sound of the wind darting over the face of the pond, the smell of the
wind itself cleansed by a mid-day rain, or scented with the piñon pine. The air
is precious to the red man. For all things share the same breath—the beasts, the
trees, the man. Like a man dying for many days, he is numb to the stench.

What is man without the beasts? If all the beasts were gone, men would die 3
from great loneliness of spirit, for whatever happens to the beasts also happens
to man. All things are connected. Whatever befalls the earth befalls the sons of
the earth.

It matters little where we pass the rest of our days; they are not many. A few 4
more hours, a few more winters, and none of the children of the great tribes that
once lived on this earth, or that roamed in small bands in the woods, will be left
to mourn the graves of a people once as powerful and hopeful as yours.

The whites, too, shall pass—perhaps sooner than other tribes. Continue to 5
contaminate your bed, and you will one night suffocate in your own waste.
When the buffalo are all slaughtered, the wild horses all tamed, the secret corners of the forest heavy with the scent of many men, and the view of the ripe
hills blotted by talking wires, where is the thicket? Gone. Where is the eagle?
Gone. And what is it to say goodbye to the swift and the hunt, the end of living
and the beginning of survival? We might understand if we knew what it was
that the white man dreams, what he describes to his children on the long winter nights, what visions he burns into their minds, so they will wish for tomorrow. But we are savages. The white man's dreams are hidden from us.

COMPREHENSION

1. What does Chief Seattle suggest is the major difference between the white man's relationship with nature and that of the red man?
2. Chief Seattle claims that perhaps the red man would understand the white man better if he understood better the "dreams" and "visions" of the white man. What does Chief Seattle suggest by these terms?
3. Chief Seattle refers to Native Americans as "savages." Why?

RHETORIC

1. The author uses a number of sensory details in describing both nature and the white man's crimes against nature. How does the eliciting of sensations help determine the relationship between writer, text, and reader?
2. The letter is written simply, with simply constructed paragraphs and sentences. What does this style suggest about the writer's voice?
3. There is a noted absence of transitional expressions in the writing, that is, such linking words as *in addition, furthermore, nevertheless, moreover*. How does this absence contribute to the directness of the writing?
4. The author uses the convention of the series, as in the following examples: "For all things share the same breath—the beasts, the trees, the man" (paragraph 2) and "When the buffalo are all slaughtered, the wild horses all tamed, the secret corners of the forest heavy with the scent of many men, and the view of the ripe hills blotted by talking wires" (paragraph 5). What is the rhetorical effect of this device?
5. Note the opening and closing sentences of the letter. How do they frame the letter? What do they suggest about one of the major themes of the letter?
6. Some scholars dispute the authenticity of the letter, attributing it to a white man who was attempting to articulate the essence of Chief Seattle's oratory in an effort to champion Native American causes. What elements of the letter resemble the rhetorical elements of a speech?

WRITING

1. Write a 250-word summary in which you compare and contrast the major differences between the white man's and the red man's perception of and relationship to nature as conceived by Chief Seattle.
2. For a research project, trace the use of the word *savage* as it has been used to describe Native Americans.
3. **WRITING AN ARGUMENT:** Argue for or against the view that the charge by Chief Seattle that the white man is contemptuous of nature is still valid today. Use at least three points to support your thesis.

www.mhhe.com/
mhreader

For more information on Chief Seattle, go to:
**More Resources > Ch. 13 Nature
& Environment**

✳

The Stone Horse

Barry Lopez

*Barry Lopez (b. 1945) was born in New York but spent much of his childhood in
Southern California. He received a BA from the University of Notre Dame and an MAT
from the University of Oregon in 1968. His early writings appeared in such magazines
as* National Geographic, Antaeus, Wilderness, Science, *and* Harper's. *He has
written both fiction and nonfiction including* Desert Notes: Reflections in the Eye of
a Raven *(1976),* Of Wolves and Men *(1978),* Arctic Dreams: Imagination and Desire in a Northern Landscape *(1986),* Crossing Open Ground *(1988),* The Rediscovery of North America *(1991),* Crow & Weasel *(1999), and an autobiography,*
About This Life *(1999). Lopez sees his role as a storyteller, someone who has the responsibility to create an atmosphere in which the wisdom of the work can reveal itself
and "make the reader feel a part of something." In the following piece, from* Crossing
Open Ground, *he focuses on an archaeological landmark to make a philosophical and
moral inquiry into the way humans regard nature.*

I

The deserts of southern California, the high, relatively cooler and wetter Mojave   1
and the hotter, dryer Sonoran to the south of it, carry the signatures of many
cultures. Prehistoric rock drawings in the Mojave's Coso Range, probably the
greatest concentration of petroglyphs in North America, are at least three thousand years old. Big-game-hunting cultures that flourished six or seven thousand years before that are known from broken spear tips, choppers, and burins
left scattered along the shores of great Pleistocene lakes, long since evaporated.
Weapons and tools discovered at China Lake may be thirty thousand years old;
and worked stone from a quarry in the Calico Mountains is, some argue, evidence that human beings were here more than 200,000 years ago.

  Because of the long-term stability of such arid environments, much of this   2
prehistoric stone evidence still lies exposed on the ground, accessible to anyone
who passes by—the studious, the acquisitive, the indifferent, the merely curious. Archaeologists do not agree on the sequence of cultural history beyond

about twelve thousand years ago, but it is clear that these broken bits of chalcedony, chert, and obsidian, like the animal drawings and geometric designs etched on walls of basalt throughout the desert, anchor the earliest threads of human history, the first record of human endeavor here.

Western man did not enter the California desert until the end of the eighteenth century, 250 years after Coronado brought his soldiers into the Zuni pueblos in a bewildered search for the cities of Cibola. The earliest appraisals of the land were cursory, hurried. People traveled *through* it, en route to Santa Fe or the California coastal settlements. Only miners tarried. In 1823 what had been Spain's became Mexico's, and in 1848 what had been Mexico's became America's; but the bare, jagged mountains and dry lake beds, the vast and uniform plains of creosote bush and yucca plants, remained as obscure as the northern Sudan until the end of the nineteenth century.

Before 1940 the tangible evidence of twentieth-century man's passage here consisted of very little—the hard tracery of travel corridors; the widely scattered, relatively insignificant evidence of mining operations; and the fair expanse of irrigated fields at the desert's periphery. In the space of a hundred years or so the wagon roads were paved, railroads were laid down, and canals and high-tension lines were built to bring water and electricity across the desert to Los Angeles from the Colorado River. The dark mouths of gold, talc, and tin mines yawned from the bony flanks of desert ranges. Dust-encrusted chemical plants stood at work on the lonely edges of dry lake beds. And crops of grapes, lettuce, dates, alfalfa, and cotton covered the Coachella and Imperial valleys, north and south of the Salton Sea, and the Palo Verde Valley along the Colorado.

These developments proceeded with little or no awareness of earlier human occupations by cultures that preceded those of the historic Indians—the Mohave, the Chemehuevi, the Quechan. (Extensive irrigation began actually to change the climate of the Sonoran Desert, and human settlements, the railroads, and farming introduced many new, successful plants into the region.)

During World War II, the American military moved into the desert in great force, to train troops and to test equipment. They found the clear weather conducive to year-round flying, the dry air and isolation very attractive. After the war, a complex of training grounds, storage facilities, and gunnery and test ranges was permanently settled on more than three million acres of military reservations. Few perceived the extent or significance of the destruction of the aboriginal sites that took place during tank maneuvers and bombing runs or in the laying out of highways, railroads, mining districts, and irrigated fields. The few who intuited that something like an American Dordogne Valley lay exposed here were (only) amateur archaeologists; even they reasoned that the desert was too vast for any of this to matter.

After World War II, people began moving out of the crowded Los Angeles basin into homes in Lucerne, Apple, and Antelope valleys in the western Mojave. They emigrated as well to a stretch of resort land at the foot of the San Jacinto Mountains that included Palm Springs, and farther out to old railroad and military towns like Twenty-nine Palms and Barstow. People also began exploring the desert, at first in military-surplus jeeps and then with a variety of all-terrain and off-road vehicles that became available in the 1960s. By the mid-1970s, the number of people using such vehicles for desert recreation had increased exponentially. Most came and went in innocent curiosity; the few who didn't wreaked a havoc all out of proportion to their numbers. The disturbance of previously isolated archaeological sites increased by an order of magnitude. Many sites were vandalized before archaeologists, themselves late to the desert, had any firm grasp of the bounds of human history in the desert. It was as though in the same moment an Aztec library had been discovered intact various lacunae had begun to appear.

The vandalism was of three sorts: the general disturbance usually caused   8 by souvenir hunters and by the curious and the oblivious; the wholesale stripping of a place by professional thieves for black-market sale and trade; and outright destruction, in which vehicles were actually used to ram and trench an area. By 1980, the Bureau of Land Management estimated that probably 35 percent of the archaeological sites in the desert had been vandalized. The destruction at some places by rifles and shotguns, or by power winches mounted on vehicles, was, if one cared for history, demoralizing to behold.

In spite of public education, land closures, and stricter law enforcement in   9 recent years, the BLM estimates that, annually, about 1 percent of the archaeological record in the desert continues to be destroyed or stolen.

II

A BLM archaeologist told me, with understandable reluctance, where to find   10 the intaglio. I spread my Automobile Club of Southern California map of Imperial County out on his desk, and he traced the route with a pink felt-tip pen. The line crossed Interstate 8 and then turned west along the Mexican border.

"You can't drive any farther than about here," he said, marking a small X.   11 "There's boulders in the wash. You walk up past them."

On a separate piece of paper he drew a route in a smaller scale that would   12 take me up the arroyo to a certain point where I was to cross back east, to another arroyo. At its head, on higher ground just to the north, I would find the horse.

"It's tough to spot unless you know it's there. Once you pick it up . . ." He   13 shook his head slowly, in a gesture of wonder at its existence.

I waited until I held his eye. I assured him I would not tell anyone else how   14 to get there. He looked at me with stoical despair, like a man who had been robbed twice, whose belief in human beings was offered without conviction.

I did not go until the following day because I wanted to see it at dawn. I ate   15 breakfast at four A.M. in El Centro and then drove south. The route was easy to follow, though the last section of road proved difficult, broken and drifted over with sand in some spots. I came to the barricade of boulders and parked. It was light enough by then to find my way over the ground with little trouble. The contours of the landscape were stark, without any masking vegetation. I worried only about rattlesnakes.

I traversed the stone plain as directed, but, in spite of the frankness of the   16
land, I came on the horse unawares. In the first moment of recognition I was
without feeling. I recalled later being startled, and that I held my breath. It was
laid out on the ground with its head to the east, three times life size. As I took in
its outline I felt a growing concentration of all my senses, as though my atten-
tiveness to the pale rose color of the morning sky and other peripheral images
had now ceased to be important. I was aware that I was straining for sound in
the windless air, and I felt the uneven pressure of the earth hard against my feet.
The horse, outlined in a standing profile on the dark ground, was as vivid before me as a bed of tulips.

I've come upon animals suddenly before, and felt a similar tension, a pre-   17
cipitate heightening of the senses. And I have felt the inexplicable but sharply
boosted intensity of a wild moment in the bush, where it is not until some minutes later that you discover the source of electricity—the warm remains of a
grizzly bear kill, or the still moist tracks of a wolverine.

But this was slightly different. I felt I had stepped into an unoccupied corri-   18
dor. I had no familiar sense of history, the temporal structure in which to think:
This horse was made by Quechan people three hundred years ago. I felt instead
a headlong rush of images: people hunting wild horses with spears on the Pleistocene veld of southern California; Cortés riding across the causeway into Mon-
tezuma's Tenochtitlán; a short-legged Comanche, astride his horse like some
sort of ferret, slashing through cavalry lines of young men who rode like farmers; a hood exploding past my face one morning in a corral in Wyoming. These
images had the weight and silence of stone.

When I released my breath, the images softened. My initial feeling, of fac-   19
ing a wild animal in a remote region, was replaced with a calm sense of antiquity. It was then that I became conscious, like an ordinary tourist, of what was
before me, and thought: this horse was probably laid out by Quechan people.
But when? I wondered. The first horses they saw, I knew, might have been those
that came north from Mexico in 1692 with Father Eusebio Kino. But Cocopa
people, I recalled, also came this far north on occasion, to fight with their neighbors, the Quechan. And *they* could have seen horses with Melchior Diaz, at the
mouth of the Colorado River in the fall of 1540. So, it could be four hundred
years old. (No one in fact knows.)

I still had not moved. I took my eyes off the horse for a moment to look   20
south over the desert plain into Mexico, to look east past its head at the bright-
ening sunrise, to situate myself. Then, finally, I brought my trailing foot slowly
forward and stood erect. Sunlight was running like a thin sheet of water over
the stony ground and it threw the horse into relief. It looked as though no hand
had ever disturbed the stones that gave it its form.

The horse had been brought to life on ground called desert pavement, a   21
tight, flat matrix of small cobbles blasted smooth by sand-laden winds. The
uniform, monochromatic blackness of the stones, a patina of iron and magne-
sium oxides called desert varnish, is caused by long-term exposure to the
sun. To make this type of low-relief ground glyph, or intaglio, the artist either

selectively turns individual stones over to their lighter side or removes areas of stone to expose the lighter soil underneath, creating a negative image. This horse, about eighteen feet from brow to rump and eight feet from withers to hoof, had been made in the latter way, and its outline was bermed at certain points with low ridges of stone a few inches high to enhance its three-dimensional qualities. (The left side of the horse was in full profile; each leg was extended at 90 degrees to the body and fully visible, as though seen in three-quarter profile.)

I was not eager to move. The moment I did I would be back in the flow of time, the horse no longer quivering in the same way before me. I did not want to feel again the sequence of quotidian events—to be drawn off into deliberation and analysis. A human being, a four-footed animal, the open land. That was all that was present—and a "thoughtless" understanding of the very old desires bearing on this particular animal: to hunt it, to render it, to fathom it, to subjugate it, to honor it, to take it as a companion. 22

What finally made me move was the light. The sun now filled the shallow basin of the horse's body. The weighted line of the stone berm created the illusion of a mane and the distinctive roundness of an equine belly. The change in definition impelled me. I moved to the left, circling past its rump, to see how the light might flesh the horse out from various points of view. I circled it completely before squatting on my haunches. Ten or fifteen minutes later I chose another view. The third time I moved, to a point near the rear hooves, I spotted a stone tool at my feet. I stared at it a long while, more in awe than disbelief, before reaching out to pick it up. I turned it over in my left palm and took it between my fingers to feel its cutting edge. It is always difficult, especially with something so portable, to rechannel the desire to steal. 23

I spent several hours with the horse. As I changed positions and as the angle of the light continued to change I noticed a number of things. The angle at which the pastern carried the hoof away from the ankle was perfect. Also, stones had been placed within the image to suggest at precisely the right spot the left shoulder above the foreleg. The line that joined thigh and hock was similarly accurate. The muzzle alone seemed distorted—but perhaps these stones had been moved by a later hand. It was an admirably accurate representation, but not what a breeder would call perfect conformation. There was the suggestion of a bowed neck and an undershot jaw, and the tail, as full as a winter coyote's, did not appear to be precisely to scale. 24

The more I thought about it, the more I felt I was looking at an individual horse, a unique combination of generic and specific detail. It was easy to imagine one of Kino's horses as a model, or a horse that ran off from one of Coronado's columns. What kind of horses would these have been? I wondered. In the sixteenth century the most sought-after horses in Europe were Spanish, the offspring of Arabian stock and Barbary horses that the Moors brought to Iberia and bred to the older, eastern European strains brought in by the Romans. The model for this horse, I speculated, could easily have been a palomino, or a descendant of horses trained for lion hunting in North Africa. 25

A few generations ago, cowboys, cavalry quartermasters, and draymen 26 would have taken this horse before me under consideration and not let up their scrutiny until they had its heritage fixed to their satisfaction. Today, the distinction between draft and harness horses is arcane knowledge, and no image may come to mind for a blue roan or a claybank horse. The loss of such refinement in everyday conversation leaves me unsettled. People praise the Eskimo's ability to distinguish among forty types of snow but forget the skill of others who routinely differentiate between overo and tobiano pintos. Such distinctions are made for the same reason. You have to do it to be able to talk clearly about the world.

For parts of two years I worked as a horse wrangler and packer in 27 Wyoming. It is dim knowledge now; I would have to think to remember if a buckskin was a kind of dun horse. And I couldn't throw a double-diamond hitch over a set of panniers—the packer's basic tie-down—without guidance. As I squatted there in the desert, however, these more personal memories seemed tenuous in comparison with the sweep of this animal in human time. My memories had no depth. I thought of the Hittite cavalry riding against the Syrians 3,500 years ago. And the first of the Chinese emperors, Ch'in Shih Huang, buried in Shensi Province in 210 B.C. with thousands of life-size horses and soldiers, a terra-cotta guardian army. What could I know of what was in the mind of whoever made this horse? Was there some racial memory of it as an animal that had once fed the artist's ancestors and then disappeared from North America? And then returned in this strange alliance with another race of men?

Certainly, whoever it was, the artist had observed the animal very closely. 28 Certainly the animal's speed had impressed him. Among the first things the Quechan would have learned from an encounter with Kino's horses was that their own long-distance runners—men who could run down mule deer—were no match for this animal.

From where I squatted I could look far out over the Mexican plain. Juan 29 Bautista de Anza passed this way in 1774, extending El Camino Real into Alta California from Sinaloa. He was followed by others, all of them astride the magical horse; *gente de razón*, the people of reason, coming into the country of *los primitivos*. The horse, like the stone animals of Egypt, urged these memories upon me. And as I drew them up from some forgotten corner of my mind— huge horses carved in the white chalk downs of southern England by an Iron Age people; Spanish horses rearing and wheeling in fear before alligators in Florida—the images seemed tethered before me. With this sense of proportion, a memory of my own—the morning I almost lost my face to a horse's hoof— now had somewhere to fit.

I rose up and began to walk slowly around the horse again. I had taken the 30 first long measure of it and was now looking for a way to depart, a new angle of light, a fading of the image itself before the rising sun, that would break its hold on me. As I circled, feeling both heady and serene at the encounter, I realized again how strangely vivid it was. It had been created on a barren bajada between two arroyos, as nondescript a place as one could imagine. The only plant life here was a few wands of ocotillo cactus. The ground beneath my shoes was so hard it wouldn't take the print of a heavy animal even after a rain. The only sounds I heard here were the voices of quail.

The archaeologist had been correct. For all its forcefulness, the horse is in- 31
conspicuous. If you don't care to see it you can walk right past it. That pleases him, I think. Unmarked on the bleak shoulder of the plain, the site signals to no one; so he wants no protective fences here, no informative plaque, to act as bea-cons. He would rather take a chance that no motorcyclist, no aimless wanderer with a flair for violence and a depth of ignorance, will ever find his way here.

The archaeologist had given me something before I left his office that now 32
seemed peculiar—an aerial photograph of the horse. It is widely believed that an aerial view of an intaglio provides a fair and accurate depiction. It does not. In the photograph the horse looks somewhat crudely constructed; from the ground it appears far more deftly rendered. The photograph is of a single mo-ment, and in that split second the horse seems vaguely impotent. I watched light pool in the intaglio at dawn; I imagine you could watch it withdraw at dusk and sense the same animation I did. In those prolonged moments its shape and so, too, its general character change—noticeably The living quality of the image, its immediacy to the eye, was brought out by the light-in-time, not, at least here, in the camera's frozen instant.

Intaglios, I thought, were never meant to be seen by gods in the sky above. 33
They were meant to be seen by people on the ground, over a long period of shifting light. This could even be true of the huge figures on the Plain of Nazca in Peru, where people could walk for the length of a day beside them. It is our own impatience that leads us to think otherwise.

This process of abstraction, almost unintentional, drew me gradually away 34
from the horse. I came to a position of attention at the edge of the sphere of its influence. With a slight bow I paid my respects to the horse, its maker, and the history of us all, and departed.

III

A short distance away I stopped the car in the middle of the road to make a few 35
notes. I could not write down what I was thinking when I was with the horse. It would have seemed disrespectful, and it would have required another kind of attention. So now I patiently drained my memory of the details it had fastened itself upon. The road I'd stopped on was adjacent to the All American Canal, the major source of water for the Imperial and Coachella valleys. The water flowed west placidly. A disjointed flock of coots, small, dark birds with white bills, was paddling against the current, foraging in the rushes.

I was peripherally aware of the birds as I wrote, the only movement in the 36
desert, and of a series of sounds from a village a half-mile away. The first sounds from this collection of ramshackle houses in a grove of cottonwoods were the distracted dawn voices of dogs. I heard them intermingled with the cries of a rooster. Later, the high-pitched voices of children calling out to each

other came disembodied through the dry desert air. Now, a little after seven, I could hear someone practicing on the trumpet, the same rough phrases played over and over. I suddenly remembered how as children we had tried to get the rhythm of a galloping horse with hands against our thighs, or by fluttering our tongues against the roofs of our mouths.

After the trumpet, the impatient calls of adults summoning children. Sun- 37 day morning. Wood smoke hung like a lens in the trees. The first car starts—a cold eight-cylinder engine, of Chrysler extraction perhaps, goosed to life, then throttled back to murmur through dual mufflers, the obbligato music of a shade-tree mechanic. The rote bark of mongrel dogs at dawn, the jagged outcries of men and women, an engine coming to life. Like a thousand villages from West Virginia to Guadalajara.

I finished my notes—where was I going to find a description of the horses 38 that came north with the conquistadors? Did their manes come forward prominently over the brow, like this one's, like the forelocks of Blackfeet and Assiniboin men in nineteenth-century paintings? I set the notes on the seat beside me.

The road followed the canal for a while and then arced north, toward Inter- 39 state 8. It was slow driving and I fell to thinking how the desert had changed since Anza had come through. New plants and animals—the MacDougall cottonwood, the English house sparrow, the chukar from India—have about them now the air of the native born. Of the native species, some—no one knows how many—are extinct. The populations of many others, especially the animals, have been sharply reduced. The idea of a desert impoverished by agricultural poisons and varmint hunters, by off-road vehicles and military operations, did not seem as disturbing to me, however, as this other horror, now that I had been those hours with the horse. The vandals, the few who crowbar rock art off the desert's walls, who dig up graves, who punish the ground that holds intaglios, are people who devour history. Their self-centered scorn, their disrespect for ideas and images beyond their ken, create the awful atmosphere of loose ends in which totalitarianism thrives, in which the past is merely curious or wrong.

I thought about the horse sitting out there on the unprotected plain. I enu- 40 merated its qualities in my mind until a sense of its vulnerability receded and it became an anchor for something else. I remembered that history, a history like this one, which ran deeper than Mexico, deeper than the Spanish, was a kind of medicine. It permitted the great breadth of human expression to reverberate, and it did not urge you to locate its apotheosis in the present.

Each of us, individuals and civilizations, has been held upside down like 41 Achilles in the River Styx. The artist mixing his colors in the dim light of Altamira; an Egyptian ruler lying still now, wrapped in his byssus, stored against time in a pyramid; the faded Dorset culture of the Arctic; the Hmong and Samburu and Walbiri of historic time; the modern nations. This great, imperfect stretch of human expression is the clarification and encouragement, the urging and the reminder, we call history. And it is inscribed everywhere in the face of the land, from the mountain passes of the Himalayas to a nameless bajada in the California desert.

Small birds rose up in the road ahead, startled, and flew off. I prayed no 42
infidel would ever find that horse.

COMPREHENSION

1. What is Lopez's primary purpose in this essay? How does its title relate to the purpose?
2. Does Lopez assume his audience has the same value position as he does? Does he assume it has a commensurate background in anthropology, history, and ecology? Explain your answer.
3. What provides Lopez with the most compelling evidence that humans have a history? Base your answer on the author's statement in paragraph 41: "This great, imperfect stretch of human expression is the clarification and encouragement, the origin and the reminder, we call history."

RHETORIC

1. How would you characterize the tone of this essay? Does the tone remain constant or does it change? Can you connect the tone with Lopez's thesis?
2. Lopez has chosen to divide his essay into three sections. Why is that decision appropriate to his strategy? How does the structure relate to the overall development of his theme?
3. While Lopez inspects the "horse," he describes it in detail. Next, he experiences a change in emotion and thinking. How does his erudition concerning history and anthropology enrich his emotional response? Explain the rhetorical progression from description to response and reflection in paragraphs 19 to 28.
4. Define the following terms: *petroglyphs* (paragraph 1), *intaglio* (paragraph 10), *arroyo* (paragraph 12), *glyph* (paragraph 21), *pastern* (paragraph 24), *conquistadors* (paragraph 38), and *totalitarianism* (paragraph 39).
5. Why does Lopez end this substantial essay with such a brief conclusion? Why has the author juxtaposed the presence of "birds" and the fear of the "infidel" in his conclusion?
6. There is a marked change in atmosphere between sections 2 and 3. What is this change? How does Lopez express it? What does it signify?
7. How does Lopez demonstrate a sense of authority on his subject matter? Does it stem naturally from the flow of his writing or is it superimposed? Explain your response.

WRITING

1. Write a descriptive/narrative essay about a personal experience in an environment that made you feel as though you were experiencing a place with

implications greater than yourself, for example, a setting in nature, a church, sports arena or stadium during a game, a concert, and so on.

2. Write an essay in which you argue for or against the public's right to gain access to archeological sites for study.

3. **WRITING AN ARGUMENT:** Argue for or against the proposition that the American system of values encourages people to despoil, destroy, or neglect historical sites and structures.

| www.mhhe.com/ **mhreader** | For more information on Barry Lopez, go to: **More Resources > Ch. 13 Nature & Environment** |

Classic and Contemporary: Questions for Comparison

1. Chief Seattle and Barry Lopez decry the destruction of nature as a physical and spiritual presence. In what ways are their stakes in this destruction the same? In what ways are they different? Does either writer have more power to effect a transformation in our attitude toward nature? Explain.

2. It has often been said that intellectual knowledge changes one's relationship with the world environment. In what ways has Lopez's "book learning" and erudition made him a different person from Chief Seattle? Base your response on the style and tone of each author.

3. Chief Seattle is literally the leader and spokesperson of a defeated nation. How does he preserve his dignity in the face of being conquered? How does he indicate to the white man that his "victory" is temporary? Lopez, on the other hand, is a successful member of his society: esteemed naturalist, award-winning writer, popular lecturer. What prevents him from reaping the benefits of his accomplishments? Explain by referring to the text.

✳

The Way to Rainy Mountain

N. Scott Momaday

Navarre Scott Momaday (b. 1934), Pulitzer Prize–winning poet, critic, and academician was born in Lawton, Oklahoma, of Kiowa ancestry. He holds degrees from the University of New Mexico and Stanford University and is a professor of English at the University of Arizona. He is the author of House Made of Dawn *(1968),* The Way to Rainy Mountain *(1969),* The Names *(1976), and other works. "I am an American In-*

dian (Kiowa), and am vitally interested in American Indian art, history and culture,"
Momaday has written. In this essay, he elevates personal experience and landscape to
the realm of poetry and tribal myth.

A single knoll rises out of the plain in Oklahoma, north and west of the Wichita   1
Range. For my people, the Kiowas, it is an old landmark, and they gave it the
name Rainy Mountain. The hardest weather in the world is there. Winter brings
blizzards, hot tornadic winds arise in the spring, and in summer the prairie is
an anvil's edge. The grass turns brittle and brown, and it cracks beneath your
feet. There are green belts along the rivers and creeks, linear groves of hickory
and pecan, willow and witch hazel. At a distance in July or August the steam-
ing foliage seems almost to writhe in fire. Great green and yellow grasshoppers
are everywhere in the tall grass, popping up like corn to sting the flesh, and tor-
toises crawl about on the red earth, going nowhere in the plenty of time. Loneliness is an aspect of the land. All things in the plain are isolate; there is no
confusion of objects in the eye, but *one* hill or *one* tree or *one* man. To look upon
that landscape in the early morning, with the sun at your back, is to lose the
sense of proportion. Your imagination comes to life, and this, you think, is
where Creation was begun.

I returned to Rainy Mountain in July. My grandmother had died in the   2
spring, and I wanted to be at her grave. She had lived to be very old and at last
infirm. Her only living daughter was with her when she died, and I was told
that in death her face was that of a child.

I like to think of her as a child. When she was born, the Kiowas were living   3
the last great moment of their history. For more than a hundred years they had
controlled the open range from the Smoky Hill River to the Red, from the head-
waters of the Canadian to the fork of the Arkansas and Cimarron. In alliance
with the Comanches, they had ruled the whole of the southern Plains. War was
their sacred business, and they were among the finest horsemen the world has
ever known. But warfare for the Kiowas was preeminently a matter of disposition rather than of survival, and they never understood the grim, unrelenting
advance of the U.S. Cavalry. When at last, divided and ill-provisioned, they
were driven onto the Staked Plains in the cold rains of autumn, they fell into
panic. In Palo Duro Canyon they abandoned their crucial stores to pillage and
had nothing then but their lives. In order to save themselves, they surrendered
to the soldiers at Fort Sill and were imprisoned in the old stone corral that now
stands as a military museum. My grandmother was spared the humiliation of
those high gray walls by eight or ten years, but she must have known from birth
the affliction of defeat, the dark brooding of old warriors.

Her name was Aho, and she belonged to the last culture to evolve in North   4
America. Her forebears came down from the high country in western Montana
nearly three centuries ago. They were a mountain people, a mysterious tribe of
hunters whose language has never been positively classified in any major
group. In the late seventeenth century they began a long migration to the south
and east. It was a journey toward the dawn, and it led to a golden age. Along

the way the Kiowas were befriended by the Crows, who gave them the culture and religion of the Plains. They acquired horses, and their ancient nomadic spirit was suddenly free of the ground. They acquired Tai-me, the sacred Sun Dance doll, from that moment the object and symbol of their worship, and so shared in the divinity of the sun. Not least, they acquired the sense of destiny, therefore courage and pride. When they entered upon the southern Plains they had been transformed. No longer were they slaves to the simple necessity of survival; they were a lordly and dangerous society of fighters and thieves, hunters and priests of the sun. According to their origin myth, they entered the world through a hollow log. From one point of view, their migration was the fruit of an old prophecy, for indeed they emerged from a sunless world.

Although my grandmother lived out her long life in the shadow of Rainy   5
Mountain, the immense landscape of the continental interior lay like memory in her blood. She could tell of the Crows, whom she had never seen, and of the Black Hills, where she had never been. I wanted to see in reality what she had seen more perfectly in the mind's eye, and traveled fifteen hundred miles to begin my pilgrimage.

Yellowstone, it seemed to me, was the top of the world, a region of deep   6
lakes and dark timber, canyons and waterfalls. But, beautiful as it is, one might have the sense of confinement there. The skyline in all directions is close at hand, the high wall of the woods and deep cleavages of shade. There is a perfect freedom in the mountains, but it belongs to the eagle and the elk, the badger and the bear. The Kiowas reckoned their stature by the distance they could see, and they were bent and blind in the wilderness.

Descending eastward, the highland meadows are a stairway to the plain. In   7
July the inland slope of the Rockies is luxuriant with flax and buckwheat, stonecrop and larkspur. The earth unfolds and the limit of the land recedes. Clusters of trees, and animals grazing far in the distance, cause the vision to reach away and wonder to build upon the mind. The sun follows a longer course in the day, and the sky is immense beyond all comparison. The great billowing clouds that sail upon it are shadows that move upon the grain like water, dividing light. Farther down, in the land of the Crows and Blackfeet, the plain is yellow. Sweet clover takes hold of the hills and bends upon itself to cover and seal the soil. There the Kiowas paused on their way; they had come to the place where they must change their lives. The sun is at home on the plains. Precisely there does it have the certain character of a god. When the Kiowas came to the land of the Crows, they could see the dark lees of the hills at dawn across the Bighorn River, the profusion of light on the grain shelves, the oldest deity ranging after the solstices. Not yet would they veer southward to the caldron of the land that lay below; they must wean their blood from the northern winter and hold the mountains a while longer in their view. They bore Tai-me in procession to the east.

A dark mist lay over the Black Hills, and the land was like iron. At the top   8
of a ridge I caught sight of Devil's Tower upthrust against the gray sky as if in the birth of time the core of the earth had broken through its crust and the mo-

tion of the world was begun. There are things in nature that engender an awful quiet in the heart of man; Devil's Tower is one of them. Two centuries ago, because they could not do otherwise, the Kiowas made a legend at the base of the rock. My grandmother said:

> Eight children were there at play, seven sisters and their brother. Suddenly the boy was struck dumb; he trembled and began to run upon his hands and feet. His fingers became claws, and his body was covered with fur. Directly there was a bear where the boy had been. The sisters were terrified; they ran, and the bear after them. They came to the stump of a great tree, and the tree spoke to them. It bade them climb upon it, and as they did so it began to rise into the air. The bear came to kill them, but they were just beyond its reach. It reared against the tree and scored the bark all around with its claws. The seven sisters were borne into the sky, and they became the stars of the Big Dipper.

From that moment, and so long as the legend lives, the Kiowas have kinsmen in the night sky. Whatever they were in the mountains, they could be no more. However tenuous their well-being, however much they had suffered and would suffer again, they had found a way out of the wilderness.

My grandmother had a reverence for the sun, a holy regard that now is all 9 but gone out of mankind. There was a wariness in her, and an ancient awe. She was a Christian in her later years, but she had come a long way about, and she never forgot her birthright. As a child she had been to the Sun Dances; she had taken part in those annual rites, and by them she had learned the restoration of her people in the presence of Tai-me. She was about seven when the last Kiowa Sun Dance was held in 1887 on the Washita River above Rainy Mountain Creek. The buffalo were gone. In order to consummate the ancient sacrifice—to impale the head of a buffalo bull upon the medicine tree—a delegation of old men journeyed into Texas, there to beg and barter for an animal from the Goodnight herd. She was ten when the Kiowas came together for the last time as a living Sun Dance culture. They could find no buffalo; they had to hang an old hide from the sacred tree. Before the dance could begin, a company of soldiers rode out from Fort Sill under orders to disperse the tribe. Forbidden without cause the essential act of their faith, having seen the wild herds slaughtered and left to rot upon the ground, the Kiowas backed away forever from the medicine tree. That was July 20, 1890, at the great bend of the Washita. My grandmother was there. Without bitterness, and for as long as she lived, she bore a vision of deicide.

Now that I can have her only in memory, I see my grandmother in the sev- 10 eral postures that were peculiar to her: standing at the wood stove on a winter morning and turning meat in a great iron skillet; sitting at the south window, bent above her beadwork, and afterwards, when her vision failed, looking down for a long time into the fold of her hands; going out upon a cane, very slowly as she did when the weight of age came upon her; praying. I remember her most often at prayer. She made long, rambling prayers out of suffering and hope, having seen many things. I was never sure that I had the right to hear, so exclusive were they of all mere custom and company. The last time I saw her she prayed standing by the side of her bed at night, naked to the waist, the light of a kerosene lamp moving upon her dark skin. Her long, black hair, always drawn and braided in the day, lay upon her shoulders and against her breasts like a shawl. I do not speak Kiowa, and I never understood her prayers, but there was something inherently sad in the sound, some merest hesitation upon the syllables of sorrow. She began in a high and descending pitch, exhausting her breath to silence; then again and again—and always the same intensity of effort, of something that is, and is not, like urgency in the human voice. Transported so in the dancing light among the shadows of her room, she seemed beyond the reach of time. But that was illusion; I think I knew then that I should not see her again.

Houses are like sentinels in the plain, old keepers of the weather watch. 11 There, in a very little while, wood takes on the appearance of great age. All colors wear soon away in the wind and rain, and then the wood is burned gray and the grain appears and the nails turn red with rust. The windowpanes are black and opaque; you imagine there is nothing within, and indeed there are many ghosts, bones given up to the land. They stand here and there against the sky, and you approach them for a longer time than you expect. They belong in the distance; it is their domain.

Once there was a lot of sound in my grandmother's house, a lot of coming 12 and going, feasting and talk. The summers there were full of excitement and reunion. The Kiowas are a summer people; they abide the cold and keep to themselves, but when the season turns and the land becomes warm and vital they cannot hold still; an old love of going returns upon them. The aged visitors who came to my grandmother's house when I was a child were made of lean and leather, and they bore themselves upright. They wore great black hats and bright ample shirts that shook in the wind. They rubbed fat upon their hair and wound their braids with strips of colored cloth. Some of them painted their, faces and carried the scars of old and cherished enmities. They were an old council of warlords, come to remind and be reminded of who they were. Their wives and daughters served them well. The women might indulge themselves; gossip was at once the mark and compensation of their servitude. They made loud and elaborate talk among themselves, full of jest and gesture, fright and false alarm. They went abroad in fringed and flowered shawls, bright beadwork and German silver. They were at home in the kitchen, and they prepared meals that were banquets.

There were frequent prayer meetings, and great nocturnal feasts. When I 13 was a child I played with my cousins outside, where the lamplight fell upon the ground and the singing of the old people rose up around us and carried away into the darkness. There were a lot of good things to eat, a lot of laughter and surprise. And afterwards, when the quiet returned, I lay down with my grandmother and could hear the frogs away by the river and feel the motion of the air.

Now there is funeral silence in the rooms, the endless wake of some final 14 word. The walls have closed in upon my grandmother's house. When I returned to it in mourning, I saw for the first time in my life how small it was. It was late at night, and there was a white moon, nearly full. I sat for a long time on the stone steps by the kitchen door. From there I could see out across the land; I could see the long row of trees by the creek, the low light upon the rolling plains, and the stars of the Big Dipper. Once I looked at the moon and caught sight of a strange thing. A cricket had perched upon the handrail, only a few inches away from me. My line of vision was such that the creature filled the moon like a fossil. It had gone there, I thought, to live and die, for there, of all places, was its small definition made whole and eternal. A warm wind rose up and purled like the longing within me.

The next morning I awoke at dawn and went out on the dirt road to Rainy 15 Mountain. It was already hot, and the grasshoppers began to fill the air. Still, it was early in the morning, and the birds sang out of the shadows. The long yellow grass on the mountain shone in the bright light, and a scissortail hied above the land. There, where it ought to be, at the end of a long and legendary way, was my grandmother's grave. Here and there on the dark stones were ancestral names. Looking back once, I saw the mountain and came away.

COMPREHENSION

1. What is the significance of Momaday's title? How does the title help explain the author's purpose?
2. Why does Momaday return to his grandmother's house and journey to her grave? In what ways was her life tied to the land?
3. List the various myths and legends the author mentions in the essay. What subjects do they treat? How are these subjects interrelated?

RHETORIC

1. What is the method of development in the first paragraph? How does the introduction, with its focus on Rainy Mountain, serve as a vehicle for the central meanings in the essay?
2. Locate and explain instances of sensory, metaphorical, and symbolic language in the essay. Why are these modes of language consistent with the subject and thesis elaborated by Momaday?
3. How does Momaday's use of abstract language affect the concrete vocabulary in the essay?

WRITING

1. Write about a person and place that, taken together, inspire a special reverence in you.
2. Momaday implies that myth—especially nature myths—is central to his life and the life of the Kiowas. What *is* myth? Do you think that myth is as strong in general American culture as it is in Kiowa culture? In what ways does it operate? How can myth sustain the individual, community, and nation? Write an analytical essay on this subject.

3. **WRITING AN ARGUMENT:** Write an essay in which you discuss Momaday's implied assertion that white Americans destroyed a way of Kiowa life
   that was rooted in reverence for nature and landscape. Determine whether
   you support this view or not.

✸ www.mhhe.com/ **mhreader**	For more information on N. Scott Momaday, go to: **More Resources > Ch. 13 Nature & Environment**

✷

The Environmental Issue from Hell

Bill McKibben

*Bill McKibben (b. 1960) was born in Palo Alto, California. After receiving a BA from
Harvard University (1982), McKibben became a staff writer for* New Yorker *magazine. McKibben's chief concern is the impact of humans on the environment and the
ways in which consumerism affects the global ecosystem. A prominent writer for the environmental movement, he has published several books, among them* The End of Nature *(1989),* The Age of Missing Information *(1992),* Long Distance: A Year of
Living Strenuously *(2000), and* Enough: Staying Human in an Engineered Age
(2003). In the essay that follows, which was published in These Times *in 2001,
McKibben argues for a new approach to global warming.*

When global warming first emerged as a potential crisis in the late 1980s, one   1
academic analyst called it "the public policy problem from hell." The years
since have only proven him more astute: Fifteen years into our understanding
of climate change, we have yet to figure out how we're going to tackle it. And
environmentalists are just as clueless as anyone else: Do we need to work on
lifestyle or on lobbying, on photovoltaics or on politics? And is there a difference? How well we handle global warming will determine what kind of century we inhabit—and indeed what kind of planet we leave behind. The issue
cuts close to home and also floats off easily into the abstract. So far it has been
the ultimate "can't get there from here" problem, but the time has come to draw
a road map—one that may help us deal with the handful of other issues on the
list of real, world-shattering problems.

Typically, when you're mounting a campaign, you look for self-interest, you   2
scare people by saying what will happen to us if we don't do something: All the
birds will die, the canyon will disappear beneath a reservoir, we will choke to
death on smog. But in the case of global warming, that doesn't exactly do the

trick, at least in the time frame we're discussing. In temperate latitudes, climate change will creep up on us. Severe storms already have grown more frequent and more damaging. The progression of seasons is less steady. Some agriculture is less reliable. But face it: Our economy is so enormous that it takes those changes in stride. Economists who work on this stuff talk about how it will shave a percentage or two off the GNP over the next few decades. And most of us live lives so divorced from the natural world that we hardly notice the changes anyway. Hotter? Turn up the air-conditioning. Stormier? Well, an enormous percentage of Americans commute from remote-controlled garage to office parking garage—it may have been some time since they got good and wet in a rainstorm. By the time the magnitude of the change is truly in our faces, it will be too late to do much about it: There's such a lag time to increased levels of carbon dioxide in the atmosphere that we need to be making the switch to solar and wind and hydrogen power right now to prevent disaster decades away. Yesterday, in fact.

So maybe we should think of global warming in a different way—as the 3 great moral crisis of our time, the equivalent of the civil rights movement of the 1960s.

Why a moral question? In the first place, no one's ever figured out a more 4 effective way to screw the marginalized and poor of this planet than climate change. Having taken their dignity, their resources, and their freedom under a variety of other schemes, we now are taking the very physical stability on which their already difficult lives depend.

Our economy can absorb these changes for a while, but consider 5 Bangladesh for a moment. In 1998 the sea level in the Bay of Bengal was higher than normal, just the sort of thing we can expect to become more frequent and severe. The waters sweeping down the Ganges and the Brahmaputra rivers from the Himalayas could not drain easily into the ocean—they backed up across the country, forcing most of its inhabitants to spend three months in thigh-deep water. The fall rice crop didn't get planted. We've seen this same kind of disaster over the past few years in Mozambique and Honduras and Venezuela and other places.

And global warming is a moral crisis, too, if you place any value on the rest 6 of creation. Coral reef researchers indicate that these spectacularly intricate ecosystems are also spectacularly vulnerable. Rising water temperatures are likely to bleach them to extinction by mid-century. In the Arctic, polar bears are 20 percent scrawnier than they were a decade ago: As pack ice melts, so does the opportunity for hunting seals. All in all, the 21st century seems poised to see extinctions at a rate not observed since the last big asteroid slammed into the planet. But this time the asteroid is us.

It's a moral question, finally, if you think we owe any debt to the future. No 7 one ever has figured out a more thoroughgoing way to strip-mine the present and degrade what comes after—all the people who will ever be related to you. Ever. No generation yet to come will ever forget us—we are the ones present at the moment when the temperature starts to spike, and so far we have not reacted. If it had been done to us, we would loathe the generation that did it, precisely as we will one day be loathed.

But trying to launch a moral campaign is no easy task. In most moral crises, 8
there is a villain—some person or class or institution that must be overcome.
Once the villain is identified, the battle can commence. But you can't really get
angry at carbon dioxide, and the people responsible for its production are, well,
us. So perhaps we need some symbols to get us started, some places to sharpen
the debate and rally ourselves to action. There are plenty to choose from: our
taste for ever bigger houses and the heating and cooling bills that come with
them, our penchant for jumping on airplanes at the drop of a hat. But if you
wanted one glaring example of our lack of balance, you could do worse than
point the finger at sport utility vehicles.

SUVs are more than mere symbols. They are a major part of the problem— 9
we emit so much more carbon dioxide now than we did a decade ago in part because our fleet of cars and trucks actually has gotten steadily less fuel efficient
for the past 10 years. If you switched today from the average American car to a
big SUV, and drove it for just one year, the difference in carbon dioxide that you
produced would be the equivalent of opening your refrigerator door and then
forgetting to close it for six years. SUVs essentially are machines for burning
fossil fuel that just happen to also move you and your stuff around.

But what makes them such a perfect symbol is the brute fact that they are 10
simply unnecessary. Go to the parking lot of the nearest suburban supermarket
and look around: The only conclusion you can draw is that to reach the grocery,
people must drive through three or four raging rivers and up the side of a
canyon. These are semi-military machines, armored trucks on a slight diet. While
they do not keep their occupants appreciably safer, they do wreck whatever they
plow into, making them the perfect metaphor for a heedless, supersized society.

That's why we need a much broader politics than the Washington lobbying 11
that's occupied the big environmental groups for the past decade. We need to
take all the brilliant and energetic strategies of local grassroots groups fighting
dumps and cleaning up rivers and apply those tactics in the national and international arenas. That's why some pastors are starting to talk with their congre-
gations about what cars to buy, and why some college seniors are passing
around petitions pledging to stay away from the Ford Explorers and Excur-
sions, and why some auto dealers have begun to notice informational picketers
outside their showrooms on Saturday mornings urging customers to think
about gas mileage when they look at cars.

The point is not that such actions by themselves—any individual actions— 12
will make any real dent in the levels of carbon dioxide pouring into our atmosphere. Even if you got 10 percent of Americans really committed to changing
their energy use, their solar homes wouldn't make much of a difference in our
national totals. But 10 percent would be enough to change the politics around
the issue, enough to pressure politicians to pass laws that would cause us all to
shift our habits. And so we need to begin to take an issue that is now the
province of technicians and turn it into a political issue, just as bus boycotts began to make public the issue of race, forcing the system to respond. That response is likely to be ugly—there are huge companies with a lot to lose, and
many people so tied in to their current ways of life that advocating change

smacks of subversion. But this has to become a political issue—and fast. The only way that may happen, short of a hideous drought or monster flood, is if it becomes a personal issue first.

COMPREHENSION

1. According to the author, what are the causes of global warming?
2. What instances of ecological disaster does the writer cite if we do not change our habits?
3. Why is a new approach to the problem of global warming needed? What approach does McKibben suggest?

RHETORIC

1. How does the writer's title capture the tone of the essay? What is McKibben's purpose in writing the essay? Does he see his readers as hostile or sympathetic to his position? How do you know?
2. How does McKibben develop his introduction? Why does he pose questions? Where does he state his claim?
3. Does McKibben make his argument through appeals to reason, emotion, ethics, or a combination of these elements? Justify your response.
4. How does the writer contend with possible objections to his position on global warming?
5. Explain the pattern of cause and effect that McKibben uses to structure his essay.
6. What varieties of evidence does the writer present to support his claim? What extended illustration does he provide, how effective is it, and why?
7. In the concluding paragraph, McKibben issues a call to action. How does the body of the essay prepare readers for this persuasive appeal?

WRITING

1. Write an essay in which you explain your own sense of the causes and effects of global warming.
2. Research your state's policy toward global warming. Present your findings in a summary essay.
3. **WRITING AN ARGUMENT:** McKibben argues that SUVs are a primary cause of wastefulness and global warming and that both moral persuasion and political activism are required to change consumer's habits. Do you agree or disagree with his assertions? Write an argumentative essay responding to this issue.

---

www.mhhe.com/
mhreader

For more information on Bill McKibben, go to:
More Resources > Ch. 13 Nature & Environment

✳

The Obligation to Endure

Rachel Carson

Rachel Carson (1907–1964) was a seminal figure in the environmental movement. Born in Pennsylvania, she awakened public consciousness to environmental issues through her writing. Her style was both literary and scientific as she described nature's riches in such books as The Sea Around Us *(1951) and* The Edge of the Sea *(1954). Her last book,* Silent Spring *(1962), aroused controversy and concern with its indictment of insecticides. In the following excerpt from that important book, Carson provides compelling evidence of the damage caused by indiscriminate use of insecticides and the danger of disturbing the earth's delicate balance.*

The history of life on earth has been a history of interaction between living 1
things and their surroundings. To a large extent, the physical form and the habits of the earth's vegetation and its animal life have been molded by the environment. Considering the whole span of earthly time, the opposite effect, in which life actually modifies its surroundings, has been relatively slight. Only within the moment of time represented by the present century has one species—man—acquired significant power to alter the nature of his world.

During the past quarter century this power has not only increased to one of 2
disturbing magnitude but it has changed in character. The most alarming of all man's assaults upon the environment is the contamination of air, earth, rivers, and sea with dangerous and even lethal materials. This pollution is for the most part irrecoverable; the chain of evil it initiates not only in the world that must support life but in living tissues is for the most part irreversible. In this now universal contamination of the environment, chemicals are the sinister and little-recognized partners of radiation in changing the very nature of the world—the very nature of its life. Strontium 90, released through nuclear explosions into the air, comes to earth in rain or drifts down as fallout, lodges in soil, enters into the grass or corn or wheat grown there, and in time takes up its abode in the bones of a human being, there to remain until his death. Similarly, chemicals sprayed on croplands or forests or gardens lie long in soil, entering into living organisms, passing from one to another in a chain of poisoning and death. Or they pass mysteriously by underground streams until they emerge and, through the alchemy of air and sunlight, combine into new forms that kill vegetation, sicken cattle, and work unknown harm on those who drink from once pure wells. As Albert Schweitzer has said, "Man can hardly even recognize the devils of his own creation."

It took hundreds of millions of years to produce the life that now inhabits 3
the earth—eons of time in which that developing and evolving and diversifying

life reached a state of adjustment and balance with its surroundings. The environment, rigorously shaping and directing the life it supported, contained elements that were hostile as well as supporting. Certain rocks gave out dangerous radiation; even within the light of the sun, from which all life draws its energy, there were shortwave radiations with power to injure. Given time—time not in years but in millennia—life adjusts, and a balance has been reached. For time is the essential ingredient; but in the modern world there is no time.

The rapidity of change and the speed with which new situations are created   4
follow the impetuous and heedless pace of man rather than the deliberate pace of nature. Radiation is no longer merely the background radiation of rocks, the bombardment of cosmic rays, the ultraviolet of the sun that have existed before there was any life on earth; radiation is now the unnatural creation of man's tampering with the atom. The chemicals to which life is asked to make its adjustment are no longer merely the calcium and silica and copper and all the rest of the minerals washed out of the rocks and carried in rivers to the sea; they are the synthetic creations of man's inventive mind, brewed in his laboratories, and having no counterparts in nature.

To adjust to these chemicals would require time on the scale that is nature's;   5
it would require not merely the years of a man's life but the life of generations. And even this, were it by some miracle possible, would be futile, for the new chemicals come from our laboratories in an endless stream; almost five hundred annually find their way into actual use in the United States alone. The figure is staggering and its implications are not easily grasped—500 new chemicals to which the bodies of men and animals are required somehow to adapt each year, chemicals totally outside the limits of biologic experience.

Among them are many that are used in man's war against nature. Since the   6
mid-1940s over 200 basic chemicals have been created for use in killing insects, weeds, rodents, and other organisms described in the modern vernacular as "pests"; and they are sold under several thousand different brand names.

These sprays, dusts, and aerosols are now applied almost universally to   7
farms, gardens, forests, and homes—nonselective chemicals that have the power to kill every insect, the "good" and the "bad," to still the song of birds and the leaping of fish in the streams, to coat the leaves with a deadly film, and to linger on in soil—all this though the intended target may be only a few weeds or insects. Can anyone believe it is possible to lay down such a barrage of poisons on the surface of the earth without making it unfit for all life? They should not be called "insecticides," but "biocides."

The whole process of spraying seems caught up in an endless spiral. Since   8
DDT was released for civilian use, a process of escalation has been going on in which ever more toxic materials must be found. This has happened because insects, in a triumphant vindication of Darwin's principle of the survival of the fittest, have evolved super races immune to the particular insecticide used, hence a deadlier one has always to be developed—and then a deadlier one than that. It has happened also because, for reasons to be described later, destructive

insects often undergo a "flareback," or resurgence, after spraying in numbers greater than before. Thus the chemical war is never won, and all life is caught in its violent crossfire.

Along with the possibility of the extinction of mankind by nuclear war, the 9 central problem of our age has therefore become the contamination of man's total environment with such substances of incredible potential for harm— substances that accumulate in the tissues of plants and animals and even pene- trate the germ cells to shatter or alter the very material of heredity upon which the shape of the future depends.

Some would-be architects of our future look toward a time when it will be 10 possible to alter the human germ plasm by design. But we may easily be doing so now by inadvertence, for many chemicals, like radiation, bring about gene mutations. It is ironic to think that man might determine his own future by something so seemingly trivial as the choice of an insect spray.

All this has been risked—for what? Future historians may well be amazed 11 by our distorted sense of proportion. How could intelligent beings seek to con- trol a few unwanted species by a method that contaminated the entire environ- ment and brought the threat of disease and death even to their own kind? Yet this is precisely what we have done. We have done it, moreover, for reasons that collapse the moment we examine them. We are told that the enormous and ex- panding use of pesticides is necessary to maintain farm production. Yet is our real problem not one of *overproduction?* Our farms, despite measures to remove acreages from production and to pay farmers *not* to produce, have yielded such a staggering excess of crops that the American taxpayer in 1962 is paying out more than one billion dollars a year as the total carrying cost of the surplus-food storage program. And is the situation helped when one branch of the Agricul- ture Department tries to reduce production while another states, as it did in 1958, "It is believed generally that reduction of crop acreages under provisions of the Soil Bank will stimulate interest in use of chemicals to obtain maximum production on the land retained in crops."

All this is not to say there is no insect problem and no need of control. I am 12 saying, rather, that control must be geared to realities, not to mythical situa- tions, and that the methods employed must be such that they do not destroy us along with the insects.

The problem whose attempted solution has brought such a train of disaster in 13 its wake is an accompaniment of our modern way of life. Long before the age of man, insects inhabited the earth—a group of extraordinarily varied and adapt- able beings. Over the course of time since man's advent, a small percentage of the more than half a million species of insects have come into conflict with hu- man welfare in two principal ways: as competitors for the food supply and as carriers of human disease.

Disease-carrying insects become important where human beings are 14 crowded together, especially under conditions where sanitation is poor, as in times of natural disaster or war or in situations of extreme poverty and deprivation. Then control of some sort becomes necessary. It is a sobering fact, however, as we shall presently see, that the method of massive chemical control has had only limited success, and also threatens to worsen the very conditions it is intended to curb.

Under primitive agricultural conditions the farmer had few insect prob- 15 lems. These arose with the intensification of agriculture—the devotion of immense acreages to a single crop. Such a system set the stage for explosive increases in specific insect populations. Single-crop farming does not take advantage of the principles by which nature works; it is agriculture as an engineer might conceive it to be. Nature has introduced great variety into the landscape, but man has displayed a passion for simplifying it. Thus he undoes the built-in checks and balances by which nature holds the species within bounds. One important natural check is a limit on the amount of suitable habitat for each species. Obviously then, an insect that lives on wheat can build up its population to much higher levels on a farm devoted to wheat than on one in which wheat is intermingled with other crops to which the insect is not adapted.

The same thing happens in other situations. A generation or more ago, the 16 towns of large areas of the United States lined their streets with the noble elm tree. Now the beauty they hopefully created is threatened with complete destruction as disease sweeps through the elms, carried by a beetle that would have only limited chance to build up large populations and to spread from tree to tree if the elms were only occasional trees in a richly diversified planting.

Another factor in the modern insect problem is one that must be viewed 17 against a background of geologic and human history: the spreading of thousands of different kinds of organisms from their native homes to invade new territories. This worldwide migration has been studied and graphically described by the British ecologist Charles Elton in his recent book *The Ecology of Invasions.* During the Cretaceous Period, some hundred million years ago, flooding seas cut many land bridges between continents and living things found themselves confined in what Elton calls "colossal separate nature reserves." There, isolated from others of their kind, they developed many new species. When some of the land masses were joined again, about 15 million years ago, these species began to move out into new territories—a movement that is not only still in progress but is now receiving considerable assistance from man.

The importation of plants is the primary agent in the modern spread of 18 species, for animals have almost invariably gone along with the plants, quarantine being a comparatively recent and not completely effective innovation. The United States Office of Plant Introduction alone has introduced almost 200,000 species and varieties of plants from all over the world. Nearly half of the 180 or so major insect enemies of plants in the United States are accidental imports from abroad, and most of them have come as hitchhikers on plants.

In new territory, out of reach of the restraining hand of the natural enemies 19 that kept down its numbers in its native land, an invading plant or animal is able to become enormously abundant. Thus it is no accident that our most troublesome insects are introduced species.

These invasions, both the naturally occurring and those dependent on hu- 20 man assistance, are likely to continue indefinitely. Quarantine and massive chemical campaigns are only extremely expensive ways of buying time. We are faced, according to Dr. Elton, "with a life-and-death need not just to find new technological means of suppressing this plant or that animal"; instead we need the basic knowledge of animal populations and their relations to their surroundings that will "promote an even balance and damp down the explosive power of outbreaks and new invasions."

Much of the necessary knowledge is now available but we do not use it. We 21 train ecologists in our universities and even employ them in our governmental agencies but we seldom take their advice. We allow the chemical death rain to fall as though there were no alternative, whereas in fact there are many, and our ingenuity could soon discover many more if given opportunity.

Have we fallen into a mesmerized state that makes us accept as inevitable 22 that which is inferior or detrimental, as though having lost the will or the vision to demand that which is good? Such thinking, in the words of the ecologist Paul Shepard, "idealizes life with only its head out of water, inches above the limits of toleration of the corruption of its own environment. . . . Why should we tolerate a diet of weak poisons, a home in insipid surroundings, a circle of acquaintances who are not quite our enemies, the noise of motors with just enough relief to prevent insanity? Who would want to live in a world which is just not quite fatal?"

Yet such a world is pressed upon us. The crusade to create a chemically ster- 23 ile, insect-free world seems to have engendered a fanatic zeal on the part of many specialists and most of the so-called control agencies. On every hand there is evidence that those engaged in spraying operations exercise a ruthless power. "The regulatory entomologists . . . function as prosecutor, judge and jury, tax assessor and collector and sheriff to enforce their own orders," said Connecticut entomologist Neely Turner. The most flagrant abuses go unchecked in both state and federal agencies.

It is not my contention that chemical insecticides must never be used. I do 24 contend that we have put poisonous and biologically potent chemicals indiscriminately into the hands of persons largely or wholly ignorant of their potentials for harm. We have subjected enormous numbers of people to contact with these poisons, without their consent and often without their knowledge. If the Bill of Rights contains no guarantee that a citizen shall be secure against lethal poisons distributed either by private individuals or by public officials, it is surely only because our forefathers, despite their considerable wisdom and foresight, could conceive of no such problem.

I contend, furthermore, that we have allowed these chemicals to be used 25 with little or no advance investigation of their effect on soil, water, wildlife, and man himself. Future generations are unlikely to condone our lack of prudent concern for the integrity of the natural world that supports all life.

There is still very limited awareness of the nature of the threat. This is an era 26 of specialists, each of whom sees his own problem and is unaware of or intolerant of the larger frame into which it fits. It is also an era dominated by industry, in which the right to make a dollar at whatever cost is seldom challenged. When the public protests, confronted with some obvious evidence of damaging results of pesticide applications, it is fed little tranquilizing pills of half truth. We urgently need an end to these false assurances, to the sugar coating of unpalatable facts. It is the public that is being asked to assume the risks that the insect controllers calculate. The public must decide whether it wishes to continue on the present road, and it can do so only when in full possession of the facts. In the words of Jean Rostand, "The obligation to endure gives us the right to know."

COMPREHENSION

1. What does Carson mean by "the obligation to endure"?
2. What reasons does the author cite for the overpopulation of insects?
3. What remedies does Carson propose?

RHETORIC

1. What tone does Carson use in her essay? Does she seem to be a subjective or an objective writer? Give specific support for your response.
2. How does the use of words such as *dangerous, evil, irrevocable,* and *sinister* help shape the reader's reaction to the piece? What emotional and ethical appeals do such words indicate?
3. Examine the ordering of ideas in paragraph 4, and consider how such an order serves to reinforce Carson's argument.
4. Paragraph 9 consists of only one sentence. What is its function in the essay's scheme?
5. Examine Carson's use of expert testimony. How does it help strengthen her thesis?
6. How effectively does the essay's conclusion help tie up Carson's points? What is the writer's intent in this final paragraph? How does she accomplish this aim?

WRITING

1. Write an essay in which you suggest solutions to the problems brought up in Carson's piece. You may want to suggest measures that the average citizen can take to eliminate the casual use of insecticides to control the insect population.
2. Write a biographical research paper on Rachel Carson that focuses on her involvement with nature and environmental issues.
3. **WRITING AN ARGUMENT:** Write an essay entitled "Insects Are Not the Problem; Humanity Is." In this essay, argue that it is humanity's greed that has caused such an imbalance in nature as to threaten the planet's survival.

www.mhhe.com/
mhreader

For more information on Rachel Carson, go to:
More Resources > Ch. 13 Nature & Environment

＊

Am I Blue?

Alice Walker

Alice Walker (b. 1944) was born in Eatonton, Georgia, attended Spelman College, and graduated from Sarah Lawrence College. Besides being a prolific novelist, short-story writer, poet, and essayist, she has also been active in the civil rights movement. She often draws on both her own history and historical records to reflect on the African American experience. Some of her well-known books are The Color Purple *(1976),* You Can't Keep a Good Woman Down *(1981),* Living in the World: Selected Writings, 1973–1987 *(1987),* The Temple of My Familiar *(1989),* By the Light of My Father's Smile *(1999), and* The Way Forward Is with a Broken Heart *(2001). In the following essay from* Living in the World, *Walker questions the distinctions commonly made between human and animal.*

For about three years my companion and I rented a small house in the country that stood on the edge of a large meadow that appeared to run from the end of our deck straight into the mountains. The mountains, however, were quite far away, and between us and them there was, in fact, a town. It was one of the many pleasant aspects of the house that you never really were aware of this.   1

It was a house of many windows, low, wide, nearly floor to ceiling in the living room, which faced the meadow, and it was from one of these that I first saw our closest neighbor, a large white horse, cropping grass, flipping its mane, and ambling about—not over the entire meadow, which stretched well out of sight of the house, but over the five or so fenced-in acres that were next to the twenty-odd that we had rented. I soon learned that the horse, whose name was Blue, belonged to a man who lived in another town, but was boarded by our neighbors next door. Occasionally, one of the children, usually a stocky teenager, but sometimes a much younger girl or boy, could be seen riding Blue. They would appear in the meadow, climb up on his back, ride furiously for ten or fifteen minutes, then get off, slap Blue on the flanks, and not be seen again for a month or more.   2

There were many apple trees in our yard, and one by the fence that Blue could almost reach. We were soon in the habit of feeding him apples, which he relished, especially because by the middle of summer the meadow grasses—so green and succulent since January—had dried out from lack of rain, and Blue   3 stumbled about munching the dried stalks half-heartedly. Sometimes he would stand very still just by the apple tree, and when one of us came out he would whinny, snort loudly, or stamp the ground. This meant, of course: I want an apple.

It was quite wonderful to pick a few apples, or collect those that had fallen to the ground overnight, and patiently hold them, one by one, up to his large, toothy mouth. I remained as thrilled as a child by his flexible dark lips, huge, cubelike teeth that crunched the apples, core and all, with such finality, and his high, broad-breasted *enormity*; beside which, I felt small indeed. When I was a child, I used to ride horses, and was especially friendly with one named Nan until the day I was riding and my brother deliberately spooked her and I was thrown, head first, against the trunk of a tree. When I came to, I was in bed and my mother was bending worriedly over me; we silently agreed that perhaps horseback riding was not the safest sport for me. Since then I have walked, and prefer walking to horseback riding—but I had forgotten the depth of feeling one could see in horses' eyes.

I was therefore unprepared for the expression in Blue's. Blue was lonely. Blue was horribly lonely and bored. I was not shocked that this should be the case; five acres to tramp by yourself, endlessly, even in the most beautiful of meadows—and his was—cannot provide many interesting events, and once the rainy season turned to dry that was about it. No, I was shocked that I had forgotten that human animals and nonhuman animals can communicate quite well; if we are brought up around animals as children we take this for granted. By the time we are adults we no longer remember. However, the animals have not changed. They are in fact *completed* creations (at least they seem to be, so much more than we) who are not likely *to* change; it is their nature to express themselves. What else are they going to express? And they do. And, generally speaking, they are ignored.

After giving Blue the apples, I would wander back to the house, aware that he was observing me. Were more apples not forthcoming then? Was that to be his sole entertainment for the day? My partner's small son had decided he wanted to learn how to piece a quilt; we worked in silence on our respective squares as I thought . . .

Well, about slavery: about white children, who were raised by black people, who knew their first all-accepting love from black women, and then, when they were twelve or so, were told they must "forget" the deep levels of communication between themselves and "mammy" that they knew. Later they would be able to relate quite calmly, "My old mammy was sold to another good family." "My old mammy was _____." Fill in the blank. Many more years later a white woman would say: "I can't understand these Negroes, these blacks. What do they want? They're so different from us."

And about the Indians, considered to be "like animals" by the "settlers" (a very benign euphemism for what they actually were), who did not understand their description as a compliment.

And about the thousands of American men who marry Japanese, Korean, Filipina, and other non-English-speaking women and of how happy they report they are, *"blissfully,"* until their brides learn to speak English, at which point the marriages tend to fall apart. What then did the men see, when they looked into the eyes of the women they married, before they could speak English? Apparently only their own reflections.

I thought of society's impatience with the young. "Why are they playing the music so loud?" Perhaps the children have listened to much of the music of oppressed people their parents danced to before they were born, with its passionate but soft cries for acceptance and love, and they have wondered why their parents failed to hear. 10

I do not know how long Blue had inhabited his five beautiful, boring acres before we moved into our house; a year after we had arrived—and had also traveled to other valleys, other cities, other worlds—he was still there. 11

But then, in our second year at the house, something happened in Blue's life. One morning, looking out the window at the fog that lay like a ribbon over the meadow, I saw another horse, a brown one, at the other end of Blue's field. Blue appeared to be afraid of it, and for several days made no attempt to go near. We went away for a week. When we returned, Blue had decided to make friends and the two horses ambled or galloped along together, and Blue did not come nearly as often to the fence underneath the apple tree. 12

When he did, bringing his new friend with him, there was a different look in his eyes. A look of independence, of self-possession, of inalienable *horse*ness. His friend eventually became pregnant. For months and months there was, it seemed to me, a mutual feeling between me and the horses of justice, of peace. I fed apples to them both. The look in Blue's eyes was one of unabashed "this is *it*ness." 13

It did not, however, last forever. One day, after a visit to the city, I went out to give Blue some apples. He stood waiting, or so I thought, though not beneath the tree. When I shook the tree and jumped back from the shower of apples, he made no move. I carried some over to him. He managed to half-crunch one. The rest he let fall to the ground. I dreaded looking into his eyes—because I had of course noticed that Brown, his partner, had gone—but I did look. If I had been born into slavery, and my partner had been sold or killed, my eyes would have looked like that. The children next door explained that Blue's partner had been "put with him" (the same expression that old people used, I had noticed, when speaking of an ancestor during slavery who had been impregnated by her owner) so that they could mate and she conceive. Since that was accomplished, she had been taken back by her owner, who lived somewhere else. 14

Will she be back? I asked. 15

They didn't know. 16

Blue was like a crazed person. Blue *was*, to me, a crazed person. He galloped furiously, as if he were being ridden, around and around his five beautiful acres. He whinnied until he couldn't. He tore at the ground with his hooves. He butted himself against his single shade tree. He looked always and always toward the road down which his partner had gone. And then, occasionally, when he came 17 up for apples, or I took apples to him, he looked at me. It was a look so piercing, so full of grief, a look so *human*, I almost laughed (I felt too sad to cry) to think there are people who do not know that animals suffer. People like me who have forgotten, and daily forget, all that animals try to tell us. "Everything you do to us will happen to you; we are your teachers, as you are ours. We are one lesson" is essentially it, I think. There are those who never once have even considered animals' rights: those who have been taught that animals actually want to be used and abused by us, as small children "love" to be frightened, or women "love" to be mutilated and raped. . . . They are the great-grandchildren of those who honestly thought, because someone taught them this: "Woman can't think" and "niggers can't faint." But most disturbing of all, in Blue's large brown eyes was a new look, more painful than the look of despair: the look of disgust with human beings, with life; the look of hatred. And it was odd what the look of hatred did. It gave him, for the first time, the look of a beast. And what that meant was that he had put up a barrier within to protect himself from further violence; all the apples in the world wouldn't change that fact.

And so Blue remained, a beautiful part of our landscape, very peaceful to 18 look at from the window, white against the grass. Once a friend came to visit and said, looking out on the soothing view: "And it *would* have to be a *white* horse; the very image of freedom." And I thought, yes, the animals are forced to become for us merely "images" of what they once so beautifully expressed. And we are used to drinking milk from containers showing "contented" cows, whose real lives we want to hear nothing about, eating eggs and drumsticks from "happy" hens, and munching hamburgers advertised by bulls of integrity who seem to command their fate.

As we talked of freedom and justice one day for all, we sat down to steaks. 19 I am eating misery, I thought, as I took the first bite. And spit it out.

COMPREHENSION

1. What is the major thesis of the essay? Is it stated explicitly in the text or does one have to infer it? Explain.
2. In paragraph 5, Walker states that animals are "*completed* creations (at least they seem to be, so much more than we) who are not likely to change." What does she mean by making this distinction between animals and humans?
3. What is the significance of the title of the essay? Does it have more than one meaning? Explain your answer.

RHETORIC

1. In paragraph 4, Walker creates a vivid description of Blue. How does she achieve this?

2. In paragraph 7, Walker makes a cognitive association between the relationship between humans and animals and the relationship between whites and blacks during slavery. Does this transition seem too abrupt, or is there a rhetorical reason for the immediate comparison? Explain.

3. Explore the other analogies she makes in paragraphs 7 and 8. Are they pertinent? What is the rhetorical effect of juxtaposing seemingly different realms to convey one central idea?

4. Walker often breaks the conventions of "college English." For example, paragraphs 8 and 9 both begin with the coordinating conjunction *and*. Paragraph 12 begins with the coordinating conjunction *but*. Paragraphs 15 and 16 are only one line each. Explain the effect of each of these rhetorical devices. Find three other unusual rhetorical strategies—either on the paragraph or sentence level—and explain their effects.

5. In paragraphs 17 and 18, Walker speeds up the tempo of her writing by beginning many of the sentences with the conjunction *and*. What is the purpose and rhetorical effect of this strategy and how does it mimic—in linguistic terms—Blue's altered emotional state?

6. Walker seems to have a profound empathy for animals, yet it is only at the end that she is repulsed by the thought of eating meat. What rhetorical strategy is she employing in the conclusion that helps bring closure to her meditation on Blue? Does it matter whether the culminating event actually occurred in her experience, or is it all right for an essayist to use poetic license for stylistic purposes?

WRITING

1. Write a personal essay in which you describe your relationship with a favorite pet. Include your observations of, responses to, and attitude toward your pet. Compare and contrast this relationship to those you have with humans.

2. Some writers have argued that it matters little if certain "nonessential" endangered species become extinct if they interfere with "human progress." Argue for or against this proposition.

3. **WRITING AN ARGUMENT:** Argue for or against one of the following practices:
   a. Hunting for the sake of the hunt.
   b. Eating meat.
   c. Keeping animals in zoos.

www.mhhe.com/
mhreader

For more information on Alice Walker, go to:
**More Resources > Ch. 13 Nature
& Environment**

✳

The Greenest Campuses:
An Idiosyncratic Guide

Noel Perrin

Noel Perrin (b. 1927) was born in New York City and worked as an editor before start-
ing a career as a college instructor at the University of North Carolina and then Dart-
mouth College, where he has taught since 1959. He has been awarded two Guggenheim
Fellowships, has been a contributor to numerous periodicals, and has authored more than
10 books. His subject matter has ranged from the scholarly, such as Dr. Bowdler's
Legacy *(1969) and* Giving Up the Gun: Japan's Reversion to the Sword, 1543–1879
(1979), to his experiences as a part-time farmer. Among the latter are First Person
Plural *(1978),* Second Person Plural *(1980),* Third Person Plural *(1983), and* Last
Person Plural *(1991). His concerns about the environment have made him a popular*
speaker on ecological issues. In the following essay, first published in The Chronicle of
Higher Education *in April 2001, Perrin creates his own "best" college guide by rank-*
ing institutions of higher learning according to their environmental awareness.

About 1,100 American colleges and universities run at least a token environ-   1
mental-studies program, and many hundreds of those programs offer well-
designed and useful courses. But only a drastically smaller number practice
even a portion of what they teach. The one exception is recycling. Nearly every
institution that has so much as one lonely environmental-studies course also
does a little halfhearted recycling. Paper and glass, usually.

There are some glorious exceptions to those rather churlish observations,   2
I'm glad to say. How many? Nobody knows. No one has yet done the necessary
research (though the National Wildlife Federation's Campus Ecology program
is planning a survey).

Certainly *U.S. News & World Report* hasn't. Look at the rankings in their an-   3
nual college issue. The magazine uses a complex formula something like this:
Institution's reputation, 25 percent; student-retention rate, 20 percent; faculty resources, 20 percent; and so on, down to alumni giving, 5 percent. The lead criterion may help explain why Harvard, Yale, and Princeton Universities so
frequently do a little dance at the top of the list.

But *U.S. News* has nothing at all to say about the degree to which a college   4
or university attempts to behave sustainably—that is, to manage its campus and
activities in ways that promote the long-term health of the planet. The magazine
is equally mum about which of the institutions it is ranking can serve as models
to society in a threatened world.

And, of course, the world is threatened. When the Royal Society in London   5
and the National Academy of Sciences in Washington issued their first-ever
joint statement, it ended like this: "The future of our planet is in the balance.

Sustainable development can be achieved, but only if irreversible degradation of the environment can be halted in time. The next 30 years may be crucial." They said that in 1992. If all those top scientists are right, we have a little more than 20 years left in which to make major changes in how we live.

All this affects colleges. I have one environmentalist friend who loves to point out to the deans and trustees she meets that if we don't make such changes, and if the irreversible degradation of earth does occur, Harvard's huge endowment and Yale's lofty reputation will count for nothing. 6

But though *U.S. News* has nothing to say, fortunately there is a fairly good grapevine in the green world. I have spent considerable time in the past two years using it like an organic cell phone. By that means I have come up with a short, idiosyncratic list of green colleges, consisting of six that are a healthy green, two that are greener still, and three that I believe are the greenest in the United States. 7

Which approved surveying techniques have I used? None at all. Some of my evidence is anecdotal, and some of my conclusions are affected by my personal beliefs, such as that electric and hybrid cars are not just a good idea, but instruments of salvation. 8

Obviously I did not examine, even casually, all 1,100 institutions. I'm sure I have missed some outstanding performers. I hope I have missed a great many. 9

Now, here are the 11, starting with **Brown University.** 10

It is generally harder for a large urban university to move toward sustainable behavior than it is for a small-town college with maybe a thousand students. But it's not impossible. Both Brown, in the heart of Providence, R.I., and Yale University (by no means an environmental leader in other respects), in the heart of New Haven, Conn., have found a country way of dealing with food waste. Pigs. Both rely on pigs. 11

For the past 10 years, Brown has been shipping nearly all of its food waste to a Rhode Island piggery. Actually, not shipping it—just leaving it out at dawn each morning. The farmer comes to the campus and gets it. Not since Ralph Waldo Emerson took food scraps out to the family pig have these creatures enjoyed such a high intellectual connection. 12

But there is a big difference in scale. Where Emerson might have one pail of slops now and then, Brown generates 700 tons of edible garbage each year. Haulage fee: $0. Tipping fee: $0. (That's the cost of dumping the garbage into huge cookers, where it is heated for the pigs.) Annual savings to Brown: about $50,000. Addition to the American food supply: many tons of ham and bacon each year. 13

Of course, Brown does far more than feed a balanced diet to a lot of pigs. That's just the most exotic (for an urban institution) of its green actions "Brown is Green" became the official motto of the university in August 1990. It was accurate then, and it remains accurate now. 14

Yale is the only other urban institution I'm aware of that supports a pig population. Much of the credit goes to Cyril May, the university's environmen- 15 tal coordinator, just as much of the credit at Brown goes to its environmental co-ordinator, Kurt Teichert.

May has managed to locate two Connecticut piggeries. The one to which he 16 sends garbage presents problems. The farmer has demanded—and received—a collection fee. And he has developed an antagonistic relationship with some of Yale's food-service people. (There are a lot of them: The campus has 16 dining facilities.) May is working on an arrangement with the second piggery. But if it falls through, he says, "I may go back on semibended knee to the other."

Yale does not make the list as a green college, for reasons you will learn 17 later in this essay. But it might in a few more years

Carleton College is an interesting example of an institution turning green 18 almost overnight. No pig slops here; the dining halls are catered by Marriott. But change is coming fast.

In the summer of 1999, Carleton appointed its first-ever environmental co- 19 ordinator, a brand-new graduate named Rachel Smit. The one-year appointment was an experiment, with a cobbled-together salary and the humble title of "fifth-year intern." The experiment worked beyond anyone's expectation.

Smit began publishing an environmental newsletter called *The Green Bean* 20 and organized a small committee of undergraduates to explore the feasibility of composting the college's food waste, an effort that will soon begin. A surprised Marriott has already found itself serving organic dinners on Earth Day.

Better yet, the college set up an environmental-advisory committee of three 21 administrators, three faculty members, and three students to review all campus projects from a green perspective. Naturally, many of those projects will be buildings, and to evaluate them, Carleton is using the Minnesota Sustainable Design Guide, itself cowritten by Richard Strong, director of facilities.

The position of fifth-year intern is now a permanent one-year position, and 22 its salary is a regular part of the budget.

What's next? If Carleton gets a grant it has applied for, there will be a mas- 23 sive increase in environmental-studies courses and faculty seminars and, says the dean of budgets, "a whole range of green campus projects under the rubric of 'participatory learning.'"

And if Carleton doesn't get the grant? Same plans, slower pace. 24

Twenty years ago, **Dartmouth College** would have been a contender for the 25 title of greenest college in America, had such a title existed. It's still fairly green. It has a large and distinguished group of faculty members who teach environmental studies, good recycling, an organic farm that was used last summer in six courses, years of experience with solar panels, and a fair number of midlevel administrators (including three in the purchasing office) who are ardent believers in sustainability.

But the college has lost ground. Most troubling is its new $50-million li- 26 brary, which has an actual anti-environmental twist: A portion of the roof requires steam from the power plant to melt snow off of it. The architect, Robert Venturi, may be famous, but he's no environmentalist.

Dartmouth is a striking example of what I shall modestly call Perrin's Law: 27
No college or university can move far toward sustainability without the active
support of at least two senior administrators. Dartmouth has no such committed senior administrators at all. It used to. James Hornig, a former dean of sciences, and Frank Smallwood, a former provost, were instrumental in creating
the environmental-studies program, back in 1970. They are now emeriti. The
current senior administrators are not in the least hostile to sustainability; they
just give a very low priority to the college's practicing what it preaches.

Emory University is probably further into the use of nonpolluting and low- 28
polluting motor vehicles than any other college in the country. According to
Eric Gaither, senior associate vice president for business affairs, 60 percent of
Emory's fleet is powered by alternative fuels. The facilities-management office
has 40 electric carts, which maintenance workers use for getting around
campus. The community-service office (security and parking) has its own
electric carts and an electric patrol vehicle. There are five electric shuttle buses
and 14 compressed–natural-gas buses on order, plus one natural-gas bus in
service.

Bill Chace, Emory's president, has a battery-charging station for electric 29
cars in his garage, and until recently an electric car to charge. Georgia Power,
which lent the car, has recalled it, but Chace hopes to get it back. Meanwhile, he
rides his bike to work most of the time.

How has Emory made such giant strides? "It's easy to do," says Gaither, 30
"when your president wants you to."

If Carleton is a model of how a small college turns green, the **University of** 31
Michigan at Ann Arbor is a model of how a big university does. Carleton is
changing pretty much as an entity, while Michigan is more like the Electoral
College—50 separate entities. The School of Natural Resources casts its six votes
for sustainability, the English department casts its 12 for humanistic studies, the
recycling coordinator casts her 1, the electric-vehicle program casts its 2, and so
on. An institution of Michigan's size changes in bits and pieces.

Some of the bits show true leadership. For example, the university is within 32
weeks of buying a modest amount of green power. It makes about half of its own
electricity (at its heating plant) and buys the other half. Five percent of that other
half soon will come from renewable sources: hydro (water power) and biomass
(so-called fuel crops, which are grown specifically to be burned for power).

The supporters of sustainability at Michigan would like to see the univer- 33
sity adopt a version of what is known as the Kyoto Protocol. The agreement,
which the United States so far has refused to sign, requires that by 2012 each nation reduce its emission of greenhouse gases to 7 percent below its 1990 figure.
Michigan's version of the protocol, at present a pipe dream, would require the
university to do what the government won't—accept that reduction as a goal.

The immediate goal of "sustainabilists" at Ann Arbor is the creation of a 34
universitywide environmental coordinator, who would work either in the president's or the provost's office.

Giants are slow, but they are also strong.   35

Tulane University has the usual programs, among green institutions, in re- 36
cycling, composting, and energy efficiency. But what sets it apart is the Tulane
Environmental Law Clinic, which is staffed by third-year law students. The director is a faculty member, and there are three law "fellows," all lawyers, who
work with the students. The clinic does legal work for environmental organizations across Louisiana and "most likely has had a greater environmental impact
than all our other efforts combined," says Elizabeth Davey, Tulane's first-ever
environmental coordinator.

At least two campuses of the **University of California** (Berkeley is not among 37
them) have taken a first and even a second step toward sustainable behavior.
First step: symbolic action, like installing a few solar panels, to produce clean
energy and to help educate students. With luck, one of those little solar arrays
might produce as much as a 20th of a percent of the electricity the university
uses. It's a start.

The two campuses are Davis and Santa Cruz, and I think Davis nudges 38
ahead of Santa Cruz. That is primarily because Davis the city and Davis the university have done something almost miraculous. They have brought car culture
at least partially under control, greatly reducing air pollution as a result.

The city has a population of about 58,000, which includes 24,000 students. 39
According to reliable estimates, there are something over 50,000 bikes in town
or on the campus, all but a few hundred owned by their riders. Most of the
bikes are used regularly on the city's 45 miles of bike paths (closed to cars) and
the 47 miles of bike lanes (cars permitted in the other lanes). The university
maintains an additional 14 miles of bike paths on its large campus

What happens on rainy days? "A surprising number continue to bike," says 40
David Takemoto-Weerts, coordinator of Davis's bicycle program.

If every American college in a suitable climate were to behave like Davis, 41
we could close a medium-sized oil refinery. Maybe we could even get rid of one
coal-fired power plant, and thus seriously improve air quality.

The **University of New Hampshire** is trying to jump straight from sym- 42
bolic gestures, like installing a handful of solar panels, to the hardest task of all
for an institution trying to become green—establishing a completely new mind-
set among students, administrators, and faculty and staff members. It may well
succeed.

Campuses that have managed to change attitudes are rare. Prescott College, 43
in Prescott, Ariz., and Sterling College, in Craftsbury Common, Vt., are rumored
to have done so, and there may be two or three others. They're not on my list—
because they're so small, because their students tend to be bright green even before they arrive, and because I have limited space.

New Hampshire has several token green projects, including a tiny solar ar- 44
ray, able to produce one kilowatt at noon on a good day. And last April it inau-
gurated the Yellow Bike Cooperative. It is much smaller than anything that
happens at Davis, where a bike rack might be a hundred yards long. But it's also

more original and more communitarian. Anyone in Durham—student, burger flipper, associate dean—can join the Yellow Bike program by paying a $5 fee.

What you get right away is a key that unlocks all 50 bikes owned by the co- 45 operative. (They are repaired and painted by student volunteers.) Want to cross campus? Just go to the nearest bike rack, unlock a Yellow, and pedal off. The goal, says Julie Newman, of the Office of Sustainability Programs, is "to greatly decrease one-person car trips on campus."

But the main thrust at New Hampshire is consciousness-raising. When the 46 subject of composting food waste came up, the university held a seminar for its food workers.

New Hampshire's striking vigor is partly the result of a special endow- 47 ment—about $12.8-million—exclusively for the sustainability office. Tom Kelly, the director, refuses to equate sustainability with greenness. Being green, in the sense of avoiding pollution and promoting reuse, is just one aspect of living sustainably, which involves "the balancing of economic viability with ecological health and human well-being," he says.

Oberlin College is an exception to Perrin's Law. The college has gotten 48 deeply into environmental behavior without the active support of two or, indeed, any senior administrators. As at Dartmouth, the top people are not hostile; they just have other priorities.

Apparently, until this year, Oberlin's environmental-studies program was 49 housed in a dreary cellar. Now it's in the $8.2-million Adam Joseph Lewis Environmental Studies Center, which is one of the most environmentally benign college buildings in the world. The money for it was raised as a result of a deal that the department chairman, David Orr, made with the administration: He could raise money for his own program, provided that he approached only people and foundations that had never shown the faintest interest in Oberlin.

It's too soon for a full report on the building. It is loaded with solar 50 panels—690 of them, covering the roof (for a diagram of the building, see www.oberlin.edu/newserv/esc/escabout.html). In about a year, data will be available on how much energy the panels have saved and whether, as Orr hopes, the center will not only make all its own power, but even export some.

Northland College, in Wisconsin, also goes way beyond tokenism. Its 51 McLean Environmental Living and Learning Center, a two-year-old residence hall for 114 students, is topped by a 120-foot wind tower that, with a good breeze coming off Lake Superior, can generate 20 kilowatts of electricity. The building also includes three arrays of solar panels. They are only token-size, generating a total of 3.2 kilowatts at most. But one array does heat most of the water for one wing of McLean, while the other three form a test project.

One test array is fixed in place—it can't be aimed. Another is like that sun- 52 flower in Blake's poem—it countest the steps of the sun. Put more prosaically, it tracks the sun across the sky each day. The third array does that and can also be tilted to get the best angle for each season of the year.

Inside the dorm is a pair of composting toilets—an experiment, to see if stu- 53 dents will use them. Because no one is forced to try the new ones if they don't want to—plenty of conventional toilets are close by—it means something when James Miller, vice president and dean of student development and enrollment, reports, "Students almost always choose the composting bathrooms."

From the start, the college's goal has been to have McLean operate so efficiently that it consumes 40 percent less outside energy than would a conventional dormitory of the same dimensions. The building didn't reach that goal in its first year; energy use dropped only 34.2 percent. But anyone dealing with a new system knows to expect bugs at the beginning. There were some at Northland, including the wind generator's being down for three months. (As I write, it's turning busily.) Dean Miller is confident that the building will meet or exceed the college's energy-efficiency goal.

There is no room here to talk about the octagonal classroom structure made of bales of straw, built largely by students. Or about the fact that Northland's grounds are pesticide- and herbicide-free.

If Oberlin is a flagrant exception to Perrin's Law, **Middlebury College** is a strong confirmation. Middlebury is unique, as far as I know, in having not only senior administrators who strongly back environmentalism, but one senior administrator right inside the program. What Michigan wants, Middlebury has.

Nan Jenks-Jay, director of environmental affairs, reports directly to the provost. She is responsible for both the teaching side and the living-sustainably side of environmentalism. Under her are an environmental coordinator, Amy Self, and an academic-program coordinator, Janet Wiseman.

The program has powerful backers, including the president, John M. McCardell Jr.; the provost and executive vice president, Ronald D. Liebowitz; and the executive vice president for facilities planning, David W. Ginevan. But everyone I talked with at Middlebury, except for the occasional student who didn't want to trouble his mind with things like returnable bottles—to say nothing of acid rain—seemed at least somewhat committed to sustainable living.

Middlebury has what I think is the oldest environmental-studies program in the country; it began back in 1965. It has the best composting program I've ever seen. And, like Northland, it is pesticide- and herbicide-free.

Let me end as I began, with Harvard, Yale, and Princeton. And with *U.S. News*'s consistently ranking them in the top five, accompanied from time to time by the California Institute of Technology, Stanford University, and the Massachusetts Institute of Technology.

What if *U.S. News* did a green ranking? What if it based the listings on one of the few bits of hard data that can be widely compared: the percentage of waste that a college recycles?

Harvard would come out okay, though hardly at the top. The university recycled 24 percent of its waste last year, thanks in considerable part to the presence of Rob Gogan, the waste manager. He hopes to achieve 28 percent this year. That's feeble compared with Brown's 35 percent, and downright puny against Middlebury's 64 percent.

But compared with Yale and Princeton, it's magnificent. Most of the information I could get from Princeton is sadly dated. It comes from the 1995 report

of the Princeton Environmental Reform Committee, whose primary recommendation was that the university hire a full-time waste manager. The university
has not yet done so. And if any administrators on the campus know the current
recycling percentage, they're not telling.

And Yale—poor Yale! It does have a figure. Among the performances of the 64
20 or so other colleges and universities whose percentages I'm aware of, only
Carnegie Mellon's is worse. Yale: 19 percent. Carnegie Mellon: 11 percent.

What should universities—and society—be shooting for? How can you 65
ask? One-hundred-percent retrieval of everything retrievable, of course.

COMPREHENSION

1. Why does Perrin call his essay an "idiosyncratic guide" when environmentalism has become a major issue in most municipalities, regions, and
   countries?
2. Is Perrin's purpose to inform, argue, or both? Does he have a clear-cut
   thesis, or does he leave it up to the reader to infer the thesis? Explain.
3. What information is Perrin's informal guide providing that is not offered in
   more conventional college rankings? Is Perrin suggesting that parents and
   students consider "green rankings" in choosing which college to apply to?
   Explain.

RHETORIC

1. What purpose might Perrin have for choosing to create a "green guide" for
   colleges when there are so many other institutions or items he could have
   selected for review, such as corporations, towns, cities, automobiles, and
   numerous household products? What makes colleges and universities a
   particularly apt target?
2. Usually, Ivy League colleges are at the top of college guide lists as most desirable. Where do they place on Perrin's list? What ironic statement is Perrin making by providing their rankings on the "green scale"? What is he
   implying about American values, particularly as they pertain to education?
3. Colleges and universities often pride themselves on the renown of their faculties. Who are the people Perrin cites as models of academic worth? Why
   has he chosen them?
4. What is the ironic purpose behind the author mentioning "Perrin's Law"
   (paragraph 27)? Is it a true "law," like the law of gravity? What body of
   knowledge is the author satirizing by invoking such a law?
5. Perrin is not didactic, since he does not recommend that other colleges
   adopt the environmental measures his model colleges have chosen. Would
   more direct advocacy on his part have strengthened his argument or weakened it, or not have had any effect? Explain.

6. In Perrin's conclusion, he changes his purpose from providing a purely informational assessment to offering a strong reprimand and recommendation. Why does he wait until the concluding paragraph to do so?

WRITING

1. Describe an environmentally friendly practice conducted at your college or university. Is it truly helpful for the environment, or is it largely symbolic?
2. Compare and contrast the academically oriented courses and programs offered at your school with what your institution actually does in the way of helping the environment. Discuss which of the two priorities is more prominent, and why.
3. **WRITING AN ARGUMENT:** Argue for or against the proposition that a magazine such as *U.S. News & World Report* should include environmental awareness and practice in its formula for assessing the rankings of colleges.

www.mhhe.com/
mhreader

For more information on Noel Perrin, go to:
More Resources > Ch. 13 Nature & Environment

✳

The Good Farmer

Barbara Kingsolver

*Barbara Kingsolver (b. 1955) was born in Annapolis, Maryland, and raised in eastern Kentucky. She studied biology at DePauw University (BA, 1977) and the University of Arizona (MS, 1981) and worked as a scientist and scientific writer before beginning her career as a writer of fiction and essays. Her highly acclaimed books include novels—*The Bean Trees *(1988),* Animal Dreams *(1990),* Pigs in Heaven *(1993), and* The Poisonwood Bible *(1998); stories—*Homeland and Other Stories *(1989); poetry—*Another America *(1992); and essays—*High Tide in Tucson: Essays from Now or Never *(1996) and* Small Wonder *(2003). The following essay provides a personal appreciation of the agrarian movement.*

Sometime around my fortieth birthday I began an earnest study of agriculture.   1
I worked quietly on this project, speaking of my new interest to almost no one because of what they might think. Specifically, they might think I was out of my mind.

Why? Because at this moment in history it's considered smart to get *out* of 2
agriculture. And because I was already embarked on a career as a writer, doing
work that many people might consider intellectual and therefore superior to
anything involving the risk of dirty fingernails. Also, as a woman in my early
40s, I conformed to no right-minded picture of an apprentice farmer. And finally, with some chagrin I'll admit that I grew up among farmers and spent the
first decades of my life plotting my escape from a place that seemed to offer me
almost no potential for economic, intellectual, or spiritual satisfaction.

It took nigh onto half a lifetime before the valuables I'd casually left behind 3
turned up in the lost and found.

The truth, though, is that I'd kept some of that treasure jingling in my pock- 4
ets all along: I'd maintained an interest in gardening always, dragging it with
me wherever I went, even into a city backyard where a neighbor who worked
the night shift insisted that her numerous nocturnal cats had every right to use
my raised vegetable beds for their litter box. (I retaliated, in my way, by getting
a rooster who indulged his right to use the hour of 6 A.M. for his personal com-
punctions.) In graduate school I studied ecology and evolutionary biology, but
the complex mathematical models of predator-prey cycles only made sense to
me when I converted them in my mind to farmstead analogies—even though,
in those days, the Ecology Department and the College of Agriculture weren't
on speaking terms. In my 20s, when I was trying hard to reinvent myself as a
person without a Kentucky accent, I often found myself nevertheless the lone
argumentative voice in social circles where "farmers" were lumped with political troglodytes and devotees of All-Star wrestling.

Once in the early 1980s, when cigarette smoking had newly and drastically 5
fallen from fashion, I stood in someone's kitchen at a party and listened to
something like a Greek chorus chanting out the reasons why tobacco should be
eliminated from the face of the earth, like smallpox. Some wild tug on my heart
made me blurt out, "But what about the tobacco farmers?"

"Why," someone asked, glaring, "should I care about tobacco farmers?" 6

I was dumbstruck. I couldn't form the words to answer: Yes, it is carcino- 7
genic, and generally grown with too many inputs, but tobacco is the last big
commodity in America that's still mostly grown on family farms, in an economy that won't let these farmers shift to another crop. If it goes extinct, so do
they.

I couldn't speak because my mind was flooded with memory, pictures, 8
scents, secret thrills. Childhood afternoons spent reading Louisa May Alcott in
a barn loft suffused with the sweet scent of aged burley. The bright, warm days
in late spring and early fall when school was functionally closed because whole
extended families were drafted to the cooperative work of setting, cutting, stripping, or hanging tobacco. The incalculable fellowship measured out in funerals,
family reunions, even bad storms or late-night calvings. The hard-muscled
pride of showing I could finally throw a bale of hay onto the truck bed myself.
(The year before, when I was 11, I'd had the less honorable job of driving the
truck.) The satisfaction of walking across the stage at high school graduation in

a county where my name and my relationship to the land were both common knowledge.

But when I was pressed, that evening in the kitchen, I didn't try to defend   9 the poor tobacco farmer. As if the deck were not already stacked against his little family enterprise, he was now tarred with the brush of evil along with the companies that bought his product, amplified its toxicity, and attempted to sell it to children. In most cases it's just the more ordinary difficulty of the small family enterprise failing to measure up to the requisite standards of profitability and efficiency. And in every case the rational arguments I might frame in its favor will carry no weight without the attendant silk purse full of memories and sighs and songs of what family farming is worth. Those values are an old currency now, accepted as legal tender almost nowhere.

I found myself that day in the jaws of an impossible argument, and I find I   10 am there still. In my professional life I've learned that as long as I write novels and nonfiction books about strictly human conventions and constructions, I'm taken seriously. But when my writing strays into that muddy territory where humans are forced to own up to our dependence on the land, I'm apt to be declared quaintly irrelevant by the small, acutely urban clique that decides in this country what will be called worthy literature. (That clique does not, fortunately, hold much sway over what people actually read.) I understand their purview, I think. I realize I'm beholden to people working in urban centers for many things I love: They publish books, invent theater, produce films and music. But if I had not been raised such a polite Southern girl, I'd offer these critics a blunt proposition: I'll go a week without attending a movie or concert, you go a week without eating food, and at the end of it we'll sit down together and renegotiate "quaintly irrelevant."

This is a conversation that needs to happen. Increasingly I feel sure of it; I   11 just don't know how to go about it when so many have completely forgotten the genuine terms of human survival. Many adults, I'm convinced, believe that food comes from grocery stores. In Wendell Berry's novel *Jayber Crow*, a farmer coming to the failing end of his long economic struggle despaired aloud, "I've wished sometimes that the sons of bitches would starve. And now I'm getting afraid they actually will."

Like that farmer, I am frustrated with the imposed acrimony between pro-   12 ducers and consumers of food, as if this were a conflict in which one could possibly choose sides. I'm tired of the presumption of a nation divided between rural and urban populations whose interests are permanently at odds, whose votes will always be cast different ways, whose hearts and minds share no common ground. This is as wrong as blight, a useless way of thinking, similar to the propaganda warning us that any environmentalist program will necessarily be anti-human. Recently a national magazine asked me to write a commentary on the great divide between "the red and the blue"—imagery taken from election-night coverage that colored a map according to the party each state elected, suggesting a clear political difference between the rural heartland and urban coasts. Sorry, I replied to the magazine editors, but I'm the wrong person to ask: I live in red, tend to think blue, and mostly vote green. If you're looking for oversimplification, skip the likes of me.

Better yet, skip the whole idea. Recall that in many of those red states, just 13 a razor's edge under half the voters likely pulled the blue lever, and vice versa—not to mention the greater numbers everywhere who didn't even show up at the polls, so far did they feel from affectionate toward any of the available options. Recall that farmers and hunters, historically, are more active environmentalists than many progressive, city-dwelling vegetarians. (And, conversely, that some of the strongest land-conservation movements on the planet were born in the midst of cities.) Recall that we all have the same requirements for oxygen and drinking water, and that we all like them clean but relentlessly pollute them. Recall that whatever lofty things you might accomplish today, you will do them only because you first ate something that grew out of dirt.

We don't much care to think of ourselves that way—as creatures whose 14 cleanest aspirations depend ultimately on the health of our dirt. But our survival as a species depends on our coming to grips with that, along with some other corollary notions, and when I entered a comfortable midlife I began to see that my kids would get to do the same someday, or not, depending on how well our species could start owning up to its habitat and its food chain. As we faced one environmental crisis after another, did our species seem to be making this connection? As we say back home, not so's you'd notice.

Our gustatory industries treat food items like spoiled little celebrities, zip- 15 ping them around the globe in luxurious air-conditioned cabins, dressing them up in gaudy outfits, spritzing them with makeup, and breaking the bank on advertising, for heaven's sake. My farm-girl heritage makes me blush and turn down tickets to that particular circus. I'd rather wed my fortunes to the sturdy gal-next-door kind of food, growing what I need or getting it from local "you pick" orchards and our farmers' market.

It has come to pass that my husband and I, in what we hope is the middle 16 of our lives, are in possession of a farm. It's not a hobby homestead, it is a farm, somewhat derelict but with good potential. It came to us with some twenty acres of good, tillable bottomland, plus timbered slopes and all the pasture we can ever use, if we're willing to claim it back from the brambles. A similar arrangement is available with the 75-year-old apple orchard. The rest of the inventory includes a hundred-year-old clapboard house, a fine old barn that smells of aged burley, a granary, poultry coops, a root cellar, and a century's store of family legends. No poisons have been applied to this land for years, and we vow none ever will be.

Our agrarian education has come in as a slow undercurrent beneath our 17 workaday lives and the rearing of our children. Only our closest friends, probably, have taken real notice of the changes in our household: that nearly all the food we put on our table, in every season, was grown in our garden or very nearby. That the animals we eat took no more from the land than they gave back to it, and led sunlit, contentedly grassy lives. Our children know how to bake bread, stretch mozzarella cheese, ride a horse, keep a flock of hens laying, help a neighbor, pack a healthy lunch, and politely decline the world's less wholesome offerings. They know the first fresh garden tomato tastes as good as it does, partly, because you've waited for it since last Thanksgiving, and that the awful ones you could have bought at the grocery in between would only subtract from this equation. This rule applies to many things beyond tomatoes. I have noticed that the very politicians who support purely market-driven economics, which favors immediate corporate gratification over long-term responsibility, also express loud concern about the morals of our nation's children and their poor capacity for self-restraint. I wonder what kind of tomatoes those men feed their kids.

I have heard people of this same political ilk declare that it is perhaps sad 18 but surely inevitable that our farms are being cut up and sold to make nice-sized lawns for suburban folks to mow, because the most immediately profitable land use must prevail in a free country. And yet I have visited countries where people are perfectly free, such as the Netherlands, where this sort of disregard for farmland is both illegal and unthinkable. Plenty of people in this country, too, seem to share a respect for land that gives us food; why else did so many friends of my youth continue farming even while the economic prospects grew doubtful? And why is it that more of them each year are following sustainable practices that defer some immediate profits in favor of the long-term health of their fields, crops, animals, and watercourses? Who are the legions of Americans who now allocate more of their household budgets to food that is organically, sustainably, and locally grown, rather than buying the cheapest products they can find? My husband and I, bearing these trends in mind, did not contemplate the profitable option of subdividing our farm and changing its use. Frankly, that seemed wrong.

It's an interesting question, how to navigate this tangled path between 19 money and morality: not a new question by any means, but one that has taken strange turns in modern times. In our nation's prevailing culture, there exists right now a considerable confusion between prosperity and success—so much so that avarice is frequently confused with a work ethic. One's patriotism and good sense may be called into doubt if one elects to earn less money or own fewer possessions than is humanly possible. The notable exception is that a person may do so for religious reasons: Christians are asked by conscience to tithe or assist the poor; Muslims do not collect interest; Catholics may respectably choose a monastic life of communal poverty; and any of us may opt out of a scheme that we feel to be discomforting to our faith. It is in this spirit that we, like you perhaps and so many others before us, have worked to rein in the free market's tyranny over our family's tiny portion of America and install values that override the profit motive. Upon doing so, we receive a greater confidence in our children's future safety and happiness. I believe we are also happier souls in the present, for what that is worth. In the darkest months I look for solace in seed catalogues and articles on pasture rotation. I sleep better at night, feeling safely connected to the things that help make a person whole. It is fair to say this has been, in some sense, a spiritual conversion.

Modern American culture is fairly empty of any suggestion that one's rela- 20
tionship to the land, to consumption and food, is a religious matter. But it's true;
the decision to attend to the health of one's habitat and food chain is a spiritual
choice. It's also a political choice, a scientific one, a personal and a convivial one.
It's not a choice between living in the country or the town; it is about understanding that every one of us, at the level of our cells and respiration, lives in
the country and is thus obliged to be mindful of the distance between ourselves
and our sustenance.

I have worlds to learn about being a good farmer. Last spring when a hard 21
frost fell upon our orchards on May 21, I felt despair at ever getting there at all.
But in any weather, I may hope to carry a good agrarian frame of mind into my
orchards and fields, my kitchen, my children's schools, my writing life, my
friendships, my grocery shopping, and the county landfill. That's the point: It
goes everywhere. It may or may not be a movement—I'll leave that to others to
say. But it does move, and it works for us.

COMPREHENSION

1. What do we learn about the author from this essay? Why does her decision
   to study agriculture and become a farmer grow out of her experience and
   training?
2. What values does Kingsolver find in the culture of tobacco farming?
3. The author speaks of "a conversation that needs to happen" (paragraph 11).
   What is the nature and substance of this conversation?

RHETORIC

1. What is the rhetorical purpose of the title? How does the thesis or claim
   grow from our understanding of the title?
2. This essay was first published in an anthology, *The Essential Agrarian Reader:
   The Future of Culture, Community, and the Land* (2003). How does Kingsolver's essay capture the spirit of this reader? What is her definition of
   agrarianism, and how does she develop it?
3. Explain the persona that Kingsolver crafts for herself in this essay. How
   does the process of self-reinvention help to organize the essay's content?
4. Kingsolver is a noted fiction writer with a sharply descriptive style of writing. Analyze paragraph 8 as an example of her descriptive ability. How do
   description and narration combine as a rhetorical strategy in this essay?
5. The author uses the comparative method to advance assertions and opinions. Where does this method appear, and what is its purpose?
6. What logical and ethical appeals does Kingsolver advance to support her
   argument?
7. How does Kingsolver construct her conclusion? How effective do you find
   it, and why?

WRITING

1. In a narrative and descriptive essay, write about some experience that made you feel close to the land, nature, or the environment. How did this experience affect you, your aspirations, and your values?
2. Research the agrarian movement, and write an essay that defines it.
3. **WRITING AN ARGUMENT:** Argue for or against the proposition that our divorce from the land keeps us from finding harmony with the world and is the cause behind economic, political, environmental, and religious strife.

✴ **www.mhhe.com/ mhreader**	For more information on Barbara Kingsolver, go to: **More Resources > Ch. 13 Nature & Environment**

CONNECTIONS FOR CRITICAL THINKING

1. Using support from the works of Lopez, Carson, Chief Seattle, and others, write a causal-analysis essay tracing our relationship to the land. To what extent have history, greed, and fear helped shape our attitude? Can this attitude be changed? How?
2. Consider the empathy and sensitivity Walker has toward animals. How do her attitude and perceptions coincide with the view expressed by Chief Seattle concerning the natural world?
3. Write a letter to the op-ed page of a newspaper objecting to a governmental ruling harmful to the environment. State the nature of the policy, its possible dangers, and your reasons for opposing it. Use support from McKibben and any of the writers in this chapter. Extra reading or research may be necessary.
4. Consider why we fear nature. Why do we consider it an enemy, an alien, something to be destroyed? How would Momaday, Walker, Lopez, and Chief Seattle respond to this question? Do you agree or disagree with them?
5. Both Lopez and Kingsolver use narration and description to explore our relationship to the land. How do they approach their subject in terms of language, attitude, and style?
6. Choose an author in this chapter whose essay, in your opinion, romanticizes nature. Compare his or her attitude with that of a writer with a more pragmatic approach to the subject. Compare the two views, and specify the elements in their writing that contribute to the overall strength of their arguments.
7. Perrin uses enumeration and illustration to structure his essay. What are some of the strengths and weaknesses of employing traditional and orderly means of presenting one's thoughts?

8. Write an essay entitled "Nature's Revenge" in which you examine the consequences of environmental abuse. Consider the short- as well as the long-term effects on the quality of life. Use support from any writer in this chapter to defend your opinion.

9. Write specifically about our relationship to other living creatures on our planet. Is it one of exploitation, cooperation, or tyranny? How does this relationship influence how we treat each other? Explore the answers to these questions in an essay. Use the works of Lopez, Walker, and Chief Seattle to support your thesis.

10. Join a newsgroup on the Web devoted to addressing a specific environmental issue, for example, atomic waste, overdevelopment, or environmental regulations and deregulations. Follow the conversation of the newsgroup for one month. Write an essay describing the chief concerns of the newsgroup members, how they address issues regarding the environment, and what specific actions they recommend or take over the course of your membership.

11. Visit the Web site of the Environmental Protection Agency. Write a report describing the agency's announcements, speeches, activities, and proposals.

12. Create your own interactive Web site focusing on the environment. Present, in its headline, this request: "In a statement of 100 words, please explain whether we are doing enough to reverse the destruction to our environment." Check back in a month, and write a report summarizing the responses.

chapter *14*

Science and Technology
What Can Science Teach Us?

Contrary to popular assumptions, contemporary science and technology are not dry subjects but rather bodies of specialized knowledge concerned with the great how and why questions of our time. In fact, we are currently in the midst of a whole series of scientific revolutions that will radically transform our lives as we enter the 21st century. The essential problem for humankind is to make sense of all this revolutionary scientific and technological knowledge, invest it with value, use it ethically, and make it serve our cultural and global needs.

As you will see in the essays assembled for this chapter, human beings are always the ultimate subject of scientific investigation. Science and mathematics attempt to understand the physical, biological, and chemical events that shape our lives. Whenever we switch on a light or turn on a computer, take an aspirin or start the car, we see that science and technology have intervened effectively in our lives. Often the specialized knowledge of science forces us to make painful decisions, and the misuse of science can have disastrous results. As Terry Tempest Williams demonstrates in her highly personal essay "The Clan of One-Breasted Women," science can have dire, unforeseen ethical implications.

The technology that arises from science affects everyday decisions as well as the larger contours of culture. Nowhere is the impact of science more apparent than in the field of biotechnology. As Dinesh D'Souza observes in his essay on the biotech revolution, science is intended to serve us, to help us with our common dilemmas. At the same time, biotechnology reminds us that despite advances, we are still mortals confronting ethical dilemmas. Even as knowledge flows from research laboratories, these mortal paradoxes tend to perplex and goad us as we seek scientific solutions to the complex problems of our era.

Science and technology as specialized bodies of knowledge can send contradictory messages because science and technology are socially constructed and reflect the contours of culture. How we manage the revolution in science—

how we harness nuclear power or battle the ravages of AIDS—will determine the health of civilization in our century.

Previewing the Chapter

As you read the essays in this chapter and respond to them in discussion and writing, consider the following questions:

- Does the author take a personal or an objective approach to the subject? What is the effect?
- What area of scientific or technological inquiry does the writer focus on?
- What scientific conflicts arise in the course of the essay?
- Is the writer a specialist, a layperson, a journalist, or a commentator? How does the background of the writer affect the tone of the essay?
- What assumptions does the author make about his or her audience? How much specialized knowledge must you bring to the essay?
- How do social issues enter into the author's presentation?
- What gender issues are raised by the author?
- How have your perceptions of the author's topic been changed or enhanced? What new knowledge have you gained? Does the writer contradict any of your assumptions or beliefs?
- Is the writer optimistic or pessimistic about the state of technology or science? How do you know?

Classic and Contemporary Images

WHERE IS SCIENCE TAKING US?

Using a Critical Perspective   Make a series of observations about each of these images. Where does your eye rest in each one? How many objects and details do you see? What reasonable inferences can you draw about the relationship of the artist who created the 15th century image to the culture and historical period? What purpose did the scientists who created and control the Hubble Space Telescope have? What purposes do the 15th century artist and 20th century scientist have in common? Argue for or against the proposition that art can actually capture the advances in science, technology, and humanity that we have experienced over time.

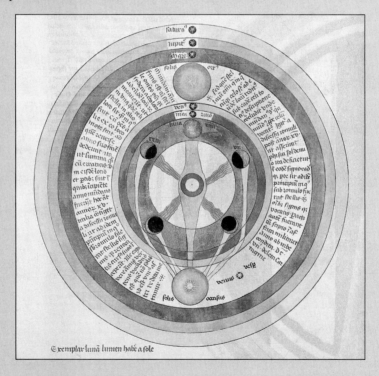

During the Renaissance in Europe, scientists such as Nicolaus Copernicus (1473–1543) and Galileo Galilei (1564–1642) revolutionized the way Europeans viewed the universe and their place in it by proving that the earth and the planets revolve around the sun, thus changing forever the worldview exemplified by the 15th century Flemish depiction of the movements of the sun and moon shown here.

Galileo's primitive telescope was a distant forerunner of the powerful Hubble
Space Telescope, launched in 1990, which is able to take photographs
of extremely distant stars and other phenomenon, such as
the gaseous pillars shown here, as it orbits the earth.

Classic and Contemporary Essays
HOW HAS NATURE EVOLVED?

Evolution seems to be more highly and hotly debated each year, insinuating itself into educational, political, scientific, and religious debates. Unfortunately, extreme positions on evolution tend to obscure what is valuable about the concept. The idea or theory of evolution is rooted in scientific creativity and the scientific method. Scientists are observers and collectors of information, and they use facts to build an explanation of the natural world and our place in it. Darwin looked carefully at nature and the physical world, as the following selection on natural selection illustrates. In his autobiography, Darwin observed that he "collected facts on a wholesale scale" before arriving at his theory of natural selection, and this inductive approach, at the heart of *The Origin of Species* (1859), typifies his method of inquiry. But does evolutionary theory explain all the facts—all of humankind's problems? Does evolution mark the progress of human civilization, or are dimensions needed to explain our relationship to the world and the world's events? Natalie Angier, one of the most lucid and persuasive contemporary science writers, tackles these questions and more in an essay prompted by the events of September 11, 2001. Basing her ideas on recent studies in evolutionary biology and other sciences, Angier presents Darwin in a new light, finding his theories generally sufficient to explain our responses to 9/11 but also suggesting that an added dimension to human behavior—beyond Darwin—is required to fully explain our struggle for existence and meaning in the modern world.

✳

Natural Selection

Charles Darwin

Charles Darwin (1809–1882) was born in England and studied medicine at Edinburgh. He also studied for the ministry at Cambridge but soon turned his interest to natural history. Through his friendship with a well-known botanist, he was given the opportunity to take a five-year cruise around the world (1831–1836) aboard the H.M.S. Beagle, *serving as a naturalist. This started Darwin on a career of accumulating and assimilating data that resulted in the formulation of his concept of evolution. He spent the remainder of his life carefully and methodically working over the information from his copious notes. He first published his findings in 1858 and a year later published his influential* Origin of Species. *This seminal work was supplemented and elaborated in many later books, including* The Descent of Man *(1871). The following selection demonstrates the methodical and meticulous method Darwin used in developing his concepts.*

In order to make it clear how, as I believe, natural selection acts, I must beg permission to give one or two imaginary illustrations. Let us take the case of a wolf, which preys on various animals, securing some by craft, some by strength, and some by fleetness; and let us suppose that the fleetest prey, a deer for instance, had from any change in the country increased in numbers, or that other prey had decreased in numbers, during that season of the year when the wolf is hardest pressed for food. I can under such circumstances see no reason to doubt that the swiftest and slimmest wolves would have the best chance of surviving, and so be preserved or selected, provided always that they retained strength to master their prey at this or at some other period of the year, when they might be compelled to prey on other animals. I can see no more reason to doubt this, than that man can improve the fleetness of his greyhounds by careful and methodical selection, or by that unconscious selection which results from each man trying to keep the best dogs without any thought of modifying the breed.

Even without any change in the proportional numbers of the animals on which our wolf preyed, a cub might be born with an innate tendency to pursue certain kinds of prey. Nor can this be thought very improbable; for we often observe great differences in the natural tendencies of our domestic animals; one cat, for instance, taking to catch rats, another mice; one cat, according to Mr. St. John, bringing home winged game, another hares or rabbits, and another hunting on marshy ground and almost nightly catching woodcocks or snipes. The tendency to catch rats rather than mice is known to be inherited. Now, if any slight innate change of habit or of structure benefited an individual wolf, it would have the best chance of surviving and of leaving offspring. Some of its young would probably inherit the same habits or structure, and by the repetition of this process, a new variety might be formed which would either supplant or coexist with the parent-form of wolf. Or, again, the wolves inhabiting a mountainous district, and those frequenting the lowlands, would naturally be forced to hunt different prey; and from the continued preservation of the individuals best fitted for the two sites, two varieties might slowly be formed. These varieties would cross and blend where they met; but to this subject of intercrossing we shall soon have to return. I may add, that, according to Mr. Pierce, there are two varieties of the wolf inhabiting the Catskill Mountains in the United States, one with a light greyhound-like form, which pursues deer, and the other more bulky, with shorter legs, which more frequently attacks the shepherd's flocks.

Let us now take a more complex case. Certain plants excrete a sweet juice, apparently for the sake of eliminating something injurious from their sap; this is effected by glands at the base of the stipules in some Leguminosae, and at the back of the leaf of the common laurel. This juice, though small in quantity, is greedily sought by insects. Let us now suppose a little sweet juice or nectar to be excreted by the inner bases of the petals of a flower. In this case insects in seeking the nectar would get dusted with pollen, and would certainly often transport the pollen from one flower to the stigma of another flower. The flowers of two distinct individuals of the same species would thus get crossed; and the act of crossing, we have good reason to believe (as will hereafter be more fully alluded to), would produce very vigorous seedlings, which consequently would have the best chance of flourishing and surviving. Some of these seedlings would probably inherit the nectar-excreting power. Those individual flowers which had the largest glands or nectaries, and which excreted most nectar, would be oftenest visited by insects, and would be oftenest crossed; and so in the long-run would gain the upper hand. Those flowers, also, which had their stamens and pistils placed, in relation to the size and habits of the particular insects which visited them, so as to favor in any degree the transportal of their pollen from flower to flower, would likewise be favored or selected. We might have taken the case of insects visiting flowers for the sake of collecting pollen instead of nectar; and as pollen is formed for the sole object of fertilization, its destruction appears a simple loss to the plant; yet if a little pollen were carried, at first occasionally and then habitually, by the pollen-devouring insects from flower to flower, and a cross thus effected, although nine-tenths of the pollen were destroyed, it might still be a great gain to the plant; and those individuals which produced more and more pollen, and had larger and larger anthers, would be selected. [3]

When our plant, by this process of the continued preservation or natural selection of more and more attractive flowers, had been rendered highly attractive to insects, they would, unintentionally on their part, regularly carry pollen from flower to flower; and that they can most effectually do this, I could easily show by many striking instances. I will give only one—not as a very striking case, but as likewise illustrating one step in the separation of the sexes of plants, presently to be alluded to. Some holly-trees bear only male flowers, which have [4] four stamens producing rather a small quantity of pollen, and a rudimentary pistil; other holly-trees bear only female flowers; these have a full-sized pistil, and four stamens with shriveled anthers, in which not a grain of pollen can be detected. Having found a female tree exactly sixty yards from a male tree, I put the stigmas of twenty flowers, taken from different branches, under the microscope, and on all, without exception, there were pollen-grains, and on some a profusion of pollen. As the wind had set for several days from the female to the male tree, the pollen could not thus have been carried. The weather had been cold and boisterous, and therefore not favorable to bees; nevertheless every female flower which I examined had been effectually fertilized by the bees, accidentally dusted with pollen, having flown from tree to tree in search of nectar. But to return to our imaginary case: as soon as the plant had been rendered so highly attractive to insects that pollen was regularly carried from flower to flower, another process might commence. No naturalist doubts the advantage of what has been called the "physiological division of labor"; hence we may believe that it would be advantageous to a plant to produce stamens alone in one flower or on one whole plant, and pistils alone in another flower or on another plant. In plants under culture and placed under new conditions of life, sometimes the male organs and sometimes the female organs become more or less impotent; now if we suppose this to occur in ever so slight a degree under nature, then as pollen is already carried regularly from flower to flower, and as a more complete separation of the sexes of our plant would be advantageous on the principle of the division of labor, individuals with this tendency more and more increased, would be continually favored or selected, until at last a complete separation of the sexes would be effected.

Let us now turn to the nectar-feeding insects in our imaginary case: we may suppose the plant of which we have been slowly increasing the nectar by continued selection, to be a common plant; and that certain insects depended in main part on its nectar for food. I could give many facts, showing how anxious bees are to save time; for instance, their habit of cutting holes and sucking the nectar at the bases of certain flowers, which they can, with a very little more trouble, enter by the mouth. Bearing such facts in mind, I can see no reason to doubt that an accidental deviation in the size and form of the body, or in the curvature and length of the proboscis, etc., far too slight to be appreciated by us, might profit a bee or other insect, so that an individual so characterized would be able to obtain its food more quickly, and so have a better chance of living and leaving descendants. Its descendants would probably inherit a tendency to a similar slight deviation of structure. The tubes of the corollas of the common red and incarnate clovers (Trifolium pratense and incarnatum) do not on a hasty glance appear to differ in length; yet the hive-bee can easily suck the nectar out of the incarnate clover, but not out of the common red clover, which is visited by humble-bees alone; so that the whole fields of the red clover offer in vain an abundant supply of precious nectar to the hive-bee. Thus it might be a great advantage to the hive-bee to have a slightly longer or differently constructed proboscis. On the other hand, I have found by experiment that the fertility of clover greatly depends on bees visiting and moving parts of the corolla, so as to push the pollen on to the stigmatic surface. Hence, again, if humble-bees were to become rare in any country, it might be a great advantage to the red clover to have a shorter or more deeply divided tube to its corolla, so that the hive-bee could visit its flowers. Thus I can understand how a flower and a bee might slowly become, either simultaneously or one after the other, modified and adapted in the most perfect manner to each other, by the continued preservation of individuals presenting mutual and slightly favorable deviations of structure.

I am well aware that this doctrine of natural selection, exemplified in the   6
above imaginary instances, is open to the same objections which were at first urged against Sir Charles Lyell's noble views on "the modern changes of the earth, as illustrative of geology"; but we now very seldom hear the action, for instance, of the coast-waves, called a trifling and insignificant cause, when applied to the excavation of gigantic valleys or to the formation of the longest lines of inland cliffs. Natural selection can act only by the preservation and accumulation of infinitesimally small inherited modifications, each profitable to the preserved being; and as modern geology has almost banished such views as the excavation of a great valley by a single diluvial wave, so will natural selection, if it be a true principle, banish the belief of the continued creation of new organic beings, or of any great and sudden modification in their structure.

COMPREHENSION

1. What does Darwin mean by the term *natural selection?*
2. What is Darwin attempting to refute by his concept of natural selection? Where in the essay is this refutation articulated?
3. Explain what Darwin means by the "physiological division of labor" (paragraph 4).
4. Define the following terms: *innate* (paragraph 2), *stamens* (paragraph 3), *pistils* (paragraph 3), *rudimentary* (paragraph 4), *incarnate* (paragraph 5), and *doctrine* (paragraph 6).

RHETORIC

1. In the introduction, Darwin makes an analogy between the needs of humans and those of nature. What is this analogy, and why is it important in devising his argument?
2. What is the tone of the essay? Consider such phrases as "beg permission" (paragraph 1) and "Let us now" (paragraph 3).
3. Darwin uses two "imaginary illustrations" in an attempt to prove his point. What are they, and why are these hypothetical illustrations more effective than real-life ones for his purpose?
4. Darwin tends to use extremely long sentences when he wishes to illustrate a process. For example, the sentence in paragraph 3 that begins "We might

have taken" is 101 words long. Deconstruct this sentence by paying special attention to its punctuation, its logical succession of clauses, and its effect on the reader of describing so many processes within its boundaries. What is the relationship between rhetorical style and its purpose?

5. Who is the implied audience for the essay? Cite specific aspects of the rhetoric that led you to your conclusion.

6. What gives Darwin his authority? Specifically, how is his authority linked to the specialized vocabulary of the essay and to the way Darwin uses language to articulate natural processes?

7. Darwin uses the argumentative technique of disarming potential critics in the final paragraph. What is the rhetorical function of this device? Does it strengthen or weaken his argument? Explain your view.

WRITING

1. Write a précis of the essay, focusing on the major points Darwin is trying to assert in his theory of natural selection.

2. **WRITING AN ARGUMENT:** In an essay, argue for or against the proposition that in order to agree with or attempt to refute Darwin's ideas of natural selection, one would have to have at least as much experience in observing nature as Darwin obviously had.

3. **WRITING AN ARGUMENT:** Argue for or against the view that Darwin's theory can have disastrous consequences for the human species if applied to politics, sociology, or economics.

---

www.mhhe.com/
mhreader

For more information on Charles Darwin, go to:
More Resources > Ch. 14 Science & Technology

✳

Of Altruism, Heroism and Nature's Gifts in the Face of Terror

Natalie Angier

Natalie Angier (b. 1958) grew up in New York City and graduated from Barnard College in 1978. She started her career a founding staff member of Discover *magazine, covering topics in biology. In 1990 she became a reporter for* The New York Times. *Her work as a* Times *science correspondent led to a Pulitzer Prize in 1991. She is also the recipient of the Lewis Thomas Award and was one of only seven journalists to receive four stars in the* Forbes Media Guide *that rated 500 reporters. Angier has also*

published in The Atlantic, Parade, Washington Monthly, *and* Reader's Digest.
Her books include The Beauty of the Beastly: New Views on the Nature of Life
(1995) and the national best seller Women: An Intimate Geography *(1999). In the
following essay, which appeared in* The New York Times *one week after the terrorist
attacks of September 11, 2001, Angier finds that evolution offers hopeful prospects for
human nature.*

For the worldless, formless, expectant citizens of tomorrow, here are some post-   1
cards of all that matters today:

Minutes after terrorists slam jet planes into the towers of the World Trade   2
Center, streams of harrowed humanity crowd the emergency stairwells, heading in two directions. While terrified employees scramble down, toward exit
doors and survival, hundreds of New York firefighters, each laden with 70 to
100 pounds of lifesaving gear, charge upward, never to be seen again.

As the last of four hijacked planes advances toward an unknown but surely   3
populated destination, passengers huddle together and plot resistance against
their captors, an act that may explain why the plane fails to reach its target,
crashing instead into an empty field outside Pittsburgh.

Hearing of the tragedy whose dimensions cannot be charted or absorbed,   4
tens of thousands of people across the nation storm their local hospitals and
blood banks, begging for the chance to give blood, something of themselves to
the hearts of the wounded—and the heart of us all—beating against the void.

Altruism and heroism. If not for these twin radiant badges of our humanity,   5
there would be no us, and we know it. And so, when their vile opposite threatened to choke us into submission last Tuesday, we rallied them in quantities so
great we surprised even ourselves.

Nothing and nobody can fully explain the source of the emotional genius   6
that has been everywhere on display. Politicians have cast it as evidence of the
indomitable spirit of a rock-solid America; pastors have given credit to a more
celestial source. And while biologists in no way claim to have discovered the
key to human nobility, they do have their own spin on the subject. The altruistic impulse, they say, is a nondenominational gift, the birthright and defining
characteristic of the human species.

As they see it, the roots of altruistic behavior far predate *Homo sapiens,* and   7
that is why it seems to flow forth so readily once tapped. Recent studies that
model group dynamics suggest that a spirit of cooperation will arise in nature
under a wide variety of circumstances.

"There's a general trend in evolutionary biology toward recognizing that   8
very often the best way to compete is to cooperate," said Dr. Barbara Smuts, a
professor of anthropology at the University of Michigan, who has published papers on the evolution of altruism. "And that, to me, is a source of some solace
and comfort."

Moreover, most biologists concur that the human capacity for language and   9
memory allows altruistic behavior—the desire to give, and to sacrifice for the
sake of others—to flourish in measure far beyond the cooperative spirit seen in
other species.

With language, they say, people can learn of individuals they have never 10 met and feel compassion for their suffering, and honor and even emulate their heroic deeds. They can also warn one another of any selfish cheaters or malign tricksters lurking in their midst.

"In a large crowd, we know who the good guys are, and we can talk about, 11 and ostracize, the bad ones," said Dr. Craig Packer, a professor of ecology and evolution at the University of Minnesota. "People are very concerned about their reputation, and that, too, can inspire us to be good."

Oh, better than good.   12

"There's a grandness in the human species that is so striking, and so pro- 13 foundly different from what we see in other animals," he added. "We are an amalgamation of families working together. This is what civilization is derived from."

At the same time, said biologists, the very conditions that encourage hero- 14 ics and selflessness can be the source of profound barbarism as well. "Moral behavior is often a within-group phenomenon," said Dr. David Sloan Wilson, a professor of biology at the State University of New York at Binghamton. "Altruism is practiced within your group, and often turned off toward members of other groups."

The desire to understand the nature of altruism has occupied evolutionary 15 thinkers since Charles Darwin, who was fascinated by the apparent existence of altruism among social insects. In ant and bee colonies, sterile female workers labor ceaselessly for their queen, and will even die for her when the nest is threatened. How could such seeming selflessness evolve, when it is exactly those individuals that are behaving altruistically that fail to breed and thereby pass their selfless genes along?

By a similar token, human soldiers who go to war often are at the beginning 16 of their reproductive potential, and many are killed before getting the chance to have children. Why don't the stay-at-homes simply outbreed the do-gooders and thus bury the altruistic impulse along with the casualties of combat?

The question of altruism was at least partly solved when the British evolu- 17 tionary theorist William Hamilton formulated the idea of inclusive fitness: the notion that individuals can enhance their reproductive success not merely by having young of their own, but by caring for their genetic relatives as well. Among social bees and ants, it turns out, the sister workers are more closely related to one another than parents normally are to their offspring; thus it behooves the workers to care more about current and potential sisters than to fret over their sterile selves.

The concept of inclusive fitness explains many brave acts observed in nature. 18 Dr. Richard Wrangham, a primatologist at Harvard, cites the example of the red colobus monkey. When they are being hunted by chimpanzees, the male monkeys are "amazingly brave," Dr. Wrangham said. "As the biggest and strongest members of their group, they undoubtedly could escape quicker than the others." Instead, the males jump to the front, confronting the chimpanzee hunters while the mothers and offspring jump to safety. Often, the much bigger chimpanzees pull the colobus soldiers off by their tails and slam them to their deaths.

Their courageousness can be explained by the fact that colobus monkeys 19 live in multimale, multifemale groups in which the males are almost always related. So in protecting the young monkeys, the adult males are defending their kin.

Yet, as biologists are learning, there is more to cooperation and generosity 20 than an investment in one's nepotistic patch of DNA. Lately, they have accrued evidence that something like group selection encourages the evolution of traits beneficial to a group, even when members of the group are not related.

In computer simulation studies, Dr. Smuts and her colleagues modeled two 21 types of group-living agents that would behave like herbivores: one that would selfishly consume all the food in a given patch before moving on, and another that would consume resources modestly rather than greedily, thus allowing local plant food to regenerate.

Researchers had assumed that cooperators could collaborate with geneti- 22 cally unrelated cooperators only if they had the cognitive capacity to know goodness when they saw it.

But the data suggested otherwise. "These models showed that under a wide 23 range of simulated environmental conditions you could get selection for prudent, cooperative behavior," Dr. Smuts said, even in the absence of cognition or kinship. "If you happened by chance to get good guys together, they remained together because they created a mutually beneficial environment."

This sort of win-win principle, she said, could explain all sorts of symbiotic 24 arrangements, even among different species—like the tendency of baboons and impalas to associate together because they use each other's warning calls.

Add to this basic mechanistic selection for cooperation the human capacity 25 to recognize and reward behaviors that strengthen the group—the tribe, the state, the church, the platoon—and selflessness thrives and multiplies. So, too, does the need for group identity. Classic so-called minimal group experiments have shown that when people are gathered together and assigned membership in arbitrary groups, called, say, the Greens and the Reds, before long the members begin expressing amity for their fellow Greens or Reds and animosity toward those of the wrong "color."

"Ancestral life frequently consisted of intergroup conflict," Dr. Wilson of 26 SUNY said. "It's part of our mental heritage."

Yet he does not see conflict as inevitable. "It's been shown pretty well that 27 where people place the boundary between us and them is extremely flexible and strategic," he said. "It's possible to widen the moral circle, and I'm optimistic enough to believe it can be done on a worldwide scale."

Ultimately, though, scientists acknowledge that the evolutionary frame- 28 work for self-sacrificing acts is overlaid by individual choice. And it is there, when individual firefighters or office workers or airplane passengers choose the altruistic path, that science gives way to wonder.

Dr. James J. Moore, a professor of anthropology at the University of Califor- 29 nia at San Diego, said he had studied many species, including many different primates. "We're the nicest species I know," he said. "To see those guys risking their lives, climbing over rubble on the chance of finding one person alive, well, you wouldn't find baboons doing that." The horrors of last week notwithstanding, he said, "the overall picture to come out about human nature is wonderful."

"For every 50 people making bomb threats now to mosques," he said, 30 "there are 500,000 people around the world behaving just the way we hoped they would, with empathy and expressions of grief. We are amazingly civilized."

True, death-defying acts of heroism may be the province of the few. For the 31 rest of us, simple humanity will do.

COMPREHENSION

1. According to Angier, how should we view the events of September 11, 2001, from the perspective of evolutionary science? What else can we learn about human nature that cannot be explained by evolutionary biology? What does the author say about Darwin?
2. Explain the relationship, as the writer presents it, between altruism and heroism. What does she mean by the "win-win principle" (paragraph 24)?
3. What experiments does the author cite to support her main ideas?

RHETORIC

1. Which paragraphs constitute the author's introduction? What strategy does she employ to start her essay? Why is this introduction especially compelling?
2. What is Angier's claim? What types of evidence does she offer to support her argument? Where does she appeal to reason, ethics, and emotion? Does she engage in refutation? Why or why not?
3. How does the author use comparison and contrast to organize this selection? How does definition serve to advance the essay and the argument?
4. Explain the writer's tone. Is she objective or subjective? Is she optimistic or pessimistic? What other attitudes toward her subject can you detect? Justify your response.
5. Where does the writer use figurative language? What is the effect?
6. Angier's concluding paragraph is brief. Do you find it effective? Why or why not?

WRITING

1. Using the data provided by Angier in her essay, and any additional research you may want to conduct, write an essay on the events of 9/11 from the perspective of evolutionary biology.

2. Write a comparison/contrast essay on altruism and heroism, explaining why they are unique to the human species. Be certain to provide sufficient examples to support your thesis.

3. **WRITING AN ARGUMENT:** In an essay, argue for or against the idea that the way we respond to crises confirms that we have evolved as a species. Provide evidence. Incorporate appeals to reason, ethics, and emotion. Refute those who might oppose your main and minor propositions.

www.mhhe.com/   For more information on Natalie Angier, go to:
mhreader   **More Resources > Ch. 14 Science & Technology**

Classic and Contemporary: Questions for Comparison

1. How do Darwin and Angier approach the subject of evolution? Do they have the same or different priorities, and why? Are they writing for the same audience? Use examples from both selections to support your response.

2. Analyze the language used in the two essays. What is similar or different about the style and diction of the two pieces? How does each use details? Is one essay more accessible to the modern reader? Why or why not?

3. How do both essays reflect the scientific method? Is one selection more "scientific" in its approach than the other? Explain your response.

✴

Nutcracker.com

David Sedaris

David Sedaris (b. 1957), who was born in Johnson City, New York, and grew up in Raleigh, North Carolina, is a well-known humorist, essayist, diarist, short-story writer, and radio commentator. After graduating from the Art Institute of Chicago in 1987, Sedaris held several temporary jobs, ranging from a cleaner of apartments to an elf in SantaLand at Macy's. His stint on National Public Radio's Morning Edition *established Sedaris as a popular if quirky humorist and led to his first collection of essays,* Barrel Fever (2000). *Termed by* Entertainment Weekly *"a crackpot in the best sense of the word," Sedaris has also written* Naked (1997), Me Talk Pretty One Day (2000), Dress Your Family in Corduroy and Denim (2004), *and other works. In this essay, Sedaris humorously explains why he is a technophobe.*

It was my father's dream that one day the people of the world would be con- 1
nected to one another through a network of blocky, refrigerator-size computers,
much like those he was helping develop at IBM. He envisioned families of the
future gathered around their mammoth terminals, ordering groceries and paying their taxes from the comfort of their own homes. A person could compose
music, design a doghouse, and . . . something more, something even better. "A
person could . . . he could . . ."

When predicting this utopia, he would eventually reach a point where 2
words failed him. His eyes would widen and sparkle at the thought of this in-
describable something more. "I mean, my God," he'd say, "just think about it."

My sisters and I preferred not to. I didn't know about them, but I was hop- 3
ing the people of the world might be united by something more interesting, like
drugs or an armed struggle against the undead. Unfortunately, my father's
team won, so computers it is. My only regret is that this had to happen during
my lifetime.

Somewhere in the back of my mind is a dim memory of standing in some 4
line holding a perforated card. I remember the cheap, slightly clinical feeling it
gave me, and recall thinking that the computer would never advance much further than this. Call me naive, but I seem to have underestimated the universal
desire to sit in a hard plastic chair and stare at a screen until your eyes cross. My
father saw it coming, but this was a future that took me completely by surprise.
There were no computers in my high school, and the first two times I attempted
college, people were still counting on their fingers and removing their shoes
when the numbers got above ten. I wasn't really aware of computers until the
mid-1980s. For some reason, I seemed to know quite a few graphic designers
whose homes and offices pleasantly stank of Spray Mount. Their floors were always collaged with stray bits of paper, and trapped flies waved for help from
the gummy killing fields of their tabletops. I had always counted on these
friends to loan me the adhesive of my choice, but then, seemingly overnight,
their Scotch tape and rubber cement were gone, replaced with odorless computers and spongy mouse pads. They had nothing left that I wanted to borrow, and
so I dropped them and fell in with a group of typesetters who ultimately betrayed me as well.

Thanks to my complete lack of office skills, I found it fairly easy to avoid di- 5
rect contact with the new technology. The indirect contact was disturbing
enough. I was still living in Chicago when I began to receive creepy Christmas
newsletters designed to look like tabloids and annual reports. Word processors
made writing fun. They did not, however, make reading fun, a point made
painfully evident by such publications as *The Herald Family Tribune* and *Wassup
with the Wexlers!*

Friends who had previously expressed no interest in torture began sending 6
letters composed to resemble Chinese take-out menus and the Dead Sea Scrolls.
Everybody had a font, and I was told that I should get one, too. The authors of
these letters shared an enthusiasm with the sort of people who now arrived at
dinner parties hoisting expensive new video cameras and suggesting that, after

dessert, we all sit down and replay the evening on TV. We, the regular people of the world, now had access to the means of production, but still I failed to see what all the fuss was about. A dopey letter is still a dopey letter, no matter how you dress it up; and there's a reason regular people don't appear on TV: we're boring.

By the early 1990s I was living in New York and working for a houseclean-   7
ing company. My job taught me that regardless of their purported virtues, computers are a pain in the ass to keep clean. The pebbled surface is a magnet for grease and dirt, and you can pretty much forget about reaming out the gaps in the keyboard. More than once I accidentally pushed a button and recoiled in terror as the blank screen came to life with exotic tropical fish or swarms of flying toasters. Equally distressing was the way people used the slanted roofs of their terminals to display framed photographs and great populations of plush and plastic creatures, which would fall behind the desk the moment I began cleaning the screen. There was never any place to plug in the vacuum, as every outlet was occupied by some member of the computer family. Cords ran wild, and everyone seemed to own one of those ominous foot-long power strips with the blinking red light that sends the message YOU MUST LEAVE US ALONE. I was more than happy to comply, and the complaints came rolling in.

Due to my general aversion to machines and a few pronounced episodes of   8
screaming, I was labeled a technophobe, a term that ranks fairly low on my scale of fightin' words. The word *phobic* has its place when properly used, but lately it's been declawed by the pompous insistence that most animosity is based upon fear rather than loathing. No credit is given for distinguishing between these two very different emotions. I fear snakes. I hate computers. My hatred is entrenched, and I nourish it daily. I'm comfortable with it, and no community outreach program will change my mind.

I hate computers for getting their own section in the *New York Times* and for   9
lengthening commercials with the mention of a Web site address. Who really wants to find out more about Procter & Gamble? Just buy the toothpaste or laundry detergent, and get on with it. I hate them for creating the word *org* and I hate them for e-mail, which isn't real mail but a variation of the pointless notes people used to pass in class. I hate computers for replacing the card catalog in the New York Public Library and I hate the way they've invaded the movies. I'm not talking about their contribution to the world of special effects. I have nothing against a well-defined mutant or full-scale alien invasion—that's *good* technology. I'm talking about their actual presence *in* any given movie. They've become like horses in a western—they may not be the main focus, but everybody seems to have one. Each tiresome new thriller includes a scene in which the hero, trapped by some version of the enemy, runs for his desk in a desperate race against time. Music swells and droplets of sweat rain down onto the keyboard as he sits at his laptop, frantically pawing for answers. It might be different if he were flagging down a passing car or trying to phone for help, but typing, in and of itself, is not an inherently dramatic activity.

I hate computers for any number of reasons, but I despise them most for 10 what they've done to my friend the typewriter. In a democratic country you'd think there would be room for both of them, but computers won't rest until I'm making my ribbons from torn shirts and brewing Wite-Out in my bathtub. Their goal is to place the IBM Selectric II beside the feather quill and chisel in the museum of antiquated writing implements. They're power hungry, and someone needs to stop them.

When told I'm like the guy still pining for his eight-track tapes, I say, "You 11 have eight-tracks? Where?" In reality I know nothing about them, yet I feel it's important to express some solidarity with others who have had the rug pulled out from beneath them. I don't care if it can count words or rearrange paragraphs at the push of a button, I don't want a computer. Unlike the faint scurry raised by fingers against a plastic computer keyboard, the smack and clatter of a typewriter suggests that you're actually building something. At the end of a miserable day, instead of grieving my virtual nothing, I can always look at my loaded wastepaper basket and tell myself that if I failed, at least I took a few trees down with me.

When forced to leave my house for an extended period of time, I take my 12 typewriter with me, and together we endure the wretchedness of passing through the X-ray scanner. The laptops roll merrily down the belt, while I'm instructed to stand aside and open my bag. To me it seems like a normal enough thing to be carrying, but the typewriter's declining popularity arouses suspicion and I wind up eliciting the sort of reaction one might expect when traveling with a cannon.

"It's a typewriter," I say. "You use it to write angry letters to airport 13 authorities."

The keys are then slapped and pounded, and I'm forced to explain that if 14 you want the words to appear, you first have to plug it in and insert a sheet of paper.

The goons shake their heads and tell me I really should be using a com- 15 puter. That's their job, to stand around in an ill-fitting uniform and tell you how you should lead your life. I'm told the exact same thing later in the evening when the bellhop knocks on my hotel door. The people whose televisions I can hear have complained about my typing, and he has come to make me stop. To hear him talk, you'd think I'd been playing the kettledrum. In the great scheme of things, the typewriter is not nearly as loud as he makes it out to be, but there's no use arguing with him. "You know," he says, "you really should be using a computer."

You have to wonder where you've gone wrong when twice a day you're of- 16 fered writing advice from men in funny hats. The harder I'm pressured to use a computer, the harder I resist. One by one, all of my friends have deserted me and fled to the dark side. "How can I write you if you don't have an e-mail address?" they ask. They talk of their B-trees and Disk Doctors and then have the nerve to complain when I discuss bowel obstructions at the dinner table.

Who needs them? I think. I figured I'd always have my family and was dev- 17
astated when my sister Amy brought home a candy-colored laptop. "I only use
it for e-mail," she said. Coming from her, these words made me physically ill.
"It's fun," she said. "People send you things. Look at this." She pushed a button, and there, on the screen, was a naked man lying facedown on a carpet. His
hair was graying and his hands were cuffed behind his doughy back. A woman
entered the room. You couldn't see her face, just her legs and feet, which were
big and mean-looking, forced into sharp-toed shoes with high, pencil-thin heels.
The man on the carpet shifted position, and when his testicles came into view,
the woman reacted as if she had seen an old balding mouse, one that she had
been trying to kill for a long time. She stomped on the man's testicles with the
toes of her shoes and then she turned around and stomped on them with the
heels. She kicked them mercilessly and, just when I thought she'd finished, she
got her second wind and started all over again.

I'd never realized that a computer could act so much like a TV set. No one 18
had ever told me that the picture could be so clear, that the cries of pain could
be heard so distinctly. This, I thought, was what my father had been envision-
ing all those years ago when words had failed him, not necessarily this scene,
but something equally capable of provoking such wonder.

"Again?" Amy pushed a button and, our faces bathed in the glow of the 19
screen, we watched the future a second time.

COMPREHENSION

1. What distinguishes the author from his father and his sister Amy?
2. Why does Sedaris hate computers? Does he actually enjoy being a "technophobe"? How do you know?
3. What sort of writer is Sedaris? Why doesn't he want to use computers to
   help him in the writing process?

RHETORIC

1. How do you interpret the title? How does the title prepare us for the tone of
   this selection?
2. Sedaris employs a personal voice in this essay. What does "I" point of view
   contribute to the selection?
3. What comic strategies does the author develop? What details stand out?
   Does comedy serve to support or undercut his claim?
4. How does Sedaris argue his case? Does he actually have a case, or a cause,
   or is his purpose simply to amuse the reader? How do you know?
5. What principle of classification appears in the essay, and how does this
   rhetorical strategy serve the author's purpose?
6. How does the writer use comparison and contrast, narration, and description to develop the essay?

7. Explain the impact and significance of the concluding scene. How does the tone alter here? What is the final effect? What is the writer's parting message to his readers?

WRITING

1. In a personal essay employing narration and description as well as analysis, describe the impact of computers on your family life. Use a comic approach to the subject.
2. Write an analysis of the elements of humor that appear in Sedaris's essay. Why is comedy an appropriate strategy for dealing with the subject of computer technology?
3. **WRITING AN ARGUMENT:** Write an essay arguing why you love or hate computers. Use ironic humor to undercut your argument—to convince readers that your opinions are actually the opposite of what you proclaim.

www.mhhe.com/
mhreader   For more information on David Sedaris, go to:
More Resources > Ch. 14 Science & Technology

✳

Entropy

K. C. Cole

K. C. Cole (b. 1946) was born in Detroit, Michigan. After receiving a BA from Colum-
bia University (1968), she worked as a freelance reporter, magazine writer, and editor
for several publications, including Saturday Review, Discover, Washington Post,
and Long Island Newsday. *Since 1994, she has been a science writer for the* Los Angeles Times *and a writer for the Exploratorium, an interactive science museum in San*
Francisco. Praised for her ability to make complex scientific concepts understandable to
a general audience, Cole has written several highly praised books, including Order in
the Universe *(1982),* The Universe and the Teacup: The Mathematics of Truth
and Beauty *(1998),* First You Build a Cloud: Reflections on Physics as a Way of
Life *(1999),* The Hole in the Universe: How Scientists Peered over the Edge of
Emptiness and Found Everything *(2001), and* Mind over Matter *(2003). In the*
following essay, Cole explains a basic law of physics and demonstrates the ways in
which it affects life on our planet.

It was about two months ago when I realized that entropy was getting the bet- 1
ter of me. On the same day my car broke down (again), my refrigerator conked
out and I learned that I needed root-canal work in my right rear tooth. The windows in the bedroom were still leaking every time it rained and my son's baby
sitter was still failing to show up every time I really needed her. My hair was
turning gray and my typewriter was wearing out. The house needed paint and
I needed glasses. My son's sneakers were developing holes and I was developing a deep sense of futility.

After all, what was the point of spending half of Saturday at the Laundro- 2
mat if the clothes were dirty all over again the following Friday?

Disorder, alas, is the natural order of things in the universe. There is even a 3
precise measure of the amount of disorder, called entropy. Unlike almost every
other physical property (motion, gravity, energy), entropy does not work both
ways. It can only increase. Once it's created it can never be destroyed. The road
to disorder is a one-way street.

Because of its unnerving irreversibility, entropy has been called the arrow of 4
time. We all understand this instinctively. Children's rooms, left on their own,
tend to get messy, not neat. Wood rots, metal rusts, people wrinkle and flowers
wither. Even mountains wear down; even the nuclei of atoms decay. In the city
we see entropy in the rundown subways and worn-out sidewalks and torn-
down buildings, in the increasing disorder of our lives. We know, without asking, what is old. If we were suddenly to see the paint jump back on an old
building, we would know that something was wrong. If we saw an egg un-
scramble itself and jump back into its shell, we would laugh in the same way
we laugh at a movie run backward.

Entropy is no laughing matter, however, because with every increase in en- 5
tropy energy is wasted and opportunity is lost. Water flowing down a mountainside can be made to do some useful work on its way. But once all the water
is at the same level it can work no more. That is entropy. When my refrigerator
was working, it kept all the cold air ordered in one part of the kitchen and
warmer air in another. Once it broke down the warm and cold mixed into a
lukewarm mess that allowed my butter to melt, my milk to rot and my frozen
vegetables to decay.

Of course the energy is not really lost, but it has diffused and dissipated 6
into a chaotic caldron of randomness that can do us no possible good. Entropy
is chaos. It is loss of purpose.

People are often upset by the entropy they seem to see in the haphazard- 7
ness of their own lives. Buffeted about like so many molecules in my tepid
kitchen, they feel that they have lost their sense of direction, that they are wasting youth and opportunity at every turn. It is easy to see entropy in marriages,
when the partners are too preoccupied to patch small things up, almost guaran-
teeing that they will fall apart. There is much entropy in the state of our country, in the relationships between nations—lost opportunities to stop the
avalanche of disorders that seems ready to swallow us all.

Entropy is not inevitable everywhere, however. Crystals and snowflakes 8 and galaxies are islands of incredibly ordered beauty in the midst of random events. If it was not for exceptions to entropy, the sky would be black and we would be able to see where the stars spend their days; it is only because air molecules in the atmosphere cluster in ordered groups that the sky is blue.

The most profound exception to entropy is the creation of life. A seed soaks 9 up some soil and some carbon and some sunshine and some water and arranges it into a rose. A seed in the womb takes some oxygen and pizza and milk and transforms it into a baby.

The catch is that it takes a lot of energy to produce a baby. It also takes en- 10 ergy to make a tree. The road to disorder is all downhill but the road to creation takes work. Though combating entropy is possible, it also has its price. That's why it seems so hard to get ourselves together, so easy to let ourselves fall apart.

Worse, creating order in one corner of the universe always creates more dis- 11 order somewhere else. We create ordered energy from oil and coal at the price of the entropy of smog.

I recently took up playing the flute again after an absence of several 12 months. As the uneven vibrations screeched through the house, my son covered his ears and said, "Mom, what's wrong with your flute?" Nothing was wrong with my flute, of course. It was my ability to play it that had atrophied, or entropied, as the case may be. The only way to stop that process was to practice every day, and sure enough my tone improved, though only at the price of constant work. Like anything else, abilities deteriorate when we stop applying our energies to them.

That's why entropy is depressing. It seems as if just breaking even is an up- 13 hill fight. There's a good reason that this should be so. The mechanics of entropy are a matter of chance. Take any ice-cold air molecule milling around my kitchen. The chances that it will wander in the direction of my refrigerator at any point are exactly 50–50. The chances that it will wander away from my refrigerator are also 50–50. But take billions of warm and cold molecules mixed together, and the chances that all the cold ones will wander toward the refrigerator and all the warm ones will wander away from it are virtually nil.

Entropy wins not because order is impossible but because there are always 14 so many more paths toward disorder than toward order. There are so many more different ways to do a sloppy job than a good one, so many more ways to make a mess than to clean it up. The obstacles and accidents in our lives almost guarantee that constant collisions will bounce us on to random paths, get us off the track. Disorder is the path of least resistance, the easy but not the inevitable road.

Like so many others, I am distressed by the entropy I see around me today. 15 I am afraid of the randomness of international events, of the lack of common purpose in the world; I am terrified that it will lead into the ultimate entropy of nuclear war. I am upset that I could not in the city where I live send my child to a public school; that people are unemployed and inflation is out of control; that tensions between sexes and races seem to be increasing again; that relationships everywhere seem to be falling apart.

Social institutions—like atoms and stars—decay if energy is not added to 16 keep them ordered. Friendships and families and economies all fall apart unless we constantly make an effort to keep them working and well oiled. And far too few people, it seems to me, are willing to contribute consistently to those efforts.

Of course, the more complex things are, the harder it is. If there were only a 17 dozen or so air molecules in my kitchen, it would be likely—if I waited a year or so—that at some point the six coldest ones would congregate inside the freezer. But the more factors in the equation—the more players in the game— the less likely it is that their paths will coincide in an orderly way. The more pieces in the puzzle, the harder it is to put back together once order is disturbed. "Irreversibility," said a physicist, "is the price we pay for complexity."

COMPREHENSION

1. What, according to the writer, is entropy? What does she mean when she calls entropy "the arrow of time" (paragraph 4)? How does entropy govern her daily life, society, nations, and the universe?
2. According to Cole, is entropy inevitable? What can we do to combat entropy?
3. What other physical laws or concepts does the author discuss in this essay? How do they relate to the theory of entropy?

RHETORIC

1. What is the author's thesis or claim? Is the thesis explicit or implicit? Explain.
2. How does Cole make this essay on a topic in physics comprehensible to a general audience? What specific strategies serve her general purpose?
3. What is the tone of this essay? How do words like *futility, tepid, loss,* and *atrophied* contribute to this tone? What other words capture the writer's attitude toward her subject?
4. How does the author develop her extended definition? What strategy does she use in the introduction? What additional rhetorical strategies does she employ, and what is the effect?
5. Would you say that the author's purpose is expository, argumentative, or a combination of both? Justify your response.

WRITING

1. Adopting Cole's personal style, explain how you contend with entropy in your daily life.

2. Take a concept from physics or any other area of science, and write an essay of definition in which you try to make this concept understandable to a general audience.
3. **WRITING AN ARGUMENT:** In a persuasive essay, either agree or disagree with Cole's implication that entropy seems ascendant in personal, social, and political (as well as physical) realms.

✳

Science, Guided by Ethics, Can Lift Up the Poor

Freeman J. Dyson

Freeman J. Dyson (b. 1923) was born in Crowthorne, England. He received a BA from Cambridge University (1945) and came to the United States in 1947 to study with Hans Bethe at Cornell University. He settled permanently in the United States in 1951 and became a naturalized citizen in 1957. Dyson was appointed professor of physics at the Institute for Advanced Study at Princeton University in 1953 and worked there until his retirement in 1994. Renowned for his research in several fields—among them mathematics, physics, astrophysics, climate studies, and nuclear engineering—Dyson also has been a professional voice advocating sanity in the application of science and technology to human affairs. His books for a general public include Disturbing the Universe *(1979),* Weapons and Hope *(1984),* Imagined Worlds *(1997), and* The Sun, the Genome, and the Internet *(1999). In 2000, Dyson was awarded the Templeton Prize for Progress in Religion; an extract from his speech, given at the Washington National Cathedral, follows.*

Throughout history, people have used technology to change the world. Our   1
technology has been of two kinds, green and gray. Green technology is seeds and plants, gardens and vineyards and orchards, domesticated horses and cows and pigs, milk and cheese, leather and wool. Gray technology is bronze and steel, spears and guns, coal and oil and electricity, automobiles and airplanes and rockets, telephones and computers. Civilization began with green technology, with agriculture and animal-breeding, 10,000 years ago. Then, beginning about 3,000 years ago, gray technology became dominant, with mining

and metallurgy and machinery. For the last 500 years, gray technology has been racing ahead and has given birth to the modern world of cities and factories and supermarkets.

The dominance of gray technology is coming to an end. During the last 50 years, we have achieved a fundamental understanding of the processes in living cells. With understanding comes the ability to exploit and control. Out of the knowledge acquired by modern biology, modern biotechnology is growing. The new green technology will give us the power, using only sunlight as a source of energy, and air and water and soil as materials, to manufacture and recycle chemicals of all kinds. Our gray technology of machines and computers will not disappear, but green technology will be moving ahead even faster. Green technology can be cleaner, more flexible and less wasteful than our existing chemical industries. A great variety of manufactured objects could be grown instead of made. Green technology could supply human needs with far less damage to the natural environment. And green technology could be a great equalizer, bringing wealth to the tropical areas of the planet, which have most of the world's sunshine, people and poverty. I am saying that green technology could do all these good things, not that green technology will do all these good things.

To make these good things happen, we need not only the new technology but the political and economic conditions that will give people all over the world a chance to use it. To make these things happen, we need a powerful push from ethics. We need a consensus of public opinion around the world that the existing gross inequalities in the distribution of wealth are intolerable. In reaching such a consensus, religions must play an essential role. Neither technology alone nor religion alone is powerful enough to bring social justice to human societies, but technology and religion working together might do the job.

We all know that green technology has a dark side, just as gray technology has a dark side. Gray technology brought us hydrogen bombs as well as telephones. Green technology brought us anthrax bombs as well as antibiotics. Besides the dangers of biological weapons, green technology brings other dangers having nothing to do with weapons. The ultimate danger of green technology comes from its power to change the nature of human beings by the application of genetic engineering to human embryos. If we allow a free market in human genes, wealthy parents will be able to buy what they consider superior genes for their babies. This could cause a splitting of humanity into hereditary castes. Within a few generations, the children of rich and poor could become separate species. Humanity would then have regressed all the way back to a society of masters and slaves. No matter how strongly we believe in the virtues of a free market economy, the free market must not extend to human genes.

I see two tremendous goods coming from biotechnology: first, the alleviation of human misery through progress in medicine, and second, the transformation of the global economy through green technology spreading wealth more equitably around the world. The two great evils to be avoided are the use of biological weapons and the corruption of human nature by buying and selling genes. I see no scientific reason why we should not achieve the good and avoid the evil. The obstacles to achieving the good are political rather than technical. Unfortunately a large number of people in many countries are strongly opposed to green technology, for reasons having little to do with the real dangers. It is important to treat the opponents with respect, to pay attention to their fears, to go gently into the new world of green technology so that neither human dignity nor religious conviction is violated. If we can go gently, we have a good chance of achieving within a hundred years the goals of ecological sustainability and social justice that green technology brings within our reach.

The great question for our time is how to make sure that the continuing scientific revolution brings benefits to everybody rather than widening the gap between rich and poor. To lift up poor countries, and poor people in rich countries, from poverty, technology is not enough. Technology must be guided and driven by ethics if it is to do more than provide new toys for the rich. Scientists and business leaders who care about social justice should join forces with environmental and religious organizations to give political clout to ethics. Science and religion should work together to abolish the gross inequalities that prevail in the modern world. That is my vision, and it is the same vision that inspired Francis Bacon 400 years ago, when he prayed that through science God would "endow the human family with new mercies." 6

COMPREHENSION

1. What does Dyson mean by "green" and "gray" technology? What has been their role in history?
2. Why, according to the author, is green technology better than gray technology for the world today?
3. What ethical issues does the author raise in this essay?

RHETORIC

1. State Dyson's claim in your own words. What evidence does he provide to substantiate his claim. Do you find this evidence convincing or not? Explain.
2. What is Dyson's purpose in contrasting green and gray technologies? Summarize these differences. How do these primary contrasts, and other comparative devices, serve to organize the essay?
3. Does the author use subject-by-subject or point-by-point comparison? Why do you think he selects the method he does?
4. Where and how does Dyson appeal to ethics in this essay?
5. Explain the functions of the writer's topic sentences. Why are they especially effective in this essay?
6. Why does Dyson inject religion into the essay in his concluding paragraph? Does this distract from the author's tone or strengthen it? Justify your answer.

WRITING

1. Using Dyson's essay as a model, write a comparison-and-contrast paper on green and gray technology. Develop your own comparative points, and provide examples and evidence to support your thesis.
2. Write a comparative paper on the creative and destructive aspects of science that confront the world today.
3. **WRITING AN ARGUMENT:** Argue for or against the proposition that what humanity needs today is a green rather than gray technological revolution. Appeal to reason and ethics as you develop your argument.

www.mhhe.com/
mhreader

For more information on Freeman J. Dyson, go to:
More Resources > Ch. 14 Science & Technology

Can We Know the Universe?
Reflections on a Grain of Salt

Carl Sagan

Carl Edward Sagan (1931–1996) received BA, BS, MA, and PhD degrees from the University of Chicago. Probably the most popular scientist in America in the 1970s and 1980s, he was the host of several television series on science and wrote a number of best-selling books on science, including The Dragons of Eden *(1977) and* Broca's Brain *(1979). The former earned him a Pulitzer Prize for general nonfiction in 1978. He also contributed hundreds of papers to scientific journals. Besides writing, Sagan served as a full-time professor at Cornell University and a visiting professor at dozens of other institutions of higher learning in the United States and abroad. He was also an activist for many philanthropic causes and served as an adviser to groups such as the Council for a Livable World Education Fund, the Children's Health Fund, and the American Committee on U.S.–Soviet Relations. Despite controversies surrounding the speculative nature of his work, Carl Sagan was one of modern science's most popular spokespersons. Sagan's philosophy may be summed up in a statement he made in a* Time *magazine interview. "We make our world significant by the courage of our questions and by the depth of our answers."*

Nothing is rich but the inexhaustible wealth of nature. She shows us only surfaces,
but she is a million fathoms deep.

—*Ralph Waldo Emerson*

Science is a way of thinking much more than it is a body of knowledge. Its goal   1
is to find out how the world works, to seek what regularities there may be, to
penetrate to the connections of things—from subnuclear particles, which may
be the constituents of all matter, to living organisms, the human social community, and thence to the cosmos as a whole. Our intuition is by no means an infal-
lible guide. Our perceptions may be distorted by training and prejudice or
merely because of the limitations of our sense organs, which, of course, perceive
directly but a small fraction of the phenomena of the world. Even so straightforward a question as whether in the absence of friction a pound of lead falls faster
than a gram of fluff was answered incorrectly by Aristotle and almost everyone
else before the time of Galileo. Science is based on experiment, on a willingness
to challenge old dogma, on an openness to see the universe as it really is. Accordingly, science sometimes requires courage—at the very least the courage to
question the conventional wisdom.

Beyond this the main trick of science is to *really* think of something: the   2
shape of clouds and their occasional sharp bottom edges at the same altitude
everywhere in the sky; the formation of a dewdrop on a leaf; the origin of a name
or a word—Shakespeare, say, or "philanthropic"; the reason for human social
customs—the incest taboo, for example; how it is that a lens in sunlight can make
paper burn; how a "walking stick" got to look so much like a twig; why the
Moon seems to follow us as we walk; what prevents us from digging a hole
down to the center of the Earth; what the definition is of "down" on a spherical
Earth; how it is possible for the body to convert yesterday's lunch into today's
muscle and sinew; or how far is up—does the universe go on forever, or if it does
not, is there any meaning to the question of what lies on the other side? Some of
these questions are pretty easy. Others, especially the last, are mysteries to which
no one even today knows the answer. They are natural questions to ask. Every
culture has posed such questions in one way or another. Almost always the proposed answers are in the nature of "Just So Stories," attempted explanations divorced from experiment, or even from careful comparative observations.

But the scientific cast of mind examines the world critically as if many alter-   3
native worlds might exist, as if other things might be here which are not. Then
we are forced to ask why what we see is present and not something else. Why
are the Sun and the Moon and the planets spheres? Why not pyramids, or
cubes, or dodecahedra? Why not irregular, jumbly shapes? Why so symmetri-
cal, worlds? If you spend any time spinning hypotheses, checking to see
whether they make sense, whether they conform to what else we know, thinking of tests you can pose to substantiate or deflate your hypotheses, you will
find yourself doing science. And as you come to practice this habit of thought
more and more you will get better and better at it. To penetrate into the heart of

the thing—even a little thing, a blade of grass, as Walt Whitman said—is to experience a kind of exhilaration that, it may be, only human beings of all the beings on this planet can feel. We are an intelligent species and the use of our intelligence quite properly gives us pleasure. In this respect the brain is like a muscle. When we think well, we feel good. Understanding is a kind of ecstasy.

But to what extent can we *really* know the universe around us? Sometimes   4
this question is posed by people who hope the answer will be in the negative, who are fearful of a universe in which everything might one day be known. And sometimes we hear pronouncements from scientists who confidently state that everything worth knowing will soon be known—or even is already known—and who paint pictures of a Dionysian or Polynesian age in which the zest for intellectual discovery has withered, to be replaced by a kind of subdued languor, the lotus eaters drinking fermented coconut milk or some other mild hallucinogen. In addition to maligning both the Polynesians, who were intrepid explorers (and whose brief respite in paradise is now sadly ending), as well as the inducements to intellectual discovery provided by some hallucinogens, this contention turns out to be trivially mistaken.

Let us approach a much more modest question: not whether we can know   5
the universe or the Milky Way Galaxy or a star or a world. Can we know, ultimately and in detail, a grain of salt? Consider one microgram of table salt, a speck just barely large enough for someone with keen eyesight to make out without a microscope. In that grain of salt there are about 10^{16} sodium and chlorine atoms. This is a 1 followed by 16 zeros, 10 million billion atoms. If we wish to know a grain of salt, we must know at least the three-dimensional positions of each of these atoms. (In fact, there is much more to be known—for example, the nature of the forces between the atoms—but we are making only a modest calculation.) Now, is this number more or less than the number of things which the brain can know?

How much *can* the brain know? There are perhaps 10^{11} neurons in the brain,   6
the circuit elements and switches that are responsible in their electrical and chemical activity for the functioning of our minds. A typical brain neuron has perhaps a thousand little wires, called dendrites, which connect it with its fellows. If, as seems likely, every bit of information in the brain corresponds to one of these connections, the total number of things knowable by the brain is no more than 10^{14}, one hundred trillion. But this number is only one percent of the number of atoms in our speck of salt.

So in this sense the universe is intractable, astonishingly immune to any hu-   7
man attempt at full knowledge. We cannot on this level understand a grain of salt, much less the universe.

But let us look a little more deeply at our microgram of salt. Salt happens to   8
be a crystal in which, except for defects in the structure of the crystal lattice, the position of every sodium and chlorine atom is predetermined. If we could shrink ourselves into this crystalline world, we would see rank upon rank of atoms in an ordered array, a regularly alternating structure—sodium, chlorine, sodium, chlorine, specifying the sheet of atoms we are standing on and all the

sheets above us and below us. An absolutely pure crystal of salt could have the position of every atom specified by something like 10 bits of information.[1] This would not strain the information-carrying capacity of the brain.

If the universe had natural laws that governed its behavior to the same de- 9 gree of regularity that determines a crystal of salt, then, of course, the universe would be knowable. Even if there were many such laws, each of considerable complexity, human beings might have the capability to understand them all. Even if such knowledge exceeded the information-carrying capacity of the brain, we might store the additional information outside our bodies—in books, for example, or in computer memories—and still, in some sense, know the universe.

Human beings are, understandably, highly motivated to find regularities, 10 natural laws. The search for rules, the only possible way to understand such a vast and complex universe, is called science. The universe forces those who live in it to understand it. Those creatures who find everyday experience a muddled jumble of events with no predictability, no regularity, are in grave peril. The universe belongs to those who, at least to some degree, have figured it out.

It is an astonishing fact that there *are* laws of nature, rules that summarize 11 conveniently—not just qualitatively but quantitatively—how the world works. We might imagine a universe in which there are no such laws, in which the 10^{80} elementary particles that make up a universe like our own behave with utter and uncompromising abandon. To understand such a universe we would need a brain at least as massive as the universe. It seems unlikely that such a universe could have life and intelligence, because beings and brains require some degree of internal stability and order. But even if in a much more random universe there were such beings with an intelligence much greater than our own, there could not be much knowledge, passion or joy.

Fortunately for us, we live in a universe that has at least important parts 12 that are knowable. Our common-sense experience and our evolutionary history have prepared us to understand something of the workaday world. When we go into other realms, however, common sense and ordinary intuition turn out to be highly unreliable guides. It is stunning that as we go close to the speed of light our mass increases indefinitely, we shrink toward zero thickness in the direction of motion, and time for us comes as near to stopping as we would like. Many people think that this is silly, and every week or two I get a letter from someone who complains to me about it. But it is a virtually certain consequence not just of experiment but also of Albert Einstein's brilliant analysis of space and time called the Special Theory of Relativity. It does not matter that these effects seem unreasonable to us. We are not in the habit of traveling close to the speed of light. The testimony of our common sense is suspect at high velocities.

---

[1]Chlorine is a deadly poison gas employed on European battlefields in World War I. Sodium is a corrosive metal which burns upon contact with water. Together they make a placid and unpoisonous material, table salt. Why each of these substances has the properties it does is a subject called chemistry, which requires more than 10 bits of information to understand. [This note is the author's.]

Or consider an isolated molecule composed of two atoms shaped some- 13
thing like a dumbbell—a molecule of salt, it might be. Such a molecule rotates
about an axis through the line connecting the two atoms. But in the world of
quantum mechanics, the realm of the very small, not all orientations of our
dumbbell molecule are possible. It might be that the molecule could be oriented
in a horizontal position, say, or in a vertical position, but not at many angles in
between. Some rotational positions are forbidden. Forbidden by what? By the
laws of nature. The universe is built in such a way as to limit, or quantize, rotation. We do not experience this directly in everyday life; we would find it startling as well as awkward in sitting-up exercises, to find arms outstretched from
the sides or pointed up to the skies permitted but many intermediate positions
forbidden. We do not live in the world of the small, on the scale of 10^{-13} cen-
timeters, in the realm where there are twelve zeros between the decimal place
and the one. Our common-sense intuitions do not count. What does count is
experiment—in this case observations from the far infrared spectra of molecules. They show molecular rotation to be quantized.

The idea that the world places restrictions on what humans might do is 14
frustrating. Why *shouldn't* we be able to have intermediate rotational positions?
Why *can't* we travel faster than the speed of light? But so far as we can tell, this
is the way the universe is constructed. Such prohibitions not only press us toward a little humility; they also make the world more knowable. Every restric-
tion corresponds to a law of nature, a regularization of the universe. The more
restrictions there are on what matter and energy can do, the more knowledge
human beings can attain. Whether in some sense the universe is ultimately
knowable depends not only on how many natural laws there are that encompass widely divergent phenomena, but also on whether we have the openness
and the intellectual capacity to understand such laws. Our formulations of the
regularities of nature are surely dependent on how the brain is built, but also,
and to a significant degree, on how the universe is built.

For myself, I like a universe that includes much that is unknown and, at the 15
same time, much that is knowable. A universe in which everything is known
would be static and dull, as boring as the heaven of some weak-minded theolo-
gians. A universe that is unknowable is no fit place for a thinking being. The
ideal universe for us is one very much like the universe we inhabit. And I
would guess that this is not really much of a coincidence.

COMPREHENSION

1. What is the thesis of the essay? In what paragraph is this thesis most clearly
   expressed?
2. Why does the author say, in paragraph 12, that in many circumstances,
   "common sense and ordinary intuition turn out to be highly unreliable
   guides"?

3. Why does Sagan say, in his conclusion, that "The ideal universe for us is one very much like the universe we inhabit"?

RHETORIC

1. What is the function of the epigram by Emerson? How does it relate to the essay proper?
2. Many of the paragraphs in the essay begin with coordinating conjunctions (a structure frowned on by many high school English teachers). What is Sagan's rhetorical purpose in using them as connecting devices?
3. What specific clues are there in the essay that Sagan's tone is one of excitement and celebration regarding science?
4. Sagan refers often to what he calls "a law of nature." Where and how in the essay does he explain, describe, or define this term?
5. The essay begins abruptly with an explanation of the concept of science. What purpose is served by diving into the subject so dramatically?
6. What is the intended effect of combining the terms *universe* and *grain of salt* in the title and subtitle? How does the author exploit this juxtaposition in his essay?
7. Examine the italicized words in the essay. Why has Sagan chosen to italicize these words? Explain.

WRITING

1. Write a personal essay in which you describe how you felt when you suddenly understood a particular topic in school that had previously eluded you.
2. For a research paper, select one of the items Sagan enumerates in paragraph 2, such as "the formation of a dewdrop on a leaf," the origin of the name *Shakespeare* or the word *philanthropic,* "the incest taboo," or "how a 'walking stick' got to look so much like a twig." Write an expository essay on your topic.
3. **WRITING AN ARGUMENT:** Argue for or against the proposition that scientific knowledge takes the mystery out of life.

---

www.mhhe.com/
mhreader

For more information on Carl Sagan, go to:
More Resources > Ch. 14 Science & Technology

✳

Staying Human

Dinesh D'Souza

Dinesh D'Souza, a leading conservative thinker, was born in Bombay, India, and came to the United States for his high school education. He graduated from Dartmouth College (BA, 1983) and subsequently wrote for several magazines, notably the National Review, *before becoming a policy analyst for the Reagan administration. His books include* Illiberal Education: The Politics of Race and Sex on Campus *(1991),* The End of Racism: Principles for a Multicultural Society *(1995),* The Virtue of Prosperity: Finding Values in an Age of Techno-Affluence *(2001), and* What's So Great about America *(2002). D'Souza has been a visiting scholar at the Hoover Institution and a research scholar at the American Enterprise Institute. In the following essay, written for the* National Review *in 2001, D'Souza offers a wide-ranging assessment of our emerging "techno-utopia."*

We are as gods, and we might as well get good at it.
                                        —*Kevin Kelly, author and techno-utopian*

The most important technological advance of recent times is not the Internet,   1
but rather the biotech revolution—which promises to give us unprecedented power to transform human nature. How should we use that power? A group of cutting-edge scientists, entrepreneurs, and intellectuals has a bold answer. This group—I call them the techno-utopians—argues that science will soon give us the means to straighten the crooked timber of humanity, and even to remake our species into something "post-human."

One of the leading techno-utopians is Lee Silver, who teaches molecular   2
biology at Princeton University. Silver reports that biotechnology is moving beyond cloning to offer us a momentous possibility: designer children. He envisions that, in the not too distant future, couples who want to have a child will review a long list of traits on a computer screen, put together combinations of "virtual children," decide on the one they want, click on the appropriate selection, and thus—in effect—design their own offspring. "Parents are going to be able to give their children . . . genes that increase athletic ability, genes that increase musical talents . . . and ultimately genes that affect cognitive abilities."

But even this, the techno-utopians say, is a relatively small step: People liv-   3
ing today can determine the genetic destiny of all future generations. Some writers, including physicist Stephen Hawking, have suggested that genetic engineering could be used to reduce human aggression, thus solving the crime problem and making war less likely. James Watson, co-discoverer of the structure of DNA, argues that if biological interventions could be used to "cure what I feel is a very serious disease—that is, stupidity—it would be a great thing for

people." Silver himself forecasts a general elevation of intellectual, athletic, temperamental, and artistic abilities so that we can over time create "a special group of mental beings" who will "trace their ancestry back to Homo sapiens," but who will be "as different from humans as humans are from the primitive worms with tiny brains that first crawled along the earth's surface."

These ideas might seem implausible, but they are taken very seriously by 4 some of the best minds in the scientific community. The confidence of the techno-utopians is based on stunning advances that have made cloning and genetic engineering feasible. In theoretical terms, biotechnology crossed a major threshold with James Watson and Francis Crick's 1953 discovery of the structure of DNA, but practical applications were slow in coming. In 1997, an obscure animal-husbandry laboratory in Scotland cloned a sheep named Dolly; today, the knowledge and the means of cloning human beings already exist, and the only question is whether we are going to do it. And why stop there? As the scientific journal *Nature* editorialized shortly after the emergence of Dolly, "The growing power of molecular genetics confronts us with future prospects of being able to change the nature of our species."

In 1999, neurobiologist Joe Tsien boosted the intelligence of mice by insert- 5 ing extra copies of a gene that enhances memory and learning; these mouse genes are virtually identical to those found in human beings. Gene therapy has already been successfully carried out in people, and now that the Human Genome Project has made possible a comprehensive understanding of the human genetic code, scientists will possess a new kind of power: the power to design our children, and even to redesign humanity itself.

The Hitler Scenario

The fact that these things are possible does not, of course, mean that they should 6 be done. As one might expect, cloning and genetic engineering are attracting criticism. The techno-utopians have not yet made their products and services available to consumers; but one can reasonably expect that a society that is anxious about eating genetically modified tomatoes is going to be vastly more anxious about a scheme to engineer our offspring and our species.

A recent book communicating that sense of outrage is Jeremy Rifkin's *The* 7 *Biotech Century.* Rifkin alleges that we are heading for a nightmarish future "where babies are genetically designed and customized in the womb, and where people are identified, stereotyped and discriminated against on the basis of their genotype." How can living beings be considered sacred, Rifkin asks, if they are treated as nothing more than "bundles of genetic information"? Biotechnology, he charges, is launching us into a new age of eugenics. In Rifkin's view, the Nazi idea of the superman is very much alive, but now in a different form: the illusion of the "perfect child."

Although Rifkin has a propensity for inflammatory rhetoric, he is raising 8 some important concerns: The new technology is unprecedented, so we should be very cautious in developing it. It poses grave risks to human health. Cloning and genetic engineering are unnatural; human beings have no right to do this to nature and to ourselves.

These criticisms meet with derision on the part of the techno-utopians. 9 Every time a major new technology is developed, they say, there are people who forecast the apocalypse. The techno-utopians point out that the new technology will deliver amazing medical benefits, including cures for genetic diseases. How can it be ethical, they ask, to withhold these technologies from people who need and want them?

Lee Silver, the biologist, is annoyed at critics such as Rifkin who keep rais- 10 ing the specter of Hitler and eugenics. "It is individuals and couples, not governments, who will seize control of these new technologies," Silver writes. The premise of the techno-utopians is that if the market produces a result, it is good. In this view, what is wrong with the old eugenics is not that it sought to eliminate defective types and produce a superior kind of being, but that it sought to do so in a coercive and collectivist way. The new advocates of biotechnology speak approvingly of what they term "free-market eugenics."

The champions of biotechnology concede that cloning and genetic engi- 11 neering should not be permitted in human beings until they are safe. But "safe," they say, does not mean "error-free"; it means safe compared with existing forms of reproduction. And they are confident that the new forms of reproduction will soon be as safe as giving birth the natural way.

The techno-utopians are also not very concerned that the availability of en- 12 hancement technologies will create two classes in society, the genetically advantaged and the genetically disadvantaged. They correctly point to the fact that two such classes exist now, even in the absence of new therapies. Physicist Freeman Dyson says that genetic enhancement might be costly at first, but won't remain permanently expensive: "Most of our socially important technologies, such as telephones, automobiles, television, and computers, began as expensive toys for the rich and afterwards became cheap enough for ordinary people."

Dyson is right that time will make genetic enhancements more widely 13 available, just as cars and TV sets are now. But the poor family still drives a secondhand Plymouth while the rich family can afford a new Porsche. This may not be highly significant when it comes to cars, because both groups can still get around fairly well. What about when it comes to genetic advantages conferred at birth? Democratic societies can live with inequalities conferred by the lottery of nature, but can they countenance the deliberate introduction of biological alterations that give some citizens a better chance to succeed than others?

The techno-utopians have not, to my knowledge, addressed this concern. 14 They emphasize instead that it is well established in law, and widely recognized in society, that parents have a right to determine what is best for their children. "There are already plenty of ways in which we design our children," remarks biologist Gregory Stock. "One of them is called piano lessons. Another is called private school." Stock's point is that engineering their children's genes is simply one more way in which parents can make their children better people.

Some people might find it weird and unnatural to fix their child in the same ₁₅ way they fix their car—but, say the techno-utopians, this is purely a function of habit. We're not used to genetic engineering, so it seems "unnatural" to us. But think about how unnatural driving a car seemed for people who previously got around on horses and in carriages. "The smallpox virus was part of the natural order," Silver wryly observes, "until it was forced into extinction by human intervention." Diseases and death are natural; life-saving surgery is unnatural.

Not Sacred after All?

Nor are the techno-utopians worried about diminishing the sanctity of human ₁₆ life because, they say, it isn't intrinsically sacred. "This is not an ethical argument but a religious one," says Silver. "There is no logic to it." Biologist David Baltimore, a Nobel laureate, argues that "statements about morally and ethically unacceptable practices" have no place in the biotechnology debate "because those are subjective grounds and therefore provide no basis for discussion." Silver and Baltimore's shared assumption is that the moralists are talking about values while they, the hard scientists, are dealing in facts.

In this view, the subjective preferences of those who seek to mystify human ₁₇ life do not square with the truths about human biology taught by science. The cells of human beings, Silver points out, are not different in their chemical makeup from the cells of horses and bacteria. If there is such a thing as human dignity, Silver argues, it derives exclusively from consciousness, from our ability to perceive and apprehend our environment. "The human mind," Silver writes, "is much more than the genes that brought it into existence." Somehow the electrochemical reactions in our brain produce consciousness, and it is this consciousness, Silver contends, that is the source of man's autonomy and power. While genes fully control the activity of all life forms, Silver writes that in human beings "master and slave have switched positions." Consciousness enables man to complete his dominance over nature by prevailing over his human nature. Silver concludes that, in a bold assertion of will, we can defeat the program of our genes, we can take over the reins of evolution, we can choose the genetic code we want for our children, and we can collectively determine the future of our species.

This triumphant note is echoed by many techno-utopians. Biotech, writes ₁₈ journalist Ronald Bailey, "will liberate future generations from today's limitations and offer them a much wider scope of freedom." Physicist Gregory Benford is even more enthusiastic: "It is as though prodigious, bountiful Nature for billions of years has tossed off variations on its themes like a careless, prolific Picasso. Now Nature finds that one of its casual creations has come back with a piercing, searching vision, and its own pictures to paint."

These are ringing statements. But do they make sense? Clearly there are ₁₉ many problems with Silver's definition of human dignity as based in consciousness. Animals are conscious; do they deserve the same dignity as human beings? Moreover, are human beings entitled to dignity only when they are conscious?

Do we lose our right to be respected, and become legitimate subjects for discarding medical experiments, when we fall asleep, or into a coma? Surely Silver would disavow these conclusions. They do, however, flow directly from his definition, which is, in fact, just as heavily freighted with values as are the statements of his opponents.

There is, behind the proclamations of scientific neutrality, an ideology that [20] needs to be spelled out, a techno-Nietzschean doctrine that proclaims: We are molecules, but molecules that know how to rebel. Our values do not derive from nature or nature's God; rather, they arise from the arbitrary force of our wills. And now our wills can make the most momentous choice ever exercised on behalf of our species: the choice to reject our human nature. Why should we remain subject to the constraints of our mortality and destiny? Wealth and technology have given us the keys to unlimited, indeed godlike, power: the dawn of the post-human era.

What is one to make of all this? In many respects, we should celebrate the [21] advent of technologies that enable us to alleviate suffering and extend life. I have no problem with genetic therapy to cure disease; I am even willing to endorse therapy that not only cures illness in patients but also prevents it from being transmitted to the next generation. Under certain circumstances, I can see the benefits of cloning. The cloning of animals can provide organs for transplant as well as animals with medicinal properties ("drugstores on the hoof"). Even human cloning seems defensible when it offers the prospect of a biological child to married couples who might not otherwise be able to have one.

Creating the Perfect Child

But there is a seduction contained in these exercises in humanitarianism: They [22] urge us to keep going, to take the next step. And when we take that step, when we start designing our children, when we start remaking human beings, I think we will have crossed a perilous frontier. Even cloning does not cross this frontier, because it merely replicates an existing genetic palate. It is unconvincing to argue, as some techno-utopians do, that giving a child a heightened genetic capacity for music or athletics or intelligence is no different from giving a child piano, swimming, or math lessons. In fact, there is a big difference. It is one thing to take a person's given nature and given capacity, and seek to develop it, and quite another to shape that person's nature in accordance with one's will.

There is no reason to object to people's attempting brain implants and so- [23] matic gene enhancements on themselves. Perhaps, in some cases, these will do some good; others may end up doing injury. But at least these people have, through their free choices, done it to themselves. The problem arises when people seek to use enhancement technologies to shape the destiny of others, and especially their children.

But, argues Lee Silver, we have the right to terminate pregnancy and control [24] our children's lives in every other way; why shouldn't parents be permitted to alter their child's genetic constitution? In the single instance of gene therapy to cure disease, I'd agree—because, in this one limited case, we can trust the parents to make a decision that there is every rational reason to believe their offspring would decide in the identical manner, were they in a position to make the choice. No child would say, "I can't believe my parents did that to me. I would have chosen to have Parkinson's disease."

But I would contend that in no other case do people have the right to bend 25 the genetic constitution of their children—or anyone else—to their will. But they might, in good conscience, be tempted to do so; and this temptation must be resisted. Indeed, it must be outlawed—because what the techno-utopians want does, in fact, represent a fundamental attack on the value of human life, and the core principle of America.

Rescuing Humanity and the American Idea

The scientific-capitalist project at the heart of the American experiment was an 26 attempted "conquest of nature." Never did the early philosophers of science, like Francis Bacon, or the American Founders conceive that this enterprise would eventually seek to conquer human nature. Their goal was to take human nature as a given, as something less elevated than the angels, and thus requiring a government characterized by separation of offices, checks and balances, limited power. At the same time, the Founders saw human nature as more elevated than that of other animals. They held that human beings have claims to dignity and rights that do not extend to animals: Human beings cannot be killed for sport or rightfully governed without their consent.

The principles of the Founders were extremely far-reaching. They called 27 into question the legitimacy of every existing government, because at the time of the American founding, no government in the world was entirely based on the consent of the governed. The ideals of the Founders even called into question their own practices, such as slavery. It took the genius of Abraham Lincoln, and the tragedy of the Civil War, to compel the enforcement of the central principle of the Declaration of Independence: that we each have an inalienable right to life, liberty, and the pursuit of happiness, and that these rights shall not be abridged without our consent.

The attempt to enhance and redesign other human beings represents a fla- 28 grant denial of this principle that is the basis of our dignity and rights. Indeed, it is a restoration of the principle underlying slavery, and the argument between the defenders and critics of genetic enhancement is identical in principle, and very nearly in form, to the argument between Stephen Douglas and Abraham Lincoln on the issue of human enslavement.

In that tempestuous exchange, which laid the groundwork for the Civil 29 War, Douglas argued for the pro-choice position. He wanted to let each new territory decide for itself whether it wanted slavery. He wanted the American people to agree to disagree on the issue. He advocated for each community a very high value: the right to self-determination.

Lincoln challenged him on the grounds that choice cannot be exercised   30
without reference to the content of the choice. How can it make sense to permit
people to choose to enslave another human being? How can self-determination
be invoked to deny others the same? A free people can disagree on many things,
but it cannot disagree on the distinction between freedom and despotism. Lincoln summarized Douglas's argument as follows: "If any one man choose to enslave another, no third man shall be allowed to object."

Lincoln's argument was based on a simple premise: "As I would not be a   31
slave, so I would not be a master." Lincoln rejects in principle the subordination
implied in the master–slave relationship. Those who want freedom for themselves, he insists, must also show themselves willing to extend it to others. At its
deepest level, Lincoln's argument is that the legitimacy of popular consent is itself dependent on a doctrine of natural rights that arises out of a specific understanding of human nature and human dignity. "Slavery," he said, "is founded
in the selfishness of man's nature-opposition to it, in his love of justice. These
principles are in eternal antagonism; and when brought into collision so fiercely
. . . convulsions must ceaselessly follow." What Lincoln is saying is that self-
interest by itself is too base a foundation for the new experiment called America. Selfishness is part of our nature, but it is not the best part of our nature. It
should be subordinated to a nobler ideal. Lincoln seeks to dedicate America to
a higher proposition: the proposition that all men are created equal. It is the denial of this truth, Lincoln warns, that will bring on the cataclysm.

Let me restate Lincoln's position for our current context. We speak of "our   32
children," but they are not really ours; we do not own them. At most, we own
ourselves. It is true that *Roe v. Wade* gives us the right to kill our unborn in the
womb. The right to abortion has been defended, both by its advocates and by
the Supreme Court, as the right of a woman to control her own body. This is not
the same as saying the woman has ownership of the fetus, that the fetus is the
woman's property. The Supreme Court has said that as long as the fetus is occupying her womb, she can treat it as an unwelcome intruder, and get rid of it.
(Even here, technology is changing the shape of the debate by moving up the
period when the fetus can survive outside the womb.) But once a woman decides to carry the pregnancy to term, she has already exercised her choice. She
has chosen to give birth to the child, which is in the process of becoming an independent human being with its own dignity and rights.

No Place for Parental Tyranny

As parents, we have been entrusted with our children, and it is our privilege and   33
responsibility to raise them as best we can. Undoubtedly we will infuse them
with our values and expectations, but even so, the good parent will respect the
child's right to follow his own path. There is something perversely restrictive
about parents who apply relentless pressure on their children to conform to their
will—to follow the same professional paths that they did, or to become the "first
doctor in the family." These efforts, however well intentioned, are a betrayal of
the true meaning of parenthood. Indeed, American culture encourages a certain

measure of adolescent rebellion against parental expectations, precisely so that young people making the transition to independence can "find themselves" and discover their own identity.

Consequently, parents have no right to treat their children as chattels; but 34 this is precisely the enterprise that is being championed by the techno-utopians. Some of these people profess to be libertarians, but they are in fact totalitarians. They speak about freedom and choice, although what they advocate is despotism and human bondage. The power they seek to exercise is not over "nature" but over other human beings.

Parents who try to design their children are in some ways more tyrannical 35 than slaveowners, who merely sought to steal the labor of their slaves. Undoubtedly some will protest that they only wish the best for their children, that they are only doing this for their own good. But the slaveowners made similar arguments, saying that they ruled the Negroes in the Negroes' own interest. The argument was as self-serving then as it is now. What makes us think that in designing our children it will be their objective good—rather than our desires and preferences—that will predominate?

The argument against slavery is that you may not tyrannize over the life and 36 freedom of another person for any reason whatsoever. Even that individual's consent cannot overturn "inalienable" rights: One does not have the right to sell oneself into slavery. This is the clear meaning of the American proposition. The object of the American Revolution that is now spreading throughout the world has always been the affirmation, not the repudiation, of human nature. The Founders envisioned technology and capitalism as providing the framework and the tools for human beings to live richer, fuller lives. They would have scorned, as we should, the preposterous view that we are the servants of our technology. They would have strenuously opposed, as we should, the effort on the part of the techno-utopians to design their offspring; to alter, improve, and perfect human nature; or to relinquish our humanity in pursuit of some post-human ideal.

Mary Shelley's 1818 novel *Frankenstein* describes a monster that is the labo- 37 ratory creation of a doctor who refuses to accept the natural limits of humanity. He wants to appropriate to himself the traditional prerogatives of the deity, such as control over human mortality. He even talks about making "a new species" with "me as its creator and source." In his rhetoric, Frankenstein sounds very much like today's techno-utopians. And, contrary to what most people think, the real monster in the novel isn't the lumbering, tragic creature; it is the doctor who creates him. This is the prophetic message of Shelley's work: In seeking to become gods, we are going to make monsters of ourselves.

COMPREHENSION

1. What is the meaning of D'Souza's title? According to the writer, what are the dangers we face in our effort to "stay human"? What must we do to retain our essential humanity?

2. What aspects of the biotech revolution does the author treat? What is his opinion of each?
3. What does the writer mean by the "Hitler Scenario" and the "American Idea"?

RHETORIC

1. What is the author's purpose? What is his tone? Does he seem reasonable or unnecessarily argumentative, objective or biased? How do you know? For what type of audience does he seem to be writing?
2. D'Souza begins his essay by introducing the ideas of such "leading techno-utopians" as Lee Silver, Stephen Hawking, James Watson and Francis Crick, and Joe Tsien. It this an effective opening strategy? Why or why not?
3. Explain the author's claim. What minor propositions does he provide? Where does he advance logical, ethical, and emotional appeals? Cite instances in which the author employs refutation to advance his argument. What faults does he find with the techno-utopians?
4. Consider the essay's section headings. How do they serve to focus the content of each section? What characterizes the progression of ideas from section to section? How does the author's decision to divide the essay into sections help him to construct his argument?
5. Does D'Souza reveal any of his own biases in this essay? Explain.
6. In the concluding paragraph, D'Souza alludes to Mary Shelley's novel *Frankenstein*. Does this allusion flow naturally from the introduction and body? Why or why not?

WRITING

1. Write a 300-word summary of this essay, transcribing all of the main aspects of D'Souza's argument.
2. Write your own survey of the biotech revolution. Refer to some of the topics and ideas mentioned by D'Souza in his essay.
3. **WRITING AN ARGUMENT:** Select one aspect of the biotech revolution—for instance, cloning or stem-cell research—and write an argumentative essay supporting or opposing developments in the field.

www.mhhe.com/ **mhreader**   For more information on Dinesh D'Souza, go to: **More Resources > Ch. 14 Science & Technology**

✳

The Clan of One-Breasted Women

Terry Tempest Williams

Terry Tempest Williams (b. 1955) is the author of many books of nonfiction, includ-
ing A Journey to Navajoland *(1984),* Coyote's Canyon *(1989),* Refuge: An Unnatural History of Family and Place *(1991),* An Unspoken Hunger *(1994), and*
Desert Quartet *(1995). Williams was identified by* Newsweek *magazine as someone*
who will have "a considerable impact on the political, economic and environmental is-
sues facing the western states in this decade." She is the recipient of a Lannan Fellow-
ship in creative nonfiction and was chosen by the periodical UTNE Reader *as a*
"visionary," one of the UTNE 100 "who could change your life." She is Naturalist-
in-Residence at the Utah Museum of Natural History in Salt Lake City. The following
essay—published in 1989 in Witness—*describes the pernicious intergenerational ef-*
fects of nuclear testing.

I belong to a Clan of One-Breasted Women. My mother, my grandmothers, and   1
six aunts have all had mastectomies. Seven are dead. The two who survive have
just completed rounds of chemotherapy and radiation.

I've had my own problems: two biopsies for breast cancer and a small tu-   2
mor between my ribs diagnosed as "a border-line malignancy."

This is my family history.   3

Most statistics tell us breast cancer is genetic, hereditary, with rising per-   4
centages attached to fatty diets, childlessness, or becoming pregnant after thirty.
What they don't say is living in Utah may be the greatest hazard of all.

We are a Mormon family with roots in Utah since 1847. The word-of-   5
wisdom, a religious doctrine of health, kept the women in my family aligned
with good foods: no coffee, no tea, tobacco, or alcohol. For the most part, these
women were finished having their babies by the time they were thirty. And only
one faced breast cancer prior to 1960. Traditionally, as a group of people, Mor-
mons have a low rate of cancer.

Is our family a cultural anomaly? The truth is we didn't think about it.   6
Those who did, usually the men, simply said, "bad genes." The women's attitude was stoic. Cancer was part of life. On February 16, 1971, the eve before my
mother's surgery, I accidentally picked up the telephone and overheard her ask
my grandmother what she could expect.

"Diane, it is one of the most spiritual experiences you will ever encounter."   7

I quietly put down the receiver.   8

Two days later, my father took my three brothers and me to the hospital to   9
visit her. She met us in the lobby in a wheelchair. No bandages were visible. I'll
never forget her radiance, the way she held herself in a purple velour robe and
how she gathered us around her.

"Children, I am fine. I want you to know I felt the arms of God around me."   10

We believed her. My father cried. Our mother, his wife, was thirty-eight   11
years old.

Two years ago, after my mother's death from cancer, my father and I were   12
having dinner together. He had just returned from St. George where his construction company was putting in natural gas lines for towns in southern Utah.
He spoke of his love for the country: the sandstoned landscape, bare-boned and
beautiful. He had just finished hiking the Kolob trail in Zion National Park. We
got caught up in reminiscing, recalling with fondness our walk up Angel's
Landing on his fiftieth birthday and the years our family had vacationed there.
This was a remembered landscape where we had been raised.

Over dessert, I shared a recurring dream of mine. I told my father that for   13
years, as long as I could remember, I saw this flash of light in the night in the
desert. That this image had so permeated my being, I could not venture south
without seeing it again, on the horizon, illuminating buttes and mesas.

"You did see it," he said.   14

"Saw what?" I asked, a bit tentative.   15

"The bomb. The cloud. We were driving home from Riverside, California.   16
You were sitting on your mother's lap. She was pregnant. In fact, I remember
the date, September 7, 1957. We had just gotten out of the Service. We were driving north, past Las Vegas. It was an hour or so before dawn, when this explosion went off. We not only heard it, but felt it. I thought the oil tanker in front of
us had blown up. We pulled over and suddenly, rising from the desert floor, we
saw it, clearly, this golden-stemmed cloud, the mushroom. The sky seemed to
vibrate with an eerie pink glow. Within a few minutes, a light ash was raining
on the car."

I stared at my father. This was new information to me.   17

"I thought you knew that," my father said. "It was a common occurrence in   18
the fifties."

It was at this moment I realized the deceit I had been living under. Children   19
growing up in the American Southwest, drinking contaminated milk from contaminated cows, even from the contaminated breasts of their mother, my
mother—members, years later, of the Clan of One-Breasted Women.

It is a well-known story in the Desert West, "The Day We Bombed Utah," or   20
perhaps, "The Years We Bombed Utah."[1] Above ground atomic testing in
Nevada took place from January 27, 1951, through July 11, 1962. Not only were
the winds blowing north, covering "low use segments of the population" with
fallout and leaving sheep dead in their tracks, but the climate was right.[2] The

---

[1] Fuller, John G., *The Day We Bombed Utah* (New York: New American Library, 1984). [This and subsequent notes in the selection are the author's.]

[2] Discussion on March 14, 1988, with Carole Gallagher, photographer and author, *Nuclear Towns:
The Secret War in the American Southwest*, published by Doubleday, Spring, 1990.

United States of the 1950s was red, white, and blue. The Korean War was raging. McCarthyism was rampant. Ike was in and the Cold War was hot. If you were against nuclear testing, you were for a Communist regime.

Much has been written about this "American nuclear tragedy." Public 21
health was secondary to national security. The Atomic Energy Commissioner, Thomas Murray said, "Gentlemen, we must not let anything interfere with this series of tests, nothing."[3]

Again and again, the American public was told by its government, in spite 22
of burns, blisters, and nausea, "It has been found that the tests may be conducted with adequate assurance of safety under conditions prevailing at the bombing reservations."[4] Assuaging public fears was simply a matter of public relations. "Your best action," an Atomic Energy Commission booklet read, "is not to be worried about fallout." A news release typical of the times stated, "We find no basis for concluding that harm to any individual has resulted from radioactive fallout."[5]

On August 30, 1979, during Jimmy Carter's presidency, a suit was filed en- 23
titled "Irene Allen vs. the United States of America." Mrs. Allen was the first to be alphabetically listed with twenty-four test cases, representative of nearly 1200 plaintiffs seeking compensation from the United States government for cancers caused from nuclear testing in Nevada.

Irene Allen lived in Hurricane, Utah. She was the mother of five children 24
and had been widowed twice. Her first husband with their two oldest boys had watched the tests from the roof of the local high school. He died of leukemia in 1956. Her second husband died of pancreatic cancer in 1978.

In a town meeting conducted by Utah Senator Orrin Hatch, shortly before 25
the suit was filed, Mrs. Allen said, "I am not blaming the government, I want you to know that, Senator Hatch. But I thought if my testimony could help in any way so this wouldn't happen again to any of the generations coming up after us . . . I am really happy to be here this day to bear testimony of this."[6]

God-fearing people. This is just one story in an anthology of thousands. 26

On May 10, 1984, Judge Bruce S. Jenkins handed down his opinion. Ten of 27
the plaintiffs were awarded damages. It was the first time a federal court had determined that nuclear tests had been the cause of cancers. For the remaining fourteen test cases, the proof of causation was not sufficient. In spite of the split decision, it was considered a landmark ruling.[7] It was not to remain so for long.

In April 1987, the 10th Circuit Court of Appeals overturned Judge Jenkins' 28
ruling on the basis that the United States was protected from suit by the legal

[3]Szasz, Ferenc M., "Downwind from the Bomb," *Nevada Historical Society Quarterly,* Fall, 1987 Vol. XXX, No. 3, p. 185.
[4]Fradkin, Philip L., *Fallout* (Tucson: University of Arizona Press, 1989), 98.
[5]Ibid., 109.
[6]Town meeting held by Senator Orrin Hatch in St. George, Utah, April 17, 1979, transcript, 26–28.
[7]Fradkin, Op. Cit., 228.

doctrine of sovereign immunity, the centuries-old idea from England in the days of absolute monarchs.[8]

In January 1988, the Supreme Court refused to review the Appeals Court decision. To our court system, it does not matter whether the United States Government was irresponsible, whether it lied to its citizens or even that citizens died from the fallout of nuclear testing. What matters is that our government is immune. "The King can do no wrong." [29]

In Mormon culture, authority is respected, obedience is revered, and independent thinking is not. I was taught as a young girl not to "make waves" or "rock the boat." [30]

"Just let it go—" my mother would say. "You know how you feel, that's what counts." [31]

For many years, I did just that—listened, observed, and quietly formed my own opinions within a culture that rarely asked questions because they had all the answers. But one by one, I watched the women in my family die common, heroic deaths. We sat in waiting rooms hoping for good news, always receiving the bad. I cared for them, bathed their scarred bodies and kept their secrets. I watched beautiful women become bald as cytoxan, cisplatin and adriamycin were injected into their veins. I held their foreheads as they vomited green-black bile and I shot them with morphine when the pain became inhuman. In the end, I witnessed their last peaceful breaths, becoming a midwife to the rebirth of their souls. But the price of obedience became too high. [32]

The fear and inability to question authority that ultimately killed rural communities in Utah during atmospheric testing of atomic weapons was the same fear I saw being held in my mother's body. Sheep. Dead sheep. The evidence is buried. [33]

I cannot prove that my mother, Diane Dixon Tempest, or my grandmothers, Lettie Romney Dixon and Kathryn Blackett Tempest, along with my aunts contracted cancer from nuclear fallout in Utah. But I can't prove they didn't. [34]

My father's memory was correct, the September blast we drove through in 1957 was part of Operation Plumbbob, one of the most intensive series of bomb tests to be initiated. The flash of light in the night in the desert I had always thought was a dream developed into a family nightmare. It took fourteen years, from 1957 to 1971, for cancer to show up in my mother—the same time, Howard L. Andrews, an authority on radioactive fallout at the National Institutes of Health, says radiation cancer requires to become evident.[9] The more I learn about what it means to be a "downwinder," the more questions I drown in. [35]

What I do know, however, is that as a Mormon woman of the fifth generation of "Latter-Day-Saints," I must question everything, even if it means losing my faith, even if it means becoming a member of a border tribe among my own [36]

[8]U.S. vs. Allen, 816 Federal Reporter, 2d/1417 (10th Circuit Court 1987), cert. denied, 108 S. CT. 694 (1988).
[9]Fradkin, Op. cit. 116.

people. Tolerating blind obedience in the name of patriotism or religion ultimately takes our lives.

When the Atomic Energy Commission described the country north of the 37 Nevada Test Site as "virtually uninhabited desert terrain," my family members were some of the "virtual uninhabitants."

One night, I dreamed women from all over the world circling a blazing fire in the 38 desert. They spoke of change, of how they hold the moon in their bellies and wax and wane with its phases. They mocked at the presumption of even-tempered beings and made promises that they would never fear the witch inside themselves. The women danced wildly as sparks broke away from the flames and entered the night sky as stars.

And they sang a song given to them by Shoshoni grandmothers: 39

Ah ne nah, nah
nin nah nah—
Ah ne nah, nah
nin nah nah—
Nyaga mutzi
oh ne nay—
Nyaga mutzi
oh ne nay—[10]

The women danced and drummed and sang for weeks, preparing them- 40 selves for what was to come. They would reclaim the desert for the sake of their children, for the sake of the land.

A few miles downwind from the fire circle, bombs were being tested. Rab- 41 bits felt the tremors. Their soft leather pads on paws and feet recognized the shaking sands while the roots of mesquite and sage were smoldering. Rocks were hot from the inside out and dust devils hummed unnaturally. And each time there was another nuclear test, ravens watched the desert heave. Stretch marks appeared. The land was losing its muscle.

The women couldn't bear it any longer. They were mothers. They had suf- 42 fered labor pains but always under the promise of birth. The red hot pains beneath the desert promised death only as each bomb became a stillborn. A contract had been broken between human beings and the land. A new contract was being drawn by the women who understood the fate of the earth as their own.

Under the cover of darkness, ten women slipped under the barbed wire 43 fence and entered the contaminated country. They were trespassing. They

---

[10]This song was sung by the Western Shoshone women as they crossed the line at the Nevada Test Site on March 18, 1988, as part of their "Reclaim the Land" action. The translation they gave was: "Consider the rabbits how gently they walk on the earth. Consider the rabbits how gently they walk on the earth. We remember them. We can walk gently also. We remember them. We can walk gently also."

walked toward the town of Mercury in moonlight, taking their cues from coyote, kit fox, antelope squirrel, and quail. They moved quietly and deliberately through the maze of Joshua trees. When a hint of daylight appeared they rested, drinking tea and sharing their rations of food. The women closed their eyes. The time had come to protest with the heart, that to deny one's genealogy with the earth was to commit treason against one's soul.

At dawn, the women draped themselves in mylar, wrapping long streamers of silver plastic around their arms to blow in the breeze. They wore clear masks that became the faces of humanity. And when they arrived on the edge of Mercury, they carried all the butterflies of a summer day in their wombs. They paused to allow their courage to settle. 44

The town which forbids pregnant women and children to enter because of radiation risks to their health was asleep. The women moved through the streets as winged messengers, twirling around each other in slow motion, peeking inside homes and watching the easy sleep of men and women. They were astonished by such stillness and periodically would utter a shrill note or low cry just to verify life. 45

The residents finally awoke to what appeared as strange apparitions. Some simply stared. Others called authorities, and in time, the women were apprehended by wary soldiers dressed in desert fatigues. They were taken to a white, square building on the other edge of Mercury. When asked who they were and why they were there, the women replied, "We are mothers and we have come to reclaim the desert for our children." 46

The soldiers arrested them. As the ten women were blindfolded and handcuffed, they began singing: 47

You can't forbid us everything
You can't forbid us to think—
You can't forbid our tears to flow
And you can't stop the songs that we sing.

The women continued to sing louder and louder, until they heard the voices of their sisters moving across the mesa. 48

Ah ne nah, nah
nin nah nah—
Ah ne nah, nah
nin nah nah—
Nyaga mutzi
oh ne nay—
Nyaga mutzi
oh ne nay—

"Call for re-enforcement," one soldier said. 49

"We have," interrupted one woman. "We have—and you have no idea of our numbers." 50

On March 18, 1988, I crossed the line at the Nevada Test Site and was arrested 51 with nine other Utahns for trespassing on military lands. They are still conducting nuclear tests in the desert. Ours was an act of civil disobedience. But as I walked toward the town of Mercury, it was more than a gesture of peace. It was a gesture on behalf of the Clan of One-Breasted Women.

As one officer cinched the handcuffs around my wrists, another frisked my 52 body. She found a pen and a pad of paper tucked inside my left boot.

"And these?" she asked sternly. 53

"Weapons," I replied. 54

Our eyes met. I smiled. She pulled the leg of my trousers back over my 55 boot.

"Step forward, please," she said as she took my arm. 56

We were booked under an afternoon sun and bussed to Tonapah, Nevada. 57 It was a two-hour ride. This was familiar country to me. The Joshua trees standing their ground had been named by my ancestors who believed they looked like prophets pointing west to the promised land. These were the same trees that bloomed each spring, flowers appearing like white flames in the Mojave. And I recalled a full moon in May when my mother and I had walked among them, flushing out mourning doves and owls.

The bus stopped short of town. We were released. The officials thought it 58 was a cruel joke to leave us stranded in the desert with no way to get home. What they didn't realize is that we were home, soul-centered and strong, women who recognized the sweet smell of sage as fuel for our spirits.

COMPREHENSION

1. What is the subject of the essay? How did you arrive at your answer?
2. The credo of the United States promotes "life, liberty, and the pursuit of happiness." In her essay, what does Williams imply is the major antagonist to this philosophy?
3. Define the following words and terms: *doctrine* (paragraph 5), *anomaly* (paragraph 6), *buttes* (paragraph 13), *mesas* (paragraph 13), *plaintiffs* (paragraph 23), and *sovereign* (paragraph 28).

RHETORIC

1. Williams often uses quotation marks to signal irony in her writing. What is the perverse irony in the following expressions: "low use segments of the population (paragraph 20)," "downwinder (paragraph 35)," and "virtually uninhabited desert terrain" (paragraph 37)?
2. Although this is a highly personal essay and reveals a profound and emotional personal experience, Williams also relies on evidence from secondary source material to support her thesis. What is the effect of using such sources in terms of the author's authority and believability?

3. The essay is divided into four segments. What is the main subject of each one?

4. Williams uses extended metaphor, personification, and comparison and contrast in paragraphs 41 and 42. Explain how she incorporates these three rhetorical devices, and their relevance to the overarching theme of the essay.

5. Williams provides a revelation to the reader in paragraph 16 that determines to a large degree the focus of her essay. What was the rhetorical purpose in waiting so long to reveal it?

6. Literary critics often state that one of the essential elements of powerful drama is some profound change that the major protagonist undergoes. What change in perspective did Williams undergo, and what was its implication for her attitudes and actions toward society?

7. Despite profound tragedy and continuous frustration in her life, Williams ends her essay on a positive note. Where is this change in tone most evident? What images reflect her ultimate triumph?

WRITING

1. In an expository essay, analyze the dangers that nuclear proliferation poses for humanity in this century.

2. For a research project, study public policy in the 1950s and 1960s regarding aboveground nuclear testing, and report on whether or not the American government foresaw that there was a good degree of certainty that some individuals would suffer dire physical infirmities as a result of such procedures.

3. **WRITING AN ARGUMENT:** Argue for or against the proposition that technology is never value-free.

---

www.mhhe.com/ **mhreader**   For more information on Terry Tempest Williams, go to:
More Resources > Ch. 14 Science & Technology

CONNECTIONS FOR CRITICAL THINKING

1. Using the essays of D'Souza, Dyson, and Williams, discuss the need for strict ethics among scientists in regards to their concern over the well-being of the general populace.
2. Compare the tone of Williams and Cole with the tone of any two male writers in this chapter. Are there any significant differences between the genders?
3. Compare the process of natural selection as advanced by Darwin with the concept of entropy as defined by Cole.
4. Search the Web for two sites: one promoting the idea of evolution, the other promoting the idea of creationism. Compare and contrast the approach of each site as well as responses made by the visitors to each site.
5. Have your class create a private chat room with screen names that do not divulge the gender of the participants. Discuss the pros and cons of computers as described by Sedaris. Have a host tally the nature of the responses, reveal the gender of the students who participated, and discuss any differences that were found between the responses of the male and female students.
6. Do a search of the Web using the keywords *cloning* and *children*. Analyze three hits to demonstrate the ethical and religious controversies raised by the subject.
7. Rent the videotape or DVD version of the television series *Cosmos,* which was based on a novel by Carl Sagan. Compare the ideas set forth in the film with those in Sagan's essay "Can We Know the Universe? Reflections on a Grain of Salt."
8. Compare and contrast the expository methods Darwin uses to explain the process of natural selection and the narrative technique Sedaris uses to describe his technophobia.
9. Using Williams's essay on the pernicious effects of technology and Angier's views on how we are "selected" to survive tragedies like 9/11 as prompts for meditation and thought on the subject, explore how technology can be either a friend or a foe of humankind.

Glossary of Terms

Abstract/concrete patterns of language reflect an author's word choice. Abstract words (for example, *wisdom, power, beauty*) refer to general ideas, qualities, or conditions. Concrete words name material objects and items associated with the five senses— words like *rock, pizza,* and *basketball.* Both abstract and concrete language are useful in communicating ideas. Generally you should not be too abstract in writing. It is best to employ concrete words, naming things that can be seen, touched, smelled, heard, or tasted in order to support generalizations, topic sentences, or more abstract ideas.

Acronym is a word formed from the first or first few letters of several words, as in OPEC (Organization of Petroleum Exporting Countries).

Action in narrative writing is the sequence of happenings or events. This movement of events may occupy just a few minutes or extend over a period of years or centuries.

Alliteration is the repetition of initial consonant sounds in words placed closely next to each other, as in "what a *t*ale of *t*error now their *t*urbulency *t*ells." Prose that is highly rhythmical or "poetic" often makes use of this method.

Allusion is a literary, biographical, or historical reference, whether real or imaginary. It is a "figure of speech" (a fresh, useful comparison) employed to illuminate an idea. A writer's prose style can be made richer through this economical method of evoking an idea or emotion, as in E. M. Forster's biblical allusion in this sentence: "Property produces men of weight, and it was a man of weight who failed to get into the Kingdom of Heaven."

Analogy is a form of comparison that uses a clear illustration to explain a difficult idea or function. It is unlike a formal comparison in that its subjects of comparison are from different categories or areas. For example, an analogy likening "division of labor" to the activity of bees in a hive makes the first concept more concrete by showing it to the reader through the figurative comparison with the bees. Analogy in exposition can involve a few sentences, a paragraph or set of paragraphs, or an entire essay. Analogies can also be used in argumentation to heighten an appeal to emotion, but they cannot actually *prove* anything.

G

Analysis   is a method of exposition in which a subject is broken up into its parts to explain their nature, function, proportion, or relationship. Analysis thus explores connections and processes within the context of a given subject. (See *causal analysis* and *process analysis*.)

Anecdote   is a brief, engaging account of some happening, often historical, biographical, or personal. As a technique in writing, anecdote is especially effective in creating interesting essay introductions and also in illuminating abstract concepts in the body of the essay.

Antecedent   in grammar refers to the word, phrase, or clause to which a pronoun refers. In writing, antecedent also refers to any happening or thing that is prior to another, or to anything that logically precedes a subject.

Antithesis   is the balancing of one idea or term against another for emphasis.

Antonym   is a word whose meaning is opposite to that of another word.

Aphorism   is a short, pointed statement expressing a general truism or an idea in an original or imaginative way. Marshall McLuhan's statement that "the medium is the message" is a well-known contemporary aphorism.

Archaic   language is vocabulary or usage that belongs to an early period and is old-fashioned today. The word *thee* for *you* is an archaism still in use in certain situations.

Archetypes   are special images or symbols that, according to Carl Jung, appeal to the total racial or cultural understanding of a people. Such images or symbols as the mother archetype, the cowboy in American film, a sacred mountain, or spring as a time of renewal tend to trigger the "collective unconscious" of the human race.

Argumentation   is a formal variety of writing that offers reasons for or against something. Its goal is to persuade or convince the reader through logical reasoning and carefully controlled emotional appeal. Argumentation as a formal mode of writing contains many properties that distinguish it from exposition. (See *assumption, deduction, evidence, induction, logic, persuasion, proposition,* and *refutation.*)

Assonance   is defined generally as likeness or rough similarity of sound. Its specific definition is a partial rhyme in which the stressed vowel sounds are alike but the consonant sounds are unlike, as in *late* and *make.* Although more common to poetry, assonance can also be detected in highly rhythmic prose.

Assumption   in argumentation is anything taken for granted or presumed to be accepted by the audience and therefore unstated. Assumptions in argumentative writing can be dangerous because the audience might not always accept the idea implicit in them. (See *begging the question.*)

Audience   is that readership toward which an author directs his or her essay. In composing essays, writers must acknowledge the nature of their expected readers— whether specialized or general, minimally educated or highly educated, sympathetic or unsympathetic toward the writer's opinions, and so forth. Failure to focus on the writer's true audience can lead to confusions in language and usage, presentation of inappropriate content, and failure to appeal to the expected reader.

Balance   in sentence structure refers to the assignment of equal treatment in the arrangement of coordinate ideas. It is often used to heighten a contrast of ideas.

Begging the question is an error or a fallacy in reasoning and argumentation in which the writer assumes as a truth something for which evidence or proof is actually needed.

Causal analysis is a form of writing that examines causes and effects of events or conditions as they relate to a specific subject. Writers can investigate the causes of a particular effect or the effects of a particular cause or combine both methods. Basically, however, causal analysis looks for connections between things and reasons behind them.

Characterization is the creation of people involved in the action. It is used especially in narrative or descriptive writing. Authors use techniques of dialogue, description, reportage, and observation in attempting to present vivid and distinctive characters.

Chronology or chronological order is the arrangement of events in the order in which they happened. Chronological order can be used in such diverse narrative situations as history, biography, scientific process, and personal account. Essays that are ordered by chronology move from one step or point to the next in time.

Cinematic technique in narration, description, and occasionally exposition is the conscious application of film art to the development of the contemporary essay. Modern writers often are aware of such film techniques as montage (the process of cutting and arranging film so that short scenes are presented in rapid succession), zoom (intense enlargement of subject), and various forms of juxtaposition, and use these methods to enhance the quality of their essays.

Classification is a form of exposition in which the writer divides a subject into categories and then groups elements in each of those categories according to their relationships with one another. Thus a writer using classification takes a topic, divides it into several major groups, and then often subdivides those groups, moving always from larger categories to smaller ones.

Cliché is an expression that once was fresh and original but has lost much of its vitality through overuse. Because terms like "as quick as a wink" and "blew her stack" are trite or common today, they should be avoided in writing.

Climactic ordering is the arrangement of a paragraph or essay so that the most important items are saved for last. The effect is to build slowly through a sequence of events or ideas to the most critical part of the composition.

Coherence is a quality in effective writing that results from the careful ordering of each sentence in a paragraph and each paragraph in the essay. If an essay is coherent, each part will grow naturally and logically from those parts that come before it. Following careful chronological, logical, spatial, or sequential order is the most natural way to achieve coherence in writing. The main devices used in achieving coherence are transitions, which help connect one thought with another.

Colloquial language is conversational language used in certain types of informal and narrative writing but rarely in essays, business writing, or research writing. Expressions like "cool," "pal," or "I can dig it" often have a place in conversational settings. However, they should be used sparingly in essay writing for special effects.

Comparison/contrast as an essay pattern treats similarities and differences between two subjects. Any useful comparison involves two items from the same class. Moreover, there must be a clear reason for the comparison or contrast. Finally, there must be a balanced treatment of the various comparative or contrasting points between the two subjects.

Conclusions are the endings of essays. Without a conclusion, an essay would be incomplete, leaving the reader with the feeling that something important has been left out. There are numerous strategies for conclusions available to writers: summarizing main points in the essay, restating the main idea, using an effective quotation to bring the essay to an end, offering the reader the climax to a series of events, returning to the beginning and echoing it, offering a solution to a problem, emphasizing the topic's significance, or setting a new frame of reference by generalizing from the main thesis. A conclusion should end the essay in a clear, convincing, or emphatic way.

Concrete (see *abstract/concrete.*)

Conflict in narrative writing is the clash or opposition of events, characters, or ideas that makes the resolution of action necessary.

Connotation/denotation are terms specifying the way a word has meaning. Connotation refers to the "shades of meaning" that a word might have because of various emotional associations it calls up for writers and readers alike. Words like *patriotism*, *pig*, and *rose* have strong connotative overtones to them. Denotation refers to the "dictionary" definition of a word—its exact meaning. Good writers understand the connotative and denotative value of words and must control the shades of meaning that many words possess.

Context is the situation surrounding a word, group of words, or sentence. Often the elements coming before or after a certain confusing or difficult construction will provide insight into the meaning or importance of that item.

Coordination in sentence structure refers to the grammatical arrangement of parts of the same order or equality in rank.

Declarative sentences make a statement or assertion.

Deduction is a form of logic that begins with a generally stated truth or principle and then offers details, examples, and reasoning to support the generalization. In other words, deduction is based on reasoning from a known principle to an unknown principle, from the general to the specific, or from a premise to a logical conclusion. (See *syllogism.*)

Definition in exposition is the extension of a word's meaning through a paragraph or an entire essay. As an extended method of explaining a word, this type of definition relies on other rhetorical methods, including detail, illustration, comparison and contrast, and anecdote.

Denotation (see *connotation/denotation.*)

Description in the prose essay is a variety of writing that uses details of sight, sound, color, smell, taste, and touch to create a word picture and to explain or illustrate an idea.

Development refers to the way a paragraph or an essay elaborates or builds upon a topic or theme. Typical development proceeds either from general illustrations to specific ones or from one generalization to another. (See *horizontal/vertical.*)

Dialogue is the reproduction of speech or conversation between two or more persons in writing. Dialogue can add concreteness and vividness to an essay and can also help reveal character. A writer who reproduces dialogue in an essay must use it for a purpose and not simply as a decorative device.

Diction   is the manner of expression in words, choice of words, or wording. Writers must choose vocabulary carefully and precisely to communicate a message and also to address an intended audience effectively; this is good diction.

Digression   is a temporary departure from the main subject in writing. Any digression in the essay must serve a purpose or be intended for a specific effect.

Discourse (forms of)   relates conventionally to the main categories of writing—narration, description, exposition and argumentation. In practice, these forms of discourse often blend or overlap. Essayists seek the ideal fusion of forms of discourse in the treatment of their subject.

Division   is that aspect of classification in which the writer divides some large subject into categories. Division helps writers split large and potentially complicated subjects into parts for orderly presentation and discussion.

Dominant impression   in description is the main impression or effect that writers attempt to create for their subject. It arises from an author's focus on a single subject and from the feelings the writer brings to that subject.

Editorialize   is to express personal opinions about the subject of the essay. An editorial tone can have a useful effect in writing, but at other times an author might want to reduce editorializing in favor of a better balanced or more objective tone.

Effect   is a term used in causal analysis to describe the outcome or expected result of a chain of happenings.

Emphasis   indicates the placement of the most important ideas in key positions in the essay. As a major principle, emphasis relates to phrases, sentences, paragraphs—the construction of the entire essay. Emphasis can be achieved by repetition, subordination, careful positioning of thesis and topic sentences, climactic ordering, comparison and contrast, and a variety of other methods.

Episodic   relates to that variety of narrative writing that develops through a series of incidents or events.

Essay   is the name given to a short prose work on a limited topic. Essays take many forms, ranging from personal narratives to critical or argumentative treatments of a subject. Normally an essay will convey the writer's personal ideas about the subject.

Etymology   is the origin and development of a word—tracing a word back as far as possible.

Evidence   is material offered to support an argument or a proposition. Typical forms of evidence are facts, details, and expert testimony.

Example   is a method of exposition in which the writer offers illustrations in order to explain a generalization or a whole thesis. (See *illustration*.)

Exclamatory sentences   in writing express surprise or strong emotion.

Expert testimony   as employed in argumentative essays and in expository essays is the use of statements by authorities to support a writer's position or idea. This method often requires careful quotation and acknowledgment of sources.

Exposition   is a major form of discourse that informs or explains. Exposition is the form of expression required in much college writing, for it provides facts and information, clarifies ideas, and establishes meaning. The primary methods of exposition are *illustration, comparison and contrast, analogy, definition, classification, causal analysis,* and *process analysis* (see entries).

Extended metaphor is a figurative comparison that is used to structure a significant part of the composition or the whole essay. (See *figurative language* and *metaphor.*)

Fable is a form of narrative containing a moral that normally appears clearly at the end.

Fallacy in argumentation is an error in logic or in the reasoning process. Fallacies occur because of vague development of ideas, lack of awareness on the part of writers of the requirements of logical reasoning, or faulty assumptions about the proposition.

Figurative language as opposed to literal language is a special approach to writing that departs from what is typically a concrete, straightforward style. It is the use of vivid, imaginative statements to illuminate or illustrate an idea. Figurative language adds freshness, meaning, and originality to a writer's style. Major figures of speech include *allusion, hyperbole, metaphor, personification,* and *simile* (see entries).

Flashback is a narrative technique in which the writer begins at some point in the action and then moves into the past in order to provide crucial information about characters and events.

Foreshadow is a technique that indicates beforehand what is to occur at a later point in the essay.

Frame in narration and description is the use of a key object or pattern—typically at the start and end of the essay—that serves as a border or structure to contain the substance of the composition.

General/specific words are the basis of writing, although it is wise in college composition to keep vocabulary as specific as possible. General words refer to broad categories and groups, whereas specific words capture with force and clarity the nature of the term. General words refer to large classes, concepts, groups, and emotions; specific words are more particular in providing meanings. The distinction between general and specific language is always a matter of degree.

Generalization is a broad idea or statement. All generalizations require particulars and illustrations to support them.

Genre is a type or form of literature—for example, short fiction, novel, poetry, drama.

Grammatical structure is a systematic description of language as it relates to the grammatical nature of a sentence.

Horizontal/vertical paragraph and essay development refers to the basic way a writer moves either from one generalization to another in a carefully related series of generalizations (horizontal) or from a generalization to a series of specific supporting examples (vertical).

Hortatory style is a variety of writing designed to encourage, give advice, or urge to good deeds.

Hyperbole is a form of figurative language that uses exaggeration to overstate a position.

Hypothesis is an unproven theory or proposition that is tentatively accepted to explain certain facts. A working hypothesis provides the basis for further investigation or argumentation.

Hypothetical examples   are illustrations in the form of assumptions that are based on the hypothesis. As such, they are conditional rather than absolute or certain facts.

Identification   as a method of exposition refers to focusing on the main subject of the essay. It involves the clear location of the subject within the context or situation of the composition.

Idiomatic language   is the language or dialect of a people, region, or class—the individual nature of a language.

Ignoring the question   in argumentation is a fallacy that involves the avoidance of the main issue by developing an entirely different one.

Illustration   is the use of one or more examples to support an idea. Illustration permits the writer to support a generalization through particulars or specifics.

Imagery   is clear, vivid description that appeals to the sense of sight, smell, touch, sound, or taste. Much imagery exists for its own sake, adding descriptive flavor to an essay. However, imagery (especially when it involves a larger pattern) can also add meaning to an essay.

Induction   is a method of logic consisting of the presentation of a series of facts, pieces of information, or instances in order to formulate or build a likely generalization. The key is to provide prior examples before reaching a logical conclusion. Consequently, as a pattern of organization in essay writing, the inductive method requires the careful presentation of relevant data and information before the conclusion is reached at the end of the paper.

Inference   involves arriving at a decision or opinion by reasoning from known facts or evidence.

Interrogative sentences   are sentences that ask or pose a question.

Introduction   is the beginning or opening of an essay. The introduction should alert the reader to the subject by identifying it, set the limits of the essay, and indicate what the thesis (or main idea) will be. Moreover, it should arouse the reader's interest in the subject. Among the devices available in the creation of good introductions are making a simple statement of thesis; giving a clear, vivid description of an important setting; posing a question or series of questions; referring to a relevant historical event; telling an anecdote; using comparison and contrast to frame the subject; using several examples to reinforce the statement of the subject; and presenting a personal attitude about a controversial issue.

Irony   is the use of language to suggest the opposite of what is stated. Writers use irony to reveal unpleasant or troublesome realities that exist in life or to poke fun at human weaknesses and foolish attitudes. In an essay there may be verbal irony, in which the result of a sequence of ideas or events is the opposite of what normally would be expected. A key to the identification of irony in an essay is our ability to detect where the author is stating the opposite of what he or she actually believes.

Issue   is the main question upon which an entire argument rests. It is the idea that the writer attempts to prove.

Jargon   is the use of special words associated with a specific area of knowledge or a particular profession. Writers who employ jargon either assume that readers know specialized terms or take care to define terms for the benefit of the audience.

Juxtaposition as a technique in writing or essay organization is the placing of elements—either similar or contrasting—close together, positioning them side by side in order to illuminate the subject.

Levels of language refer to the kinds of language used in speaking and writing. Basically there are three main levels of language—formal, informal, and colloquial. Formal English, used in writing or speech, is the type of English employed to address special groups and professional people. Informal English is the sort of writing found in newspapers, magazines, books, and essays. It is popular English for an educated audience but still more formal than conversational English. Finally, colloquial English is spoken (and occasionally written) English used in conversations with friends, employees, and peer group members; it is characterized by the use of slang, idioms, ordinary language, and loose sentence structure.

Linear order in paragraph development means the clear line of movement from one point to another.

Listing is a simple technique of illustration in which facts or examples are used in order to support a topic or generalization.

Logic as applied to essay writing is correct reasoning based on induction or deduction. The logical basis of an essay must offer reasonable criteria or principles of thought, present these principles in an orderly manner, avoid faults in reasoning, and result in a complete and satisfactory outcome in the reasoning process.

Metaphor is a type of figurative language in which an item from one category is compared briefly and imaginatively with an item from another area. Writers use such implied comparisons to assign meaning in a fresh, vivid, and concrete way.

Metonymy is a figure of language in which a thing is not designated by its own name but by another associated with or suggested by it, as in "The Supreme Court has decided" (meaning that the judges of the Supreme Court have decided).

Mood is the creation of atmosphere in descriptive writing.

Motif in an essay is any series of components that can be detected as a pattern. For example, a particular detail, idea, or image can be elaborated upon or designed to form a pattern or motif in the essay.

Myth in literature is a traditional story or series of events explaining some basic phenomenon of nature; the origin of humanity; or the customs, institutions, and religious rites of a people. Myth often relates to the exploits of gods, goddesses, and heroes.

Narration as a form of essay writing is the presentation of a story in order to illustrate an idea.

Non sequitur in argumentation is a conclusion or inference that does not follow from the premises or evidence on which it is based. The non sequitur thus is a type of logical fallacy.

Objective/subjective writing refers to the attitude that writers take toward their subject. When writers are objective, they try not to report their personal feelings about the subject; they attempt to be detached, impersonal, and unbiased. Conversely, subjective writing reveals an author's personal attitudes and emotions. For many varieties of college writing, such as business or laboratory reports, term papers, and literary analyses, it is best to be as objective as possible. But for many personal essays in composition courses, the subjective touch is fine. In the hands of skilled writers, the objective and subjective tones often blend.

Onomatopoeia   is the formation of a word by imitating the natural sound associated with the object or action, as in *buzz* or *click.*

Order   is the arrangement of information or materials in an essay. The most common ordering techniques are *chronological order* (time in sequence); *spatial order* (the arrangement of descriptive details); *process order* (a step-by-step approach to an activity); *deductive order* (a thesis followed by information to support it); and *inductive order* (evidence and examples first, followed by the thesis in the form of a conclusion). Some rhetorical patterns, such as comparison and contrast, classification, and argumentation, require other ordering methods. Writers should select those ordering principles that permit them to present materials clearly.

Overstatement   is an extravagant or exaggerated claim or statement.

Paradox   is a statement that seems to be contradictory but actually contains an element of truth.

Paragraph   is a unit in an essay that serves to present and examine one aspect of a topic. Composed normally of a group of sentences (one-sentence paragraphs can be used for emphasis or special effect), the paragraph elaborates an idea within the larger framework of the essay and the thesis unifying it.

Parallelism   is a variety of sentence structure in which there is balance or coordination in the presentation of elements. "I came, I saw, I conquered" is a standard example of parallelism, presenting both pronouns and verbs in a coordinated manner. Parallelism can appear in a sentence, a group of sentences, or an entire paragraph.

Paraphrase   as a literary method is the process of rewording the thought or meaning expressed in something that has been said or written before.

Parenthetical   refers to giving qualifying information or explanation. This information normally is marked off or placed within parentheses.

Parody   is ridiculing the language or style of another writer or composer. In parody, a serious subject tends to be treated in a nonsensical manner.

Periphrasis   is the use of many words where one or a few would do; it is a roundabout way of speaking or writing.

Persona   is the role or characterization that writers occasionally create for themselves in a personal narrative.

Personification   is giving an object, a thing, or an idea lifelike or human characteristics, as in the common reference to a car as "she." Like all forms of figurative language, personification adds freshness to description and makes ideas vivid by setting up striking comparisons.

Persuasion   is the form of discourse, related to argumentation, that attempts basically to move a person to action or to influence an audience toward a particular belief.

Point of view   is the angle from which a writer tells a story. Many personal and informal essays take the *first-person* (or "I") point of view, which is natural and fitting for essays in which the author wants to speak in a familiar way to the reader. On the other hand, the *third-person* point of view ("he," "she," "it," "they") distances the reader somewhat from the writer. The third-person point of view is useful in essays in which the writers are not talking exclusively about themselves, but about other people, ideas, and events.

Post hoc, ergo propter hoc in logic is the fallacy of thinking that a happening that follows another must be its result. It arises from a confusion about the logical causal relationship.

Process analysis is a pattern of writing that explains in a step-by-step way how something is done, how it is put together, how it works, or how it occurs. The subject can be a mechanical device, a product, an idea, a natural phenomenon, or a historical sequence. However, in all varieties of process analysis, the writer traces all important steps, from beginning to end.

Progression is the forward movement or succession of acts, events, or ideas presented in an essay.

Proportion refers to the relative emphasis and length given to an event, an idea, a time, or a topic within the whole essay. Basically, in terms of proportion the writer gives more emphasis to a major element than to a minor one.

Proposition is the main point of an argumentative essay—the statement to be defended, proved, or upheld. It is like a *thesis* (see entry) except that it presents an idea that is debatable or can be disputed. The *major proposition* is the main argumentative point; *minor propositions* are the reasons given to support or prove the issue.

Purpose is what the writer wants to accomplish in an essay. Writers having a clear purpose will know the proper style, language, tone, and materials to utilize in designing an effective essay.

Refutation in argumentation is a method by which you recognize and deal effectively with the arguments of your opponents. Your own argument will be stronger if you refute—prove false or wrong—all opposing arguments.

Repetition is a simple method of achieving emphasis by repeating a word, a phrase, or an idea.

Rhetoric is the art of using words effectively in speaking or writing. It is also the art of literary composition, particularly in prose, including both figures of speech and such strategies as *comparison and contrast, definition,* and *analysis.*

Rhetorical question is a question asked only to emphasize a point, introduce a topic, or provoke thought, but not to elicit an answer.

Rhythm in prose writing is a regular recurrence of elements or features in sentences, creating a patterned emphasis, balance, or contrast.

Sarcasm is a sneering or taunting attitude in writing, designed to hurt by evaluating or criticizing. Basically, sarcasm is a heavy-handed form of *irony* (see entry). Writers should try to avoid sarcastic writing and to use more acceptable varieties of irony and satire to criticize their subject.

Satire is the humorous or critical treatment of a subject in order to expose the subject's vices, follies, stupidities, and so forth. The intention of such satire is to reform by exposing the subject to comedy or ridicule.

Sensory language is language that appeals to any of the five senses—sight, sound, touch, taste, or smell.

Sentimentality   in prose writing is the excessive display of emotion, whether intended or unintended. Because sentimentality can distort the true nature of a situation or an idea, writers should use it cautiously, or not at all.

Series   as a technique in prose is the presentation of several items, often concrete details or similar parts of grammar such as verbs or adjectives, in rapid sequence.

Setting   in narrative and descriptive writing is the time, place, environment, background, or surroundings established by an author.

Simile   is a figurative comparison using *like* or *as*.

Slang   is a kind of language that uses racy or colorful expressions associated more often with speech than with writing. It is colloquial English and should be used in essay writing only to reproduce dialogue or to create a special effect.

Spatial order   in descriptive writing is the careful arrangement of details or materials in space—for example, from left to right, top to bottom, or near to far.

Specific words   (see *general/specific words.*)

Statistics   are facts or data of a numerical kind, assembled and tabulated to present significant information about a given subject. As a technique of illustration, statistics can be useful in analysis and argumentation.

Style   is the specific or characteristic manner of expression, execution, construction, or design of an author. As a manner or mode of expression in language, it is the unique way each writer handles ideas. There are numerous stylistic categories—literary, formal, argumentative, satiric—but ultimately, no two writers have the same style.

Subjective   (see *objective/subjective.*)

Subordination   in sentence structure is the placing of a relatively less important idea in an inferior grammatical position to the main idea. It is the designation of a minor clause that is dependent upon a major clause.

Syllogism   is an argument or form of reasoning in which two statements or premises are made and a logical conclusion drawn from them. As such, it is a form of deductive logic—reasoning from the general to the particular. The *major premise* presents a quality of class ("All writers are mortal"). The *minor premise* states that a particular subject is a member of that class ("Ernest Hemingway was a writer."). The conclusion states that the qualities of the class and the member of the class are the same. ("Hemingway was mortal.").

Symbol   is something—normally a concrete image—that exists in itself but also stands for something else or has greater meaning. As a variety of figurative language, the symbol can be a strong feature in an essay, operating to add depth of meaning and even to unify the composition.

Synonym   is a word that means roughly the same as another word. In practice, few words are exactly alike in meaning. Careful writers use synonyms to vary word choice without ever moving too far from the shade of meaning intended.

Theme   is the central idea in an essay; it is also termed the *thesis*. Everything in an essay should support the theme in one way or another.

Thesis   is the main idea in an essay. The *thesis sentence*, appearing only in the essay (normally somewhere in the first paragraph) serves to convey the main idea to the reader in a clear and emphatic manner.

Tone   is the writer's attitude toward his or her subject or material. An essay writer's tone may be objective, subjective, comic, ironic, nostalgic, critical, or a reflection of numerous other attitudes. Tone is the voice that writers give to an essay.

Topic sentence   is the main idea that a paragraph develops. Not all paragraphs contain topic sentences; often the topic is implied.

Transition   is the linking of ideas in sentences, paragraphs, and larger segments of an essay in order to achieve *coherence* (see entry). Among the most common techniques to achieve smooth transitions are (1) repeating a key word or phrase; (2) using a pronoun to refer to a key word or phrase; (3) relying on traditional connectives such as *thus, however, moreover, for example, therefore, finally,* and *in conclusion;* (4) using parallel structure (see *parallelism*); and (5) creating a sentence or paragraph that serves as a bridge from one part of an essay to another. Transition is best achieved when a writer presents ideas and details carefully and in logical order.

Understatement   is a method of making a weaker statement than is warranted by truth, accuracy, or importance.

Unity   is a feature in an essay whereby all material relates to a central concept and contributes to the meaning of the whole. To achieve a unified effect in an essay, the writer must design an effective introduction and conclusion, maintain consistent tone or point of view, develop middle paragraphs in a coherent manner, and above all stick to the subject, never permitting unimportant or irrelevant elements to enter.

Usage   is the way in which a word, phrase, or sentence is used to express a particular idea; it is the customary manner of using a given language in speaking or writing.

Vertical   (see *horizontal/vertical.*)

Voice   is the way you express your ideas to the reader, the tone you take in addressing your audience. Voice reflects your attitude toward both your subject and your readers. (See *tone.*)

Credits

Photos

p. 20: Rosenthal/AP/Wide World; **p. 21**: Thomas Franklin/The Bergen Record/Corbis SABA; **p. 132**: The Prado Museum, Spain/Art Resource; **p. 133**: Eddie Adams/AP/Wide World Photos; **p. 165**: Courtesy, fedworld.gov; **pp. 167, 168**: Courtesy, Auraria Library; **p. 201**: MANDALAY ENT/BALTIMORE PICS/The Kobal Collection; **p. 202**: The Everett Collection; **p. 254**: Oberlin College Archives, Oberlin, Ohio; **p. 255**: Tom Stewart/Corbis/Stock Market; **p. 296**: Museum of Fine Art, Ghent, Belgium/The Bridgeman Art Library; **p. 297**: AP/Wide World Photos; **p. 352**: National Park Service: Statue of Liberty National Monument; **p. 353**: Reforma Archivo/AP/Wide World Photos; **p. 404**: The Granger Collection, New York; **p. 405**: AP/Wide World Photos; **p. 456**: Detroit Institute of Arts/The Bridgeman Art Library. Gift of Edsel Ford; **p. 457**: George Haling/Photo Researchers; **p. 512**: Photofest; **p. 513**: HBO/The Kobal Collection; **p. 568**: Musee Rodin/Bridgeman Art Collection; **p. 569**: Rabbit, 1986, Stainless steel, 41″ × 19″ × 11⅞″ © Jeff Koons; **p. 618**: Angels: Messengers of the Gods/Thames & Hudson/Indian Museum, Calcutta; **p. 619**: Gillian Darley/EDIFICE; **p. 668**: Corbis; **p. 669**: Gary Bramnick/AP/Wide World Photos; **p. 673**: Bibliotheque Royale, Brussels/The Bridgeman Art Library; **p. 726**: Smithsonian American Art Museum/Art Resource; **p. 727**: Damian Dovarganes/AP/Wide World Photos; **p. 780**: Musee Condee, Chantilly, France/The Bridgeman Art Library; **p. 781**: Space Telescope Science Institute/Science Photo Library/Photo Researchers

Color Insert Photos

p. 1: Association of American Publishers; **p. 2**: Estate of Ernest S. Klempner; **p. 3**: Courtesy, Sean John; **p. 4**: AP/Wide World Photos; **p. 5**: Courtesy, www.amendforarnold.com; **p. 6**: AP/Wide World Photos; **p. 7**: Courtesy, Wendy's International; **p. 8**: Irwin Thompson/The Dallas Morning News/AP/Wide World Photos

Text

Adler, Mortimer, J. "How to Mark a Book" by Mortimer J. Adler. Reprinted by permission of the author.

Alvarez, Julia "Writing Matters" by Julia Alvarez from *Something to Declare,* Algonquin Books of Chapel Hill, 1998. First appeared in a different version in *The Writer,* September 1998. Copyright © 1982, 1998 by Julia Alvarez. Reprinted by permission of Susan Bergholz Literary Services, New York. All rights reserved.

C

Angier, Natalie "Why Men Don't Last: Self-Destruction as a Way of Life" by Natalie Angier from *The New York Times*, February 17, 1999. Copyright © 1999 The New York Times Co. Reprinted with permission.

Angier, Natalie "Of Altruism, Heroism and Nature's Gifts in the Face of Terror" by Natalie Angier from *The New York Times*, November 18, 2001. Copyright © 2001 by The New York Times Co. Reprinted with permission.

Appiah, K. Anthony "The Multicultural Mistake" by K. Anthony Appiah from *The New York Review of Books*, October 9, 1997. Copyright © 1997 NYREV, Inc. Reprinted with permission from The New York Review of Books.

Atwood, Margaret "The Female Body" by Margaret Atwood. Copyright © O. W. Toad Ltd., 1992. Reprinted from *Good Bones* by Margaret Atwood with permission of Coach House Press.

Baldwin, James "Stranger in the Village" from *Notes of a Native Son* by James Baldwin. Copyright © 1955, renewed 1983, by James Baldwin. Reprinted by permission of Beacon Press, Boston.

Barry, Dave "Red, White & Beer" from *Dave Barry's Greatest Hits* by Dave Barry. Copyright © 1988 by Dave Barry. Reprinted by permission of the author.

Bordo, Susan "The Globalization of Eating Disorders" by Susan Bordo. Copyright © Susan Bordo, Otis A. Singletary Professor of the Humanities, University of Kentucky. Reprinted by permission of the author.

Brooks, David "Love, Internet Style" by David Brooks from *The New York Times*, November 8, 2003. Copyright © 2003 The New York Times Co. Reprinted with permission.

Carson, Clayborne "Two Cheers for *Brown vs. Board of Education*" by Clayborne Carson from *Journal of American History*, June, 2004: pp. 26–31. Copyright © Organization of American Historians. Reprinted with permission.

Carson, Rachel L. "The Obligation to Endure" from *Silent Spring* by Rachel Carson. Copyright © 1962 by Rachel L. Carson. Copyright renewed 1990 by Roger Christie. Reprinted by permission of Houghton Mifflin Company. All rights reserved.

Carter, Stephen L. Excerpts from *The Culture of Disbelief: How American Law and Politics Trivialize Religious Devotion* by Stephen L. Carter. Copyright © 1993 by Stephen L. Carter. Reprinted by permission of Basic Books, a member of Perseus Books, L.L.C.

Catton, Bruce "Grant and Lee: A Study of Contrasts" by Bruce Catton from *The American Story*, edited by Earl Schenck Miers. Reprinted with permission of William Catton.

Cofer, Judith Ortiz "The Myth of the Latin Woman: I Just Met a Girl Named Maria" from *The Latin Deli: Prose & Poetry* by Judith Ortiz Cofer. Copyright © 1993 by Judith Ortiz Cofer. Reprinted by permission of the University of Georgia Press.

Cole, K. C. "Entropy" by K. C. Cole from *The New York Times*, March 18, 1982. Copyright © 1982 by The New York Times Co. Reprinted with permission.

Coles, Robert "I Listen to My Parents and Wonder What They Believe" by Robert Coles. Originally published in *Redbook* magazine, February 1980. Reprinted by permission of the author.

Collins, Billy "American Sonnet" from *Questions about Angels* by Billy Collins. Copyright © 1991. Reprinted by permission of the University of Pittsburgh Press; "Lines Composed over Three Thousand Miles from Tintern Abbey" from *Picnic, Lightning* by Billy Collins. Copyright © 1998. Reprinted by permission of the University of Pittsburgh Press.

Dahlke, Laura Johnson "Plath's 'Lady Lazarus'" by Laura Johnson Dahlke from *The Explicator*, 60:4 (2002): 234–236. Copyright © 2002. Reprinted with permission of the Helen Dwight Reid Educational Foundation. Published by Heldref Publications, 1319 Eighteenth Street, NW, Washington, DC 20036-1802.

Didion, Joan "Marrying Absurd" from *Slouching Towards Bethlehem* by Joan Didion. Copyright © 1966, 1968, renewed 1996 by Joan Didion. Reprinted by permission of Farrar, Straus and Giroux, LLC.

Dillard, From *An American Childhood* by Annie Dillard, pp. 110–117. Copyright © 1987 by Annie Dillard. Reprinted by permission of HarperCollins Publishers, Inc.; From *The Writing Life* by Annie Dillard, pp. 3–5. Copyright © 1989 by Annie Dillard. Reprinted by permission of HarperCollins, Inc.

Dove, Rita "Loose Ends" excerpted from *The Poet's World* by Rita Dove. The Library of Congress, copyright © 1995 by Rita Dove. Reprinted by permission of the author.

D'Souza, Dinesh "Staying Human" by Dinesh D'Souza from *National Review,* January 22, 2001. Reprinted with permission from the author.

Dyson, Esther "Cyberspace: If You Don't Love It, Leave It" by Esther Dyson from *The New York Times,* July 16, 1995. Copyright © 2001 The New York Times Co. Reprinted with permission.

Dyson, Freeman J. "Science, Guided by Ethics, Can Lift Up the Poor" by Freeman Dyson. Speech given at the Washington National Cathedral for Templeton Prize for Progress in Religion, 2000. Reprinted with permission of the author.

Ebbeling, Cara B., Kelly B. Sinclair, Mark A. Pereira, Erica Garcia-Lago, Henry A. Feldman, David S. Ludwig "Compensation for Energy Intake from Fast Food Among Overweight and Lean Adolescents" by Cara B. Ebbeling et al. from *Journal of the American Medical Association,* June 16, 2004, Volume 291, No. 23, pp. 2828–2833. Reprinted by permission of The American Medical Association.

Ehrenreich, Barbara Excerpt from "Scrubbing in Maine" from *Nickel and Dimed: On (Not) Getting By in America* by Barbara Ehrenreich. Copyright © 2001 by Barbara Ehrenreich. Reprinted by permission of Henry Holt and Company, LLC.

Elbow, Peter "Freewriting" from *Writing Without Teachers* by Peter Elbow. Copyright © 1973, 1998 by Peter Elbow. Used by permission of Oxford University Press, Inc.

Etzioni, Amitai "Parenting as an Industry" from *The Spirit of Community* by Amitai Etzioni. Copyright © 1993 by Amitai Etzioni. Used by permission of Crown Publishers, a division of Random House, Inc.

Forster, Edward Morgan "Not Looking at Pictures" from *Two Cheers for Democracy* by Edward Morgan Forster. Copyright © 1951 by E.M. Forster and renewed 1979 by Donald Parry. Reprinted by permission of Harcourt, Inc. and The Provost and Scholars of King's College, Cambridge and The Society of Authors as the Literary Representatives of the Estate of E. M. Forster.

Friedman, Thomas L. "Prologue: The Super-Story" from *Longitudes and Attitudes* by Thomas L. Friedman. Copyright © 2002 by Thomas L. Friedman. Reprinted by permission of Farrar, Straus and Giroux, LLC.

Gates, Henry Louis Jr. "2 Live Crew Decoded" by Henry Louis Gates Jr. from *The New York Times,* June 19, 1990. Copyright © 1990 The New York Times Co. Reprinted with permission.; "Delusions of Grandeur" by Henry Louis Gates, Jr. Copyright © 1991 by Henry Louis Gates, Jr. Originally published in *Sports Illustrated.* Reprinted by permission of the author.

Gelernter, David "Unplugged: The Myth of Computers in the Classroom" by David Gelernter. Copyright © 1994 by David Gelernter. Reprinted by permission of *The New Republic.*

Gitlin, Todd "Supersaturation, or, The Media Torrent and Disposable Feeling" from *Media Unlimited: How the Torrent of Images and Sounds Overwhelms* by Todd Gitlin. Copyright © 2001 by Todd Gitlin. Reprinted by permission of Henry Holt and Company, LLC.

Glasser, Ronald J. "We Are Not Immune" by Ronald J. Glasser. Copyright © 2004 by *Harper's Magazine.* All rights reserved. Reproduced from the July issue by special permission.

Goodall, Jane "A Question of Ethics" by Jane Goodall from *Newsweek,* May 7, 2001. Reprinted by permission of Jane Goodall, DBE, founder of the Jane Goodall Institute and United Nations Messenger of Peace.

Goodman, Ellen "I Worked Hard for That Furrowed Brow" by Ellen Goodman from *Boston Globe,* April 2, 2002. Copyright © 2002 The Washington Post Writers Group. Reprinted with permission.

Gould, Jon B. "Playing with Fire: The Civil Liberties Implications of September 11" by Jon B. Gould from *Public Administration Review,* 62:1, September 2002, pp. 74–80. Reprinted by permission of Blackwell Publishing.

Gould, Stephen Jay "The Terrifying Normalcy of AIDS" by Stephen Jay Gould. Reprinted by permission of the author.

Hall, Edward T. "The Arab World" from *The Hidden Dimension* by Edward T. Hall. Copyright © 1966, 1982 by Edward T. Hall. Used by permission of Doubleday, a division of Random House, Inc.

Havel, Vaclav, "The Divine Revolution, Lifting the Iron Curtain of the Spirit," *Civilization Magazine,* July/August 1998.

Hogan, Linda "Hearing Voices" by Linda Hogan. Copyright © 1991 Linda Hogan. Reprinted by permission of the author.

Orwell, George "Politics and the English Language" reprinted from *Shooting an Elephant and Other Essays* by George Orwell. Copyright © 1946 by Sonia Brownwell Orwell and renewed 1974 by Sonia Orwell. Reprinted by permission of Harcourt, Inc., and Bill Hamilton as the Literary Executor of the Estate of the late Sonia Brownell Orwell and Secker and Warburg Ltd.

Perrin, Noel "The Greenest Campuses: An Idiosyncratic Guide" by Noel Perrin. Copyright © 2001 by Noel Perrin. Originally appeared in *Chronicle of Higher Education.* Reprinted by permission of the author.

Pogrebin, Letty Cottin "Superstitious Minds" by Letty Cottin Pogrebin. Copyright © 1988. Reprinted by permission of *Ms. Magazine.*

Posner, Richard A. "Security versus Civil Liberties" by Richard A. Posner. First published in *The Atlantic Monthly,* December 2001. Reprinted by permission of the author.

Quindlen, Anna "Sex Ed" from *Living Out Loud* by Anna Quindlen. Copyright © 1987 by Anna Quindlen. Used by permission of Random House, Inc.; "Men at Work" from *Thinking Out Loud* by Anna Quindlen. Copyright © 1993 by Anna Quindlen. Used by permission of Random House, Inc.; "A New Look, and Old Battle" by Anna Quindlen. Copyright © 2001 Anna Quindlen. Reprinted by permission of International Creative Management, Inc.

Reed, Ishmael "America: The Multinational Society" from *Writin' Is Fightin': Forty-Three Years of Boxing on Paper* by Ishmael Reed. Copyright © Ishmael Reed. Reprinted by permission of Lowenstein Associates.; "The Patriot Act of the 18th Century" by Ishmael Reed from *Time Magazine,* July 5, 2004. Copyright © 2004 TIME Inc. Reprinted by permission.

Reich, Robert "Why the Rich Are Getting Richer, and the Poor, Poorer" from *The Work of Nations* by Robert Reich. Copyright © 1991 by Robert Reich. Used by permission of Alfred A. Knopf, a division of Random House, Inc.

Rodriguez, Richard "Los Pobres" by Richard Rodriguez. Copyright © 1981 by Richard Rodriguez. Originally appeared in *New West.* Reprinted by permission of Georges Borchardt, Inc., on behalf of the author.; "The Lonely, Good Company of Books" by Richard Rodriguez. Copyright © 1981 by Richard Rodriguez. Reprinted by permission of Georges Borchardt, Inc., on behalf of the author.; "Family Values" by Richard Rodriguez. Copyright © 1992 by Richard Rodriguez. Originally appeared in *The Los Angeles Times.* Reprinted by permission of Georges Borchardt, Inc., on behalf of the author.

Rorem, Ned "The Beatles" from *A Ned Rorem Reader* by Ned Rorem. Copyright © 2001 by Ned Rorem. Reprinted by permission of Yale University Press.

Rosenzweig, Paul "Face Facts: Patriot Act Aids Security, Not Abuse" by Paul Rosenzweig. From *Christian Science Monitor,* July 29, 2004. Copyright © 2004 by Christian Science Publishing Society. Reproduced with permission of Christian Science Publishing Society via Copyright Clearance Center.

Rushdie, Salman "Not About Islam?" from *Step Across This Line* by Salman Rushdie. Copyright © 2002 by Salman Rushdie. Used by permission of Random House, Inc.

Sabin, Heloisa "Animal Research Saves Lives" by Heloisa Sabin. Originally appeared in *The Wall Street Journal,* October 18, 1995. Reprinted by permission of Americans for Medical Progress on behalf of the author.

Sagan, Carl "Can We Know the Universe? Reflections on a Grain of Salt" from *Broca's Brain* by Carl Sagan. Copyright © 1979 by Carl Sagan. Copyright © 2005 by the Estate of Carl Sagan. Reprinted with permission.

Scarry, Elaine "Acts of Resistance" by Elaine Scarry from *Harper's,* May 2004. For the full version of this essay, please see *Boston Review,* February/March 2004. Reprinted with permission of the author.

Schlesinger, Arthur M. Jr. "The Cult of Ethnicity." *Time,* 7/8/81. Copyright © 1981 by Arthur M. Schlesinger, Jr. Reprinted by permission of the author.

Sedaris, David "Nutcracker.com" from *Me Talk Pretty One Day* by David Sedaris. Copyright © 2000 by David Sedaris. Used by permission of Little, Brown and Co., Inc.

Selzer, Richard "Sarcophagus" from *Confessions of a Knife* by Richard Selzer. Copyright © 1979 by David Goldman and Janet Selzer (New York: William Morrow). Reprinted by permission of Georges Borchardt, Inc. on behalf of the author.

Sen, Amartya "A World Not Neatly Divided" by Amartya Sen from *The New York Times,* November 23, 2001. Copyright © 2001 The New York Times Co. Reprinted with permission.

Index

I